Transforming E-Business Practices and Applications:
Emerging Technologies and Concepts

In Lee
Western Illinois University, USA

Information Science REFERENCE

INFORMATION SCIENCE REFERENCE

Hershey · New York

Director of Editorial Content:	Kristin Klinger
Senior Managing Editor:	Jamie Snavely
Assistant Managing Editor:	Michael Brehm
Publishing Assistant:	Sean Woznicki
Typesetter:	Kurt Smith, Sean Woznicki
Cover Design:	Lisa Tosheff
Printed at:	Yurchak Printing Inc.

Published in the United States of America by
Information Science Reference (an imprint of IGI Global)
701 E. Chocolate Avenue
Hershey PA 17033
Tel: 717-533-8845
Fax: 717-533-8661
E-mail: cust@igi-global.com
Web site: http://www.igi-global.com/reference

Library of Congress Cataloging-in-Publication Data

Transforming e-business practices and applications : emerging technologies and concepts / In Lee, editor.
p. cm.

Includes bibliographical references and index.
Summary: "This book presents an integrated view of the latest issues and technologies evolving from business transactions and support"--Provided by publisher.

ISBN 978-1-60566-910-6 (hardcover) -- ISBN 978-1-60566-911-3 (ebook) 1. Electronic commerce. 2. Technological innovations--Economic aspects. I. Lee, In, 1958-
HF5548.32.T725 2009
658.8'72--dc22

2009044465

British Cataloguing in Publication Data
A Cataloguing in Publication record for this book is available from the British Library.

Editorial Advisory Board

Table of Contents

Detailed Table of Contents

Shopping online has emerged as one of the most popular Internet applications, providing a plethora of purchasing opportunities for consumers and sales challenges for retailers. The aim of this chapter is to shed further light on the past and present status of the e-consumer phenomenon, by looking into online shopping behaviour and by examining the major reasons for being motivated or being de-motivated from buying online, focusing on the trust element. Building on that analysis, the possible future status of e-consumer behaviour is presented via an examination of ubiquitous retailing, which denotes the next stage of that retail revolution.

Consumers' lack of trust is identified as one of the greatest barriers inhibiting business-to-consumer (B2C) e-commerce. This may be partly attributable to the lack of face-to-face interpersonal exchanges that support trust behavior in conventional commerce. It was proposed that initial trust may be built by simulating face-to-face interaction. To investigate this, an extensive laboratory-based experiment was conducted to assess the initial trust in consumers using four online vendors' websites with a variety of still and video images of sales personnel, both Western and Saudi Arabian. Initial trust was found to be enhanced for websites employing photographs and video clips compared to control websites lacking such images; also the effect of culture was stronger in the Saudi Arabian setting when using Saudi photos rather than Western photos.

This study examines the driving forces of customer loyalty in the broadband market in Hong Kong. The authors developed and empirically tested a model to examine the antecedents of customer loyalty towards Internet service providers (ISPs) in Hong Kong. Structural equation modeling (SEM) was used to evaluate the proposed model. A total of 737 valid returns were obtained through a questionnaire survey. The results show that customer satisfaction, switching cost, and price perception are antecedents that lead directly to customer loyalty, with customer satisfaction exerting the greatest influence. Although research found that service quality significantly influences customer satisfaction, which in turn leads to customer loyalty, a direct relationship between service quality and customer loyalty was not found. Results also reveal that corporate image is not related to customer loyalty. This empirical investigation suggests that investing huge resources in building corporate image can indeed be a risky strategy for ISPs.

Gathering customer data over the Internet is largely limited to collecting the responses to a set of easily answerable questions, such as Yes/No questions and Likert scale questions. These data are then analyzed to identify customer trends or other items of interest to management. The data can be useful, but key to their usage is the application of suitable mathematical tools. Traditionally little more than standard statistics has been used in the analysis of ordinal, or category, data. This can be inaccurate and in some cases, misleading. This chapter introduces measures of agreement and dissent to the field of e-business analysis and shows how ordinal data can be analyzed in more meaningful ways.

To facilitate online consumer decision making, a number of e-commerce businesses are augmenting their Web site features. The Web-based decision support for consumers is often provided by eliciting consumer preferences directly or indirectly to generate a set of product recommendations. The software that captures consumer preferences and provides recommendations is called a Web-based decision support system (WebDSS). It is important for WebDSS to support consumers' decision strategies. These decision strategies could be compensatory or non-compensatory in nature. Based on a synthesis of previous research, the authors argue that compensatory based WebDSS are perceived by consumers to be better than non-compensatory WebDSS in terms of decision quality, satisfaction, effort, and confidence. This chapter presents a study that examined the level of online support provided to the consumers' execution of compensatory and non-compensatory strategies. The results based on investigating 375 e-commerce websites indicate that moderate levels of support exists for consumers to implement non-compensatory choice strategies, and virtually no support exists for executing multi-attribute based compensatory choice strategies. The results of this study suggest that there is an opportunity in waiting for e-commerce businesses to implement compensatory WebDSS to improve the decision making capabilities of their consumers.

Many case examples in the mobile market have shown that the success of mobile services is difficult to predict. Different factors either boost or hinder the adoption of mobile services. The present chapter has covered earlier research on mobile service adoption and utilized a unique handset-based research platform in collecting data from 548 Finnish smartphone users in 2006. The main research goal is to understand the process of mobile service adoption by extracting new kinds of data straight from handsets. In addition to descriptive results, a path analysis model is developed that models the mobile service adoption process. The chapter finds that user intentions have a strong impact on consequent adoption of the service, expectedly. What is more, perceived hedonic benefits from the service are the strongest factor driving intentions to use the service. The perceived technical capability to use the service and the role of the surrounding social network do not drive the intentions of early-adopter users. Interestingly multimedia service adoption is driven by more capable (new) handsets, and mobile web browsing benefits significantly from block or flat-rate (instead of usage-based) pricing plans for transmitted data. The chapter develops several indices that measure time-varying characteristics of mobile services. Calculated indices for a set of mobile services suggest that the studied mobile services are currently experiencing different phases in their life cycle.

The increasing utilization of mobile commerce technologies in e-business raises the question of their use in supply chain integration and management. This chapter presents a multiple case study investigation of the adoption of mobile technology in the supply chain. A technology-organization-environment framework of the contextual influences on technological innovation adoption is used to inform an analysis of three companies' adoption and use of mobile data solutions for sales automation, freight tracking and service support. Analysis of the three case studies found that the relative advantage of the technological innovation and the information intensity of the company were the most important factors influencing adoption. Other factors that appeared to influence adoption included the compatibility of the technology with the company's business approach, the presence of top management support and the degree of organizational readiness. Environmental factors such as competition within the industry or business partner influence seemed less influential for these pioneers of mobile technology use in supply-side activities.

The recent growth of business-to-consumer (B2C) Internet auctions challenges researchers to develop empirically-sound explanations of critical factors that allow merchants to earn price premiums in these auctions. The absence of a comprehensive model of Internet auctions led the authors of this chapter to conduct an exploratory study to elucidate and rank critical factors that lead to price premiums in Internet auctions. The authors employ Classification and Regression Trees (CART), a decision-tree induction technique, to analyze data collected in a field study of eBay auctions. Analysis yields decision trees that visually depict noteworthy factors that may lead to price premiums and that indicate the relative importance of these factorsShipping cost, reputation, initial bid price, and auction ending time were found to be the factors most predictive of price premiums in B2C Internet auctions.

In recent years, there has been an increased interest in the types of online auction. Yet many auctions with fixed-end times are experiencing "sniping" or submission of bids in the final minute of an auction. Late bidding deprives rivals of the ability of seeing one's bid and undercutting it. Late bidding facilitates colludes or independent pricing well above that predicated by auction mechanism. This chapter aims to propose and implement a new type of online auction called Least and Unique Price (LUP). In the LUP auction, the winner will be the bidder who submits the least and unique price. Moreover, late bidding and specific closing time also be overcome by the LUP auction. In addition, this chapter presents the practical implementation of the proposed auction. In order to evaluate the proposed auction a comparative analysis of different auction types and the proposed one has been done.

Internet auctions have gained widespread appeal as an efficient and effective means of buying and selling goods and services. This study examines buyer behavior on eBay, one of the most well-known Internet auction Web sites. eBay's auction format is similar to that of a second-price, hard-close auction, which gives a rational participant an incentive to submit a bid that is equal to his or her maximum willingness to pay. But while traditional second-price, hard-close auctions assume that participants have reliable information about their own and other bidders' reservation prices, eBay participants usually do not. This raises the possibility that eBay participants may adapt their bidding strategies and not actually bid

organizations must pursue in order to facilitate a successful transformation from traditional brick-and-mortar organizations to click-and-brick business models. Diffusion theory is used to demonstrate how these success factors create business value within an organization. The research model is fully grounded in information technology business value and productivity literature (e.g., Kauffman & Kriebel (1988), Mahmood, Gemoets, Hall, & Lopez (2008) Mahmood & Mann (1993), and Zhu (2004)). The manuscript utilizes an existing sample set consisting of a population of more than 550 company executives who are successfully implementing e-commerce strategies. The research examines constructs found in the literature and focuses on two importance dimensions of creating business value through e-commerce strategies: IT alignment to organizational strategies (ITOrS) and the quality and effectiveness of existing online systems (OnSQE). Critical success factors for e-commerce business success were found to include ITOrS (IT alignment to organizational strategies), IOrSA (Quality and effectiveness of online systems, OnSE (Online systems efficiency), and OnSQE (Online systems quality and effectiveness. The research produces empirical evidence that strategic decision making concerning implementation of e-commerce technologies and alignment with top management strategic planning is critical to the success of creating business value for e-commerce enabled organizations. The manuscript concludes with limitations of the research and implications for future research studies.

Businesses invest in developing information systems resources to gain competitive advantages. Literature has demonstrated the requirement of strategic alignment in converting these competitive advantages into sustained superior business performance. The knowledge of information systems strategic orientation and its impact on business performance will enable these businesses to fine tune their strategic information systems applications portfolio in achieving required strategic alignment. This chapter describes a research study that focuses on the information systems strategic orientation of small businesses and investigates its relationship with their perceived business performance. The organizational impact of adoption of the initial stages of electronic business development is also examined. The data were collected from small businesses on nine strategy areas, through mail survey. The result reveals three multifaceted dimensions of information systems strategic orientation. These dimensions of strategic orientation have significantly influenced their business performance. This phenomenon is explained with a model named Linear Strategic Alignment Model. For the adopters of Web presence, all these three dimensions remain significant in explaining their business performance.

This research discusses Nigerian financial firms' perspectives on key determinants of e-business deployment. It explores possible differences that exist among financial firms using in-house e-business capabilities and those that outsource their e-business capabilities. This chapter contributes to the few pieces of literature on e-business experiences among firms operating in Africa, particularly Nigeria.

Today, Web services are of interest to both academia and industry. However, little has so far been accomplished in terms of design and development methods to assist those who are responsible for specifying and deploying applications based on Web services. For this purpose, the authors developed in this chapter a method based on Context and Policy for Web Services known as CP4WS. In this method, policies manage various aspects related to Web services such as participation in composition scenarios and adjustment in response to environmental changes, and context provides the necessary information that permits for instance to trigger the appropriate policies and to regulate the interactions between Web services with respect to the current state of the environment. CP4WS consists of several steps such as the identification of user needs and the behavioral specification of Web services. Each step has a specific graphical notation that facilitates the representation, description, and validation of the composition operations of Web services. A case study that illustrates and highlights the use and originality of CP4WS, respectively, is provided in this chapter.

Recent years have seen a dramatic increase in business processes and research in distributed computing environments. Applications today can be composed of very heterogeneous components: some involve having the user in the loop; some deal with streaming data; while some require high-performance resources for their execution. This chapter examines the performance of a series of process-based models for the development of e-Business using enterprise software applications. Merging management technology in workflow systems is a critical step to provide service-oriented architecture and on-demand business. The authors propose a value-oriented process technique as a strategic alignment to improve investment value. The framework focuses on the guidelines for traditional users to identify the structural conflicts in integrating web services. A comparative study of workflow models for intra-and inter-organizational process control is presented. This chapter identifies the current progress in the adaptability in the design of process models coupled with structural changes of workflow views. The study provides a resource list of successful implementations for practitioners in organizational management. The research highlights the motivation of market facilitation, expert sharing and collaboration that enable commercial applications to support complex heterogeneous, autonomous and distributed information systems.

Preface

Emerging web technologies and e-business models are changing the ways we do businesses in the global economy. The technology-driven changes are complex and pervasive and are taking place in all parts of our society, ranging from the corporate competition, globalization, education, life style, entertainment, customer relationship, to technological innovations. To stay ahead of the competition and to seize opportunities, firms need to assess the potentials and limitations of these web technologies and develop the business models that fit corporate strategies. This paper reviews emerging e-business concepts and technologies, discusses relevant issues and implications, and stimulates future research efforts.

INTRODUCTION

Since the dot.com bust in the early 2000, e-business field has experienced rapid business restructuring internally and externally. At the same time, web technologies have advanced at an unprecedented rate. The growth of e-business has been phenomenal in terms of sales volume and the number of corporate and individual adopters. For several years before 2006, growth rates exceeded 25% each year. Forrester forecasted that online sales will continue to grow to $204 billion in 2008 (BuzReport, 2008). It is expected that in 2009 online sales will reach $235 billion, by 2010 it will reach $267 billion, and by 2012 online sales are expected to reach $334 billion. Though still growing faster than that of offline sales, the growth rate is declining as the market becomes more saturated with online shoppers. Obstacles to online retail sales growth include trust, security, and product availability.

As hyper-competition and new technologies pose new opportunities and challenges, firms face increased pressures from stakeholders to create e-business values. They are trying to assess how innovations in the e-business will affect existing business processes and competition in order to develop new business models and applications that will take advantage of the new technologies. For example, advancements of wireless technologies have extended the reach of e-business to mobile business environments. As a growing number of customers utilize mobile devices to exchange information and to conduct business transactions, firms are competing to provide the most value-added, innovative, convenient mobile services for their customers.

Although a large volume of literature exists in the area of e-business, further research is critical as firms are challenged by newly emerging concepts and technologies. The purposes of this study are: (1) to survey literature on emerging e-business concepts and technologies for sustainable business practices and applications, and (2) to discuss implications for e-business practices and future research direction.

EMERGING CONCEPTS AND THEORIES IN E-BUSINESS

In this section, emerging theories and concepts in B2B, B2C, and m-commerce areas will be discussed.

Business-to-Consumer (B2C) E-Commerce

B2B technologies create different business value for firms depending on how firms utilize them (Lee et al., 2003). "Basic B2B e-commerce" is used primarily for the inter-firm exchange of electronic documents. In contrast, "collaborative B2B e-commerce" is used to create new inter-firm collaborations with business partners. A survey result from a grocery industry showed that the real source of performance improvement in the B2B e-commerce comes not from the basic B2B e-commerce, but from the collaborative B2B e-commerce. Current B2B e-marketplaces actively embrace inter-firm collaborations and supply chain integration among partners as well as electronic business transactions and data interchange.

B2B e-commerce was also investigated from a global perspective (Raisinghani et al, 2005; Unhelkar, 2005). As the economic activities are more globalized, existing international information systems have been embracing web-based e-business technologies. Global B2B e-commerce provides potential to reduce global business transaction costs, to lower entry barriers to a global market, and to enhance global business collaborations. Gibbs et al. (2003) observed that B2B e-commerce is likely to be driven by global forces, whereas B2C seems to be more of a local phenomenon. The difference is attributable to the fact that B2B e-commerce is driven by global economy and multinational companies that push the adoption of B2B e-commerce to their global suppliers, customers, and subsidiaries. On the other hand, the expansion of B2C to global markets encountered many obstacles such as language barriers, national law, accounting practices, and cultural differences.

As e-business business models and technologies develop, the supply chain dimension of e-business, called e-supply chain, has become one of the most discussed topics in the various industries (Kim et al, 2005; Nath & Angeles, 2005). E-supply chain emerged as a promising alternative to traditional supply chain and dramatically changed the way procurement is conducted. E-supply chain also opens up new opportunities for small and medium-sized enterprises (SMEs), considering the traditional proprietary EDI technology was not available for most SMEs. A study of organizational barriers to e-supply chain integration revealed that (1) internal barriers impeded e-integration more than either upstream supplier barriers or downstream customer barriers and (2) e-integration has a positive effect on the performance (Frohlich, 2002). These barriers to implementation and organizational issues need to be addressed to realize potential benefits.

E-supply chain enhances supply chain flexibility much needed in times of hypercompetitive business environments, sophisticated customer preferences, and widespread use of disruptive technologies. Based on a survey result of supply chain relationships in the IT industry, Gosain, et al. (2004) proposed two e-supply chain design principles: (1) modular design of interconnected processes and structured data connectivity, and (2) deep coordination-related knowledge. They also suggested that sharing a broad range of information with partners is counterproductive to supply chain flexibility, and that organizations should instead enhance information quality.

The influence of the past experience in EDI facilitates B2B adoption (Angeles et al, 2001). Implementing B2B requires certain technological knowledge, so that the previous orientation of firms towards EDI-related systems proves to be important in its development, thanks to the generation of affinity and the simplification of the learning process. Experience with systems such as EDI or Internet has a direct influence on the use of e-CRM and there is also a direct and positive transmission of knowledge from e-CRM to B2B (Ortega et al, 2008). The knowledge accumulated from using a technological innovation can be considered an important organizational knowledge asset, which allows firms to obtain a number of benefits as a result of deploying systems that are complementary.

A new generation of intermediaries relies on new information technologies such as the semantic web, rule-based triggers, and knowledge-based constraint maintenance systems (Orman, 2008). The author discusses in detail three types of intermediaries with respect to their use of information technologies: data-based intermediaries, trigger-based intermediaries, and constraint-based intermediaries. The interdependence and fragmentation of the intermediaries are likely to increase with the development of new web technologies such as Web services and semantic web. New intermediaries emerge on top of the existing intermediaries and create new services. Web services are likely to encourage proliferation of intermediary services, because they enable the intermediaries to share Web services using standardized information exchange formats such as web service description languages, web service directories, and ontologies, and XML (Waldfogel & Chen, 2006).

Business-to-Consumer (B2C) E-Commerce

B2C e-commerce created many successful e-commerce startups with unique business models such as eBay, Amazon.com, and expedia.com. B2C e-commerce is characterized by intense competition, low market entry barriers, and a low degree of customer loyalty which in part are the reason behind the demise of numerous B2C e-commerce startups in the late 1990s and early 2000s. Due to the rapid growth of the Internet population and online sales, most traditional retailers have been transformed into click and mortar organizations by establishing the B2C ecommerce web sites. For these B2C e-commerce organizations, understanding online consumer behavior is one of the most important tasks for their business success.

To understand online consumer behavior, a number of researchers have applied grounded theories and investigated web site characteristics, motivations, facilitators, inhibitors, trust, attitude, intention, and loyalty which underlie individual acceptance of B2C e-commerce applications (Kulviwat et al, 2006; Jih, 2007). Most of these empirical studies adapted Theory of Planned Behavior (TPB), Technology Acceptance Model (TAM), and SERVQUAL which have been widely used in IT adoption studies. These studies showed that consumer's acceptance of e-commerce is influenced by subjective norm, perceived usefulness, perceived site quality, risk, vendor quality, security, reliability, assurance, privacy, and user's web experience.

As B2C e-commerce technologies advance and consumers gain more online experience, e-services have drawn attention from researchers and practitioners (Gefen & Straub, 2003; Baida et al, 2007). E-services enhance consumers' online shopping experience, and include Internet radio, web-based decision support, personalization, e-payment, online inquiry, electronic document sharing, and e-product support services. One of the major drawbacks of e-services is that e-services frequently lack the social presence. Findings of a free simulation experiment showed the positive influence of social presence on trust and its ultimate contribution to online purchase intentions of e-services (Gefen & Straub, 2003). Another study indicated service convenience, web site service quality, and risk are significant factors affecting consumers' satisfaction level, which in turn affects intention to use e-services (Zhang & Prybutok, 2005).

Researchers in information systems and marketing have been focusing on investigating the effectiveness of Web-based decision support systems (WebDSS) in providing accurate and satisfying choices for customers (Gudigantala et al, 2008). Overwhelming evidence supports the notion that compensatory WebDSS are better than non-compensatory WebDSS in terms of decision quality, satisfaction, effort, and confidence. However, their investigation of 375 U.S. based company websites shows that though moderate levels of support exists for consumers to implement non-compensatory choice strategies, virtually no support exists for executing multi-attribute based compensatory choice strategies. The results help managers in providing more support for compensatory based decision strategies, and at the same time raise the question of the lack of popularity of compensatory WebDSS.

Developing web quality index and measuring web quality can help stakeholders understand and improve e-business web sites. A number of measurement instruments have been developed to measure the quality of web sites including WebQual, eTailQ, Sitequal, and NetQual. WebQual metrics have been developed to evaluate the quality of web sites (Barnes & Vidgen, 2002). WebQual metrics have been developed and validated through various e-business applications. Some of the dimensions include web site usability, information quality, and service interaction quality. WebQual is grounded in the perceptions of web site users, and analysis of data collected from users is used to guide web site design. eTailQ was developed by utilizing online and offline focus groups, along with a classification task and an online survey of a panel of consumers (Wolfinbarger & Gilly, 2003). The instrument consists of four factors (a) site design that includes navigation, search for information, product selection, order process, and personalization, (b) customer service including online assistance, response to customers' emails, ease of returning items, empathy, and reactivity, (c) reliability/respect for commitments refers to adequate description, presentation, and delivery of products or services ordered at the promised quality level, and (d) security/privacy reflected by security of payments and confidentiality of personal data. Sitequal consists of nine items reflecting 4 dimensions (a) ease of use and capacity to obtain information, (b) design and creativity of site with multimedia content and colors, (c) speed of the order process and reactivity to consumers' requests, and (d) security of financial and personal information (Yoo & Donthu, 2001).

NetQual includes 18 items distributed along 5 dimensions (a) quality and quantity of information available, (b) ease of site use, (c) design or aesthetic aspect of the site, (d) reliability or respect for commitment, (e) security/confidentiality of personal and financial data (Bressolles, 2006; Bressolles & Nantel, 2008). The dimensions retained refer to the functional characteristics of both the site and transaction. To compare the above-mentioned four instruments, the authors conducted a laboratory experiment involving two Canadian Websites in travel and online insurance. The results show NetQual best fits the data and offers the highest explanatory power.

Mobile Commerce (M-Commerce)

Mobile commerce (m-commerce) ushered in a new wave of e-business evolution fueled by the increasing use of mobile devices such as cell phones and handheld devices. Although m-commerce can be generally considered to be an extension of electronic commerce, it has a number of unique characteristics and business models, as it embraces many emerging technologies such as mobile or handheld computing devices and wireless technologies. During the past several years, m-commerce has significantly changed the way organizations do business, and entered into the center of the e-business research and practices. Most m-commerce adopters are individuals who play the dual roles of technology user and service consumer.

Coupled with the rapid advances in the wireless communications technologies, m-commerce has enormous potential to become a dominant form of market mechanism. However, with m-commerce still in its infancy, we still need to explore opportunities and challenges posed by m-commerce, and identify the appropriate business models and business strategies for the success of m-commerce. In addition, the wireless networking infrastructure, W3C standards, and the open and global wireless application protocol (WAP) specification need to be fully established to take advantage of m-commerce opportunities. The success of any mobile application requires a full understanding of application types, user requirements, technological constraints, and market potentials.

The technology behind m-commerce and the products and services available were examined (Frolick and Chen, 2004). They also examined the benefits and challenges of m-commerce, as well as the issues to be addressed when considering the implementation of m-commerce solutions. A number of studies addressed consumer perception and loyalty on m-commerce (Mahatanankoon et al., 2005; Mallat & Dahlberg, 2005; Jih, 2007). These studies reported that m-commerce consumer behaviors are similar to those of e-commerce, indicating that customer intention to use m-commerce is also affected by trust, habit, and customer satisfaction. Kim et al, (2007) adopted the theory of consumer choice and decision making from economics and marketing research, and developed the Value-based Adoption Model (VAM), and explained customers' m-commerce adoption from the value maximization perspective. The findings demonstrated that consumers' perception of the value of m-commerce is a principal determinant of adoption intention. Organizational issues, key attributes in developing m-commerce, and the driving and impeding forces of m-commerce were also investigated (Bai et al., 2005).

The unique characteristics of telecommunication markets along with the increasing trend for global e-business have led to a growing need for cross-national studies on m-commerce. Implications of m-commerce from a customer perspective across different countries were studied (Kim et al., 2004). Results from the online survey in the three countries yielded the following findings. First, adoption patterns for m-commerce were different across three countries. Second, customers' perceptions of the importance of the m-commerce and their willingness to pay for m-commerce services differed significantly across different countries. Finally, customers' preferred services in m-commerce also differed across the three countries.

A number of studies have examined the potential for mobile commerce to be applied to supply chain management (SCM) (Shankar & O'Driscoll, 2002; Kalakota et al., 2003; Lau et al., 2006). However, there are few empirical studies that focus on the adoption and implementation of mobile commerce in the supply chain activities of companies (Doolin & Al Haj Ali, 2008). In order to close the knowledge gap in the mobile supply chain, the authors present a multiple case study investigation of the adoption of mobile technology in the supply chain. A technology-organization-environment framework of the contextual influences on technological innovation adoption was used to analyze companies' adoption and use of mobile data solutions for sales automation, freight tracking and service support. Analysis of the three case studies found that the relative advantage of the technological innovation and the information intensity of the company were the most important factors influencing adoption. Other factors that appeared to influence adoption included the compatibility of the technology with the company's business approach, the presence of top management support, and the degree of organizational readiness.

Mobile services are an emerging m-commerce application which has evolved from mere communication oriented services (circuit-switched voice, text messaging, voice mailbox) to multimedia, content retrieval, browsing and other advanced services. Many case studies in the mobile services area have indicated that the success of mobile services is difficult to predict, because it is difficult to pinpoint the reasons behind successes and failures as many issues affect the adoption of a particular mobile service (Verkasalo, 2008). The reasons can be categorized into two main categories: technological/business strategic and behavioral (Pedersen, 2001).

Verkasalo (2008) utilized a newly developed handset-based mobile end-user research platform and investigated the process of mobile service adoption in an attempt to understand drivers and bottlenecks for service adoption. Based on data extracted straight from handsets of 548 Finnish panelists, the author develops a path analysis model to explain mobile service adoption contingent on a given set of explanatory variables. His study finds that user intentions have a strong impact on consequent adoption of the service. Perceived hedonic benefits from the service are the strongest factor that drives user's intentions to use the service. Multimedia services are strongly driven by newer, more capable handsets and mobile Internet browsing benefits significantly from block or flat-rate (instead of usage-based) pricing plans for transmitted data.

EMERGING TECHNOLOGIES

In this section, Web services, semantic web, and ubiquitous computing technologies will be discussed.

Web Services

According to the World Wide Web Consortium (2008), "Web services provide a standard means of interoperating between different software applications, running on a variety of platforms and/or frameworks. Web services can be combined in a loosely coupled way in order to accelerates application development and integration inside and outside the enterprise (Bose & Sugumaran, 2006). Programs providing simple services can interact with each other in order to deliver sophisticated added-value services. Extensible Markup Language (XML) plays a central role in the development of Web services by providing a data interchange format that is independent of programming languages and operating systems.

The Web Services are based on a core set of enabling technologies, including XML, SOAP, WSDL, and UDDI, which reflect the work of researchers and consultants from a variety of companies and industry organizations. Web services technologies are emerging as the platform that can universally standardize the communication of applications for automating both the provider and consumer ends of e-business transactions. In order to connect systems, business partners, and customers cost-effectively, Web services let programs invoke requests to other programs over the Internet via open protocols and standards businesses.

Due to the significant potential benefits, many popular Web sites such as Google and Amazon.com are boosting their traffic through Web services. Web services can drive down costs in application development by achieving automated code generation, reuse, and interoperability. Web services have shown promising results such as greater development productivity gains and easier and faster integration with trading partners. However, despite the rapid development in the Web services area, many issues including information security still remain to be resolved in the context of e-commerce (Misra et al., 2007; Yau et al., 2007).

The lack of design and development methods is an issue which received little attention from the research community (Maamar and Benslimane, 2008). To close a gap in this area, the authors develop CP4WS that stands for Context and Policy for Web Services. CP4WS is a context-based and policy-driven method for designing and developing composite Web services. Policies manage various aspects related to Web services like participation in composition and adjustment due to changes in the environment, while context provides the necessary information that triggers the appropriate policies and regulates the interactions between Web services according to the current state of the environment. CP4WS consists of five steps: user needs identification and specification, Web services orchestration, Web services contextualization, Web services behavior specification, and Web services deployment. Each step has a specific graphical notation that facilitates the representation, description, and validation of the composition operations of Web services.

Semantic Web

The Semantic Web has drawn attention from both industry and academia (Cannoy & Iyer, 2007; Joo et al., 2007). The semantic web is an evolving extension of the World Wide Web in which information and services on the web are rendered as means for computers and people to work in cooperation (Berners-Lee et al, 2001). The semantic web comprises the standards and tools of XML, XML Schema, Resource Description Framework (RDF), RDF Schema, and Web Ontology Language (OWL). Using semantic web technologies for e-business tasks such as product search or content integration requires ontologies for products and services. Ontologies can be used to describe products and services so that agents can advertise and discover them according to a semantic specification of functionality as well as other parameters such as cost, security, etc. (Trastour et al., 2003). The semantic web can make e-commerce interactions more expressive and flexible by standardization of ontologies, message content, and message protocols (Li & Ling, 2007). .

One important semantic web application area is Web services. Evidence shows that semantic web services are mandatory components of the semantic web, primarily because entities are more willing to expose functionality than data in business settings (Hepp, 2006). Semantic web services aim to describe and implement web services so as to make them more accessible to automated agents. Semantic web services can support a service description language that can be used to enable an intelligent agent to behave more like a human user in locating suitable Web services.

One of the major problems faced by the semantic web is heterogeneity which causes interoperability among domain ontologies. This is a significant problem particularly for ontologies of similar domains. The current mediation techniques focus mainly on syntactic matching. A hybrid ontology mediation algorithm for the semantic web combines both semantic and syntactic matching technique and proves to be a better solution to this problem (Muthaiyah & Kerschberg, 2008). The authors provide empirical evidence with hypothesis tests and also provide several new measures such as relevance, reliability and precision to validate our approach.

While the development of Web services and semantic web has been impressive, numerous research opportunities exist. For example, semantic web technologies can help firms to improve internal and external data integration, knowledge discovery, knowledge management, and service-oriented architectures. To realize these benefits, research and practitioner communities need to collaborate on the standard and tool development to encourage the exploration and exploitation of the semantic web technologies, and service-oriented application development to enhance inter-firm data sharing and collaboration. Because web servers typically contain proprietary information from multiple sources, security control over server access is essential. The best security measures for this purpose are yet to be determined.

Ubiquitous Computing and U-Commerce

Ubiquitous computing is a paradigm shift where technology becomes virtually invisible in our lives. The advancement of new technologies such as radio frequency identification (RFID) and sensor networks has initiated a trend towards ubiquitous computing, which is also called "anytime, anywhere" computing (Lyytinen et al., 2004). In a ubiquitous computing environment, computing devices, applications, networks, and data will be fully integrated and merged (Junglas & Watson, 2006). Due to the "anytime, anywhere" pervasive computing, organizational activities become more nomadic. The ubiquitous computing environment will make possible new forms of organizing, communicating, working and living. However, ubiquitous computing systems create new risks to security and privacy.

One important technology for ubiquitous computing is radio frequency identification (RFID) technology. Like other IT value measures, RFID business value includes lead time reduction, productivity improvement, cost reduction, increased revenue, customer satisfaction, competitive advantage, inventory reduction, and other metrics of performance (Michael and McCathie, 2005; Angeles, 2007; Veeramani et al., 2008). A recent survey shows that the cost of the tags and hardware, and the availability of these components are the main issues hampering the widespread adoption of the technology by suppliers (Vijayaraman and Osyk, 2006). Many organizations take a "wait and see" stance and hope to learn more from the early adopters, since the suggested benefits of RFID are still uncertain while RFID technology requires significant up-front investment (Reyes and Jaska, 2007).

Supply chain RFID is an emerging application that has attracted a lot of attention from researchers and practitioners in the U.S., Europe, and Asia (Soon and Gutiérrez, 2008). RFID allows automatic identification and data capture using radio waves, a tag, and a reader. The tag can store more product data than traditional barcodes (Jones et al, 2004). The tag contains product data in the form of Electronic Product Code (EPC), a global RFID-based item identification system developed by the Auto-ID Center. Product data the RFID tag stores include product ID, production location, production date, and shipping container ID. RFID technology enables supply chains to easily and inexpensively collect and share information, thus enhancing supply chain visibility. The enhanced supply chain visibility leads to reduced stock-out, lower labor costs, reduced transaction costs, and improved inventory management in their supply chains (Twist, 2005).

In addition to data storage and information sharing capability, RFID improves information quality significantly. Managers may not use information provided from supply chain partners if they do not have confidence in information quality, and furthermore will not share their own information with their partners. While RFID technology is known to provide more accurate, current, and reliable information to supply chain partners than the traditional barcode technology, which leads to a better collaboration among supply chain partners, challenges such as false read, data overload, real-time acquisition of data, data security, and privacy must be dealt with (Bose & Lam, 2008).

Ubiquitous computing has enabled a new paradigm of commerce which goes beyond any traditional commerce (Junglas & Watson, 2006). This type of commerce is called "ubiquitous commerce", or simply "u-commerce", and is expected to have a great impact on the businesses. Despite the promising future of ubiquitous computing and the tremendous benefits it can bring to customers, customers' privacy concerns appear to be the biggest obstacle and social issue (Asif & Mandviwalla, 2005). The advancement of technologies embedded and used in the u-commerce environment raises concerns of customers because their personal information can not only be constantly accessed and continuously tracked, but can also be easily disseminated and possibly used in ways unknown to them (Gunther & Spiekermann,

2005). From the technical point of view, research need to address the technical features of successful ubiquitous computing applications. From organizational and behavioral point of view, user interface design, perceptions, satisfaction, and privacy issues need to be addressed.

CONCLUSION

Business organizations are in the midst of the turbulent global economy, and face unprecedented challenges and opportunities. For some firms, the new technologies will enable them to increase competitive advantage if they successfully align their business strategies with the new technology-enabled business models. Other firms, which cannot see the opportunities, may suffer significantly from the missed opportunities and eroded competitive advantage.

Emerging e-business concepts and technologies for sustainable business practices and applications which are to be drawn from the recent studies include: (1) previous orientation of some firms towards EDI-related systems proves to be important in the B2B development; (2) m-commerce has enormous potential to become a dominant form of market mechanism; (3) perceived hedonic benefits from the mobile service are the strongest factor that drives user's intentions to use the service; (4) Web services have shown promising results such as greater development productivity gains and easier and faster integration with trading partners. However, despite the rapid development in the Web services area, many issues including information security still remain to be resolved in the context of e-commerce; (5) supply chain RFID is an emerging application that has attracted a lot of attention from researchers and practitioners. These observations should be taken into account in the design of sustainable business practices and applications

In Lee
Editor, International Journal of E-Business Research (IJEBR)

REFERENCES

Angeles, R. (2007). An empirical study of the anticipated consumer response to RFID product item tagging. *Industrial Management & Data System, 107*(4), 461-483.

Angeles, R., Corritore, C., Choton, S. and Nath, R. (2001). Success factors for domestic and international electronic data interchange (EDI) implementation for US firms.*Information & Management*, 21, 329-347.

Asif, Z., & Mandviwalla, M. (2005). Integrating the supply chain with RFID, a technical and business analysis. *Communications of the Association for Information Systems*, 15, 393-427.

Bai, L., Chou, D.C., Yen, D.C., & Lin, B. (2005). Mobile commerce: Its market analyses. *International Journal of Mobile Communications, 3*(1), 66 - 81.

Baida, Z., Gordijn, J., Akkermans, H., Sæle, H., & Morch, A. (2007). How e-services satisfy customer needs: A software-aided reasoning. In I. Lee, *E-Business Innovation and Process Management* (pp. 198-233). Hershey, PA: CyberTech Publishing.

Barnes, S.J., & Vidgen, R.T. (2002). An integrative approach to the assessment of e-commerce quality, *Journal of Electronic Commerce Research, 3*(3), 114-127.

Berners-Lee, T., Hendler, J., & Lassila, O. (2001). The Semantic Web. *Scientific American Magazine.* May 17.

Bose, I., & Lam, C.Y. (2008). Facing the challenges of RFID data management. *International Journal of Information Systems and Supply Chain Management, 1*(4), 1-19.

Bose, R., & Sugumaran, V. (2006). Challenges for deploying web services-based e-business systems in SMEs. *International Journal of E-Business Research, 2*(1), 1-18.

Bressolles, G. (2006). La qualité de service électronique : NetQual. Proposition d'une échelle de mesure appliquée aux sites marchands et effets modérateurs [Electronic service quality: NetQual – Proposition of a measurement scale to commercial Websites and moderating effects]. *Recherche et Applications en Marketing, 21*(3), 19-45.

Bressolles, G., & Nantel, J. (2008). The measurement of electronic service quality: Improvements and application. *International Journal of E-Business Research, 4*(3), 1-19.

BuzReport. (2008). Retrieved April 15, 2008, from http://www.bizreport.com/2008/02/forrester_online_shopping_growth_to_slow.html

Cannoy, S., & Iyer, L. (2007). Semantic web standards and ontologies in the medical sciences and healthcare. In A. Salam & J. Stevens (Eds.), *Semantic Web Technologies and E-Business: Toward the Integrated Virtual Organization and Business Process Automation* (pp. 405-420). Hershey, PA: IGI Publishing.

Doolin, B., & Al Haj Ali, E. (2008). Adoption of Mobile Technology in the Supply Chain: An Exploratory Cross-Case Analysis. *International Journal of E-Business Research, 4*(4), 1-15.

Frohlich, M.T. (2002). E-integration in the supply chain: barriers and performance. *Decision Sciences, 33*(4), 537-556.

Frolick, M.N., & Chen, L.D. (2004). Assessing m-commerce opportunities. *Information Systems Management, 21*(2), 53 - 61.

Gefen, D., & D Straub, D. (2003). Managing user trust in B2C e-services. *e-Service Journal, 2*(2), 7-24.

Gibbs, J., Kraemer, K.L., & Dedrick, J. (2003). Environment and policy factors shaping global e-commerce diffusion: a cross-country comparison. *The Information Society, 19*(1), 5-18.

Gosain, S., Malhotra, A., & El Sawy, O.A. (2004). Coordinating for flexibility in e-business supply chains. *Journal of Management Information Systems, 21*(3), 7 - 45.

Gudigantala, N., Song, J., & Jones, D.R. (2008). How well do e-commerce web sites support compensatory and non-compensatory decision strategies? An exploratory study. *International Journal of E-Business Research, 4*(4), 43-57.

Gunther, O., & Spiekermann, S. (2005). RFID and the perception of control: the consumer's view. *Communications of the ACM, 48*(9), 73-76.

Hepp, M. (2006). Semantic Web and semantic Web services: father and son or indivisible twins? *IEEE Internet Computing, 10*(2), 85-88.

Jih, Wen-Jang (2007). Effects of consumer-perceived convenience on shopping intention in mobile commerce: An empirical study. *International Journal of E-Business Research, 3*(4), 33-48.

Jones, P., Clarke-Hill, C., Shears, P., Comfort D., & and Hillier, D. (2004). Radio frequency identification in the UK: opportunities and challenges. *International Journal of Retail and Distribution Management, 32*(3), 164-171.

Joo, J., Lee, S., & Jeong, Y. (2007). Application of semantic web based on the domain-specific ontology for global km. In A. Salam & J. Stevens (Eds.), *Semantic Web Technologies and E-Business: Toward the Integrated Virtual Organization and Business Process Automation* (pp. 287-309), Hershey, PA: IGI Publishing.

Junglas, I. A., & Watson, R. T. (2006). The u-constructs: four information drives. *Communications of the Associations of Information Systems*, 17, 569-592.

Kalakota, R., Robinson, M., & Gundepudi, P. (2003). Mobile applications for adaptive supply chains: A landscape analysis. In E.-P. Lim & K. Siau (Eds.), *Advances in Mobile Commerce Technologies* (pp. 298-311). Hershey, PA: Idea Group Publishing.

Kim, K.C., Im, I. & Kang, M.S. (2005). The effects of IT on supply chain management in the automobile industry. In N. Shin, *Strategies for Generating E-Business Returns on Investment* (pp. 86-101). Hershey, PA: Idea Group Publishing.

Kim, H.W., Chan, H.C., & Gupta, S. (2007). Value-based Adoption of Mobile Internet: An empirical investigation. *Decision Support Systems*, *43*(1), 111-126.

Kim, J., Lee, I., Lee, Y., Choi, B., Hong, S.-J., Tam, K.Y., Naruse, K., & Maeda, Y. (2004). Exploring e-business implications of the mobile internet: a cross-national survey in Hong Kong, Japan and Korea. *International Journal of Mobile Communications*, *2*(1), 1-21.

Kulviwat, S., Thaku, R., & Guo, C. (2006). an exploratory study of consumer adoption of online shipping: mediating effect of online purchase intention. *International Journal of E-Business Research, 2*(2), 68-82.

Lau, H.C.W., Lee, C.K.M., Ho, G.T.S., Ip, W.H., Chan, F.T.S., & Ip, R.W.L. (2006). M-commerce to support the implementation of a responsive supply chain network. *Supply Chain Management, 11*(2), 169-178.

Lee, S.C., Pak, B.Y., & Lee, H.G. (2003). Business value of B2B electronic commerce: the critical role of inter-firm collaboration. *Electronic Commerce Research and Applications*, *2*(4), 350-361.

Li, C., & Ling, T. (2007). A Basis for the semantic web and e-business: efficient organization of ontology languages and ontologies. In A. Salam; J. Stevens, *Semantic Web Technologies and E-Business: Toward the Integrated Virtual Organization and Business Process Automation* (pp. 212-235), Hershey, PA: IGI Publishing.

Lyytinen, K., & Yoo, Y. (2002). Research commentary: the next wave of nomadic computing. *Information Systems Research, 13*(4), 377-388.

Maamar, Z, & Benslimane, D. (2008). A context-based and policy-driven method to design and develop composite Web services. *International Journal of E-Business Research, 4*(3), 40-63.

Mahatanankoon, P., Wen, H.J., & B Lim, B. (2005). Consumer-based m-commerce: exploring consumer perception of mobile applications. *Computer Standards & Interfaces*, *27*(4), 347-357.

Mallat, N., & Dahlberg, T. (2005). Consumer and merchant adoption of mobile payment solutions. In T. Saarinen, M. Tinnila & A. Tseng, *Managing Business in a Multi-Channel World: Success Factors for E-Business* (pp. 32-50). Hershey, PA: Idea Group Publishing.

Michael K., & McCathie, L. (2005). The pros and cons of RFID in supply chain management. In *Proceedings of the International Conference on Mobile Business*, (pp. 623-629).

Misra, S.C., Kumar, V., & Kumar, U. (2007). An approach for intentional modeling of Web services security risk assessment. In G. Radhamani & G.S.V.R.K. Rao (Eds.), *Web Services Security and E-Business* (pp. 363-379). Hershey, PA: IGI Publishing.

Muthaiyah, S., & Kerschberg, L. (2008). A hybrid ontology mediation approach for the semantic web. *International Journal of E-Business Research, 4*(4), 79-91.

Nath, R., & Angeles, R. (2005). Relationships between supply characteristics and buyer-supplier coupling in e-procurement: An empirical analysis. *International Journal of E-Business Research, 1*(2), 40-55.

Orman, L.V. (2008). Knowledge-Based Intermediaries. *International Journal of E-Business Research, 4*(2), 1-13.

Ortega, B.H., Marinez, J.J., De Hoyos, M.J.M. (2008). The role of information technology knowledge in B2B development. *International Journal of E-Business Research, 4*(1), 40-54.

Pedersen, PE. (2001). An adoption framework for mobile commerce. *Proceedings of the 1st. IFIP Conference of E-Commerce*, Zürich, Switzerland, October 3-5, 2001.

Raisinghani, M.S., Melemez, T., Zou, L., Paslowski, C., Kimvidze, I., Taha, S., & Simons, K. (2005). E-business models in B2B: Process based categorization and analysis of B2B models. *International Journal of E-Business Research, 1*(1), 16-36.

Reyes, P.M., & Jaska, P. (2007). Is RFID right for your organization or application? *Management Research News*, *30*(8), 570-580.

Shankar, V., & O'Driscoll, T. (2002). How wireless networks are reshaping the supply chain. *Supply Chain Management Review*, *6*(4), 44-51.

Soon, C.-B., & Gutiérrez, J.A. (2008). Effects of the RFID Mandate on Supply Chain Management. *The Journal of Theoretical and Applied Electronic Commerce Research.* 3(1), 81-91.

Trastour, D., Bartolini, C., & Preist, C. (2003). Semantic Web support for the business-to-business e-commerce pre-contractual lifecycle. Computer Networks: The International Journal of Computer and Telecommunications Networking, *42*(5), 661 - 673.

Twist, D.C., (2005). The impact of radio frequency identification on supply chain facilities. *Journal of Facilities Management*, *3*(3), 226-239.

Unhelkar, B. (2005). Global e-business alliances: the socio-cultural perspectives, influence, and mitigation. In Y.-C. Lan, *Global Information Society: Operating Information Systems in a Dynamic Global Business Environment* (pp. 94-112), Hershey, PA: Idea Group Publishing.

Veeramani, D., Tang, J., & Alfonso Gutierrez, A. (2008). A Framework for Assessing the Value of RFID Implementation by Tier-One Suppliers to Major Retailers. *The Journal of Theoretical and Applied Electronic Commerce Research, 3*(1), 55-70.

Verkasalo, H. (2008). Dynamics of Mobile Service Adoption. *International Journal of E-Business Research, 4*(3), 40-63.

Vijayaraman, B.S., & Osyk, B.A. (2006). An empirical study of RFID implementation in the warehousing industry. *The International Journal of Logistics Management, 17*(1), 6-20, 2006.

Waldfogel, J., & Chen, L. (2006). Does information undermine brand? Information intermediary use and preference for branded Web retailers. *Journal of Industrial Economics, 54*(4), 425-449.

Wolfinbarger, M., & Gilly, M. C. (2003). eTailQ: Dimensionalizing, measuring and predicting eTail quality. *Journal of Retailing, 79*(3), 183-198.

(W3C) Extensible Markup Language. (2008). Retrieved June 20, 2008, from http://www.w3.org/XML

World Wide Web Consortium (W3C) Web Services Activity Statement. (2008). Retrieved April 12, 2008, from http://www.w3.org/2002/ws/Activity

Yau, W.-C., Rao, G.S.V., & Krishna, R. (2007). Web services security in e-business: attacks and countermeasures. In G. Radhamani & G.S.V.R.K. Rao (Eds.), *Web Services Security and E-Business* (pp. 165-183). Hershey, PA: IGI Publishing.

Yoo, B., & Donthu, N. (2001). Developing a scale to measure the perceived quality of internet shopping sites (SITEQUAL). *Quarterly Journal of Electronic Commerce, 2*(1), 31-47.

Zhang, X. & Prybutok, V.R. (2005). A consumer perspective of e-service quality. *IEEE Transactions on Engineering Management, 52*(4), 461- 477.

Chapter 1
E-Consumer Behaviour:
Past, Present and Future Trajectories of an Evolving Retail Revolution

M. Bourlakis
Brunel University, UK

S. Papagiannidis
Newcastle University, UK

Helen Fox
Newcastle University, UK

ABSTRACT

Shopping online has emerged as one of the most popular Internet applications, providing a plethora of purchasing opportunities for consumers and sales challenges for retailers. The aim of this paper is to shed further light on the past and present status of the e-consumer phenomenon, by looking into online shopping behaviour and by examining the major reasons for being motivated or being de-motivated from buying online, focusing on the trust element. Building on that analysis, the possible future status of e-consumer behaviour is presented via an examination of ubiquitous retailing, which denotes the next stage of that retail revolution.

INTRODUCTION

Shopping online has emerged as one of the most popular Internet applications. Initially, the selling focus was on durable, non-food items such as books, but nowadays almost any product can be traded on-line. It is not surprising then that the major retailers have capitalised on that selling format. For example, in the U.K., Tesco was the first grocery retailer to launch this facility in 1996.

The paper examines the past, present and future status of e-consumer behaviour and aims to shed further light on that phenomenon. The next section analyses the past status of the e-consumer behaviour concept by discussing findings from the literature followed by a subsequent discussion on the present status of the phenomenon. The last sections provide our views on the future state of e-consumer behaviour, by presenting ubiquitous retailing as a possible evolution of retailing, before drawing relevant conclusions.

INTERNET CONSUMER SHOPPING: PAST STATUS

Rowley (1998) states that the Internet shopping experience has become a challenge for Internet retailers that need to ensure success at each stage (Figure 1).

For the first two stages, the website can be adapted for each different consumer, allowing them to have their own home page for specific needs and wants. Their shopping habits can be recorded, which helps in making the selection and ordering a quicker experience. For the Internet medium to be attractive to retailers, there are a number of issues associated with delivery, distribution, and relationships in the supply chain that will need to be pre-considered.

Focusing on the demographic element of the E-consumer, Mintel (2000, 2003) reports that the UK Internet user is predominately male, aged 20 – 30 and has an AB socio-economic background. Gender is believed to influence the extent and pattern of participation in web activities and Rodger and Harris (2003) found that women were less emotionally satisfied with Internet shopping than men. Specifically, females expressed lower emotional gratification with Internet shopping and are more sceptical of online shopping than males, perhaps because that emotional bond with the retailer is not evident in a virtual environment (Rodger and Harris, 2003). Men reported greater trust in Internet shopping, and perceived the Internet as a more convenient shopping outlet than did women. Overall, emotion and trust are the two critical determinants of consumer shopping attitudes and behaviour. In a similar vein, Girard *et al.* (2003) illustrated how shopping orientation and demographics have differential roles to play, based on the type of product purchased on the Internet. They also believe that gender, education, and household income revealed strong influences on preferences for shopping online whilst convenience is another key reason for purchasing online (Mayer, 2002;

Figure 1. The stages of the Internet shopping experience (Source: Rowley 1998)

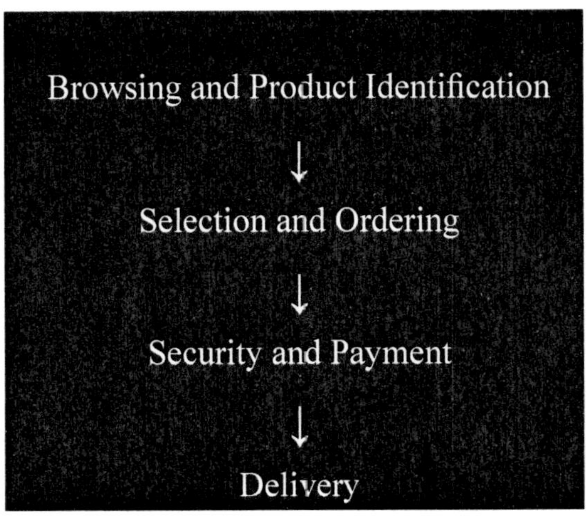

Phau and Poon, 2000; Poon, 2000; Seybold, 2002; Shim *et al.*, 2001; Teo, 2002; Thomas, 2003). A useful categorisation of the key influential factors for Internet shopping is developed by Shim *et al.* (2001) including transaction services (related to security, product guarantees, safety, privacy, and service), convenience (which relates to overall speed of Internet shopping and freedom from hassles), sensory experiences (which includes the social, personalising, and recreational experiences of shopping) and merchandise (product information, comparative shopping opportunities, and variety of merchandise choice). The attitude toward Internet shopping encompasses specific attributes related to transaction services (Shim *et al.*, 2001).

Similarly, there have been many studies indicating reasons for abstaining from the Internet. Anon [A] (2002) states that, in principle, web retailing is not a cheap option, compared to traditional retailing as the cost of packaging, distribution, stock management, and record keeping is likely to be higher for distance selling compared to traditional retailing. Anon [B] (2002) cites security problems, the lack of trust in Internet retailers, the lack of Internet knowledge and finally, the long delivery time for goods. Jones and Vijayasarathy (1998) suggest that individuals have unfavourable perceptions of Internet shopping security as they are wary of giving credit card details over the Internet and Rowley (1996, 1998) argues that businesses should provide alternative arrangements. For example, consumers should be able to make arrangements using phone, fax, or post, should use tokens on different sites, should apply encryption for their credit card numbers and should use electronic cash by withdrawing 'digital money' from an Internet bank and stored on the hard disk.

Focusing on the 'risk' element, Forsythe and Shi (2003) analyse the types of perceived risk and demographics on online shopping behaviour that contains six types of perceived risk. These are the financial risk, the product performance risk, the financial risk, the social risk, the psychological risk, the physical risk, and time / convenience risk (Forsythe and Shi, 2003). Product performance risk is defined as the loss incurred when a brand or product does not perform as expected. Financial risk is defined as a net loss of money to a customer. Psychological risk may refer to disappointment, frustration, and shame experience if one's personal information is disclosed. Time / convenience risk may refer to the loss of time and inconvenience incurred due to difficulty of navigation and / or submitting orders, finding appropriate web sites, or delays receiving products. Social risk involves fears of isolation from people and not receiving the pleasure whilst shopping. Physical risk involves not being able to use the senses, such as touch and smell.

In order to alleviate these risks, trust is required. This becomes a *prerequisite* for fostering and nurturing online shopping relationships. Aiming for that, Lee and Turban (2001) propose a model for developing consumer trust during Internet shopping and identify specific 'trust building' constructs. These include the following: trustworthiness of the Internet merchant (ability, integrity, benevolence), trustworthiness of the Internet shopping medium (technical competence, reliability, medium understanding) and other contextual factors (effectiveness of 3rd party certification and effectiveness of security infrastructure). They also elucidate on the key parameters that may affect consumer trust in Internet shopping, including credit card loss assurance policies, product warranty policies, policy on returned merchandise, availability of escrow service, ability to schedule human customer service sessions and, ability of user-friendly, reliable, efficient storefront interfaces with animated characteristics (Lee and Turban, 2001). Rowley (1998) also states that to attract the Internet shopper, the Internet retailer needs to focus on the speed of transaction, convenience, selection, and price. Online shopping offers retailers the opportunity to gain new customers notwithstanding the given opportunities to improve customer loyalty (Roberts *et al.*, 2003).

To conclude, there is a range of factors that affect e-consumer shopping either positively (motivating factors) or negatively (inhibiting factors) and subsequently, they can motivate or de-motivate shoppers. Table 1 synthesises the previous discussion providing a literature review thematic chart for the motivating and inhibiting consumer-related factors during online shopping plus possible areas for further improvement.

FURTHER COMMENTS ON THE PRESENT STATUS

The previous analysis examined academic material published till 2004 and hence it will be worth considering whether any progress has been made on the issues raised in Table 1. Aiming to address this point, the current section analyses relevant material published in professional associations' magazines, government reports and practitioners' journals from 2005 onwards.

Table 1. The E-Consumer: A literature review thematic chart

Central Theme	Author(s)	Year	Major Findings
Motivating factors for online shopping	Girard *et al.*	2003	- Shopping orientation and demographics have differential roles to play, based on the type of product purchased on the Internet - Convenience and recreational shoppers were the dominant orientations that influence consumers' preferences for shopping online, and this influence varied with the product types - Gender, education, and household income revealed strong influence on preferences for shopping online
	Lee, Turban	2001	- The parameters that may affect consumer trust in Internet shopping include: 1) credit card loss assurance policies, 2) product warranty policies, 3) policy on returned merchandise, 4) availability of escrow service, 5) ability to schedule customer service sessions, 6) ability of user-friendly storefront interfaces with animated characteristics
	Morganosky, Cude	2000	- The majority cited convenience and time saving as their primary motivation for buying groceries online - Shopping online appears to be the most advanced leading edge technology in grocery shopping - Online grocery shoppers seem to recognise and value differences between the online grocery shopping experience and the in-store shopping experience
	Phau, Poon	2000	- Online marketing should be perceived as having five components: 1) promotions, 2) one-to-one contact, 3) closing, 4) transaction, 5) fulfilment - Internet shopping is generally still unfamiliar to most Internet users - Expensive goods, such as automobiles, jewellery and stereo systems are not ready for web selling. The monetary risks involved in buying these products are too great. These products also require more than visual inspection - People shop online mainly because of convenience
	Poon	2000	- Industry sector dimension, actual experience of competitive advantage, and quality information support are key to Internet commerce benefit - For those retailers who have been online for about 2 years, there is evidence that Internet commerce has been providing benefit to its adopters
	Rodgers, Harris	2003	- Emotion, trust, and convenience are three critical determinants of people's shopping attitudes and behaviour - Females express lower emotional gratification with e-shopping because of their inclination toward left-hemisphere processing - Men reported greater trust in Internet shopping, and perceived the Internet as a more convenient shopping outlet than did women

continued on following page

Specifically, there are still security concerns relating to Internet transactions amongst consumers and in 2006 it was estimated that almost $2 billion would be lost (Finextra, 2006) in existing and potential e-commerce sales. That survey, executed by Gartner with 5000 online US adults, found that almost half of these (46%) had concerns about theft of information and Internet data breaches. Similarly, the research indicated that 70% of the surveyed online consumers do not trust unknown companies and delete suspicious emails from unrecognised sources (Finextra, 2006). Another survey echoed the above and illustrated further how online consumers were concerned about cyber crime and are still concerned about online privacy and security (NewsBlaze, 2007). Unavoidably, these concerns lead to changes in online behaviour with online consumers preferring to shop from recognised retailers which have already worked on how to increase online consumer confidence.

Therefore, it can be reasonably suggested that trust is still of pivotal importance (and a prerequisite) as it has always been for online transactions. However, the most successful online firms, including retail firms, have made the trust element the key differentiator for their online strategies compared to other online firms, which continued with the same online customer practices and subsequently, have achieved mediocre results (Riegelberger, 2006). For example, the successful online firms have developed detailed online trust-building strategies (see also Lee and Turban, 2001) with their customers including a range of 'embedding' strategies such as temporal embedding, social embedding and institutional embedding ones. Firstly, temporal embedding implies developing a strong relationship with a customer by signalling the firm's long-term plans, goals and investments or even by being attached to a well-established brand. It is not surprising then that many successful online retailers are

Table 1. continued

Inhibiting factors for online shopping and other concerns	Anon [B]	2002	- The main concern for not purchasing online is security - Another reason for not purchasing online is that people believe it is more enjoyable and easier to buy goods and services in a store - Books, followed by Music (CDs) are the most popular products purchased online
	Forsythe, Shi	2003	- Internet represents a fundamentally different environment for retailing from traditional retailing media - Internet browsers appear to be much more concerned than shoppers with the risk associated with Internet shopping. Risk perception was much greater among browsers than those who shop on the Internet
	Koyuncu, Bhattacharya	2004	- Individuals are inclined to increase their shopping from the Internet since online shopping provides better prices, and allows individuals to shop more quickly than other shopping alternatives - Individuals tend to shop less from the Internet because online shopping requires longer delivery time for items bought online, and payment involves risk
	Rowley	1998	- The major problems with Internet shopping are transaction problems / concerns, lack of credit card security, difficulty in locating products / services, poor product quality / insufficient information, technical problems in software / slow interface - For Internet retailing to be attractive to retailers, there are a number of issues associated with delivery, distribution, and relationships in the supply chain that will need to be satisfactorily resolved
	Teo	2002	- Main deterrents to purchasing online have been customers' preference to examine products, the need to possess a credit card, and security concerns

continued on following page

Table 1. continued

Areas for improvement for online shopping	Anon [A]	2002	- Retailers need to address some of the concerns about credit card security, since most online sales will involve credit card payments - If you are trying to sell a product that is obtainable readily from the local supermarket or high street, you are less likely to succeed than if you are selling something rare, different, or unusual - Web retailing is not a cheap option, compared to traditional retailing. The cost of packaging, distribution, stock management, and record keeping are likely to be higher for distance selling compared to traditional retailing
	Fram, Grady	1997	- To improve shopping, respondents of this survey most often asked for easier processes for locating products / services, along with improved visuals / graphics - Women (in sample) find little difference between the quality of goods found in local stores / catalogues and those sold on the net
	Jones, Wi-jayasarathy	1998	- Electronic shopping has the potential to radically alter the structure of consumer in-home shopping behaviour - Security is an issue; channels may not be secure, and credit card numbers might be intercepted and then misused
	Mayer	2002	- Consumers cite fear that credit card details will be stolen - Although certain things have improved, e.g. delivery, there is still substantial room for improvement - Key information is often lacking, e.g. failure to disclose whether a product is in stock - Returning goods and getting a refund was often problematic
	Raphel	1996	- Supermarkets are now using the Internet because 'that's where the customers are' - Customers are cautious about using the Internet for shopping. They are concerned about security
	Roberts *et al.*	2003	- Significant number of people will stick to conventional supermarkets for grocery shopping because of the nature of social contacts and the need to touch and feel the products - Main driving force for groceries online is convenience and time saving, but consumers are sensitive to product prices, delivery charges, and Internet access costs - Deepest concern with Internet shopping is security
	Rowley	1996	- The Internet is useful tool for marketing (depending upon the audience that you try to reach), but is far less effective in generating sales - Security is a major issue with four methods overcoming this; prior arrangements, tokens, encryption, and electronic cash - The future of successful e-tailing will be about rediscovering the fundamental principles of why people really buy
	Vijayasara-thy	2003	- There is a need to profile the online shopper using more sophisticated psychographic measures such as shopping orientations, rather than relying solely on demographics - Results suggest that shopper segments (community, home, and apathetic) derived from shopping orientations differ when online

the ones which also enjoy a traditional / physical retail presence and hence, are able to transfer their positive brand image from physical retailing to online retailing. This can be partially explained by the fact that an established, trusted brand does "carry considerable weighting in consumer decisions about online shopping" (Brown *et al.*, 2007, p. vi). Also, offering customer incentives can only strengthen that relationship with consumers. Under social embedding strategies, online firms capitalise on their positive recommendations and positive word-of-mouth communications from friends and family members and these firms could also use the Internet to disseminate reputation information, as is the case with the major online retailer, Amazon, with its affiliate programme (Riegelberger, 2006). Lastly, institutional embedding strategies include the development of trust programmes via the use of industry associations and regulatory programmes. However, Riegelberger (2006) argues that using trust as a differentiation point by online firms will soon erode as online consumer expectations are on the increase and will soon become a 'must do' tool. An example of such an increase in online expectations is apparent when online consumers engage in online window shopping. Based on a survey carried out by Scan-Alert examining the behaviour of more than 7 million online shoppers, it was found that the average online consumer requires over 19 hours to make their first purchase on a website following a first visit (Leonard, 2005). That increased length of time indicates that current Internet shopping involves consumers spending a considerable amount of time cross-checking and comparing websites or a 'catalog of catalogs' before making a final decision (ScanAlert, 2005). It also illustrates how online retailing is an ultra competitive business and far more competitive than the traditional / physical retailing (Leonard, 2005). Nevertheless, there is little evidence that price is the principal criterion of online shopping. However, providing to the consumer the ability to check prices online and get the best possible price is, and will always be, a very attractive selling point (Brown *et al.*, 2007).

Brown *et al.* (2007) examined the demographic element of the E-consumer and reported that differences between demographic groups in the UK are gradually decreasing. This was explained by the fact that specific consumer groups are catching up, including older people and lower socioeconomic groups. In addition, Brown *et al.* (2007) noted that, although online shopping is increasing its popularity to both men and women, the purchasing behaviour between them does vary.

Following the previous discussion, we can conclude that some progress has been made; however, there is still a long way to go with a plethora of issues being unresolved. One of these is the further enhancement of trust and reduction of risk during online shopping. Apart from the industry, the latter has also attracted the attention of research funding bodies as witnessed by, *inter alia*, the considerable funding given by the Data Information Fusion Defence Technology Centre to De Montfort University in the UK (Net4now, 2007). It is anticipated that further developments will emerge from these initiatives notwithstanding the rapidly evolving and dynamic field we are dealing with. The next part provides a synopsis of these possible future developments and scenarios by making the appropriate interconnections with the previous discussion.

A FUTURISTIC SCENARIO: WHEN IS MY FRIDGE FINALLY GOING TO ORDER MILK?

Jones and Wijayasarathy (1998) suggest that electronic shopping has the potential to radically alter the structure of shopping behaviour. Also, Forsythe and Shi (2003) believe that the Internet represents a fundamentally different environment for retailing from traditional retailing media. However, any radical transformations would have required new approaches to consumer behaviour and attitude and one could argue that improvements and electronic evolutions of existing shopping mechanisms are not enough. For example, Fletcher (1999) believes that grocery home shopping will never be more than a niche market (only 4% of all food shopping is done from home). This argument finds support from Roberts *et al.* (2003), who suggest that a significant number of people will stick to conventional supermarkets for grocery shopping, because of the nature of social contacts and the need to touch and feel the products. Although it is difficult, especially when taking into consideration the fluid nature of the networked environment, to predict what may happen in the future, one could draw suggestions from one aspect of the evolution of computing, that of ubiquitous computing. Ubiquitous computing encourages the seamless integration of technology in the environment, allowing users to interact with it naturally.

Ubiquitous retailing could be an application of such an approach to interacting with technology. The pervasive nature of the interaction allows users to radically alter the mechanisms of ordering goods. As everything could potentially be transformed into a point of sale, the consumer would be constantly surrounded by spending opportunities that are accessible without having to visit a web site, login, add the products to the shopping cart and then checkout. Established relationships, coupled with semi-automated ordering mechanisms, could significantly alter the shopping experience. The convenience factor could be easily further strengthened, if one allows the environment itself to assume some control of the shopping. If a light bulb is burnt out, then the chandelier could order one by itself. To some extent, we are already looking at this phenomenon: mobile commerce and location-based services.

Mobile phones and other mobile devices like PDAs have gradually become powerful enough to provide an additional online conduit, which has only recently started to look attractive for wider commercial applications. Mobile marketing, micro payments and location based services are among the first applications to reach such devices, allowing the consumer to interact with near-by points of sales. For example, a consumer can buy refreshments from a vending machine and pay using his mobile phone or get the latest offers from nearby shops.

In scenarios like the above, the customer is still largely in charge of the transaction and purchasing decision. In the future, this may change, as pervasive computing gradually finds its way in the environment and a wider-range of purchasing opportunities become a reality. At that point, rule-based purchasing may become an attractive proposition for consumers, who could program the points-of-sale to automate purchasing based on certain conditions (e.g. the chandelier would order light bulbs only when a third of them were burnt out). Grocery shopping is ideal for this kind of purchasing. Most items can be restocked with minimum associate risk: not much is lost if you end up ordering a bit more milk and bread than you needed.

When it comes to trust, it is difficult to hypothesise whether ubiquitous retailing will increase or decrease customers' trust when shopping online, as this will depend on how each consumer uses the technology. The time and convenience elements of buying online may be further enhanced, as customers will have a plethora of opportunities within their environment to complete purchases, without being confined within the narrow boundaries of desktop computers. In addition, although browser security may not be an issue any more (as there will be no browsers as we know them today), security and privacy implications may be even more complicated. If securing one channel is as difficult as it has been, one can only start to imagine the implications of securing and monitoring so many points of sale!

Such automated purchasing, based on rules, will have a number of significant implications for both the consumers and the retailers. Goods may be classified as commodity items whose purchasing could be delegated to the technology and items that the consumer feels require personal attention when purchasing. For the first type of items impulse buying may suddenly become a thing of the past, as the consumer does not need to worry about having enough milk in the fridge again. Which products end-up in these two categories will depend on the consumer and his special needs and requirements. This will probably result in consumers being positioned between the two emerging extremes: those who would not mind automating as many of their purchases as possible and those who would prefer the 'traditional shopping' and engage in every step. There have been signs of such patterns already. For example, Seybold (2002) suggests that during shopping, consumers are disappointed to lose the opportunity to touch and feel the products and to make 'impulse buys'. These customers would probably not allow all their shopping to be automated, even if it were very convenient.

From the retailers' point of view, ubiquitous retailing may spark a chain reaction of changes, as they will not get to engage with the customer in the same way. If the retailer's role becomes that of a supplier automatically filling the consumer's shopping basket and delivering the goods, when is the retailer going to build a relationship with the customer? The answer may be that pervasive retailing and especially automated shopping will end up being the best profiling method ever! Data mining techniques could potentially generate very detailed customer profiles. In order to take advantage of these, retailers would need to rethink their customer relationship management strategies and how they market their products to the consumer. Whether convenience justifies such extreme profiling, at least with today's standards, and whether balance between privacy and ease of purchasing can be achieved, is something still to be seen. Considering that via the use of these techniques the retailer is gaining consumers' trust we recommend that retailers should formulate customer relationship management programs building on this emerging relationship with consumers. Based on the arguments posed in this paper, we also propose a conceptual framework that depicts the retail revolution in online retailing, where trust enjoys a different status depending on the stage of that revolution (see Figure 2).

Apart from the above, traditional ordering processes will also be significantly affected. Issues like the ease of navigation of a web site will not be featuring in the lists of consumer concerns, simply because there will be no need to visit a web site to order. Ubiquitous retailing, by definition, will aim to seamlessly integrate the point of sale with the environment; ease of use will have to be an intrinsic characteristic of the new systems. The specialisation of point of sales to perform well-defined purchasing would also allow for speedier transactions, enhancing the convenience factor. In fact, transacting models themselves may be significantly affected. Instead of performing one-off transactions for many items, goods may be purchased one at a time or placed on temporary shopping carts, either on the consumer's or the retailer's side. A balance will then be required between the processing-periods of such carts (Figure 3), which could be time-based (e.g. once a day), cost-based (e.g. when the items' total cost reaches a predefined amount) or urgency-based (e.g. I need a light bulb now!). An example of such an approach can be seen

Figure 2. Trust and online retail revolution

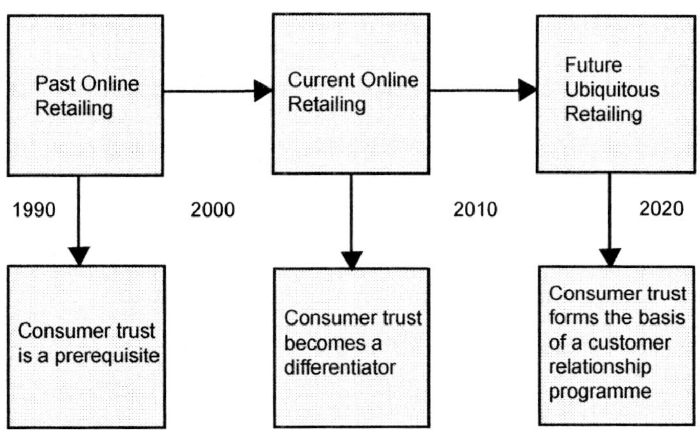

Figure 3. Ubiquitous retailing transaction modes

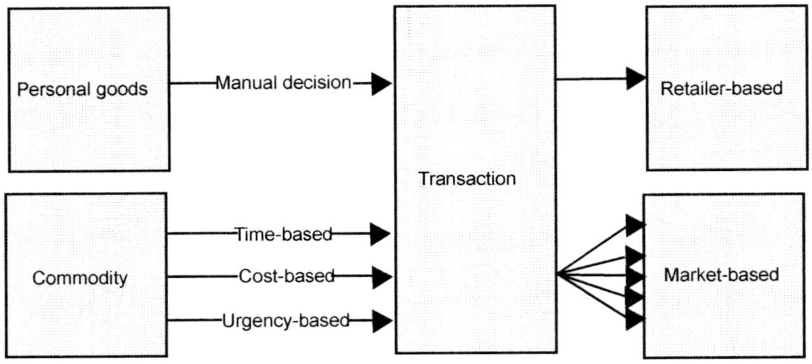

in the 'Intelligent shelves' case (Metro, 2006), which guarantees that customers no longer face empty shelves. The products placed in the system are equipped with Smart Chips that contain information relevant to the product itself or its logistic processes. A RFID reader integrated into the shelf automatically recognizes when an item is removed by reading the product information. Should stocks diminish, the system can take the necessary action.

Shopping carts may be 'retailer-based' or 'market-based' (Figure 3). In the first scenario the customer and the automated agents place the items in one cart that is to be undertaken by one retailer. In the second, and perhaps more interesting scenario, a number of pre-defined retailers get to fulfil the order, based on their offered price. As the cart will be automatically processed by the customer's purchasing agent, the best price will be selected among the prices offered by the different retailers, which could spark 'price wars' among them! In such a networked environment, in which transactions will be performed by automated agents, fine tuning these agents would have a significant impact on retailers' adopted strategies.

Finally, pervasive and ubiquitous shopping could help convergence in online shopping. Traditionally customers usually order products that are of low to medium value, e.g. books or DVDs, but not very expensive items e.g. a house. They would not buy few low value items (e.g. a few pieces of fruit or a box of washing powder) from the online supermarket either, as it is easier to buy these from a nearby store and any potential saving will be lost in the now relatively high delivery cost, compared to the product's cost. Automated purchasing could result in economies of scale, extending the cost boundaries within which online retailing currently operates, and reduce the impact that the distance from the nearest store may have on the decision to buy online.

CONCLUSION

The current paper has suggested a conceptual framework illustrating an online retail revolution in connection with consumer trust and e-consumer behaviour. Whether these changes justify a claim that there has been a radical transformation in retailing or that these changes were just the natural evolution of retailing, due to technological advances, is something that can be debated. Either way, online retailing is here to stay; although difficult to predict with what form and shape. The last section presented a potentially revolutionary scenario, that of ubiquitous retailing, signs of which we are already manifesting themselves through mobile commerce.

The authors consider this revolution to be very useful to business managers and other stakeholders such as consumer bodies and associations. For example, retail managers should pay further attention to the trust element and should still differentiate their strategies by developing 'trust-building' programs with consumers. By doing so, they will be able to move to the next stage, which is the formulation of customer relationship management programmes, which were found to be very appropriate for future ubiquitous retailing. Consumer bodies and associations should also take into consideration the proposed arguments, especially the one related to the future state of the revolution. We believe that they will find these points extremely beneficial in order to protect and guarantee the welfare and well-being of their members and to minimise their possible exploitation.

Last but not least, further work is required to confirm / disconfirm our findings, including further empirical research work with retail managers and consumers. The recommended revolutionary process in connection with the trust element requires further attention whilst the possible application of this revolution to other online firms, not just retailers, could form another investigation.

REFERENCES

Anon [A] (2002). Converting Web users to Web shoppers. *Strategic Direction* 18(4) 10–12.

Anon [B] (2002). Internet users weigh in worldwide: a new report finds that even though more Internet users plan to buy online this year, security is the highest concern for Web shoppers worldwide. *Direct Marketing* 65(6), 28–36.

Brown, D., Oleksik, G., & Bisdee, D. (2007). Consumer attitudes review, Internet shopping – Annexe E. *Prepared for the Office of Fair Trading, Crown Copyright.*

Finextra (2006). Security fears scare off US customers from online banking, shopping. Retrieved July 17, 2007, from: http://www.finextra.com/fullstory.asp?id=16204

Fletcher, W. (1999). Buying the groceries via home shopping is not a fresh concept. *Marketing* (May), 16–17.

Forsythe, S.M., & Shi, B. (2003). Consumer patronage and risk perceptions in Internet shopping. *Journal of Business Research* 56, 867–875.

Fram, E.H., & Grady, D.B. (1997). Internet shoppers: is there a surfer gender gap?. *Direct Marketing* 59 (9), 46 – 51.

Girard, T. Korgaonkar, P. , & Silverblatt, R. (2003). Relationship of type of product, shopping orientations and demographics with preference for shopping on the Internet. *Journal of Business and Psychology* 18(1), 101–120.

Jones, J.M., & Vijayasaratgy L.R. (1998). Internet consumer catalog shopping: findings from an exploratory study and directions for future research. *Internet Research: Electronic Networking Applications and Policy* 8(4) 322 – 330.

Koyuncu, C., & Bhattacharya, (2004). The impacts of quickness, price, payment risk, and delivery issues on on-line shopping. *Journal of Socio-Economics 33(2), 241-251.*

Lee, M.K.O., & Turban, E. (2001). A trust model for consumer Internet shopping. *International Journal of Electronic Commerce* 6(1), 75–91.

Leonard, K. (2005) .It's not shopping cart abandonment, it's comparison shopping. Retrieved July 17, 2007, from Internet Retailer Web site: http://www.internetretailer.com.

Mayer, R.N. (2002). Shopping from a list: international studies of consumer online experiences. *Journal of Consumer Affairs* 36(1), 115–127.

Metro Stores (2006). *Innovative technologies at the METRO Group.* Retrieved July 28, 2007, from: http://www.future-store.org/servlet/PB/show/1008042/off-Publikationen-Innovativ-eTechnologien_engl.pdf.

Mintel (2000). *UK vs. US Online Shopping.* Mintel, London.

Mintel (2003). *Home Shopping, UK.* Mintel, London

Morganosky, M.A., & Cude, B.J. (2000). Consumer response to online grocery shopping *International Journal of Retail and Distribution Management* 28(1), 17–26.

Net4now (2007). *De Montfort to make technology trustworthy.* Retrieved July 17, 2007, from: http://www.net4now.com/isp_news/news_article.asp?News_ID=3883.

News Blaze (2007). *Online privacy and security: The fear factor.* Retrieved July 17, 2007, from: http://newsblaze.com/story/20060314050502000001.mwir/newsblaze/LEGALLAW/Legal

Phau, I., & Poon, S.M. (2000). Factors influencing the types of products and services purchased over the Internet. *Internet Research: Electronic Networking Applications and Policy* 10(2), 102–113.

Poon, S. (2000). Business environment and Internet commerce benefit – a small business perspective. *European Journal of Information Systems* 9, 72 – 81.

Raphel, M. (1996). How supermarkets capture customers with their net. *Direct Marketing* 59(1), 14–17.

Riegelsberger, J. (2006). *Big brands must actively build trust on line.* Retrieved July 17, 2007, from: http://www.e-consultancy.com/about/press-centre/e-consultancy-press-releases.asp?id=1492.

Roberts, M. Xu, M.X., & Mettos, N. (2003). Internet Shopping: the supermarket and consumer perceptions. *Journal of Electronic Commerce in Organizations* 1 (2) 32–43.

Rodgers, S. and Harris, M.A. (2003). Gender and e-commerce an exploratory study. Journal of Advertising Research 43(3), 322–330.

Rowley, J. (1996). Retailing and shopping on the Internet. *International Journal of Retail and Distribution Management* 24(3), 26–38.

Rowley, J. (1998). Internet food retailing: the UK in context. *British Food Journal* 100 (2), 85–95.

ScanAlert (2005). *New report shows online consumers window shopping for days before buying.* Retrieved July 17, 2007 from: http://www.marketwire.com/2.0/release.do?id=661502&sourceType=1.

Seybold, P. (2002). *Shopping online at Tesco.* Retrieved October 10, 2007 from: Business Line, Financial Daily from the Hindu Group of Publications Web Site:

http://www.blonnet.com/catalyst/2002/03/07/stories/2002030700120200.htm.

Shim, S. Eastlick, M.A. Lotz, S.L., & Warrington (2001). An online pre-purchase intentions model: the role of intentions to search. *Journal of Retailing* 77, 397–416.

Teo, T.S.H. (2002). Attitudes towards online shopping and the Internet. *Behaviour and Information Technology* 21(4), 209–271.

Thomas, D. (2003). E-shopping continues to boom. *Computer Weekly* 27(February), 3.

Vijayasarathy, L.R. (2003). Psychographic Profiling of Online Shopper. *Information Science Publishing* 1(3), 48-74.

This work was previously published in the International Journal of E-Business Research, Vol. 4, Issue 3, edited by I. Lee, pp. 64-76, copyright 2008 by IGI Publishing (an imprint of IGI Global).

Chapter 2
Putting the Human Back into e-Business:
Building Consumer Initial Trust through the Use of Media-Rich Social Cues on e-Commerce Websites

Khalid Aldiri
University of Bradford, UK

Dave Hobbs
University of Bradford, UK

Rami Qahwaji
University of Bradford, UK

ABSTRACT

Consumers' lack of trust is identified as one of the greatest barriers inhibiting business-to-consumer (B2C) e-commerce. This may be partly attributable to the lack of face-to-face interpersonal exchanges that support trust behavior in conventional commerce. It was proposed that initial trust may be built by simulating face-to-face interaction. To investigate this, an extensive laboratory-based experiment was conducted to assess the initial trust in consumers using four online vendors' websites with a variety of still and video images of sales personnel, both Western and Saudi Arabian. Initial trust was found to be enhanced for websites employing photographs and video clips compared to control websites lacking such images; also the effect of culture was stronger in the Saudi Arabian setting when using Saudi photos rather than Western photos.

INTRODUCTION

The rapid advance of the Internet and global information technology has changed the way many people view shopping and undertake daily transactions. The Internet allows the purchasing of goods and services at almost any time and from almost any location. Consumers also have more choices and they are able to research the right product and make quick comparisons, taking other people's experiences into account through the use of elec-

DOI: 10.4018/978-1-60566-910-6.ch002

tronic communities. Despite these advantages that together have attracted immense interest in the Internet, numerous studies have demonstrated, with empirical evidence, that the penetration rate of Internet shopping remains low (Aldiri, Hobbs, Qahwaji, 2007). This penetration is commonly explained by the lack of trust in this new shopping mode. Lack of trust represents a significant factor inhibiting online purchases and is one of the critical roadblocks to success in e-commerce (Aldiri, Hobbs, Qahwaji, 2007).

Consumer trust may be even more important in electronic transactions than it is in traditional transactions. Characteristically, Internet transactions are blind, borderless, can occur twenty-four hours a day and seven days a week, and are non-instantaneous.

Also, in e-commerce there is a distinct lack of rules and traditions regulating this field. Furthermore, online transactions lack the assurance provided in traditional settings through formal proceedings and receipts.

Indeed, in contrast to face-to-face commerce and to other applications of trust, which encompass a wide range of emotions involving various types of social interaction with humans, there are typically no interpersonal interactions in e-commerce, neither direct nor implied. Such interactions, or even cues relating to them, are notably missing from e-commerce websites. Online shopping may be viewed as lacking human warmth and sociability.

Consequently, trust in an Internet business is focused much more on transaction processes, in contrast to that of traditional transactions involving brick-and-mortar stores where trust tends to be focused on face-to-face personal relationships. Since trust is likely to play an essential role in online transactions, it is important to identify the antecedents of a consumer's trust in the context of an Internet transaction so that consumers can feel relaxed and confident.

Research indicates that human beings like to reduce their social uncertainty. In other words, they seek ways to recognize, predict, and occasionally attempt to control the behaviour of other people (Kim & Tadisina, 2005). When social uncertainty cannot be reduced through regulations and traditions, people resort to trust and, to a lesser degree, to familiarity as a major social complexity reduction method (Zeng, Zeng, & Guo, 2005).

Trust has thus become a central concern of human computer interaction (HCI) research.

Trust is crucial wherever risk, uncertainty or interdependence are present (Gillespie, 2000). These conditions increase in many settings, and without doubt exist in the relationship between e-commerce vendors and customers (Gefen & Straub, 2004).

Research Objectives and Organization of the Paper

Online trust is a growing area of research and in itself is a substantial field. However, one aspect deserves specific mention, namely, how website design conveys trustworthiness to users (Egger, 2001; Feng, Lazar, & Preece, 2004; Kim, Ferrin, & Rao, 2003; Nielsen, 2004). This approach offers imperative practical guidance for web designers, who want to improve the interface usability of a specific e-commerce vendor.

Specific design elements have been found to have an effect on online trustworthiness including ease of navigation (Cheskin & Sapient, 1999), good use of visual design elements (Kim & Tadisina, 2005), an overall professional look of the website (Belanger, Hiller, & Smith, 2002; Kim & Tadisina, 2005), ease of carrying out transactions (Nielsen, 2004), and appropriate and useful content to the target audience (Shelat & Egger, 2002).

This research is founded on the belief that investigating and establishing design guidelines for increasing website trustworthiness is essential, and thus it aims to investigate how to increase the perceived trustworthiness of a vendor website by an individual user.

With regard to this aspect, attractive people in media cues (photo, video clip, and audio) have been utilized extensively by the marketing industry to stimulate affective responses from consumers. In contrast, in the field of Business-to-Consumer (B2C) e-commerce they have been used rarely, even in recent years; although preliminary usage of media cues can be seen represented by a smiling photo of a man or women in certain vendor websites. Research into the effects of social cues (or interpersonal cues) represented in media cues on B2C e-commerce trust is scarce and the findings contradictory. Hence, this study investigates this element. In addition, this study goes further to test the effects of cultural and non cultural social cues, and multiple forms of media cues on trust, based on the perceptions of different groups of users in order to investigate the effects of media cues; all with respect to vendor website trustworthiness.

After introducing the general problem of online trust, above, this section now introduces the set of specific problems to be addressed by this research in the form of research objectives. They are structured into empirical and methodological objectives.

Research Objective 1 (Empirical)

To investigate the effect of social presence cues (interpersonal cues) on user trust. Another aspect under investigation in the current online trust debate is the role of social presence cues in the formation of user online trust. A limited amount of research has paid attention to these practices in the B2C e-commerce trust literature, while in the marketing industry it has been widely implemented. Addressing new practices in B2C e-commerce interface design, this study aims to investigate the effect of rich representations of interpersonal cues (given in facial photos, video clips representations) embedded in web pages on the element of trust in B2C e commerce vendors.

Research Objective 2 (Empirical)

To investigate the effects of cultural interpersonal cues represented in multi media cues (facial photos, video clips representations) on trust in B2C e commerce vendors. Web visitors and shoppers increasingly represent a multicultural community. The advent of the web as a new global communication medium raises new issues and challenges about designing standardized global websites or localized websites for different countries. A major obstacle for companies extending their e-business globally is to understand the needs of global customers and overcome the cultural barriers on the website. However, to date little research has systematically examined web preferences of users related to a variety of online design characteristics. Even in the academic literature few studies have attempted to study website cultural preference issues, and they have not produced any generalizable empirical evidence in support of adapting websites to different cultures.

Research Objective 3 (Empirical)

This research was concerned with investigating trust as a dependent variable and other variables as independent variables, as well as how the depth of a user's exploration of a site (exploration depth) influences the effect of a media cue on user trust.

Research Objective 4 (Methodological)

On a methodological level, the study aimed to overcome part of the limitations of conventionally used trust questionnaires free from financial risk by introducing trust questionnaires that were elicited under conditions of financial risk.

Finally, this paper is organized as follows: Section 2.0 examines previous work related to e-commerce and trust and reviews the research hypotheses. Section 3.0 outlines the methodology of the study, while the data analysis is presented in

Section 4.0. The results are discussed in Section 5.0 and in Section 6.0 conclusions are drawn.

THEORETICAL BACKGROUND AND HYPOTHESES

Trust

Trust has widely conflicting conceptual definitions; the literature on trust in general, and on trust in e-commerce in particular, is in a state of confusion (McKnight & Chervany, 2001). The confusion in trust terminology is expressed in terms of two kinds of problems. First, similar concepts are given different names and second, the same terms are used for different concepts (Krauter, Kaluscha, & Fladnitzer, 2006). There is no universally accepted scholarly definition of trust (Corritore, Kracher, & Wiedenbeck, 2003).

Summarizing the different definitions of trust across various research disciplines, it can be concluded that all trust definitions address one or more of the following perspectives (Krauter et al., 2006): 1) context characteristics, 2) trustor properties, and 3) characteristics of the trusted object. Many definitions also address the interaction or relationships between two or all three of these perspectives. The key concepts of most of the trust definitions are risk (Mayer, Davis, & Schoorman, 1995), vulnerability (Mayer et al., 1995), expectation (Baier, 1986), and confidence (Lewicki & Bunker, 1995). These factors are gathered by the following trust definition, which represent the most common definition of online trust (Mayer et al., 1995):

Trust is the willingness to depend upon another party and be vulnerable to the actions of this other party based on the assumption that the trusted party will deliver without taking advantage of the situation. p.712

This research adopts this contemporary definition since it is widely recognized and the most frequently cited (Rousseau, Sitkin, Burt, & Camerer, 1998); also it can be seen as the lowest common denominator of HCI trust research (Corritore, Kracher, & Wiedenbeck, 2003), it embeds two critical components of trust: confident expectations and a willingness to be vulnerable.

As it is very abstract, this definition can be seen as the most common in the human computer interaction (HCI) trust research (Corritore et al., 2003). It defines trust as an internal state or attitude which entails cognitive and affective aspects (Corritore et al., 2003).

However, trust as an internal state, is different from trusting action (e.g. buying online, entering credit card details, relying on advice) or cooperation (e.g. in a team setting), which are observable behaviors (Corritore et al., 2003). Also trust is not the same as trustworthiness; trust is the act of a trustor while perceived trustworthiness is a characteristic of someone or something that is the object of trust (Corritore et al., 2003; Serva, Benamati, & Fuller, 2005).

In the e-commerce context, trust has been defined rather narrowly (e.g. Bhattacherjee, 2002, defined trust as trusting beliefs) or more broadly (e.g. McKnigh, Choudhury, & Kacmar, 2002, who defined trust as trusting beliefs and trusting intentions). Trusting beliefs (i.e. perceptions of web vendor attributes), and trusting intentions (i.e. intentions to engage in trust-related behaviors with a web vendor) are based on the theory of reasoned action (TRA) which was introduced by (Ajzen & Fishbein, 1980) to analyze the psychological processes that reflect observed relationships among beliefs, attitudes, intentions, and behaviors. The theory asserts that intention to perform behavior is determined by the individual's attitude toward the behavior, and a person's attitude is affected by his/her beliefs.

Since trust can be seen as a belief, confidence, sentiment, or expectation about an exchange partner's intention and/or likely behavior, we believe that it is posited to be directly related to the attitudes toward purchasing from a vendor and indirectly related to consumers' willingness to buy through purchasing attitudes. This is how we define trust in this paper; this is consistent with how trust has been previously defined in the literature (Gefen, 2002).

The Varying Dimensionality of Trust

To investigate how trust is formed in commercial relational exchanges that take place within electronic environments, a review of the literature identified a body of works that provides the necessary theoretical background to aid understanding of this complex context.

Gefen et al. (2003) summarized the conceptualizations of trust from prior research as:

- A set of distinct beliefs related to the trusted party, consisting of integrity, benevolence, and ability.
- A general belief or trusting intention that another party could be trusted.
- A feeling of confidence and security in the caring response of the other party.
- A combination of these factors.

Another different perspective of trust was introduced by Krauter et al. (2006) who reviewed the different constructions of trust across various research disciplines and concluded that all trust definitions address one or more of the following perspectives: 1) trustor properties; 2) characteristics of the trusted object; and 3) context characteristics. Some studies also address the interaction or relationship between two or all three of these perspectives. In this study we adopt this classification since it cover all the trust interaction parties.

PROPERTIES OF THE TRUSTOR

The properties of the trustor perspective deal with elements that characterize the trustor. From the perspective of the individual trustor many trust constructions can be categorized into different conceptual types, such as attitudes, beliefs, intentions, behaviours, and dispositions. Under this perspective, constructs are partly affective (or emotional) and cognitive (or rational) determined and are embedded in a process which may result in trusting behaviour, which is the behavioural demonstration of trust.

Under this perspective, most of the trust researchers proposed two main components of trust: cognitive or rational, and affective or emotional (e.g., (Lewis & Weigert, 1985; Luhmann, 1979; Shneiderman, 2000). More emphasis and discussion will now be given with regard to this important point since it forms the basis for the subject of this study.

Feng, Lazar, Preece (2004) argue that the affective component of trust has been to some extent ignored in Human Computer Interaction (HCI) trust research, in particular in the domain of e-commerce, where research is largely focused on factors that contribute to cognitive trust, such as security, error absence, trust seals. For retail transactions, cognitive factors can increase the confidence that a transaction will be successfully completed. Affective factors can increase the feelings of attraction and trust.

The distinction between cognitive and affective trust received some support in an empirical study on long-term work relationships (McAllister, 1995), and strong support in a face-to-face study that manipulated affective trust (Johnson-George & Swap, 1982). However, only a few studies have examined affective-oriented factors in an e-commerce (B2C) environment.

This oversight of the affective element of trust does not reflect the increasing consideration that is given to affective processes in multiple disciplines (Riegelsberger et al., 2005): e.g. in marketing (Kotler, 2006), psychology (Zajonc, 1980), and sociology (Giddens, 1991). Marketing and advertising research are the fields that have the greatest power in investigating affective elements in decision-making processes. In these disciplines it is well-known that a consumer's purchase decision depends on rational process and on positive affective reactions (Kotler, 2006).

This perspective also considers trust beliefs and trust intentions, since they are based on the trustor's cognitive and affective perceptions of the trust warranting attributes of the object of trust. In an online context trusting beliefs include the online consumer's beliefs and expectancies about the trust object and includes all related characteristics (Kim et al. 2004). Trust intention is the extent to which the online consumer is willing to depend on, or intends to depend on, the mediated party in a given situation on the Internet even though he/she cannot control this party (Kim et al. 2004).

The above studies argue that both trusting beliefs and trusting intentions must be present for trust to exist. It can be seen that trust beliefs are a necessary but insufficient condition for trust to exist, because increasing trusting beliefs will not always have a corresponding positive effect on trust intentions.

CHARACTERISTICS OF THE TRUSTED PARTY

With respect to the characteristics of the trusted party, several scholars have in recent years offered trust definitions that highlight trust related attributes and characteristics of the trusted party. Researchers now tend to include specific characteristics of the trusted party, that cab be clustered into three dimensions: ability, benevolence and integrity (McKnight et al., 2002, Fogg et al. 2001). These characteristics or attributes of the trusted party are often referred to interchangeably as elements, antecedents, underlying dimensions or determinants of online trust (Wang & Emurian, 2005). 'Ability' reflect consumers' confidence that the firm has the skills necessary to perform the job (Mayer et al. 1995); 'benevolence' reflect confidence that the firm has a positive orientation toward its consumers beyond an 'egocentric profit motive' (Mayer et al. 1995); and 'integrity' reflect confidence that the firm adheres to a set of moral principles or professional standards that guide its interactions with customers. These dimensions are related, yet distinct.

Context Characteristic

The context characteristic perspective addresses the relationship between trust and risk and the fact that trust only exists in an uncertain and risky environment, since trust would not be needed if actions could be undertaken with complete certainty and no risk. In both online and offline transactions the trustor must lack information regarding the behaviour or characteristics of the object of trust (i.e. uncertainty) and there must be something that the trustor could lose if the trust is violated (i.e. risk). Thus, the relationship between risk and trust is reciprocal: risk creates an opportunity for trust, which leads to risk taking.

Online Trust in the HCI Literature

Based on the accumulated information and understanding of online trust as reviewed in the previous section, online merchants can take many effective actions to build and enhance merchant-customer relationships before, during, and/or after any online transaction, which can increase consumer trust. While those effective actions might be useful in retaining customers, online merchants need more efficient ways to communicate their trustworthiness to first-time visitors (Initial trust) and to convert them into consumers.

Initial trust is very important since web-based companies must rely on their websites to represent them and to show their new customers that they are reliable and trustworthy. Customers visiting a website for the first time may rely heavily on website cues to form their initial trust beliefs, but they still lack the important information that develops through interaction over time (McKnight et al., 2002). Another important reason for establishing customer trust through the first website visit is the possibility of low switching costs for new customers compared with switching costs for expert customers (Reichheld & Schefter, 2000). Since the web contains vast numbers of alternatives, either in vendors or in products, it has become very easy for customers to switch to a different online store after their first visit. So it has become an essential process for online vendors to try to establish and retain a trusting relationship with first time customers. Therefore, we believe that a thorough examination of the factors that promote initial trust in new customers of online vendors is critical.

Moreover, the physical appearance of business buildings and facilities, together with direct contact with company sales employees, which have been found to have effects on consumer trust in a conventional business context, are missing in the online environment. Online merchants depend mainly on their website to attract customers and to communicate with them. Therefore, applying trust-inducing features to the websites of online merchants can be an effective technique for enhancing online trust (Wang & Emurian, 2005).

Many researchers in HCI have embarked upon studying trust in an online context. Some researchers have focused on the effect of computer errors on trust; while others have examined the cues that may affect trust. These cues range from design and interface elements (e.g. Nielsen, 2004; Fogg, 2002), to perceived website credibility (e.g. Fogg et al., 2001), to the extent to which the technology is perceived and responded to as a social actor (e.g. Nass et al., 1999).

However, research on the use of media cues in websites is contradictory, with some studies finding such media cues to be positive (Nielsen et al., 2000; Fogg et al., 2001a; Steinbruck et al., 2002), while others finding them to be neutral or even negative (Riegelsberger & Sasse, 2001).

Social Presence and Online Trust

The trust phenomenon represents an ongoing interpersonal interaction with another party, whether a person or organization, over an extended period of time. Indeed, trust is typically built gradually through extensive ongoing interactions that enable individuals to create reliable expectations of what the other party may do (Luhmann, 1979). This social context is a key characteristic of trust and trust-building behaviour, since trust, in general, is built through constructive interactions with other people (Luhmann, 1979). So, the perception of a high degree of social presence, implying direct or indirect human contact in the relationship should, arguably, contribute to the building of trust. Extending this logic implies that another way in which trust in an e-commerce website may be built is through embedding the website interface with a high social presence, such as the perception that there is a medium of communication that represents personal, sociable, and sensitive human contact (Gefen & Straub, 2004).

Social presence has been defined as the extent to which a medium allows users to experience others as being psychologically present (Fulk, Schmitz, & Power, 1987). Social presence theory by (Short, Williams, & Christie, 1976) describes how the social context affects medium use; they see social presence as a quality inherent in a communication medium. Some researchers characterize the social presence of a medium as its capacity to transmit information about facial expressions, posture, dress and non-verbal cues (Short et al., 1976). Others focus on the psychological connection, where social presence is concerned with warmth and whether it conveys a feeling of human contact or sociability (Yoo & Alavi, 2001). While others still, focus on its close relationship to information richness theory (Straub, 1994), which concentrates on the interactivity of the media. Related to media information richness theory (Straub, 1994), social presence theory considers the extent to which a medium enables a communicator to experience communication partners as being psychologically present (Short et al., 1976). High social presence is typically found in face-to-face communication. However, medium richness can vary depending on circumstances (Zmud, Lind, & Young, 1990).

A significant difference between online and offline shopping environments is that the latter involve a wide range of social interactions with humans through multiple sensory channels. Online shopping, on the other hand, primarily tends towards reducing the users' affective or emotional factors through functional and performance based website design. As such, online shopping may be viewed as lacking human sociability, since it is more impersonal and automated than traditional offline commerce. Social Responses to Computer Technology (SRCT) research paradigm proposes that individual interactions with computers and other communication technologies are fundamentally social and natural, and people interact with computers and mediated stimuli using the same social attitudes and behaviours that they apply to other people. The CASA (Computers Are Social Actors) paradigm by (Nass, Moon, & Carney, 1999) initiated this field of research and uses theories and experiments derived from psychology, sociology, and communication to develop and validate theories. Granted, online shopping websites typically involve no actual interaction with other people, however, website interface features have been suggested to impact the perception of social presence cues, also known as interpersonal cues (discussed in more detail in the next section), that can be embedded in different ways. Higher perceived social presence cues in a website may increase online trust through their effect on increased electronic communication since communication is a necessary element of constructive interaction (Gefen & Straub, 2004). Trust may increase when the trusted party shows behaviour or other indicators in accordance with one's expectations; the perception that the vendor is embodying a high degree of social presence cues in the website should increase consumer trust to the degree that such indications are expected (Luhmann, 1979).

Social presence cues can convey a sense of personal, sociable and sensitive human contact, so too should multimedia websites. Indeed, recently, many online shopping websites have used interface features and embedded social presence cues. Advertising research has long relied on imagery of 'friendly faces' to build a positive attitude towards products.

To summarise, trust can only occur in a social context; in the absence of other people, trust is not relevant (Fukuyama, 2000).

Social Presence Cues (Interpersonal Cues)

Social presence cues, also known as interpersonal cues, are the signals that make a person aware of the presence of other people (Short et al., 1976). They are non-verbal and para-verbal communication methods and suggest participant appearance each time a person interacts with another. They may be intentional or unintentional and are part of the rapid stream of communication that passes between two interacting individuals. Although there are a number of definitions of social presence cues (interpersonal cues), in the broadest sense they entail communication that transcends the bare elements of the written or spoken word. The interpretation of this communication has been shown to have a central effect on participants' perceptions of the event and can give information about an individual's background, motivations and emotional attitudes. They augment spoken messages by helping people express their feelings or thoughts through the use of their bodies, their facial expressions, and their tone of voice and so on (Mehrabian & Epstein, 1972).

In many social encounters, individuals wish to develop positive relationships with others. They do so in part by adopting interaction behaviours that signal that they are warm, friendly and approachable (Andersen, 1985). This argument is consistent with intuitive expectations about why individuals might perceive another person positively. Most people tend to feel positive about someone who smiles, is expressive, appears relaxed, addresses them by name, asks them questions, and discloses information through personal anecdotes. Psychological studies have concluded that more than 65 percent of the information exchanged during a face-to-face interaction is expressed through non-verbal means. Fromkin and Rodman (2007) suggest that up to 90 per cent of the communicative process takes place non-verbally. The characteristics of the individuals involved and their response to coding and decoding signals govern the role of social presence cues in any exchange. According to Birdwhistell (1970), non-verbal cues can pass information from one person to the other and can integrate the communication process. Also, non-verbal cues can guide the turn-taking process (Whittaker & O'Connail, 1997) and play an important role in both impression formation (i.e., the process of getting to know the other person) and building interpersonal relationships (Short et al., 1976; Walther, 1992, 1993).

Three specific variables can be identified as impacting on the nature of non-verbal communication during the exchange; gender, culture and personal traits (Kim, 1996). Social presence cues (interpersonal cues) are of relevance in the debate on trust as they can be interpreted as signals for trustworthiness. A trustor can form an impression of perceived trustworthiness of a trustee from the interpersonal cues he/she perceives in a face-to-face situation; they work as inferences (Steinbruck, Schaumburg, Kruger, & Duda, 2002). Interpersonal cues lead to instant impression formation and thus have an immediate impact on affective trust.

The different types of cues differ in their reliability for trust assessments, and in how they are affected by transmission over media (Riegelsberger, Sasse, & McCarthy, 2005).

There are many classifications of interpersonal cues (Hinton, 1993); the most common types are:

1. **Paralanguage** - The vocal cues that accompany spoken language.
2. **Kinesics** - Body movements.
3. **Occulesics** - Eye behavior.
4. **Appearance/artifacts**.
5. **Proxemics** - The non-verbal study of space and distance.
6. **Haptics** - The non-verbal communication study of touch.
7. **Olfactics** - The non-verbal communication study of smell.
8. **Chronomics** - The non-verbal communication study of time.
9. **Facial expressions**.

Online Trust and Media Cue

Existing technologies allow for various representations of interpersonal cues that are embodied in different kinds of media cues (e.g. photo, audio, video, embodied agent) to be integrated into one platform.

The reduction in the number of interpersonal cues when interacting online is seen as one of the reasons for the lack of trust online (Shneiderman, 2000). This assumption is supported by many studies (e.g. Hassanein & Head, 2004). These studies found that richer representations result in higher awareness, preference and interpersonal focus. The most commonly used example of visual interpersonal cues is the smile - this has been identified as powerful in stimulating immediate affective responses, and can form a basis for affective trust. Thus, website interface elements such as photographs, video clips, or synthetic voices can create some level of social presence that may enhance the level of affective trust compared to website interfaces lacking such elements, and can therefore be taken as an indication of the trustworthiness of the trustee.

There is much research on trust in e-commerce vendors, while in contrast very few have specifically tested the effect of interpersonal cues on trust (Al-Diri, Hobbs, & Qahwaji, 2006). Some existing studies focused on the effect of synthetic interpersonal cues in avatars (embodied agents); others investigated the effect of displaying facial photos of humans on e-commerce sites. All of these studies tested users' trust either in the form of quantitative questionnaires or with qualitative interviews, using two mock-ups of an e-commerce website, one with and one without a photo or avatar (Bickmore & Picard, 2005).

VISUAL MEDIA CUE

Visual media cue has been studied in different disciplines, such as the mass media field for a long time. Graber (1989) discusses the value of visual information: "*Research has shown that audiences report visual content more accurately than verbal content and retention rates are much higher for visual information*" (p. 149).

In addition to helping people access their memories for associations, visual media cues can enable people to directly process information and thus make fewer errors in understanding the information (Graber, 1990). In her study on how visual media cues contribute to learning from the news, Graber (1991) found that visual media cues contribute to story comprehension and information gain by adding reality and clarification the story, and providing emotional support. Basil (1994) suggests that although people process audio and written words in common ways, we probably process visual media cues differently; visual media cues take fewer mental resources to process – visual media cue processing is more automatic.

These studies on the effects of visual media cues in traditional mass media support the usefulness of visual media cues in giving strength to messages. Because the website is used to communicate messages and is very capable of presenting visual information, it seems very reasonable to apply these findings to website design.

In summary, rich media representations result in higher levels of trust in mediated interactions. This view is supported by researches that suggest richer representations give more interpersonal cues that may lead to increased e-commerce vendor trustworthiness.

Video

In general, video is considered to be a rich media cue and thus is considered to afford high social presence cues as it transmits many visual and audio cues. In a study by (Swerts, Krahmer, Barkhuysen, & van de Laar, 2003), on the effect of social presence cues on the detection of speaker uncertainty, they found the best discriminative ability for video with audio, compared to video only or audio only. Brosig, Ockenfels, & Weimann, (2002) found that video with audio reached levels of cooperation that were close to those reached in face-to-face communication, even though they were reached after a longer time than in face-to-face communication.

Horn, Olson, & Karasik, (2002) in a study on lie detection in job interviews found a better performance in discriminating lies from truthful statements in high quality video than in audio only. The study also compared low resolution video and low frame rate video. They found that low resolution video, which suppresses detailed facial cues, gives a performance similar to that of high-quality video and has a good performance in lie detection. (Horn et al., 2002) attribute this result to the reduction in truth bias in the absence of recognizable facial cues and the fact that the presence of facial cues may lead to a trusting reaction.

A study by (Mulken, Andre, & Müller, 1999) investigated trust in advice. This study varied the representation of an advisor (video, embodied agent, audio and text only) and the quality of the advice. Hence, the effect of media representation could be compared to the effect of advice quality. The study found a preference for high quality advice in all representations, but only insignificant indication of a positive effect for video on the behavioural measures. Overall, the presence of more facial cues may lead to an overly positive trusting reaction.

Photos

Research on the use of personal photos in websites is limited in number and and the outcomes are contradictory, with some studies finding such images to be positive cues (Nielsen, 2004); (Fogg, 2002); (Steinbruck et al., 2002), while others found them to be neutral (Riegelsberger, Sasse, & McCarthy, 2002). Urban, Fareena, & Qualls, (1999) found that screen-sized facial photographs of shopping assistants embedded into a shopping website interface led to a wide range of reactions as some users liked it, whilst others considered it unnecessary. In the Fogg et al. (2001) study on the credibility of online news articles, they found that photos of authors increased credibility. Riegelsberger, Sasse, & McCarthy (2003) found that virtual re-embedding had a positive effect on user trust for medium experienced shoppers. Highly experienced Internet users, as well as consumers with a high level of distrust towards online-vendors, benefited little from the provision of social cues in the interface (Riegelsberger & Sasse, 2002). Steinbruck et al. (2002), in an experimental study, investigated whether adding a photo of an employee to the homepage of an online-banking site increased user trust in the bank – they found a positive effect on trust. These findings provide additional support for the notion that just the presence of interpersonal cues can positively increase trust.

As a result of the foregoing it was hypothesized that:

H-1: Subjects differ significantly on their rating of trust belief and trust intention across vendor's websites.

H-2-a: The first rating of a vendor's websites trustworthiness will result for those presenting video clips.

H-2-b: The second rating of a vendor's websites trustworthiness will result for those presenting photos

H-2-c: The third rating of a vendor's websites trustworthiness will result for those without photos.

Website Design and Culture

The creation of virtual organizations brings specific consequences for communication (El-Shinnawy & Markus, 1998). Specifically, non-face to face communication becomes more important as technology shrinks the world, bringing multiple cultures into virtual relationships and increasing global communication and business opportunities.

The source of online shoppers is progressively becoming more global and represents a multicultural community. In 2004, there were 736.6 million people with Internet access (Global Reach, 2004) and hence access to online consumer products. Of those Internet users the primary language was English (57.4%), followed by Chinese (12.2%), Japanese (9.5%), Spanish (8%) and German (7%), with the United States accounting for 66% of e-revenues. In 2003, this share decreased to 39% as other regions expanded their Internet activities, including buying online (International Data Corporation).

However, Hofstede, (2005) describe culture as the collective programming of the mind which distinguishes the members of one group from another. At its most basic level, culture is conceptualized as shared symbols, norms, and values in a social collective, such as a nation (Hofstede, 2005). According to (Matsumoto, 2000), culture is characterized as the degree to which people share attributes, values, beliefs and behaviours.

Culture has been related to a wide range of consumer preferences in non-Internet settings including attitudes toward advertising (de Mooij, 2006), brand loyalty (Deshpande, Hoyer, & Donthu, 1986), consumer values, consumption patterns, and perceived risk. The results of the studies suggest culture does have a large potential influence on consumers.

Links between communication differences and cultural value differences have been found across various countries (Cheng & Schweitzer, 1996). Several researchers, therefore, have emphasized the use of country-specific cultural value appeals when developing international advertising campaigns (Albers-Miller & Gelb, 1996); (Zhang & Gelb, 1996). This communication strategy has been found to be more persuasive and appealing than standardized advertising (Gregory & Munch, 1997). There are many ways by which culture can be reflected. Some forms of representations include symbols, rituals, behaviours, values and communication values (Hofstede, 2005).

However, studies on the effectiveness of communication using culturally appropriate appeals in advertising and marketing have only dealt with traditional print and broadcast media.

Culture has implications in Internet settings as well, and is proposed to affect consumer trust (Jarvenpaa & Tractinsky, 1999), Internet diffusion (Ferle, Edwards, & Mizuno, 2002), Internet marketing (Tian & Emery, 2002), website development (Kang & Corbitt, 2001), and web interface acceptance and preferences (Okayazaki & Rivas, 2002).

Understanding how to build trust for diverse consumers in electronic markets is imperative (Gommans, Krishan, & Scheddold, 2001); (Grewal, Munger, Iyer, & Levy, 2003); (Jarvenpaa & Tractinsky, 1999); (McKnight & Chervany, 2001b); (Yoon, 2002), and despite shifting figures and anticipated large numbers of web users from diverse cultures, to date little research has examined the effects of cultural website preferences on online trust. While, in the communication literature, one topic that is frequently studied is the link between adaptation and local cultural values (e.g. (Albers-Miller & Gelb, 1996).

THE IMPORTANCE OF DESIGNING A CULTURAL WEBSITE

Despite an anticipated large number of consumers from multiple cultures, few studies have examined web preferences of users related to design characteristics across cultures. (Chen & Dhillon, 2003) noted that, the website is a way a firm communicates with its customers. Therefore its appearance and structure encourage or discourage a consumer's purchase intentions.

In order to gain market advantage, firms may think it enough to just translate language, currency, date, time formats, etc. However, researchers, such as (Del Galdo & Neilson, 1996), find these methods inadequate. To meet the needs of the diverse market, it is necessary to localize Internet sites for the target market (Del Galdo & Neilson, 1996); (Sun, 2001).

Previous studies relating to cross-cultural issues on websites have explored website characteristics such as interactivity, site quality, navigation, and their perception by international consumers and found cross-cultural differences (Fock, 2000); (Robbins & Stylianou, 2003); (Tsikriktsis, 2002). Some studies have investigated how various theories like the theory of planned behaviour and the theory of flow can explain cross-cultural consumer behaviour on the web environment (Luna, Peracchio, & de Juan, 2002). Cyr and Trevor-Smith, (2004) found statistically significant characteristics in website design for municipal websites across cultures.

Simon (2001) examined cultural differences related to website satisfaction and website perception among the residents of Asia, Europe, Latin and South America, and North America based on Hofstede's model. The study found different preferences for colours and navigation.

A study by Almoajel (2007) introduced a model that defines the relationship between trust among Middle Eastern consumers and their level of comprehension (understanding) for B2C e-commerce website content that is written in a language other than their native language; the study showed a significant effect of culture on trust among the study participants.

Sun (2001) study examined some cultural elements in web design and focused on language, pictures and graphics, colours, and page layout and found culture to be an important design consideration that increases usability of multilingual websites. All of these studies point to one general conclusion – the website needs to be adapted to the culture of its target consumers.

Singh et al (2003) employed content analysis of 40 American-based firms to compare their domestic and foreign website differences. Significant differences in cultural characteristics were found for all major categories tested. This study concluded that the web is not a culturally neutral medium.

However, website interface characteristics such as colour, screen images were found have different preferences by users across cultures (Del Galdo & Nielsen, 1996; Marcus & Gould, 2000).

There are many ways that culture can be reflected. Some forms of representations include symbols, rituals, behaviours, values and communication values (Hofstede, 2005). Cultural elements represent major variables in determining the acceptability and usability of a website. To launch a website successfully, subtle cultural nuances and cross-cultural communication issues must be considered. (Badre, 2000) refers to the merging of cultural values and website usability as 'culturability' – when cultural elements are considered in website design and are expected to directly affect the way a user interacts with the website.

To achieve cultural sensitivity, many multilingual websites use culturability to close the distance between local users and corporations, and to localize websites on the cultural level. Specific culturability signifies a cultural affiliation and denotes a conventionalized use of the features in the website. Such cultural affiliation can be seen in the frequent use of certain items, such as a national symbol, colour or particular spatial organization in website design.

With respect to online trust, an effective website design can engage and attract online consumer trustworthiness (Fogg, 2002; Fogg et al., 2001); (Nielsen, 2004) which is also considered central to trust development (Egger, 2001). (Fogg, 2002) found a high percentage of consumers responding to a survey assessed the trust of websites based on overall visual design.

Design elements considered often include architecture of the information, familiarity of metaphors, transparency of terminology, ease of access, symbols and graphics and photographs as a way to build trust (Steinbruck et al., 2002). Also, colour preferences, site features (links, maps, search functions, page layout), language and content (Egger, 2001) were important. Winn and Beck (2002) described the persuasive power of design elements on an e-commerce website, and offered guidelines to website designers that appeal to user's logic and emotions and were considered trustworthy.

Building trust on the web requires user interface characteristics appropriate for culturally diverse audiences (Hillier, 2003; Marcus & Gould, 2000; Robbins & Stylianou, 2003). The Cheskin Research Group (1999, 2000) conducted work on trust in e-commerce. They identified six fundamental forms for the communication of trustworthiness on the web: brand, navigation, fulfilment, presentation, up-to-date technology and the logos that guarantee security.

This research adopts the view of Shannon (2000) that the goal of localizing user website interfaces is to provide a technologically, linguistically and culturally neutral platform from which to launch global e-commerce initiatives while allowing a framework that incorporates local content and functionality. This involves enhancing the website to fit the target users in different nations (Alvarez, Kasday, & Steven, 1998; Lagon, 2000).

Cultural Photo as a Symbol

Symbols are an important element denoting culture (Marcus & Gould, 2000). Symbols are 'metaphors' denoting actions of the user (Barber & Badre, 2001), they vary and may represent a wide range of features (Fernandes, 1995). One important form of symbols is multimedia relating to culture which few researchers have examined. Marcus & Gould, (2000) found site examples that suggest strong use of graphics and multimedia is more likely when the website designer is from a culture that values material goods and is more assertive (close to masculinity in Hofstede's dimension). Multimedia elements might include streaming video, sound and animation.

Simon (2001) addressed the issue of a communication-based approach by using conceptualizations of website satisfaction and perception. In a web environment, perception was defined as the degree to which participants felt the site was appropriate for their home country based on three key variables of media perception – social presence (i.e., transmission of information rich in socioeconomic content), communication effectiveness, and communication interface.

Visual features on a website are increasingly important for supporting information as well as interaction between consumers and e-vendors. Consumers are interested in information, representing style and interaction with web usage depending on the amount of visual components. Visual components support user understanding of the website.

Visual web design aids user enthusiasm towards the site and is likely to help them engage with a website when browsing. This can be attained by the use of icons, symbols, or other website interface tools (Pullman, 1998). Based on the numerous researches presented in the above sections, it is possible to take the view that cultural cues given in a website interface will increase user trust because they increase cultural presence.

Hence the following additional research hypotheses were proposed:

H-3: Across websites including human portraits there will be significant statistical differences in trustworthiness between websites with local social presence and websites with foreign social presence.

H-4: Saudi subjects will trust a website with Saudi social presence (photo) more than a website with Western social presence (photo).

EXPERIMENTAL METHODOLOGY

This study was designed as a one-factor experiment manipulating three levels of website social presence cues (or interpersonal cues). Each of the four specially-designed websites displayed the same products but each represented different vendors. Only the media cues were manipulated on the sites. Thus, the study attempted to investigate and examine the effects of the interpersonal cues that can be manipulated by facial photographs, video clips, and culture as control variables, which used photos of Saudi and Western people when forming the initial trust toward online vendors. In addition the study set out to measure some auxiliary parameters.

Experiment Participants

The experiment was conducted in one of the most famous IT training institutes in Riyadh, Kingdom of Saudi Arabia (KSA). This institute was established some 20 years ago and is renowned for the diversity of participants entering its gates.

Without question, local knowledge, experience and familiarity of its set up and prestige make this a hub for students and workers enrolling from far afield. For this reason, there is a strong evidence to suggest that the sample and applicability of findings discussed in this study will be a fair and representative reflection of the country at large and not purely for Riyadh. However, prior knowledge of the background of those sampled provided unequivocal evidence of the broad base of findings observed.

The number of participants totalled 72 students. Since online consumers are generally younger and more highly educated than conventional customers, student samples are close to the online consumer population (McKnight et al., 2002) and therefore are representative of online shoppers (Gefen & Straub, 2004).

Experiment Material

An initial survey on the most popular online products was carried out and found that laptops came first in popularity. The stimulus experiment material used in this research was carefully selected, having noted that previous studies in online trust used mock-ups of shopping sites to test the effect of interpersonal cues (e.g. (Steinbruck et al., 2002); (Riegelsberger et al., 2002). This experiment used semi-functional copies of existing vendors' sites, chosen in consultation with the four most famous reviewer business sites; BizRate.com, ResellerRating.com, Price Grabber.com, and Epinion.com to facilitate the task of rating online shopping sites. These services aggregate feedback from customers of e-commerce vendors based on post service and handling of privacy and security, which represents an aspect of vendor trustworthiness. Western shopping sites were selected as they constitute a realistic scenario with relatively high risk, due to the vendor and the users being in two different countries. The selection was based on the rating of high trustworthiness of the vendors and the number of reviewers of the selected site. Also we made a usability test for the four vendors' websites to check the usability index for each website and to make sure that all selected websites had almost the same usability criteria. Also a usability test was performed for the four vendors' websites to check the usability index for each website and to make sure that all selected websites had almost the same usability criteria. To do that, five PhD students in computer science were asked to perform usability assessment tests for all four vendor websites by using the checklist developed by (Keevil, 1998). The checklist has over 200 questions in five categories, and even though it is not very recent, it is still valid and robust. The results showed that the four websites had almost the same usability index.

Semi-functional copies of the website including the homepage and some subsequent layers depending on the available links in each layer were designed, so that participants were able to browse and search general information about the site. Also any certification or reputation seals that were present on some pages were removed.

The media cues (photos and the video clip) were selected by another five professionals in computing and business, who were asked to rate the photos and select the most appropriate based on a realistic image of a customer service representative. The media cues (photo or video clip) were placed in appropriate places in the first page of the site showing the selected product (without deleting or hiding anything from the page itself). This page was connected to the entire website so the subject could browse and search the site. In addition, each media cue was presented on each vendor website based on a predefined website experiment scenario display. Since these media cues represent customer representatives the video clip says the following statements: "Welcome to our website and it is hope that you like this website and if there is anything that we can do to help with please feel free to contact us." This media cue representation ranged from one seconds to five seconds in length, and was streamed with

Figure 1. Snap shot of the experimental websites.

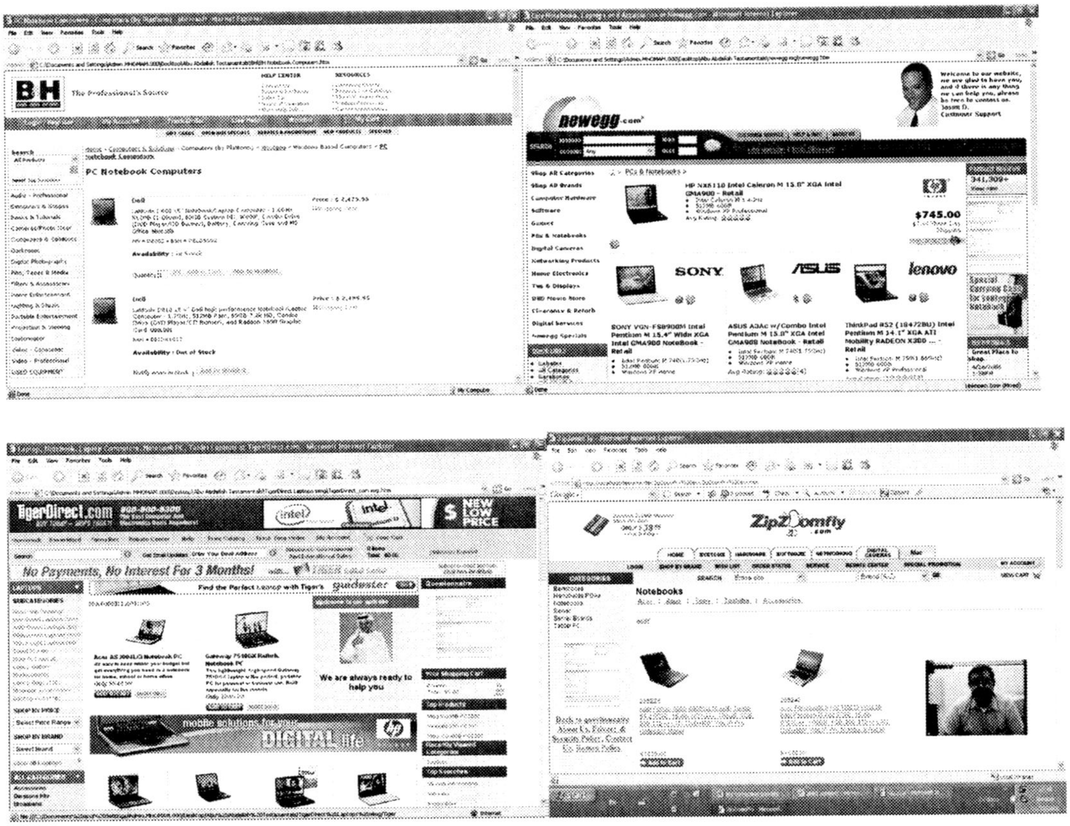

Windows Media Encoder (350 kbps, 320x240). Audio was encoded with 22.050 kHz, 16 bit, mono. See figure 1.

Data Collection

The research methodology of studies on trust in e-commerce can be categorized into qualitative approaches, using either semi-structured interviews (e.g. (Egger, 2001), or qualitative interviews in conjunction with user evaluation sessions (e.g. (Witkowski, Neville, & Pitt, 2003). The second category entails trust questionnaires in an experiment environment, in this approach the majority of studies used an experiential survey, i.e. participants were asked to navigate to a specified or self-selected Internet company and had to perform several predefined tasks and afterwards report on their impressions by filling out a questionnaire (e.g. (Jarvenpaa, Tractinsky, & Vitale, 2000). A second group of studies that used this approach, applied a "basic" survey approach, i.e. subjects were administered a questionnaire or they were pointed to an on-line-questionnaire form without previously visiting any e-commerce website (e.g.(Bhattacherjee, 2002). The third category involves social dilemma games, studies in this category measure players' rate of cooperation and defection to infer media effects on trust and trustworthy behavior (e.g. (Brosig et al., 2002).

From a clear review of these approaches, there are distinct advantages and disadvantages. For this study it was decided to use the hypothesis-testing laboratory and experimental quantitative approach. This approach allows a high level of experimental control for independent variables while keeping resource requirements relatively low.

All experimental tasks during this research experiment were performed in a computer laboratory. The research instrument to measure the constructs of interest was developed by adapting existing measures from the literature to the current research context. All items were scored on a five-point Likert-type scale ranging from (1) strongly disagree to (5) strongly agree (See Appendix-1). As the experiment was conducted in an Arabic speaking country, the questionnaire, originally written in English, was translated into Arabic by a bilingual person whose native language is Arabic. The Arabic questionnaire was then translated back into English by another bilingual person. This English version was then compared with the original version and no items were found to deviate significantly in terms of language. This process was conducted not only because it can prevent any distortions in meaning across cultures, but also because it can enhance the translation quality. The study question items consisted of five sections. The first section included the basic demographic characteristics such as age, gender, education level and Internet experience. In the second section, respondents were asked to answer questions on their online purchasing experience, logos of the four experiment websites were presented in this section, and participants were asked, as a filtered question, whether they had seen or purchased from any of these websites, the reason for this was to make sure of the validity of implementing the research interest of initial trust. In the third section, respondents were tested for their trust propensity or 'disposition to trust' an individual trait, defined as a "general willingness based on extended socialization to depend on others" (McKnight & Chervany, 2001). The questionnaire items were selected based on their high reliability and validity to discriminate the construct and achieve a high Cronbach alpha value; this was adapted from (Teo & Liu, 2005) and (Gefen, 2000). In the fourth section, respondents were asked to answer a question adapted from (McKnigh et al., 2002) regarding their belief towards the Internet as a new medium or as a new environment (i.e. how confident they feel in the Internet system). This item was adapted based on high value of reliability and validity - its Cronbach alpha value was 94%. The fifth section tested respondents on their trust and trust intentions toward the e-commerce vendor. The questionnaire items reflected the most common trust belief dimensions, which are ability, integrity, and benevolence. Also it measured trust intentions - intentions to engage in trust-related behaviors with the web vendor. All items in this section were adapted from Kammerer (2000), since his questionnaire items effectively discriminate trustworthy from less trustworthy online vendors. Also, Kammerer (2000) reports excellent reliability scores with Cronbach alpha values of 0.97%.

Experimental Procedure and Tasks

Previous online trust research has been criticized for relying on measuring trust without inducing any form of risk (Riegelsberger, Sasse, & McCarthy, 2003), which represents an important key related to trust. To overcome these criticisms, this study induced financial risk in a laboratory situation. While it does not fully represent a real-world risk, nevertheless, it allowed combining a laboratory setting with some element of real-world risk by informing participants that the experiment website trustworthiness had been assessed and rated by independent business reviewer sites and one of their tasks was to identify the trustworthiness of each shopping site; those whose rating matched the real trustworthiness rate would be entered into a lucky draw with prizes such as a laptop to be offered in a random draw conducted at the end of the study. In this way we induced a slightly higher level of risk thus, increasing the realism of the experiment and encouraging participation.

At the beginning of the experiment a brief introduction and the total estimated time that it would take were given. Then participants started the first part of the experiment by opening the first experiment page, which includes the experiment objectives and instructions, then they started filling out sets of the first four questionnaires sections, that extracted some demographic characteristics, online purchasing experience, disposition to trust, and their belief towards the Internet as a new medium or as a new environment. In the second part each subject was asked to look at four websites and browse them. This involved looking at the website, opening the available links and website layers, looking at the laptop brands and specifications and features, and so on, and then evaluating the e-commerce vendor using the online vendor trust questionnaire. This process was repeated for all four websites in the set.

However, in order to avoid predictability and the effect of learning within the scenarios (the possibility that participants exposed to a particular interface or task would do better the next time they were asked to use it), within the experiment software a module with the capability to control the display order scenario of the four websites was included so that the four experimental websites were presented in a different order for each participant. Also, one of the experiment software design package capabilities was to display the media cues within the four websites in a different order, so each media cue was displayed in each vendor website.

When subjects finished, they were asked to move on to the third part which comprised one task. In this task each subject was asked to search the first website seen (displayed automatically) and imagine that they had enough money to buy a laptop that would serve their needs for the next two years. The subjects were required to indicate the model, price and certain product specifications of their final laptop choice. Once the participants had found their products and completed their responses on the instruction form, they were asked to fill out the vendor trust questionnaire again.

Sample Size

Choosing a representative sample size in any study can be a question of debate. Obviously, the larger the sample the better the interpretations that can be derived. However, with obvious limitations of time, cost and resources at one's disposal, there is a clear balance, on the one side of the information required in making the studies meaningful and constraints described above. The chosen sample size selected (n=72) was derived using statistical sample size calculations based on adequate power to drive the hypothesis under consideration (Ferguson, 1989; Shirley, Stanley, & Daniel, 2004). The power of a statistical test in being able to detect differences between the different groups is a function of the sample size, the significance of the test (commonly 5%) and the likely differences between the groups that we expected to observe. Where the difference was expected to be smaller, a larger sample size would have been needed to observe the change. On these grounds, while 72 may not seem to be a considerable amount, it is nevertheless adequate from a statistical consideration.

DATA ANALYSIS

All data analysis was conducted using the SPSS Windows software package. A total of 72 subjects participated in this study, all male and between the ages of 18-25 and 26-35 respectively; most (79.2%) held bachelor degrees. As expected, this group was 'Internet-savvy' with over 39% of the respondents spending between 6 and 10 hours online per week. On average, the majority made at least one online purchase per week and 28% of the respondents spent 2000SR (1$=3.75SR) or more per online purchase.

Table 1. Friedman test for Trust Belief and Trust Intention between the Four Websites

	Trust Belief			Trust Intention
	Ability	Integrity	Benevolence	
N	72	72	72	72
Friedman test Sig at 5%	Yes	Yes	Yes	Yes

Testing the Research Hypotheses

Data analysis for this experiment was performed using a Friedman test. This non-parametric test represent an appropriate statistical test that is normally used in this situation (Peter & Smeeton, 2007a; Sheskin, 2004; Shirley et al., 2004) and was deemed much suitable as the data complied to the assumptions of non-normality. The two main statistical procedures to check the data normality are either by plotting the data in the shape of histogram for example and check whether the data fit a normal distribution curve or not, and the second procedure is to use a formal statistical test like for example Shapiro-Wilk Test (Peter & Smeeton, 2007; Sheskin, 2004; Shirley et al., 2004). Both of these statistical procedures used to confirm the non-normality of the data.

This test is an alternative to the repeated measures ANOVA test, when the assumption of normality or equality of variance is not met. This, like many non-parametric tests, uses the ranks of the data rather than their raw values to calculate the statistic. This test was chosen because it is suitable for comparing variant vendor websites embedded with different media cues when the relative data sample has some characteristics also presented in this context, including the following (Peter & Smeeton, 2007a); (Sheskin, 2004):

- The experimental data is not parametric, i.e., the dependent variables are not normally distributed.
- The number of compared websites is two and more. In our context, we typically compared several vendors websites embedded with different media cues.
- The same subjects were used to observe different manipulation in vendors' websites.

To test the first hypothesis (**H-1**), the Friedman test was used. This test was computed for trust belief and trust intention of all websites to see if there were any significant statistical differences between the subjects' answers with regard to the trustworthiness of the four websites. Two types of data were used in the analysis, the first was the mean value of each subject answer, and the second was the actual subject answer regarding each questionnaire item. Results showed the subjects differed significantly on their rating of their initial trust and trust intention regarding the four vendors' websites owing to the overall statistical significance (P =.000) at 0.05 levels using the two kinds of data; so, the first hypothesis was supported see table 1.

In order to test the second hypothesis (**H-2-a, b, c**) it was necessary to conduct comparisons contrasting specific conditions with one another. Wilcoxon matched-pairs signed-ranks test was recommended by (Sheskin, 2004) and (Peter & Smeeton, 2001) for conducting the three pairwise comparisons (video clip, photo, no photo) with regard to trust belief and trust intention. Table 2 contains the results.

After reviewing the three analyses, it was found that all pairwise comparisons were significant, with the exception of the video-photo website comparison in trust intention. Subjects rated the trust belief and trust intention for the photo website the highest, the video clip website next highest, and the no photo website as the lowest. Thus, the second hypothesis was only partially supported,

Table 2. Results of Wilcoxon Signed Rank Test

Website	Trust type	Rank	Wilcoxon Signed Rank Test Asymp.Sig
With photo	Trust Belief	1	Yes
	Trust Intention	1	Yes
With video clip	Trust Belief	2	Yes
	Trust Intention	2	No
With no photo	Trust Belief	3	Yes

Table 3. Friedman test for Trust Belief and Trust Intention between the Two Websites

	Trust Belief			Trust Intention
	Ability	Integrity	Benevolence	
N	72	72	72	72
Friedman test Sig at 5%	Yes	Yes	No	Yes

since the vendor website utilizing the video clip came second rather than the expected first position. A possible explanation for this unexpected result is that the video clip was not recorded to professional standards.

The same procedure followed when testing the first and the second hypothesis was also used to test the **H-3** and the **H-4** hypotheses, but in this case between two vendor websites only, (the website with the Saudi photo and the website with the Western photo). The Friedman matched samples results are shown in Table 3. The analysis showed that subjects differed significantly on their rating of their trust belief (ability and integrity, although not for benevolence dimension), and also for trust intention in respect of the two vendor websites (overall statistical significance of p=.000 at 0.05 levels using the two kinds of data); so the third hypothesis was fully supported (see Table 3).

With respect to the fourth hypothesis (**H-4**), the Wilcoxon matched-pairs signed-ranks test was conducted for comparison, contrasting specific conditions with one another with regard to trust belief and trust intention. Table 4 contains the results.

From Wilcoxon test analyses, it can be said that pairwise comparison was significant. Subjects rated the trust belief and trust intention for the website using a Saudi photo as the highest; the website with a Western photo next. So the fourth hypothesis was supported.

Analysis of the Exploration Depth

In terms of investigating the factor depth of a users' exploration of a site (exploration depth)

Table 4. Results of Wilcoxon Signed Rank Test

Website	Trust type	Rank	Wilcoxon Signed Rank Test Asymp.Sig
With Saudi photo	Trust Belief	1	Yes
	Trust Intention	1	Yes
With Western photo	Trust Belief	2	Yes
	Trust Intention	2	Yes

which influences the effect of media-subject on user trust (Research Goal 3), a superficial exploration of the homepage and additional pages with no task browsing was compared to a subsequent in-depth exploration of a vendor's site, beyond the homepage with predefined task browsing. Hence, each participant in this scenario saw the first vendor's site twice: first only the homepage and the additional pages that were linked from it, but with a general browsing task (superficial exploration); then the homepage and additional pages, but with the specific task of imaginary purchasing - an in-depth exploration. The introduction of the exploration depth was used for several reasons. Most trust models indicated that browsing time is an essential factor affecting trust. Initial trust in the first encounter depends on factors other than behavioral trust that emerges from a long-standing relationship. Applying this consideration to the initial trust that can be measured in a laboratory experiment, it can be proposed that the level of trust will be based upon factors other than initial trust after a detailed exploration of the site. A media cue might have a strong effect based on a quick look, but might lose its significance once more information (e.g. from a site's security and privacy policies) is obtainable. On the other hand, it can be expected that the first impression (e.g. influenced by a media cue) can color further information processing and thus have a long-term effect (Baron & Byrne, 2004).

The same analysis techniques were used in investigating the data. The analysis confirmed the same finding, that there was a statistical significance between trust of the four vendors' websites, where the websites with facial photo were ranked first and then those with a video clip. Also the vendor website with a Saudi photo was trusted more (statistically significant) than that with a Western photo. See Tables 5, 6, 7, and 8.

Finally, many nonparametric correlation tests were conducted to see whether there were any significant relationships between the trust belief, trust intention and participants' trust disposition, system assurance, age, education level, or Internet usage. Results showed no statistically significant differences between any of these variables.

Discussion

This experiment investigated the initial trust in e-commerce vendors. It researched the effect of adding media cues (video clip, facial photo) to the homepage of an e-commerce vendor and the effect of interpersonal cues (or social presence cues) that are implicit in media cue on users' trust in that vendor. It further investigated the effect of the culture represented by a facial photo of a

Table 5. Friedman test for Trust Belief and Trust Intention between the Four Websites

	Trust Belief			Trust Intention
	Ability	Integrity	Benevolence	
N	72	72	72	72
Friedman test Sig at 5%	Yes	Yes	Yes	Yes

Table 6. Friedman test for Trust Belief and Trust Intention between the Two Websites

	Trust Belief			Trust Intention
	Ability	Integrity	Benevolence	
N	72	72	72	72
Friedman test Sig at 5%	Yes	Yes	No	Yes

Table 7. Results of Wilcoxon Signed Rank Test

Website	Trust type	Rank	Wilcoxon Signed Rank Test Asymp.Sig
With photo	Trust Belief	1	Yes
	Trust Intention	1	Yes
With video clip	Trust Belief	2	Yes
	Trust Intention	2	Yes
With no photo	Trust Belief	3	Yes
	Trust Intention	3	Yes

Table 8. Results of Wilcoxon Signed Rank Test

Website	Trust type	Rank	Wilcoxon Signed Rank Test Asymp.Sig
With Saudi photo	Trust Belief	1	Yes
	Trust Intention	1	Yes
With Western photo	Trust Belief	2	Yes
	Trust Intention	2	Yes

Western and Saudi man added to the homepage of an e-commerce vender, on user trust in that vendor. This study also investigated the influences and the effect of the interaction between media cues and the depth of a users' exploration of a website (exploration depth) on e-vendor website trustworthiness, by comparing a superficial exploration of the homepage and additional pages with no task browsing to a subsequent in-depth exploration of a vendor's site beyond the homepage with predefined task browsing. On a methodological level, the study aimed to overcome part of the limitations of conventionally used trust questionnaires conducted under conditions of no financial risk by introducing trust questionnaires that were elicited under financial risk.

Media Cue Effects

The first question addressed by this experiment considered whether interpersonal cues (social presence cues), represented by media cues (facial photo, video clip), perceived by users in the website interface had an effect on the trustworthiness of e-commerce vendors (H1).

In contrast to earlier studies that investigated the effect of only one kind of media cue (i.e. photo) on single e-commerce website trustworthiness, this experiment investigated the effect of multiple kinds of media cue (video clip, and photo) across four different e-commerce websites. Earlier studies results regarding this subject were contradictory since some studies showed that user trust perceptions can be influenced by adding one photo to one homepage (e.g., Nielsen, 2004; Fogg, et al. 2001; Fogg, 2002; Steinbruck, et al. 2002), while other studies found a moderate or negative effect (Riegelsberger & Sasse 2002); the aim of this experiment therefore was to identify whether the strategy of adding different media cues (either video, photo) could be beneficial when tested across several vendors' websites and when trust is measured under experiment laboratory financial risk. Also this experiment, in contrast to some previous studies on user trust in e-commerce, used copies of existing e-commerce websites whose trustworthiness was known to the researcher. Vendor trustworthiness was incorporated in the form of customer ratings, which had been aggregated by Biz Rate, Reseller Rating, Price Grabber, and Epinion.

The analysis found positive effects of adding media cues on the study participants and this effect caused significant differences in perceptions of vendors' trustworthiness. It was found that media cues can manipulate and bias users' judgment towards vendors' trustworthiness based upon an inspection of surface cues. Thus, this experiment demonstrated the importance of interpersonal cues, represented by media cues, as trust cues that can be perceived by users. This finding is consist with and supports earlier studies (e.g. Nielsen, 2004; Fogg et al., 2001; Fogg, 2002; Steinbruck et al. 2002), but contradicts study results from Riegelsberger & Sasse (2002).

From the perspective of an e-commerce vendor, this effect of a media cue is important. Users arriving at a site homepage will decide, based upon such first impressions, whether to explore the site in more depth or leave it and go instead to a competitor's site. The positive influence of a media cue can thus help e-commerce vendors to overcome the barrier in the process of converting a visitor to a customer. The findings of this experiment underline the importance of the interface as a communicator of trustworthiness.

Culture Effects

Online trust research to date has been mostly limited to a Western context, particularly focusing on Western culture. However, the trust theories and mechanisms developed in the Western context might not apply for other societies, especially since culture may affect the antecedents of trust. For example, (Sako & Helper, 2002) noted that there were differences in trust perceptions between subjects from two countries in terms of the level of trust, the way in which trust was conceptualized, and the way in which it was formed. Thus, there is a need to re-examine the notion of trust and identify its determinants in the context of different markets and cultures.

In contrast to other studies on the effect of adding a Western facial photo on e-commerce vender website trustworthiness, this experiment took a different path by adding the culture factor as an independent variable; it investigated the effect of different cultures' photos (a Western facial photo and a Saudi facial photo) across different e-commerce websites' trustworthiness.

As proposed, but not examined by earlier research, the analysis of vendors' website trustworthiness indicates that there were statistically significant differences between perceived websites trustworthiness with a Western facial photo and a Saudi facial photo. This means that the culture factor has a positive manipulation effect in the participants' perceptions during their trust assessment process. This finding can be interpreted as cultural media cues providing participants with situational cues that help them to absorb information within a familiar context. They create a virtual landscape with its own landmarks where participants feel at home.

These findings are consonant with prior work (e.g., Cyr & Trevor-Smith, 2004; Yoon, 2002; Simon, 2001; Lee, Kim, & Moon, 2000; Marcus & Gould, 2000; Fernandes, 1995); that the website requires interface characteristics suitable for culturally diverse audiences. Various preferences exist among Saudis for media cue kinds. Differences also exist in terms of the media cue through which Saudi website visitors prefer to interact. Of note, greater presence of cultural media cue was represented by the Saudi facial photo. One possible explanation for this difference is that Japan is considered a high-context culture, where additional cultural tools in media cues, beyond the regularly used Western media cue format, are preferred. This finding was consistent during the superficial exploration task, the-depth exploration task, and in the preference rank.

In summary, this research provides statistically significant evidence to support other work in website design – design preferences differ across cultures. It also supports the general call for localization of website content and provides some directions related to specific cultural preferences in website design.

The implications of these results are significant when exploring how various elements of web design must be considered in the context of culture. In addition, there are implications for accessibility to increasingly larger non-English-speaking populations on the Internet. It is expected that when websites are appropriate and culturally sensitive, then users will have increased access to content and enhanced user experiences.

Local vendors simply reach their local customers through local mass media, but for e-vendors these methods are not valid. Thus, vendors' websites do not target only one segment of consumers: instead, they target all global consumers. Therefore, it becomes critical for online firms to develop culturally adapted international websites that adapt to the needs of their international audience.

In-Depth Exploration Effects

This study also investigated the influences and the effect of the interaction between media cues and the depth of a users' exploration of a website (exploration depth) on e-vendor website trustworthiness. This was done by comparing a superficial exploration of the homepage and additional pages with no task browsing to a subsequent in-depth exploration of a vendor's site beyond the homepage with predefined task browsing. Exploration depth was included as an additional within-subject variable. Hence, each participant (in this scenario) saw the first vendor twice: first only the homepage and additional pages that were linked from it, but with a general browsing task (superficial exploration), followed by the homepage and additional pages that were linked from it, but with specific task (purchasing imagination) (in-depth exploration). The aim was to investigate how the exploration depth influences the effect of a media cue on user trust of a vendor website.

The introduction of the exploration depth was used for many reasons. Most trust models indicated that browsing time is an essential factor affecting trust. Initial trust in first encounters depends on factors other than behavioural trust that emerges from a long-standing relationship. Applying this consideration to the initial trust that can be measured in a laboratory experiment, it can be proposed that the level of trust after superficial exploration will be based upon factors other than initial trust after a detailed exploration of the site. A media cue might have a strong effect based on a quick look, but might lose its significance once more information (e.g. from a site's security and privacy policies) is obtainable. On the other hand, it can be expected that the first impression (e.g. influenced by a media cue) can colour further information processing and thus have a long-term effect (Baron & Byrne, 2008).

An interesting and important result is that the effect of trust was improved once users had explored vendors' sites in more depth, i.e. once they had looked for the purpose of a specific task in pages beyond the homepage. It can be interpreted that the cues that carried reliable information about a vendors' trustworthiness were perceived, and media cues could easily influence this decision. Hence, a media cue has a long-term effect on user trust and it biases the perception of other site elements (Baron & Byrne, 2008). Rather, the supplementary information perceived on additional pages improved the relative impact of the media cue. While a study by Riegelsberger and Sasse (2002) found that the effects of a facial photo diminished beyond the first page or the home page. This difference can be interpreted by the different methods of investigation.

From a practitioner point of view this study can benefit an e-commerce vendor, as Internet users who open an e-commerce website homepage choose either to stay or to quickly close it and go to another one based upon first impressions. Attractive first impressions can be seen as essential, but they are not enough for keeping customers.

CONCLUSION

This experiment addressed the main research goals of the study. First, it investigated a media cue's (facial photo or video clip) ability to influence user trust in respect of e-commerce vendors, based on surface cues that were implemented in the media cue. This goal is of high relevance, because this strategy is now being used by e-commerce vendors (especially facial photographs, but not yet video) in an attempt to increase trust and attract customers. Secondly, this research tested the effect on user trust of adding a facial photo from two different cultures (Western and Saudi) to an e-commerce vendor's homepage. It thus focused on the symbolic use of interpersonal cues. This goal, despite its importance for the development of trust in e-commerce, has not been addressed in previous research. Thirdly, the experiment was concerned with investigating how time, as an independent variable, represented by the depth of a users' exploration of a site (exploration depth), influences the effect of a media cue on user trust by comparing a superficial exploration of the homepage and additional pages with no task browsing to a subsequent in-depth exploration of a vendor's site beyond the homepage with predefined task browsing. Fourthly, most of the previous studies tested the effects of adding one photo to a mock-up of one e-commerce site. This experiment aimed to overcome this limitation by testing several photos on several semi-functional copies of existing vendors' sites. In addition, this experiment introduced a method for measuring trust that required participants to make decisions under conditions of financial risk.

This experiment found that media cues in the interface are indeed able to affect a vendor's trustworthiness based on the surface cues it contains.

A clear picture emerged regarding the effect of photos from different cultures. The experimental results found a positive effect of the media cues in the two stages of exploration - a superficial exploration and an in-depth exploration of a vendors' website. These visceral reactions, however, appear to color the subsequent processing of signals, as their effect improved with the perception of additional signals from an in-depth exploration.

From a methodology point of view, this experiment verified that financial risk can be used in a laboratory experiment to enhance the validity of trust research. Analyzing users' behavior leads to a relevant outcome to e-commerce vendors since they are concerned with users' purchase decisions.

With respect to investigating the influence of the auxiliary variables on trust, such as trust disposition, system assurance, age, education level, and Internet usage, the experiment did not find any significant evidence that could support any kind of influence.

Finally, based on the findings of the experiment it is suggested that web designers and e-commerce vendors should keep in mind the following recommendations when introducing e-commerce applications in Middle Eastern countries in general, and in Saudi Arabia in particular:

There is a significant effect of a media cue in B2C e-commerce websites. Users arriving at a site homepage will decide, based upon their first impressions, whether to explore the site in more depth or leave it and go instead to a competitor's site. The positive, attractive impressions of a media cue can thus help e-commerce vendors in the process of converting a visitor to a customer. The findings of this experiment underline the importance of the interface as a communicator of trustworthiness.

In B2C e-commerce applications it is very important to take cultural aspects into consideration when designing an e-commerce website. It is expected that when websites are appropriate and culturally sensitive, then users will have increased access to content and enhanced user experiences. Selecting photos of customer service representatives that relate to targeted markets represent an easy way to enhance the vendors' trustworthiness.

REFERENCES

Ajzen, I., & Fishbein, M. (1980). *Understanding attitudes and predicting social behavior*. Englewood Cliffs, NJ: Prentice-Hall.

Al-Diri, K., Hobbs, D., & Qahwaji, R. (2006). *Effects of Social Presence and Cultural Representation on Websites for Promoting Trust in B2C E-Commerce - A Saudi Arabian Study.* Paper presented at the Proceedings of the IADIS e-Commerce 2006 International Conference, Barcelona, Spain.

Al-Diri, K., Hobbs, D., & Qahwaji, R. (2007). *E-Commerce: An exploratory study of on-line shopping in Saudi Arabia.* Paper presented at the Proceedings of the Saudi Innovation Conference (SiC) 2007, Newcastle upon Tyne, UK.

Al Moajel, N. (2007). *Fostering Trust for Middle-Eastern B2C e-Commerce "An Investigation of the Effects of Voice, Accent and Culture an English e-commerce website on Trust perception among Middle Eastern Consumers".* PhD thesis in Bradford University, Bradford, UK.

Albers-Miller, N. D., & Gelb, B. D. (1996). Business advertising appeals as a mirror of cultural dimensions: a study of eleven countries. *Journal of Advertising, 25*(4), 57–70.

Badre, A. (2001). The Effects of Cross Cultural Interface Design Orientation on World Wide Web User Performance. *Georgia Institute of Technology.*

Bagozzi, R. P., Wong, N., Abe, S., & Bergami, M. (2000). Cultural and situational contingencies and the theory of reasoned action: Application to fast food restaurant consumption. *Journal of Consumer Psychology, 9*(2), 97–106.

Baier, A. (1986). Trust and Antitrust. *Ethics, 96*, 231–260.

Barber, W., & Badre, A. N. (1998). *Culturability: The merging of culture and usability.* Paper presented at the 4th Conference on Human Factors and the Web, Baskin, Ridge, New Jersey, US.

Belanger, F., Hiller, J., & Smith, W. (2002). Trustworthiness in electronic commerce: the role of privacy, security, and site attributes. *The Journal of Strategic Information Systems, 11*(3-4), 245–270.

Bhattacherjee, A. (2002). Individual trust in online firms: Scale development and initial test. *Journal of Management Information Systems, 19*(1), 211.

Bickmore, T. W., & Picard, R. W. (2005). Establishing and maintaining long-term human-computer relationships. *ACM Transactions on Computer-Human Interaction, 12*(2), 293–327.

Brosig, J., Ockenfels, A., & Weimann, J. (2002). The effects of communication media on cooperation. *German Economic Review, 4*(2), 217–241.

Chen, S. C., & Dhillon, G. S. (2003). Interpreting Dimensions of Consumer Trust in E-Commerce. *Information Technology and Management, 4*(2-3), 303.

Cheng, H., & Schweitzer, J. C. (1996). Cultural values reflected in Chinese and U.S television commercials. *Journal of Advertising Research*, (May/June): 27–45.

Cheskin, & Sapient. (1999). *eCommerce Trust Study*, from http://www.studioarchetype.com/cheskin

Chow, S., & Holden, R. (1997). Toward an understanding of loyalty the moderating role of trust. *Journal of Managerial Issues, 9*(3), 275–298.

Corritore, C. L., Kracher, B., & Wiedenbeck, S. (2003). On-line trust: concepts, evolving themes, a model. *International Journal of Human-Computer Studies, 58*(6), 737–758.

Cyr, D., & Trevor-Smith, H. (2004). Localization of web design: An empirical comparison of German, Japanese, and United States web site characteristics. *Journal of the American Society for Information Science and Technology, 55*(13), 1199–1206.

De Mooij, M. (2006). *Global marketing and advertising: Understanding cultural paradoxes*: Sage Publications: London.

Del Galdo, E., & Neilson, J. (1996). *International user interfaces*. New York: John Wiley & Sons.

Deshpande, R., Hoyer, W. D., & Donthu, N. (1986). The intensity of ethnic affiliations: A study of the sociology of Hispanic consumption. *The Journal of Consumer Research, 13*, 214–220.

Egger, F. N. (2001, June 27-29). *Affective design of e-commerce user interface: How to maximize perceived trustworthiness*. Paper presented at the the International Conference on Affective Human Factors Design, Singapore.

El-Shinnawy, M., & Markus, M. (1998). Acceptance of communication media in organizations: richness or features? *Professional Communication . IEEE Transactions on, 41*(4), 242–253.

Ernst, & Young. (2002). *Gulf Cooperation Council eCommerce Study*: Gulf Cooperation Council.

Feng, J., Lazar, J., & Preece, J. (2004). Empathy and online interpersonal trust: A fragile relationship. *Behaviour & Information Technology, 23*(2), 97–106.

Ferguson, G. A. (1989). *Statistical analysis in psychology and education* (5 ed.): New York; London: McGraw-Hill

Ferle, C., Edwards, S., & Mizuno, Y. (2002). Internet diffusion in Japan: Cultural consideration. *Journal of Advertising Research*, (March-April): 65–79.

Fernandes, T. (1995, May). *Global User Interface Design*. Paper presented at the ACM Conference on Human Factors in Computing Systems—CHI '95 Mosaic of Creativity, New York.

Fock, H. (2000). *Cultural influences on marketing communication on the World Wide Web*. Paper presented at the Multicultural Marketing Conference, Hong Kong, September.

Fogg, B. J. (2002). *Stanford Guidelines for Web Credibility A Research Summary from the Stanford Persuasive Technology Lab*, from http://www.webcredibility.org/guidelines

Fogg, B. J., Marshall, J., Kameda, T., Solomon, J., Rangnekar, A., Boyd, J., et al. (2001). *Web credibility research: A method for online experiments and early study results*. Paper presented at the CHI '01 extended abstracts on Human factors in computing systems, Seattle, WA.

Fromkin, V., & Rodman, J. (1983). An Introduction to Language. New York, NY: CBS College Publishing.

Fulk, J., Schmitz, J., & Power, G. J. (1987). A Social Information Processing Model of Media Use in Organizations. *Communication Research, 14*, 520–552.

Gefen, D. (2000). E-commerce: the role of familiarity and trust. *Omega: The International Journal of Management Science, 28*(6), 725–737.

Gefen, D. (2002). Reflections on the dimensions of trust and trustworthiness among online consumers. *The Data Base for Advances in Information Systems, 33*(3), 38.

Gefen, D., & Straub, W. D. (2004). Consumer trust in B2C e-Commerce and the importance of social presence: experiments in e-Products and e-Services. *Omega.Oxford*, *32*(6), 407.

Giddens, A. (1991). *The Consequences of Modernity*. Cambridge, UK: Polity Press.

Gillespie, T. (2000). Trust and Risk in Internet Commerce. *Library Journal*, *125*(10), 186.

Gommans, M., Krishan, K. S., & Scheddold, K. B. (2001). From brand loyalty to e-loyalty: A conceptual framework. *Journal of Economic and Social Research*, *3*(1), 43–58.

Gregory, G., & Munch, J. (1997). Cultural values in international advertising: An examination of familial norms and roles in Mexico. *Psychology and Marketing*, *14*(2), 99–119.

Grewal, G., Munger, J. L., Iyer, G. R., & Levy, J. (2003). The influence of Internet retailing factors on price expectations. *Psychology and Marketing*, *20*(6), 477–493.

Hassanein, K., & Head, M. M. (2004). Manipulating social presence through the Web interface and its impact on consumer attitude towards online shopping. Paper presented at the McMaster eBusiness Research Centre (MeRC)Working Paper Series, Hamilton, ON, Canada.

Hillier, M. (2003). The role of cultural context in multilingual website usability. *Electronic Commerce Research and Applications*, *2*, 2–14.

Hofstede, G. (2005). *Cultures and Organizations*: Harper Collins Business, London.

Horn, D. B., Olson, J. S., & Karasik, L. (2002). The effects of spatial and temporal video distortion on lie detection performance. *Conference on Human Factors in Computing Systems - Proceedings*, 714-715.

Jarvenpaa, S. L., & Tractinsky, N. (1999). Consumer trust in an internet store a cross-cultural validation. *Journal of Computer-Mediated Communication*, *5*(2), 1–35.

Jarvenpaa, S. L., Tractinsky, N., & Vitale, M. (2000). Consumer trust in an Internet store. *Information Technology and Management*, *1*(1-2), 45–71.

Johnson-George, C., & Swap, W. C. (1982). Measurement of specific interpersonal trust: Construction and validation of a scale to assess trust in a specific other. *Journal of Personality and Social Psychology*, *43*(6), 1306–1307.

Jones, M. A. (1999). Entertaining shopping experiences: An exploratory investigation. *Journal of Retailing and Consumer Services*, 6.

Kacen, J., & Lee, J. A. (2002). The influence of culture on consumer impulsive buying behavior. *Journal of Consumer Psychology*, *12*(2), 163–176.

Kammerer, M. (2000). Die Bestimmung von Vertrauen in Internetangebote. Zurich: Lizenziatsarbeit der Philisophischen Fakultaet der Universitaet Zurich.

Kang, K. S., & Corbitt, B. (2001). Effectiveness of graphical components in web site e-commerce application-A cultural perspective. *The Electronic Journal Systems in Developing Countries.*

Keevil, B. (1998, September 24 to 26, 1998). *Measuring the Usability Index of Your Web Site.* Paper presented at the the annual Special Interest Group on Systems Documentation conference SIGDOC98, Quebec City, Province of Quebec, Canada.

Kim, D. J., Ferrin, D. L., & Rao, H. R. (2003). A Study of the Effect of Consumer Trust on Consumer Expectations and Satisfaction: The Korean Experience. *Proceedings of the ACM Conference on Electronic Commerce, 5*, 310-315.

Kim, E., & Tadisina, S. (2005). Factors Impacting Customers' Initial Trust in E-Businesses: An Empirical Study. *Proceedings of the Hawaii International Conference on System Sciences, Big Island, HI, January 2005*, 1-10.

Kotler, P. (2002). *Marketing Management*. Englewood Cliffs, NJ: Prentice Hall.

Krauter, S. G., Kaluscha, E. A., & Fladnitzer, M. (2006). *Perspectives of Online Trust and Similar Constructs –A Conceptual Clarification*. Paper presented at the ICEC'06, Fredericton, Canada.

Kumar, N., & Benbasat, I. (2002). Para-social presence and communication capabilities of a Web site: A theoretical perspective. *e-Service Journal, 1*(3).

Lewicki, R. J., & Bunker, B. B. (1995). Trust in Relationships: A Model of Trust Development and Decline. In I. B. B. B. J. Z. Rubin (Ed.), *Conflict, cooperation and justice* (pp. 133-173). San Francisco.

Lewis, J. D., & Weigert, A. (1985). Trust as a social reality. *Social Forces, 63*(4), 967–985.

Luhmann, N. (1979). *Trust and power*: Wiley, London.

Luna, D., Peracchio, L. A., & de Juan, M. D. (2002). Cross-cultural and cognitive aspects of web site navigation. *Journal of the Academy of Marketing Science, 30*(4), 397–410.

Marcus, A., & Gould, E. W. (2000). Cultural dimensions and global web user interface design. *Interaction*, (July/August): 33–46.

Matsumoto, D. (2000). *People: Psychology from a Cultural Perspective*. CA: Brookes/Cole.

Mayer, R. C., Davis, J. H., & Schoorman, F. D. (1995). An integrative model of organizational trust. *Academy of Management Review, 3*(20), 709–734.

McAllister, D. J. (1995). Affect- and cognition-based trust as foundations for interpersonal cooperation in organizations. *Academy of Management Journal, 38*(1), 24–59.

McKnigh, D. H., Choudhury, V., & Kacmar, C. (2002). Developing and validating trust measures for e-commerce: An integrative typology. *Information Systems Research, 13*(3), 334.

McKnight, D., & Chervany, N. (2001). What trust means in e-commerce customer relationships: An interdisciplinary conceptual typology. *International Journal of Electronic Commerce, 6*(2), 35–59.

McKnight, D. H., Choudhury, V., & Kacmar, C. (2002). The impact of initial consumer trust on Intentions to transact with a web site: a trust building model. *The Journal of Strategic Information Systems, 11*(3/4), 297–323.

McKnight, D. H., Cummings, L. L., & Chervany, N. L. (1998). Initial trust formation in new organizational relationships. *AMR, 3*(23), 472–490.

Mehrabian, A., & Epstein, N. (1972). A measure of emotional empathy. *Journal of Personality, 40*, 525–543.

Nass, C., Moon, C., Fogg, Y., Reeves, B. J., & Dryer, D. C. (1996). Can computer personalities be human personalities? *International Journal of Human-Computer Studies, 43*, 223–239.

Nass, C., Moon, Y., & Carney, P. (1999). Are respondents polite to computers? Social desirability and direct responses to computers. *Journal of Applied Social Psychology, 5*(29), 1093–1110.

Nielsen, J. (1996). *Top Ten Mistakes in Web Design*, from http://www.sun.com/columns/alertbox/9605.html. date of download.02-08-2006

Nielsen, J. (2004). *Designing for Web Usability*. Indianapolis: New Riders Publications.

Peter, S., & Smeeton, N. C. (2007). *Applied Non-parametric Statistical Methods*: CRC Press.

Reichheld, F. F., & Schefter, P. (2000). E-loyalty: your secret weapon on the web. *Harvard Business Review, 78*(4), 105–113.

Riegelsberger, J., & Sasse, M. A. (2002). *Face it - photos don't make a web site trustworthy.* Paper presented at the CHI '02 extended abstracts on Human factors in computing systems, Minneapolis, Minnesota, USA.

Riegelsberger, J., Sasse, M. A., & McCarthy, J. D. (2002). *Eye-catcher or Blind Spot? The Effect of Photographs of Faces on E-commerce Sites.* Paper presented at the Proceedings of the 2nd IFIP Conference on E-commerce, E-business, E-government (i3e), Boston: Kluwer.

Riegelsberger, J., Sasse, M. A., & McCarthy, J. D. (2003). *Shiny happy people building trust? Photos on e-commerce websites and consumer trust.* Paper presented at the Proceedings of CHI 2003, New York,US.

Riegelsberger, J., Sasse, M. A., & McCarthy, J. D. (2005). *Do people trust their eyes more than ears?: media bias in detecting cues of expertise.* Paper presented at the CHI '05 extended abstracts on Human factors in computing systems, Portland, OR, USA.

Riegelsberger, J., Sasse, M. A., & McCarthy, J. D. (2005). *Do people trust their eyes more than ears?: media bias in detecting cues of expertise.* Paper presented at the CHI '05 extended abstracts on Human factors in computing systems, Portland, OR, USA.

Robbins, S., & Stylianou, A. (2003). Global corporate web sites: an empirical investigation of content and design. *Information & Management, 40*(3), 205–212.

Rousseau, D. M., Sitkin, S. B., Butt, R. S., & Camerer, C. (1998). Not so different after all: a cross-discipline view of trust. *Academy of Management Review, 23*(3), 393–404.

Serva, M. A., Benamati, J., & Fuller, M. A. (2005). Trustworthiness in B2C E-Commerce: An Examination of Alternative Models. *The Data Base for Advances in Information Systems, 36*(3), 89.

Shelat, B., & Egger, F. (2002). *What makes people trust online gambling sites?* Paper presented at the Conference on Human Factors in Computing Systems.

Sheskin, D. J. (2004). *Handbook of Parametric and Nonparametric Statistical Procedures* (2004 ed.): CRC Press.

Shirley, D., Stanley, W., & Daniel, C. (2004). *Statistics for research* (3 ed.): New York; Chichester: Wiley-Interscience

Shneiderman, B. (2000). Designing trust into online experiences. *Communications of the ACM, 43*(12), 57–59.

Short, J., Williams, E., & Christie, B. (1976). *The Social Psychology of Telecommunications.* London: Wiley.

Simon, S. J. (2001). The impact of culture and gender on web sites: An empirical study. *The Data Base for Advances in Information Systems, 1*(32), 18–37.

Singh, N., Xhao, H., & Hu, X. (2003). Cultural adaptation on the Web: A study of American companies' domestic and Chinese websites. *Journal of Global Information Management, 11*(3), 63–80.

Steinbruck, U., Schaumburg, H., Kruger, T., & Duda, S. (2002). *A picture says more than a thousand words photographs as trust builders in e-commerce websites.* Paper presented at the Conference on Human Factors in Computing Systems.

Straub, D. W. (1994). The effect of culture on IT diffusion: E-mail and FAX in Japan and the U.S. *Information Systems Research*, *5*, 23–47.

Sun, H. (2001). Building a culturally-competent corporate web site: An explanatory study of cultural markers in multilingual web design. SIGDOC, 1(October 21-24), 95-102.

Swerts, M., Krahmer, E., Barkhuysen, P., & van de Laar, L. (2003). *Audiovisual cues to uncertainty.* Paper presented at the ISCA Workshopon Error Handling in Spoken Dialog Systems, Switzerland.

Teo, T. S. H., & Liu, J. (2005). Consumer trust in e-commerce in the United States, Singapore and China. *Omega*, *35*(1), 22–38.

Tian, R. G., & Emery, C. (2002). Cross-cultural issues in Internet marketing. *Journal of American Academy of Business*(March), 217-224.

UNCTAD. (2003). *United Nations Conference on Trade and Development, E-commerce and Development Report.* Paper presented at the xviii.

UNCTAD. *(2004).* United Nations Conference on Trade and Development E-commerce and Development Report 2004.

Urban, G. L., Fareena, S., & Qualls, W. (1999). *Design and Evaluation of a trust based advisor on the internet*. Retrieved 20-January, 2006, from http://ebusiness.mit.edu/research/papers/123%20 Urban,%20Trust%20Based%20Advisor.pdf. date of download.11-10-2006

Van Mulken, S., Andre, E., & Müller, J. (1999). An empirical study on the trustworthiness of life-like interface agents. In (pp. 152-156). Mahwah, NJ: Lawrence Erlbaum.

Wallis, G. (2006). Internet spending: measurement and recent trends. *Economic Trends, 628*(MARCH,2006).

Wang, Y., & Emurian, H. (2005). An overview of online trust: Concepts, elements, and implications. *Computers in Human Behavior*, *21*(1), 105–125.

Witkowski, M., Neville, B., & Pitt, J. (2003). Agent mediated retailing in the connected local community. *Interacting with Computers*, *15*(1), 5–32.

Yoo, Y., & Alavi, M. (2002). Media and group cohesion: Relative influences on social presence, task participation, and group consensus. *MIS Quarterly*, *25*, 371–390.

Zajonc, R. (1980). Feeling and thinking: Preferences need no inferences. *The American Psychologist*, *35*(2), 151–175.

Zeng, X., Zeng, J., & Guo, Q. (2005). *Research of trust on B2C electronic commerce.* Paper presented at the the Proceedings of the 7th international conference on Electronic commerce, Xi'an, China.

Zhang, X., & Zhang, Q. (2005). *Online trust forming mechanism: approaches and an integrated model.* Paper presented at the Proceedings of the 7th international conference on Electronic commerce, Xi'an, China.

Zhang, Y., & Gelb, B. (1996). Matching Advertising Appeals to Culture: The Influence of Products' Use Conditions. *Journal of Advertising*, *25*(3).

Zmud, R. O., Lind, M., & Young, F. (1990). An attribute space for organizational communication channels. *Information Systems Research*, *14*, 440–457.

Chapter 3
The Driving Forces of Customer Loyalty:
A Study of Internet Service Providers in Hong Kong

T.C.E. Cheng
The Hong Kong Polytechnic University, Hong Kong

L.C.F. Lai
The Hong Kong Polytechnic University, Hong Kong

A.C.L. Yeung
The Hong Kong Polytechnic University, Hong Kong

ABSTRACT

In this study we examine the driving forces of customer loyalty in the broadband market in Hong Kong. We developed and empirically tested a model to examine the antecedents of customer loyalty towards Internet service providers (ISPs) in Hong Kong. Structural equation modeling (SEM) was used to evaluate the proposed model. A total of 737 valid returns were obtained through a questionnaire survey. The results show that customer satisfaction, switching cost, and price perception are antecedents that lead directly to customer loyalty, with customer satisfaction exerting the greatest influence. Although we found that service quality significantly influences customer satisfaction, which in turn leads to customer loyalty, we did not find a direct relationship between service quality and customer loyalty. Our results also reveal that corporate image is not related to customer loyalty. Our empirical investigation suggests that investing huge resources in building corporate image can indeed be a risky strategy for ISPs.

INTRODUCTION

Due to a recent significant surge in the number of ISPs, the broadband market in Hong Kong has become very crowded, leading to fierce price competition, which has eventually resulted in the elimination of many ISPs from the market. From 2001 to 2006, the number of ISPs in Hong Kong dropped from 258 to 181. As the broadband market matures, the focus of ISPs has shifted from customer acquisition to customer retention. In March 2006, there were around 2.6 million Internet users, including both broadband and narrowband users, representing a 39% penetration rate in Hong Kong. About 64% of these users access through the broadband Internet (Office of the Telecommunications Authority, 2006). These figures establish Hong Kong as one of the most Internet-connected cities in the Asian-Pacific region.

The significance of customer loyalty cannot be overemphasized because it relates closely to the continued survival, as well as the future growth, of companies. For a company to maintain a stable profit level when the market reaches the saturation point, a defensive strategy aiming at retaining existing customers is more important than an offensive one, which targets at expanding the size of the overall market by inducing potential customers to subscribe to its services (Ahmad & Buttle, 2002; Fornell, 1992).

Previous studies on customer loyalty focused on customer satisfaction and switching barriers (Dick & Basu, 1994; Gerpott, Rams, & Schindler, 2001; Lee & Cunningham, 2001). These studies have found that customers experiencing a high level of satisfaction are likely to remain with their existing service providers and maintain their service subscriptions. Switching barriers, on the other hand, play a moderating role in the relationship between customer satisfaction and customer loyalty (Colgate & Lang, 2001; Lee & Cunningham, 2001). Researchers in this area have further elaborated on the linkages between price factors and perceived value (e.g., Grewal, Monroe, & Krishnan, 1998), as well as between price and customer loyalty (e.g., Voss, Parasuraman, & Grewal, 1998). In addition, the marketing literature supports the general notion that pricing factors affect the price perceptions of customers, which in turn contribute to customer loyalty (Reichheld, 1996).

By using SEM, this study empirically analyzes whether customer satisfaction, switching cost, price perception, and corporate image are antecedents of customer loyalty in the context of the ISP market in Hong Kong. We also seek to identify elements of service quality as antecedents of satisfaction, and their levels of impact on satisfaction, and to ascertain whether service quality is a direct antecedent of customer loyalty. We examine the degree to which switching cost and price perception account for the variations in the strength of consumer loyalty to ISPs. Finally, we test if corporate image has any impact on customers' loyalty to their present ISPs.

THEORETICAL BACKGROUND AND HYPOTHESIS DEVELOPMENT

Customer loyalty is a purchase behavior, which, unlike customer satisfaction, is an attitude (Griffin, 1996). Customer loyalty is concerned with the likelihood of a customer returning, making business referrals, providing strong word of mouth, as well as offering references and publicity (Bowen & Shoemaker, 1998). Loyal customers are less likely to switch to competitors in view of a given price inducement, and they make more purchases compared to less loyal customers (Baldinger & Rubinson, 1996). Although most research on loyalty has focused on frequently purchased package goods (i.e., brand loyalty), the loyalty concept is also important for industrial goods (i.e., vendor loyalty), services (i.e., service loyalty), and retail establishments (i.e., store loyalty) (Dick & Basu, 1994). As evidenced in the previous discussions, customer loyalty has been generally described as occurring when customers repeatedly purchase goods or services over time, have word of mouth, and make referrals to other customers.

Antecedents of Customer Loyalty

One of the major factors found to affect customer loyalty is customer satisfaction. Halstead, Hartman, and Schmidt (1994) considered customer satisfaction as an affective response that focuses on product performance against some prepurchase standard during or after consumption. Mano and Oliver (1993) referred to satisfaction as an attitude or evaluative judgment varying along the hedonic continuum focusing on the product, which is evaluated after consumption. Fornell (1992) identified satisfaction as an overall evaluation based on the total purchase and consumption experience of the target product, or service performance compared with prepurchase expectations over time. Oliver (1997, 1999) regarded satisfaction as a fulfillment response or judgment on a product or service, which is evaluated for one-time or ongoing consumption.

Service quality can be defined as the result of the comparison between a customer's expectations on a service and their perception of the way the service has been delivered (Gronroos, 1984; Lehtinen & Lehtinen, 1982; Lewis & Booms, 1983; Parasuraman, Zeithaml, & Berry, 1985, 1988, 1994). Perceived service quality is usually measured by two dimensions, namely *process* quality and *output* quality. Parasuraman et al. (1985, 1988, 1994) developed the 22-item SERVQUAL instrument, which has been widely used to measure service quality in many industries, such as banking (Mukherjee & Nath, 2005), health care (Choi, Lee, Kim, & Lee, 2005), and airport service (Fodness & Murray, 2007). The SERVQUAL instrument assesses the overall service quality by comparing service expectation and actual performance, in terms of five generic dimensions, namely, tangibles, reliability, responsiveness, assurance, and empathy.

When consumers switch service providers, they will incur various costs ranging from the time spent in gathering information about potential alternatives to the benefits forfeited due to termination of the existing service. Patterson and Smith (2003) defined switching cost as the perception of the magnitude of the additional cost incurred to terminate a relationship and to secure an alternative one. Selnes (1993) defined switching cost as the technical, financial, and psychological factors that make it difficult or expensive for a customer to change brands.

Corporate image is defined as the overall impression about a company formed on the minds of the public (Barich & Kotler, 1991; Dichter, 1985; Kotler, 1982). It relates to the different physical and behavioral attributes of a company, such as business name, logo, corporate values, tradition, ideology, and the impression of quality communicated by a customer to a potential customer (i.e., word of mouth). The building of corporate image is a lengthy process. The sensory process starts with ideas, feelings, and previous experience with a company that are retrieved from memory and transformed into a mental image (Yuille & Catchpole, 1977). Past studies have suggested that a host of factors, including advertising, public relations, physical image, word of mouth, and customer's actual experience with the goods and services, influence the corporate image of a company in the mind of a customer (Normann, 1991).

Researchers (e.g., Slater, 1997) and consultants (e.g., Gale, 1994) have recommended that companies should adjust their strategies to retain customers in order to achieve superior customer value delivery as customer value incorporates both the costs and benefits of staying with a company. As such, customers' perceived value is considered as a strong driver of customer retention. Nevertheless, some important questions about the role of price in services have remained unanswered. One is whether price perception has a direct effect on overall customer loyalty. If so, it is essential for companies to actively manage their customers' price perceptions because of their impact on value perceptions. Another question is about the formation of price perception in services. Answers to these questions can help clarify the measurement and management of price perception.

Conceptual Model and Hypotheses

We propose a conceptual model that theorizes the relationships among consumer loyalty, service quality, customer satisfaction, switching cost, and corporate image as shown in Figure 1. In what follows, we justify the postulated relationships in the model and formulate several hypotheses to test the model.

Service Quality and Customer Satisfaction

Service quality researchers refer to satisfaction as a transaction-specific evaluation, and to quality as an overall evaluation based on a whole set of cumulative evaluations. Parasuraman et al. (1994) recommended examining service quality and satisfaction, and their causal link, from both transaction-specific and global perspectives. In the context of the ISP business, which mainly hinges on the ongoing relationship between a customer and their service provider, the cumulative-specific perspective is more suitable to view this ongoing relationship. Moreover, service quality is usually considered as an antecedent of customer satisfaction in the ISP business. Therefore, we hypothesize that

H1: *Perceived service quality is positively related to customer satisfaction.*

Customer Satisfaction and Customer Loyalty

The marketing literature suggests that customer loyalty can be defined in two distinct ways, namely the "behavioral approach" and the "attitude approach" (Jacoby & Kyner, 1973). From the behavioral perspective, customer loyalty is identified as the actual repurchase behavior of a customer (Cunningham, 1961). In contrast, the attitude-based perspective refers to customer loyalty as the intention to repurchase (Fournier

Figure 1. Theoretical framework

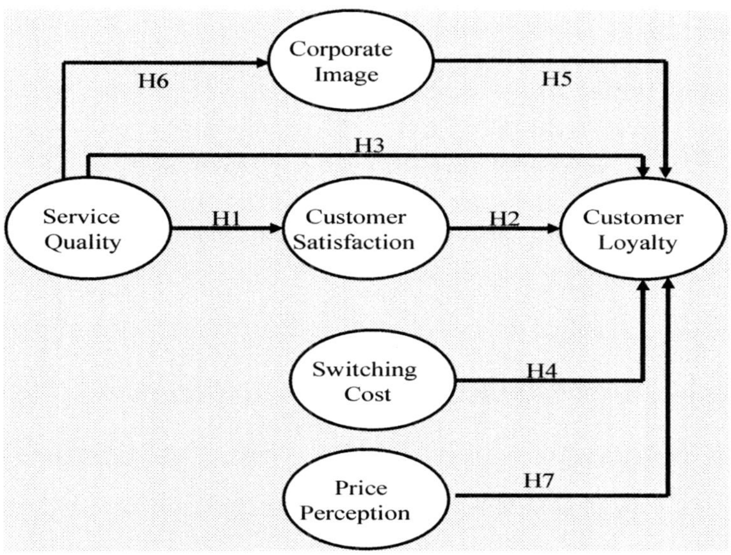

& Yao, 1997). Ajzen and Fishbein (1977) argued that attitude and behavior are consistent in most situations, and that attitude is a strong predictor of future behavior. Thus, Dick and Basu (1994) developed a model that integrates both approaches to study loyalty. Moreover, marketing researchers have investigated the relationships between customer loyalty and different variables, for example, switching cost, which are considered as significant antecedents of customer satisfaction (Bearden & Teel, 1983; Berne, 1997; Bloemer & Kasper, 1993, 1995; Bloemer & Lemmink, 1992; Boulding, Kalra, Staelin, & Zeithaml, 1993; Cronin & Taylor, 1992; Fornell, 1992; Kasper, 1988; LaBarbera & Mazursky, 1983; Oliva, Oliver, & MacMillan, 1992; Oliver, 1999). Research based on the American Customer Satisfaction Index supported empirically that customer loyalty is positively related to customer satisfaction (Fornell, Johnson, Anderson, Cha, & Bryant, 1996). Chiou (2004) obtained this result in his study of the ISP industry, too. Thus, we propose the following hypothesis:

H2: *Customer satisfaction is positively related to customer loyalty.*

Service Quality and Customer Loyalty

The *cognitive evaluation-emotional response-behavioral intention* link explains conceptually how customers form their behavioral intentions. Many studies have also identified a direct positive link between service quality perception and customer behavioral intention (e.g., Boulding et al., 1993; Ranaweera & Neely, 2003; Zeithaml, Berry, & Parasuraman, 1996).

Researchers have attempted to measure the effect of service quality perception on retention. Cronin, Brady, and Hult (2000) found that there exist direct, linear effects of service quality perception, customer satisfaction, and value, on behavioral intention in their large-scale survey of six industries. Particularly, their findings show that service quality perception has a much greater impact than price on determining value. Therefore, the researchers concluded that service customers may consider service quality more important than the cost of acquiring their services. These results are generally consistent with the earlier studies reported previously. However, the study by Cronin and Taylor (1992) showed that using either the SERVQUAL instrument or the SERVPERF instrument to measure service quality fails to confirm the service quality perception—customer behavioral intention link. Using alternative measures of service quality, they found that only satisfaction determines repurchase intention. However, Cronin and Taylor cautioned that their results do not mean that "service quality fails to affect purchase intentions."

Furthermore, some past studies attempting to link customer satisfaction (a similar construct to service quality perception) with customer retention in the retail sector, which is characterized by few or no switching barriers, have established a significant non-linear relationship between these two constructs (e.g., Jones & Sasser 1995; Mittal & Kamakura, 2001). Therefore, a non-linear association between service quality perception and customer retention is also plausible. However, to echo major past research findings, we hypothesize a linear association between service quality perception and customer retention as follows:

H3: *Perceived service quality is positively related to customer loyalty.*

Switching Cost and Customer Loyalty

Switching cost is referred to as the cost incurred by a customer who switches from an existing service provider to a new service provider. The switching cost includes time, money, and psychological cost (Dick & Basu, 1994). It also contains the perceived risks of potential losses perceived by customers at switching, such as losses of a financial, performance-related, social, psychological, and safety-related nature (Murray, 1991). In the ISP environment, when switching cost is high, customers tend to continue using their ISPs' broadband services. The reason is that switching incurs risk (Anton Martin, Garrido Samaniego, & Rodriguez Escudero, 1998; Klemperer, 1995; Ruyter, Wetzels, & Bloemer, 1996; Selnes, 1993; Wernerfelt, 1991). Therefore, we have the following hypothesis:

H4: *Perceived switching cost is positively related to customer loyalty.*

Corporate Image and Customer Loyalty

Corporate image is regarded as the portrait projected by a firm in the mind of its customers. It is the result of an aggregation process that incorporates a range of information used by customers to form a perception of the firm, based on their own previous experience or on the information they acquire from other sources, such as advertising and word of mouth. Corporate image may further establish and affect customer loyalty (Andreassen & Lindestad, 1998; Kandampully & Suhartanto, 2000; Nguyen & Leblanc, 2001). We therefore hypothesize that

H5: *Corporate image is positively related to customer loyalty.*

Service Quality and Corporate Image

Bitner (1992) proposed that cues from the physical environment, which is an important element of service quality, are one of the means that can effectively convey a firm's purposes and image to its customers. Gronroos (1984) argued that corporate image is built mainly by service quality, in terms of both technical quality and functional quality of services. In a study of the airline industry, Ostrowski, O'Brien, and Gordon (1993) concluded that "positive experience over time (following several good experiences) would ultimately lead to positive image and preference." More recent studies have shown that service quality is considered to be partly responsible for the resulting corporate image (Nguyen & LeBlanc 1998; Zins, 2001). Thus, we postulate the following hypothesis:

H6: *Perceived service quality is positively related to corporate image.*

Price Perception and Customer Loyalty

Limited research has been undertaken to investigate the linkage between price perception and customer loyalty (Ranaweera & Neely, 2003; Varki & Colgate 2001). Ranaweera and Neely (2003) showed that price perception has a direct linear relationship with customer loyalty in the telecommunications sector. We believe that such a relationship may be more explicit in the ISP environment in Hong Kong, where there is fierce price competition. Hence, we formulate the following hypothesis:

H7: *Price perception is positively related to customer loyalty.*

RESEARCH METHODOLOGY

Sample

This study targets customers of Internet services in Hong Kong. We collaborated with a local marketing research company to conduct a large-scale questionnaire survey of users of Internet services in Hong Kong. We randomly e-mailed 100,000 invitations to users of Internet services captured in the database of the marketing company to participate in our survey.

Data Collection Procedure

We conducted a pilot study with 20 ISP users to assess the relevance of the indicators to the corresponding constructs and the clarity of the instructions for completing the questionnaire. Upon completing the pilot study, we made minor changes to the questionnaire in order to improve its validity and readability. The questionnaire was developed in English and translated into Chinese. To ensure its face validity, the questionnaire was reviewed by industry practitioners and scholars. Based on their evaluations, corrections and amendments were made. As suggested by Ueltschy, Laroche, Eggert, and Bindl (2007), some measures of both customer satisfaction and service quality may be non-equivalent across cultures, which would limit their usage across borders. When measurement scales are created in one country and then translated for use in another, the interpretation and connotation of certain terms may negatively impact their applicability. In order to minimize the cultural effect, a pretest involving exploratory interviews with users and experts of ISPs was conducted to ensure the questionnaire was relevant and clear to the respondents with the Chinese culture.

With the assistance of a marketing research firm in Hong Kong, a total of 100,000 e-mail invitations were sent randomly to Internet users within the company's database. They were asked to participate in our survey by clicking the hyperlink included in our e-mail invitations. Out of 100,000 invitations sent out, 3,247 recipients opened the e-mail. Once a respondent completed the questionnaire, their answers were automatically entered into our database. We received 856 completed questionnaires. However, 119 questionnaires were not answered by ISP users, so only 737 usable returns were obtained from 3,247 recipients who had opened our invitation e-mails, yielding an effective response rate of 22.7%.

The respondents varied in demographics and background. In general, they are mature and well educated. Table 1 summarizes the respondent characteristics. The relatively high educational and income profile of the respondents indicates that they may not be a perfect representative sample of Internet users in Hong Kong. Though the samples were selected on a random basis, it is difficult to avoid the potential bias that respondents with high educational background and incomes are more likely to response.

Non-response bias was evaluated by following Armstrong and Terry's (1977) suggested approach. We used the mid-point of the data collection period to distinguish early and late respondents. Seventy-seven percent of the responses were from early respondents while the remaining 23% were from late respondents. Applying the independent sample *t*-test, we compared the responses of the early and late respondents. We observed no significant differences in the answers ($p < 0.05$) between the early and late respondents, which suggests that non-response bias did not appear to be a problem in our study.

Measurement and Operationalization of Constructs

To develop the instrument for our study, we based our efforts on an extensive review of the relevant literature. We subsequently revised some of the items of the instrument, taking into consideration the findings of the pilot study and comments from some experienced researchers. We list in Table 2 the complete instrument that was included in our survey. Hereafter we discuss the measurement and operationalization of each of the constructs embedded in our conceptual model.

- **Service quality:** We measured service quality using the SERVPERF instrument developed by Cronin and Taylor (1992). SERVPERF is a 22-item scale consisting of five dimensions, namely, reliability, responsiveness, assurance, empathy, and tangibles. It treats service quality as dis-

Table 1. Summary of respondent characteristics

Gender	Percentage	
Male	69%	
Female	31%	

Age	Percentage	Cumulative Percentage
15 or below	0%	0%
16 – 25	12%	12%
26 – 35	35%	47%
36 – 45	34%	81%
46 or above	19%	100%

Education	Percentage	Cumulative Percentage
Primary school	0%	0%
Secondary school	14%	14%
Post secondary	14%	28%
Tertiary	72%	100%

Income level	Percentage	Cumulative Percentage
Below HK$5,000 / mth	3%	3%
HK$5,000 – 9,999 / mth	11%	14%
HK$10,000 – 14,999 / mth	19%	33%
HK$15,000 – 19,999 / mth	14%	47%
HK$20,000 – 29,999 / mth	17%	64%
HK$30,000 / mth or above	36%	100%

confirmation between expectation and performance. The perception data relative to a respondent's expectation are collected directly. Each respondent was asked to rate each item of service quality on a five-point scale, anchored at 1 = strongly agree and 5 = strongly disagree. The Cronbach alpha was 0.84, indicating high reliability for this construct.

- **Customer satisfaction:** We adopted Oliver's (1980) instrument to assess customer satisfaction. We asked respondents to evaluate their satisfaction with the decision to choose their ISPs, their belief of making a right decision, and their overall satisfaction with their ISPs. Respondents were invited to rate the indicators on a five-point, Likert-type scale, anchored at 1 = strongly agree and 5 = strongly disagree. The Cronbach alpha was 0.95, indicating very high construct reliability.
- **Customer loyalty:** We measured the attitude aspect of customer loyalty, which is a common means of assessing this latent construct as recommended by Berne (1997). We used "change to another ISP," "continuity in us-

Table 2. Questionnaire and its measurement properties

Service Quality (Cronbach's α = 0.87, AVE=0.68)	
SQ1*	My ISP has up-to-date equipment (e.g., Modem)
SQ1*	My ISP's physical facilities are visually appealing (e.g., Design of the stores)
SQ1*	My ISP's customer service staff are well-dressed and appear neat.
SQ1*	The appearance of the physical facilities of my ISP is in keeping with the type of other ISPs.
SQ2	When my ISP promises to do something by a certain time, it does so.
SQ2	When I have problems, my ISP takes corrective action without delay.
SQ2	My ISP is dependable (e.g., High connection speed, high availability of network).
SQ2	My ISP customer service staff make an effort to explain things in a simple way.
SQ2	My ISP keeps its records accurately.
SQ3*	My ISP does not tell customers exactly when services will be performed.
SQ3*	It is difficult to contact my ISP whenever necessary.
SQ3*	My ISP's customer service staff are not always willing to help customers.
SQ3*	My ISP's customer service staff are too busy to respond to customer requests promptly.
SQ4	I can trust my ISP's customer service staff.
SQ4	I feel safe in my transactions with my ISP's customer service staff.
SQ4	My ISP's customer service staff are polite.
SQ4	My ISP's customer service staff get adequate support from their firm to do their jobs well.
SQ5	My ISP keeps me informed of things that I need to get the best use of the service.
SQ5	My ISP's customer service staff give me personal attention.
SQ5	My ISP's customer service staff understand my needs best.
SQ5	My ISP has my best interests at heart.
SQ5	I find the operating hours of my ISP convenient.
Customer satisfaction (Cronbach's α = 0.95, AVE=0.87)	
S1	I am happy about my decision to choose this ISP.
S2	I believe that I did the right thing when I chose this ISP.
S3	Overall, I am satisfied with this ISP.
Corporate Image (Cronbach's α = 0.90, AVE=0.75)	
IMA1	I have always had a good impression of my ISP.
IMA2	In my opinion, my ISP has a good image in the minds of customers.
IMA3	I believe that my ISP has a better image than its competitors.
Price Perception (Cronbach's α = 0.88, AVE=0.80)	
PP1	The prices charged by my ISP are reasonable.
PP2	My ISP's services are value-for-money.

Responses to the following questions ranged from 1 = strongly agree to 5 = strongly disagree.

**Deleted item*

continued on following page

Table 2. continued

Switching Cost (Cronbach's α = 0.95, AVE=0.54)	
SC1	To change to another ISP involves investing time in searching for information about other ISPs.
SC2*	To change to another ISP involves the sacrifice of existing benefits and privileges accumulated with my existing ISP.
SC3	To change to another ISP incurs a risk in choosing another ISP that might turn out not to satisfy me.
Customer Loyalty (Cronbach's α = 0.84, AVE=0.64)	
L1	I will not change to another ISP because I value my ISP.
L2	I will continue to use my ISP within the next 12 months.
L3	I would always recommend my ISP to someone who seeks my advice.

Responses to the following questions ranged from 1 = strongly agree to 5 = strongly disagree.
**Deleted item*

ing the ISP," and "recommending the ISP to others" as indicators for this construct. Respondents were requested to rate these indicators on a five-point, Likert-type scale, anchored at 1 = strongly agree and 5 = strongly disagree. The Cronbach alpha was 0.84, indicating high construct reliability.

- **Switching cost:** We adopted the typology proposed by Vilagines (1994) to measure switching cost. Particularly, we focused on assessing switching cost by the time required to search for information about other ISPs, the effort involved in deciding on another ISP, and the risk of making a mistake with the switch. A five-point, Likert-type scale ranging from 1 = strongly agree to 5 = strongly disagree was used. The Cronbach alpha was 0.65, suggesting moderate and marginally acceptable construct reliability.
- **Corporate image:** We followed Nguyen and LeBlanc's (2001) suggestion to assess the construct of corporate image. We measured this latent construct by good impression, good image in the minds of customers, and better image than competitors. Respondents were asked to rate the indicators on a five-point, Likert-type scale, anchored at 1 = strongly agree to 5 = strongly disagree. The Cronbach alpha was 0.90, indicating high construct reliability.
- **Price perception:** We measured price perception by two questions. One refers to the "reasonableness of price," which was used in Ranaweera and Neely's (2003) study. It captures the way in which price is perceived relative to that of competitors. Another question concerns "value for money," which was used in Varki and Colgate's (2001) study. It reflects the relative standing of one's service provider in terms of price. A five-point, Likert-type scale ranging from 1 = strongly agree to 5 = strongly disagree was used. The Cronbach alpha was 0.88, suggesting high construct reliability.

DATA ANALYSIS AND RESULTS

We applied SEM to examine our proposed model, using analysis of moment structures (AMOS). We followed Anderson and Gerbing's (1982) two-step approach, whereby we estimated the measurement model prior to estimating the structural model. To

Table 3. Reliability and convergent validity of the final measurement model

Indicators	Description	R^2	*t*-value	Alpha
SQ2	Reliability	.745	-[a]	0.8683
SQ4	Assurance	.610	18.37	
SQ5	Empathy	.560	17.19	
IMA1	I have good impression of my ISP	.817	-	0.9003
IMA2	Customers have good image	.774	24.21	
IMA3	Better image than competitors	.669	21.01	
S1	Happy with my decision	.882	-	0.9526
S2	I did the right thing	.837	32.09	
S3	I am satisfied with my ISP	.893	36.01	
SC1	Changing ISPs involves investing time	.407	-	0.6933
SC3	Risk of choosing a bad ISP	.691	4.39	
PP1	Charge is responsible	.608	-	0.8750
PP2	Value for money	.994	23.70	
L1	I will not change to other ISP	.627	-	0.8426
L2	Continue to use 12 months	.570	14.99	
L3	Will recommend to others	.723	17.09	

[a] *Indicates a parameter fixed at 1.0 in the original solution.*

[b] *Fit indices:* $\chi^2 = 218$ $(p = 0.000)$, $df = 95$, $\chi^2 / df = 2.295$, $NNFI = 0.954$, $CFI = 0.974$

avoid sample bias, we randomly divided the 737 usable responses into two groups. The first group of 368 responses was used to test the measurement model, while the second group of 369 responses was used to test the structural model. In what follows, we present and discuss the results of the measurement model analysis, structural model analysis, and hypothesis testing.

Measurement Model Results

We assessed the convergent and discriminant validity of the scales by the methods outlined in Bollen (1989) and Chau (1997). Convergent validity is assessed by the significance of the *t*-values of the item loadings. In addition, it would be difficult to justify a proposed item for a latent construct in research if its reliability is less than 0.50, because in that case 50% of its variance is error variance. It is common to drop the worst performing item from its respective scale and to re-estimate the parameter values, if any item exhibits an R^2 value below 0.50. This may require several iterations and the goal is to produce an acceptable model that maximizes performance for a given sample. Table 3 summarizes the reliability (R^2 values) and convergent validity (*t*-values) of the final measurement model. All R^2 values were greater than the 0.50 threshold level, and all the item loadings of the constructs were significant, with *t*-values of at least 4.39 ($p < 0.01$). These results provide sufficient evidence of reliability and convergent validity of the constructs examined in our study.

Evidence of discriminant validity of a construct is present if the average variance extracted (AVE) of the construct is greater than its squared correlations with other constructs (Fornell & Larcker,

1981). The AVEs of service quality, customer satisfaction, corporate image, switching cost, price perception, and customer loyalty were 0.684, 0.870, 0.753, 0.543, 0.801, and 0.641, respectively. They were all larger than the squared correlations between any target construct and other constructs, which ranged from 0.001 to 0.516 (Table 4). The elements on the diagonal are all larger than the off-diagonal elements in Table 4. The largest squared correlation between two different constructs (off-diagonal) was 0.516 and the smallest AVE (on the diagonal) was 0.543. These results provide sufficient evidence of discriminant validity of the constructs included in our study.

Structural Model Result and Hypothesis Testing

Table 5 presents the overall model fit and the results of testing of each of the research hypotheses using the second group of 369 responses. The results of the structural model indicate an adequate fit: $\chi^2 = 308$ ($p = 0.000$), df = 96, χ^2 / df = 3.208, NNFI = 0.933, CFI = 0.947, and RMSEA = 0.077. The structural model meets all the criteria for fit measures except the χ^2 / df value of 3.2, which was marginally higher than the acceptable value of 3.0. This demonstrates that the structural model fits the data very well.

Table 4. Summary of AVEs and squared correlations

	Construct	1	2	3	4	5	6
1	Service Quality	**0.684**					
2	Customer Satisfaction	0.212	**0.870**				
3	Corporate Image	0.516	0.170	**0.753**			
4	Switching Cost	0.028	0.001	0.015	**0.543**		
5	Price Perception	0.408	0.011	0.038	0.020	**0.801**	
6	Customer Loyalty	0.044	0.347	0.016	0.045	0.022	**0.641**

[a] *AVE on the diagonal*

[b] *Squared correlation off the diagonal*

Table 5. Overall model fit and tests of research hypotheses

Causal path	Hypothesis	Expected sign	Path coefficient	R^2	*t*-value	*p*-value	($p \leq 0.05$)
SQ -> Satisfaction	H1	+	0.856	0.733	17.33	***	Significant
Satisfaction -> Loyalty	H2	+	0.726	0.527	6.55	***	Significant
SQ -> Loyalty	H3	+	0.109	0.012	-0.68	0.611	Insignificant
Switching cost -> Loyalty	H4	+	0.176	0.031	3.99	***	Significant
Image -> Loyalty	H5	+	0.134	0.018	1.48	0.168	Insignificant
SQ -> Image	H6	+	0.827	0.684	15.10	***	Significant
Price -> Loyalty	H7	+	0.130	0.017	2.57	***	Significant

Note: $\chi^2 = 308$ ($p = 0.000$), $df = 96$, $\chi^2 / df = 3.208$, $NNFI = 0.933$, $CFI = 0.947$, $RMSEA = 0.077$

Figure 2 displays the results of hypothesis testing. All hypothesized relationships, except H3 and H5, were highly significant at $p = 0.05$. The estimate of the standardized path coefficient indicates that the linkage between service quality and customer satisfaction is highly significant (H1 was supported: path coefficient = 0.86, t = 17.333, p = 0.000). The estimate of the standardized path coefficient shows that customer satisfaction affects customer loyalty substantially (H2 was supported: path coefficient = 0.73, t = 6.55, p = 0.000). The estimate of the standardized path coefficient displays that the link between switching cost and customer loyalty is significant (H4 was supported: path coefficient = 0.18, t = 3.99, p = 0.000). The estimate of the standardized path coefficient indicates that the linkage between service quality and corporate image is significant (H6 was supported: path coefficient = 0.83, t = 15.1, p = 0.000). The estimate of the standardized path coefficient displays that price perceptions affects customer loyalty greatly (H7 was supported: path coefficient = 0.13, t = 2.57, p = 0.000).

DISCUSSIONS AND CONCLUSION

Our findings show that service quality and customer satisfaction explained 73% and 53% of the variance of customer loyalty, respectively, suggesting that service quality and customer satisfaction have significant effects on customer loyalty. Moreover, as shown by the path estimates in Table 5, the relationship between service quality and customer loyalty mediated by customer satisfaction is much stronger than the one without the mediation effect of customer satisfaction. This supports the notion that customer satisfaction is a mediator in the link between service quality and customer loyalty.

The findings suggest that the reliability, assurance, and empathy dimensions of service quality are significant predictors of customer satisfaction. The reliability dimension is related to the connection speed and availability of the network, which form the core of user experience. The assurance dimension includes security of the transactions and trustworthiness of the customer service staff. They are essential to provide online services and transactions. Given the 24-hour operation of ISP users, the empathy dimension, which includes

Figure 2. Hypothesized model and its path estimates (p < 0.05)

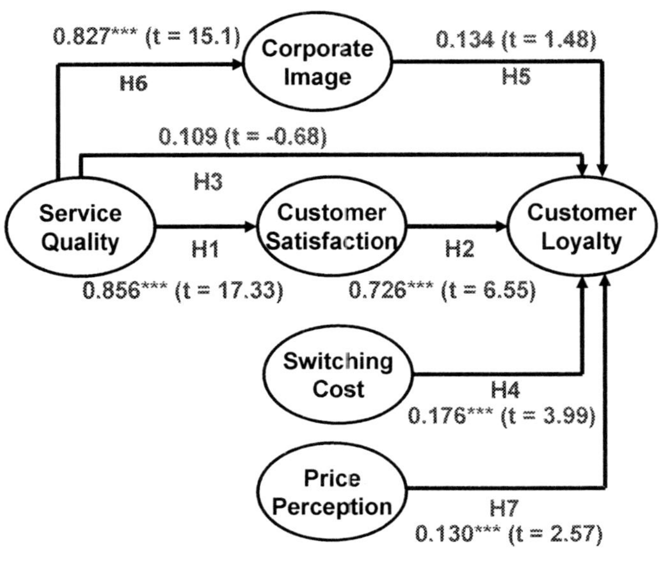

convenient operating hours with support, is crucial. In addition, due to the complexity of IT issues, the empathy dimension reflects the extent to which customer service staff understand customers' needs. This enables service staff to assist customers easier and better.

It is interesting to note that switching cost and price perception only explained 3.1% and 1.7% of the variance of customer loyalty, respectively. These findings suggest that ISP users are not price sensitive and switching cost is not a substantial barrier to ISP users to consider switching. In addition, our results show that corporate image is not significantly related to customer loyalty either. This is because ISP users are rational, and therefore advertising and image building campaigns have a relatively limited impact on users' retention behaviors.

In a competitive market, it may be more difficult to recruit new customers than to retain existing customers. Quite often, profits generated from loyal customers increase as the relationships between service providers and customers grow in strength and intensity. Customer loyalty is considered as an effective way to long-term profitability in both business-to-business and business-to-consumer exchange relationships (Reichheld, 1996). Thus, companies have shifted their marketing focus from pure satisfaction generation to loyalty cultivation (Reichheld, 2001). They are more committed to creating and maintaining effective customer retention programs (Bolton, Kannan, & Bramlett, 2000), especially in regard to service subscriptions in the service industry such as the ISP sector. Many ISPs have expended great effort on devising competitive loyalty programs to retain their customers. Therefore, examining the factors that influence consumer loyalty intention is helpful for companies to design more effective customer retention strategies.

In addition, service firms have the tendency to invest heavily in building their corporate images. It is widely accepted that corporate image has the ability to instill loyalty in customers (Nguyen & LeBlanc, 1998; Zins, 2001). However, our study reveals that, in a competitive ISP market, the impact of corporate image on customer loyalty is not really significant. This suggests that it is not advisable for ISPs to channel substantial resources to establishing their corporate images with a view to retaining customers.

REFERENCES

Ahmad, R., & Buttle, F. (2002). Customer retention management: A reflection of theory and practice. *Marketing Intelligence and Planning, 20*(3), 149-161.

Ajzen, I., & Fishbein, M. (1977). Attitude-behavior relations: A theoretical analysis and review of empirical research. *Psychological Bulletin, 84*(5), 888-918.

Anderson, J. C., & Gerbing, D. W. (1982). Some methods for respecifying measurement models to obtain unidimensional construct measurement. *Journal of Marketing Research, 19,* 453-460.

Andreassen, T. W., & Lindestad, B. (1998). The effect of corporate image in the formation of customer loyalty. *Journal of Service Research, 1*(1), 82-92.

Anton Martin, C., Garrido Samaniego, M. J., & Rodriguez Escudero, A. I. (1998). El Grado de Elaboracion del Proceso de Eleccion de Compra como Refuerzo de la Lealtad de Marca. X Encuentro de Profesores Universitarios de Marketing, Santander (pp. 435-448).

Armstrong, J. S., & Terry, S. O. (1997). Estimating nonresponse bias in mail surveys. *Journal of Marketing Research, 14,* 396-402.

Baldinger, A. L., & Rubinson, J. (1996). Brand loyalty: The link between attitude and behavior. *Journal of Advertising Research, 36*(6), 22-34.

Barich, H., & Kotler, P. (1991). A framework for marketing image management. *Sloan Management Review, 32*(2), 94-104.

Bearden, W. O., & Teel, J. E. (1983). Selected determinants of consumer satisfaction and complaint reports. *Journal of Marketing Research, 20,* 21-28.

Berne, C. (1997). Modelizacion de la Poscompra: Satisfaccion y Lealtad. In J. M. Mugica Grijalva & S. Ruiz de Maya (Eds), *El Comportamiento del Consumidor,* Cap. 5, Ariel Economia, Barcelona, 163-180.

Bitner, M. J. (1992). Servicescapes: The impact of physical surroundings on customers and employees. *Journal of Marketing, 56*(2), 57-71.

Bloemer, J. M., & Kasper, H. (1993). Brand loyalty and brand satisfaction: The case of buying audio cassettes anew in the Netherlands. In *Proceedings of the 22nd European Academy Conference,* Barcelona, Spain.

Bloemer, J. M., & Kasper, H. (1995). The complex relationship between consumer satisfaction and brand loyalty. *Journal of Economic Psychology, 16*(2), 311-329.

Bloemer, J. M., & Lemmink, J. G. (1992). The importance of customer satisfaction in explaining brand and dealer loyalty. *Journal of Marketing Management, 8*(4), 351-364.

Bollen, K. A. (1989). *Structural equations with latent variables.* New York: Wiley.

Bolton, R. N., Kannan, P. K., & Bramlett, M. D. (2000). Implications of loyalty program membership and service experiences for customer retention and value. *Journal of Academy of Marketing Science, 28*(1), 95-108.

Boulding, W., Kalra, A., Staelin, R., & Zeithaml, V. A. (1993). A dynamic process model of service quality: From expectations to behavioral intentions. *Journal of Marketing Research, 30,* 7-27.

Bowen, J. T., & Shoemaker, S. (1998). Loyalty: A strategic commitment. *Cornell and Hotel Restaurant Administration Quarterly, 39*(1), 12-25.

Chau, Y. K. (1997). Reexamining a model for evaluating information center success using a structural equation modeling approach. *Decision Science, 28*(2), 309-334.

Chiou, J. S. (2004). The antecedents of customers' loyalty toward Internet service providers. *Information and Management, 41*(6), 685-695.

Choi, K. S., Lee, H., Kim, C., & Lee, S. (2005). The service quality dimensions and patient satisfaction relationships in South Korea: Comparisons across gender, age and types of service. *Journal of Services Marketing, 19*(3), 140-149.

Colgate, M., & Lang, B. (2001). Switching barriers in consumer markets: An investigation of the financial services industry. *Journal of Consumer Marketing, 18*(4/5), 332-347.

Cronin, J. J., Brady, M. K., & Hult, G. T. M. (2000). Assessing the effects of quality, value and customer satisfaction on consumer behavioral intentions in service environments. *Journal of Retailing, 76*(2), 193-218.

Cronin, J. J., Jr., & Taylor, S. A. (1992). Measuring service quality: A re-examination and extension. *Journal of Marketing, 56*(3), 55-68.

Cunningham, R. M. (1961). Customer loyalty to store and brand. *Harvard Business Review, 39*(6), 127-137.

Dichter, E. (1985). What's an image? *Journal of Consumer Marketing, 2*(1), 75-81.

Dick, A. S., & Basu, K. (1994). Customer loyalty: Toward an integrated conceptual framework. *Journal of the Academy of Marketing Science, 22*(2), 99-113.

Fodness, D., & Murray, B. (2007). Passengers' expectations of airport service quality. *Journal of Services Marketing, 21*(7), 492-506.

Fornell, C. (1992). A national customer satisfaction barometer: The Swedish experience. *Journal of Marketing, 56*(1), 6-21.

Fornell, C., Johnson, M. D., Anderson, E. W., Cha, J., & Bryant, B. E. (1996). The American customer satisfaction index: Nature, purpose, and findings. *Journal of Marketing, 60*(4), 7-18.

Fornell, C., & Larcker, D. F. (1981). Evaluating structural equations models with unobservable variables and measurement error. *Journal of Marking Research, 18,* 39-50.

Fournier, S., & Yao, J. L. (1997). Reviving brand loyalty: A conceptualization within the framework of consumer-brand relationships. *International Journal of Research in Marketing, 14*(5), 451-472.

Gale, B. T. (1994). *Managing customer value.* New York: The Free Press.

Gerpott, T., Rams, W., & Schindler, A. (2001). Customer retention, loyalty, and satisfaction in the German mobile cellular telecommunications market. *Telecommunication Policy, 25*(4), 249-269.

Grewal, D., Monroe, K. B., & Krishnan, R. (1998). The effects of price comparison advertising on buyers' perceptions of acquisition value and transaction value. *Journal of Marketing, 62*(2), 46-59.

Griffin, J. (1996). The Internet's expanding role in building customer loyalty. *Direct Marketing, 59*(7), 50-53.

Gronroos, C. (1984). A service quality model and its market implications. *European Journal of Marketing, 18*(4), 36-44.

Halstead, D., Hartman, D., & Schmidt, S. L. (1994). Multisource effects on the satisfaction formation process. *Journal of the Academy of Marketing Science, 22,* 114-129.

Jacoby, J., & Kyner, D. B. (1973). Brand loyalty vs repeat purchasing behavior. *Journal of Marketing Research, 10,* 1-9.

Jones, T. O., & Sasser, W. E. (1995). Why satisfied customers defect? *Harvard Business Review, 73*(6), 88-99.

Kandampully, J., & Suhartanto, D. (2000). Customer loyalty in the hotel industry: The role of customer satisfaction and image. *International Journal of Contemporary Hospitality Management, 12*(6), 346-351.

Kasper, J. D. (1988). On problem perception, dissatisfaction and brand loyalty. *Journal of Economic Psychology, 9*(3), 387-397.

Klemperer, P. (1995). Competition when consumers have switching costs: An overview with applications to industrial organization macroeconomics and international trade. *Review of Economics Studies, 62,* 515-539.

Kotler, P. (1982). *Marketing for nonprofit organizations* (2nd ed.). Englewood Cliffs, NJ: Prentice Hall.

LaBarbera, P. A., & Mazursky, D. (1983). A longitudinal assessment of consumer satisfaction/dissatisfaction. *Journal of Marketing Research, 20,* 393-404.

Lee, M., & Cunningham, L. F. (2001). A cost/benefit approach to understanding service loyalty. *Journal of Services Marketing, 15*(2), 113-130.

Lehtinen, U., & Lehtinen, J. R. (1982). *Service quality—A study of dimensions.* Working Paper. Service Management Institute, Helsinki, Finland.

Lewis, R. C., & Booms, B. H. (1983). The marketing aspects of service quality. In L. L. Berry, G. Shostack, & G. Upah (Eds), *Emerging perspectives in service marketing* (pp. 99-107). Chicago: American Marketing Association.

Mano, H., & Oliver, R. L. (1993). Assessing the dimensionality and structure of the consumption experience: Evaluation, feeling, and satisfaction. *Journal of Consumer Research, 20*(3), 451-66.

Mittal, V., & Kamakura, W. A. (2001). Satisfaction, repurchase intent and repurchase behavior: Investigating the moderating effect of customer characteristics. *Journal of Marketing Research, 38,* 131-142.

Mukherjee, A., & Nath, P. (2005). An empirical assessment of comparative approaches to services quality measurement. *Journal of Services Marketing, 19*(3), 174-84.

Murray, K. B. (1991). A test of services marketing theory: Consumer information acquisition activities. *Journal of Marketing, 55*(1), 10-25.

Nguyen, N., & LeBlanc, C. (1998). The mediating role of corporate image on customers' retention decisions: An investigation in financial services. *International Journal of Bank Marketing, 16*(2), 52-65.

Nguyen, N., & LeBlanc, G. (2001). Corporate image and corporate reputation in customers' retention decisions in services. *Journal of Retailing and Consumer Services, 8*(4), 227-236.

Normann, R. (1991). *Service management: Strategy and leadership in service business.* New York: John Wiley and Sons.

Office of the Telecommunications Authority Web site. Retrieved from www.ofta.gov.hk.

Oliva, T. A., Oliver, R. L., & MacMillan, I. C. (1992). A catastrophe model for developing services satisfaction strategies. *Journal of Marketing, 56*(3), 83-95.

Oliver, R. L. (1980). A cognitive model of the antecedents and consequences of satisfaction decisions. *Journal of Marketing Research, 17,* 460-469.

Oliver, R. L. (1997). *Satisfaction—A behavioral perspective on the consumer.* New York: McGraw-Hill.

Oliver, R. L. (1999). Whence consumer loyalty? *Journal of Marketing, 63*(4), 33-44.

Ostrowski, P. L., O'Brien, T. V., & Gordon, G. L. (1993). Service quality and customer loyalty in the commercial airline industry. *Journal of Travel Research, 32,* 16-24.

Parasuraman, A., Zeithaml, V. A., & Berry, L. L. (1985). A conceptual model of service quality and its implication for future research. *Journal of Marketing, 49*(2), 41-50.

Parasuraman, A., Zeithaml, V. A., & Berry, L. L. (1988). SERVQUAL: A multiple-item scale for measuring consumer perceptions of service quality. *Journal of Retailing, 64*(1), 12-40.

Parasuraman, A., Zeithaml, V. A., & Berry, L. L. (1994). Reassessment of expectations as a comparison standard in measuring service quality: Implications for further research. *Journal of Marketing, 58*(1), 111-124.

Patterson, P. G., & Smith, T. (2003). A cross-cultural study of switching barriers and propensity to stay with service providers. *Journal of Retailing, 79*(2), 107-120.

Ranaweera, C., & Neely, A. (2003). Some moderating effects on the service quality-customer retention link. *International Journal of Operations and Production Management, 23*(2), 230-248.

Reichheld, F. (1996). *The loyalty effect.* Boston: Harvard Business School Press.

Reichheld, F. (2001). Lead for loyalty. *Harvard Business Review, 79*(7), 76-84.

Ruyter, K., Wetzels, M., & Bloemer, J. (1996). On the relationship between perceived service quality and behavioral intentions: A cross-sectional perspective. In J. Beracs, A. Bauer, & J. Simons

(Eds), *Proceedings 25th EMAC Conference*, Budapest, Hungary.

Selnes, F. (1993). An examination of the effect of product performance on brand reputation, satisfaction and loyalty. *European Journal of Marketing, 27*(9), 19-35.

Slater, S. F. (1977). Developing a customer value-based theory of the firm. *Journal of the Academy of Marketing Science, 25*(2), 162-167.

Ueltschy, L. C., Laroche, M., Eggert, A., & Bindl, U. (2007). Service quality and satisfaction: An international comparison of professional services perceptions. *Journal of Services Marketing, 21*(6), 410-423.

Varki, S., & Colgate, M. (2001). The role of price perceptions in an integrated model of behavioural intentions. *Journal of Service Research, 3*(3), 232-240.

Vilagines. (1994). *Marketing Relocional: Como Obtener Clientes Leales y Rentables*. Ed., Gestion S. A., Barcelona.

Voss, G., Parasuraman, A., & Grewal, D. (1998). The role of price and quality perceptions in pre-purchase and postpurchase evaluation of services. *Journal of Marketing, 62*(4), 46-61.

Wernerfelt, B. (1991). Brand loyalty and market equilibrium. *Marketing Science, 10*(3), 229-245.

Yuille, J. C., & Catchpole, M. J. (1977). The role of imagery in models of cognition. *Journal of Mental Imagery, 1*(1), 171-180.

Zeithaml, V., Berry, L. L., & Parasuraman, A. (1996). The behavioral consequences of service quality. *Journal of Marketing, 60*(2), 31-46.

Zins, A. H. (2001). Relative attitudes and commitment in customer loyalty models—Some experiences in the commercial airline industry. *International Journal of Service Industry Management, 12*(3), 269-294.

This work was previously published in the International Journal of E-Business Research, Vol. 4, Issue 4, edited by I. Lee, pp. 26-42, copyright 2008 by IGI Publishing (an imprint of IGI Global).

Chapter 4
E-Business Decision Making by Agreement

William J. Tastle
Ithaca College, USA

Mark J. Wierman
Creighton University, USA

ABSTRACT

Gathering customer data over the Internet is largely limited to collecting the responses to a set of easily answerable questions, such as Yes/No questions and Likert scale questions. These data are then analyzed to identify customer trends or other items of interest to management. The data can be useful, but key to their usage is the application of suitable mathematical tools. Traditionally little more than standard statistics has been used in the analysis of ordinal, or category, data. This can be inaccurate and in some cases, misleading. This paper introduces measures of agreement and dissent to the field of eBusiness analysis and shows how ordinal data can be analyzed in more meaningful ways.

INTRODUCTION

Gathering data from customers is a common activity and much research has gone into design and planning (Parsons, 2007; Solomon, 2001), improving response rates (Cook, Heath, & Thompson, 2000; Kaplowitz, Hadlock, & Levine, 2004; Schmidt, Calantone, Griffin, & Montoya-Weiss, 2005), the study of privacy and ethics (Couper, 2000), mode of questionnaire delivery (Denscombe, 2006), the effect of subject lines of survey responses (Porter & Whitcomb, 2005), and the analysis of Web usage using traditional statistics (Korgaonkar & Wolin, 1999; Stanton, 1998), but little has been written about the evolution of ordinal scale survey results, typical of Likert or Likert-like scale surveys. Acknowledging that getting respondents to answer surveys, either paper or digital, can be a challenge, and once the data is collected the effort to squeeze as much information from the data as possible begins.

Traditionally, data analysis is well founded in statistics, even though the same underpinnings of statistics recognize that there are limits to this branch of mathematics. Statistics are at home when dealing with ratio or interval data (Tastle & Wierman, 2006a), but once the scale shifts to ordered categories the use of statistics is circumspect, for what does it mean to say the average of "warm" and "hot" is reported as "warm-and-a-half" (Jamieson, 2004). Ordinal scales of measurement typically consist of ordered category hierarchies such as: strongly agree (SA), agree (A), neither agree nor disagree (N), disagree (D), and strongly disagree (SD); very cold, cold, cool, tepid, warm, hot, and very hot. The instrument typically used to collect this kind of data is called the Likert scale, though there are variations of this scale such as Likert-like, Likert-type, and ordered response scales. Researchers utilize this kind of instrument to collect data that cannot be ascertained using traditional measures, for the data being collected are feelings, perceptions, sensations, emotions, impressions, sentiments, opinions, passions, or the like. Unfortunately, the application of standard statistics to these data can be improper (Cohen, Manion, & Morrison, 2000; Jamieson, 2004; Pell, 2005). This article looks at the different kinds of scales and presents a new measure for analyzing ordinal scale data.

The identification of consensus in a group environment was the motivation for the original research into ways of assessing ordinal data. The authors sought to identify some mathematical way by which a discussion leader could be guided towards getting a group of discussants to arrive at consensus as quickly as possible. The consensus measure can be easily applied to situations whereby a quick survey of perceptions of discussants to one statement is taken. Given the statement "The group has arrived at consensus" the discussants would check either SA, A, N, D, or SD. The resulting calculation of consensus could guide the leader in the direction of conversation or to determine if there is sufficient agreement to move forward. The authors have expanded on this idea to identify the group agreement with a targeted category, such as SA, on a data collection instrument. It would be nice to know if, in response to some survey statement on a matter of critical importance to the organization, the overall percentage of agreement for each Likert category, not just the mode category. Notice we do not use the mean, for the meaning of the average of two ordered categories is not clear, that is, the average of *acceptable* and *unacceptable* is *acceptable-and-a-half*, or so the interval and ration scale mathematics tells us. Also, standard deviation is based on the presence of at least an interval scale, so its use on ordinal scales is suspect at least, and invalid at most. The dissent measure gives a result that is much easier to interpret and carries more intuitive meaning. In this article we focus on the agreement measure and how it can be used to foster a group agreement assessment that is especially important when a business is largely limited to Internet activities and must rely on survey-type data for assessments that might typically be ascertained through an in-person sales force.

BACKGROUND

We begin with a discussion of the meaning of *consensus*, for it plays a critical role in the analysis and interpretation of **ordinal** data that is collected using Internet-based survey forms, and then conclude this section with a discussion of other works.

It is common for a group of well-intentioned individuals, engaged in purposeful dialogue, to utilize the concept of consensus in making decisions, especially when it is important to maintain some sort of collegiality. In America there exists a set of rules used by most boards and organizations as the arbiter of the structure for group discussions and it is called Robert's Rules of Order. While Robert's Rules are effective, it usually results in someone or some group *losing* in the resulting decision if the leader or chair calls for a vote having *sensed* that most are in agreement. Such feelings may be incorrect. Although consensus building is a typical method used in decision making, few measures exist that allow for the easy determination of the degree to which a group is nearing the point of agreement. When dealing with Internet-based surveys, the ordinal data collected must be analyzed to determine the level of consensus or agreement of the respondents with respect to the questions or issues raised. The purpose of this article is to show a mathematical measure (Tastle & Wierman, 2005, 2006a, 2006b, in press-a, in press-b, in press-c; Wierman & Tastle, 2005) that is intuitive, satisfies the requirements of a measure of consensus, and is easy to apply to the analysis of ordinal surveys.

The survey analysis requires finding some means by which the consensus of the respondents to an ordinal survey can be identified, understood, and compared. As a number of business and political analysts have pointed out in the past, there are problems associated with determining consensus in a group or by survey; the problems are similar. If a too-strict requirement of consensus is asserted, it is possible for a minority group to hold a veto power over decisions. Conversely, a too-loose requirement permits the domination of the minority by the majority, an equally undesirable outcome. It is entirely possible for a decision by consensus to take an extremely long time to occur, and thus may be inappropriate for urgent matters such as decisions involving strategic policy or competitive advantage. Sometimes, consensus decision making encourages groupthink, a situation in which people modify their opinions to reflect what they believe others want them to think. This can lead to a situation in which a group makes a decision that none of the members individually support and may lead to a few dominant individuals making all decisions. Fortunately, survey respondents are not impacted by this problem. Finally, consensus decision making may fail when there simply is no agreement possible, a problem that is theoretically possible when half of the survey respondents select *strongly agree* and the other half select *strongly disagree*. Does such a possibility justify reporting a *neutral* category being the average category chosen? Even if including a standard deviation in the resulting report (for a five category Likert scale, such a standard deviation is 2), there is an expectation and visual image by most readers of values scattered around a mean value of neutral, clearly an erroneous expectation.

Consensus (n.d.) has two common meanings. One is a general agreement among the members of a given group or community; the other is as a theory and practice of getting such agreements. Many discussions focus on whether agreement needs to be unanimous and even dictionary definitions of consensus vary. These discussions miss the point of consensus, which is not a voting system but a taking seriously of everyone's input, and a trust in each person's discretion in follow-up action. In consensus, people who wish to take up some action want to hear those who oppose it because they do not wish to impose, and they trust that the ensuing conversation will benefit everyone. Action despite opposition will be rare and done with attention to minimize damage to relationships. In a sense, consensus simply refers to how any group of people who value liberty might work together.

To capture how someone *feels* towards an issue under discussion, some mechanism must be used by which that person may express his/her opinions, but in a manner such that the data can be quantified. The Likert scale easily fulfills this requirement. Unfortunately, the Likert scale has no interval property. To *solve* this problem some have advocated placing numbers next to the linguistic labels, that is, strongly agree = 1, in an effort to force an interval. This does not work; the presence of an interval means that a respondent has carefully reviewed the available data (or searched his/her mind for a proper feeling) and has evidence that 2.1 is too high and 1.9 is too low, so the choice of 2 is checked. Forcing the presence of numbers does not change an ordinal scale to an interval scale. It remains simply a set of ordered categories and the use of ratio and interval scale mathematics is not conceptually sound when analyzing ordered categories, though the results are accepted as accurate. We propose another way of analyzing ordinal data, and it has great potential in e-business as we attempt to gather as much information as possible out of available data.

There is substantial work on the ranking of discrete data (Chamberlin, Cohen, & Coombs, 1984; Murphy & Martin, 2003) and every good statistics text has a section devoted to Kendall, Spearman, and Cayley rankings (Murphy & Martin, 2003), and sometimes the Hamming and Euclidian distances. Ranking is a means by which items in a collection can be evaluated such that any two items can be compared to see which should be placed higher in the ranking. Hence it is easy to see that presidential candidates can be ranked, as can the top golfers, the National Football League, or World Soccer teams, the flavors of ice cream, and attributes of a product. Unfortunately, we sometimes confuse ordinal ranking with ordinal measures. An ordinal ranking is the assignment of a unique ordinal number to all items in a collection. An ordinal measure is the assignment of a degree of acceptability, desirability, favor, discernment, and so forth to each single attribute. To ask a subset of Internet customers to rank the products in order of desirability is quite different from asking them to assess their agreement that property X is an important quality of product Y. In the latter ranking the customers merely need the list of products and a space next to each item into which their number value can be placed. In the former example, some ordinal scale is provided to which the customer will check a response. For example, in response to the statement "it is important for product Y to have property X" the ordinal scale might be *strongly agree*, *agree*, *neither agree nor disagree*, *disagree*, or *strongly disagree*. It is obviously not useful to take a set of responses to these Likert attributes and attempt to forcibly rank order them, and that is the purpose of this article: to show a new method by which such data can be evaluated from a perspective of group agreement. Using ranks, there is a winner and a looser! Using this novel method of assessing group agreement, each Likert category has a degree of agreement.

Davies (2005) investigated the combination of a fully anchored Likert scale with a numerical rating scale and found that by providing visual cues yielded a more discriminating result in which respondents more consistently applied their ratings. Applying our method to his data yielded identical results without the need for any other visual cues. The method presented here is computationally easy to apply and gives consistent results. We do, however, acknowledge that this work is still in-process, and much more must be done before the measure becomes main stream (see the Conclusion). We hope that readers will build upon our efforts.

METHOD

Let us assume that a data set of Likert scale responses has been collected by means of an Internet survey. The data are represented by a listing of numbers, each one from 1 to 5 representing the standard Likert categories from strongly agree to strongly disagree. We can apply standard statistics to this listing, but a more conceptually accurate method is offered on some reflections on the properties needed to analyze these data.

We postulate that the following set of rules must be satisfied before any measure can be considered a viable solution to the Likert-scale problem:

1. For a given (even) number of **n** individuals participating in a survey on some matter of interest, if **n/2** select the strongly disagree category and the other **n/2** select the strongly

agree category, the group is considered to have **no** consensus. This is called the **diametric opposition** quality.

2. If all the participants classify themselves in the same category, regardless of the label given that category, then the group is considered to be in consensus.
3. If the mix of participants is such that **n/2 + 1** survey respondents assign themselves to any one category, the degree of consensus must be greater than 0, for the balance in the group is no longer equal.

Conversely, dissention requires the following set of rules be satisfied:

1. For a given (even) number of **n** individuals participating in a survey on some matter of interest, if n/2 select strongly disagree and the remaining **n/2** respondents select strongly agree, the group is considered to have **maximum dissention**.
2. If all the respondents classify themselves in the same category, regardless of the label given that category, then the dissention is considered to be **zero**.
3. If the mix of respondents is such that n/2 + 1 respondents assign themselves to any one category, the degree of dissention must be less than maximal.

Consensus and dissention are inverse functions of shared group feelings towards an issue. This *feeling* can be captured through a Likert scale that measures the extent to which a person agrees or disagrees with the statement under investigation. The most common scale is 1 to 5. Often the scale will be 1 = strongly agree, 2 = agree, 3 = not sure, 4 = disagree, and 5 = strongly disagree. Other number assignments can be made, such as: -2 = strongly agree, -1 = agree, 0 = not sure, 1 = disagree, and 2 = strongly disagree, or 0.0 = strongly agree, 0.25 = agree, 0.50 = neutral, and so forth. Likert scales can also be from two to nine categories in width. The issues of scale, symmetry, selection of clusters, and ordinal vs. interval data are not addressed here, but Munshi (1990) has produced a very nice article that describes these aspects in straightforward terms. A rather complete bibliography can also be found there.

THE CONSENSUS AND DISSENTION MEASURES

The properties of a consensus measure is defined (Tastle & Wierman, 2005) as:

$$Cns(X) = 1 + \sum_{i=1}^{n} p_i \log_2\left(1 - \frac{|X_i - \mu_X|}{d_X}\right) \quad (1)$$

where the random variable X represents the Likert scale values, X_i is the particular Likert attribute value, p_i is the probability associated with each X_i, d_X is the width of X, and $E(X) = \sum_{i=1}^{n} p_i X_i = \mu_X$ is the mean of X. This measure adequately fulfills the previous rules as evidenced by the following *illustrations*.

The mirror image of consensus is dissention and has the following form:

$$Dnt(\mathbf{X}) = -\sum_{i=1}^{n} p_i \log_2\left(1 - \frac{|X_i - \mu|}{d_X}\right) \quad (2)$$

In other words, $Cns = 1 - Dnt$ and $Dnt = 1 - Cns$. One of the interpretations of the dissent measure is that of dispersion. If the frequency distribution is balanced on the extreme categories of the Likert scale, for example at *strongly agree* and *strongly disagree*, the dispersion is maximized at 1 (and the consensus is zero). As the frequency distribution approaches the assignment of all probability to a single category, the dispersion approaches 0 (and the consensus approaches one). This is the essence of the consensus measure: the more the respondent assignments are tightly clustered around one category, the higher the consensus and the less the dissent. This dispersion is always a value in the unit interval, [0..1].

Let us assume that we have a five-attribute Likert scale: strongly agree (SA), agree (A), neutral (N), disagree (D), and strongly disagree (SD). Let us further assign a numerical scale of SA = 1, A = 2, N = 3, D = 4, and SD = 5. Then X = {1, 2, 3, 4, 5}, $X_1 = 1$, and so forth, $d_x = 5 - 1 = 4$. Using an arbitrary number of random integer values to populate the scale, the following table denotes the required properties.

Table 1 contains data on eight aspects: the first column is simply an index of the rows, columns SA through SD denote the frequencies assigned to the Likert scale attributes (for comparison purposes all frequencies sum to 12), the expected mean for the attribute values, the standard deviation for the attribute values, Cns and Cns% are the consensus values in decimal and rounded percent, and Dnt and Dnt% are the dissension values in decimal and rounded percent (Cns = 1 - Dnt). Row 1 shows a maximum amount of dissent in consensus since n/2 observations are reflected in each of the extreme attributes. As a point of interest, had the n/2 values been associated with agree and disagree, the consensus would have been 3.0 and the standard deviation 1.0, since these attributes are closer to each other. Rows 2 through 9 show a convergence of opinion moving towards agree. An examination of the mean column shows a modest fluctuation of the values but, in general, a movement of value from neutral (3) to agree (2). This is supported by the StdDev column as the values continue to converge towards 0 as the values surrounding the attributes merge. The consensus shows continuous movement towards 1; it is arguably easier to associate the consensus as a percent to easily visualize the movement towards a consensus. Conversely, one can monitor the dissent from total presence (row 1) to total absence (row 10). Finally, row 10 shows the attribute values firmly in one category. The mean is trivially at 2, the StdDev is now zero, consensus is complete at 100%, and dissent does not exist.

The proof that this measure of ordinal data satisfies the rules listed previously is found in Wierman and Tastle (2005).

THE AGREEMENT MEASURE

Consensus (Equation 1) can become agreement (Equation 3) when the mean μ_X is replaced with

Table 1. Illustration of ten sets of values ranging from the most extreme (row 1) to the most concentrated (row 10). Calculations of the mean, standard deviation, consensus, consensus as a percent, dissent, and dissent as a percent, are shown.

	SA	A	N	D	SD	Mean	St Dev	Cns	Cns%	Dnt	Dnt%
1	6	0	0	0	6	3.0	2	0	0%	1	100%
2	6	0	0	1	5	2.917	1.93	0.049	5%	0.951	95%
3	6	0	0	2	4	2.833	1.86	0.097	10%	0.903	90%
4	5	1	0	2	4	2.917	1.80	0.146	15%	0.854	85%
5	5	1	2	4	0	2.417	1.32	0.425	43%	0.575	58%
6	1	5	2	4	0	2.750	1.01	0.605	61%	0.395	40%
7	1	5	4	2	0	2.583	0.86	0.675	68%	0.325	33%
8	0	6	5	1	0	2.583	0.64	0.758	76%	0.242	24%
9	0	9	3	0	0	2.250	.043	0.896	90%	0.104	11%
10	0	12	0	0	0	2.0	0	1	100%	0	0%

some target value, τ, and we divide by twice the width, $2d_x$, in the denominator. The target, τ, is usually some desired value identified by the manager. For our purposes let us assume that the desired response is *strongly agree*. Since that is the first category in our Likert item, it is assigned a numerical value of 1. Hence, in response to the declarative statement "Customer service is exceptionally good," we desire for our survey respondents to strongly agree with this statement, that is, the target is $\tau=1$. This measure is called *agreement* to distinguish it from measures that use an unspecified target such as the mean, median, or mode. Equation 3 shows τ in place of μ and an expanded width. Doubling the width prevents the equation from exploding when extreme values are reflected in the frequency distribution. We have found the agreement function to work especially well in practice and, for this current work, have limited ourselves to the $2d_x$ denominator. For the most part, either consensus or agreement will work very well, but it is necessary to be consistent in their use. It should also be mentioned that consensus, dissent, and agreement are invariant with respect to linear transformations of the random variable X.

$$Agr(X|\tau) = 1 + \sum_{i=1}^{n} p_i \log_2\left(1 - \frac{|X_i - \tau|}{2d_X}\right) \quad (3)$$

While targeting provides a novel way of measuring distance from a desired goal, it assumes that all elements of the assessment are equally important.

Table 2 shows the Table 1 Likert data and the mean, and adds the agreement measures for each of the Likert categories. The usefulness of the measure becomes evident with practice. For example, the first row shows a higher agreement for target 3 (neutral) than for the other values, and the extreme values, 1 (SA) and 5 (SD), have the smallest measure of agreement. At first, this seems counterintuitive, but the agreement measure is actually mid-way between consensus and consent for SA and SD. Table 1 shows a consensus for the entire first row having the value of 0. With respect to the entire distribution of categories, there is no *consensus* whatsoever. However, with respect to the neutral category there is a 50% *agreement*. When the set of surveys support the extreme categories, there is a de facto agreement on the middle category (neutral in this case) but the level of agreement is certainly not 0, nor is it 100%. It

Table 2. Illustration of the same ten sets of values ranging from the most extreme (row 1) to the most concentrated (row 10). Calculations of the mean and agreement values for each category, that is, Agt(1) is read as "agreement with respect to strongly agree as the target."

	SA	A	N	D	SD	Mean	Agt(1)	Agt(2)	Agt(3)	Agt(4)	Agt(5)
1	6	0	0	0	6	3.0	0.500	0.565	0.585	0.565	0.500
2	6	0	0	1	5	2.917	0.527	0.587	0.603	0.581	0.484
3	6	0	0	2	4	2.833	0.554	0.608	0.622	0.597	0.468
4	5	1	0	2	4	2.917	0.538	0.625	0.641	0.619	0.495
5	5	1	2	4	0	2.417	0.689	0.749	0.747	0.651	0.393
6	1	5	2	4	0	2.750	0.625	0.813	0.821	0.738	0.501
7	1	5	4	2	0	2.583	0.668	0.851	0.853	0.706	0.464
8	0	6	5	1	0	2.583	0.674	0.885	0.888	0.712	0.472
9	0	9	3	0	0	2.250	0.752	0.952	0.856	0.641	0.388
10	0	12	0	0	0	2.0	0.807	1.000	0.807	0.585	0.322

is logical that some middle value is appropriate, like 50%. Looking down to the values in row 7 we note that there are more respondents who have selected A than any other category. The mean is 2.58, which indicates that the average is almost between agree and neutral, perhaps agree-and-a-half. There is a 68% consensus on the part of the respondents with a dispersion of about 33%. The respondent values are becoming clustered, but what is the data telling us? The agreement with a target of neural has the greatest value, 0.853, which is interpreted as an 85.3% agreement of the respondents for the neutral category. The four who selected neutral are not the deciders of the agreement but rather, the five plus 2 that surround it. Agreement takes into account all the data in the distribution.

Finally, examination of row 10 shows complete consensus for the overall distribution as would be expected with all respondents having selected the same category, and the agreement with respect to agree is also 1.0. It is also evident that there is some agreement with the contiguous categories, like 80% agreement with both strongly agree and neutral, and even a modest level of agreement with strongly disagree of 32%. The absence of data from one or more categories does not mean an absence of agreement. All agreement values are shown in Figure 1.

CONCLUSION

Data collected through the Internet can be analyzed in many statistically proper ways, but the fundamental premise of the presence of an interval or ratio scale is absent from ordinal data. This method is a new way of examining ordered ordinal data, is very intuitive, and requires little effort in calculating. The authors have used a spreadsheet to perform the calculations. The category that is most targeted by the survey respondents can be identified, and the degree of overall consensus with respect to the frequency distribution is interpreted as a percentage of the whole. Also, the measure of dissent is an indicator of dispersion. A visual representation of the proximity of categories can show confusion (multiple categories with close

Figure 1. The measures of agreement for all 10 rows of data shown in Table 2

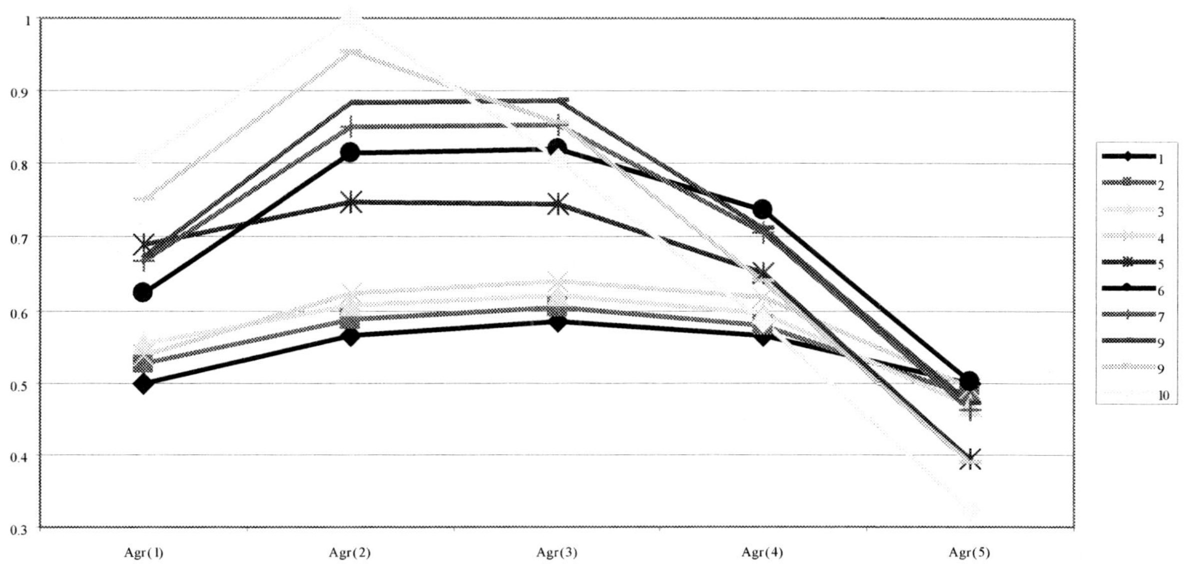

agreement measures, or a mandate for a particular category (steep drop-off in agreement measure for categories on either side). Combined, these measures permit the examination of data from any entirely new perspective.

Other work in the identification of measures to assist group decision making and analysis of non-interval or ratio scale data offers considerable potential for the interested researcher. What follows is an incomplete listing of potential opportunity:

- The authors think that a statistical change in Likert categorical importance, assuming a normal distribution of potential responses, can be approximated by the normal measure of significance. However, it has not yet been proven.
- The Likert scale demands the selection of only one category, but what about the possibility of degrees of categorical selection, as would be found in a typical fuzzy set.
- Our preliminary investigations suggest that the presence or absence of an interval does not change the result of the measure. Thus, we can be very confident in the resulting choice of the selected category, but we have not yet undertaken this research.
- We seek to identify a measure of consistency; preliminary attempts using covariance do not satisfy our intuition as to the properties we require of such a measure.
- Lastly, nominal measures are reputed to have only one valid statistical measure, that of the mode. We have found, only in the most preliminarily way, that a measure of dispersion can be validly calculated for nominal measures, but much work remains before an article can be written.

Using these measures and the family of measures that could develop from them, analysis of nominal and ordinal data might be able to move past the traditional statistical approach

ACKNOWLEDGMENT

The authors are grateful to the insightful comments from the anonymous reviewers. We wish we could thank them in person, but that would invalidate the peer review process so we thank them publicly.

REFERENCES

Chamberlin, J. R., Cohen, J. L., & Coombs, C. H. (1984). Social choice observed: Five presidential elections of the American Psychological Association. *The Journal of Politics, 46*(2), 479-502.

Cohen, L., Manion, L., & Morrison, K. (2000). *Research methods in education* (5th ed.). London: Routledge Falmer.

Consensus. (n.d.). Retrieved from http://essentialfacts.com/primary/ethics/Consensus.html

Cook, C., Heath, F., & Thompson, R. (2000). A meta-analysis of response rates in Web- or Internet-based surveys. *Education and Psychological Measurement, 60*(6), 821-836.

Couper, M. (2000). Web surveys: A review of issues and approaches. *The Public Opinion Quarterly, 64*(4), 464-494.

Denscombe, M. (2006). Web-based questionnaires and the mode effect. *Social Science Computer Review, 24*(2), 246-254.

Jamieson, S. (2004). Likert scales: How to (ab)use them. *Medical Education, 38,* 1217-1218.

Kaplowitz, M., Hadlock, T., & Levine, R. (2004). A comparison of Web and mail survey response rates. *Public Opinion Quarterly, 68*(1), 94-101.

Korgaonkar, P., & Wolin, L. (1999). A multivariate analysis of Web usage. *Journal of Advertising Research, 39*(2), 53-68.

Munshi, J. (1990). A method for constructing Likert scales. Retrieved June 2004, from http://www.munshi.4t.com

Murphy, T. B., & Martin, D. (2003). Mixtures of distance-based models for ranking data. *Computational Statistics & Data Analysis, 41,* 645-655.

Parsons, C. (2007). Web-based surveys: Best practices based on the research literature. *Visitor Studies, 10*(1), 1064-5578. Retrieved July 15, 2007, from http://www.informaworld.com/10.1080/10645570701263404

Pell, G. (2005). Use and misuse of Likert scales. *Medical Education, 39,* 970.

Porter, S., & Whitcomb, M. (2005). E-mail subject lines and their effect on Web survey viewing and response. *Social Science Computer Review, 23*(3), 380-387.

Schmidt, J., Calantone, R., Griffin, A., & Montoya-Weiss, M. (2005). Do certified mail third-wave follow-ups really boost response rates and quality? *Marketing Letters, 16*(2), 129-141.

Solomon, D. (2001). Conducting Web-based surveys. *Practical Assessment Research and Evaluation, 7*(19).

Stanton, J. (1998). An empirical assessment of data collection using the Internet. *Personal Psychology, 51*(3), 709-725.

Tastle, W. J., & Wierman, M. J. (2005). Consensus and dissention: A new measure of agreement. *North American Fuzzy Information Processing Society (NAFIPS) Conference,* Ann Arbor, MI.

Tastle, W. J., & Wierman, M. J. (2006a). An information theoretic measure for the evaluation of ordinal scale data. *Behavioral Research Methods, 3,* 487-494.

Tastle, W. J., & Wierman, M. J. (2006b). Consensus and dissension: A new measure of agreement. In *NAFIPS 2006,* Montreal, Canada.

Tastle, W. J., & Wierman, M. J. (in press-a). Consensus: A new measure of ordinal dispersion measure. *International Journal of Approximate Reasoning.*

Tastle, W. J., & Wierman, M. J. (in press-b). The development of agreement measures: From general to targeted. *International Journal of Approximate Reasoning.*

Tastle, W. J., & Wierman, M. J. (in press-c). Determining risk assessment using the weighted ordinal agreement measure. *Journal of Homeland Security.*

Wierman, M. J., & Tastle, W. J. (2005). Consensus and dissention: Theory and properties. *North American Fuzzy Information Processing Society (NAFIPS) Conference,* Ann Arbor, MI.

This work was previously published in the International Journal of E-Business Research, Vol. 4, Issue 4, edited by I. Lee, pp. 16-25, copyright 2008 by IGI Publishing (an imprint of IGI Global).

Chapter 5
Transforming Consumer Decision Making in E-Commerce:
A Case for Compensatory Decision Aids

Naveen Gudigantala
Texas Tech University, USA

Jaeki Song
Texas Tech University, USA

Donald R. Jones
Texas Tech University, USA

ABSTRACT

To facilitate online consumer decision making, a number of e-commerce businesses are augmenting their Web site features. The Web-based decision support for consumers is often provided by eliciting consumer preferences directly or indirectly to generate a set of product recommendations. The software that captures consumer preferences and provides recommendations is called a Web-based decision support system (WebDSS). It is important for WebDSS to support consumers' decision strategies. These decision strategies could be compensatory or non-compensatory in nature. Based on a synthesis of previous research, the authors argue that compensatory based WebDSS are perceived by consumers to be better than non-compensatory WebDSS in terms of decision quality, satisfaction, effort, and confidence. This chapter presents a study that examined the level of online support provided to the consumers' execution of compensatory and non-compensatory strategies. The results based on investigating 375 e-commerce websites indicate that moderate levels of support exists for consumers to implement non-compensatory choice strategies, and virtually no support exists for executing multi-attribute based compensatory choice strategies. The results of this study suggest that there is an opportunity in waiting for e-commerce businesses to implement compensatory WebDSS to improve the decision making capabilities of their consumers.

DOI: 10.4018/978-1-60566-910-6.ch005

INTRODUCTION

The last decade has witnessed a substantial growth in both Internet penetration and e-commerce activities. As of 2006, the Internet is used by 73% of all American adults (Pew Internet and American Life, 2006). Similarly, according to a report from Forrester research, sales from e-commerce activities are expected to reach $331 billion by 2010 with online sales expecting to account for 13% of total retail sales in 2010, up from 7% in 2004.[1] The growth of e-commerce has resulted in electronic markets offering a wide variety of product choices, elaborate product related information, and great convenience for consumers. Consequently, ever greater numbers of individuals are interacting with online environments to search for product related information and to buy products and services (Xiao & Benbasat, 2007). In fact, searching for product or service related information was the next most popular activity on the Internet in 2003 after email or instant messaging (US Department of Commerce Report, 2004). These statistics suggest that the growth in Internet penetration and e-commerce resulted in increased consumer reliance on the Internet for a variety of decision making processes ranging from searching for products, comparing them, and often resulting in making a final purchase.

Although increased access to information has been a blessing to consumers, the online environment has also resulted in an overabundance of information (Haubl & Murray, 2003). For instance, a search for products on Google shopping reveals that there are more than 3000 options available for a 42 inch LCD television and more than 6000 options available for women's handbags.[2] This amount of information is guaranteed to overwhelm the limited information processing capabilities of human beings (Simon, 1955). Therefore, many web retailers are incorporating **web-based decision support systems** (WebDSS from here on) to assist consumers with their decision making process (Grenci & Todd, 2002). Web-based decision support systems capture individual user preferences for products either explicitly or implicitly, and provide recommendations based on such preferences (Xiao & Benbasat, 2007). WebDSS have the potential to ease consumers' information overload and to reduce search complexity in addition to improving their decision quality (Haubl & Trifts, 2000).

Improving consumer decision making in online environments has been the subject of interest for researchers in a number disciplines. Researchers from computer science, library sciences, social psychology, marketing, management, and information systems have been making important contributions to this area of research. Consequently, the array of decision support tools implemented on e-commerce websites is known with different terminology although they all refer to the same tool to be used by the consumers. Examples include intelligent agents, electronic product **recommendation agents**, recommendation systems, and web-based decision support systems. In their extensive review of electronic recommendation agents, Xiao and Benbasat (2007) categorized recommendation agents (RA) into three types. The first type of recommendation agents includes content-filtering, collaborative-filtering, and hybrid agents. The second type includes feature-based and need-based recommendation agents. Finally, the third type of recommendation agents includes compensatory and non-compensatory based systems.

We consider only compensatory and non-compensatory WebDSS in this chapter. We present a synthesis of literature concerning the effectiveness of implementing compensatory versus non-compensatory DSS, and then examine whether or not such findings have made their way into the design of commercial websites. We do so by examining the level of consumer support provided on commercial websites to execute compensatory or non-compensatory strategies. We believe that understanding the reality of the extent to which e-commerce websites support compensatory

Table 1. Four apartment alternatives

Alternatives	Attributes				
	Rent ($)	A/C	Covered Parking	Washer and Dryer	Dish Washer
A	350	Yes	No	No	No
B	400	Yes	No	Yes	No
C	450	Yes	No	Yes	Yes
D	500	Yes	Yes	Yes	Yes

and non-compensatory strategies is important for several reasons. From a practical standpoint, if we find that a relatively smaller fraction of websites provide compensatory based support despite a previously well-supported finding that such support is normatively better, then that would highlight an opportunity for the web retailers to increase the support levels to their customers. From a theoretical standpoint, such finding would raise further questions concerning the factors affecting the implementation of non-compensatory and compensatory WebDSS.

The rest of the chapter is organized as follows. We first present backgrounds concerning WebDSS types and research concerning the effectiveness of compensatory versus non-compensatory DSS. We then describe the methodology used for the study and present the results. We conclude with a discussion on managerial implications, directions for future research, and conclusion.

BACKGROUND: TYPES OF WEB-BASED DECISION SUPPORT SYSTEMS

One of the common implementations of WebDSS use filtering based methods. Content-filtering WebDSS consider users' most desired attributes and provide recommendations accordingly. Some of the commercial implementations of content filtering WebDSS include Active Buyers Guide and MySimon (Xiao & Benbasat, 2007). Collaborative-filtering WebDSS use the suggestions provided by like-minded consumers to provide recommendations (Ansari, Essegaier, & Kohli, 2000). Amazon, CD Now provide collaborative filtering WebDSS on their websites.

On the other hand, WebDSS can also be implemented using decision strategies. Research investigating the decision strategies used by individual decision makers has a long history. Much of the knowledge acquired from the research domain of traditional decision support systems is now guiding the research that examines the effectiveness of WebDSS implementation using different decision strategies. The scope of the study presented in this chapter is limited to studying the commercial implementation of compensatory and non-compensatory WebDSS. The following section provides an in-depth treatment of the related concepts.

Decision strategies refer to the rules employed by individuals to arrive at decisions (Hogarth 1987, p. 72). The decision strategies can be classified into compensatory and non-compensatory decision strategies. These strategies are discussed using an example of renting an apartment (See Table 1).

Non-Compensatory WebDSS

A **non-compensatory WebDSS** implements one of the many non compensatory decision strategies. The use of a **non-compensatory strategy** avoids confronting the conflicts inherent in the choice situation and does not allow the decision maker to trade off a low value on one attribute against a high value on another attribute (Hogarth, 1987). In the apartment rental example, if a university student decides that covered parking represents the most important attribute, options A, B, and C are immediately eliminated. Irrespective of the attractiveness of options A, B, and C, the use of one attribute to make the decision results in their elimination.

There are many examples of non-compensatory strategies. The *conjunctive* strategy sets cut off points for certain attributes, and the options that do not meet all of these thresholds are eliminated. The use of the *disjunctive* strategy allows an option to remain under consideration so long as either of two attributes has a value that meets its specified cutoff. Executing an *elimination-by-aspects* strategy (EBA) requires selecting an attribute at every stage and eliminating the options that do not include such aspect. The process continues until a winner is selected. The *satisficing* strategy compares each attribute value with a pre-determined cut-off level and the option that fails to meet the cut-off level is rejected (Hogarth, 1987).

Computer support for non-compensatory strategies focuses on allowing the consumer to have some control over viewing and manipulating features of interest. For example, a website could allow the consumer to specify the value of an attribute (e.g., enter text for searching a web site or choosing from a list of criteria) to help select a set of products for further investigation. Even more support is given when the consumer can sort products based on the value of an attribute (as in sorting by price, weight, or category). Nonetheless, many websites do not give even this level of support, forcing the user to "drill down" in each product to find its features and their values.

In general, the use of non-compensatory WebDSS results in presentation of options that meet the predetermined thresholds set by the decision maker. However, the use of non-compensatory WebDSS by decision makers may not always yield the recommendations suggested by normative decision making models. First, non-compensatory WebDSS do not consider a consumer's preference function for multiple attributes. Second, at times, the use of non-compensatory WebDSS may result in the elimination of some options based on criterion set on one attribute, though such options may be very attractive on the other attributes.

Compensatory WebDSS

A **compensatory WebDSS** implements one of the many compensatory decision strategies on a website. The use of a **compensatory strategy** mandates confronting the conflicts inherent in the choice situation, and allows the decision maker to trade-off a low value on one attribute against a high value on another attribute (Hogarth, 1987). Using the example presented in Table 1, if a university student decides to rent an apartment, the use of a compensatory strategy would facilitate balancing the lack of some apartment features of option A against a high value (low rent) of another apartment feature.

Compensatory strategies enable desirable values of one attribute of a product or service to "compensate" for undesirable values of another. Compensatory strategies require computations or judgmental assessments that combine multiple variables for each product being considered. An example of a compensatory strategy is a *weighted-additive* strategy in which a weight (relative importance) is assigned to each attribute and multiplied by its value. These products are summed to provide a score for each product. Then the product with the highest score is selected. Variations of the weighted-additive strategy use judgmental assessment of these tradeoffs rather than an actual calculation. Compensatory strategies are more accurate than non compensatory strategies when potential tradeoffs exist between variables, such as when the consumer would say, "I would rather have a convertible, but not if I have to pay too much extra to get it." Unfortunately, compensatory strategies are so difficult that they are rarely used except for small sets of products or if some sort of computer-based assistance is provided (Johnson & Payne 1985; Todd & Benbasat, 1994).

Computer support for compensatory strategies would also make it easier for consumers to see several attributes for several products at a time, such as when a website displays an array of rows of products with columns for various attributes. Such an array makes it easier for the consumer to identify and assess the tradeoffs inherent in compensatory strategies. Some websites provide the display of such an array. Support for compensatory strategies could also enable the selection of products based on multiple features. Still further support would be to assist the consumer in evaluating tradeoffs by providing some sort of scoring computation of the values of various attributes. Such a "model" would best serve the consumer if the consumer were allowed to say which attributes were important and how much weight to give each attribute. Finally the consumer would need a way to easily find out which products or services obtained the highest scores, such as having the system sort or select products based on scores provided by the model.

Partial support for compensatory strategies can come from allowing the consumer to see external ratings of products (such as Consumer Reports). External ratings have the advantages of being an overall assessment of multiple criteria (i.e., they are "compensatory") and of being objective. They have the disadvantage that they are based on someone else's assumption about the relative importance of various criteria. They do not capture the consumer's own preferences.

Edwards and Fasolo (2001) note that "the idea of a procedure for making important decisions that does not depend heavily on human inputs seems unlikely as well as unattractive. Selection, training, and elicitation of responses from the person...become crucial" (p. 588). Compensatory WebDSS are specifically designed to implement such an idea. As opposed to a non-compensatory WebDSS, compensatory WebDSS need to draw on the processing capacity and storage abilities of the computers to implement the normative algorithms such as multi attribute utility theory, Bayesian nets, and subjective expected utility theory that are otherwise very difficult for the unaided decision maker to implement (Larrick, 2004). Compensatory WebDSS execute algorithms in the background and also perform consistency checks on user provided weights making the decision tools more appealing to end users (Larrick, 2004).

Research Comparing the Effectiveness of Compensatory versus Non-Compensatory WebDSS

How effective are compensatory WebDSS compared to non-compensatory WebDSS? We address this question in this section based on empirical results from five studies that compared compensatory and non-compensatory WebDSS. The results are summarized in Table 3. The compensatory WebDSS used in these studies elicit weights from users on attributes and present alternatives with final scores based on expected values. The non-compensatory WebDSS used were based on any of the conjunctive, disjunctive, EBA, or satisficing strategies. The explanations of the variables used to compare the two types of WebDSS are provided in Table 2.

Satisfaction

Previous research suggests that users experience more satisfaction with the use of compensatory WebDSS compared to that of non-compensatory WebDSS (Fasolo, McClelland, & Lange, 2005; Olson and Widing, 2002; Periera 2001; Song, Jones, & Gudigantala, 2007). Widing and Talarzyk (1993) found that no such differences exist between the two formats. However they used only one item to measure satisfaction. The result by Fasolo et al. (2005) concerning the effectivness of compensatory WebDSS is noteworthy because their research study employed an alternative set

Table 2. Variables used for comparing the effectiveness of Compensatory and Non-Compensatory WebDSS

Variable	Explanation
Satisfaction	The user's satisfaction with the WebDSS in supporting the decision making process
Decision Quality/Accuracy	Preference matching: The extent to which the choice selected by the user matches her stated preferences Product switching: Once a purchase decision is made, will the user change his mind and switch to another choice given a chance?
Effort	The amount of cognitive resources exerted by the user in processing the information to arrive at the choice
Confidence	The degree to which a user has confidence in WebDSS's recommendations
Decision Time	The time taken to arrive at the final choice

with negative inter-attribute correlations. Inter-attribute correlation is obtained by calculating the average correlation among the all the pair of attributes. With positive inter-attribute correlations (friendly environment), the alternative that is favorable on one attribute tends to be favorable on others. With negative inter-attribute correlations (unfriendly environment), the more attractive level of one attribute is associated with less attractive level on the other. Alternatives characterized by negative inter-attribute correlations such as those between cost and quality, are common and lower consumer confidence and satisfaction (Fasolo et al., 2005). Therefore, consumers require more support when dealing with alternatives with negative inter-attribute correlations. Hence, existing evidence overwhelmingly supports the notion that compensatory WebDSS contribute to better satisfaction ratings compared to those of non-compensatory WebDSS.

Decision Quality

Decision quality has been measured in various ways in literature. Preference matching measures the extent to which the choice selected by the user matches her stated preferences (Pereira, 2001) where as product switching measures the extent to which the user is likely to change his mind and switch to alternative choice given that he already made a purchase decision (Widing & Talarzyk, 1993). The existing evidence strongly suggests that compensatory WebDSS provide better **decision quality** to users compared to non-compensatory

Table 3. Studies that compared the effectiveness of compensatory and non-compensatory WebDSS

	Fasolo et al. (2005)	Song et al. (2007)	Olson and widing (2002)	Periera (2001)	Widing and Talarzyk (1993)
Satisfaction	Compensatory WebDSS are better	Compensatory WebDSS are better	Compensatory WebDSS are better	Compensatory WebDSS are better	No Difference
Decision Quality/ Accuracy	Compensatory WebDSS are better	Compensatory WebDSS are better	Compensatory WebDSS are better	Compensatory WebDSS are better	Compensatory WebDSS are better
Effort	Compensatory WebDSS are better	Compensatory WebDSS are better		Compensatory WebDSS are better	
Confidence				Compensatory WebDSS are better	
Decision Time			No Difference		Non-Compensatory WebDSS took more time

WebDSS (Fasolo et al., 2005; Olson and Widing, 2002; Periera 2001; Song et al., 2007; Widing & Talarzyk, 1993).

Effort

The existing evidence supports the idea that non-compensatory WebDSS requires more effort from users than that of compensatory WebDSS (Fasolo et al., 2005; Periera 2001; Song et al., 2007).

Confidence

Research by Pereira (2001) found out that users felt more confident when making choices with compensatory WebDSS as opposed to a non-compensatory WebDSS.

Decision Time

While research by Widing & Talarzyk (1993) suggests that non-compensatory DSS took more time to arrive at a choice, Olson & Widing (2002) found that no such differences exist. Therefore, the existing evidence is inconclusive whether the use of compensatory WebDSS saves time.

Therefore, overwhelming evidence supports the notion that compensatory WebDSS are better than non-compensatory WebDSS on important variables such as satisfaction, decision quality, effort and confidence. In addition, even in the absence of decision tools implementing compensatory decision strategies, websites that facilitate the comparison of alternatives contribute to an increased use of compensatory strategies by users (Jedetski, Adelman, & Yeo, 2002). However, research also suggests that non-compensatory WebDSS support consumer decision making better than websites that just provide products by alphabetical order (Widing & Talarzyk, 1993). When technology makes compensatory strategies easier, consumers are more likely to use them (Todd & Benbasat, 1994). Moreover, consumers are more likely to use ratings based on their own weightings of the features, rather than someone else's preferences (Jones & Brown, 2003). Hence, based on the empirical results that non-compensatory WebDSS are better than websites providing product recommendations in alphabetical order, and compensatory WebDSS are better than non-compensatory WebDSS, we conducted a study to investigate how well the commercial websites provide support to compensatory and non-compensatory strategies. In the following sections we provide the methodology and results of the study.

METHODOLOGY

Research on evaluation of web-based information systems in the context of e-commerce has been on-going. Previous work has examined e-commerce websites based on concordance analysis (Jinling & Guoping, 2005) and the factors that influence web-based information systems success (Garrity et al., 2005). Garrity et al.(2005) found that decision support satisfaction plays an important role in web-based information system success. Given this finding and the evidence that compensatory and non-compensatory WebDSS provide varying levels of decision support satisfaction (with compensatory WebDSS providing more satisfaction), we set out to evaluate retail websites based on three criteria:

(1) Does the website have useful features commonly found on competitors' websites?
(2) How much support does the website give to consumers' non-compensatory decision strategies?
(3) How much support does the website give to compensatory strategies in a way that captures consumers' own preferences and weightings of product features?

Selection of Data Source

In identifying business firms used in this study (as opposed to personal or very small operations), we used the "Business and Company Center" database[3]. This is one of the most comprehensive web-based business databases available today that offers extensive information on hundreds of thousands of companies worldwide. We focused on retail and service industry based on four-digit industry codes. The scope of our analysis is restricted to the U.S. based companies. The database contains approximately 2600 U.S. companies out of approximately 7600 world wide companies. We selected about 25% of the U.S. companies by selecting every fourth company from the database consisting of 2600 companies.

Questionnaire Preparation and Pilot Testing

The first part of the data collection procedure involved preparation of the questionnaire to be used by the researcher evaluating a web site. The questionnaire elicited information concerning different kinds of decision support provided by web retailers. The questionnaire was intended to capture information concerning the support provided to help users locate, evaluate, and compare products. In addition, information concerning the provision of a multi-attribute model that would elicit user preferences as well as the provision of others ratings' about products was captured. The information concerning the communication of privacy policy was also captured. The questionnaire is included in Appendix A. In a nutshell, the questionnaire was intended to capture website support for executing non-compensatory strategies, compensatory strategies, product related information, and security and privacy based information.

A pilot study was conducted to refine the questionnaire and to examine the agreement about the information collected. The three authors individually visited 30 websites and gathered data. The authors found 90% agreement on the data collected. Based on discussions about the sources of the few disagreements, further revisions were made to the questionnaire to remove ambiguities in the questions.

Data Collection

We selected about 25% of the 2600 U.S. companies, and one of the authors visited 610 websites to collect data[4]. In order to reduce the possible bias in the sample, we avoided any duplication, such as companies listed in multiple SIC industries. In addition, the URL's of many companies are out of date, and some URL's represented replications of those that were already considered for the study. Finally, the data set consisted of complete observations for 375 business firms operating on the Web. Out of the 375 websites, 310 were retail websites and 65 were service websites. The retail industry contained websites on merchandize stores, apparel and accessory stores, furniture, house hold appliances, electronics etc., where as the service industry consisted of websites on hotels and motels, rooming and boarding houses, sporting and recreational camps, RV parks, and software services etc. The collected data provided rich description of the typical features, their level of support for consumers' non-compensatory strategies, and their level of support for consumers' compensatory strategies and preferences.

To give further assurance of accuracy and validity of data collection, a second author randomly gathered data about some companies in the sample to compare to the other author's data collection. There was almost perfect agreement between the two authors.

Results

Our overall findings are displayed graphically in Figure 1. Typical web site features are shown first. Of our sample of 375 web sites, all give general

Figure 1. The percentage of web-retailers' web sites investigated (375 total) having various web site features, including features that would support consumers' decision strategies and preferences.

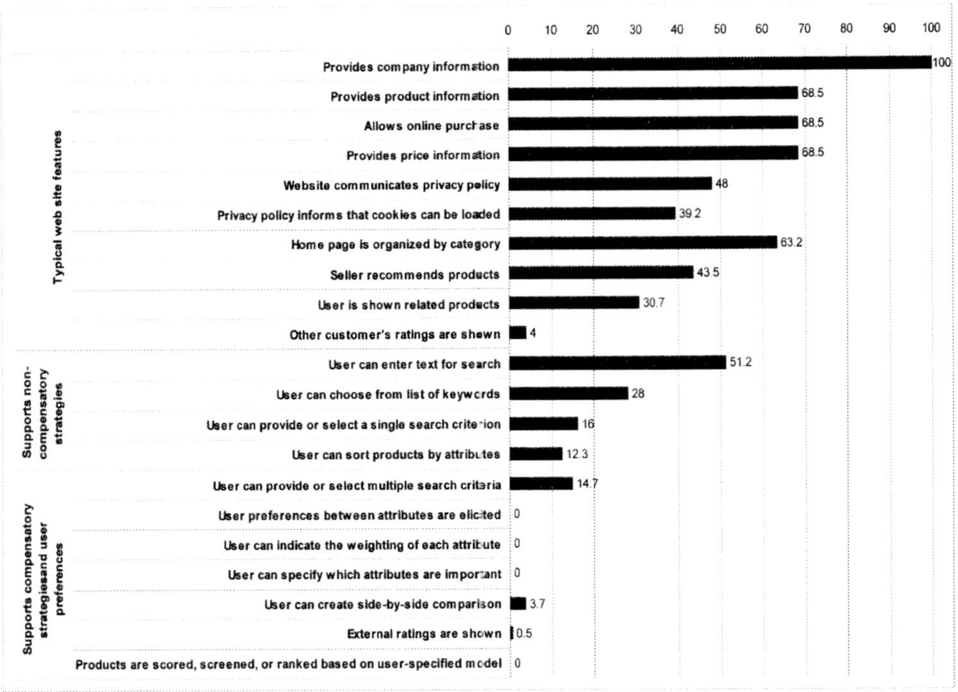

company information and about two-thirds (68.5 percent) support online purchasing of products or services. Most of the websites that support online purchases display the privacy policy and inform that cookies can be loaded to the consumer's computer. Most of the websites that support online purchases also enable consumers to find specific categories, which facilitates consumers' search. About half of the websites recommend products in some way, about a third show related products. Only four percent of the websites surveyed show other customers' ratings.

In the middle of the graphic, the results are shown for features that would be helpful to consumers desiring to execute non-compensatory strategies. Most of the websites that supported selling had at least one feature that would enable the consumer to find products based on a certain criterion, such as entering text for a search, choosing from a list of keywords, or providing a single search criterion. Nonetheless, only 12.3 percent of the websites enable the sorting of products based on an attribute value.

At the bottom of the graphic, the results are shown for features that would be helpful to consumers desiring to execute compensatory strategies. When we considered the support for compensatory strategies that incorporated consumer preferences, we found almost no support. Just 14.7 percent of the websites supported searches based on multiple criteria. Only 3.7 percent displayed side-by-side comparison. Only one-half of one percent showed external ratings of products or services. NONE of the websites assisted the consumers by allowing the users to give weights of attributes or specify which weights are important. NONE of the websites provided for scoring based on user-specified models.

To gain further insight into the breakdown of the websites in our sample, we subdivided our sample two ways: retail versus service, and sales volume above or below average. These results are shown in Table 4. Inspection of these breakdowns reveals several patterns. First, the typical website features are provided more often for retail products than for services.

Service industry websites are more prone to just give company information and not try to sell directly on the website. On the other hand, company size did not appear to affect the extent of online selling, perhaps because there are few financial or technological barriers to a small business that wants to begin selling on the Internet. The larger companies appear to attempt to market their products somewhat more by recommending products, showing related products and showing other customer ratings.

Retailers of products more frequently allowed users to enter text for a search, while service companies more frequently allowed a choice of keywords or provision of a single search criterion. Since these features are merely different ways of achieving the same objective, we do not see sellers of products or services as dominating in supporting ways of specifying criteria. For the few websites that supported sorting of products by attributes, this feature was more frequently provided by retailers of products than by service firms. The sort feature was also more frequently provided by large firms than small firms.

For compensatory strategies, the main result is that websites gave little support at all. For some reason service firms gave more support in searching multiple criteria than sellers of products. Of the few websites showing side-by-side comparisons, all were retailers of products (rather than services) and most were large companies. External ratings were all of products rather than services. This may be due to a lack of available external ratings of services.

MANAGERIAL IMPLICATIONS

A recent Forrester research survey found that out of all consumers who use **recommendation features** of websites, seventy percent find the recommendations useful, and close to one-third of customers provided with product recommendations are likely to purchase products from such recommendations.[5] Similarly, a survey by Choice stream found that 69% of respondents who spent more than $1,000 online in 2007 were more likely to shop with personalized recommendations than those that do not.[6] These surveys suggest that there is a clear value to having WebDSS that provide recommendations consistent with shoppers preferences.

The main finding of our investigation of e-commerce websites is a complete absence of support for consumers' compensatory strategies based on their own preferences. Given the results of academic research that compensatory WebDSS provide better decision quality, satisfaction, and confidence to consumer, and reduce effort an opportunity is waiting for managers to start looking for ways to implement such tools.

The purpose of a decision support system is to help a customer pick the best possible choice in all situations. The use of non-compensatory DSS is not associated with better decision quality (Fasolo et al., 2005). However, managers have to make sure that compensatory WebDSS are easy to use. Most of the compensatory WebDSS implemented in research experiments typically have two screens. In the real world, as the number of screens used to capture consumer preferences increases, the longer it takes for customers to make a decision. Such design may discourage users. Therefore, to the extent that compensatory WebDSS are easy to use, they are likely to be used by consumers.

The execution of compensatory strategies requires users to submit weights to attributes and then the DSS recommends products with highest expected values. But, how does a user know what algorithm is being used to come up with the results? Therefore, it is recommended that managers provide information concerning how the final scores (expected values) are calculated from the user supplied weights.

Table 4. Survey of WebDSS Attributes

Attributes	All (N=375)	Industry		Sales Volume	
		Retail (N=310)	Service (N=65)	Above (N=188)	Below (N=187)
Typical Web Site Features					
Provides company information	100.0 (%)	100.0(%)	100.0(%)	100.0(%)	100.0(%)
Provides product information	68.5	72.3	50.8	69.7	67.4
Allows online purchase	68.5	72.3	50.8	69.7	67.4
Provides price information	68.5	72.3	50.8	69.7	67.4
Website communicates privacy policy	48.0	51.7	29.2	59.6	36.4
Privacy policy informs that cookies can be loaded	39.2	42.6	23.1	53.7	24.6
Home page is organized by category	63.2	66.1	49.2	63.8	62.6
Seller recommends products	43.5	47.7	23.1	50.0	36.9
User is shown related products	30.7	36.8	1.5	39.9	21.4
Other customers' ratings are shown	4.0	4.8	0.0	7.4	0.5
Web Site Features Supportive of Non-Compensatory Decision Strategies					
User can enter text for search	51.2	60.0	9.2	51.6	50.8
User can choose from list of keywords	28.0	25.5	40.0	30.9	25.1
User can provide or select a single search criterion	19.2	15.2	38.5	20.7	17.6
User can sort products by attributes	12.3	13.5	6.2	16.0	8.6

continued on the following page

Table 4. continued

	Web Site Features Supportive of Compensatory Decision Strategies or User Preferences				
User can provide or select a multiple search criterion	14.7	10.0	36.9	13.8	15.5
User preferences between attributes are elicited	0.0	0.0	0.0	0.0	0.0
User can indicate the weighting to each attribute	0.0	0.0	0.0	0.0	0.0
User can specify which attributes are important	0.0	0.0	0.0	0.0	0.0
User can create side-by-side comparison	3.7	4.5	0.0	5.3	2.1
External ratings are shown	0.5	0.6	0.0	0.5	0.5
Products are scored, screened, and ranked based on user specified model	0.0	0.0	0.0	0.0	0.0

It is also possible that the lack of expertise and developmental costs may influence managers not to implement compensatory WebDSS. We believe that the extent to which the benefits of implementing such WebDSS outweigh the costs implies that it would be a worthwhile proposition for managers to consider developing compensatory based decision support tools.

Directions for Future Research

While our study results showed absence of support for executing compensatory strategies in e-commerce websites based on consumer preferences, with some additional research, we found some third party web sites that provided such support. Examples of such third party sites include My product advisor (http://www.myproductadvisor.com), Select smart (http://www.selectsmart.com), and Yahoo! shopping smart sort computer and electronic recommendations (http://shopping.yahoo.com/smartsort). Future research could investigate two research questions. First, what are the factors that inhibit e-commerce websites from providing support for compensatory based strategies based on consumer preferences? Second, what are the implications for e-commerce websites with third party web sites providing such support when consumers expect such support from the web retailers themselves?

A second area of research could look into the issues surrounding consumers' adoption of **decision technology** implemented to support individuals' decision making processes. Research shows that less than 10% of home users visit shop bots (Montgomery, Hosanagar, Krishnan, & Clay, 2004). Therefore, future research could look into various factors that would improve the consumer adoption of decision technology. Furthermore, additional research is needed to understand how individual differences in decision makers affect adoption and usage of decision technology on e-commerce web sites.

The present survey considers only compensatory and non-compensatory based systems, and the results suggest that an important gap exists between theory and practice. Future studies could conduct similar kinds of studies to investigate how well e-commerce websites provide support concerning content, collaborative, and hybrid WebDSS as well as the feature and need based WebDSS. It is our hope that as with our study, important insights could be brought out by conducting studies that investigate the extent of website support concerning other types of WebDSS.

Compensatory decision tools that are implemented in the experiments may face challenges when extended to the real world. For example, most of the compensatory WebDSS designed in experiments contain all the attribute values for a given alternative set. However, in the real world, attributes values may be missing for some alternatives, and therefore computing expected values for such alternatives could be problematic. Therefore, future research could look at the effects of missing information on consumer choices in online decision support environments.

Future research could also look at measuring the monetary benefit to an organization implementing a web-based decision support tool on its website. The existing research so far has focused on decision outcome variables such as satisfaction, decision quality, effort etc. Of interest to managers could be whether improved WebDSS tools augment the user's willingness to purchase.

CONCLUSION

Research conducted by decision scientists over the last few decades has examined how decisions are made and have identified several decision strategies individuals use to make decisions. These decision strategies are compensatory and non-compensatory in nature. After the advent of Internet and the subsequent growth of e-commerce market, most websites are implementing web-based decision support tools to help consumer make their choices. One category of web-based decision tools uses decision strategies to provide consumer support. In this chapter, we focused on website support for executing consumers' compensatory and non-compensatory strategies. By synthesizing the existing literature concerning the effectiveness of implementing compensatory versus non-compensatory WebDSS, we found that a majority of the evidence favors implementing compensatory WebDSS. If compensatory WebDSS are so effective, one would expect to observe e-commerce websites increasing the level of support for executing consumers' compensatory strategies. Based on a study of 375 U.S. company websites, we found that very little support exists for features that support compensatory strategies (such as side-by-side comparison of alternatives) and no support exists for executing compensatory strategies based on consumer preferences.

We also note several limitations of our study. As far as we are aware, there is no study that explored how well websites provide support for compensatory and non-compensatory based strategies. Though it is problematic to generalize the findings of U.S. based companies to companies world wide, a future study could look into how well such strategies are supported in websites worldwide. Secondly, choosing 25% of U.S. based companies is purely arbitrary. However, we believe that the results of our study are representative of the current situation on e-commerce websites. For example, Fasolo et al. (2005) state that "although we have no precise data to support it, we are under the impression that real World Wide Web compensatory sites are having rougher and shorter lives than non-compensatory sites....We have anecdotal evidence that transparency and length might be a reason for the lack of success of compensatory ones." (p. 341).

The results of the study presented in this chapter open up an opportunity for managers to start providing more support for compensatory based decision strategies, and at the same time beg the question of the lack of popularity of such tools. A number of potential reasons have been presented and a host of research questions have been raised. It is our hope this attempt fuels further research in improving the design of WebDSS and finding factors that affect the adoption of WebDSS, ultimately contributing to the benefit of both the websites and users.

REFERENCES

Ansari, A., Essegaier, S., & Kohli, R. (2000). Internet Recommendation Systems. *JMR, Journal of Marketing Research*, *37*(3), 363–375. doi:10.1509/jmkr.37.3.363.18779

Edwards, W., & Fasolo, B. (2001). Decision Technology. *Annual Review of Psychology*, *52*(1), 581–606. doi:10.1146/annurev.psych.52.1.581

Fasolo, B., McClelland, G. H., & Lange, K. A. (2005). The effect of site design and interattribute correlations on interactive web-based decisions. In C.P. Haugtvedt, K. Machleit, & R. Yalch (Eds.), *Online consumer psychology: understanding and influencing behavior in the virtual world* (pp. 325-344). Lawrence Erlbaum Associates, Inc.

Garrity, E. J., Glassberg, B., Kim, Y. J., Sanders, G. L., & Shin, S. K. (2005). An experimental investigation of Web-based information systems success in the context of electronic commerce. *Decision Support Systems*, *39*(3), 485–503. doi:10.1016/j.dss.2004.06.015

Grenci, R. T., & Todd, P. A. (2002). Solutions-Driven Marketing. *Communications of the ACM*, *45*(2), 64–71. doi:10.1145/504729.504730

Haubl, G., & Trifts, V. (2000). Consumer decision making in online shopping environments: The effects of interactive decision aids. *Marketing Science*, *19*(1), 14–21. doi:10.1287/mksc.19.1.4.15178

Hauble, G., & Murray, K. (2003). Preference Construction and Persistence in Digital Marketplaces: The Role of Electronic Recommendation Agents. *Journal of Consumer Psychology*, *13*(1), 75–91. doi:10.1207/153276603768344807

Hogarth, R. (1987). *Judgment and Choice* (2nd ed). New York: John Wiley and Sons.

Jedetski, J., Adelman, L., & Yeo, C. (2002). How Web Site Decision Technology Affects Consumers. *IEEE Internet Computing*, *6*(2), 72–79. doi:10.1109/4236.991446

Jinling, C., & Guoping, X. (2005). Comprehensive Evaluation of E-commerce Websites Based on Concordance Analysis. In *Proceedings of the 2005 IEEE International Conference on e-Business Engineering* (pp. 179-182).

Johnson, E. J., & Payne, J. W. (1985). Effort and accuracy in choice. *Management Science*, *31*(4), 394–414. doi:10.1287/mnsc.31.4.395

Jones, D. R., & Brown, D. (2003). The division of labor between human and computer in the presence of decision support system advice. *Decision Support Systems*, *33*(4), 375–388. doi:10.1016/S0167-9236(02)00005-2

Larrick, R. P. (2004). Debiasing. In D.J. Koehler & N. Harvey (Eds.), *Blackwell Handbook of Judgment and Decision Making*. Oxford: Blackwell Publishing.

Montgomery, A. L., Hosanagar, K., Krishnan, R., & Clay, K. B. (2004). Designing a Better Shopbot. *Management Science*, *50*(2), 189–206. doi:10.1287/mnsc.1030.0151

Olson, E. L., & Widing, R. E. (2002). Are interactive decision aids better than passive decision aids? A comparison with implications for information providers on the Internet. *Journal of Interactive Marketing*, *16*(2), 22–33. doi:10.1002/dir.10007

Pereira, R. E. (2001). Influence of query-based decision aids on consumer decision making in electronic commerce. *Information Resources Management Journal*, *14*(1), 31–48.

Pew Internet and American Life. Internet Penetration and Impact (2006). Retrieved November 9, 2007 from http://www.pewinternet.org/PPF/r/182/report_display.asp

Simon, H. A. (1955). A Behavioral Model of Rational Choice. *The Quarterly Journal of Economics*, *69*(1), 99–118. doi:10.2307/1884852

Song, J., Jones, D., & Gudigantala, N. (2007). The Effect of Incorporating Compensatory Choice Strategies in Web-based Consumer Decision Support Systems. *Decision Support Systems*, *43*(2), 359–374. doi:10.1016/j.dss.2006.10.007

Todd, P., & Benbasat, I. (1994). The influence of decision aids on choice strategies: An experimental analysis of the role of cognitive effort. *Organizational Behavior and Human Decision Processes*, *60*(1), 36–65. doi:10.1006/obhd.1994.1074

U.S. Department of Commerce Report. (2004). *A Nation Online, Entering the Broadband Age*. Retrieved November 9, 2007 from http://www.ntia.doc.gov/reports/anol

Widing, R. E., & Talarzyk, W. W. (1993). Electronic Information Systems for Consumers: An Evaluation of Computer-Assisted Formats in Multiple Decision Environments. *JMR, Journal of Marketing Research*, *30*(2), 125–141. doi:10.2307/3172823

Xiao, B., & Benbasat, I. (2007). E-Commerce Product Recommendation Agents: Use, Characteristics, and Impact. *MIS Quarterly*, *31*(1), 137–209.

ENDNOTES

[1] http://www.forrester.com/Research/Document/Excerpt/0,7211,34576,00.html

[2] http://www.google.com/prdhp?hl=en&tab=wf

[3] Please visit http://www.galegroup.com/pdf/facts/bcrc.pdf to find more about this database

[4] The questionnaire captures general details, support for user to locate a product, evaluate individual products, support in terms of others ratings, support to compare products, support for multi-attribute models, and information about cookies. The only place where the researcher's perceptions could bias the results is the section on support provided to user to select a specific product. This part is not used in the analysis. The rest of the variables are binary in nature. For example, a website can provide a key-word based search or not. Similarly, a website can let the users pick important attributes or not, weight the attributes or not. Therefore, we believe that what is needed from a data collector is general observation skills and since perceptions are not recorded, we believe that use of one of the authors to collect data is reasonable.

[5] http://www.ecommerce-guide.com/resources/market_research/article.php/3721731

[6] http://www.choicestream.com/pdf/2007_ChoiceStream_Personalization_Survey_FINAL.pdf

APPENDIX A

URL: ____________________________________

SIC Code: ___________________ Preparer __________

Name of Business ____
Date__________

Types of Products Offered ____

Circle all that apply: shows company info, shows product info, shows prices, allows online purchase

Support That Helps User Locate a Product

Y N Home page is organized by category to assist with product search
Y N User can enter text for search
Y N User can choose from list of keywords for search
Y N User can provide or select a single search criterion (e.g., homes with 3 bedrooms, < $200,000)
Y N User can provide or select multiple search criteria
Y N User is shown related products

Support That Helps User Evaluate Individual Products

BA A AA Products are described in detail (Below average, average, above average)
BA A AA Products are shown in high quality pictures

Special features (pictures): ____

Support That Provides User with Others' Ratings of a Specific Product

Y N Other customers' ratings or comments are shown for products
Y N External ratings (e.g. Consumer Reports ratings) are shown for products

Source: ____

Y N Seller recommends some products (e.g., "best value")

Verbiage: ____

Support That Helps User Compare Products

Y N User can sort products by an attribute: ____

—

Y N User can create side-by-side comparison of products on a single web page

Support That Creates Multi-Attribute Model of Elicited User Preferences

Y N User can specify which attributes are important and system picks products for user to review

Explain: ____

Y N User preferences between attributes are elicted by system (e.g., providing user with pairs of product attributes and asking user which is more important).

Y N User can indicate how much weight should be given to each attribute.

Y N Products are scored, screened, or ranked (indicate which) based on multi-attribute model of user preferences

Explain: ____

System Informs of Cookies in Privacy Policy

Y N Website communicates a privacy policy

Y N Privacy policy informs that cookies might be loaded onto user's computer

Other Type of Support

Please describe in detail any other type of decision support provided for the consumer

____

____

Chapter 6
Modeling the Adoption of Mobile Services

Hannu Verkasalo
Helsinki University of Technology, Finland

ABSTRACT

Many case examples in the mobile market have shown that the success of mobile services is difficult to predict. Different factors either boost or hinder the adoption of mobile services. The present chapter has covered earlier research on mobile service adoption and utilized a unique handset-based research platform in collecting data from 548 Finnish smartphone users in 2006. The main research goal is to understand the process of mobile service adoption by extracting new kinds of data straight from handsets. In addition to descriptive results, a path analysis model is developed that models the mobile service adoption process. The chapter finds that user intentions have a strong impact on consequent adoption of the service, expectedly. What is more, perceived hedonic benefits from the service are the strongest factor driving intentions to use the service. The perceived technical capability to use the service and the role of the surrounding social network do not drive the intentions of early-adopter users. Interestingly multimedia service adoption is driven by more capable (new) handsets, and mobile web browsing benefits significantly from block or flat-rate (instead of usage-based) pricing plans for transmitted data. The chapter develops several indices that measure time-varying characteristics of mobile services. Calculated indices for a set of mobile services suggest that the studied mobile services are currently experiencing different phases in their life cycle.

1. INTRODUCTION

Mobile services have evolved quite a lot from mere communication oriented services (circuit-switched voice, text messaging, voice mailbox) to today's multimedia, content retrieval, browsing and other advanced services. The **mobile Internet** (see Funk 2004) is emerging and the IP-based service delivery is likely to hit the mobile mass market domain very soon. Overlay networks existing already in the Internet (Clark et al. 2006) may have spill-over effects to the mobile industry. The mobile Internet scenario

DOI: 10.4018/978-1-60566-910-6.ch006

contrasts sharply with the dominant, vertically-oriented way of doing mobile business (see e.g. Karlson et al. 2003, Verkasalo 2007a and Vesa 2005). The emergence of the mobile Internet is driven by the wide-scale adoption of **smartphones** (i.e. converged devices) along with improvements in both cellular (GSM and 3G) and alternative (e.g. WiFi) radio networks. In terms of data services the same service evolution trends have been seen in the "wired" Internet earlier that can be seen in the mobile domain today. For example, the movement from messaging data services to static content (Web) and further to multimedia streaming can already be seen in mobile service studies (Verkasalo 2007b).

Amidst the rapid evolution of the mobile industry many commercial service failures have taken place. Only a small number of services can be considered successful after the exception of SMS (short messaging service) that was a killer success in Europe in the 90s. Third party instant messaging clients providing connectivity to Internet instant messaging (IM) communities, ringtone/wallpaper downloads and mobile webmail have appeared in the market and succeeded pretty well. However, at the same time many disputed services, typically pushed to the market by handset vendors and operators, have not been widely adopted and revenue have been mediocre or poor. Some of these services include MMS (multimedia messaging service), WAP (wireless application protocol), operator-standardized IM, VoIP (voice over Internet protocol) and video calls.

It is difficult to pinpoint the reasons behind successes and failures. Typically not one but many issues affect the adoption of a particular mobile service. The reasons can be categorized into two main categories. First of all, a commercial/technical perspective includes issues that relate to marketing, positioning, developing, implementing, delivering and timing of the mobile service. These factors include e.g. demand forecasting, pricing, positioning of the service in the service provider's service portfolio, promotional activities, creation of end-user awareness, service quality management, and strategic push of the service in the value-chain (i.e. distribution management). These factors are called as *technological* or *business strategic* in Pedersen (2001). Second, the end-user perspective deals with end-user related factors driving or inhibiting service adoption. This perspective is called as *behavioral* in Pedersen (2001). Factors under this perspective include e.g. service usability, social pressure, network externalities, contextual environment, consumption choices, the user's motivation and technical capabilities. The first perspective deals more with the producer side of the market whereas the second perspective deals with the demand side of the market. Drivers and bottlenecks for service adoption might emerge in either domain.

Even though many potential factors explaining successes and failures of mobile services can be identified, it is often difficult to test hypotheses in practice. No suitable empirical research approaches have existed earlier to provide actual usage data to study the dynamics of **mobile service adoption**. Accurate data from end-users can be nowadays acquired with a **handset-based mobile end-user research platform** that was introduced in Verkasalo & Hämmäinen (2007). The new platform provides accurate usage statistics along with flexible tools to deploy questionnaire studies. The present article attempts to provide descriptive results on mobile service adoption with data from Finland 2006. In addition, a path analysis model is built explaining the main drivers and bottlenecks of mobile service adoption based on empirical usage data and questionnaire studies.

2. EARLIER RESEARCH

2.1. Theoretical Models Explaining Technology Adoption

The adoption research can generally be divided into four categories:

- Diffusion research (market focus)
- Adoption approach (individual user focus)
- Gratification research (needs of users focus)
- Domestication research (consequence of adoption focus)

The **diffusion research** focuses on the market-level phenomena, and studies the diffusion of technology in the whole market. Adoption research, on the other hand, includes many statistical frameworks, and it considers individual users as a focal research object. Adoption research focuses on the micro-level process in adopting technologies. Gratification research contributes by analyzing the different kinds of benefits users seek from new technologies, and domestication research analyzes the role of new technologies in integrating to the every day life of people. Although this article mainly applies the statistical models introduced in the adoption research, elements from other research approaches are also applied in building the framework introduced in chapter 3.3. Therefore all these approaches are discussed now in detail.

First, Rogers (1962) introduced the idea of "*diffusion of innovations*", and approached the adoption process in diffusion terms. In his research the adoption process follows a bell curve, suggesting that different kinds of people adopt new technologies at different pace. Only the most technology enthusiastic people adopt new products/services at first, the mass market being more cautious and thus adopting slower. The late-adopters are the most technology averse people, typically purchasing new technology only when it is inevitable. Rogers' research still serves as the general background for many kinds of adoption research.

The diffusion research can be used in studying the emergence of new services. In Rogers (1995) it is argued that adoption is initiated by a new technology, after which the social setting and communication channels boost the diffusion. Rogers' theory has laid ground for many other research frameworks, for example Christensen's (1997) theory of "disruptive technologies" that take over dominant technologies by having a disruptive diffusion path. Time is the core factor in Rogers' idea of technology diffusion, as adoption (penetration of service) follows different patterns at different points of time after its introduction. Diffusion research best conforms to studies interested in the adoption process on the whole market level. The present article largely deals with the demand side dynamics, assuming mostly that the technology is available and some other factors determine if usage takes place or not. Therefore the diffusion approach is not the best one.

The second approach is the **adoption perspective**, in which an individual user standpoint is taken. Each individual makes her own decisions in considering whether to try the service or not. As one possible adoption framework, a technology adoption model provides understanding on a given technology by modeling the process through which the end-user adapts to the technology (see Constat 2001). The framework presents how individuals go through the processes of awareness, engagement, activation, purchase and usage. The key idea is the adoption DNA involved in each process. The adoption DNA reflects various factors, for example the individual's perceived utility from using the technology, societal norms and perceived difficulty of using the technology. In the technology adoption model this adoption DNA then determines how individuals adopt the technology in question, for example whether they adopt it quickly or slowly (or do they adopt it at all). This framework suggests that the usage is determined by a complex model of social, cultural, as well as individual factors.

A wide domain of research stems from adoption of information systems science, utilizing theoretical models developed a couple of decades ago. Earlier models attempting to explain adoption of technologies (particularly information system technologies and this is why the approach is sometimes called as IS adoption science) include the *theory of reasoned action*, *theory of planned behavior* and *technology acceptance model*.

A theory of reasoned action (Fishbein & Ajzen 1975) is based on the individual's attitude towards the action and subjective norm of the action (expected behavior of others in response to the individual's action). Together these determine the behavioral intention to use the technology. Ajzen later expanded the model; in **a theory of planned behavior** (Ajzen 1985; 1991 a third concept exists, namely the perceived behavioral control. This reflects the difficulty of performing the action. These models communicate that technology adoption depends both on the individual's own perceived benefit of performing the action and the social norm driven by people around the individual. All in all, these frameworks suggest that usage patterns not only depend on the individual's own capabilities and interests, but also on the sociological environment and norms in the culture.

Davis (1989) follows similar logic in his framework - **a technology acceptance model** (TAM). He distinguishes two concepts. First, the perceived usefulness reflects the expected benefits from using a certain technology. Second, the perceived ease of use reflects pretty much the same thing as the perceived behavioral control in the theory of planned behavior, i.e. how difficult it is to use the technology. In predicting information technology adoption the TAM model developed by Davis is the most used framework, and by 2000 more than 400 journal articles had cited the two original TAM articles (Venkatesh & Davis 2000). Almost all the information technology adoption articles that will be discussed later in this article stem from the TAM model. Despite its popularity, further development of the model has taken place. For example, the original model as projected below suggested that perceived usefulness and perceived ease-of-use mediate all external factors (e.g. demographics), though this is not always the case (see e.g. Burton-Jones & Hubona 2005). The original model typically explains 40% of usage intentions and 30% of actual use (see e.g. Venkatesh & Davis 2000 and Meister & Compeau 2002). Applications of the framework might in the best case achieve better explanatory power.

Holistic adoption models (see e.g. Pedersen and Thorbjørnsen 2003) deployed earlier in the mobile context and stemming from the TAM model are close to the research approach applied in developing the theoretical framework for this article. Holistic behavioral models typically utilize statistical methods such as structural equation modeling (SEM), and they derive from theoretical models illustrated above (particularly Davis' research). In Nysveen et al. (2005a) the adoption approach utilizing variations of the TAM model with SEM analysis is called as information systems research, as most studies utilizing structural equation models from the adoption perspective deal with the adoption of ICT services and systems. The research done by e.g. Pedersen and Thorbjørnsen (2003) is used as a basis for the model of this article. The next chapter will discuss the theoretical approaches to study mobile services in detail. Because of e.g. increasing context-specific nature and various ubiquitous characteristics (Heinonen and Pura 2006; Rask and Dholakia 2001) mobile services should be considered carefully when applying earlier information systems adoption models.

The other approaches to study the adoption of technology include a uses and gratifications research approach (see e.g. Leung and Wei 2000; Höflich and Rössler 2001) and domestication perspective (Haddon 2001; Ling 2001; Skog 2002). The former approach (**gratification research**) deals with the gratifications that users look for when using mobile services. These gratifications can be either utilitarian (business value or direct utility) or hedonic (entertainment oriented) (Flanagin and Metzger 2001). The latter approach (**domestication research**) has close linkage to sociology, anthropology and ethnology. Sometimes domestication research does not focus on individual adoption only, but extends to the adoption from the cultural or societal point of view (see e.g. Ling and Yttri 2002). Domestication research attempts to tackle the consequences of service usage and the integration of the technology/services in the customer's every day life (Pedersen 2005). These two frameworks reflect softer scientific disciplines in studying mobile services than the diffusion or adoption perspectives. All approaches share same concepts and ideas, and they are thus not totally separate from each other.

2.2. Emergence of Mobile Services

Little research on the emergence of mobile services exists. This chapter briefly describes some of the main outcomes of earlier research, particularly in the light of this article. Many of the focal services studied in this article have just been introduced to the market, and no research on them is available. However, this chapter attempts to emphasize generic observations in the adoption of mobile services that have relevance also with new emerging services.

The **mobile Internet** is defined in this article to consist of new packet-switched mobile data services. The first mobile data services were hyped quite a lot in the public in the late 90s, but in practice the mobile Internet has not kicked off yet (Saarikoski 2006). Particularly the WAP technology and mobile email can be considered failures (Sigurdson 2001). Many reasons are suggested as possible bottlenecks, from pricing (see e.g. Gao et al. 2002) to general difficulties in terminal configuration (Verkasalo 2007b). These reasons were found as bottlenecks in mobile payment adoption, too (Mallat 2006b). MMS messaging experienced similar disappointing customer adoption rates at first than WAP, email and electronic payments are experiencing right now. In Japan NTT DoCoMo has achieved satisfactory demand for mobile Internet services, and Japan can be considered as the world's leading mobile market in many other dimensions, too (Saarikoski 2006; Minges 2005). The key difference to Western operators is that NTT DoCoMo has bundled service interfaces directly to the handset and actively pushed new value-added services. Consequently both the customer awareness has been increased and ease of adoption significantly improved.

The Internet service expansion to the mobile domain could serve as a prospective spark that could significantly push the extent of mobile data service usage. The reasons for the potential impact of Internet services can be divided into two. First of all, the pricing/commercial models of Internet services are different from operator-based mobile services (in terms of pricing, for example, Internet services are free of charge, or sold in flat-rate bundles). Second, the spill-over effects and value already embedded in Internet services (e.g. instant messaging communities, the variety of WWW-based content services, webmail) serve as a strong force if suitable mobile access technologies (e.g. the network access and adequate terminals for the usability's sake) can be deployed with low price. Earlier research has pinpointed the need for alternative radio access methods to fixed Internet service access, the key alternative being mobile cellular connectivity (Kearney 2001). These two factors (commercial/pricing models and mobile extension of the Internet) are evident in the final conclusions of Aarnio et al. (2002), as they state on the future prospects of mobile services that "*... prices must come down to overcome the critical mass threshold and in most cases the services should be integrated to the Internet...*".

As people are currently claiming technical challenges and prices for the low realized actual usage of POP/IMAP (post office protocol, internet message access protocol) mobile email (Verkasalo 2007b), webmail services accessible with new mobile web browsers hold a lot of potential to dominate all dedicated email clients and POP/IMAP-based complicated configurations. In addition, a well implemented mobile WWW access opens doors also to other mobile services, such as centralized calendar services, news portals, electronic banking - even office applications (in the vein of Google's web-based office tools). Similar potential for the mobile Internet expansion exists also in voice (from telecom-dominated voice to mobile VoIP), instant messaging (from SMS to mobile instant messaging networks) and multimedia streaming (from offline content playback to online multimedia content like mobile YouTube), to mention a few examples. The open Internet world allows for rapid and flexible network-edge based service innovation and engineering.

Niina Mallat (2006a) illustrates some interesting characteristics of mobile services in her dissertation. According to her studies, several successes and failures could be identified in the mobile payment service sector. Mallat concludes her dissertation with the note that usage situation and context are important in explaining the adoption of mobile electronic payments. The same can be generalized to the wider mobile domain. In studying mobile service adoption the special characteristics of mobile services – the freedom of context and location – should be internalized in the framework is possible. Mallat criticizes earlier mobile service adoption research for not taking contextual factors that well into account. Also Hejden et al. (2005) emphasized the importance of context in influencing the perceived value of services.

In tackling the drivers and bottlenecks of mobile services, Aarnio et al. (2002) studied 1 553 Finnish respondents and found five user segments from the sample that resembled classical adoption categories. Their paper studied both traditional Internet and new mobile services. The authors emphasize the important role of prices in mobile service adoption. Furthermore, they suggest that social norm should not be overlooked in mobile services. However, Nysveen et al. (2005a) suggest that different mobile services depend on explanatory factors differently. For example, some services are more dependent on social pressure/norms than others. For example in the research of Pedersen and Thorbjørnsen (2003) the social norm did not explain that much of the variance.

Pedersen and Thorbjørnsen (2003) developed a structural equation model incorporating motivational, attitudinal, social and resource-related influences on the intention to use mobile services. Their model explained about 62-75% of the variance in the dependent variable (intention to use). In all of their case studies (focusing on different services) they found that both extrinsic (utilitarian) and derived (i.e. expressiveness) motivations have an important role to play, whereas intrinsic (entertainment-oriented) motivations are less explanatory. Pedersen (2005) continues by suggesting that many of the factors suggested in earlier ICT adoption research work fine in the context of mobile services, but also many new dimensions should be taken into account (such as expectations and subjective norm). Pedersen (2005) argues that by extending the decomposed theory of planned behavior (Taylor & Todd 1995) with elements from domestication research the explanatory power of the models in mobile services can be increased. To criticize existing statistics models Nysveen et al. (2005b) argue that gender could have moderating effects in the adoption of mobile services, though the original TAM model suggests that all background variables should be fully mediated by the included explanatory factors. All in all, in statistically evaluating the determinants of usage, different services should be treated individually and one should be careful with cross-service generalizations.

Some mobile services are more goal-directed (utilitarian) instead of experimental (hedonic) (Nysveen et al. 2005a). Pedersen and Thorbjørnsen (2003) acknowledge the same. Also Wong and Hiew (2005) argue for the emphasis on the expected benefits from mobile services. According to their study, satisfaction of the end-user market is the most important factor determining success, and therefore e.g. strong technology push is not the way in which mobile services should be deployed in the market. According to Wong and Hiew, value-added to the end-user and direct benefits from using the service are the key drivers particularly among younger, early-adopter, customers. Hejden et al. (2005) comply by concluding that utilitarian and hedonic values have some correlation but it is particularly utilitarian value that has a significant impact on the intention to use the services whereas hedonic values relate to more ad-hoc use cases. The utilitarian importance might have, however, derived from the market context at the time of the study in Hejden et al (2005), as many new mobile services emerging today are actually quite entertainment instead of business-oriented. Most new mobile services are therefore targeted at consumers instead of business customers. Hejden et al. (2005) nevertheless found that perceived risks from the service negatively drive the service's utilitarian value, whereas hedonic value is not affected by the perceived risks (e.g. the service is justified to be more difficult to use if it relates to "killing time"). Some papers (e.g. Tseng et al. 2007) emphasize that in addition to direct value also network externalities matter in analyzing the benefits of electronic services. In mobile instant messaging, for example, the value embedded in the buddy network affects the utility from using mobile instant messaging.

Pedersen (2005) notes that adoption studies in the mobile domain might have differing objects of research. He distinguishes between users (Green et al. 2001; Bakalis et al. 1997), services (see Verkasalo & Hämmäinen 2007; Kim 2001; Pedersen et al. 2001) and terminals (see Chuang et al. 2001; Skog 2000). In this article the approach is mainly from the user point of view, and the process through which end-users adopt services is the focal research objective.

3. EMPIRICAL ANALYSIS

3.1. Research Method and Dataset

A pioneering mobile end-user research platform was utilized in acquiring data for this article. The new mobile end-user research platform is based on a developed Symbian/S60 smartphone client that observes all kinds of usage actions taking place in the handset. Usage-levels stamps on any application, network or user interface level action is logged with accurate context-specific information (e.g. time). This data is sent to centralized servers every night for analysis purposes. Research is conducted in panels lasting typically 2-3 months, and a typical panel includes hundreds of interested customers which in the panel become panelists. All panelists participating in these study panels sign a contract and they are aware of the research process. Usage data is complemented with various WWW-based questionnaires through which to acquire data on issues that are not usage-related (e.g. motivations and attitudes). Though the accuracy and scope of acquired data is a clear contribution in the world of end-user research, the challenges include e.g. biased end-user domain (**early-adopter users**) and sample size (being still in the range of 400-1000 panelists). The research method has been used already in various papers, such as (Verkasalo 2006a, Verkasalo 2006b, Verkasalo & Hämmäinen 2006, Verkasalo 2007a, Verkasalo 2007b, Verkasalo 2007c and Verkasalo 2007d). For more information on the research method, see Verkasalo & Hämmäinen (2007).

In acquiring data for this article SMS recruitment messages were sent to 28 000 Finnish consumer subscribers who owned a Nokia S60 device. 1 071 (3.8%) customers visited the recruitment site and answered the beginning questionnaire. Out of the registered panelists 695 (65%) managed to generate at least three active weeks of smartphone usage data. The rest either did not manage to install the research client or then quit the panel study earlier than was supposed. Some people might have used the smartphone with the research client installed as a secondary phone, and thus they were excluded, too. Out of the active panelists 548 (79%) answered the final questionnaire, which was more comprehensive than the beginning questionnaire. In addition to demographics, the beginning and final questionnaires covered various background questions related to user motivation, usage patterns and opinions on usability and performance of mobile services.

The details of the panel are introduced in Verkasalo (2007b). In brief, most panelists were young male consumer customers. The panel consisted of early-adopter customers, as in earlier handset-based service studies (see Verkasalo 2005). 37% of panelists considered themselves either experienced or very experienced smartphone users, whereas only 19% considered themselves as beginners. 56% considered themselves as normal in terms of smartphone usage patterns.

The questions asked from the panelists in background questionnaires are listed in Appendix A. The whole dataset used in this article consists of a questionnaire that was filled in by the panelists, together with a comprehensive set of aggregated usage-based measurements that reflect objectively panelists' actual behavior during the panel. The background questionnaire was used in obtaining data on the factors that have been found important earlier in studying technology adoption (see particularly Pedersen's research). Because of respondents' limited time, only one question was asked for each of the identified factors. Earlier statistical adoption studies have had more comprehensive questionnaires, and therefore they have the possibility to use both the measurement and structural parts of structural equation modeling. In this article only the path model is estimated. The acquired data is collected for the services that were identified as being of great interest at that point of time (in 2006):

- WEB/WAP services, Internet browsing
- Email
- Instant messaging / chat
- MMS messaging
- Gaming
- Internet telephony
- P2P (point-to-point) file sharing
- User-created content creation, blogging
- Streaming/Internet multimedia
- Offline multimedia playback

For each of these services two key usage-level variables were calculated. First, a simple Boolean variable was calculated indicating whether a particular user had used a particular service or not during the panel Second, a usage frequency variable was calculated measuring the share of days when a particular service was used out all active days spent in the panel. If a service is in constant use, this variable ranges from 50% (meaning usage every second day) to 100% (usage every day). If usage is less frequent, the usage frequency variable takes lower values. This variable well reflects the extent of use, and it is restricted to take values in the range of 0% to 100%. This variable is used as an indicator of usage in the estimated path model later in this article.

3.2. Descriptive Statistics

Figure 1 depicts the users' own opinion about themselves in terms of technical experience with smartphones and estimated average minutes of smartphone service usage per day. More experienced users spend time with smartphone devices more than less experienced users by observation. People with stronger technical interests and capabilities should be more likely to explore new services on the one hand and more likely to learn using them and adopting the service into every day use, on the other hand. Therefore the observed results on total smartphone usage time are quite expected. Because no data on user experience was available for all of the panelists, the total smartphone usage minutes per day is used a proxy for user experience in path modeling later in this article.

Next the focus is on particular services for which a comprehensive dataset was collected combining both questionnaires and actual usage data. Figures below projects the identified key

Figure 1. Correlation of technical orientation and extent of handset use

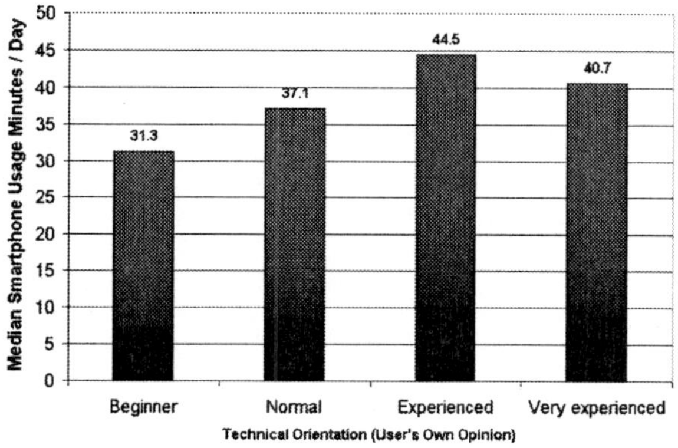

smartphone services in two dimensions: the share of panelists who intended to use the service before the panel (*intention index*), and the share of panelists who actually used them (*usage index*). The projected diagonal line depicts the points in which the service's all users who have intentions to use the service actually used the service during the panel. In some cases it, however, might be that some users who did not intend to use the service actually used it anyways. The services that fall off the line have some bottlenecks in the adoption process because not all of the interested panelists used them. Conversely, those services that are above the line have been adopted by more panelists than who actually intended to do so.

WEB/WAP browsing, MMS, offline multimedia and gaming have experienced usage approximately according to intentions. In other words, approximately those who intended to use the services also were able to do so. Therefore these services can be considered as sort of successes in this study. More interestingly, there are many services that lie relatively far away from the diagonal line. These include e.g. instant messaging, mobile email (both embedded email clients and webmail included) and streaming multimedia. People had intentions to use these services, but for a reason or another usage did not realize. Some bottlenecks for adoption remain for some services, whereas for some services bottlenecks are significantly lower. All services that received high usage indices (located in the upper right hand corner of Figure 2) have been available in the market for longer than those that received low usage indices. Clearly service maturity reflects in the results.

Figure 3 demonstrates how intentions correlate with service adoption. Those who stated they have strong interest towards the service had significantly higher adoption rates. Strong correlation between intentions and the consequent adoption process exists. As was found above, WEB/WAP browsing, MMS, gaming and offline multimedia playback catch a wide domain of those users who originally had intentions towards the particular service. On the other hand, only 60% of those who had a very strong interest to use mobile email and 33% of those who had a very strong interest to use various streaming multimedia services actually did use these services. There are many arguments to explain why certain services experience good adoption whereas some others do not. Some of these reasons include the technical difficulty to configure/use the service, immaturity of the service, unavailability of necessarily technical help, problems in the marketing / communication functions and suboptimal pricing.

The figures above demonstrated that there exists a gap between user intention and actual usage. This is called here as the "adoption gap", as it

Figure 2. User intentions and actual use of services

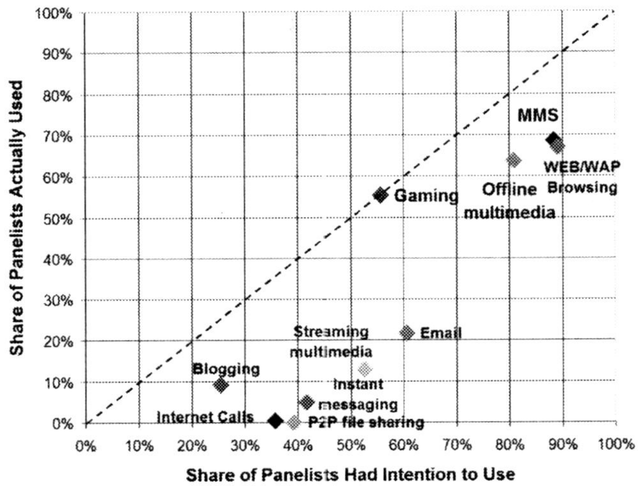

reflects the problems in moving from intentions to actual usage. This article develops a service *adoption index* measuring the extent of this gap. The index can be calculated by measuring the share of panelists who actually used the service out of those who intended to use it. For services that experience little bottlenecks in adoption, this should be close to 100%, whereas for services with significant bottlenecks this should be closer to 0%. Graphically all services located close to the diagonal line have high service adoption indices. The advantage of this index is that it can be used when comparing services with different customer awareness (e.g. different diffusion levels of original intention to use), as effectively the measure is normalized based to the number of people who intended to use the service (instead of all panelists).

In addition, there exists a gap between intended usage (reflecting current short term interest to explore the service) and expected future service diffusion (reflecting future expectations for the service, i.e. is the service going to succeed in the mobile market by extending the "wired" version of the service). A background question was asked with regards to users' expectations of different mo-

Figure 3. User intentions and actual service usage

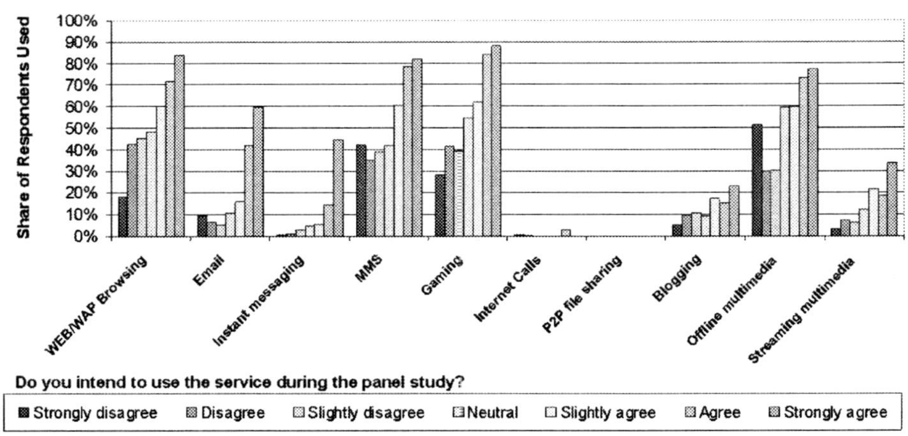

bile services succeeding in the mobile market by replacing their use of the service in other platforms (e.g. desktop computers or MP3 players). This is called as an attitude towards the service variable in this article as it reflects positive expectations for the mobile version of the service. Rather than reflecting general attitude it reflects positive long-term perception of the service eventually being used by the panelist herself. *Attitude index* measures the share of panelists having a positive attitude towards the service. Further, *timing index* measures the share of panelists having positive intentions to use the service during the panel (in the short term) out of those panelists having a positive (long-term) attitude towards the service. This variable is very useful, as it normalizes the number of people intending to use the service by the number of people having long-term interest (i.e. attitude) towards the service. Therefore the timing index can be used in evaluating whether the market is ready to adopt a particular service right now, given that positive attitudes and therefore eventual latent demand for the service exist. The timing index is high if a relatively high share of interested panelists are likely to adopt a given service in the short-term, and low if service adoption is for a reason or another (e.g. unavailability of the service) not acute right now.

Table 1 summarizes the calculated indices for the identified services. By looking at the attitude index, it can be said that people have positive attitudes (future expectations for mobile services replacing their use of corresponding services with other devices) towards most services, the most so with regards to offline multimedia playback, MMS (replacing corresponding multimedia messaging clients in desktop computers, such as some IM clients) and mobile email. The second column (intention index), however, tells that most people have short-term interest to use only browsing, MMS, games and offline multimedia services. Indeed, the timing index is higher than 50% for these services. They are likely to generate usage right now. Some other services (having high at-

Table 1. Measured service indices

	Attitude Index	Intention Index	Timing Index	Usage Index	Adoption Index	
WEB/WAP Browsing	73%	88%	92%	69%	72%	
Email	87%	61%	65%	22%	31%	
Instant messaging	79%	42%	42%	5%	10%	
MMS	94%	89%	90%	67%	70%	
Gaming	68%	56%	58%	55%	71%	
Internet Calls	83%	36%	34%	0%	0%	
P2P file sharing	63%	39%	41%	0%	0%	
Blogging	76%	25%	26%	9%	13%	
Offline multimedia	95%	81%	82%	64%	70%	
Streaming multimedia	85%	53%	52%	13%	19%	

Attitude index	Share of panelist thinking that the mobile service has replaced / will replace corresponding existing use of the service (i.e. computers)
Intention index	Share of panelists having an intention to use the service in the near future
Timing index	Share of panelists intending to use the service in the near future out of those having a (long-term) positive attitude
Usage index	Share of panelists having used the service during the panel study
Adoption index	Share of panelists having actually used the service out of those who intended to do so

titude indices) might be successes further in the future according to the panelists, an example being e.g. mobile blogging which has a rather positive attitude index but low intention and consequently timing index. Also mobile VoIP (Internet calls) have promising prospects, as many people have positive long-term attitudes towards the service though no short-term interest (intention) to use the service exists. The timing index therefore reflects current propensity of taking service into use, given that there is a positive attitude towards the service. The fourth column projects the actual share of panelists having used the service during the panel, and finally adoption index in the fifth column states that only offline multimedia, browsing, MMS and gaming have been adopted without significant bottlenecks. All of these five indices can be measured over time. Longitudinal comparisons would provide visualizations on the development of services over their individual adoption curves.

The panelists themselves have opinions with regards to the key bottlenecks for new mobile services not to actually succeed in the market. All together 71 panelists were asked randomly (utilizing the pop-up questionnaire functionality in the research platform) why they think new mobile services do not survive in the market. Each respondent could pick only one answer, and the results are projected in Figure 4.

According to the panelists it is still pricing that has the most influence on actual usage. In other words, even though people have interests, too high a price level or suboptimal pricing structure might prevent them from using a particular service. The second most important thing the respondents identified as a bottleneck is technical implementation. Many services (according to the panelists particularly handset-embedded mobile email) are simply too difficult to use or configure, and this serves as a bottleneck. The article now turns to studying the adoption process in a more detailed manner by utilizing path modeling and comprehensive background data.

3.3. Path Analysis Model

Path analysis is used as a tool in verifying the theoretical hypotheses of this article. Path analysis is an extension of multiple regression. Wright (1921; 1934; 1960) was first to utilize path analysis in empirically studying direct and indirect effects

Figure 4. Reasons for the failure of new mobile services

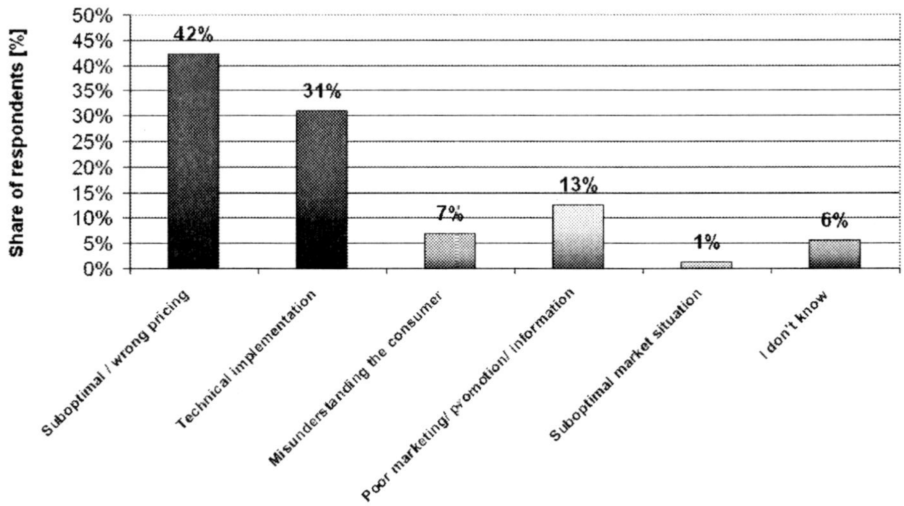

of theoretical models. In essence, path analysis extends multiple regression by including a number of equations instead of only one (Schumacker & Lomax 2004).

Pedersen and Thorbjørnsen (2003) is used as a basis for the theoretical model developed here. Therefore factors perceived enjoyment, perceived usefulness, social push (called in other contexts as subjective norm) and behavioral control (i.e. perceived ease of use) are included to have either direct or indirect (mediated by attitude) effect on intention to use. The perceived expressiveness variable is not included in this model to keep it simpler. After all, it has not been included in all of the earlier TAM models. For the attitude variable a question was asked with regards to a particular service having already replaced or replacing potentially in the future the existing use of the service with e.g. computers or MP3 players. If people have a positive attitude towards the service, they answer positively to this question. Each factor is measured by a single question, listed in Appendix A. Due to limited data (and consequent unobserved uncertainty) the developed path model does not work that well by itself (Figure 5). However, the path models estimated for each of the services can still be compared against each other in analyzing the role of each factor in the context of a cross-service study. The original TAM model assumes that demographics (such as gender and age) and many other background variables are totally mediated out by the included independent factors in the extended TAM model. This simplifies the analysis.

Because of the variety of data available not only related to background variables but also on actual usage, the theoretical model is extended to include a factor reflecting actual usage (usage frequency of the service). Actual use is depicted to depend on intention to use, technical capability of the phone (the handset features serve either as a bottleneck or enabler), data pricing (lower marginal data prices should push services that generate data charging records), monetary pool available for smartphone usage (more monetary resources should have a positive impact on service usage that

Figure 5. Developed theoretical path model of mobile service adoption

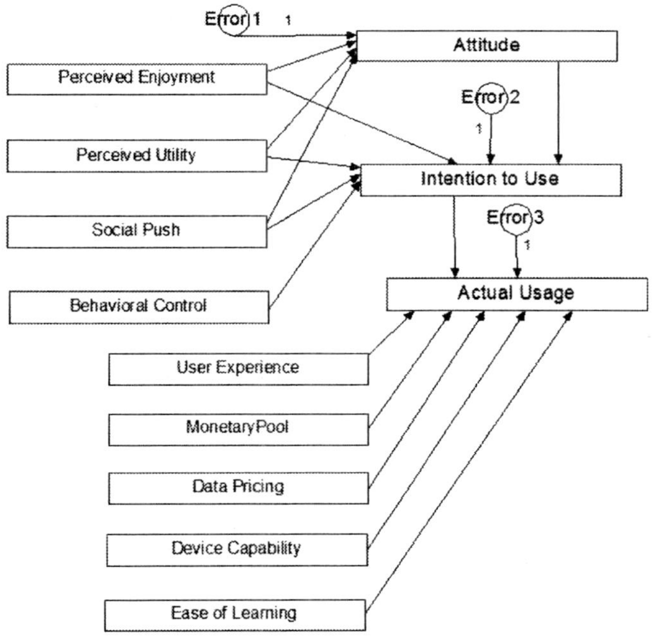

generate charging records) and user experience (more experienced users should be more likely to be capable of using the service). The technical capability of the phone is a boolean variable taking value 1 if the handset is based on Nokia S60 3rd edition platform (all new Nokia smartphones) and 0 otherwise (older Nokia S60 handset). Data pricing takes values 0 (no data plan, expensive usage-based charging), 1 (block-priced data plan, relatively cheaper to use data services) and 2 (flat-rate, unlimited usage with zero marginal cost). Monetary pool is proxied through ABPU (average billing per user) rates of subscribers, hypothesizing that high average bills of the past reflect the user's mobile service budget. User experience is here communicated through estimated average minutes of smartphone usage per day (that has strong correlation to user's experience according to a smaller sample of data).

One more variable is included to reflect a positive perceived easiness of learning and ease of getting help (i.e. kind of facilitating conditions in actually using the service) for the service. This is projected to have a direct impact on actual use. The behavioral control variable introduced earlier better reflects the expected difficulty of use thus having implications in the initial intention to use, whereas the ease of learning component is hypothesized to be relevant if the initial interest towards the service exists and the problem is how to actually use the service given positive intentions.

The final model to be estimated for each service separately is projected in the figure above; see Appendix A for the description of data. Variables monetary pool, user experience and data pricing have strong correlation according to correlation matrixes, and these correlations are allowed for in estimating the path model.

Based on earlier research (Nysveen et al. 2005a), utilitarian and hedonic services should differ in the impacts corresponding end-user perceptions have on intention to use. In addition, Hejden et al. (2005) suggests that expected technical challenges should have a stronger negative impact on intention to use for utilitarian rather than for hedonic services.

All in all, the following hypotheses are laid down for the model:

1. Perceived enjoyment has both direct and indirect (attitude) impact on intention
2. Perceived usefulness has both direct and indirect (attitude) impact on intention
3. Social push has both direct and indirect (attitude) impact on intention
4. Behavioral control has direct impact on intention
5. Ease of learning has direct impact on actual use
6. Intention to use, technical capability of the phone and user experience have direct impacts on actual use
7. Data pricing has a direct impact on actual usage for services that generate data charging records
8. Higher monetary pool for mobile service usage has a direct impact on actual usage in services that generate charging records
9. Higher handset capability has a positive impact on actual use
10. Utilitarian services should be driven more by perceived usefulness and hedonic services should be drive more by perceived enjoyment
11. Poor behavioral control (i.e. technical difficulty) is justified for hedonic services but nor for utilitarian ones

The models are estimated with AMOS (add-on package for SEM in SPSS), using the maximum likelihood approach. Because there are missing data points (not all people answered the questionnaires), only panelists with full data are included. All in all, 426 panelists are included in the path estimation.

3.4. Results of Path Analysis

The estimated path analysis diagrams for the chosen services are depicted in Appendix B. The figures include path coefficients (above the paths) and for dependent variables the share of variance explained by the model (above the box representing the variable). Some model fit indices are reported, too. First, a fit index CMIN/DF is included - the relative chi-square that represents how much the fit of data has been reduced because one or more paths have been dropped from the full model. Generally if this index is more than 3, too many paths have been dropped. The second included fit index is GFI (goodness of fit index). This tells how much of the variance in the sample variance-covariance matrix is accounted for by the model. Generally this should be above 0.9. Finally RMSEA (root mean square error of approximation) estimates lack of fit if compared to the full model. 0.08 or below is considered generally as adequate fit.

By looking at the results in Appendix B, the first observation is that the fit indices are quite poor. In other words, the developed model is perhaps not the best one to explain the involved complex relationships. GFI indices range from 0.82 (IM and email) to 0.86 (browsing). The second observation is that the model quite poorly explains the dependent variables under interest. The variance explained for the intention to use variable range from 0.05 (blogging) to 0.24 (gaming and MMS). The variance explained for the actual usage variable range from 0.01 (blogging) to 0.27 (gaming). Lots of uncertainty (e.g. non-internalized variables) exists that would explain intentions and actual usage, and the model therefore cannot depict all of the variance in the dependent variables. Almost all of the variables included in the framework model the demand-side, and therefore the missing (explanatory) variables are probably related to supply-side factors (e.g. service push, pricing and availability).

Without going into the details of each service (the estimated path models can be found from Appendix B), the table below summarizes the statistically significant path coefficients for each service.

It is surprising that perceived enjoyment has very strong impact on attitude for all services. People who think the enjoyment value is high (for any service) have high expectations that they are going to replace existing use of the service with the mobile version of the service (this is here called considered positive attitude). Perceived enjoyment positively (and statistically significantly) drives intentions directly, too. It can be generalized that the advanced mobile service market in Finland is building on expected hedonic benefits even in services which have been generally considered business-oriented (email).

Further, utilitarian benefits are important, too. The impact towards attitude is positive for all services expect for streaming multimedia, though the relationships are not as strong as with hedonic benefits. Interestingly the positive forward-looking attitude towards mobile instant messaging is strongly pushed by perceived utilitarian benefits. This confirms that instant messaging is a viable communication channel rather than a service meant for entertainment purposes solely. The direct impact of perceived utility on intentions is not that strong as the impact of perceived enjoyment (hedonic benefits), but the model nevertheless confirms that the intentions to use browsing, MMS, email (strong effect), blogging and IM are at least partly driven by utility-related expectations. Expectedly perceived utility does not drive offline multimedia playback, gaming or streaming multimedia usage intentions, because they are quite entertainment-oriented by nature.

Social push is all about the impact of supportive/driving social context with regards to usage attitude and/or intentions. Social push has positive impacts particularly on email and instant messaging. This can be explained by the fact that these services are currently available, and the marginal impact of somebody in the close social neighborhood helping or recommending the service might be significant to drive attitude towards possible future use. The direct effect of the social push on intention to use is not very strong for any of the other services, communicating the fact that these early-adopter users are quite independent.

Behavioral control measures the user perception of herself being capable of using the service and to overcome technical challenges. Particularly for browsing and gaming this factor is a significant driver of intentions, whereas for most other services it has a positive though less significant effect. The statistical significance for the path originating from behavior control is generally not that strong. This might be due to the fact that Hejden et al. (2005) hypothesized that for hedonic services behavioral control does not have that strong an impact. For MMS there is no relation at all, as MMS is already well integrated into smartphones and thus there is little variance in end-user behavioral control (see Table 2).

The TAM model suggests that attitude towards the service should drive intentions. The attitude variable included in this research is geared towards current or future role of the mobile service possibly replacing existing use of the service with other devices (e.g. laptops or MP3 players). Therefore this variable is not reflecting general attitude in the purest sense. There is a statistically significant relation between attitude and intention only for mobile email. Positive mobile email attitude levels correlate strongly with positive intentions. People who think mobile email is or will be an important part of their smartphone usage in the future also have strong short-term intentions of using the service. For many other services the effect of attitude on intention is statistically insignificant. This can be explained by the fact that people generally perceive that these new advanced services are unavailable / technically poor, and therefore they lack short-term intentions of use though there is a positive general attitude. Another argument is relatively stronger effects of direct effects of background variables. In other words, rather than for many background factors having an indirect impact via positive attitude that the service would replace something (having a flavor that something else is sacrificed), for example perceived enjoyment or utility directly drive intentions and these relations are very strong.

When moving to actual use of services, it becomes clear that the variables hypothesized to have direct effect on realized usage are important. User experience (proxy through average usage minutes per day) has statistically significant relationships with actual usage of browsing, MMS, offline multimedia playback and gaming. These are services users have probably used already earlier with their smartphones, and therefore experience variable strongly drives usage. For some very new services such as email, IM or streaming the experience variable does not have a statistically significant effect. This might be because earlier experience relates to services that have been in the market already for some time (on which experience can accumulate), whereas experience does not help with regards to services that are just about to hit the market or are available later in the future.

Monetary pool has a strong effect on MMS usage, which has strong correlation with other mature service (such as voice and SMS) usage. In other words, higher consumption levels on mobile telephony do not drive any of the truly new service categories. First, some services are simply free and they do not depend on consumption levels or available monetary pool but on e.g. perceived enjoyment instead. Second, still most of the ABPU (average billing per user) rate consists of voice and SMS service charging, and therefore it does not reflect data service prices/consumption that well. All of the services that require network connectivity in this research are based on IP connectivity and are therefore independent of circuit-switched service pricing.

Data pricing (the lower marginal cost of using data services) has the most significant relationship expectedly with browsing usage. Browsing is currently the most important data (Internet) service. There is also weaker (but still statistically significant) effect on email usage. However, for

Table 2. Statistical significance of estimated paths

Path	Browsing	MMS	Email	IM	Offline multimedia	Streaming multimedia	Gaming	Blogging
Perceived Enjoyment -> Attitude	***	***	***	***	***	***	***	***
Perceived Utility -> Attitude	**	**	**	***	**		**	**
Social Push -> Attitude	*		**	**	*			
Bhavioral Control -> Intention to Use	**		*	*	*	*	***	*
Perceived Enjoyment -> Intention to Use	***	***	***	***	***	***	***	
Prceived Utility -> Intention to Use	*	*	***	*				***
Social Push -> Intention to Use				*	*			
Atitude -> Intention to Use			*		**			
User Experience -> Usage	***	**			**		***	
Monetary Pool -> Usage		**			#			
Data Pricing -> Usage	***		**					
Device Capability -> Usage			***		***	**		
Ease of Learning -> Usage								
Intention to Use -> Usage	***	***	***	***	***	**	***	

*** $p \leq 0.001$ (positive relation)
** $p \leq 0.01$ (positive relation)
* $p \leq 0.05$ (positive relation)
# $p \leq 0.05$ (negative relation)

other data services data pricing does not have an effect. This is partly due to little actual usage realized for other than browsing data service.

Interestingly device capability has the biggest influence on email and multimedia functionalities. The impact on multimedia is not a surprise, as if nothing else, new handsets include at least better multimedia functionalities (camera chips, memory capacity, MP3 player functionalities, better displays for movie/video playback) than older ones. The marginal email usage is coming from webmail oriented email usage. This is partly driven by better browsers included in newer handsets together with new high-resolution displays. This explains the path from device capability to email usage.

Ease of learning variable does not have an impact on actual usage at all. These early-adopter users, after all, all know that there is help available for example in the Internet. On the other hand, they also tend to play around with the handset by themselves rather than think that they need somebody else for help, not to talk about manuals that are shipped with new handsets.

Finally, the most important variable in the model (intention to use) has statistically significant effects on all service usage expect for blogging. As concluded already earlier with descriptive statistics, people who have strong intentions to use the service tend to adopt the service (use it in practice) with a higher probability than those who do not have intentions. Blogging has not been used by that many, and therefore intentions cannot correlate with usage. Most blogging usage comes actually from Nokia's preinstalled Lifeblog application. Because of this it can be that those who explored blogging merely randomly launched Lifeblog, even though they did not have clear intentions. All in all, usage intentions are still the most important driver of advanced mobile service usage, therefore suggesting that the needs of end-users are the key issue in bringing new mobile services to the market instead of push-centric service introduction strategies.

A couple of reasons exist for the general poor fit of the estimated models. First, these are **advanced services**, and there are probably many random factors that lead into a usage event. The service must be in the market available for panelists and the technical performance/accessibility has to be in order. These are some of the variables the model cannot internalize, and this explains the poor share of variance explained for usage-level variables. Similarly intentions originate from many other factors (in advanced services) than those included in the model. For example, promotional events and availability of the service certainly do affect the short-term intention to use a service. These variables are not included in the model. In many cases these reasons explain the poor fit of the model. Indicative results suggest that the depicted theoretical model is not the optimal one. For example mobile blogging is a rather new service on which customers have very little or limited knowledge on. For this service most paths are insignificant and model fit is very poor. In addition, it is difficult to explain the variance of dependent variables if these variables have no or very little variance. Gaming is a well-known service and the market-level factors (e.g. operator competition, network performance, pricing, promotional events) have less impact. Therefore the theoretical adoption model receives much better results for gaming than for blogging, for example.

4. CONCLUSION

The article utilized a newly developed **handset-based mobile service research platform** in deploying questionnaires and measuring actual usage of mobile services. This data is collected for ten emerging mobile services that are about to hit the mobile market as of 2006. The data is analyzed with **descriptive statistics** and a specialized **path model** developed particularly for this study (though originating from earlier information systems adoption science). In addition, various indices are developed that communicate the characteristics of mobile services. This study differs from other handset-based research in its specific focus on the adoption process and associated measurements, utilizing also questionnaires.

Descriptive statistics reveal that for most services not all mobile service demand (i.e. usage intentions) is fulfilled, i.e. not all panelists interested in services are actually using them. Therefore some bottlenecks for adoption definitely exist. Some of these bottlenecks originate from supply-side challenges (e.g. unavailability of the service, marketing failures), and some are related to demand-side problems (e.g. technical difficulties of taking the service into use). However, those strongly intending to use the service, ceteris paribus, expectedly have higher probabilities of adopting the service in practice.

The article developed five indices that can be measured over time and therefore utilized in longitudinal analysis. These indices communicate the general attitude towards mobile services (**attitude index**), short-term intentions of usage (**intention index**), the current propensity of taking services into use (**timing index**), the extent of service usage (**usage index**) and the probability of using the service given intentions exist (**adoption index**). These indices communicate that panelists on average have positive attitudes towards mobile versions of the services included in the study, but only matured mobile services (introduction more than a year ago) experience positive short-term intentions of usage and consequent adoption. Different services are clearly experiencing different phases of their life-cycle at this point of time. The developed indices can be utilized in projecting the phase of diffusion for each service emerging in the market.

The estimated path models reveal that perceived hedonic (enjoyment) benefit is the strongest driver of both attitude and intention towards the service. This is in contrast to Pedersen and Thorbjørnsen (2003) who found that utility-centric benefits drive intention to use mobile services. In the estimated models of this article perceived utility drives intention to use only in the case of mobile email. In general, many of the newly developed mobile services should be considered as generating hedonic rather than business/utilitarian value to end-users. The social setting around panelists or expected technical difficulties do not explain intentions to use. This is most likely due to panelists in the dataset being early-adopters (quite independent and technically advanced users). User experience with smartphones explains actual use of services that have already been in the market for some time, but experience of the user does not explain the adoption of the most recent services. All in all, supply-side dynamics might be more relevant when studying why certain services do not succeed among the very high-end of the end-user domain, as the technical front was found to have little explanatory power in the adoption process. Pricing of data traffic has a strong effect only on browsing and email as they are the most visible mobile Internet services currently. Multimedia services, on the other hand, benefit from higher capability of handsets. The most important thing driving actual use of services, however, is the intention of end-users to use the service, and therefore user needs should be looked upon in the future instead of technology push strategies.

The estimated path models do not have very good fit. Instead of using standardized sets of questions and structural equation modeling in verifying the theoretical model, a more simplified path modeling exercise was chosen due to limitations in the panel study implementation. Improvement of the theoretical model remains as a future work task in addition to the acquisition of more comprehensive dataset supporting not only path but also measurement part of the SEM analysis. One of the key conclusions of this article is, however, that the internalized variables of the estimated adoption model poorly explain the variance observed in intention and usage-level variables. Therefore most advanced mobile services have special characteristics and external sources of variance that were not observed in this analysis. These likely originate from the supply side of the market. Nevertheless, the demonstrated fitting of empirical data with path analysis was found to be an efficient method to compare the role of different variables in explaining the adoption of emerging mobile services.

From the economic perspective a profitable mobile service business entails two necessary conditions. On the one hand, one has to attract a reasonable base of customers. On the other hand, one should make the most money out of that customer base. This article has indicated several implications with regards to the first condition. First, the perceived hedonic benefit (perceived enjoyment of using the service, entertainment value, possibility to "kill time") is the biggest source of user interest in emerging mobile services. Second, intentions of users clearly drive realized usage. Therefore lots of effort should be put on generating intentions among the potential customer population. The probability of mobile services replacing already existing services (long-term attitude), social network effects and perceived ease of use do not play that big a role in driving intentions. Rather, promotion and smooth availability of the service (that were not internalized in the model) might play a significant role. Third, new handsets push particularly multimedia services and some services embedded in the Internet, such as webmail and WWW browsing. Fourth, data pricing plans (being block-priced or totally flat-rate) drive the usage of the most important data services (e.g. browsing). When asked from the users, technical problems with new services together with suboptimal pricing are the major causes of non-realized usage. The statistical model emphasized particularly the latter.

Some mobile services are truly successful in the market, whereas some others are hyped quite a lot but few people actually adopt them. A need exists to acquire data on user opinions and service usage in order to understand the actual dynamics of mobile service adoption. This understanding is valuable in e.g. better commercializing mobile services and in improving mobile customer relationship management processes (for mobile CRM see Liljander et al. 2007). This article showed that new handset-based data can help in better understanding the mobile service adoption process. The research differed from other handset-based research in focusing on the modeling of the adoption process, and combining efficiently questionnaire data with usage data. The article introduced many new indices that reflect the different sides of the adoption process, and additionally demonstrated that different services can be compared against each other with the handset-based research approach. Future handset-based research should deploy wider questionnaires to make the statistical procedure of path modeling more reliable. In particular, the measurement part of the analysis should be deployed before estimating the structural part of the path model. In addition, a wider perspective towards the adoption process should be taken. For example, potential long-term usage should be better modeled against potential short-term usage, and reasons why explorative usage sometimes does not lead to sustainable and repetitive usage should be looked upon.

REFERENCES

Aarnio, A., Enkenberg, A., Heikkilä, J., & Hirvola, S. (2002). Adoption and Use of Mobile Serivces - Empirical Evidence from a Finnish Survey. In *Proceedings of the 35th Hawaii International Conference on System Sciences 2002* (pp. 1454–1463).

Ajzen, I. (1985). From Intentions to Actions: A Theory of Planned Behavior. In J. Kuhl & J. Beckmann (Eds.), *Action Control: From Cognition to Behavior*. New York: Springer.

Ajzen, I. (1991). The theory of planned behavior. *Organizational Behavior and Human Decision Processes*, *50*, 179–211. doi:10.1016/0749-5978(91)90020-T

Bakalis, S., Abeln, M., & Mante-Meijer, E. (1998) *Adoption and use of mobile telephony in Europe*. Cost 248 Report. Telia, Farsta, Sweden.

Burton-Jones, A., & Hubona, G. S. (2005, Spring). Individual Difference and Usage Behavior: Revisiting a Technology Acceptance Model Assumption. *The Data Base for Advances in Information Systems*, *36*(2), 75.

Christensen, C. M. (1997). *The Innovator's Dilemma*. Harvard Business School Press.

Chuang, M. C., Chang, C. C., & Hsu, S. H. (2001). Perceptual factors underlying user preferences toward product form of mobile phones. *International Journal of Industrial Ergonomics*, *27*, 247–258. doi:10.1016/S0169-8141(00)00054-8

Clark, D. D., Lehr, W., Bauer, S. J., Faratin, P., Sami, R., & Wroclawski, J. (2006). Overlay Networks and Future of the Internet. *Journal of Communications and Strategies*, *3*(63), 1–21.

Constat. 2001. ConStat's Technology Adoption Model. The Framework for the WISDM Program (White Paper). Retrieved November 26, 2006 from http://www.constat.com/wisdm/ConStat_Tech_Adoption_Model.pdf

Davis, F. D. (1989). Perceived Usefulness, Perceived Ease of Use, and User Acceptance of Information Technology. *MIS Quarterly*, *13*, 319–340. doi:10.2307/249008

Davis, F. D., Bagozzi, R. P., & Warshaw, P. R. (1989). User Acceptance of Computer Technology: A Comparison of Two Theoretical Models. *Management Science*, *35*, 982–1003. doi:10.1287/mnsc.35.8.982

Fishbein, M., & Ajzen, I. (1975). *Belief, attitude, intention and behavior: an introduction to theory and research.* Reading, MA: Addison-Wesley.

Flanagin, A. J., & Metzger, M. J. (2001). Internet use in the contemporary media environment. *Human Communication Research, 27*, 153–181. doi:10.1093/hcr/27.1.153

Funk, J. L. (2004). *Mobile Disruption - the technologies and applications driving the mobile internet.* Wiley.

Gao, M., Hyytinen, A., & Toivanen, O. (2006). Demand for and Pricing of Mobile Internet: Evidence from a Real-World Pricing Experiment. *Helsinki Center of Economic Research Discussion Paper No. 122* (October 2006).

Green, N., Harper, R. H. R., Murtagh, G., & Cooper, G. (2001). Configuring the mobile user: sociological and industry views. *Personal and Ubiquitous Computing, 5*, 146–156. doi:10.1007/s007790170017

Haddon, L. (2001). *Domestication and mobile telephony.* Presented at the conference "Machines that Become Us," Rutgers University, April 18-19, New Jersey.

Heijden van der, H. & Ogertschnig, M. & van der Gaast, L. (2005). Effects of Context Relevance and Perceived Risk on User Acceptance of Mobile Information Services. *ECIS 2005.*

Heinonen, K., & Pura, M. (2006). *Classifying Mobile Services.* Paper Presented at Helsinki Mobility Roundtable, 1-2 June 2006, Helsinki.

Höflich, J. R., & Rössler, P. (2001). Mobile schriftliche Kommunikation oder: E-Mail für das Handy. *Medien & Kommunikationswissenschaft, 49*, 437–461.

Karlson, B., Bri, A., Link, J., Lönnqvist, P., & Norling, C. (2003). *Wireless Foresight: Scenarios of the Mobile World in 2015.* John Wiley and Sons.

Kearney, A. T. (2001). *Satisfying the experienced on-line shopper: Global e-shopping survey.* London: A.T. Kearney Ltd., Inc.

Kim, J. (2001). *Analyzing mobile Internet users.* Paper Presented at the CHI 2001 Workshop on mobile communications.

Leung, L., & Wei, R. (2000). More than just talk on the move: Uses and gratifications of the cellular phone. *J&MC Quarterly, 77*, 308–320.

Liljander, V., Polsa, P., & Forsberg, K. (2007). Do Mobile CRM Services Appeal to Loyalty Program Customers? *International Journal of E-Business Research, 3*(2), 24.

Ling, R. (2001). *It is 'in.' It doesn't matter if you need it or not, just that you have it. Fashion and the domestication of the mobile telephone among teens in Norway.* Oslo: Working Paper, Telenor R&D, Norway.

Ling, R., & Yttri, B. (2002). Hyper-coordination via mobile phone in Norway. In J. E. Katz & M. Aakhus (Eds.), *Perpetual contact.* New York: Cambridge University Press.

Mallat, N. (2006a). *Consumer and merchant adoption of mobile payments.* Doctoral dissertation. Helsinki School of Economics.

Mallat, N. (2006b). *Exploring Consumer Adoption of Mobile Payments.* Paper Presented at Helsinki Mobility Roundtable, 1-2 June 2006, Helsinki.

Meister, D. B., & Compeau, D. R. (2002). Infusion of Innovation Adoption: An Individual Perspective. In *Proceedings of the ASAC, Winnipeg, Manitoba.*

Minges, M. (2005). Is the Internet Mobile? Measurements from the Asia-Pacific region. *Telecommunications Policy, 29*, 113–125. doi:10.1016/j.telpol.2004.11.002

Nokia Telecommunications. (1999). *The Demand for Mobile Value Added Services: A Market Study*, Retrieved from http://www.nokia.com/press/background/pdf/study-vas.pdf

Nysveen, H., Pedersen, P. E., & Thorbjørnsen, H. (2005a). Intentions to Use Mobile Services: Antecedents and Cross-Service Comparisons. *Journal of the Academy of Marketing Science*, *33*(3), 330–346. doi:10.1177/0092070305276149

Nysveen, H., Pedersen, P. E., & Thorbjørnsen, H. (2005b). Explaining intention to use mobile chat services: Moderating effects of gender. *Journal of Consumer Marketing*, *22*(5), 247–256. doi:10.1108/07363760510611671

Pedersen, P. E. (2001). An adoption framework for mobile commerce. In *Proceedings of the 1st. IFIP Conference of E-Commerce*, Zürich, Switzerland, October 3-5, 2001.

Pedersen, P. E. (2005). Adoption of mobile Internet services: An exploratory study of mobile commerce early adopters. *Journal of Organizational Computing and Electronic Commerce*, *15*(3), 203–221. doi:10.1207/s15327744joce1503_2

Pedersen, P. E., Methlie, L. B., & Thorbjørnsen, H. (2001). *Understanding mobile commerce end-user adoption: a triangulation perspective and suggestions for an exploratory service evaluation framework*. Paper presented at HICSS-35, Hawaii, US, Jan 7-10, 2002

Pedersen, P. E., & Thorbjørnsen, H. (2003). *Adoption of Mobile Services*. Model development and Cross-Service study. Norwegian School of Economics and Business Administration.

Rask, M., & Dholakia, N. (2001). Next to the customer's hart and wallet: Frameworks for exploring the emerging m-commerce arena. *AMA Winter Marketing Educator's Conference*, 12, 372-378.

Rogers, E. M. (1962). *Diffusion of innovation*. New York, The Free Press.

Rogers, E. M. (1995). *Diffusion of innovations*. (4th ed. New York: The Free Press.

Saarikoski, V. (2006). *The Odyssey of the Mobile Inter*net. Doctoral Dissertation. University of Oulu, Finland.

Schumacker, R. E., & Lomax, R. G. (2004). *A Beginner's Guide to Structural Equation Modeling* (2nd ed.). Mahwah, NJ: Lawrence Erlbaum Associates.

Sigurdson, J. (2001). *WAP OFF - Origin, Failure and Future*. Prepared for the Japanese-European Technology Studies (JETS), October 9, 2001, University of Edinburgh.

Skog, B. (2000). The mobile phone as symbolic capital in teenage cultures (in Norwegian). In R. Ling & K. Thrane (Eds.), *Social consequences of mobile telephony* (in Norwegian), Proceedings from a seminar on society, children, and mobile telephony, Telenor FoU - Notat nr. 38/2000.

Skog, B. (2002). Mobiles and the Norwegian teen: identity, gender and class. In J. E. Katz & M. Aakhus (Eds.), *Perpetual contact*. New York: Cambridge University Press.

Taylor, S., & Todd, P. A. (1995). Understanding information technology usage: a test of competing models. *Information Systems Research*, *6*, 144–176. doi:10.1287/isre.6.2.144

Tseng, F. C., Teng, C. I., & Chiang, D. M. (2007). Delivering Superior Customer Perceived Value in the Context of Network Effects. *International Journal of E-Business Research*, *3*(1), 41.

Venkatesh, V., & Davis, F. D. (2000). A Theoretical Extension of the Technology Acceptance Model: Four Longitudinal Field Studies. *Management Science*, *46*(2), 186–204. doi:10.1287/mnsc.46.2.186.11926

Verkasalo, H. (2005). *Handset-Based Monitoring of Mobile Customer Behavior*. Master's Thesis Series. Networking Laboratory. Department of Electrical and Telecommunications Engineering. Helsinki University of Technology.

Verkasalo, H. (2006b). *Empirical Observations on the Emergence of Mobile Multimedia Services in the U.S. and Europe*. Paper presented at the 5th International Conference on Mobile and Ubiquitous Multimedia. December 4-6, 2006, Stanford University, California, 2006.

Verkasalo, H. (2007a). A Cross-Country Comparison of Mobile Service and Handset Usage. Licentiate's thesis, Helsinki University of Technology, Networking Laboratory, Finland.

Verkasalo, H. (2007b). Empirical Insights on the Evolution of the Finnish Mobile Market. Paper presented at Conference on Telecommunication Techno-Economics (CTTE) 2007, 14-15 June 2007, Helsinki, Finland.

Verkasalo, H. (2007c). *Empirical Findings on the Mobile Internet and E-Commerce*. Paper presented at 20th Bled e Conference: eMergence: Merging and Emerging Technologies, Processes, and Institutions. Bled, Slovenia, June 4 - 6, 2007.

Verkasalo, H. (2007d). *Contextual Usage-Level Analysis of Mobile Services*. Accepted for publication at The 4th Annual International Conference on Mobile and Ubiquitous Systems: Computing, Networking and Services (MOBIQUITOUS 2007). August 6-10, 2007 - Philadelphia, PA, USA.

Verkasalo, H., & Hämmäinen, H. (2006). *Handset-Based Monitoring of Mobile Subscribers*. Paper presented at Mobility Roundtable, 1-2 June, 2006, Helsinki, Finland.

Verkasalo, H., & Hämmäinen, H. (2007). A Handset-Based Platform for Measuring Mobile Service Usage. *INFO: The Journal of Policy, Regulation and Strategy, 9*(1).

Verkasalo, H. (2006a). *Mobile Data Service Evolution - Empirical Implications from Europe and the USA*. Paper presented at the 3rd International CICT Conference, November 30 - December 1, 2006, Denmark, 2006.

Vesa, J. (2005). *Mobile Services in the Networked Economy*. Hershey, PA: IRM Press.

Wong, C. C., & Hiew, P. L. (2005). Correlations between factors affecting the diffusion of mobile entertainment in Malaysia. *ICEC, 2005*, 615–621. doi:10.1145/1089551.1089661

Wright, S. (1921). Correlation and causation. *Journal of Agricultural Research, 20*, 557–585.

Wright, S. (1934). The method of path coefficients. *Annals of Mathematical Statistics, 5*, 161–215. doi:10.1214/aoms/1177732676

Wright, S. (1960). Path coefficients and path regression: Alternative or complementary concepts? *Biometrics, 16*, 189–202. doi:10.2307/2527551

APPENDIX A. QUESTIONS ASKED IN THE BACKGROUND QUESTIONNAIRE

Background factors are recoded into the following scale:

- ABPU... 1 (small) - 5 (big) (asked before the panel)
- Work... 0 (not) - 1 (yes) (asked before the panel)
- Age... 0 (young) - 1 (old) (asked before the panel)
- Gender... 0 (women) - 1 (men) (asked before the panel)
- Device_Capability... 0 (old device = lower capability) - 1 (new device = higher capability) (asked before the panel)
- Data_Pricing... 0 (usage-based), 1 (block-priced), 2(flat) (asked before the panel)
- Experience (of smartphone usage in the user's own opinion)... 1 = Beginner, 2 = Normal, 3 = Experienced, 4 = Very experienced (asked before the panel)
- Panel_Days... Number of days observed active in the panel (derived from usage data)
- Handset_Usage_Activity... Number of application activations during an average day (derived from usage data)
- Smartphone_Usage_Day... Minutes of smartphone usage on an average day (derived from usage data)

Questions are asked in relation to the following mobile services:

- WEB/WAP services, Internet browsing
- Email
- Instant messaging / chat
- MMS messaging
- Gaming
- Internet telephony
- P2P file sharing
- User-created content creation, blogging
- Streaming/Internet multimedia
- Offline multimedia

1. I intend to use the following services during the next couple of months… (intention to use) (asked before the panel)
2. Mobile versions of the following services are going to replace / have already replaced the use of those services with other devices (such as PCs, MP3 players, Radio)... (attitude) (asked after the panel)
3. The use of the following services generates enjoyment, pleasure and entertainment to me... (enjoyment value) (asked after the panel)
4. The use of the following services increases my work or study related productivity and performance... (utilitarian value) (asked after the panel)
5. I do not have / would not have significant technical difficulties in using the following services... (behavioral control) (asked after the panel)

6. It is easy for me to learn and develop my skills in using the following services... (ease of learning) (asked after the panel)
7. People around me (e.g. friends or family) have recommended the use or have helped me in using the following services... (social push) (asked after the panel)

Likert scale answers:

1 = Strongly disagree, 2 = Disagree, 3 = Slightly disagree, 4 = Neutral, 5 = Slightly agree, 6 = Agree, 7 = Strongly agree

Appendix B. Results of the Path Analysis (Figures 6, 7, 8, 9, 10, 11, 12, and 13)

Figure 6.

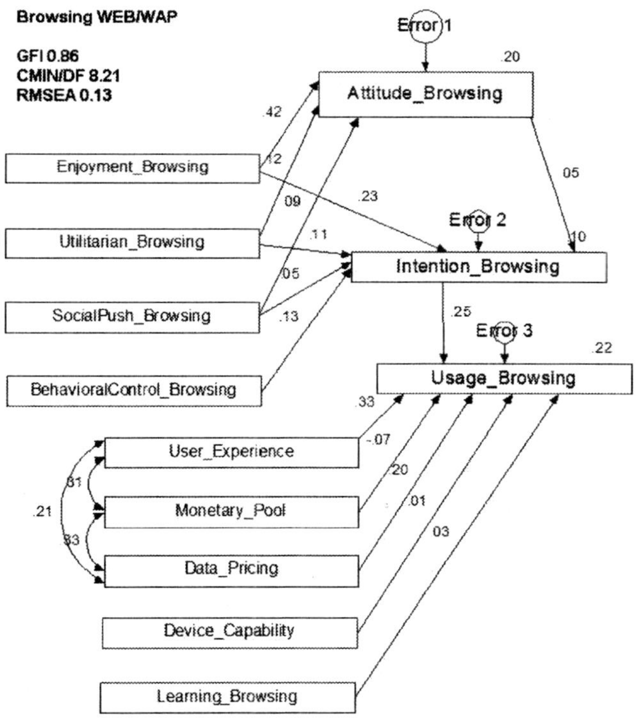

Figure 7.

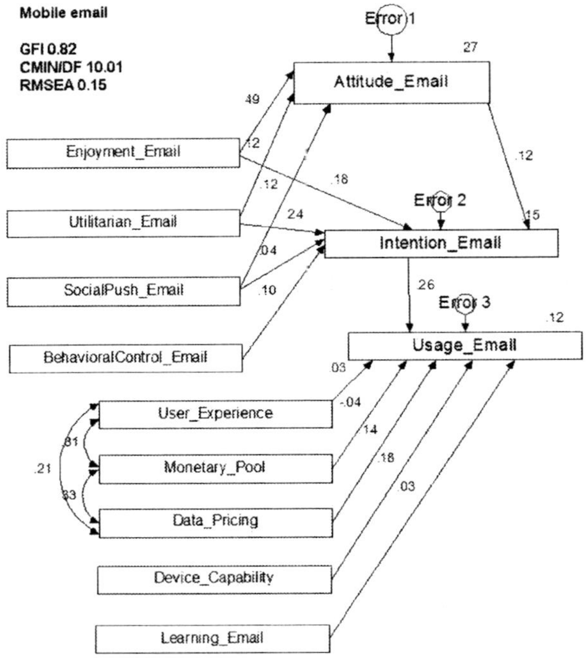

Figure 8.

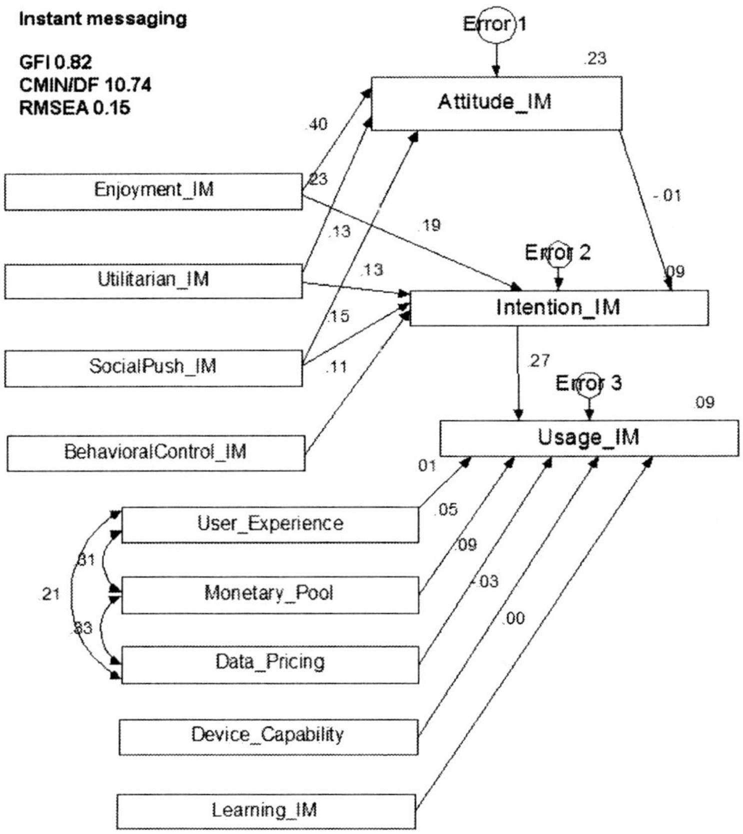

Figure 9.

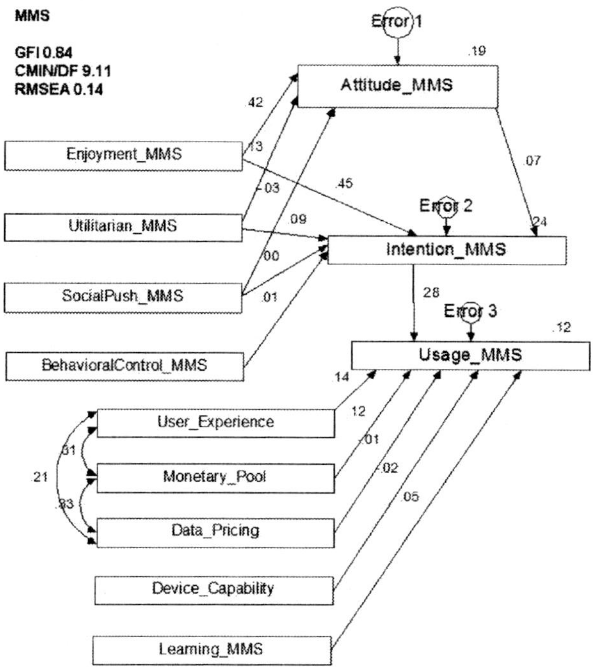

Figure 10.

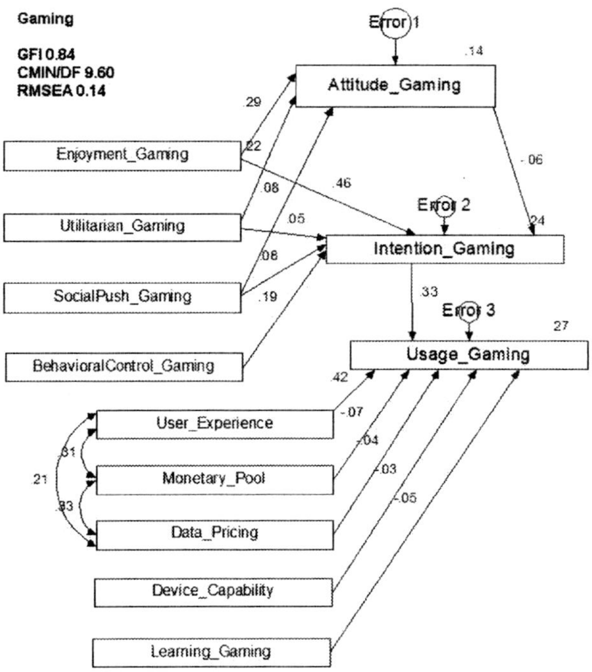

Figure 11.

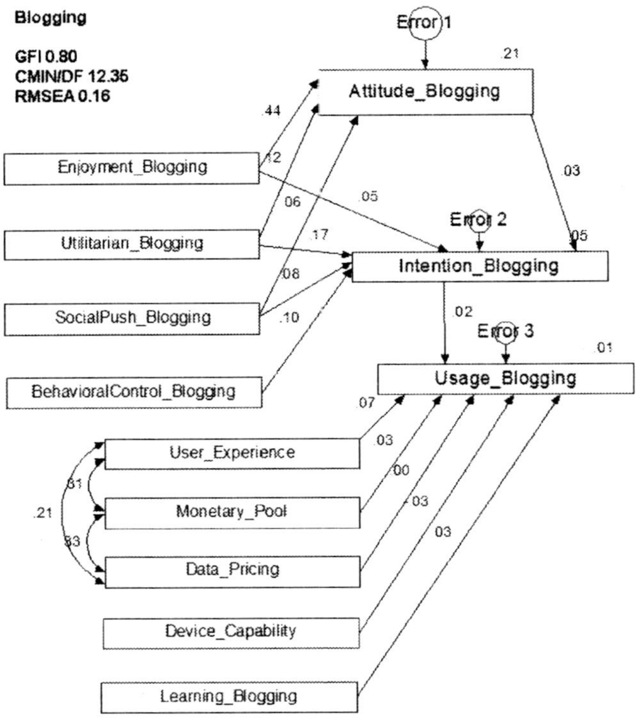

Figure 12.

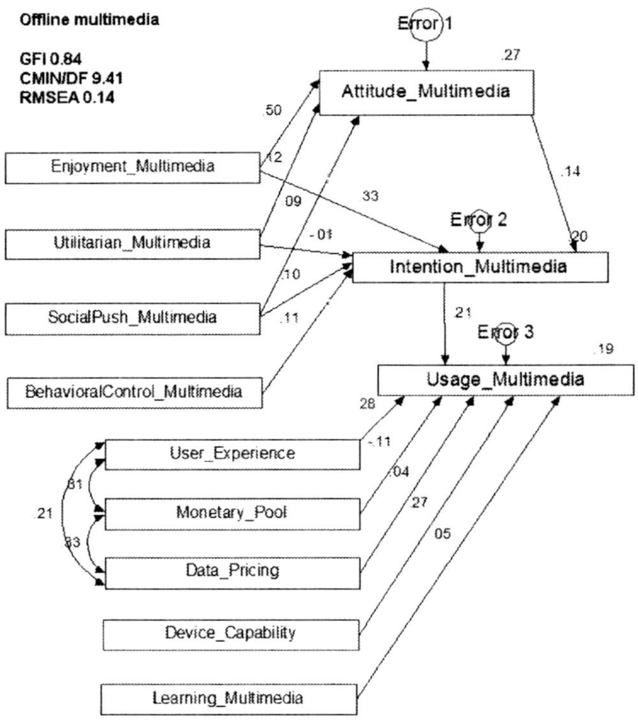

Figure 13.

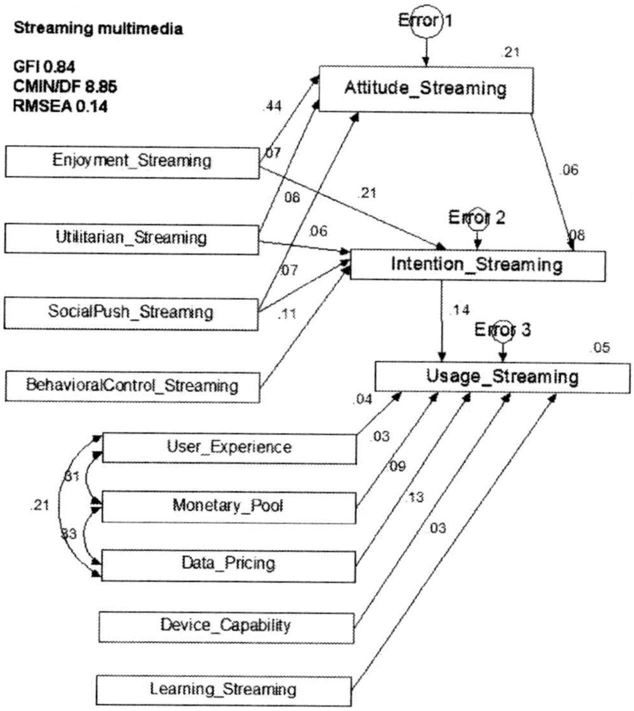
Streaming multimedia
GFI 0.84
CMIN/DF 8.85
RMSEA 0.14
Error 1
Attitude_Streaming
Enjoyment_Streaming
Utilitarian_Streaming
SocialPush_Streaming
BehavioralControl_Streaming
Error 2
Intention_Streaming
Error 3
Usage_Streaming
User_Experience
Monetary_Pool
Data_Pricing
Device_Capability
Learning_Streaming

Chapter 7
Mobile Technology Adoption in the Supply Chain

Bill Doolin
Auckland University of Technology, New Zealand

Eman Ibrahim Al Haj Ali
United Arab Emirates University, UAE

ABSTRACT

The increasing utilization of ***mobile commerce*** *technologies in e-business raises the question of their use in* ***supply chain*** *integration and management. This chapter presents a multiple case study investigation of the adoption of* ***mobile technology*** *in the* ***supply chain****. A technology-organization-environment framework of the contextual influences on technological* ***innovation adoption*** *is used to inform an analysis of three companies' adoption and use of* ***mobile data solutions*** *for sales automation, freight tracking and service support. Analysis of the three case studies found that the relative advantage of the technological innovation and the information intensity of the company were the most important factors influencing adoption. Other factors that appeared to influence adoption included the compatibility of the technology with the company's business approach, the presence of top management support and the degree of organizational readiness. Environmental factors such as competition within the industry or business partner influence seemed less influential for these pioneers of* ***mobile technology*** *use in supply-side activities.*

INTRODUCTION

Supply chain management (SCM) can be defined as "the process of managing relationships, information, and materials flow across enterprise borders to deliver enhanced customer service and economic value" (Mentzer et al., 2001, p. 10). Information technology (IT) is pervasive in SCM (Russell & Hoag, 2004), and with the development of electronic commerce it is playing an increasingly strategic role as **supply chain** activities are conducted, linked and integrated electronically (Bhatt & Emdad, 2001; Kalakota & Robinson, 2001). Companies are seeking to gain competitive advantage and create responsiveness to markets by adopting IT that enables them to utilize and manage information and knowledge

DOI: 10.4018/978-1-60566-910-6.ch007

within and across the extended enterprise (Lau et al., 2006; Porter & Millar, 1985). Of relevance to this chapter is the relatively recent but rapid development of **mobile commerce** (Kalakota & Robinson, 2002; Kumar & Zahn, 2003) and its application to SCM.

Mobile commerce is the conduct of electronic commerce through mobile or handheld computing devices (e.g. mobile phones, PDAs and tablet PCs), using wireless technologies and telecommunication networks (Siau et al., 2003). Such mobile technologies facilitate communication, Internet access, data exchange and transactional capabilities largely independent of time and location. The result is increased real-time interaction between companies, employees and **supply chain** partners, including customers (Kalakota & Robinson, 2002), enhancing operational efficiency and providing new opportunities for customer service (Kumar & Zahn, 2003; Shankar & O'Driscoll, 2002).

A number of studies have examined the potential for **mobile commerce** to be applied to SCM. Mobile technologies are envisaged to have the most impact in areas of SCM such as e-procurement, materials handling, warehousing, inventory management, logistics and fulfilment, asset tracking, sales and field force automation, and dispatch management (Alanen & Autio, 2003; Benou & Bitos, 2008; Kalakota & Robinson, 2002; Kalakota et al., 2003; Lau et al., 2006; Rangone & Renga, 2006; Ruhi & Turel, 2006; Shankar & O'Driscoll, 2002). For example, it has been argued that mobile applications integrated with a company's enterprise systems can provide greater visibility into **supply chain** operations, leading to real-time order status information and more responsive service management (Kalakota et al., 2003). When deployed to mobile employees such as sales representatives or technical field service teams, mobile technologies can automate data collection, deliver necessary information to employees wherever their location, and reduce the time needed to update data from the field for the rest of the company, resulting in improved workforce productivity, process efficiency, data accuracy and service quality (Rangone & Renga, 2006).

The idea that **mobile commerce** can transform SCM is reflected in the development of concepts such as 'untethered' (Shankar & O'Driscoll, 2002), 'adaptive' (Kalakota et al., 2003) and 'responsive' (Lau et al., 2006) **supply chains**. However, there are few empirical studies that that focus on the adoption and implementation of **mobile commerce** in the **supply chain** activities of companies. Those that do have tended to report on financially modest or relatively simple applications that support mobile activities (operational mobility) rather than the mobile transmission of data (transmission mobility) (Rangone & Renga, 2006). In contrast, this chapter examines the adoption of more complex mobile applications that support transmission mobility as well as operational mobility, and that integrate with existing company information systems and have the potential to change operating procedures and activities.

The aim of the chapter is to provide an empirical analysis of why organizations adopt **mobile commerce** technologies in the **supply chain**. Since the organizational adoption of **mobile commerce** is not well understood, we use an exploratory approach grounded in qualitative data to provide an analysis of three New Zealand companies' development and use of **mobile data solutions**. We treat these as innovations and draw on the IT **innovation adoption** literature to inform our analysis. The next section summarizes this literature and presents a conceptual framework based on technological, organizational and environmental factors influencing the **innovation adoption** decision. We then outline the research method used in the study before presenting our analysis of the three case studies. The final part of the chapter synthesizes some conclusions from the cross-case comparison, and discusses the implications for research and practice in this area.

ORGANIZATIONAL ADOPTION OF INFORMATION TECHNOLOGY INNOVATIONS

There is a long-standing interest in the adoption of IT innovations in the study of information systems (Kwon & Zmud, 1987). In this chapter we are concerned with the primary adoption of an innovation by an organisation (Frambach & Schillewaert, 2002; Gallivan, 2001; Russell & Hoag, 2004), rather than its secondary adoption by individuals in the organisation (e.g. Compeau & Higgins, 1995; Karahanna et al., 1999; Moore & Benbasat, 1991; Taylor & Todd, 1995; Venkatesh & Davis, 2000). By organizational adoption of an innovation we mean a process beginning with initial awareness and evaluation of a new technology or product, followed by a decision to purchase and implement the innovation, and finally its acceptance or assimilation within the organization (Frambach & Schillewaert, 2002).

Researchers have utilized a number of approaches in attempting to explain why organisations adopt IT-related innovations. Probably the most common approach used is one based around the identification of a set of contingency factors that collectively explain the **innovation adoption** decision or outcome (Allen, 2000; Fichman, 2004; Frambach & Schillewaert, 2002; Jeyaraj et al., 2006; Kwon & Zmud, 1987; Prescott & Conger, 1995).

Many contingency or factor studies of IT **innovation adoption** tend to follow a **'technology-organization-environment' model** pioneered by DePietro, Wiarda and Fleischer (1990). Underlying this model is the assumption that **innovation adoption** is complex and context-sensitive (Wolfe, 1994), and that almost all factors influencing the organisational adoption of an innovation can be classified as either characteristics of the innovation itself, the organisation, or its environment (Flanagin, 2000; Frambach & Schillewaert, 2002; Kwon & Zmud, 1987). The number of empirical studies following this approach provides support for its usefulness (e.g. Al-Qirim, 2005; Chau & Hui, 2001; Dedrick & West, 2004; Elliot, 2002; Henriksen, 2006; Kuan & Chau, 2001; Premkumar et al., 1997; Teo et al., 1997), and following calls to extend this framework to other innovation domains (Chau & Tam, 1997; Thong, 1999; Zhu et al., 2003), we have used it to organise our exploratory study of the contextual influences on the organizational adoption of **mobile commerce** technologies in the **supply chain**.

As noted above, the **technology-organization-environment model** proposes that organizational **innovation adoption** is influenced by three elements of context: the perceived attributes of the technology, organisational characteristics and environmental conditions. Prior studies of **innovation adoption** have identified a complex and rich group of potentially relevant factors within each of these three elements – too many for a single study to examine (Frambach & Schillewaert, 2002; Russell & Hoag, 2004). The adoption model we use in this study is shown in Figure 1. It includes three high-level factors for each contextual element, which we believe have an influence on organizational adoption of **mobile commerce** technologies in the **supply chain**. Each factor is discussed below.

('+' and '-' indicate a positive or negative influence on adoption, respectively)

Technology Attributes

Technological factors focus on how characteristics and features of the technology or innovation itself influence adoption (DePietro et al., 1990). Frambach & Schillewaert (2002) note that organizational decision-makers' perceptions of an innovation affect their evaluation of and propensity to adopt a new technology or product. In their meta-analysis, Tornatzky & Klein (1982) found that three perceived attributes were consistently associated with **innovation adoption** behaviours: the relative advantage of an innovation over its predecessor, its compatibility with the organiza-

Figure 1. Contextual influences on mobile commerce technology adoption

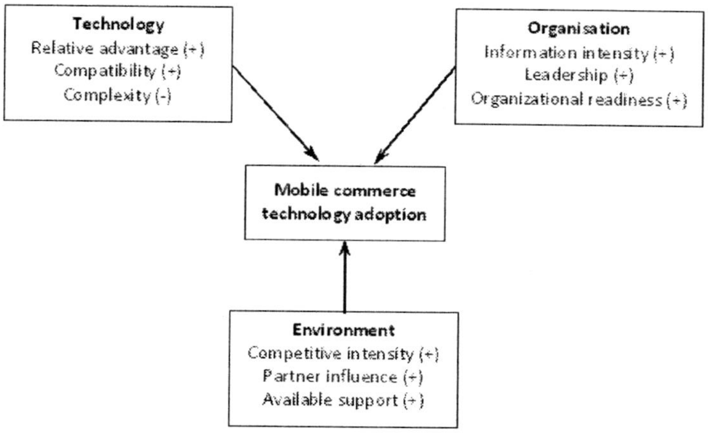

tion's needs and existing systems, and its complexity to understand and use (Rogers, 2003). Potential adopters typically evaluate the *relative advantage* of a technological innovation in terms of whether the costs of adoption are outweighed by the benefits likely to be received (Premkumar et al., 1994). This evaluation may be in economic or in more subjective terms; what is important is that an organization perceives the new technology as advantageous in comparison with existing or alternative technologies. For example, the Internet is generally perceived to offer a range of benefits in relation to electronically-mediated communication, information gathering, marketing and value-added services. As such it can be perceived as representing a significant relative advantage over traditional tools for performing these operations (Mehrtens et al., 2001; Poon & Swatman, 1998; Tan & Teo, 1998; Teo et al., 1997). Relative advantage may depend on how satisfied the organization is with their existing technological solution (Chau & Tam, 1997, 2000).

The more *compatible* a technological innovation, the less changes or adjustments will needed and the lower the possible level of resistance to the technology when it is adopted (Teo et al., 1997). With respect to IT-related innovations, some authors distinguish between organizational compatibility and technological compatibility. The former focuses attention on congruence with organizational values and operating practices. For example, Russell & Hoag (2004) suggest that **supply chain** innovations that are designed to assure **supply chain** reliability are more likely to be adopted by organizations with cultures that focus on reliability rather than price competitiveness. Doolin et al. (2003) discuss an organization in which Internet retailing was compatible with the existing sales infrastructure and experience acquired through the organization's other distribution channels. Flanagin (2000) found that some organizations adopted IT innovations at an early stage that they considered compatible with their perceived industry leadership or reputation. Technological compatibility reflects the ability of the adopting organization to successfully integrate the new technology with its existing IT infrastructure and legacy systems (Dedrick & West, 2004; Jones & Beatty, 1998; Premkumar et al., 1994).

Complexity has been found to influence organizational IT **innovation adoption** decisions in a number of studies (Bouchard, 1993; Elliot, 2002; Grover, 1993; Russell & Hoag, 2004). Not only the complexity of the technological innovation, but also the number and difficulty of processes and activities involved in its adoption, may negatively influence adoption (Niederman, 1998). For example, the establishment of traditional EDI relationships with multiple trading partners or involving multiple transaction types potentially involves dealing with a divergent range of hardware, software, protocols, telecommunications links and business practices (Ramamurthy et al., 1999).

Organizational Characteristics

Factors related to the organizational context that may facilitate or inhibit adoption and implementation of an innovation are usually defined in terms of various characteristics of the organization, its employees and available internal resources (DePietro et al., 1990; Zhu et al., 2003). Thong (1999) suggests that organisations that are more *information-intensive* in their products or services are more likely to adopt IT innovations based on their greater potential for strategic use for IT and perception of IT as a source of competitive advantage. This may be reflected in the organization's view of itself as innovative towards IT (Dedrick & West, 2004; Teo et al., 1997). Certainly, the centrality or strategic importance of IT to the organization's business and operations has been found to correlate with IT-related **innovation adoption** (Chwelos et al., 2001; Dedrick & West, 2004; Grover, 1993), as has the level of IT use or sophistication of the organization (Flanagin, 2000; Mehrtens et al., 2001).

It is well-established that an organization's *leadership* influences IT adoption decisions (Russell & Hoag, 2004). For example, senior management's willingness to innovate and explore the possibilities of new technologies has been found to facilitate adoption of innovations (Al-Qirim, 2003; Elliot, 2002; Thong, 1999). In a study of established retailing firms, Doolin et al. (2003) found that a conservative management attitude to the Internet typically led to a cautious approach to Internet channel development. In particular, the commitment, involvement and support of senior managers with an **innovation adoption** decision can provide direction, ensure adequate resources are made available, and signal the importance of the adoption (Grover, 1993; Premkumar & Potter, 1995; Premkumar & Ramamurthy, 1995; Premkumar et al., 1997; Ramamurthy et al., 1999; Russell & Hoag, 2004; Teo et al., 1997). Other members of the organization (often IT professionals) may act as internal champions for an innovation, raising awareness of the innovation and its benefits with managers and potential users (Grover, 1993; Mehrtens et al., 2001; Premkumar & Potter, 1995; Russell & Hoag, 2004).

The *readiness* of an organization to adopt and introduce a technological innovation relates to the existence of adequate financial, human and technical resources within the organization (Chau & Hui, 2001; Chwelos et al., 2001; Dedrick & West, 2004; Henriksen, 2006; Iacovou et al., 1995; Wymer & Regan, 2005). Of particular importance is the level of internal technical expertise available to implement the new technology (Chau & Tam, 1997, 2000; Doolin et al., 2003; King & Gribbins, 2002; Kuan & Chau, 2001; Mehrtens et al., 2001; Premkumar & Potter, 1995; Zhu et al., 2003). Adoption may depend not just on the level of skills and knowledge among the organization's IT professionals, but also on that of other employees, whose ability and confidence to operate IT-related innovations affect both the human capital available for a new technology's adoption and its acceptance within the organization (Fillis et al., 2004; Thong, 1999).

Environmental Conditions

The environmental context constitutes the arena in which adopting organizations conduct their business (DePietro et al., 1990). Within this context, influence on **innovation adoption** decisions may be expressed through the competitive pressure in an industry, relationships with business partners, and access to resources and support from government, vendors or standards. The higher the *competitive intensity* in an industry, the stronger the pressure on an organization to adopt innovations in order to gain or maintain competitiveness (Chwelos et al., 2001; Grover, 1993; Kuan & Chau, 2001; Premkumar & Ramamurthy, 1995; Premkumar et al., 1997; Ramamurthy et al., 1999). Competition also leads to environmental uncertainty, increasing the propensity for **innovation adoption** (Chau & Tam, 1997; Thong, 1999). Zhu et al. (2003) found a significant association between competitive pressure and e-business adoption.

Business or trading *partner influence*, whether supportive or coercive, can motivate an organization to adopt an innovation (Chau & Hui, 2001; Chwelos et al., 2001; Henriksen, 2006). For example, external pressure from a trading partner has been found to be an important predictor of EDI adoption (Bouchard, 1993; Iacovou et al., 1995). Where organizations have established trading relationships, the adoption decision may be mutual and built on trust (Premkumar et al., 1997; Ramamurthy et al., 1999). The readiness (or not) of business partners may also be a significant facilitator in adoption of some IT innovations (Chwelos et al., 2001; Zhu et al., 2003).

The perceived level of *available support* from vendors (Al-Qirim, 2005; Chau & Hui, 2001), government (Damsgaard & Lyytinen, 2000; Elliot, 2002) or third parties (Dedrick & West, 2004; Doolin et al., 2003) for an IT innovation and its implementation is sometimes an important influence on an organization's adoption decision. Perceived support may also relate to infrastructural support for the use of an innovation. For example, a lack of standards may act as a barrier to the diffusion of a relatively complex IT innovation, such as EDI (Bouchard, 1993; Damsgaard & Lyytinen, 2000).

METHOD

The research objective was to provide an empirical exploration of why organizations might adopt **mobile commerce** technologies in the **supply chain**. Although prior studies of organizational adoption of IT innovations have been conducted, there are few empirical studies that address the adoption of **mobile commerce** in organizational **supply chain** activities. As a consequence, our understanding of this technological innovation context is relatively undeveloped. The intention of this study was to increase this understanding through the application of a preliminary model of **mobile commerce** technology adoption to three case studies of organizations that had adopted **mobile data solution**s in their **supply chain**s.

We used a case study approach because of the exploratory nature of the research and the relative lack of a strong theoretical base for understanding supply-side **mobile commerce** technology adoption (Benbasat et al., 1987). Further, a case study approach facilitates our focus on the contextual conditions of **mobile commerce** technology adoption (Yin, 2003). As Yin (2003, p. 13) suggests, case studies are useful for investigating "a contemporary phenomenon within its real-life context, especially when the boundaries between phenomenon and context are not clearly evident". Our primary source of data was semi-structured interviews conducted during 2004 with key informants in three New Zealand companies that had developed and used **mobile data solution**s. The qualitative data generated from the interviews were supplemented with secondary data sources, including publicly available information on the companies and their activities.

The interviews were based on a common set of questions designed to elicit information on the company and its operations, its use of IT, the decision to adopt mobile technologies, the perceived benefits of the technology, factors facilitating or inhibiting adoption, the implementation process, and any implications of adoption for the company.

Table 1. Case studies

	FoodCo	FreightCo	PowerCo
Business	Food manufacturing and marketing	Freight, logistics and warehousing	Electricity network and distribution
Company size	900 employees	1200 employees	280 employees
Turnover	NZ$220 million	NZ$890 million	NZ$870 million
IT team	4 employees	20 employees	20 employees
Application	Mobile sales automation	Mobile freight tracking	Mobile service support
Interviewees	IT Manager Systems Administrator Commercial Manager	IT Manager Logistics Manager Stock Controller	IS Manager Customer Service Manager

The interviews were audio-taped and transcribed for qualitative data analysis. This involved both within-case and cross-case thematic analysis organized around the theoretical propositions identified above (Yin, 2003). Table 1 summarises the interviews and background details of the three case study companies.

In the following sections we present our analysis of the three case studies of **mobile technology** adoption. The analysis of each case is structured around a brief description of the company and the **mobile data solution** studied, followed by a discussion of the three types of contextual influence identified in the research model outlined in Figure 1: attributes of the technology itself, organizational characteristics, and wider environmental or industry conditions. Selected quotes from the interviews are used to illustrate the analysis.

MOBILE SALES AUTOMATION AT FOODCO

FoodCo is a New Zealand subsidiary of a multinational food company. It manufactures and distributes a range of over two hundred grocery product lines to several thousand retailers, relying on fifty mobile sales representatives as its contact point with customers. The management of stock from its three production plants to its large customer base is an ongoing challenge for the company, with speed and efficiency in order taking and fulfilment perceived as essential to maintaining customer satisfaction. FoodCo has a small IT department of four people, mainly for routine maintenance of the company's information systems.

FoodCo was a pioneer in New Zealand in the use of bar code scanners to capture order information at the point of customer contact and the transmission of this data to its sales office, first by dial-up modem over a landline and then by car phone over a cellular phone network. In 1999, the company decided to upgrade its system and outsourced development of a customised **mobile data solution** used by the sales force via laptop computers. This system has been progressively updated since then both in terms of software and hardware. The major motivation for the adoption of a mobile sales automation technology was "to move key strokes out of the office into the field" (IT Manager).

Battery-operated mobile units based on a tablet PC are used by the sales force to download updated product information, customer information, sales promotions and territory management information. Stock levels and replenishment dates are also available to sales representatives in the field. The mobile units allow sales representatives to write directly into the screen in interacting with the **mobile data solution**. Inputted order and invoicing information is transferred to the company's sales office where the information is processed via the company's ERP system and the required goods are dispatched as quickly as possible. Customer information and in-store negotiated promotion details can also be updated in real time via the system. The **mobile data solution** includes a shelf management function that enables sales representatives to capture shelf stocking information for managing product placement in large customers such as supermarkets. A sales effort screen provides a range of information on sales targets and volumes. Sales representatives can track their performance at product level throughout a month and compare it with previous months' performances. Data is transmitted over a **GPRS wireless network**, although the units also have built-in modems for use with a landline and infrared ports for use with mobile phones if alternative data transmission mechanisms are needed.

Technology Attributes

FoodCo clearly perceives a relative advantage in their **mobile data solution**: "The benefits have certainly been there and pretty much delivered to our expectations" (Commercial Manager). The **mobile data solution** effectively automates the sales process, eliminating the paper work which sales representatives were previously doing. Lightweight tablet PCs have replaced the "huge, big briefcases of paper" previously carried by sales representatives (Systems Administrator). The added information and functionality provided by their **mobile data solution** enables FoodCo's sales representatives to undertake promotion management, conduct in-store deals and manage customer relationships on a one-to-one, real-time basis. This was seen as enabling a shift in their role: "We see the [mobile] unit becoming even less an order entry unit and much more of a business management tool" (IT Manager).

The **mobile data solution** has enabled FoodCo to improve the efficiency of its order processing and logistics. Timely receipt of sales orders means that planning associated with warehouse picking and truck delivery loads can begin earlier: "We are becoming more and more focused in that area of getting that whole process more and more efficient. And having the orders coming in effectively within five minutes of them being taken into [the ERP system], ready to be picked, has been beneficial to us" (Systems Administrator). The **mobile data solution** is also considered to be a source of competitive advantage through the way that it integrates and synchronizes information regarding customers, products and distribution, enabling the company to manage its key customer accounts more efficiently: "Historically we were very good at transactions and you've got good competitive advantage by being able to transact better than anybody else. But now it's not about transactions, it's about knowledge management" (IT Manager).

The current tablet PC technology is considered to be a significant improvement over previous units in terms of weight, screen size and processing power. While some transmission and coverage issues had been experienced with the cellular network originally used to transmit the data, data is now transmitted over a **GPRS wireless network** selected because of its continuous availability, connection stability, high speed and its relatively cheap (data-driven) rates. Ironically, the speed and efficiency of the wireless transmission led to an unintended increase in projected data costs as sales representatives began transmitting data after every sales call (until reined in).

In terms of its compatibility, FoodCo's **mobile data solution** matched the business approach of the company in a number of ways. For example, the units allow sales representatives to manage customer relationships with key accounts in person rather than from head office. Similarly, sales representatives take a proactive role with small retailers: "It's all about presence in the marketplace and being there in front of them and actually influencing buying patterns" (Systems Administrator). The **mobile data solution** was also compatible with the IT infrastructure and approach used by FoodCo. The existence of the company's ERP system and the simultaneous rollout of its sales and distribution modules provided the necessary complementary technology for the **mobile data solution** to function effectively.

Extensive training was required to upskill the sales force in using both the mobile computer units and the extended range of functionality. The tradeoff of the more powerful, large-screened tablet PC units was their complexity, which made them more prone to breakdown and damage when dropped or mishandled. In addition, the **mobile data solution** project grew in size and complexity, creating some difficulties in coordination between the various departments involved in its use: "I think the biggest thing was that it ended up bigger than it was ever planned to be ... Sometimes what you find is that when you revisit it that a lot of the facility there isn't being used to its capability" (IT Manager).

Organizational Characteristics

The adoption of **mobile technology** for sales automation reflects both FoodCo's history of IT use (including sales automation) and its innovative attitude towards IT. FoodCo had been actively monitoring and developing the e-business side of its operations since 1999: "[FoodCo] has always been at the front of deploying that kind of technology to the market ... We tend to pick up the new technologies quickly if we can see there's a clear business input" (Systems Administrator). The small IT department within FoodCo actively looks for ways to utilize new and innovative IT in the company's operations. However, the decision to explore new technological options in sales automation was a strategic one taken by FoodCo's senior management. According to the IT Manager, "That type of leadership has always been there ... The current management is very, very supportive."

However, the unwillingness of some sales representatives to embrace the new technology initially slowed adoption and use of the **mobile data solution** within the company. A number of representatives lacked computer literacy, were reluctant to change established ways of doing things, or were reluctant to utilize the new functionality in front of customers in case they showed their inadequacy. As the Systems Administrator explained, "Some of our reps have been with the company for a long time ... and putting a computer in front of them was terribly daunting." However, with time and training this barrier was overcome, with many of these representatives becoming advocates for using the new technology.

Environmental Conditions

FoodCo perceive themselves as leaders in their industry, particularly in gaining competitive advantage through the innovative use of IT for knowledge management. In relation to their use of **mobile technology**, "We were seen to be again, you know, market leading and out there doing things at the forefront basically" (Systems Administrator). FoodCo's largest customers, major supermarket chains, were beginning to move their suppliers to electronic ordering and invoicing, and FoodCo's significant investment in sales automation technology meant that they were well-perceived by these key customers. The proactive contact and support provided by FoodCo's **GPRS wireless network** provider was mentioned in our interviews as positively influencing the company's adoption of a wireless data solution.

MOBILE FREIGHT TRACKING AT FREIGHTCO

FreightCo is a **supply chain** logistics provider with operations in New Zealand, Australia, Asia and the United States. The company offers a full range of logistics services, including managed warehousing, domestic distribution and international air and sea freight operations, linked with information technology and systems. FreightCo operates a nationwide fleet of more than 600 vehicles in New Zealand servicing some 8,000 customers. FreightCo coordinates its distribution operation through a centralized database supplied with real-time freight tracking data from delivery drivers in the field. FreightCo has an IT team of some twenty people working on systems maintenance and IT innovations, although it tends to outsource much its development work.

The original motivation for deploying a mobile freight tracking system was to "get even more satisfaction to the customers and get in that customer focus" (IT Manager). Drivers scan the barcode of each piece of freight on delivery using a lightweight handheld device with an inbuilt scanner. A consignment note, the date, time and location of delivery, the driver's identity and the recipient's name is uploaded to the company's central database, where that information is made available via a Website to customers, who can track the movement and status of their freight consignment in real time. The delivery information is also used as the basis for payment of the owner-drivers. New job information or updates flow back to the driver's handheld unit from FreightCo's administrative centre. FreightCo was a pioneer in using systems such as this, transmitting data over a third-party operated trunk radio network via radio telephones in the delivery trucks since 1992. In 2004, FreightCo commenced transmitting data over a GRPS wireless network.

Technology Attributes

At FreightCo, the **mobile data solution** implemented for freight tracking had the effect of removing the need for paperwork and reducing the administrative workload on the distribution fleet drivers (an estimated time saving of up to one hour per day per driver). These efficiency gains enabled the company to process a tenfold increase in consignment notes over a ten year period: "Basically we're piling through the freight, or the paperwork about the freight, in a much more efficient manner … The piles of paperwork that we would have had would have been enormous" (IT Manager). The automated system also decreases the chance of errors, improves the timeliness of information, and increases the speed at which information becomes available to customers: "[It] gave us the advantage of managing our network much better, in such a way that we knew where the freight was much better, we knew what our timing was, we knew we could monitor when things went wrong." (IT Manager)

FreightCo sees information and technology as central to its business of providing "intelligent" logistics solutions for its customers. It perceives technology to be the key differentiator in the logistics industry, and sees its ability to provide real-time information across the **supply chain** to customers as a competitive advantage: "It meant that we had much more to sell. I think we were already the premium provider out there, but it kept us the premium provider. Having been ahead of the technology, like we were, enabled us to continue to charge higher prices" (IT Manager).

The use of a **GPRS wireless network** for data transmission was seen by FreightCo as superior to the previous trunk radio network used, as it increased the amount of data that could be sent at from a mobile unit any one time (including, for example, customer signatures captured directly on the screen of the handheld devices) and also the overall data transmission capacity available to the company's distribution fleets. As the IT Manager observed: "[GPRS] was becoming a necessity … The more trucks we put on, the more delays we were getting with the data backing up and not coming through … [GPRS] seems to be unlimited."

The **mobile data solution** for freight tracking is compatible with FreightCo's business model and desire for technology leadership: "We've always had this fundamental business model of being the best … Although many companies may have said, 'Well, what's the benefit of … having the mobile communications today?', we didn't look at it like that" (IT Manager). Going mobile also allowed the company to cope with the huge growth that it experienced and continues to experience as a result of its business strategy.

Organizational Characteristics

As a company, FreightCo is proactive in keeping its IT capability ahead of the business in order to respond to new challenges in the business environment: "We wanted to take ideas to customers before they required it of us, so you know we wanted to be very forward thinking" (IT Manager). IT is essential in linking together and managing the company's range of logistics services. Expenditure on IT is high. Around 2004, the company was spending some NZ$10 million per year on software development, telecommunications and equipment. The IT department actively seeks "innovative solutions and ideas" (IT Manager). While adoption of the new **mobile technology** was initially IT-driven, FreightCo's management was quick to see the benefits and supported the innovation. As the company's IT Manager recounted: "We just had a belief that it would be better and we talked directly to the owners of the business and they thought it would be better and away we went."

Initially, the owner-driver contractors who comprise FreightCo's distribution fleets resisted accepting the new technology. The required expenditure on new technology may have been one reason for this, although FreightCo did subsidize half the cost of purchasing the handheld units: "There was a lot of resistance by the drivers ... Resistance to change and technology. Yeah, they didn't want to do it ... Therefore we have to have a good business case to sell them" (IT Manager). However, when FreightCo more recently acquired a competitor's fleet, the newly arrived owner-drivers were generally receptive to using the new **mobile data solution**. The IT Manager suggested that this was because of the benefits to drivers were evident by then, and associated with "a company that has succeeded rather than the previous company which was dying."

Environmental Conditions

The most important environmental influence on FreightCo's adoption of **mobile technology** was the competitive intensity of the logistics industry in which the company operates. As noted earlier, FreightCo's use of information provides them with a perceived competitive advantage: "We wanted to be ahead of the competition like we always are" (IT Manager). The availability and benefits of a supported GPRS network were acknowledged by FreightCo's IT Manager: "There's just going to be an exponential expansion ... and you've got networks that are prepared to invest the money in it."

MOBILE SERVICE SUPPORT AT POWERCO

PowerCo is a large electricity distribution company with over 600,000 customers. Good customer service in the form of reliable power supply is important to the company, so response times to the over 20,000 emergency callouts a year the company experiences are critical. PowerCo uses field crews from outsourced contractors to maintain and fix its electricity network. Around 2001, the company "identified the fact that we needed to get real time information back from the field, we needed to get more accurate information out to the field" (Customer Services Manager) in order to improve the response process. In 2003, after extensive piloting and field testing, PowerCo implemented a **mobile data solution** purchased from an overseas vendor and then customized for the company by predominantly outsourced developers. The company has an in-house IT team of some twenty staff, mostly working on systems maintenance.

PowerCo's **mobile data solution** facilitates efficient and effective response to field service requests. When a fault is reported to PowerCo's call centre or detected by the company's network management system, details are sent to a field crew's handheld PDA (personal digital assistant) via a secure GPRS network using a Bluetooth wireless-capable mobile phone as a modem. Crews can upload information on the job status, fault location, work required and billing in real time from the field. Data is captured once, and automatically updated on PowerCo's central information systems, including its customer relationship management (CRM) system and its geographical information system (GIS). The CRM system allows customer contact representatives to access real-time information in order to accurately and quickly answer customer queries or claims. Service requests are logged against actual network assets and fault location data is uploaded from the field to the GIS, which facilitates monitoring, management and long-term planning of PowerCo's networks.

Technology Attributes

The **mobile data solution** was perceived as better than the previous system based on two-way radios and various paper-based forms, and its benefits matched PowerCo's expectations: "The rate of return on the investment [was]… I would say about 18 months" (IS Manager). Invoices are now created automatically from data relevant to a service request entered in the field as the job progresses. This reduces the need for administrative data entry, decreasing costs and speeding up the invoicing process. Other benefits included a reduction in data duplication or redundancy, with a consequential decrease in the risk of errors in data entry: "So the main drive is reducing paper, data quality and only capturing data once" (IS Manager). The efficiency of the emergency response process also improved markedly using the **mobile data solution**, with faster response times and more accurate information sent to and from field crews: "We were collecting data at the call centre but it was never making it to the guys in the field … Now, everything gets passed through … so the sort of level of accuracy of information that the guys in the field are getting is much higher" (Customer Services Manager). In addition, the GPRS network used by PowerCo provides much wider coverage than the network they previously used.

The information provided via the **mobile data solution** has enabled the call centre to deal with customers complaints efficiently and effectively, and to keep them informed of progress in a timely manner. Because information is updated from the field in real-time and made accessible to the call centre operators: "We know when they're [field crew] on-site. We know when they've restored power. We know that the job has been completed … We can follow up all the details … It's made a huge difference to us in terms of resolving customer complaints because all the information is actually there" (Customer Services Manager).

This use of accurate, real-time information to maintain continuous power supply and improve customer service is consistent with PowerCo's role as a network provider of critical energy services. The **mobile data solution** also matches PowerCo's business needs in relation to their range of service processes. By not being device-driven, the technological solution can be applied in other company services, such as electricity connections: "At the end of the day, the solution can use multiple, you know, types of devices and I think that was another thing that was really important to us because we knew that not all of our processes could be supported by one device" (Customer Service Manager).

Aspects of the complexity of the **mobile data solution** did become issues in various ways for PowerCo. For example, the limited battery life of the PDAs (which often stay docked in the field crews' vehicles in order to remain powered) and the range of the Bluetooth wireless connection between the PDA and the mobile phone modem (about ten meters) effectively shape the crews' use of the technology. PowerCo's IS Manager described how aspects of the **mobile data solution** were designed to cope with crews periodically moving out of coverage. The crews are able to continue to work with the application offline, updating the job status and then uploading the data when they come back within range. Screen layout and sequence on the PDAs was also modified to enhance the application's operability in field conditions.

In fact, the **mobile data solution** was deliberately developed in a way that accommodated the conditions and characteristics of field crews, who were consulted extensively. As the IS Manager recounted: "[The development company] supplied most of the developers and it was young people … [They design] so you can just click it like this, and you do it like this. It might be flashy but it's not always practical … And then I arranged for them to go out with a field crew and their whole attitude changed. They suddenly started to think like the field crew and not just like a developer." Nevertheless, some aspects of the **mobile data solution** remain complex for the field crews to use: "The guys struggle a little bit with the GIS stuff and it's been quite a big learning curve for them, but they're getting there" (Customer Service Manager).

Organizational Characteristics

PowerCo has invested significantly in adopting new technology. It generates, on a daily basis, large volumes of multidimensional and interrelated asset, customer, financial and operational data, which is compiled and displayed in a number of formats to allow users to select and drill into various areas for information. Business intelligence provides information analysis and distribution, data visualization and spatial analysis for decision making and planning: "We're … an IT focused [company] and we believe in IT solutions too. And it was most definitely a business decision that we needed to, that we wanted to go down that track [in adopting **mobile technology**]" (Customer Service Manager).

PowerCo's IT team takes a reactive approach to IT solutions for the company, focusing on supporting business requirements rather than 'pushing' technology: "We're really in there to try and understand the business needs before we even talk systems" (IS Manager). The impetus for the adoption of **mobile technology** was from top management: "It was top down. It was a benefit that our executives and the contractors' executives saw. And so, like, everybody's using wireless despatching in field crews and we should actually also be using it" (IS Manager).

PowerCo uses outsourced field crews, which meant that the contractors had to be convinced to adopt and use the new **mobile data solution**, including taking responsibility for maintaining the **mobile technology** itself: "We've provided a certain number of the devices to start with but then from then on they've got to buy their own, they've got to support their own hardware, that type of thing. So we had to sell it into them as well" (Customer Services Manager). However, PowerCo provided them with training. Project team members would go into the field with the field crews, "holding their hands" as they used the **mobile technology**: "You have to break the habit of what they would normally do. So you just need to have someone sitting in there and saying, 'No, don't write that down on a piece of paper, and don't write that in your diary, you need to do it in here'" (Customer Services Manager).

The field crews generally accepted and used the new mobile units, despite management's concern that the modern 'white collar' technology might be perceived as out of place in the 'blue collar' field environment and that the field crews would "struggle with it." In fact, although it was technology that most of the crews had not experienced before, "They picked it up pretty quickly. Yeah, they're not unintelligent people … I think we thought that we'd have more problems teaching them than sort of we did" (Customer Services Manager). The field crews selected to participate in piloting the system actually refused to return the units at the end of the pilot, wanting to continue using them, and placing unforeseen demands on the company's resources as they continued supporting the pilot while developing the full **mobile data solution**.

Environmental Conditions

The outsourced contractors who supply the field crews are an important business partner for PowerCo. As noted earlier, the contractors' senior management apparently played a role in identifying the potential benefits of using wireless technology for dispatching field crews, and recognized that at some stage they would need to adopt it: "I think they were quite pleased that we made the choice to actually roll it out, that they didn't have to do something themselves ... I think they were pretty supportive. They could see the end result should be beneficial for their business" (Customer Service Manager).

Maintaining "robust connections" between the handheld PDA units and the **GPRS wireless network**, remains problematic according to PowerCo's IS Manager. The company initially used wireless cards in the PDAs to access the GPRS network, but experienced a high level of disconnections, hence the shift to using dedicated mobile phones as modems. However, there were still problems with disconnections, which appeared to be related to the standard that handles communication between the GPRS network and the mobile application: "That standard is still a grey area. It's not just related to [our application]; we are also talking to other people in the industry and we've found that they lose a lot of connections ... Bit annoying, but we working with [network and application providers] to resolve it" (IS Manager).

Support from the original application vendor also became an issue, as while the application worked satisfactorily on the original handheld units used, it did not necessarily do so on the latest technology purchased by the contractor users: "We're having some problems with newer technology, getting it to be able to support the software ... That's been another issue to stop us rolling it [the **mobile data solution**] out wider, because there's been changes of device and [the vendor] hasn't necessarily kept up with that side" (Customer Service Manager).

DISCUSSION

Table 2 summarizes the findings of our cross-case analysis of the adoption of **mobile data solutions** in the three case studies.

The perceived *relative advantage* of the **mobile data solutions** appeared to play an important role in the successful adoption of **mobile commerce** technology in all three companies. The benefits achieved through the use of **mobile technology** in these three companies relate variously to (a) administrative efficiency, in the form of paperwork reduction and time savings; (b) improved data accuracy and timeliness; (c) improved operational efficiency in **supply chain** operations; (d) enhanced roles for company users of the **mobile technology**; and (e) competitive advantage. The *compatibility* of the **mobile data solution** with a company's business approach or existing processes was also a common factor across the three cases, particularly in that it matched the focus on customer service observed in all three companies. The *complexity* of the **mobile data solutions** only appeared relevant in two of the case studies, where it was perceived to increase the level of user training required in those companies.

In terms of the organizational context, *information intensity* emerged as a common factor influencing adoption. All three companies are

Table 2. Summary of innovation adoption findings

	FoodCo	FreightCo	PowerCo
Technology Attributes			
Relative advantage	• Information integration and synchronization a source of competitive advantage • "Manage the business within the supermarket rather than just take an order"	• Providing real-time information to customers is a competitive advantage • "The piles of paperwork … would have been enormous"	• "Reducing paper, [improving] data quality and only capturing data once" • "It's made a huge difference to us in terms of resolving customer complaints"
Compatibility	• "It's all about presence in the marketplace"	• Freight tracking system is a good fit with the company's focus on customer service	• "The end result is that customers spend less time in the dark"
Complexity	• "It ended up bigger than it was ever planned to be"	• Not significant	• Aspects of the mobile data solution had to be modified for field conditions • "The guys struggle a little bit with the GIS stuff"
Organizational Characteristics			
Information intensity	• "We tend to pick up the new technologies quickly if we can see there's a clear business input"	• If a company's IT capability stays ahead of the business, the business will always be prepared for new challenges	• "We're an IT focused [company] and we believe in IT solutions"
Leadership	• "The current management is very, very supportive"	• Management were quick to see the benefits and supported the innovation	• Adoption of mobile technology was initially a top-down decision
Technical readiness	• IT team actively scans the technological environment	• IT team actively seeks "innovative solutions and ideas"	• "We're really in there to try and understand the business needs before we even talk systems"
User readiness	• "Putting a computer in front of [some of] them was terribly daunting"	• "There was a lot of resistance by the drivers … to change and technology"	• "It's just technology that they're not used to" • "You have to break the habit of what they would normally do"
Environmental Conditions			
Competitive intensity	• "We were seen to be again you know market leading"	• "We wanted to be ahead of the competition"	• Not significant
Partner influence	• Major supermarket chains were beginning to move their suppliers to electronic ordering and invoicing	• Not significant	• "The contractors … were quite pleased that we made the choice to actually roll it out, that they didn't have to do something themselves"
Available support	• Proactive support from wireless network provider	• Not significant	• Continually changing hardware technology requires vendor support for software compatibility • Mobile device to wireless network communication standard problematic

information-intensive in that information processing is an important part of their business and that IT is integral in managing customer services. The importance of this factor was reflected in the history of IT use in all three companies and their proactive and innovative attitude towards IT, and e-business in particular. *Leadership*, in the form of top management support for the **innovation adoption**, was also a common theme across all three case studies. Even where the initial awareness of the innovation was not management-driven, management adopted a supportive attitude to the business use of new technology. Examination of the *organizational readiness* of the companies to adopt their **mobile data solutions** revealed mixed findings. On one hand, the role played by at least two of the companies' IT teams in actively seeking innovative uses for IT appeared to be a positive influence on adoption of the **mobile commerce** technology. On the other hand, the lack of readiness of some intended users of the **mobile data solutions** to embrace the new technology tended to slow adoption or increase the time and training needed to achieve deployment of the solutions.

Although we expected wider environmental or industry conditions to play a significant role in shaping **innovation adoption** decisions in the three case studies, overall they seemed to have less effect than technology attributes or organizational characteristics and their occurrence varied across the case studies. In the case of two of the companies, FoodCo and FreightCo, *industry competitiveness* was reflected primarily in each company's desire to be market leaders through the use of IT. The influence of business *partners* was also a finding in two of the companies, with some of FoodCo's major customers innovating with electronic transactions themselves, and PowerCo's sub-contractors providing support for the innovation based on their recognition of the benefits of the mobile dispatch technology. Third-party *support* was taken into account in the development and implementation of a **mobile data solution** in two of the companies, although in different ways. FoodCo received proactive support from its wireless network provider, while PowerCo found itself reliant on vendor support because of changing or problematic technology. In both cases, while the availability of appropriate support was a factor in the adoption experience, it did not seem to be a direct consideration in terms of the adoption decision itself.

CONCLUSION

This chapter has presented an exploratory empirical study into why organizations adopt **mobile commerce** technologies in the **supply chain**. Since the organizational adoption of **mobile commerce** is not well understood, the findings presented here are likely to be useful for both researchers and practitioners interested in the application of **mobile technology** in this area. However, further research that refined or expanded the research model used in this study would advance our understanding of the complex processes involved in organizational adoption of **mobile commerce** innovations.

The evidence from the three case studies demonstrates that the **innovation adoption** model presented in the chapter can help explain why companies do or do not adopt **mobile data solutions** in a **supply chain** context. In particular, the relative advantage offered by the **mobile technology** and the information intensity of the companies appeared to be the most relevant contextual influences on adoption in all three companies. Compatibility, leadership and organizational readiness also arose across all three case studies in our analysis. With respect to the latter factor, an interesting distinction emerged between the positive influence of 'technical readiness', which reflected the expertise of and role played by each company's IT team, and the negative influence of 'user readiness', which reflected the lack of familiarity with and even resistance to the new technology shown by some users in each case study. Interestingly, the industry environment did not seem to play an important role in adoption of the **mobile data solution**s in the three case studies. This may reflect the pioneering status of the three companies in their respective industries in New Zealand with respect to the use of **mobile commerce** technology in the **supply chain**.

Based on our exploratory analysis of the three case studies we believe there is a need for further research to refine and test the propositions developed in our model. This could take four forms. First, larger scale survey research could be used to statistically confirm such propositions at a more general level. Second, studies carried out in different organizational or industry settings and for different types of **mobile commerce** innovations would potentially increase the applicability of the model. Third, the contextual factors used in our model were selected for their perceived relevance to **supply chain** applications of **mobile technology**. There is a need for other potentially relevant factors to be explored in the context of this type of technological **innovation adoption**. Examples include the possible role of factors such as the trialability or observability of an innovation, or the influence of network externalities or a critical mass of adopters, on the organizational **innovation adoption** decision (Doolin & Troshani, 2007). Finally, our exploratory case study approach does not enable us to reliably assess the degree of influence on the organizational **innovation adoption** process of the various factors in our model. Further research is needed to investigate the relative influence of different factors in **mobile commerce** technology adoption.

By identifying the factors that appear to influence organizational **innovation adoption** of **mobile commerce** technologies in the **supply chain**, we have developed a model that represents a preliminary step towards understanding this phenomenon. As Frambach & Schillewaert (2002) note, such models are of use to practitioners, including both technology suppliers and organizational managers, in marketing innovations to organizations and in gaining acceptance of innovations within organizations.

For **supply chain** or IT practitioners, the issue that came through strongly in the analysis of the three case studies was that adopting and implementing a **mobile data solution** involves more than automating existing processes. This is not a new finding, as business process redesign has often been associated with leveraging the benefits of IT innovations (e.g. Ahadi, 2004; Ndede-Amadi, 2004; Venkatraman, 1994). With respect to **mobile commerce** technologies, the mobility, localisation and immediacy aspects of the technology provide opportunities for process redesign which imply new ways of doing things for users. Addressing the latter involves extensive and carefully thought out training, but also recognition that changing existing user behaviours may be necessary.

REFERENCES

Ahadi, H. R. (2004). An examination of the role of organizational enablers in business process reengineering and the impact of information technology. *Information Resources Management Journal*, *17*(4), 1–19.

Al-Qirim, N. (2003). E-commerce in the aerial mapping industry: a New Zealand case study. *Journal of Systems & Information Technology*, *7*(1), 67–92.

Al-Qirim, N. (2005). An empirical investigation of an e-commerce adoption-capability model in small businesses in New Zealand. *Electronic Markets*, *15*(4), 418–437. doi:10.1080/10196780500303136

Alanen, J., & Autio, E. (2003). Mobile business services: A strategic perspective. In B. E. Mennecke & T. J. Strader (Eds.), *Mobile Commerce: Technology, Theory and Applications* (pp. 162-184). Hershey, PA: Idea Group Publishing.

Allen, J. P. (2000). Information systems as technological innovation. *Information Technology & People*, *13*(3), 210–221. doi:10.1108/09593840010377644

Benbasat, I., Goldstein, D. K., & Mead, M. (1987). The case research strategy in studies of information systems. *MIS Quarterly*, *11*(3), 369–386. doi:10.2307/248684

Benou, P., & Bitos, V. (2008). Developing mobile commerce applications. *Journal of Electronic Commerce in Organizations*, *6*(1), 63–78.

Bhatt, G. D., & Emdad, A. F. (2001). An analysis of the virtual value chain in electronic commerce. *Logistics Information Management*, *14*(1/2), 78–84. doi:10.1108/09576050110362465

Bouchard, L. (1993). Decision criteria in the adoption of EDI. In J. I. DeGross, R. P. Bostrom & D. Robey (Eds.), *Proceedings of the Fourteenth International Conference on Information Systems* (pp. 365-376). Orlando FL: ICIS.

Chau, P. Y. K., & Hui, K. L. (2001). Determinants of small business EDI adoption: an empirical investigation. *Journal of Organizational Computing and Electronic Commerce*, *11*(4), 229–252. doi:10.1207/S15327744JOCE1104_02

Chau, P. Y. K., & Tam, K. Y. (1997). Factors affecting the adoption of open systems: an exploratory study. *MIS Quarterly*, *21*(1), 1–24. doi:10.2307/249740

Chau, P. Y. K., & Tam, K. Y. (2000). Organizational adoption of open systems: a 'technology-push, need-pull' perspective. *Information & Management*, *37*(5), 229–239. doi:10.1016/S0378-7206(99)00050-6

Chwelos, P., Benbasat, I., & Dexter, A. S. (2001). Research report: empirical test of an EDI adoption model. *Information Systems Research*, *12*(3), 304–321. doi:10.1287/isre.12.3.304.9708

Compeau, D. R., & Higgins, C. A. (1995). Computer self-efficacy: development of a measure and initial test. *MIS Quarterly*, *19*(2), 189–211. doi:10.2307/249688

Damsgaard, J., & Lyytinen, K. (2000). The dynamics of factors explaining EDI diffusion in Hong Kong in the late 1990s. In J. Thong, P. Chau & K. Y. Tam (Eds.), *Proceedings of the Fourth Pacific Asia Conference on Information Systems* (pp. 1061-1074). Hong Kong: AIS.

Dedrick, J., & West, J. (2004). An exploratory study into open source platform adoption. In *Proceedings of the 37th Hawaii International Conference on System Sciences* (pp. 1-10). Hawaii: IEEE.

DePietro, R., Wiarda, E., & Fleischer, M. (1990). The context for change: organization, technology, and environment. In L. G. Tornatzky & M. Fleischer (Eds.), *The Processes of Technological Innovation* (pp. 151-175). Lexington MA: Lexington Books.

Doolin, B., McLeod, L., McQueen, B., & Watton, M. (2003). Internet strategies for established retailers: four New Zealand case studies. *Journal of Information Technology Cases and Applications*, *5*(4), 3–20.

Doolin, B., & Troshani, I. (2007). Organizational adoption of XBRL. *Electronic Markets*, *17*(3), 199–209. doi:10.1080/10196780701503195

Elliot, S. (2002). Research model and theoretical implications. In S. Elliot (Ed.), *Electronic Commerce B2C Strategies and Models* (pp. 291-326). Brisbane: Wiley.

Fichman, R. G. (2004). Going beyond the dominant paradigm for information technology innovation research: emerging concepts and methods. *Journal of the Association for Information Systems*, *5*(8), 314–355.

Fillis, I., Johannson, U., & Wagner, B. (2004). Factors impacting on e-business adoption and development in the smaller firm. *International Journal of Entrepreneurial Behaviour & Research, 10*(3), 178–191. doi:10.1108/13552550410536762

Flanagin, A. J. (2000). Social pressures on organizational Website adoption. *Human Communication Research, 26*(4), 618–646. doi:10.1111/j.1468-2958.2000.tb00771.x

Frambach, R. T., & Schillewaert, N. (2002). Organizational innovation adoption: a multi-level framework of determinants and opportunities for future research. *Journal of Business Research, 55*(2), 163–176. doi:10.1016/S0148-2963(00)00152-1

Gallivan, M. J. (2001). Organizational adoption and assimilation of complex technological innovations: development and application of a new framework. *The Data Base for Advances in Information Systems, 32*(3), 51–85.

Grover, V. (1993). An empirically derived model for the adoption of customer-based interorganizational systems. *Decision Sciences, 24*(3), 603–640. doi:10.1111/j.1540-5915.1993.tb01295.x

Henriksen, H. Z. (2006). Motivators for IOS adoption in Denmark. *Journal of Electronic Commerce in Organizations, 4*(2), 25–39.

Iacovou, C. L., Benbasat, I., & Dexter, A. S. (1995). Electronic data interchange and small organizations: adoption and impact of technology. *MIS Quarterly, 19*(4), 465–485. doi:10.2307/249629

Jeyaraj, A., Rottman, J. W., & Lacity, M. C. (2006). A review of the predictors, linkages, and biases in IT innovation adoption research. *Journal of Information Technology, 21*(1), 1–23. doi:10.1057/palgrave.jit.2000056

Jones, M. C., & Beatty, R. C. (1998). Towards the development of measures of perceived benefits and compatibility of EDI: a comparative assessment of competing first order factor models. *European Journal of Information Systems, 7*(3), 210–220. doi:10.1057/palgrave.ejis.3000299

Kalakota, R., & Robinson, M. (2001). *e-Business 2.0: Roadmap for Success* (2nd ed.). Boston, MA: Addison-Wesley.

Kalakota, R., & Robinson, M. (2002). *M-Business: The Race to Mobility*. New York: McGraw-Hill.

Kalakota, R., Robinson, M., & Gundepudi, P. (2003). Mobile applications for adaptive supply chains: A landscape analysis. In E.-P. Lim & K. Siau (Eds.), *Advances in Mobile Commerce Technologies* (pp. 298-311). Hershey, PA: Idea Group Publishing.

Karahanna, E., Straub, D. W., & Chervany, N. L. (1999). Information technology adoption across time: a cross-sectional comparison of pre-adoption and post-adoption beliefs. *MIS Quarterly, 23*(2), 183–213. doi:10.2307/249751

King, R. C., & Gribbins, M. L. (2002). Internet technology adoption as an organisational event: an exploratory study across industries. In *Proceedings of the 35th Annual Hawaii International Conference on System Sciences* (Vol. 7, pp. 201). Washington DC: IEEE Computer Society.

Kuan, K. K. Y., & Chau, P. Y. K. (2001). A perception-based model for EDI adoption in small businesses using a technology-organization-environment framework. *Information & Management, 38*(8), 507–521. doi:10.1016/S0378-7206(01)00073-8

Kumar, S., & Zahn, C. (2003). Mobile communications: Evolution and impact on business operations. *Technovation, 23*(6), 515–520. doi:10.1016/S0166-4972(02)00120-7

Kwon, T. H., & Zmud, R. W. (1987). Unifying the fragmented models of information systems implementation. In R. J. Boland & R. Hirschheim (Eds.), *Critical Issues in Information Systems Research* (pp. 227-251). New York: Wiley.

Lau, H. C. W., Lee, C. K. M., Ho, G. T. S., Ip, W. H., Chan, F. T. S., & Ip, R. W. L. (2006). M-commerce to support the implementation of a responsive supply chain network. *Supply Chain Management, 11*(2), 169–178. doi:10.1108/13598540610652564

Mehrtens, J., Cragg, P. B., & Mills, A. M. (2001). A model of Internet adoption by SMEs. *Information & Management, 39*(3), 165–176. doi:10.1016/S0378-7206(01)00086-6

Mentzer, J. T., DeWitt, W., Keebler, J. S., Min, S., Nix, N. W., & Smith, C. D. (2001). Defining supply chain management. *Journal of Business Logistics, 22*(2), 1–25.

Moore, G. C., & Benbasat, I. (1991). Development of an instrument to measure the perceptions of adopting an information technology innovation. *Information Systems Research, 2*(3), 192–222. doi:10.1287/isre.2.3.192

Ndede-Amadi, A. A. (2004). What strategic alignment, process redesign, enterprise resource planning, and e-commerce have in common: enterprise-wide computing. *Business Process Management Journal, 10*(2), 184–199. doi:10.1108/14637150410530253

Niederman, F. (1998). The diffusion of electronic data interchange technology. In T. J. Larsen & E. McGuire (Eds.), *Information Systems Innovation and Diffusion: Issues and Directions* (pp. 141-160). London: Idea Group.

Poon, S., & Swatman, P. M. C. (1998). A combined-method study of small business Internet commerce. *International Journal of Electronic Commerce, 2*(3), 31–46.

Porter, M. E., & Millar, V. E. (1985). How information gives you competitive advantage. *Harvard Business Review, 63*(4), 149–160.

Premkumar, G., & Potter, M. (1995). Adoption of computer aided software engineering (CASE) technology: an innovation adoption perspective. *The Data Base for Advances in Information Systems, 26*(2/3), 105–124.

Premkumar, G., & Ramamurthy, K. (1995). The role of interorganizational and organizational factors on the decision mode on for adoption of interorganizational systems. *Decision Sciences, 26*(3), 303–336. doi:10.1111/j.1540-5915.1995.tb01431.x

Premkumar, G., Ramamurthy, K., & Crum, M. (1997). Determinants of EDI in the transportation industry. *European Journal of Information Systems, 6*(2), 107–121. doi:10.1057/palgrave.ejis.3000260

Premkumar, G., Ramamurthy, K., & Nilakanta, S. (1994). Implementation of electronic data interchange: an innovation diffusion perspective. *Journal of Management Information Systems, 11*(2), 157–186.

Prescott, M. B., & Conger, S. A. (1995). Information technology innovations: a classification by IT locus of impact and research approach. *The Data Base for Advances in Information Systems, 26*(2/3), 20–41.

Ramamurthy, K., Premkumar, G., & Crum, M. R. (1999). Organizational and interorganizational determinants of EDI diffusion and organizational performance: a causal model. *Journal of Organizational Computing and Electronic Commerce, 9*(4), 253–285. doi:10.1207/S153277440904_2

Rangone, A., & Renga, F. M. (2006). B2e mobile internet: An exploratory study of Italian applications. *Business Process Management Journal, 12*(3), 330–343. doi:10.1108/14637150610667999

Rogers, E. M. (2003). *Diffusion of Innovations* (5th ed.). New York: Free Press.

Ruhi, U., & Turel, O. (2006). Enabling the glass pipeline: the infusion of mobile technology applications in supply chain management. In Y. Lan & B. Unhelkar (Eds.), *Global Integrated Supply Chain Systems* (pp. 291-309). Hershey, PA: Information Science Publishing.

Russell, D. M., & Hoag, A. M. (2004). People and information technology in the supply chain: social and organizational influences on adoption. *International Journal of Physical Distribution & Logistics Management, 34*(2), 102–122. doi:10.1108/09600030410526914

Shankar, V., & O'Driscoll, T. (2002). How wireless networks are reshaping the supply chain. *Supply Chain Management Review*, *6*(4), 44–51.

Siau, K., Lim, E.-P., & Shen, Z. (2003). Mobile commerce: Current states and future trends. In E.-P. Lim & K. Siau (Eds.), *Advances in Mobile Commerce Technologies* (pp. 1-17). Hershey, PA: Idea Group Publishing.

Tan, M., & Teo, T. S. H. (1998). Factors influencing the adoption of the Internet. *International Journal of Electronic Commerce*, *2*(3), 5–18.

Taylor, S., & Todd, P. A. (1995). Understanding information technology usage: a test of competing models. *Information Systems Research*, *6*(2), 144–176. doi:10.1287/isre.6.2.144

Teo, T. S. H., Tan, M., & Buk, W. K. (1997). A contingency model of Internet adoption in Singapore. *International Journal of Electronic Commerce*, *2*(2), 95–118.

Thong, J. Y. L. (1999). An integrated model of information systems adoption in small businesses. *Journal of Management Information Systems*, *15*(4), 187–214.

Tornatzky, L. G., & Klein, K. J. (1982). Innovation characteristics and innovation adoption implementation: a meta-analysis of findings. *IEEE Transactions on Engineering Management*, *EM-29*(1), 28–45.

Venkatesh, V., & Davis, F. D. (2000). A theoretical extension of the Technology Acceptance Model: four longitudinal field studies. *Management Science*, *46*(2), 186–204. doi:10.1287/mnsc.46.2.186.11926

Venkatraman, N. (1994). IT-enabled business transformation: from automation to business scope redefinition. *Sloan Management Review*, *35*(2), 73–87.

Wolfe, R. A. (1994). Organisational innovation: review, critique and suggested research directions. *Journal of Management Studies*, *31*(3), 405–431. doi:10.1111/j.1467-6486.1994.tb00624.x

Wymer, S. A., & Regan, E. A. (2005). Factors influencing e-commerce adoption and use by small and medium businesses. *Electronic Markets*, *15*(4), 438–453. doi:10.1080/10196780500303151

Yin, R. K. (2003). *Case Study Research: Design and Methods* (3rd ed.). Thousand Oaks, CA: Sage.

Zhu, K., Kraemer, K., & Xu, S. (2003). Electronic business adoption by European firms: a cross-country assessment of the facilitators and inhibitors. *European Journal of Information Systems*, *12*(4), 251–268. doi:10.1057/palgrave.ejis.3000475

Chapter 8
Exploring Decision Rules for Sellers in Business-to-Consumer (B2C) Internet Auctions

Jeff Baker
Texas Tech University, USA

Jaeki Song
Texas Tech University, USA

ABSTRACT

The recent growth of business-to-consumer (B2C) Internet auctions challenges researchers to develop empirically-sound explanations of critical factors that allow merchants to earn price premiums in these auctions. The absence of a comprehensive model of Internet auctions leads us to conduct an exploratory study to elucidate and rank critical factors that lead to price premiums in Internet auctions. We employ Classification and Regression Trees (CART), a decision-tree induction technique, to analyze data collected in a field study of eBay auctions. Our analysis yields decision trees that visually depict noteworthy factors that may lead to price premiums and that indicate the relative importance of these factors. We find shipping cost, reputation, initial bid price, and auction ending time as the factors most predictive of price premiums in B2C Internet auctions.

INTRODUCTION

Over the past decade, Internet auctions have grown from a mere curiosity to a major focus of both researchers and businesses. In their early days, Internet auctions were dominated by individuals selling collectibles such as antiques, celebrity memorabilia, stamps, toys, coins, and trading cards; the vast majority of transactions were consumer-to-consumer (C2C) (Lucking-Reiley, 2000a). More recently, researchers have noted the growth of business-to-business (B2B) and business-to-consumer (B2C) auctions (Bapna, Goes, & Gupta, 2001). In B2C auctions, large merchants such as Dell, Disney, Home Depot, IBM, Motorola, Sears, Sun Microsystems, and Sharper Image have been able to use Internet auctions to sell excess inventory for greater profit than they would receive from using a liquidator (Dholakia, 2005b; Gentry, 2003; Grow, 2002; Vogelstein, Boyle, Lewis, & Kirkpatrick, 2004). As further evidence of the growth of B2C Internet auctions, by the first quarter of 2006, Internet auctioneer eBay alone hosted approximately 383,000 eBay stores worldwide, including 171,000 on Web sites other than their U.S. Web site (eBay, 2006). As firms continue to make extensive use of Internet auctions, the interest in developing sound guidelines for businesses as well as developing theory to advance research will likely continue to grow as well.

While many studies have examined the factors that determine an auction item's final bid price, the number of bids an item receives, whether a sale is completed, or the revenue earned by a seller, the examination of price premiums (above-average final bid prices) is relatively understudied. In economics, price premiums are defined as prices that yield above-average profits (Klein & Leffler, 1981; Shapiro, 1983). Price premiums within the Internet auction context have been defined as *"the monetary amount above the average price received by multiple sellers for a certain matching product"* (Ba & Pavlou, 2002, pp. 247-248). Restated, a number of auctions exist where sellers have earned above-average prices, or price premiums, on the items they have auctioned. In this study, we compare the group of auctions that have achieved above-average prices with those that have not, to observe significant differences. To our knowledge, only two studies have previously examined price premiums (Ba & Pavlou, 2002; Pavlou, 2002). Since it is only by maximizing revenue and profit that a firm can remain viable in the marketplace (Seth & Thomas, 1994), an increased focus on how businesses that rely upon Internet auctions can earn price premiums may prove beneficial. The focus on price premiums is the first contribution of this study. As we investigate price premiums, we examine many of the independent variables that have been considered in previous studies to determine if they are also predictive of price premiums. The second contribution is the application of CART analysis to Internet auctions as a tool to generate decision rules. CART analysis is a tree-based method of recursive partitioning for explaining or predicting a response to order variables by significance (Brieman, Friedman, Olshen, & Stone, 1984). It generates decision trees and decision rules that may be used as guidelines (by sellers in Internet auctions, in this case). While electronic commerce research has demonstrated that CART analysis can be used to improve one-to-one Internet marketing (Kim, Lee, Shaw, Chang, & Nelson, 2001), CART has not yet been applied to Internet auctions. Thus, our study is, to our knowledge, the first to use a statistically-based decision making technique to demonstrate how sellers can use quantitative data to decide how to sell products in B2C Internet auctions. The third and final contribution of this study is the examination (by CART analysis) of variables that have been found (generally by multiple-regression analysis) to be determinants of auction outcome in previous studies. This confirmation of variables identified as critical factors in other types of analysis is the third contribution of this study.

The article will be organized as follows. We begin by reviewing literature on auctions, including relevant research on both traditional auctions as well as Internet auctions. Next, we present literature on machine-learning techniques that enable the induction of decision trees. Following the literature review, we discuss our methods, including our dataset, variables, and our research design. Specifically, we describe the collection and analysis of field data from Internet auctioneer eBay. We then present the results of our analysis. Following the presentation of our results, we discuss our findings and note the implications of our study. Finally, we conclude by briefly noting the limitations of our study and directions for future research.

LITERATURE REVIEW

Literature pertinent to this study will be selectively drawn from two areas of research. Given that one of the objectives of this study is to investigate factors enabling sellers to earn price premiums in Internet auctions, the first area from which we draw theory is that of auction literature. An additional objective—namely, describing a technique for developing decision rules for sellers in Internet auctions—leads us to the second area of research that is pertinent to the present study: decision-tree induction techniques.

Auctions

Auctions have been described as "*a market institution with an explicit set of rules determining resource allocation and prices on the basis of bids from the market participants*" (McAfee & McMillan, 1987, p. 701). A vast amount of research addresses the topic of auctions. Numerous surveys of auction literature can be found (Engelbrecht-Wiggans, 1980; Klemperer, 1999, 2000; Krishna, 2002; McAfee & McMillan, 1987; Milgrom, 1985, 1986; Rothkopf & Harstad, 1994; Wilson, 1987), including a bibliography of earlier literature (Stark & Rothkopf, 1979) and a review of experimental auction literature (Kagel, 1995).

Auction Mechanisms and Auction Theory

Auction mechanisms are generally categorized as: (1) English or ascending-price auctions; (2) Dutch or descending-price auctions; (3) first-price sealed-bid auctions; or (4) second-price sealed bid or Vickrey auctions (McAfee & McMillan, 1987). A thorough description of these mechanisms can be found in the recent work of Lucking-Reiley (2000a). Internet auctions on eBay, the point of data collection for this study, have been described by scholars as a hybrid of the English and second-price auctions (Lucking-Reiley, 2000a, 2000b; Ward & Clark, 2002; Wilcox, 2000). Researchers assert that eBay uses a hybrid auction type on the grounds that the presence of a proxy-bidding mechanism ensures that a winning bidder will pay only one increment more than the second-highest bidder's price. Since this study examines only auctions of the hybrid eBay type, a discussion of how various types of auction mechanisms impact auction outcome is beyond the scope of the present study.

Auction theory is often centered around or developed in response to the seminal work of William Vickrey (1961), who described the Independent Private Values Model (IPV). In this model, each bidder formulates a valuation for the item being auctioned without an awareness of competing bidders' valuations. Even if valuations were shared among all bidders, each individual bidder's valuation would be unaffected by the additional information that competing bidders' valuations would provide. In this way, the bidder's value is independent of the influence of competing bidders and is privately determined. In contrast, the Common Values Model (CV) posits that the value of the item being auctioned is common to all bidders, but incomplete information causes each bidder to formulate a valuation for the item that falls either above or below the common value (Rothkopf, 1969; Wilson, 1969). If it is assumed that bidders' valuations are normally distributed about the common value, the winner of the auction is the bidder with the valuation that is farthest above the common value. This person incurs the "winner's curse," because he or she has likely overpaid for the item. An integrative approach, referred to as the Affiliated Values Model (AV), explains that bidder valuations depend upon the bidder's personal preferences, the preferences of others, and the intrinsic qualities of the item being sold (Milgrom & Weber, 1982). Bidders' valuations are described as affiliated because a high valuation by one bidder makes a high valuation by other bidders more likely (Milgrom & Weber, 1982). The AV model is a more general conceptualization of the valuation of items in auctions than the IPV or CV models; both the IPV and CV models can be understood as special cases of the more general AV model (McAfee & McMillan, 1987). Recent studies of Internet auctions rely upon and explicitly mention the merits of the AV model (Dholakia & Soltysinski, 2001; Gilkeson & Reynolds, 2003; Segev, Beam, & Shanthikumar, 2001; Wilcox, 2000). These studies empirically validate the AV model in Internet auctions by demonstrating that bidders may be influenced not only by their own valuation of the item, but also by the behavior of other bidders.

Internet Auctions

Internet auctions have a relatively brief history. Among the earliest electronic auctions were the auctioning of pigs in Singapore (Neo, 1992) and flowers in Holland (van Heck & van Damme, 1997) conducted over a LAN. Auctions on the Internet, conducted via newsgroups and e-mail discussion lists, were the next major development in the Internet auction timeline (Lucking-Reiley, 1999, 2000a). The explosion in popularity of Internet auctions, however, did not begin until the 1995 launches of U.S. Web sites Onsale and eBay (Lucking-Reiley, 2000a). By 1999, there were an estimated 200 auction sites on the Internet (Crockett, 1999). The continued growth of Internet auctions is demonstrated by the performance of international industry leader eBay, a company that operates auction Web sites in 24 countries, includes over 180 million registered users, and generated US$ 4.552 billion in sales in 2005 (eBay, 2006). International competition includes firms such as QXL.com in Europe, Taobao.com in Asia, and MercadoLibre in Latin America. Following Möllenberg (2004, pp. 360-371), we will define Internet auctions to mean *virtual marketplaces relying on Internet services (such as the World Wide Web) and Internet protocols to conduct auctions.*

In spite of the relatively short history of Internet auctions, they have begun to draw interest not only from economists, but also from researchers in marketing, information systems, and computer science (see Appendix A for a selective listing of recent studies in each of these disciplines). The general questions that many of these studies seek to answer are, "What is the optimal way to auction an item?" or "How is the marketplace changing as a result of Internet auctions?" or "What factors should be considered when buying or selling in an Internet auction?" We will generally limit our discussion of Internet auctions to empirical studies that deal with variables that are under the control of the seller (rather than variables under the control of the other two parties to the auction transaction, the auctioneer and the bidder). Since this study focuses on developing decision rules for sellers in single-item B2C Internet auctions, we will reserve exploration of multi-unit auctions and buyer behavior for other researchers. To organize the list of variables that have been investigated in previous studies, we introduce the categories of: (1) selling information, (2) seller information, (3) product information, and (4) delivery information. We will define and discuss each of these categories in turn.

Selling information includes general information about the auction and the terms of an item's sale. The initial bid price, the availability of a buy-now option, the auction duration, and the auction's ending time are included as selling information variables. Table 1 contains a list of these variables, their definitions, and a list of studies in which they have been investigated. There have been a number of important findings in this area. It has been observed that an item's final bid price can be significantly affected by its initial bid price (Brint, 2003). Bidders have been found to sometimes ignore a buy-now option even when

Table 1. Previous empirical studies measuring selling information variables

Variable	Description	Source
Initial Bid Price	Starting bid price	(Gilkeson & Reynolds, 2003; McDonald & Slawson, 2002; Standifird, 2001; Standifird, Roelofs, & Durham, 2004)
Buy-Now Option	Presence or absence of option for bidder to end auction early by purchasing at a seller-determined fixed price (eBay's Buy-it-Now option)	(Standifird, Roelofs, & Durham, 2004)
Auction Duration	Length of auction in days	(Dholakia & Soltysinski, 2001; Gilkeson & Reynolds, 2003; McDonald & Slawson, 2002; Mehta, 2002; Standifird, 2001; Standifird, Roelofs, & Durham, 2004; Subramaniam, Mittal, & Inman, 2004)
Auction Ending Time	Time of day auction ends	(Dholakia & Soltysinski, 2001; Gilkeson & Reynolds, 2003; McDonald & Slawson, 2002; Mehta, 2002; Standifird, 2001)

Variable	Description	Source
Number of Positive Feedback Ratings	Total number of eBay positive feedback ratings	(Ba & Pavlou, 2002; McDonald & Slawson, 2002; Standifird, 2001)
Number of Negative Feedback Ratings	Total number of eBay negative feedback ratings	(Ba & Pavlou, 2002; McDonald & Slawson, 2002; Standifird, 2001)
Product Information Variables		
Number of Pictures	Number of pictures	(Ottaway, Bruneau, & Evans, 2003)
Number of Bids	Total number of bids submitted for item	(Dholakia, 2005b; Dholakia & Soltysinski, 2001; Gilkeson & Reynolds, 2003; McDonald & Slawson, 2002; Standifird, 2001; Subramaniam, Mittal, & Inman, 2004; Wilcox, 2000)
Delivery Information Variables		
Availability of Expedited Delivery	Availability of express delivery	
Availability of International Delivery	Possibility to Deliver Internationally	
Shipping Cost	Amount of shipping and handling charges	(Gilkeson & Reynolds, 2003; McDonald & Slawson, 2002)

buy-now prices are set below prevailing market prices (Standifird, Roelofs, & Durham, 2004). Setting a buy-now price may, however, enhance revenue for sellers (Budish & Takeyama, 2001) in some situations. The time of day or week that an auction ends, and the duration of an auction are frequently used as either control variables or dependent variables (Bruce, Haruvy, & Rao, 2004; Dholakia & Soltysinski, 2001; Gilkeson & Reynolds, 2003; McDonald & Slawson, 2002; Standifird, 2001; Standifird, Roelofs, & Durham, 2004; Subramaniam, Mittal, & Inman, 2004), but have not, to our knowledge, been conclusively linked to higher closing prices.

Seller information is defined as the various facets of the seller's feedback rating. The ease with which buyers are able to provide feedback has made a seller's feedback rating one of the most significant predictors of auction closing price. Feedback mechanisms can help sellers earn higher prices (Bruce, Haruvy, & Rao, 2004; McDonald & Slawson, 2002; Ottaway, Bruneau, & Evans, 2003) and have been shown in one previous study to play a role in generating price premiums for reputable sellers (Ba & Pavlou, 2002). The number of positive feedback ratings and the number of negative feedback ratings are included as seller information variables in this study (see Table 1). We investigate both positive as well as negative feedback, because it has been found that positive and negative feedback have an asymmetrical effect upon the final bid price. Specifically, positive feedback is mildly influential in determining final bid price, while negative feedback is highly influential (Standifird, 2001). Thus, it has been clearly demonstrated that seller information is also an important subset of variables to examine when researching Internet auctions.

Product information refers to the information provided by the seller or by other bidders about the item being auctioned. Frequently, product information is measured by recording the number of pictures of an item and the number of bids which an item receives (see Table 1). One study has explained that pictures of an item being auctioned on the Internet may affect information processing and ultimately the item's final closing price (Ottaway, Bruneau, & Evans, 2003). Another found that detailed descriptions of the item were significant predictors of a completed sale (Gilkeson & Reynolds, 2003)[1]. Other researchers have included product description as a control variable in their studies (Bruce, Haruvy, & Rao, 2004; Dholakia & Soltysinski, 2001; Gilkeson & Reynolds, 2003; Standifird, Roelofs, & Durham, 2004), giving at least informal credence to the notion that product information, such as pictures of an item, can influence an item's final closing price. Finally, the number of bids and the number of bidders has been shown to be factors leading to higher closing prices (Dholakia & Soltysinski, 2001; Gilkeson & Reynolds, 2003; Wilcox, 2000). Following the lead of these scholars, and in order to reach a more definitive conclusion regarding the possible impact of product description on auction prices, we also include product information in our analysis of Internet auctions.

Finally, ***delivery information*** simply refers to the cost of shipping and to the available delivery options. The availability of expedited delivery, international delivery, and the item's shipping cost are included here as variables (see Table 1). Relatively few researchers have included this subset of variables within their models. However, one study argues that high seller reputation and delivery efficiency may covary (McDonald & Slawson, 2002), while another includes shipping cost as a control variable (Gilkeson & Reynolds, 2003). We introduce the examination of international delivery because we believe that, with the increasing level of international activity in Internet retailing and Internet auctions, international shipping will become more important to sellers wishing to ensure the largest possible set of potential bidders. To gain a more complete perspective on all factors impacting auction prices, we will include each of the aforementioned delivery attributes in our analysis.

Recent scholarly commentary identifies three approaches that researchers have taken in their studies of Internet auctions: (1) concept discovery, which explores new phenomena; (2) process explanation, which seeks to provide an economic, psychological, or social explanation for behavior; and (3) theory deepening, which uses electronic markets to develop and test theories (Dholakia, 2005a). It has been noted that concept discovery and process explanation have received the majority of researchers' attention, while theory-deepening approaches are relatively few in number (Dholakia, 2005a). In the absence of established theory, continued exploratory work such as this study seems warranted.

While the foregoing findings from Internet auction research are noteworthy in their own right, they have a limited usefulness even when taken in sum. Without being able to ascertain which variables will provide the greatest benefit relative to other variables, businesses are left without guidance for generating price premiums in Internet auctions. In light of this need, we will capitalize upon previous work in a novel way. Rather than simply searching among the myriad attributes of an Internet auction to find those that are predictive of the final closing price, we propose a descriptive model based upon empirical data which ranks the attributes of Internet auctions by their importance. A classification and regression tree will be produced which can be used to guide businesspeople who are making decisions regarding how to auction their products in B2C auctions. At this point, we will turn our attention to decision-tree induction, a technique capable of producing decision rules for sellers.

Decision-Tree Induction Techniques

Decision rules, or rules of classification, can be deduced from data by using various machine-learning techniques (Tsai & Koehler, 1993). Information gained by analyzing data with these inductive learning techniques can be represented in various forms, including mathematical statements, logical expressions, formal grammar, decision trees, graphs, and networks (Kim, Lee, Shaw, Chang, & Nelson, 2001). Decision trees are essentially visual presentations of sets of nested if-then statements. One advantage of using decision trees is that they depict rules that can be readily expressed in words, thus facilitating comprehension by decision-makers (Kim, Lee, Shaw, Chang, & Nelson, 2001).

Several algorithms for building decision trees exist; they include CART (Classification and Regression Trees), QUEST (Quick, Unbiased and Efficient Statistical Tree), SLIQ (Supervised Learning In Quest), CHAID (Chi-squared Automatic Interaction Detector), IC (Interval Classifier), ID3, and C5.0 (Agarwal, Arning, Bollinger, Mehta, Shcafer, & Srikant, 1996; Mehta, Agarwal, & Rissanen, 1996; Quinlan, 1990). While decision-tree induction allows data analysts to deduce decision rules for both continuous and discrete variables, not all algorithms are equally well-suited for use with both types of variables. For instance, CHAID and C5.0 are restricted to the analysis of categorical variables only (Berry & Linoff, 1997; Zanakis & Becerra-Fernandez, 2005). CART, on the other hand, can analyze either categorical or continuous variables. Classification-tree analysis can be used for categorical criterion[2] variables; regression-tree analysis is used for continuous criterion variables (Brieman, Friedman, Olshen, & Stone, 1984). Because of this characteristic of the CART algorithm, and because we intend to make binary splits of our dataset into price premium and non-price premium groups at each node, CART is ideally suited to our study. We now turn to a brief description of the CART decision-tree induction process.

Classification- and Regression-Tree Analysis (CART) is a nonparametric procedure that determines the optimal decision tree for classifying observations on the basis of a large number of predictive variables (Brieman, Friedman, Olshen, & Stone, 1984). CART recursively splits a dataset into non-overlapping subgroups based upon the independent variables until splitting is no longer possible (Kim, Lee, Shaw, Chang, & Nelson, 2001). One of the principal advantages of CART is that it tends to be less-biased than other data analysis methods (Lhose, Biolsi, Walker, & Reuter, 1994; Sorensen, Miller, & Ooi, 2000; Zanakis & Becerra-Fernandez, 2005). For instance, multiple discriminant analysis (MDA) and LOGIT methodologies need to satisfy the assumption of multivariate normality for independent variables; in addition, MDA requires that the groups' covariance structure be equal. Thus, if the variables follow some distribution other than the multivariate normal distribution, MDA and LOGIT will give biased results. The assumptions of multivariate normality and equal covariance can be easily violated in empirical datasets; biased classification can result. In such a situation, CART is preferable because it rests upon more realistic, less-frequently violated assumptions. CART assumes only that the groups are discrete, non-overlapping, and identifiable (Brieman, Friedman, Olshen, & Stone, 1984). Thus, CART is a data analysis technique that may be well-suited to real-world electronic commerce datasets. Now that some of the merits of CART have been described, we turn to an explanation of the process of decision-tree induction with CART.

The decision-tree induction technique begins as a dataset is subdivided into *N* sub-datasets. *N-1* subsets are used as training datasets, and the remaining dataset is used to test the model. The first training dataset is analyzed to find the single most important independent variable for classifying the observations into two groups. CART thus makes its most significant split first, at the root node (Berry & Linoff, 1997; Zanakis & Becerra-Fernandez, 2005). Each subgroup is then examined again with the algorithm to find the next-most important variable for classifying observations. After this partition, the process continues until only inconsequential variables remain (Berry & Linoff, 1997). The possibility of erroneously classifying some observations is computed by summing the predictive error rate at each split (Zanakis & Becerra-Fernandez, 2005). At this point, the tree is "pruned" to remove branches that inflate the error rate without providing substantial improvements in predictive power (Berry & Linoff, 1997). After the decision tree is generated from the first training dataset, the subsequent training datasets are analyzed to refine the tree. This process is known as cross-validation. Analysis of the training datasets thus generates a decision tree—a predictive model for classifying observations. Finally, the test dataset is analyzed to verify that the decision tree generated using the training dataset accurately classifies the remainder of the data as well.

To our knowledge, the use of decision-tree induction techniques to analyze Internet auction data and generate decision rules has not been undertaken. The application of the decision-tree analysis technique to Internet auction data may help to unify and bring coherence to the disparate extant findings in Internet auction research. It may also provide perspective on the relative importance of the numerous factors that have been proven to significantly impact auction outcome.

METHOD

We present the following analysis in order to answer questions about the variables enabling merchants to earn price premiums in Internet auctions and also to describe the decision rules for these variables.

Sample

Data was collected over a one-month period in 2005 from eBay's U.S. Web site. Data from international industry leader eBay has been frequently used as the point of data collection for studies of Internet auctions (Ba & Pavlou, 2002; Brint, 2003; Bruce, Haruvy, & Rao, 2004; Dholakia, 2005b; Dholakia & Soltysinski, 2001; Gilkeson & Reynolds, 2003; Standifird, Roelofs, & Durham, 2004; Ward & Clark, 2002; Wilcox, 2000). Data from eBay is used for three reasons. First, eBay data is often used because the realism of such data is often preferable to data collected in an experimentally-controlled laboratory setting. Field experiments with auctions present an obvious trade-off between experimental control and realism (List & Lucking-Reiley, 2000). Laboratory experiments of auctions have been criticized on the grounds that subjects' behavior in an artificial laboratory environment may not be exactly the same as it would be in real-world conditions (Lucking-Reiley, 1999). It has been argued that experimental subjects have no incentive to develop optimal bidding strategies or apply experience gained from bidding (Ward & Clark, 2002). Collection of data from a field setting reduces questions regarding its generalizability to the marketplace. For these reasons, our goal of developing a guideline for selling in Internet auctions that is both descriptive and prescriptive leads us to follow the precedent of these researchers in using field data rather than experimental data.

The second reason that researchers often use eBay data is simply that eBay continues to be the Internet auctioneer of choice. EBay continues to lead the industry because of the circular effect of high seller volume eliciting high bidder interest, which in turn motivates sellers to continue to utilize eBay (Wingfield, 2001). Thus, eBay provides substantial numbers of auctions to observe and numerous points of measurement.

The third and final reason for the use of eBay data is that eBay is the largest and most international of the Internet auctioneers. Their auction mechanism and terminology are used more widely than any other auctioneer's. Thus, in an endeavor to provide the most generalizable results, we have selected eBay as the point of data collection for this study.

The items examined in this study are a DVD movie (404 auctions) and a popular MP3 player (366 auctions). All DVD auctions were for the same, new, identically-packaged movie title (the popular animated feature "The Incredibles"), and all MP3 player auctions were for the same, new, first-quality, identically-packaged model of the device (the 4 GB Apple iPod). All items were described as "new," "never-used," "new in box," or "brand new." We included these items to sample a reasonably-broad spectrum of items, ranging from inexpensive (DVD) to relatively expensive (MP3 player). We collected data during a three-week window of time to guard against effects due to changes in the market price (due to the release of new versions of the products, or due to a reduction in cost in fixed-price markets). Additionally, these items were examined because their value should not change with the fortunes of a team or individual (as sports collectibles or celebrity memorabilia might). Finally, the high sales volume of these items facilitates data collection.

Variables

The variables for this study are those listed and defined earlier in Table 1. As we noted earlier, variables studied in previous research as predictors of auction outcome can be classified into four categories: selling information, seller information, product information, and delivery information. In addition, the dependent variable of interest is final bid price. We define final bid price as the highest bid submitted for a given item.

Measurement of Variables

Table 2 reports our coding scheme for the variables in the Internet auction. Table 3 reports the descriptive statistics of the data for 404 DVD auctions and 366 MP3 player auctions.

Research Design

This study uses CART to determine the most important variables that sellers should consider to earn price premiums. The reader will recall first, that price premiums have been defined as *"the monetary amount above the average price received by multiple sellers for a certain matching product"* (Ba & Pavlou, 2002, pp. 247-248) and second, that CART is a nonparametric procedure that determines the optimal decision tree for classifying observations on the basis of a large number of predictive variables (Brieman, Friedman, Olshen, & Stone, 1984).

We perform two analyses with CART: classification-tree analysis and regression-tree analysis. We first use final bid price as the criterion variable for classification-tree analysis. The classification-tree algorithm identifies the predictors that best separate our data into categories where an auction

Table 2. Data coding scheme

Variables	Coding
Criterion (Dependent) Variable:	
Final Bid Price	Continuous: dollars and cents
Independent Variables:	
Selling Information Variables	
(1) Initial Bid Price	Continuous: dollars and cents
(2) Buy-Now Option	Binary: 0—not available, 1—available
(3) Auction Duration	Continuous: duration of auction in days
(4) Auction Ending Time	Categorical: 1: Weekday before 4 PM 2: Weekday after 4 PM 3: Weekend before 4 PM 4: Weekend after 4 PM
Seller Information Variables	
(5) Number of Positive Feedback Ratings	Continuous: number of positive ratings
(6) Number of Negative Feedback Ratings	Continuous: number of negative ratings
Product Information Variables	
(7) Number of Pictures	Continuous: number of pictures
(8) Number of Bids	Continuous: total number of bids submitted
Delivery Service Information Variables	
(9) Availability of Expedited Delivery	Binary: 0—not available, 1—available
(10) Availability of International Delivery	Binary: 0—not available, 1—available
(11) Shipping Cost	Continuous: dollars and cents

yields a price premium (denoted in subsequent figures as PP) or fails to yield a price premium (denoted as NPP). Second, we use number of bids as a criterion variable for regression-tree analysis. We use number of bids as criterion variable because the number of bids is highly and directly correlated with the final bid price. Thus, the results should be substantially similar to those in the classification-tree analysis.

In the tree-building process, CART requires that the user select a computational method for validating the tree. CART provides cross validation in which the dataset is divided into *N* sub-datasets. *N-1* subsets are used as training datasets, and the remaining dataset is used for testing the model. To validate our trees, we use 10-fold cross-validation, a procedure in which nine subsets are used as a training sample, and one subset is used as a test sample (Steinberg & Colla, 1997). In the10-fold cross-validation process, the data are divided into approximately 10 equal subsets, where subsets are determined by random sampling on the criterion variable, and the tree-growing process is repeated 10 times.

Table 3. Descriptive statistics

	DVD Movie (N=404)		MP3 Player (N=366)	
	Mean	**Std. Dev.**	**Mean**	**Std. Dev.**
Criterion (Dependent) Variable:				
Final Bid Price	9.74	2.94	187.58	19.34
Independent Continuous variables:				
(1) Initial Bid Price	4.58	3.89	34.98	69.39
(3) Auction Duration	4.47	2.20	2.93	2.05
(5) Number of Positive Feedback Ratings	849	2616	2374	3176
(6) Number of Negative Feedback Ratings	5.70	9.35	20.04	32.38
(7) Number of Pictures	0.58	0.56	2.58	1.71
(8) Number of Bids	6.14	4.23	23.28	12.60
(11) Shipping Cost	4.20	1.07	16.32	5.50
Independent Categorical Variables				
	Frequencies		**Frequencies**	
(2) Buy-Now option	No: 384, Yes: 20		No: 357, Yes: 9	
(4) Auction Ending Time				
Weekday Morning	98		164	
Weekday Afternoon	153		48	
Weekend Morning	50		20	
Weekend Afternoon	103		134	
(9) Availability of Expedited Delivery	No: 354, Yes: 50		No: 256, Yes: 110	
(10) Availability of International Delivery	No: 187, Yes: 217		No: 130, Yes: 236	

RESULTS

Figure 1 and Figure 2 demonstrate the decision trees of the two different data sets that have been induced using CART analysis. Table 4 summarizes the decision rules derived from the trees. In the first dataset (the DVD movie), the classification-tree analysis shows that if the initial bid price is greater than $9.63, then the final bid price is above the average final bidding price. In other words, when sellers set the initial bid price at a level greater than $9.63, these sellers earn a price premium (PP). This result (shown both at the top of Figure 1 and also as Rule 1 in Table 4) shows that the initial bid price significantly impacts the final bid price (a finding consistent with earlier regression-based Internet auction studies). The remaining rules pertain to situations in which the sellers' initial bid price is less than $9.63. These rules together show that shipping cost, positive feedback, and the auction ending time are important determinants of price premiums. Rule 2 (see both Figure 1 and Table 4) shows that if the shipping cost is less than or equal to $3.97, a price premium is earned. Rules 3 through 5 indicate that, in the situation of relatively-high shipping cost, positive feedback and an ending time during the PM hours of the weekday are predictors of price premiums. In Rule 3, if the seller has less than 549.5 positive feedback ratings, the seller fails to

Figure 1. Decision tree for DVD movie

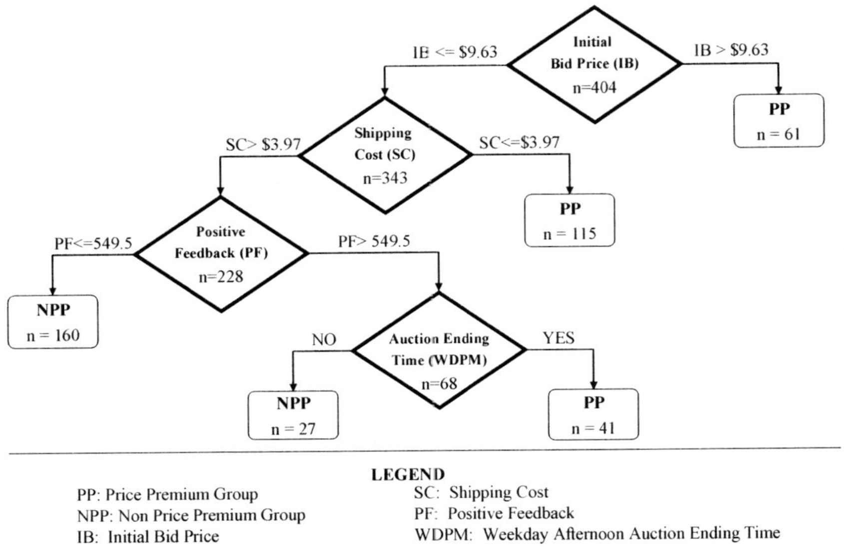

Figure 2. Decision tree for MP3 player

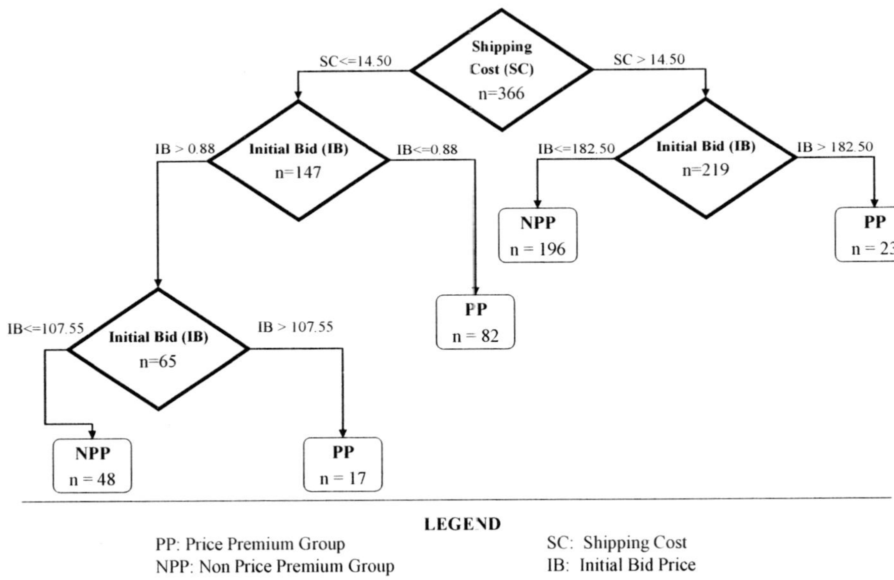

earn a price premium. Rule 4 and Rule 5 together show that if the seller has more than 549.5 positive feedback ratings, the ending time of the auction plays an important role. If the auction does not end during the PM hours of a weekday, the items fail to earn a price premium.

The fit statistics for the model are as follows. Resubstitution relative cost for the optimal tree is 0.428 and its complexity is 0.006. In addition, error rates for misclassification based on the criterion variable are 0.24 for price premiums and 0.23 for non-price premium auctions. In other words, 24% of observations are misclassified to NPP and 23% of observations are misclassified to PP.

In regression-tree analysis, the initial bid price and shipping cost are the most important

Table 4. Decision rules based on CART analysis for DVD dataset (N=404)

	Rule for Terminal Node
Classification-Tree Analysis: Dependent Variable = Final Bid Price	
Rule 1	IF Initial Bid Price (IB) > $9.63, THEN Price Premium Group (PP).
Rule 2	IF Initial Bid Price (IB) <= $9.63 AND Shipping Cost (SC) <= $3.97, THEN Price Premium Group (PP).
Rule 3	IF Initial Bid Price (IB) <= $9.63 AND Shipping Cost (SC) > $3.97 AND Positive Feedback (PF) <= 549.5, THEN Non Price Premium Group (NPP).
Rule 4	IF Initial Bid Price (IB) <= $9.63 AND Shipping Cost (SC) > $3.97 AND Positive Feedback (PF) > 549.5 AND Ending Time ≠ Weekday afternoon, THEN Price Premium Group (PP).
Rule 5	IF Initial Bid Price (IB) <= $9.63 AND Shipping Cost (SC) > $3.97 AND Positive Feedback (PF) > 549.5 AND Ending Time = Weekday afternoon, THEN Non Price Premium Group (NPP).
Regression-Tree Analysis: Dependent Variable = Number of Bids	
Rule 1	IF IB <= $ 3.15, THEN Average Number of Bids (AVG-NB) = 9.29
Rule 2	IF IB > $3.15 AND SC > $3.9, THEN Average Number of Bids (AVG-NB) = 2.93.
Rule 3	IF IB > $3.15 AND IB <= $9.25 AND SC <= $3.9, THEN Average Number of Bids (AVG-NB) = 6.28
Rule 4	IF IB > $9.25 AND SC <= $3.9, THEN Average Number of Bids (AVG-NB) = 2.73

predictors, just as in classification-tree analysis (see the bottom half of Table 4 for regression-tree analysis). Rule 1 shows that if the initial bid price is less than or equal to $3.15, the average number of bids is 9.29. If the initial bid price is set above $3.15 and the shipping cost is greater than $3.90, then the average number of bids is 2.93 (Rule 2). If, however, the initial bid price is set above $3.15 but the shipping cost is less than $3.90, then the average number of bids is 6.28 (Rule 3). Finally, if the initial bid price is greater than $9.25 and the shipping cost is less than or equal to $3.90, then the average number of bids is 2.73. In regression analysis, CART does not report a misclassification rate.

In the MP3 player dataset, the first rule shows that if the shipping cost (SC) is greater than $14.50 and the initial bid price (IB) is less than or equal to $182.50, then no price premium (NPP) is earned (see Figure 2 and Table 5, Rule 1). If, however, the shipping cost (SC) is less than or equal to $14.50 and the initial bid price is greater than $182.50, then a price premium (PP) is earned (Rule 2). These results show that the bidders were significantly influenced by shipping cost and the initial bid price. The remaining rules pertain to situations in which the shipping cost is less than or equal to $14.50. Rule 3 shows that if the shipping cost is less than or equal to $14.50, and the seller sets the initial bid less than or equal to $0.88, then a price premium was earned. In Rule 4, if the shipping cost is less than or equal to $14.50 and if the seller set the initial bid between $0.88 and $107.55, then the seller failed to earn a price premium. Rule 5 shows that if the shipping cost is less than or equal to $14.50 and if the seller set the initial bid price higher than $107.55, then a price premium was earned. In this dataset, resubstitution relative cost for the optimal tree is 0.348 and its complexity is 0.008. In addition, error rates for the NPP group and the PP group are 0.06 and 0.27, respectively.

In regression-tree analysis, the initial bid price and positive feedback ratings are the most important predictors, a slightly different result than in the classification-tree analysis. Rule 1 shows that

Table 5. Decision rules based on CART analysis for MP3 player dataset (N=366)

	Rule for Terminal Node
Classification-Tree Analysis: Dependent Variable = Final Bid Price	
Rule 1	IF SC > $14.50 AND IB < = $182.50, THEN NPP
Rule 2	IF SC > $14.50 AND IB > $182.50, THEN PP
Rule 3	IF SC <= $14.50 AND IB <= $0.88, THEN PP
Rule 4	IF SC <= $14.50 AND IB > $0.88 AND IB <= $107.55, THEN NPP
Rule 5	IF SC <= $14.50 AND IB > $107.55, THEN PP
Regression-Tree Analysis: Dependent Variable = Number of Bids	
Rule 1	IF IB > $135.40, THEN Average Number of Bids (AVG-NB) = 2.27
Rule 2	IF IB >= $85.00 AND IB < $135.40, THEN Average Number of Bids (AVG-NB) = 14.44
Rule 3	IF IB >= $7.50 AND IB < 85.40, THEN Average Number of Bids (AVG-NB) = 21.34
Rule 4	IF IB < $7.50 AND PF > 2635.5, THEN Average Number of Bids (AVG-NB) = 25.74

if the initial bid price is greater than $135.40, the average number of bids is 2.27. If the initial bid price is above $85.00 and below $135.40, then the average number of bids is 14.44 (Rule 2). If the initial bid price is between $7.50 and $85.00, then the average number of bids is 21.34 (Rule 3). Finally, if the seller sets the initial bid price lower than $7.50, positive feedback ratings plays an important role. Specifically, if the initial bid price is lower than $7.50 and positive feedback ratings are greater than 2635.5, then the average number of bids is 25.74.

Based on the rules derived from CART analysis, initial bid price, shipping cost, and positive feedback appear as important variables to determine the final bid price for both products.

CONCLUDING REMARKS

Implications for Research

This study contributes to the literature in several ways. This study, with its broad examination of variables from earlier research, is a step towards a comprehensive theoretical understanding of B2C Internet auctions. Exploratory work to this point has identified significant variables, and here all variables have been considered concurrently to find the ones with the greatest relative import. Thus, this study may provide guidance as researchers begin to develop definitive lists of variables that impact the outcome of Internet auctions. The creation of such definitive lists has been identified as a necessary preliminary step to developing theory (Ba & Pavlou, 2002; Pavlou, 2002; Weick, 1995; Whetten, 1989).

More specifically, we have been able to identify three independent variables—shipping cost, initial bid price, and reputation—that may play a larger role in Internet B2C auctions than was previously realized. In particular, the identification of shipping cost as a primary determinant of price premiums has not been previously reported by researchers. Our findings lead us to the conclusion that shipping cost is the single most important factor in earning price premiums in Internet auctions. This is somewhat surprising in light of the fact that so few researchers have considered shipping cost as a part of their models. Positive feedback was also found to be a critical factor in the ability to earn price premiums. This finding is in line with recent studies that place seller reputation in a place of great importance in Internet auctions (Ba & Pavlou, 2002; Bruce, Haruvy, & Rao, 2004; McDonald & Slawson, 2002; Ottaway, Bruneau, & Evans, 2003). Initial bid price was found to be a critical factor in our study, but has not been found to be significant in others (Gilkeson & Reynolds, 2003). We argue that initial bid price is significant because a low initial bid price attracts the greatest number of possible bidders. Sellers desire to have the largest possible number of bidders, because having more bidders helps to ensure that an item does not remain unpurchased. Because of the inconclusive history of this variable, further examination is warranted. Finally, ending time has been proposed as a significant factor in popular literature (Ribeiro, 2004; Witt, 2005) but, to our knowledge, has not been conclusively linked to auction outcome until this study found a weekday afternoon end time to be a significant predictor of the ability to earn a price premium.

We also note that the CART analysis performed here produces results that are similar to the vast majority of existing empirical work on Internet auctions. The majority of previous work has used some form of regression analysis to reach its conclusions; CART corroborates these results using a different methodology. CART has been used in e-commerce research (Kim, Lee, Shaw, Chang, & Nelson, 2001), but has not, to our knowledge, been applied to the specific topic of Internet auctions. One of the strengths of CART is that it is a non-parametric technique, which means that no assumptions are made regarding the distributions of the predictor variables. Normal, non-normal, skewed, categorical, and ordinal variables can be included in CART analyses. These conditions may be present in the datasets collected by Internet auction researchers, thus making CART a potentially useful analytic tool. For these reasons, the application of CART in Internet auction research is a new methodological contribution.

Implications for Practice

Internet Auctions have become a popular sales and marketing channel for businesses seeking to enhance profits. Offering auction services, selling by auction, and adopting the appropriate auction-pricing policy may increase the attractiveness of a Web site, reduce the inventory cost of slow-selling products, reduce transaction costs, and provide valuable insight into customer preferences. In some cases, retailers have realized revenue gains of 50% or more on excess inventory sold through Internet auctions rather than through liquidators (Gentry, 2003).

For the benefit of practitioners, we have demonstrated the use of a relatively new tool, CART Analysis, which can be used to investigate auctions. Businesses can use this tool, available in a number of data mining software packages, to glean insights from their data about how to most effectively list items for auction. Useful outputs of this analysis technique are the decision tree, which gives a fairly intuitive visual representation of the critical variables which are identified, and the table of decision rules, which demonstrates the priority of the critical factors. CART can provide complex, exact models that include decision rules for all variables considered in a given analysis. Here, we have attempted to strike a balance between simplicity and detail by demonstrating a simpler decision tree that indicates only the most important variables. Outputs are simple to interpret (particularly for individuals with little statistical training in regression techniques), provide precise results, and suggest logical, sequential decisions to practitioners. Also, we note that CART analysis is essentially a data-driven method; it is a tool that can be useful even when the analyst has little experience in selling the specific product for which he or she may be developing decision rules. Finally, as noted in the previous section, CART does not require that the data be normally distributed, thus making it a viable option in situations where techniques with more stringent assumptions will not work.

More practically, based upon our findings, merchants desiring to utilize Internet auctions should consider competing with other sellers on the bases of shipping cost, reputation, initial bid price, and auction ending time. We urge caution, however, in applying these findings to products that differ from the ones used in our study. The specific decision rules generated by our analysis may not be widely generalizable to other products (or even to other movie titles or other MP3 players). Investigations of other types of products may generate different decision rules. We encourage sellers to investigate each of the variables listed earlier in Table 1 to find the factors that are most important for the particular product being auctioned. Nevertheless, the growing corpus of literature on Internet auctions indicates that shipping cost, reputation, initial bid price, and auction ending time may be important in the sale of many other types of items as well (e.g., Brint, 2003; Gilkeson & Reynolds, 2003; McDonald & Slawson, 2002; Standifird, 2001).

Limitations

One limitation of this study is the use of a "greedy" classification algorithm. Classification trees use what is known as a "greedy" algorithm to determine splits in the dataset (Harrison, 1997). Some have criticized the use of classification trees because of the use of "greedy" algorithms, instead arguing that splits should be made based upon two or more levels at once (TwoCrows, 1999). Essentially, the "greedy" algorithm executes its task without considering the impact that any split may have on subsequent splits (TwoCrows, 1999). Other criticisms come from researchers who have expressed a desire for classification systems that make multivariate rather than univariate splits (TwoCrows, 1999). These debates are outside the scope of this study. However, it is worth noting that in spite of these theoretical issues, classification trees continue to be widely used and trusted in data mining applications by researchers and software developers.

Another limitation of our study is that we have analyzed data on only two products. While the fact that we have only examined a DVD movie and an MP3 player does not diminish the fact that CART analysis has been demonstrated as a useful technique in Internet auction research and practice, it is a limitation from the standpoint of identifying significant independent variables. A broader selection of products and different types of analysis might generate more broadly-applicable guidelines for Internet auction retailers. For instance, both DVDs and MP3 players are small, easily transportable, internationally-used products. It is conceivable that different products would generate decision trees with different critical factors. Thus, while our specific decision rules may not be widely applicable, the technique that we have demonstrated is.

Similarly, we have collected data from only one Internet auctioneer. While eBay has a larger international presence than any other Internet auctioneer, we remind readers that caution should be used when applying the findings here to auctions conducted in different contexts.

A different type of data analysis might also present additional useful findings. Other approaches for analyzing Internet auction datasets include, but are not limited to, binary logistic regression (with the auction's final closing price as a binary dependent variable, for instance) and multiple regression (perhaps with number of bids as a dependent variable). Using the same data from which the decision trees were generated, both binary logistic regression and multiple regression identify predictors very similar to those identified in CART analysis (interested readers may consult Appendix B for these results). These analyses identify significant independent variables, but do not generate sequential decision rules. Thus, it is likely that such techniques may be of greater interest to researchers than to practitioners. One final point with regard to alternate data analysis techniques deserves mention. The reader will recall from earlier discussion that regression techniques have more stringent assumptions than CART, and thus regression may not be appropriate for analysis of all datasets.

Future Research

One opportunity for future research is the investigation of auctions of different types of products. DVDs and MP3 players represent only a small fraction of the myriad items that are auctioned on the Internet. It is conceivable that auctions of different types of items may yield different decision rules. An examination of how the type of product impacts auction outcome may be a fruitful area of inquiry for researchers.

Future work will also need to delve into the motivations for bidder, seller, and auctioneer behavior. This is necessary in order to more completely explain the significance of findings in Internet auction studies. Some researchers have examined the roles that bidder experience (Ward & Clark, 2002; Wilcox, 2000), bidder strategy (Bapna, Goes, & Gupta, 2003; Easley & Tenorio, 2004), bidder acceptance of technology (Stafford & Stern, 2002), and bidder motivation (Cameron & Galloway, 2005; Standifird, Roelofs, & Durham, 2004) play in Internet auctions. As this study has assisted in the development of sellers' decision rules and strategies for Internet auctions, we believe future research should continue to examine bidders' decision rules and strategies as well. Future work could integrate findings from these two streams of Internet auction research, fitting the seller model of behavior to the bidder model. The ultimate goal should be the development of a comprehensive model of buyer and seller characteristics, motivation, and behavior.

REFERENCES

Agarwal, R., Arning, A., Bollinger, T., Mehta, M., Shcafer, J., & Srikant, R. (1996). *The quest data mining system.* Paper presented at the 2nd International Conference on Knowledge Discovery in Databases and Data Mining, Portland, OR.

Ba, S., & Pavlou, P. A. (2002). Evidence of the effect of trust building technology in electronic markets: Price premiums and buyer behavior. *MIS Quarterly, 26*(3), 243-269.

Ba, S., Whinston, A. B., & Zhang, H. (2003). Building trust in online auction markets through an economic incentive mechanism. *Decision Support Systems, 35*(3), 273-286.

Bapna, R., Goes, P., & Gupta, A. (2000). A theoretical and empirical investigation of multi-item on-line auctions. *Information Technology and Management, 1*(1-2), 1-23.

Bapna, R., Goes, P., & Gupta, A. (2001). Insights and analyses of online auctions. *Association for Computing Machinery. Communications of the ACM, 44*(11), 42-50.

Bapna, R., Goes, P., & Gupta, A. (2003). Analysis and design of business-to-consumer online auctions. *Management Science, 49*(1), 85-102.

Bapna, R., Goes, P., Gupta, A., & Karuga, G. (2002). Optimal design of the online auction channel: Analytical, empirical, and computational insights. *Decision Sciences, 33*(4), 557-578.

Berry, M., & Linoff, G. (1997). *Data mining techniques for marketing, sales, and customer support.* New York: John Wiley & Sons.

Breiman, L., Friedman, J., Olshen, R., & Stone, C. (1984). *Classification and regression trees.* Belmont, CA: Wadsworth International Group.

Brint, A. T. (2003). Investigating buyer and seller strategies in online auctions. *The Journal of the Operational Research Society, 54*(11), 1177-1188.

Bruce, N., Haruvy, E., & Rao, R. (2004). Seller rating, price, and default in online auctions. *Journal of Interactive Marketing, 18*(4), 37-51.

Budish, E., & Takeyama, L. N. (2001). Buy prices in online auctions: Irrationality on the Internet? *Economics Letters, 72*(3), 325-334.

Cameron, D., & Galloway, A. (2005). Consumer motivations and concerns in online auctions. *International Journal of Consumer Studies, 29*(3), 181-193.

Chong, B., & Wong, M. (2005). Crafting an effective customer retention strategy: A review of halo effect on customer satisfaction in online auctions. *International Journal of Management and Enterprise Development, 2*(1), 12-26.

Crockett, R. O. (1999). Going, going...richer. *Business Week* (3659), EB16.

Dholakia, U. M. (2005a). Concept discovery, process explanation, and theory deepening in e-marketing research: The case of online auctions. *Marketing Theory, 5*(1), 117-124.

Dholakia, U. M. (2005b). The usefulness of bidders' reputation ratings to sellers in online auctions. *Journal of Interactive Marketing, 19*(1), 31-41.

Dholakia, U. M., & Soltysinski, K. (2001). Coveted or overlooked? The psychology of bidding for comparable listings in digital auctions. *Marketing Letters, 12*(3), 225-237.

Ding, M., Elishaberg, J., Huber, J., & Saini, R. (2005). Emotional bidders - An analytical and experimental examination of consumers' behavior in a Priceline-like reverse auction. *Management Science, 51*(3), 352-365.

Easley, R. F., & Tenorio, R. (2004). Jump bidding strategies in Internet auctions. *Management Science, 50*(10), 1407-1420.

eBay (2006). eBay Inc. announces fourth quarter and full year 2005 financial results. Retrieved April 3, 2006, from www.investor.ebay.com/news/Q105/ebay0420-2932ld.pdf

Engelbrecht-Wiggans, R. (1980). Auctions and bidding models: A survey. *Management Science, 26*(2), 119-143.

Geng, X., Stinchcombe, M., & Whinston, A. B. (2001). Radically new product introduction using on-line auctions. *International Journal of Electronic Commerce, 5*(3), 169-189.

Gentry, C. R. (2003). Recycled returns. *Chain Store Age, 79*(1), 72-73.

Gilkeson, J. H., & Reynolds, K. (2003). Determinants of Internet auction success and closing price: An exploratory study. *Psychology & Marketing, 20*(6), 537-566.

Gregg, D., & Walczak, S. (2003). E-commerce auction agents and online-auction dynamics. *Electronic Markets, 13*(3), 242-251.

Grow, B. (2002). Excess inventory? EBay to the rescue. *Business Week*, 8.

Harrison, J. G. (1997). *Enhancements to the data mining process.* Unpublished Ph.D., Stanford University, CA.

Hu, X., Lin, Z., Whinston, A. B., & Zhang, H. (2004). Hope or hype: On the viability of escrow services as trusted third parties in online auction environments. *Information Systems Research, 15*(3), 236-250.

Kagel, J. H. (1995). A survey of experimental research. In J. H. Kagel & A. E. Roth (Eds.), *The handbook of experimental economics* (pp. 501-586). Princeton, NJ: Princeton University Press.

Kannan, P. K., & Kopalle, P. (2001). Dynamic pricing on the Internet: Importance and implications for consumer behavior. *International Journal of Electronic Commerce, 5*(3), 63-83.

Kim, J. W., Lee, B. H., Shaw, M. J., Chang, H. -L., & Nelson, M. (2001). Application of decision-tree induction techniques to personalized advertisements on Internet storefronts. *International Journal of Electronic Commerce, 5*(3), 45-62.

Klein, B., & Leffler, K. B. (1981). The role of market forces in assuring contractual performance. *The Journal of Political Economy, 89*(4), 615-641.

Klemperer, P. (1999). Auction theory: A guide to the literature. *Journal of Economic Surveys, 13*(3), 227-286.

Klemperer, P. (Ed.). (2000). *The economic theory of auctions*. Cheltenham, UK: Edward Elgar.

Krishna, V. (2002). *Auction theory.* San Diego, CA: Academic Press.

Lhose, G. L., Biolsi, K., Walker, N., & Reuter, H. H. (1994). A classification of visual representations. *Communications of the ACM, 37*(12), 36-49.

List, J. A., & Lucking-Reiley, D. (2000). Demand reduction in multiunit auctions: Evidence from a sportscard field experiment. *The American Economic Review, 90*(4), 961-982.

Lucking-Reiley, D. (1999). Using field experiments to test equivalence between auction formats: Magic on the Internet. *The American Economic Review, 89*(5), 1063-1081.

Lucking-Reiley, D. (2000a). Auctions on the Internet: What's being auctioned, and how? *The Journal of Industrial Economics, 48*(3), 227-253.

Lucking-Reiley, D. (2000b). Vickrey auctions in practice: From nineteenth-century philately to twenty-first-century e-commerce. *The Journal of Economic Perspectives, 14*(3), 183-193.

McAfee, R. P., & McMillan, J. (1987). Auctions and bidding. *Journal of Economic Literature, 25*(2), 699-739.

McDonald, C. G., & Slawson, V. C. (2002). Reputation in an Internet auction market. *Economic Inquiry, 40*(4), 633-650.

Mehta, K. (2002). *Understanding revenue drivers of electronic auctions.* Unpublished Ph.D., University of Illinois at Chicago.

Mehta, M., Agarwal, R., & Rissanen, J. (1996). *SLIQ: A fast, scalable classifier for data mining.* Paper presented at the Fifth International Conference on Extending Database Technology, Avignon, France.

Milgrom, P. (1985). The economics of competitive bidding: A selective survey. In L. Hurwicz, D. Schmeidler, and H. Sonnenschein (Ed.), *Social goals and social organization.* Cambridge: Cambridge University Press.

Milgrom, P. (1986). Auction theory. In T. E. Bewley (Ed.), *Advances in economic theory.* Cambridge: Cambridge University Press.

Milgrom, P. R., & Weber, R. J. (1982). A theory of auctions and competitive bidding. *Econometrica, 50*(5), 1089-1122.

Neo, B. S. (1992). The implementation of an electronic market for pig trading in Singapore. *Journal of Strategic Management Systems, 1*(5), 278-288.

Oh, W. (2002). C2C versus B2C: A comparison of the winner's curse in two types of electronic auctions. *International Journal of Electronic Commerce, 6*(4), 115-138.

Ottaway, T. A., Bruneau, C. L., & Evans, G. E. (2003). The impact of auction item image and buyer/seller feedback rating on electronic auctions. *The Journal of Computer Information Systems, 43*(3), 56-60.

Pavlou, P. A. (2002). Trustworthiness as a source of competitive advantage in online auction markets. *Academy of Management Proceedings*, A1-A7.

Porter, R., & Shoham, Y. (2004). On cheating in sealed bid auctions. *Decision Support Systems,* (39), 41-54.

Quinlan, J. R. (1990). Induction of decision trees. In J. W. Chavlik, W. Jude & T. G. Dietterich (Eds.), *Readings in machine learning* (pp. 57-69). San Mateo, CA: Morgan Kauffman.

Ribeiro, J. (2004, March 2). Researcher compiles auction data for online merchants. Retrieved June 7, 2005, from http://www.computerworld.com/managementtopics/ebusiness/story/0,10801,90668,00.html

Rothkopf, M. H. (1969). A model of rational competitive bidding. *Management Science, 15*(7), 362-374.

Rothkopf, M. H., & Harstad, R. M. (1994). Modeling competitive bidding: A critical essay. *Management Science, 40*(3), 364-384.

Segev, A., Beam, C., & Shanthikumar, J. G. (2001). Optimal design of Internet-based auctions. *Information Technology and Management, 2*(2), 121-163.

Seth, A., & Thomas, H. (1994). Theories of the firm: Implications for strategy research. *Journal of Management Studies, 31*, 165-191.

Shapiro, C. (1983). Premiums for high quality products as returns to reputations. *The Quarterly Journal of Economics, 98*(4), 659-680.

Sinha, A. R., & Greenleaf, E. A. (2000). The impact of discrete bidding and bidder aggressiveness on sellers' strategies in open English auctions: Reserves and covert shilling. *Marketing Science*, (19), 244-266.

Sorensen, E. H., Miller, K. L., & Ooi, C. K. (2000). The decision tree approach to stock selection. *Journal of Portfolio Management, 27*(1), 42-52.

Stafford, M. R., & Stern, B. (2002). Consumer bidding behavior on Internet auction sites. *International Journal of Electronic Commerce, 7*(1), 135-150.

Standifird, S. S. (2001). Reputation and e-commerce: EBay auctions and the asymmetrical impact of positive and negative ratings. *Journal of Management, 27*(3), 279-295.

Standifird, S. S., Roelofs, M. R., & Durham, Y. (2004). The impact of eBay's buy-it-now function on bidder behavior. *International Journal of Electronic Commerce, 9*(2), 167-176.

Stark, R., & Rothkopf, M. H. (1979). Competitive bidding: A comprehensive bibliography. *Operations Research, 27*(2), 364-391.

Steinberg, D., & Colla, P. (1997). *CART: Interface and documentation.* San Diego, CA: Salford Systems.

Subramaniam, R., Mittal, V., & Inman, J. J. (2004). Revenue equivalence in online auctions: The case of Yahoo and eBay. In P. T. L. Popkowski Leszczyc (Ed.), *Advances in consumer research* (Vol. 31, pp. 1-2).

Tsai, L. H., & Koehler, G. J. (1993). The accuracy of concepts learned from induction. *Decision Support Systems, 10*(2), 161-172.

TwoCrows (1999). *Introduction to data mining and knowledge discovery.* Potomac, MD: Two Crows Corporation.

van Heck, E., & van Damme, E. (1997). *New entrants and the role of IT: The tele-flower auctions in the Netherlands.* Paper presented at the 30th HICSS, Hawaii.

Vickrey, W. (1961). Counterspeculation, auctions, and competitive sealed tenders. *Journal of Finance, 16*(1), 8-37.

Vogelstein, F., Boyle, M., Lewis, P., & Kirkpatrick, D. (2004). 10 tech trends to bet on. *Fortune, 149*(4), 75-81.

Ward, S. G., & Clark, J. M. (2002). Bidding behavior in on-line auctions: An examination of the eBay Pokemon card market. *International Journal of Electronic Commerce, 6*(4), 139-155.

Weick, K. (1995). What theory is not, theorizing is. *Administrative Science Quarterly, 40*(3), 385-390.

Whetten, D. (1989). What constitutes a theoretical contribution. *Academy of Management Review, 14*(4), 490-495.

Wilcox, R. T. (2000). Experts and amateurs: The role of experience in Internet auctions. *Marketing Letters, 11*(4), 363-374.

Wilson, R. B. (1969). Communications to the editor: Competitive bidding with disparate information. *Management Science, 15*(7), 446-449.

Wilson, R. B. (1987). Auction theory. In J. Eatwell, M. Milgate & P. Newman (Eds.), *The new Palgrave: A dictionary of economic theory.* London: Macmillan.

Wingfield, N. (2001). EBay watch: Corporate sellers put the online auctioneer on even faster track --- Goods from IBM, Disney help dot-com pioneer post a surge in profits --- Why mom and pop are mad. *Wall Street Journal,* A.1.

Witt, L. (2005). Building sales on eBay. Retrieved May 10, 2005, from http://www.fortune.com/fortune/smallbusiness/answercentral/0,15704,601929

Zanakis, S., & Becerra-Fernandez, I. (2005). Competitiveness of nations: A knowledge discovery examination. *European Journal of Operational Research, 166*(2), 185-211.

ENDNOTES

1 It should be noted, however, that this study examined the sale of sterling silver flatware, including pieces manufactured in the 1890's. This study may only demonstrate the fact that sales of collectible items will likely be disproportionately affected by the quality of the item and the level of detail in its description.

2 In classification trees, the dependent variable is often referred to as the "criterion variable." We will adopt this usage.

APPENDIX A

Table A.1. Academic disciplines investigating Internet auctions and citations of recent studies

Economics	Marketing	Information Systems	Computer Science
(Budish & Takeyama, 2001; Easley & Tenorio, 2004; Lucking-Reiley, 2000a; Lucking-Reiley, 2000b; McDonald & Slawson, 2002; Sinha & Greenleaf, 2000; Standifird, 2001; Standifird, Roelofs, & Durham, 2004; Wilcox, 2000)	(Bruce, Haruvy, & Rao, 2004; Chong & Wong, 2005; Dholakia, 2005b; Dholakia & Soltysinski, 2001; Ding, Elishaberg, Huber, & Saini, 2005; Geng, Stinchcombe, & Whinston, 2001; Gilkeson & Reynolds, 2003; Kannan & Kopalle, 2001; Stafford & Stern, 2002; Subramaniam, Mittal, & Inman, 2004)	(Ba & Pavlou, 2002; Ba, Whinston, & Zhang, 2003; Bapna, Goes, & Gupta, 2000; Bapna, Goes, & Gupta, 2001; Bapna, Goes, & Gupta, 2003; Bapna, Goes, Gupta, & Karuga, 2002; Gregg & Walczak, 2003; Hu, Lin, Whinston, & Zhang, 2004; Oh, 2002; Pavlou, 2002; Segev, Beam, & Shanthikumar, 2001; Ward & Clark, 2002)	(Ottaway, Bruneau, & Evans, 2003; Porter & Shoham, 2004)

APPENDIX B

Table B.1. Binary logistic and multiple regression analyses

Variables	DVD		MP3 Player	
Criterion (Dependent) Variable	Final Bid	Number of Bids	Final Bid	Number of Bids
Independent Variables:				
Constant	1.972	10.47	-1.49	26.09
Initial Bid Price	0.027**	-0.89***	0.01***	-0.13***
Shipping Cost	-0.783	-0.57***	-0.03	0.20**
Buy-Now Option	2.331***	4.06***	-0.51	2.64
Ending Time: WDM	-0.421**	0.23	0.65	1.18
Ending Time: WKM	-1.288**	-0.57	-1.02	-4.17*
Ending Time: WKA	0.641	0.03	-0.82***	0.11
Auction Duration	0.101	0.09	0.06	-0.05
Log (Positive Feedback)	0.317**	0.39***	0.13*	-0.10
Number of Negative Feedback Ratings	-0.265*	-0.50***	0.00	-0.01
Number of Pictures	-0.196	-0.10	0.19**	-0.23
Expedited Delivery	-0.569	0.99	-1.30***	-0.03
International Delivery	-0.433*	-0.26	0.31	0.04
Log-likelihood ratio	-130.82***		86.52***	
F-Value		25.73***		32.23***
Adjusted R^2		55.2%		51.4%

$^{*}p<0.10$ $^{**}p<0.05$ $^{***}p<0.01$

This work was previously published in the International Journal of E-Business Research, Vol. 4, Issue 1, edited by I. Lee, pp. 1-21, copyright 2008 by IGI Publishing (an imprint of IGI Global).

Chapter 9
An Implementation of a New Type of Online Auction

M. A. Otair
Arab Academy for Banking and Financial Sciences, the League of Arab States

Ezz Hattab
Arab Academy for Banking and Financial Sciences, the League of Arab States

ABSTRACT

In recent years, there has been an increased interest in the types of online auction. Yet many auctions with fixed-end times are experiencing "sniping" or submission of bids in the final minute of an auction. Late bidding deprives rivals of the ability of seeing one's bid and undercutting it. Late bidding facilitates colludes or independent pricing well above that predicated by auction mechanism. This article aims to propose and implement a new type of online auction called Least and Unique Price (LUP). In the LUP auction, the winner will be the bidder who submits the least and unique price. Moreover, late bidding and specific closing time also be overcome by the LUP auction. In addition, this article presents the practical implementation of the proposed auction. In order to evaluate the proposed auction a comparative analysis of different auction types and the proposed one has been done.

INTRODUCTION

Each year, millions of Internet users utilize online auctions to purchase and sell various products. Despite the recent economic slowdown, consumers and businesses continue to spend money online. Due to the explosive popularity of the Internet, the use of electronic commerce as revenue-generating businesses has also increased. Government, corporations, as well as businesses are keen to find ways of efficiently allocating their resources with the use of auction type mechanisms (Okamoto, 2002). There are several auction types used on the market today are: the English auction, the Dutch auction, and second price sealed bid auction (Yuen, Sung, & Wong, 2003).

An online auction is an indirect sale transaction type of e-commerce. An online auction increases competition among vendors in addition to broaden the set of potential consumers (Richard &Halstead, 2004). Online auction has become very popular, even though there are some aspects of an online auction that its electronic version cannot encompass, such as seeing and touching the real product. An auction is a public sale in which the price is determined by bidding, and the item is sold to the highest bidder. To participate in an auction means to bid to obtain an item (Turban, King, Lee, & Viehland, 2004). For example, in the English auction, the person who offers the highest bid wins the right to purchase the item at that price. Online auctions have been widely used throughout the world in order to set prices of such unique items such as fine are and antiquities.

Various researchers (Parente, Venkataraman, Fizel, & Milett, 2001; Gao, 2005; Orson & Donald, 2004) have pointed out that, there are a variety of auction formats and many characteristics that define auctions. These formats and factors are valid for online auctions. The number of bidders and pattern of bidding is determined by the rules of the auction and its surrounding environment. Relevant environmental factors include the type of the good being sold, risk preferences of bidders, the time frame of the auction, and the available information concerning the bidding process.

One auction characteristic that may influence auction success is auction time of the bidding period. Most auctions are initiated with advanced notice of a specific closing time. The fixed end time poses an incentive problem – the early bid serves no benefit to the bidder but reveals information to his rivals (Parente et al., 2001).

Bapna, Goes, and Gupta (2001) report that, the major problem is that of a bidder hiding as until near the end of the auction, to conceal its true interests/values from opponents. Example: "bid-sniping" in eBay (auction last a week, but all meaningful bids occur in the last 5 minutes). One of the possible remedies for the snipers problem is an auction "Overtimes" which can restore the desirable bidding properties of reverse auctions. An overtime or extension to the auction (Parente et al., 2001) is invoked if any bidding occurs in a designated final phase of the auction (e.g., bid in the last 3 minutes). The overtime may iterate if late bidding continues in the previous overtime. The additional time allows bidders the opportunity to react to "snipers" and minimizes the potential for pricing rings. However, Chafferry (2004) noticed that the disadvantage of overtime is that it obligates serious bidders to return to the auction at closing time and remain through subsequent extension periods.

In addition to propose a new type of online auction and resolve the sniping problem, this article aims to understand the business of online auction that which type of business model that it's under, how it works, types of it, advantages and disadvantages of it, payment options and the trust that exist between customers on the Web with the proposed auction.

This article is organized as follows: in section 2, background on online auctions is presented, including information on the varieties, benefits, and uses online auctions. Next in section 3, the LUP auction is discussed, which is the proposed auction. Implementation of the proposed auction is then proposed in section 4. Section 5 introduces the comparative analysis between the proposed auction and different auction types. Finally, conclusions and future researches on online auction are discussed.

GENERAL AUCTION TYPES

Online auctions are rapidly gaining popularity (Reginaldo, Braga, & Masiero, 2001). Online auction systems are classified base on different aspects such as: bid definition, specification of auction rules, and definition of transactions. Several reasons motivate their use, among which are the ability to increase the number of potential customers, the low prices for the consumers, the low transaction costs, and the short period of time in which they can be done (Heck & Vervest, 1998; Kumar & Feldman, 1998).

Auctions are of many types, classified according to the method of submission of bids, method used in deciding who the winner is, the final price paid by the winner, and for valuing the object being auctioned. Some examples of different types of auction(traditional and electronic) are as follows: Sealed bid auctions, English/ascending auctions, Dutch/descending auctions, First price auctions, Second price auction, Common value/objective value auctions, and Private value/subjective value auctions. There are also variations and combinations of these designs.

Types of Online Auctions

There exist many different forms of auctions that have evolved to suit differing needs and applications. Figure 1 gives taxonomy of the types of electronic auctions that will be discussed in this section. This section also provides a review of the some variants of the main types electronic auction schemes that have been proposed.

Online auctions can be categorized according to the number of sellers and the style by which bidders submit bids. An online auction that has only one seller and many buyers (bidders) is called a single auction. While an online auction with many sellers and many buyers (i.e., a securities market) is called a double auction. Bidding in an auction in general can typically be of two types: sealed or open. In a sealed bid auction, secret bids (not known to the other bidders) are submitted to the Auctioneer. Once the bidding period finishes, the bids are opened and the winner of the auction is determined according to some publicly known rule (e.g., the highest bidder wins as in English auction). In an open bid auction, bids are known to all participants (bidders) during the bidding period.

1. English Auction

English auction is also called Open-outcry or Ascending-price. The auctioneer stars the bidding process with a minimum price for the product. The auction participants bid increasing the price until no more bids happen or some condition to be satisfied. Then, the highest price wins. An English auction is the most well known type of auction and is widely used for real estate.

Some researchers proposed many modifications on English auction mechanisms (Stubblebine & Syverson, 1999; Omote & Miyaji, 2001). Stubblebine and Syverson (1999) present a first price

Figure 1. Taxonomy of electronic auction schemes-this figure taken from (Trevathan, 2005)

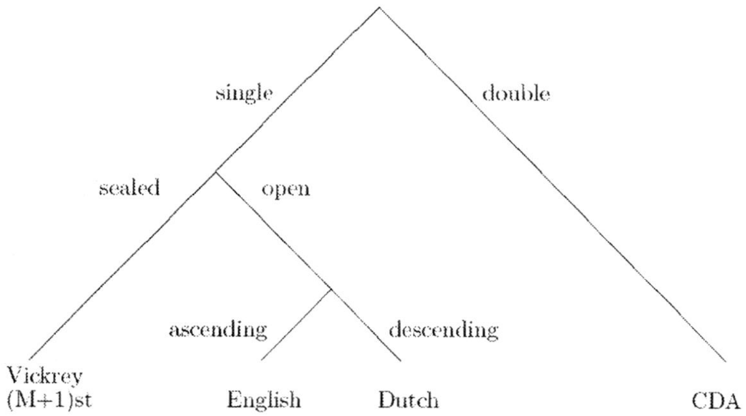

English auction. The scheme attempts to stop the Auctioneer from skewing its clock or selectively blocking bids. Bids are committed using secret bid commitment. This allows the Auctioneer to commit to a bid before he knows who it is from or what the bid amount is. After the Auctioneer has committed to the bid submission, the bidder can reveal the key. The Auctioneer must commit to the bids received at regular intervals. Anyone can verify the Auctioneer's actions by obtaining a notarized (time-stamped) version of the bid history from an online Notary.

Omote and Miyagi (2001) present an English auction scheme using a group signature. In this scheme there are two managers responsible for conducting a series of k auctions. The Registration Manager secretly knows the correspondence of the bidder's identity and registration key. The Registration Manager works as an identity escrow agency. The Auction Manager hosts the auction and prepares bidder's keys in each round. Bidders must acquire a new key prior to each auction in order to submit bids.

2. Dutch auction

Alternatively an auction can be descending if the reimbursement price is bid down instead of up. In a Dutch auction, the seller progressively lowers his/her offer until a buyer accepts that price. Dutch auctions have the natural property of concealing the losing bids. So far, all auctions discussed have been single in the sense that there is only one seller.

Sako (1999) proposes a first price, sealed bid auction which takes a Dutch-style approach to opening bids. Bidders choose their bids from a price list and send this information (encrypted) to a set of Auctioneers. The Auctioneers open (decrypt) the bids from highest to lowest in the price list until a winner is found. As soon as a winner is found, the Auctioneers refrain from opening any further bids (i.e., they do not decrypt any bids of a lesser value in the price list). This has the effect of concealing the values of losing bids (as in a Dutch auction). The auction outcome and the privacy of losing bids is ensured unless a threshold t, of the Auctioneers collude.

3. Vickery Auction or Uniform Second Price

A Vickrey auction (Vickrey, 1961) is a second price, sealed bid auction. The bidder with the highest bid wins and they must pay an amount corresponding to the second highest bid. An English auction is a first price, open bid auction. The bid amounts are known to all bidders, and they attempt to outbid each other. The highest bidder (for a given time-out period) wins and must pay an amount equal to the winning bid. English and Vickrey auctions are considered as ascending due to the bidders bidding up the reimbursement price.

Franklin and Reiter (1996) present what can be considered the first formal attempt at constructing a secure auctioning system. They propose a second price, sealed bid auction scheme which distributes the role of the Auctioneer across several servers. The auction process is considered safe as long as a threshold of the Auctioneers is not corrupt. Bids are submitted using a digital cash protocol. Bidders split their bid using a secret sharing scheme and then send their shares to the appropriate servers. At the close of bidding the servers multicast their shares to each other and jointly compute the result. Each server owns a share of the winner's digital coin signed with a verifiable signature sharing scheme. The signature prevents a single Auctioneer from altering a bid or throwing an auction to a single bidder. Their system further enables the bids to be backed by escrowing financial commitments of the bidders.

Kikuchi, Harkavy, and Tygar (1998) describe a set of protocols for performing sealed-bid electronic auctions which preserve the privacy of the submitted bids using a form of secure distributed computation. Bids are never revealed to any party, even after the auction is completed. Both first-price and second-price (Vickrey) auctions are supported, and the computational costs of the methods are low enough to allow their use in many real-world auction situations.

Naor, Pinkas, and Sumner (1999) propose a second-price, sealed bid auction. This scheme uses two Auctioneer servers each owned by a separate entity. These Auctioneers are referred to as the Auction Issuer and the Auction Manager, respectively. Auction Manager is responsible for conducting the auction. Auction Issuer aids Auction Manager in determining the winner. The Auctioneers communicate using an oblivious transfer protocol which effectively prevents each party from learning too much information about the bidders and the bids they submit. This scheme (and its variants) have thus far proved to be more efficient then threshold trust schemes.

Current online auctions use open-cry, English type auctions that involve multiple rounds of biddings. Studies by Roth and Ockenfels (2001) have shown that 20% of all bidders made their last bids in the last hour before closing eBay (14% in the last 10 minutes). Also 68% of auctions on eBay had the last bid received in the last hour of bidding (55% during last 10 minutes). These findings show that current Internet auctions are highly concentrated near the end. This is not a desirable feature, especially for the Internet where in people use different connection services to access the network. Current Internet auctions require bidders to be online at the last few minutes of the bidding stage to win the auction. This feature is also appealing because it solves the problem of sniping-as "Overtime" technique- wherein bidders wait for the last seconds to bid.

4. Continuous Double Auction

In a Continuous Double Auction (CDA) there are many buyers (bidders) and sellers who endlessly trade a single product or service. Bids are matched and cleared over time according to the auction rules. CDAs are typically open bid. The best known example of a CDA is a stock or securities market such as the New York Stock Exchange (NYSE).

Wang and Leung (2004) discuss security and anonymity issues for CDAs used in Internet retail markets. They propose a scheme that involves three servers to conduct the auction, but only requires trust between two of the servers. In order to participate in the auction, a bidder must first create a pseudonym that is blindly signed by one of the servers. This pseudonym is authenticated and signed by the second server and the bidder is issued a certificate to participate in the auction. To place a bid, a bidder forms his/her bid and presents this with his/her certificate to the third server whom conducts the auction. Bids are posted on a public bulletin board and are cleared according to the auction's matching strategy. When there is a dispute, the two servers used during registration can reveal the identity of a bidder.

5. Limbo auction [www.41414.com]

In most Limbo auctions, bidders can initiate a bid online and then play on their cell phone. So, if they want to be a member of the Limbo community they must have a mobile phone on any phone networks that share their text messaging revenue with Limbo to play. Moreover, Limbo works only on the some U.S. carriers.

The lowest unique bidder is the person who has bid the lowest amount (low bid in cents) that no other person has bid. When two people place the same bid, that number is no longer unique. When a person has a unique bid but there are others that are lower, then it's not the lowest. To win a Limbo Auction, the bidder must be the one with the lowest AND unique bid at the auction's end. If the bid price that submitted from any bidder is unique, but later someone bids the same amount as him, he gets bumped! Then, he will each need to try and find the next lowest unique number. Once you're a winner, you must wait for 120 days before you/they can win again.

Every bidder can know the status of the auction through sending a text message to know who the current winner is and what is the remaining time to close the auction? So, the sniping problem is increased. Limbo auctions are charge a fee (usually 99 cents) for each bid. Bidders do, of course, pay the cost of sending and receiving text messages based on their cell phone service plan. Most plans charge between 4 and 10 cents per message, but some are now charging up to 15 cents. Each Limbo Auction has a maximum number of bids they can place each day. It typically ranges between 10 and 30 bids, which increases the difficulty of the game.

6. Unique auction [www.UniqueAuction.com]

Unique Auction is a Web-based auction providing the opportunity for person-to-person bidding in an effort to accumulate new products from trusted sources. On Unique Auction the maximum price for any product is 20% of the retail price. For example, a product with a retail price of $1,000 would be sold under the Unique Auction hammer for a maximum price of only $200. The unique site ensures that no bids can exceed the $200 maximum price threshold and the highest unique bidder under $200 wins. Each bid price placed has an administration cost of $1 for members and $2 for nonmembers. The Unique auction will be closed if the time of the auction is terminated, or the number of bids on the item has reached reserve bids amount. Each item has a reserved number of bids, or a fixed number of bids that need to be met before an official winner can be declared for the auction.

Unique Auction is an online auction that is based on value and entertainment with countless opportunities. The chances of winning depend on the bidder. Also, unique auction suffers in some cases from the sniping problem.

Benefits of Online Auctions

For sellers, the benefit of auctions consists of access to an international customer base to sell merchandise quickly, easily, at low cost and sometimes premium prices. Even old damaged, returned, and refurbished merchandise can be unloaded. Furthermore, new markets are established and a company's customer base can be enlarged (Richard & Halstead, 2004).

For buyers, online auctions represent 24-hour availability 7 days a week, instant gratification, and interactivity. This interactivity sometimes allows customers more control over price and product/service customization. Increased learning about a product category may also be a positive outcome (Poel & Leunis, 1999). Consumer search costs can be reduced dramatically even while the number of available options increases. Adding to both convenience and savings (Machlis, 1998). Industrial buyers can save 6-15% or more by opening up their purchasing processes to online auctions (Woolley, 1998).

LEAST AND UNIQUE PRICE AUCTION (LUP)

Erica Klarreich (2003) showed that "Anyone who has bid in more than a handful of eBay online auctions has probably run into the phenomenon called (sniping), in which bidders place their bids in the last few seconds of an auction, leaving rivals no time to respond." This is not good behavior and not fair action, so all bidders will be frustrated, and possibly they will not return to the auction once again.

Barnes and Vidgen (2001) discuss how online auction become a main stream business, it is crucial for auction type developers to find ways to discourage collusive, entry-deterring and predatory behavior, as well as to prevent suboptimal pricing of items and many other problems. Even though there are many solutions for these problems, there is no technique or mechanism associated with any auction type to solve all of these problems. The attractive feature with the proposed auction LUP is the ability to overcome many of these problems especially the problem of sniping and it appeals to people with a betting nature, people brought up with online games. Similar to other online auctions in the pursuit for cheering fun and playing, LUP Auction is rightly distinct in some very unusual mechanisms.

For each type of online auctions, both reverse and forward auctions can be set: *Reverse* auction which is an auction to buy, and *Forward* auction is an auction to sell (Kauffman & Wood, 2005), whereas the proposed auction can be considered as Forward auction.

How the Proposed Auction Work?

To join an auction you have to register online and it's absolutely free if you are a buyer. However, you have to pay a certain amount of money to the LUP auction if you are a seller. You as buyer can select the item that you want and start bidding until the closing time of the auction. The seller will set the acceptable price (or market price). Bids are placed and accepted online, and result of your bid (if accepted or no) will be seen on the screen immediately after you have submitted it. Therefore, you can track only the name of the current winner right on your computer. The most interesting thing in LUP is the closing mechanism, where the auction will close when the number of placed bids plus the least and unique bid price are greater or equal to the market price.

The Steps of LUP Auction

The processes of the LUP auction can be achieved as follows:

1. The seller puts up an item for sale and specifies an acceptable price which is the desired price or retail price (may be closed or equals to the market price).
2. Item is the posted on the LUP auction site
3. Set the *AMOUNT* of this item to *Null*
4. Set the current *WinnerName* to Null, and the *WinnerBid* is set equal to acceptable price as initial value (the aim of this step is to make the first coming bid- after that- is the winner bid (i.e., to make the first bid is the least)).
5. Accept a new bid (i.e., the name of the bidder with his bid). (Note: bidders can not place a bid with cents, and each bidder must pay one dollar for each submitted bid, also it must be greater than or equals to one dollar).
6. The bidder's account will be decreased by one dollar (or by 0.5 dollar if the bidder is a golden member-as you will see in subsection 4.1-)
7. *If* the bidder's account greater than or equal to submitted bid *Then* go to step 8 *Else* go to step 5 (i.e., there is no enough money in his account, and in this case the system may show apologizing).
8. Increase the *AMOUNT* by one dollar (or by 0.5 dollar if the bidder is a golden member).
9. The bidder name with his bid will insert into the database (for the settlement purposes).
10. *If* the submitted bid is less than the current winner bid *Then* go to step 11 *Else* go to step 5.
11. *If* the submitted bid is unique *Then* go to step 12 *Else* reset the *WinnerName* to Null and go to step 5.
12. Show the current bidder name as the current *WinnerName* without showing his submitted bid (only, it will be saved into the database). Therefore, the bidders can track the name of the current winner on their computers.
13. *If* the *AMOUNT* + *current bid* is greater than or equal to the *acceptable* price *Then* close the auction and the least and unique bid is displayed with the name of bidder -with his submitted bid- who submits this bid, *Else* go to step 5.

At the end of the auction, the winner will be the bidder who submits the least and unique bid price. Also the winner must pay the winner bid plus the number of his bids during the auction life. In addition, each bidder who joined with the auction (not winner) must pay an amount which is equal to the number of his submitted bids (all these data extracted from the database). Under this framework, any seller can maximize the expected revenue from the auction. In the other side any winner will take an item with very low price. LUP auction charges an entry fee for each bid to cover the true market value of the product. With this technique, the seller can get full price of the item (i.e., true market value) while the winner buys it with very low price.

The LUP auction also can be implemented on very expensive items such as new cars, apartments, and so forth. However, in these cases normally an entry fee must be greater than one dollar, for example, it must be $50 or $100 for each submitted bid (it depends on the cost of that item). Any company as well can utilize the LUP auction for its customers by set a proper policy or strategy. For instance, an airline company can put 50 networked PCs in any hall in an airport with installed LUP software, and the passengers can participate in the LUP auction to win very cheap ticket before the flight. After that, each bidder must pay number of dollars equal to the number of bids that he done (this company may use this technique as marketing strategy).

The big advantage of the LUP auction is there are no snipers, because there is a lack of information about the auction period and the current winner bid price. In other words, the closing time of the LUP auction is not known and may be get very little time, may be minutes or possibly just a few seconds without showing the intermediate winner bid prices.

The expected time for completing an auction and the amount of signaling messages needed to conduct an auction are important consideration factors. These two elements are not considered in classical auctioning models. For Internet-based auctions, in particular those available on wireless accesses, the issue of communication cost cannot be ignored (Yuen, Sung & Wong, 2003). From this perspective, the LUP auctioning scheme intuitively has advantage over the other auctioning scheme.

Table 1. The four scenarios of the LUP auction

Submitted Price	Unique	Not Unique
Least	*1. Accepted*	*2. Not Accepted*
Not Least	*3. Not Accepted*	*4. Not Accepted*

The Scenarios of the LUP Auction

Once the bidder enters his bid for the selected item, he can see his bid if it fails or no. The submitted bid must be the least and unique. There are four scenarios will be as in the following table:

- **First scenario:** If the submitted bid is less than the bid of the current winner (i.e., Least and Unique), then the name of the bidder will be displayed as new winner (because the bid must be the least).
- **Second scenario:** If the submitted bid is the least but not unique (i.e., it is equal to the bid of the current winner- Least but not Unique-), then the name of the current bidder will be removed (because the bid must be unique) and the auction without winner. So, the first coming bid after that will be the current winner.
- **Third scenario:** If the submitted bid is unique but it is greater than the bid of the current winner (Unique but not the least), then the bid is not accepted (because the bid must be the least).
- **Fourth scenario:** If the submitted bid is neither unique nor less than the bid of the current winner (Not unique and not the

least), then the bid is not accepted (because the bid must be the least and unique).

AN IMPLEMENTATION OF LUP AUCTION

The following procedures are very important to implement the LUP auction and to make it practical auction:

Registration

Any customer interested in using the LUP auction must first register with the system by filling out the following registration forms in three steps (The personal information is protected by LUP's privacy policy and encrypted):

- **The first step:** each customer must fill out his personal and contacts (shipped) information. As shown in Appendix-A1 he must provide his real name, postal address and e-mail address, and a desired username and password.
- **The second step:** each customer enters his User ID (if it is not used because it must be unique) and Password. Also, he must agree with the conditions and policies. Appendix-A2 shows this step. In this form he submits these information; first name, last name, two addresses, city, province/state, country, postal/ZIP code, phone number, and any valid e-mail address is required in order to activate his account.
- **The third step:** The aim of this step is to confirm identity of the customer. Even though the registration process is free, however, a credit card or swift code of the transferred money or the password of issued card by LUP auction is required to confirm his identity and to eliminate any nonserious bidders. For instance, this step helps keep LUP a safe place to buy. Moreover, his credit card will be debited by $20 as credit for him to enable him to participate in any LUP auction. All these information will be entered as shown in Appendix-A3. All registered customers must deposit a certain amount of money or some other security with the LUP auction registration time. The money is transferred to a bank under control of the LUP auction. When the bidder is bidding for products, the number of the bids and the current bid price placed by a customer may never exceed the money available on his account. Once the registration process is completed, the customer becomes a member of the LUP auction. Furthermore, customer can use prepaid cards (issued by the LUP auction site with several values) or by transfer amount of money to anyone of several bank accounts of the LUP auction (from anywhere). That is, to make it worthwhile for larger community.

Each bid placed has a participation cost of $0.5 for golden members and $1 for nongolden members. Golden Membership cuts the $1 for each bid fee in half by pay golden membership fee which costs $10 monthly or $100 yearly. The golden member can return these amounts in one or two auctions (if he submits 20 bids in any auction, he will save $10).

Login

A member of the LUP auction services who wants to make use of the system must first login to the system using his username and password. Once logged, the member may choose from one of the following possibilities:

- Start a new auction.
- Browse the current auction.
- Consult the history of other members, or
- Deposit or withdraw money from his account.

Participating and Bidding in an Auction

To bid on an item, a participant simply has to enter the amount of the bid. A valid bid (with no cents) must satisfy the following requirements:

- The remaining amount in the bidder account that wants to place a bid price must be less than the number of all his bids plus the new bid price. This requirement ensures that a member is always in the position to pay for all items he placed bids on.
- The auction has not closed.

The bidding process in LUP auction is very simple as shown in Figure 2. Any bidder can participate in the LUP auction after logging in, using submit his bid, and clicking the command button which is called bid now. After that, the bidder will see the results of his bid and whether his bid price will be rejected or accepted (i.e., he will be the current or the final winner, if the bid is the last bid when the auction is being closed). In addition, any bidder can recharge his account at any time.

Closing an Auction

The time of closure of an auction depends on the number of bids and the acceptable price of the product. As shown, the auction will close when *the number of all bids* **plus** *the least and unique bid* price are greater than or equal to the acceptable price.

If the auction closes and ends successfully, then the participant having placed the least and unique bid wins the auction. The money is withdrawn from the accounts of winning participant and the other participants and deposited on the account of the LUP auction.

COMPARATIVE ANALYSIS

In order to have a better evaluation of the performance of the LUP auction a comparative analysis with five common auction types (English, Vick-

Figure 2. The interface of the bidding process in LUP auction

ery, Dutch, Limbo, and Unique) has been done. The following table shows that the comparative analysis has been done through 10 criterions. Although all these criterions are qualitative, any auction type would be affected (Positively or negatively) by them. Table 2 shows the whole comparative analysis.

The descriptions of the above criterions are listed below:

1. **Number of bids for each auction:** In each auction type there are number of bids that submitted by all bidders. In Dutch auction there is only one bid which is the first bids (N/A). In English, Vickery, and Limbo; the number of bids are unlimited (i.e., any bidder can submit any number of bids while the auction is opened). Whereas, in LUP and Unique auctions the number of bids are limited (because, the auction will close when the number of bids exceed the unit price).
2. **Revenue for sellers:** In the first three auction types mentioned above, the revenue for the sellers is medium because the closing of auction depends on the time of the auction (Meaning that, the desired or the reserve price may be not taken by the seller because the time of the auction is end).
3. **Price for buyers:** In English and Vickery auctions, the prices for buyers are medium because it uses ascending bids, then the winner who submits the highest or the second highest price. In Dutch auction the winner will be the first buyer who submits the first bid, so the buyer may not take the items with the lowest prices. While in the last three types –Limbo, Unique, and LUP—all winners will take the prices with the lowest prices (in many cases the winner can take the item with less than 10-20% of the market price).
4. **Avoiding of snipers:** All auction types try to resolve the snipers problem, however they can not resolved or terminate it. While in the LUP auction, snipers will not be found later on (because there is no person knows when the auction will close).
5. **Time of auction:** All auction types are limited time. In other words, each auction has beginning and ending time except the Unique and LUP auctions.
6. **Interactivity:** In any Web application, the interactivity is a very important issue. As you seen from the previous table, only the English, Unique, and LUP provide the interactivity with the bidders.
7. **Luck:** Each bidder needs to have skills to win any item in English auction. Whereas, bidders in Vickery and Dutch auctions need not to have many skills because the whole process of bidding is hidden. In Limbo, Unique, and LUP auctions combine skill

Table 2. Comparative analysis between LUP and common five auctions

Auction Type / Criteria	English	Vickery	Dutch	Limbo	Unique	LUP
No. of bids for each auction	Un-Limited	Un-Limited	N/A	Un-Limited	Limited	Limited
Revenue for Sellers	Mid	Mid	Mid	High	High	High
Price for Buyers	Mid	Mid	Mid	Low	Low	Low
Avoidance of sniping	Low	N/A	N/A	Mid	Low	High
Time of Auction	Limited	Limited	Limited	Limited	Limited or Dynamic	Dynamic
Interactivity	High	Low	N/A	Mid	High	High
Luck	Low	High	High	Mid	Mid	Mid
Open/Secret	Open	Secret	Open	Semi-Secret	Semi-Secret	Semi-Secret
Funnily (Attractively)	Mid	Mid	Low	High	High	High
Cost for each bid	Low	Mid	Mid	High	High	Mid

with luck, so no strategy can guarantee success.

8. **Open/secret:** Auction can be classified as: open (public) or secret (sealed). Auction will be open when every bidder can see another bidder who submits the bids and his price, meaning that all the bids placed are completely visible to everyone. While in secret auction every bidders does not know any thing about any other bidders or their prices. English and Dutch are open, whereas Vickery is secret. Limbo, Unique and LUP are semisecret because all bidders see part of the bidding process such as the name of the current winner as in the LUP auction.
9. **Funnily (attractively):** Although the online auction is the most attractive application in online applications, however, the Limbo, Unique, and LUP are more attractive and fun than other types. Because the bidding process is performed in a very funny way.
10. **Cost for each bid:** This is cost for each bid on the item. In English auction the bidder can submit any number of bids with no additional cost (i.e., for free), whereas bidders in the remaining auctions pay bidding fee for each time he place a bid. Bidders in LUP auction as mentioned above have to pay $1 for each bid, which is a medium cost when compared with Limbo and Unique auctions.

Even though the above criteria showed that the Unique and LUP auctions have many useful features, however, each bidder thinks that the auction which he uses is the best one.

CONCLUSION

The LUP is a descending price auction and requires bidders to submit bids as the price for the product being auctioned decreases. The LUP is also semisealed bid auction, because just the name of the current winner is shown, without showing his bid price. The LUP is often suited to situations where the number of bidders is relatively large and items are not expensive such as mobile phones. Moreover, as any descending bid auction, the LUP promotes very low price, mitigating the winner's curse, so that prices are lower than the true market values from buyer perspective.

The LUP auction will bring in bidders that previously did not want to spend time sitting in front of the PC during the bidding process. From this point of view, the proposed auction type is more attractive than other auction types. Moreover, the problem of sniping was resolved using the proposed auction.

In most types of online auction, the winner will not pay more than the market price of the product or the service. So, the revenue for the seller will be very low in many cases. In contrast, in the LUP auction the winner will take the product with very low price and the seller will determine the acceptable price and then the revenue for him will be very high.

FUTURE WORK

The future work will make the LUP auction as one of the real online auctions. Bidders of LUP auction can use their credit cards or the prepaid cards (issued by the LUP auction site with several values) or by transfer amount of money to the account bank of the LUP auction (from anywhere). That is, to make it worthwhile for larger community.

REFERENCES

Bapna, R., Goes, P., & Gupta, A. (2001). Comparative analysis of multi-item online auctions: Evidence from the laboratory. *Decision Support System*.

Barnes, S., & Vidgen, R. (2001). Assessing the quality of auction Web sites. In *Proceedings of the 34th Annual Hawaii International Conference on System Sciences* (Vol. 7).

Chaffey, D. (2004). *E-business and e-commerce management* (2nd ed.). Prentice Hall.

Franklin, M., & Reiter, M. (1996). The design and implementation of a secure auction service. *In IEEE Transactions on Software Engineering, 22*(5), 302-312.

Gao, S. (2005). Designing a limited supply online auction. In *Proceedings of The Best Conference Ever '05,* Vancouver.

Heck, E., & Vervest, P. (1998). How should CIOs deal with Web-based auctions?. *Communications of the ACM, 41*(7).

Kumar, M., & Feldman, S. (1998). Internet auctions. In *Proceedings of the 3rd USENIX Workshop on Electronic Commerce.*

Kikuchi, H., Harkavy, M., & Tygar, J. (1998). *Electronic auctions with private bids.* In *Proceedings of the 3rd USENIX Workshop on Electronic Commerce,* (pp. 61-73).

Klarreich, E. (2003). Sold to the latest bidder. *SIAM News, 36*(2).

Kauffman, R., & Wood, C. (2005). The effects of shilling on final bid prices in online auctions. *Journal of Electronic Commerce Research and Applications.*

Machlis, S. (1998). Online auction services thriving. *Computerworld, 32*(24).

Naor, M., Pinkas, B., & Sumner, R. (1999). Privacy preserving auctions and mechanism design. In *The 1st ACM Conference on Electronic Commerce.*

Orson, L., & Donald, J. (2004). Critical success factors for online auction Web sites. *Bulletin of applied computing and information technology* (Vol. 2).

Okamoto, T. (2002). Analysis and design of an online "discreet" Vikery auction protocol. *Technical Project.*

Omote, K., & Miyaji, A. (2001). A practical English auction with one-time registration. *In Proceedings of the Sixth Australiasian Conference on Information Security and Privacy,* (pp. 221-234).

Poel, D., & Leunis, J. (1999). Consumer acceptance of the Internet as a channel of distribution. *Journal of Business Research, 45.*

Parente, D., Venkataraman, R., Fizel, J., & Milett, I. (2001). B2B online reverse auctions: What's new?. *Decision Line.*

Reginaldo, R., Braga, R., & Masiero, P. (2001). A pattern language for online auctions management. In *Proceedings of the PLoP 2001 Conference.*

Richard, C., & Halstead, D. (2004). Characteristics and Internet marketing strategies of online auction sellers. *International Journal of Internet Marketing and Advertising, 1*(1).

Roth, A., & Ockenfels, A. (2001). Last-minute bidding and the rules for ending second-price auctions: Evidence from eBay and Amazon auctions on the Internet. *American Economic Review.*

Sako, K. (1999). Universal verifiable auction protocol which hides losing bids. In *Proceedings of the SCIS'99,* (pp. 35-39).

Stubblebine, S., & Syverson, P. (1999). Fair online auctions without special trusted parties. In *Proceedings of the Third Annual Conference on Financial Cryptography,* (pp. 230-240).

Trevathan, J. (2005). *Electronic auctions literature review* (JCU Tech. Rep.).

Turban, E., King, D., Lee, J., & Viehland, D. (2004). *Electronic commerce: A managerial perspective.* Pearson Education.

Vickrey, W. (1961). Counter speculation, auctions and sealed tenders. *Journal of Finance*, *16*, 8-37.

Viswanathan, K., Boyd, C., & Dawson, E. (2000). A three phased schema for sealed bid auction system design. In *ACISP*.

Wang, C., & Leung, H. (2004). Anonymity and security in continuous double auctions for Internet retails market. In *Proceedings of the 37th Hawaii International Conference on System Sciences*.

Woolley, S. (1998). E-muscle. *Forbes*, *161*(5).

Yuen, W., Sung, W., & Wong, W. (2003). Optimal price decremental strategy for Dutch auctions.

APPENDIX-A1

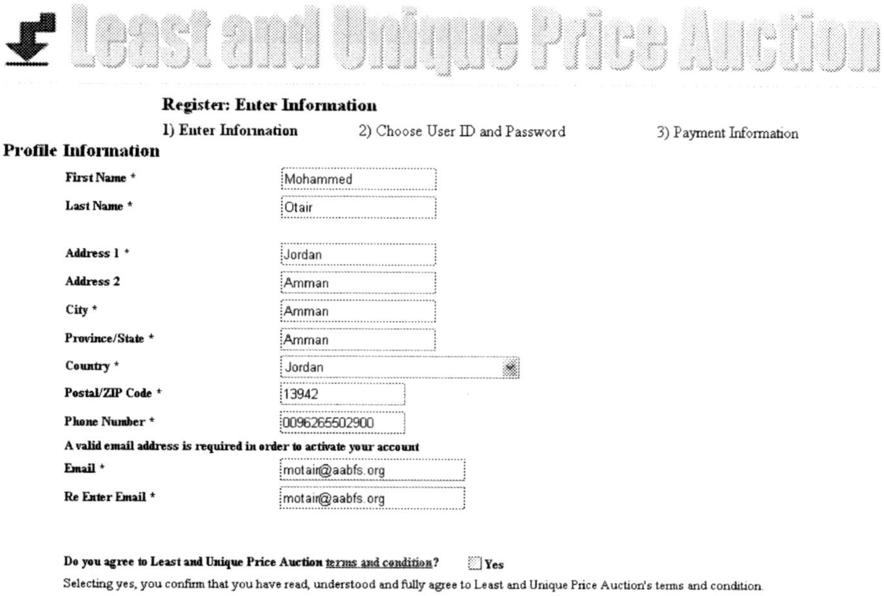

Least and Unique Price Auction

Register: Enter Information

1) Enter Information 2) Choose User ID and Password 3) Payment Information

Profile Information

First Name *	Mohammed
Last Name *	Otair
Address 1 *	Jordan
Address 2	Amman
City *	Amman
Province/State *	Amman
Country *	Jordan
Postal/ZIP Code *	13942
Phone Number *	0096265502900

A valid email address is required in order to activate your account

Email *	motair@aabfs.org
Re Enter Email *	motair@aabfs.org

Do you agree to Least and Unique Price Auction terms and condition? ☐ Yes

Selecting yes, you confirm that you have read, understood and fully agree to Least and Unique Price Auction's terms and condition.

APPENDIX-A2

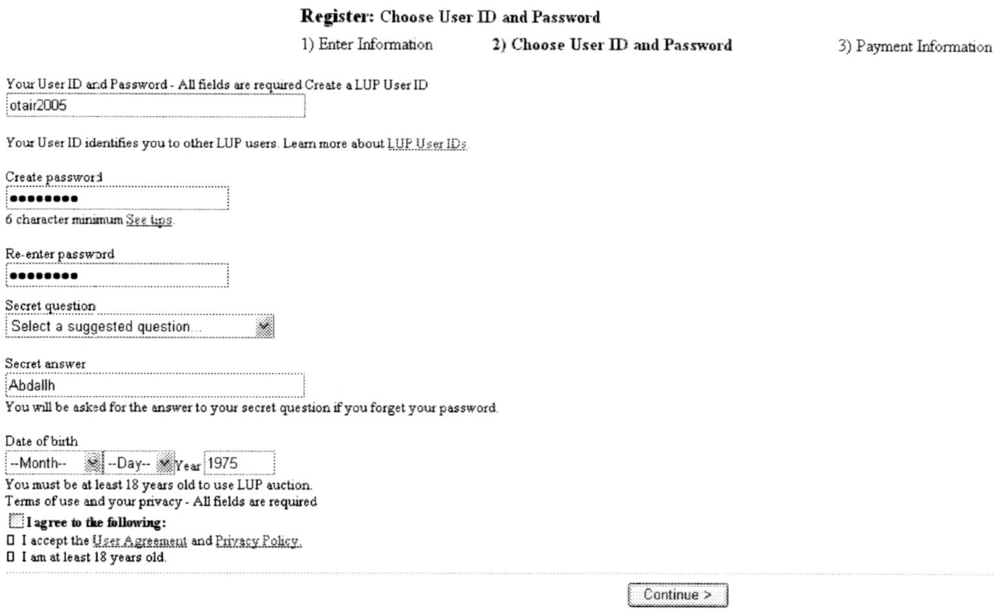

Register: Choose User ID and Password

1) Enter Information **2) Choose User ID and Password** 3) Payment Information

Your User ID and Password - All fields are required Create a LUP User ID

otair2005

Your User ID identifies you to other LUP users. Learn more about LUP User IDs.

Create password

••••••••

6 character minimum See tips.

Re-enter password

••••••••

Secret question

Select a suggested question...

Secret answer

Abdallh

You will be asked for the answer to your secret question if you forget your password.

Date of birth

--Month-- --Day-- Year 1975

You must be at least 18 years old to use LUP auction.

Terms of use and your privacy - All fields are required

☐ **I agree to the following:**

- I accept the User Agreement and Privacy Policy.
- I am at least 18 years old.

Continue >

APPENDIX-A3

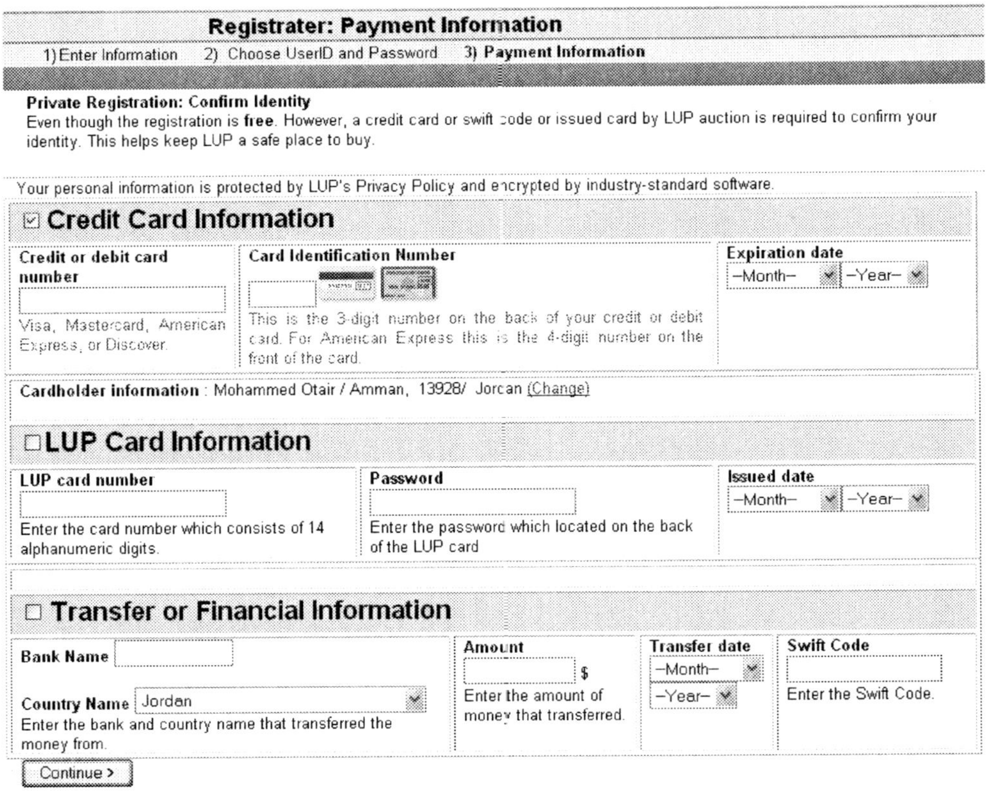

This work was previously published in the International Journal of E-Business Research, Vol. 4, Issue 2, edited by I. Lee, pp. 88-102, copyright 2008 by IGI Publishing (an imprint of IGI Global).

Chapter 10
Nibbling, Sniping, and the Role of Uncertainty in Second-Price, Hard-Close Internet Auctions:
Empirical Evidence from eBay

Daniel Friesner
Gonzaga University, USA

Carl S. Bozman
Gonzaga University, USA

Matthew Q. McPherson
Gonzaga University, USA

ABSTRACT

Internet auctions have gained widespread appeal as an efficient and effective means of buying and selling goods and services. This study examines buyer behavior on eBay, one of the most well-known Internet auction Web sites. eBay's auction format is similar to that of a second-price, hard-close auction, which gives a rational participant an incentive to submit a bid that is equal to his or her maximum willingness to pay. But while traditional second-price, hard-close auctions assume that participants have reliable information about their own and other bidders' reservation prices, eBay participants usually do not. This raises the possibility that eBay participants may adapt their bidding strategies and not actually bid their reservation prices because of increased uncertainty. In this article, we empirically examine the bidding patterns of online auction participants and compare our findings to the behavior of bidders in more conventional auction settings.

INTRODUCTION

Over the past decade, Internet auctions have gained widespread appeal as an efficient and effective means of buying and selling goods and services (Stafford & Stern, 2002). These auctions provide a market that brings together a large number of participants, a large selection of goods and services to be exchanged, and a more flexible time frame within which to conduct transactions. Internet auctions have, as a consequence, become a multibillion dollar industry where a broad range of products, from raw materials to used consumer goods, are regularly bought and sold (Anonymous, 2004; Baatz, 1999).

Conducting an auction using an electronic medium necessitates that the rules for participation differ somewhat from more traditional auction formats. For example, auctions on eBay have a specific time frame within which participants are able to bid, and the value of the highest bid is displayed at any given point in time. These characteristics, in conjunction with other Internet auction attributes, either individually or in combination, allow bidders to employ a series of unique strategies in an attempt to gain an advantage over rivals. Sniping and nibbling are two such commonly-employed strategies.

Sniping occurs when a bidder with a very high reservation value waits until the final moments of a hard-close auction to submit a bid. By waiting until the last moment to submit a bid, this individual may win the auction and do so at a price that is significantly below his or her reservation value by tendering a bid that is only marginally higher than the existing high bid. Nibbling is the strategy employed when a bidder is unsure about the value of the good or service being auctioned and uses an incremental process to approximately deduce the value of the good or service. In addition, nibbling may be used to determine the maximum willingness to pay of the current high bidder.

Nibblers bid one increment above the highest current bid in the auction, and then will wait to see if someone else outbids them. If they are not outbid, then they win the auction at a price very close to what at least one other person was willing to pay. If they are outbid, they may automatically place a new bid that is, again, one increment above the current high bid. Nibblers repeat this process until either they are certain that they have met their reservation value or the auction is complete. Nibblers are more generally classified within the context of a larger group of auction participants known as incremental bidders, which include all participants who place multiple, responsive bids over the course of the auction.

Using the online auction environment as a setting to develop, test, and extend knowledge regarding electronic commerce and its influence on consumer behavior is an important research effort (Dholakia, 2005a,b; Kauffman & Walden, 2001). The characteristics of Internet auctions have implications for both business managers and policy-makers, because bidder strategies can be interpreted as either unethical or reducing the efficiency of an auction's outcome (Gardner, 2003; Marcoux, 2003). As such, it is of interest to empirically characterize the impacts of these strategies on auction outcomes, particularly for those auctions with high public visibility or those used frequently by the public.

LITERATURE REVIEW

Recent auction literature has focused on the impact of auction ending rules and sniping on the efficiency of Internet auctions. For example, Roth and Ockenfels (2002) examined eBay and Amazon auctions for both antiques and computers and find that the format for ending the auction has a significant impact on both the amount of sniping and the auctions' subsequent outcomes. Ockenfels and Roth (2002) also examined the impact that artificial bidding agents have on the amount of sniping in eBay auctions. Bajari and Hortacsu (2003) collected a sample of data on eBay coin auctions and estimated how various characteristics of bidder behavior, such as sniping, impact the efficiency of auction outcomes.

Others have examined the role that experience, rationality, and risk tolerance play in bidding behavior. Wilcox (2000) found that individuals with higher levels of experience in online auctions were more likely to employ strategies consistent with those predicted by traditional auction theory. However, he also found that some experienced players continued to employ strategies inconsistent with the theory.

Kamins, Dreze, and Folkes (2004) found that final (winning) auction bids were significantly different depending on whether the auction imposed a high reference price (e.g., regular price, suggested retail price, etc.) or a minimum bid constraint. This implies that perceived valuations about the product being auctioned (which is a function of experience and risk tolerance, among other factors) impact a bidder's optimal strategy. In addition, Ariely and Simonson (2003) found that high starting prices may influence a bidder's value judgment about good, which in turn may influence the final winning bid.

Similarly, McDonald and Slawson (2002) found that auctions with sellers who had a strong reputation (whether positive or negative) induced bidders to behave differently than in similar markets where the seller was anonymous. Their conclusion was that bidders base their expected product valuation and subsequent bidding strategies, in part, on their perceptions of the sellers' reliability. Dholakia and Simonson (2005) find that when sellers encourage bidders to compare prices, the winners in auctions with explicit reference points tended to bid later, submit fewer bids, snipe, and avoid multiple simultaneous auctions.

As indicated by Dholakia (2005), the most important type of research in online auctions is theory deepening, using the online environment as the setting or context to develop, elaborate on, and test general marketing and consumer behavior theory. Our contribution to the growing Internet auction literature is to empirically examine the relationship between the uncertainty in an auction and the incentive of participants to nibble and snipe. To do so, we randomly select auctions conducted on eBay across two types of goods: one that exhibits a significant degree of uncertainty about the product's value (used cars), and one where there is significantly less uncertainty about the product's value (certified coins), and look for mean differences in the amount and intensity of nibbling and sniping across each type of auction. In addition, we compare the strategies used for each product group to optimize behavior in traditional second-price, hard-close auctions.

HYPOTHESIS DEVELOPMENT

The formats by which Internet auctions are conducted vary almost as much as the number of products being purchased and sold. For example, auctions may be classified as "first price" when the winning bidder submits the highest bid (or the lowest if bidding to provide a good or service) and pays a price equal to that bid, or "second price" when the winning bidder submits the highest bid, but pays the second-highest price for the product or service. Concomitantly, auctions may be classified as "private value" when each individual has her own independent valuation for the product or service being auctioned, or as a "common value" auction when bidders' valuations are interdependent or the value of the product/service being auctioned is unknown. Dholakia and Soltysinski (2001) posit a more general classification, known as an "affiliated value" auction, which encompasses the common- and private-value auctions as special cases. This auction type allows for varying degrees of correlation among the multiple bidders' valuations. That is, there may be some degree of benefit from observing others' behavior, but not as much as in the common-value case.

Auctions may also be classified depending on the rules for ending the auction. A "hard-close" auction is one that is concluded at a specific date and time, while an "automatic extension" auction ends when there is a predefined length of time between consecutive bids. Clearly, these definitions are neither mutually exclusive nor collectively exhaustive. As a result, auctioneers have the ability to "mix and match" various auction formats, possibly tailoring the auction to the nature of the good being sold. EBay, for example, uses a second-price, hard-close auction format, while Amazon.com (eBay's major competitor) uses a second-price, automatic-extension format (Bajari & Hortacsu, 2003; Ockenfels & Roth, 2002; Roth & Ockenfels, 2002; Wilcox, 2000).

Conventional auction theory (and its suggestions for optimal bidding strategies) is based on a specific set of assumptions governing the amount of information available to participants, both about the rules governing the auction as well as the motivations and value judgments of the other bidders. Thus, as the amount of information or rules governing the auction change, so do the optimal bidding strategies. Moreover, as the optimal strategies change, so do the relative efficiencies of the auctions being compared. For example, under a first-price, private-value, automatic-extension auction with perfect information, rational bidders have an incentive to submit a bid that is less than their maximum willingness to pay (also known as their "reservation price"). However, if this same auction were conducted in a second-price format (holding all other auction features constant), a rational bidder has an incentive to bid her reservation price. Because a rational bidder in the aforementioned second-price auction submits a bid that is closer to her willingness to pay than in the first-price auction, the second-price auction is said to be more "efficient" than its first-price counterpart (Gardner, 2003).

The rules governing each type of auction force participants to face different types of information uncertainties, and (as in the case of any risky consumption activity) bidders utilize different strategies to reduce the impact of uncertainty (Bauer, 1960; Celsi, Rose, & Leigh, 1993; Puto, Patton, & King, 1985). In eBay auctions, a bidder has perfect information about when the auction will close. But the hard-close auction also allows players (particularly with high reservation values) to mask their reservation values from the other participants by sniping (Roth & Ockenfels, 2002). Conversely, participants may attempt to deduce other bidders' reservation values through nibbling (or other incremental bidding) strategies (Roth & Ockenfels, 2002; Wilcox 2000).

The former analyses examined differences in bidding behavior across individuals in different auctions for the same good being auctioned. A related line of analysis compares differences in bidding strategies within the same general auction format for different goods (each with potentially widely-divergent values). Gilkeson and Reynolds (2003), for example, examined eBay auctions for flatware and found that an auction's opening price (relative to the perceived value of the good in question) has a significant impact on the auction's outcomes. Brint (2003) examined the relationship between the amount of available price information and bidding behavior. Using eBay auctions for three categories of goods, one with detailed, readily-available price guides (UK gold coins), one with partial price information (Wisden cricket books), and one with no published guides (Esso Football tokens), Brint (2003) concluded that setting a moderately-high starting price benefits a seller, especially for items with no real price guide. In addition, Brint found that bidders who delay their bids as late as possible can significantly improve their chance of winning the auction.

In this study, we combine ideas from both strands of the literature. From the former, there is evidence that nibbling is a bidding strategy that is most likely to occur in an auction format where there is a greater degree of uncertainty about the value of the good being auctioned. The latter set of studies argues that, even within the same auction format, differences in the goods being auctioned (each with its unique level of uncertain value) will induce bidders to behave differently.

EBay incorporates auction rules that are consistent with the second-price, hard-close format (Bajari & Hortacsu, 2003; Heyman, Orhun, & Ariely, 2004; Wilcox, 2000). The observed behavior of eBay bidders', however, is less than rational and inconsistent with the predictions of traditional auction theory (Standifird, Roelofs, & Durham, 2004; Ward & Clark, 2002). For example, bidder valuations have been found to be dependent on the absolute amount and number of bids from other bidders as well as the information supplied by sellers. Herd behavior has occurred in eBay auctions, where bidders gravitate toward items with more bids and ignore auctions of equivalent items or items of equal or superior value (Dholakia, Basuroy, & Soltysinski, 2002; Dholakia, & Soltysinski, 2001). This type of behavior is accentuated in circumstances of greater uncertainty. That is, herd behavior is greater whenever buyers or sellers are less experienced and auctions are varied in the quality of information available.

This finding is consistent with other studies which examine the role experience, rationality, and risk tolerance play in bidding behavior. Wilcox (2000) found that individuals with higher levels of experience in online auctions were more likely to employ strategies similar to those predicted by traditional auction theory. However, he also found that some experienced bidders continued to employ strategies inconsistent with efficient auction outcomes. Auction sellers that had a strong reputation, whether positive or negative, also have been shown to induce discrepant bidder behavior (Dholakia, 2005; McDonald & Slawson, 2002). Auction participants base their product valuations and subsequent bidding strategies, in part, on their perceptions of sellers' reliability.

Inefficient auction outcomes are expected in circumstances where bidder perceptions diverge and various risk reduction strategies are employed (Sandholm, 2000). In eBay auctions, a bidder only has perfect information about when the auction will close. In this instance, the hard-close auction format provides a bidder with the incentive, particularly anyone with a high reservation value, to mask his or her own willingness to pay from other bidders. Not surprisingly, many auction participants attempt to deduce each other's reservation values through incremental bidding strategies like nibbling (Roth & Ockenfels, 2002; Wilcox, 2000).

Nibbling, or incremental bidding in general, is essentially an information gathering technique (Marcoux, 2003; Ockenfels & Roth, 2002). Auction participants use nibbling when: (1) they are unsure of the value of the object being auctioned; (2) they are unsure about how their willingness to pay compares to the other participants'; or (3) some combination of (1) and (2). Thus, it stands to reason that, in auctions where there is more uncertainty about the good being auctioned or where the number of participants is high, there should be a higher occurrence of nibbling. The value of nibbling as a means of reducing uncertainty should also be greater for more expensive items (Oh, 2002). More specifically, information search behavior is expected to be greater among bidders participating in an Internet auction for a more expensive product. The first two hypotheses examine the information search behavior proposed in the preceding nibbling discussion.

Hypothesis 1: *The amount of nibbling will be greater for goods of less certain value.*

Hypothesis 2: *The intensity of nibbling will be greater for goods of less certain value.*

An association between nibbling and sniping is also expected within the same general auction format. As Roth and Ockenfels (2002) note, participants are more likely to snipe when the incentive to nibble is reduced. Specifically, we expect a negative correlation between sniping and nibbling (both in terms of the amount and intensity of sniping and nibbling) in an auction for a product with a more certain value. Because bidders have a substantial amount of information about the value of the good being auctioned, there is less incentive to nibble. Concomitantly, the incentive to snipe may be increased, especially if the current maximum bid is significantly below its market value, because bidders may be able to win the auction at a price significantly below the good's market value by not revealing their valuations to other bidders.

Reference prices supplied by sellers reduce uncertainty and have been shown to influence bidders' perceived value judgments about a good as well as their bidding strategy (Ariely & Simonson, 2003; Dholakia & Simonson, 2005; Gilkeson & Reynolds, 2003). These findings are reliable whether or not the seller proposed a high reference price (e.g., regular price, suggested retail price, etc.) or a minimum bid constraint (Kamins, Dreze & Folkes, 2004). This result is also consistent with Brint (2003), who discusses the benefits of waiting as long as possible to place bids for goods with readily-available pricing guides.

We expect, as a consequence, a positive association between nibbling and sniping in auctions where the value of the product being auctioned is less certain. In these auctions, we anticipate a greater frequency (and intensity) of nibbling, as bidders have a strong incentive to deduce the value of the item being auctioned (either their own valuation, or the valuation of the bidder with the highest willingness to pay). At the same time, because the product's value is uncertain, bidders are likely to have a wide range of initial product valuations. In this situation, bidders with high initial valuations are likely to attempt to conceal their reservation price (and thereby win the auction at a price which is less than their reservation value) by sniping. Thus, in an auction whose product's value is less certain, we would expect sniping and nibbling to occur in tandem, thereby creating a positive correlation between these variables. Hypothesis 3 and 4 examine whether perceptions of value lead to discrepant nibbling and sniping behavior within second-price, hard-close auctions.

Hypothesis 3: *The amount (and intensity) of nibbling is inversely related to the amount (and intensity) of sniping for goods of more certain value.*

Hypothesis 4: *The amount (and intensity) of nibbling is positively related to the amount (and intensity) of sniping for goods of less certain value.*

DATA

Our dataset consists of an interval random sample taken from completed eBay auctions, where the unit of analysis is a single auction. This is chosen because, in order to determine the presence and intensity of sniping and nibbling, it was necessary to construct counts of nibbling and snipers within each auction, making the auction itself the smallest possible unit of analysis. We collected information on two distinct types of product categories. The first set of auctions contains information on professionally-graded, U.S. coins (Numismatic Guaranty Corporation, Numistrust Corporation, or the Professional Coin Grading Service). There are numerous pricing guides for graded coins, with the most-commonly-used dealer reference being the Coin Dealer Newsletter, or Greysheet. The Greysheet is published weekly in abbreviated form. The monthly issue includes dealer bid and ask prices for every type of U.S. coin, and separate prices for each major grading service, based on the standards of that particular service. In the absence of overt fraud, the readily-available market value of this type of good leaves little doubt as to its true valuation.

In contrast, the second set of auctions contains information on used automobiles. Although there are numerous pricing guides for used automobiles, the lack of a third-party grading service adds a great deal of uncertainty to the true value of the good.

METHOD

For each auction, information was collected on a number of relevant variables. The number of bids per auction and the number of bidders per auction were collected as a baseline measure of activity. To measure sniping activity, we collected the number of bids in the last minute of an auction (Bajari & Hortacsu, 2003). We also measured sniping intensity by calculating the portion of bids that were placed in the last minute of an auction.

Following Marcoux (2003), we defined nibbling as incremental bidding, usually one bid increment above the current price, which continues until the nibbler's last bid exceeds the reservation price of the top bidder. Automatic bidding agents in eBay result in many multiple bids being categorized as nibbles. This outcome was deemed inconsequential since bidders were still provided with additional information regardless of the absolute value of their incremental bid.

Given this definition, we were able to calculate several empirical measures of nibbling, including the number of nibblers in an auction and the number of times within an auction that an individual practiced nibbling. To measure the intensity of nibbling, we calculated the proportion of bids that were nibbling bids as well as proportion of bidders in an auction that nibbled.

While the data available from eBay provides a number of useful variables for our analysis, it is not an exhaustive source of information. Unfortunately, variables not provided by eBay are needed to perform more sophisticated analyses (for example, actual shipping costs which influence the true price of the item paid by the winning bidder), such as maximum likelihood regression techniques. Failure to include these omitted variables would result in biased and inconsistent estimates. As a result, we adopt a more parsimonious approach of using analysis of variance and correlation analysis techniques. An additional concern is that several of our variables are likely to be non-normally distributed, which precludes the use of parametric hypothesis tests. Instead, we test for mean (and distributional) differences in the number and intensity of nibbling across auction groups using a nonparametric test (the Mann-Whitney U-Test). We measure the association between the number and intensity of nibbling and the number and intensity of sniping by calculating correlation coefficients between our proxies for sniping behavior and our nibbling variables. Consistent with our prior discussion, we calculate these correlations in nonparametric (Spearman) fashion. We also conduct hypothesis tests to determine whether each correlation coefficient is significantly positive or negative.

RESULTS

Table 1a contains descriptive statistics for coin auction data. On average, each auction consisted of approximately five bidders, who submitted approximately seven bids. In addition, each auction contained, at the mean, slightly more than four nibblers, who nibble 4.7 times per auction. The proportion of bidders who nibbled is 0.685, and the proportion of bids that are nibbles is 0.567. During the last minute of the auction, the average number of bidders is 0.8, the average number of bids is 0.975, the proportion of all bids is 0.226, and the proportion of bidders is 0.229.

Table 1a also provides some information about the distribution of responses. A comparison of mean and median values indicates that several of the variables, including the number of bids, the number of bidders, and the number of nibbles and nibblers are likely normally distributed, since the mean and median values for each variable are similar in magnitude. However, the mean and median values for our proportional variables (bidders who nibble, bids that are nibbles, and bids

Table 1a. Descriptive statistics Panel A: Coin auction data[a)]

Variable	Mean	Std. Dev.	First Quartile	Median	Third Quartile
Number of Bids	7.025	4.605	3.250	6.500	10
Number of Bidders	5.000	3.258	2	4.500	7
Number of Nibblers	4.050	3.178	1.250	3	6
Number of Nibbles	4.700	3.716	1.250	4	7
Number of Bids per Bidder	1.475	0.776	1	1.225	1.607
Proportion of Bidders who Nibble	0.685	0.301	0.525	0.800	0.875
Proportion of Bids that are Nibbles	0.567	0.281	0.400	0.667	0.750
Number of Bids in Last Minute	0.975	1.097	0	1	2
Number of Bidders in Last Minute	0.800	0.911	0	1	1
Proportion of Bids in Last Minute	0.226	0.306	0	0.083	0.458
Proportion of Bidders in Last Minute	0.229	0.298	0	0.134	0.383

[a]$\underline{n} = 49$

Table 1b. Descriptive statistics Panel B: Automobile auction data

Variable	Mean	Std. Dev.	First Quartile	Median	Third Quartile
Number of Bids	25.915	13.336	18	26	35
Number of Bidders	12.085	6.463	6	13	16
Number of Nibblers	10.787	6.196	5	12	15
Number of Nibbles	16.213	9.484	8	17	23
Number of Bids per Bidder	2.253	0.900	1.688	2.059	2.667
Proportion of Bidders who Nibble	0.849	0.184	0.750	0.880	1
Proportion of Bids that are Nibbles	0.599	0.176	0.500	0.633	0.724
Number of Bids in Last Minute	0.660	0.939	0	0	1
Number of Bidders in Last Minute	0.596	0.798	0	0	1
Proportion of Bids in Last Minute	0.025	0.043	0	0	0.041
Proportion of Bidders in Last Minute	0.042	0.060	0	0	0.077

[b]$\underline{n} = 49$

and bidders in the last minute of the auction) are somewhat different, implying that non-normality may be an issue.

Table 1b collects the descriptive statistics for car auctions. At the mean, each auction contained 12 bidders placing a total of 26 bids. In addition each auction contained nearly 11 nibblers, who nibbled 16 times per auction. The proportion of bidders who nibble is 0.849, and the proportion of bids that are nibbles is 0.599. During the last minute of the auction, the number of bidders is 0.596, the number of bids is 0.660, the proportion of bids is 0.025, and the proportion of bidders is 0.042.

The empirical distributions for each of our variables in Table 1b are analogous to those for the coin auctions summarized in Table 1a. Comparisons of the mean and median values indicate

Table 2. Nonparametric (Mann-Whitney u-test) hypothesis tests

Variable	M-W Test	Z-Stat	Prob.
Number of Bids	192.500	-6.372	0.000
Number of Bidders	332.000	-5.189	0.000
Number of Nibblers	344.500	-5.083	0.000
Number of Nibbles	275.000	-5.671	0.000
Number of Bids per Bidder	345.000	-5.085	0.000
Proportion of Bidders who Nibble	578.500	-3.090	0.002
Proportion of Bids that are Nibbles	909.000	-0.265	0.791
Number of Bids in Last Minute	796.000	-1.342	0.180
Number of Bidders in Last Minute	827.500	-1.055	0.292
Proportion of Bids in Last Minute	647.000	-2.704	0.007
Proportion of Bidders in Last Minute	651.000	-2.667	0.008

that a number of variables are likely to be non-normally distributed. As a result, it is necessary to employ nonparametric analysis to control for this possibility.

Comparing Tables 1a and 1b, we note differences in nibbling activity across the two types of auctions. When evaluated at the sample mean, the automobile auctions have a higher number of bids and bidders, nibbles and nibblers, and a higher proportion of bids that are nibbles and bidders that nibble. With the exception of bids that are nibbles, these differences also hold true at the sample median as well. These statistics lend support to hypothesis H_1, as the automobile auctions (which exhibit more uncertainty) have both higher numbers and intensities of nibbling. However, it remains to be seen from the subsequent hypothesis tests whether these differences are, in fact, statistically significant.

We can also make inferences about sniping activity across the two auction formats. The automobile auctions exhibit lower mean and median values for both our sniping activity and intensity variables. The latter implies that, when we control for size of the auction, there is less sniping and more nibbling in the more uncertain market. One possible explanation is that these auctions also exhibit more (and more intense) nibbling, which may lead to higher bid prices, and drive some of the potential snipers (who may not have extremely-high reservation values) out of the market before they have a chance to snipe.

Our hypothesis tests are contained in Tables 2 through 4. Table 2 contains the Mann-Whitney test used to address hypotheses 1 and 2, while Tables 3 and 4 contain the correlation coefficient analysis used to address hypotheses 3 and 4.

The results in Table 2 provide clear evidence that the graded coin and automobile auction data show different nibbling behavior for all of our measures ($p < 0.01$), except the proportion of bids that are nibbles. Specifically, the automobile auction data contained significantly more nibblers, nibbles, and portion of bidders who nibble. The automobile auctions also exhibited significantly more participation, as evidenced by the significantly higher number of bidders and bids per auction.

As a corollary to these findings, Table 2 also provides evidence about differences in sniping activity across the two product categories. With regard to the total amount of sniping, the samples are not statistically significantly different, but are different in the proportion of bids and bidders in the last minute ($p < 0.01$). This supports our earlier argument that the amount and intensity of sniping differs within the same auction format. The auctions whose product has a more certain value (certified coins) exhibited more sniping behavior.

Tables 3 and 4 present our analysis of hypotheses 3 and 4. Analysis of the coin auction data in Table 3 indicate that there is no significant correlation ($p < 0.05$) between the amount of sniping and the amount or intensity of nibbling in the coin data. However, there is a significant correlation between the intensity of sniping behavior and both the amount and intensity of nibbling. Specifically, the intensity of sniping behavior (when measured as both the proportion of bids in the last minute of the auction as well as the proportion of bidders who snipe) is negatively correlated ($p < 0.05$) with the number of nibbles, the number of nibblers, and the portion of bids that are nibbles.

In direct contrast, the results for the automobile auctions demonstrate a significant positive correlation between the amount of sniping and the amount of nibbling, as well as the intensity of sniping and the intensity of nibbling. Specifically, the amount of sniping, as measured by the

Table 3. Nonparametric (Spearman) correlations for graded coin auctions

	Number of Bids in Last Minute		Number of Bidders in Last Minute	
	Spearman		Spearman	
Variable	**Correlation**	**Prob.**	**Correlation**	**Prob.**
Number of Bids	0.072	0.330	0.085	0.301
Number of Bidders	0.139	0.197	0.168	0.151
Number of Nibblers	0.078	0.317	0.109	0.253
Number of Nibbles	0.101	0.269	0.121	0.229
Number of Bids per Bidder	-0.061	0.354	-0.151	0.176
Proportion of Bidders who Nibble	-0.197	0.111	-0.180	0.133
Proportion of Bids that are Nibbles	0.149	0.180	0.177	0.138
	Proportion of Bids in Last Minute		**Proportion of Bidders in Last Minute**	
	Spearman		**Spearman**	
Variable	**Correlation**	**Prob.**	**Correlation**	**Prob.**
Number of Bids	-0.299	0.031	-0.286	0.037
Number of Bidders	-0.228	0.079	-0.261	0.052
Number of Nibblers	-0.281	0.040	-0.314	0.024
Number of Nibbles	-0.271	0.045	-0.284	0.038
Number of Bids per Bidder	-0.250	0.060	-0.225	0.082
Proportion of Bidders who Nibble	-0.468	0.001	-0.491	0.001
Proportion of Bids that are Nibbles	-0.168	0.151	-0.204	0.103

Note: Probability values apply to a one-tailed test

number of sniping bids and the number of snipers, is significantly and positively correlated ($p < 0.05$) with the number of nibbles, the number of nibblers, and the proportion of bids that are nibbles. The intensity of sniping, as measured by the proportion of bids in the last minute and the proportion of bidders who snipe in the last minute, is significantly and positively associated with the number of nibbles, the number of nibblers, and the proportion of bids that are nibbles.

When the results contained in Tables 3 and 4 are taken together, our results extend and clarify Ockenfels and Roth's (2002) assertion that participants resort to sniping more frequently in circumstances of higher uncertainty in an effort to avoid the bidding wars that may occur when several participants nibble in that same auction. In auctions with less uncertainty regarding the value of the good which is being auctioned, the number of nibblers and the intensity of their nibbling have a negative impact on sniping. This is consistent with the idea that when the true value of the good has little uncertainty, it is optimal to conceal your reservation price.

Table 4. Nonparametric (Spearman) correlations for automobile auction

	Number of Bids in Last Minute		**Number of Bidders in Last Minute**	
	Spearman		**Spearman**	
Variable	**Correlation**	**Prob.**	**Correlation**	**Prob.**
Number of Bids	0.486	0.001	0.481	0.001
Number of Bidders	0.528	0.000	0.536	0.000
Number of Nibblers	0.509	0.000	0.507	0.000
Number of Nibbles	0.568	0.000	0.574	0.000
Number of Bids per Bidder	-0.156	0.148	-0.167	0.132
Proportion of Bidders who Nibble	0.218	0.070	0.183	0.109
Proportion of Bids that are Nibbles	0.438	0.001	0.457	0.001
	Proportion of Bids in Last Minute		**Proportion of Bidders in Last Minute**	
	Spearman		**Spearman**	
Variable	**Correlation**	**Prob.**	**Correlation**	**Prob.**
Number of Bids	0.336	0.011	0.357	0.007
Number of Bidders	0.420	0.002	0.374	0.005
Number of Nibblers	0.410	0.002	0.358	0.007
Number of Nibbles	0.437	0.001	0.442	0.001
Number of Bids per Bidder	-0.220	0.069	-0.134	0.184
Proportion of Bidders who Nibble	0.178	0.116	0.159	0.143
Proportion of Bids that are Nibbles	0.407	0.003	0.365	0.006

Note: Probability values apply to a one-tail test

CONCLUSION

We randomly selected items from completed auctions on eBay for two types of goods: one that exhibits a significant degree of uncertainty about the product's value (used cars); and one where there is significantly less uncertainty about the product's value (certified coins). Using nonparametric (Mann-Whitney) analysis of variance and Spearman correlation analysis, we test hypotheses based on bidder behavior in the same auction format, but across different product groups. Our results indicate that auctions with a high amount of uncertain value exhibit significantly more nibbling. Specifically, more nibbling occurred both in terms of the number (and proportion) of participants who use a nibbling strategy, as well as the number (and proportion) of incremental bids that are submitted over the course of the auction. Auction participants were also more prone to engage in sniping in auctions with more uncertain value. Moreover, sniping occurred most often in high-risk auctions when several other participants attempted nibbling strategies. This is consistent with the findings of Ockenfels and Roth (2002), which suggest that many bidders use sniping to avoid getting into bidding wars. For lower-risk auctions, where the value of the good is known with more certainty, we find that auction participants are less prone to nibble, and thereby withhold information concerning their reservation price, engaging in sniping behavior.

It is also interesting to note that, even though participants did not have reliable information about other participants' reservation prices, coin auction participants' behavior deviated only slightly from that predicted in conventional second-bid, hard-close auctions. Our analysis indicates that there may be little benefit from observing others' bidding behavior in this circumstance. However, in car auctions, where there was a large amount of uncertainty concerning the good being purchased, auction participants' behavior deviated significantly from that predicted for traditional second-bid, hard-close auctions. Our results are consistent with the idea that bidders engage in nibbling and sniping to overcome a lack of information concerning the valuation of the product. In this case, the lack of reliable knowledge concerning the reservation price of other participants may have induced alternative bidding strategies (nibbling and sniping) in an attempt to overcome information deficiencies.

While our findings provide an interesting clarification of Internet auction theory, our study is preliminary in nature and could be extended in a number of ways. First, the spectrum of goods included in the study could be expanded in order to give a clearer picture of how uncertainty influences online bidding behavior. In addition, it would be interesting to study the ramifications of other online auction rules, such as those employed by Amazon.com, on the behavior of auction participants.

Another potential limitation of our study is the use of Spearman correlations. While correlations are a parsimonious technique, an alternative might be the use of maximum likelihood-based regression techniques. In this case, many other factors would be needed to control for omitted variables. If important control variables are not included, our results would be biased and inconsistent. Most of the control variables needed to perform this type of analysis are not available through e-Bay or other online auction providers. We leave these topics for future research.

REFERENCES

Anonymous (2004). At the drop of a hammer. *Economist, 371*(8375), 10.

Ariely, D., & Simonson, I. (2003). Buying, bidding, playing, or competing? Value assessment and decision dynamics in online auctions. *Journal of Consumer Psychology, 13*(1&2), 113-123.

Baatz, E. (1999). Online auctions start to pick up steam. *Purchasing, 127*(6), 46-56.

Bajari, P., & Hortacsu, A. (2003). The winner's curse, reserve prices, and endogenous entry: Empirical insights from eBay auctions. *Rand Journal of Economics, 34*(2), 329-355.

Bauer, R. (1960). Consumer behavior and risk taking. In R. Hancock (Ed.), *Proceedings of the American Marketing Association* (pp. 389-398).

Brint, A. T. (2003). Investigating buyer and seller strategies in online auctions. *Journal of the Operational Research Society, 54*(11), 1177-1197.

Celsi, R., Rose, R., & Leigh, T. (1993). An exploration of high-risk leisure consumption through skydiving. *Journal of Consumer Research, 20*(1), 1-23.

Dholakia, U. (2005). Concept discovery, process explanation, and theory deepening in e-marketing research: The case of online auctions. *Marketing Theory, 5*(1), 117-124.

Dholakia, U. (2005). The usefulness of bidders' reputation ratings to sellers in online auctions. *Journal of Interactive Marketing, 19*(1), 31-40.

Dholakia, U., Basuroy, S., & Soltysinski, K. (2002). Auction or agent (or both)? A study of moderators of the herding bias in digital auctions. *International Journal of Research in Marketing, 19*(2), 115-130.

Dholakia, U., & Simonson, I. (2005). The effect of explicit reference points on consumer choice and online bidding behavior. *Marketing Science, 24*(2), 206-217.

Dholakia, U., & Soltysinski, K. (2001). Coveted or overlooked? The psychology of bidding for comparable listings in digital auctions. *Marketing Letters, 12*(3), 225-237.

Gardner, R. (2003). *Games for business and economics, 2nd ed.* Hoboken, NJ: John Wiley and Sons Publishers.

Gilkeson, J., & Reynolds, K. (2003). Determinants of Internet auction success and closing price: An exploration study. *Psychology and Marketing, 20*(6), 537-566.

Heyman, J. E., Orhun, Y., & Ariely, D. (2004). Auction fever: The effect of opponents and quasi-endowment on product evaluations. *Journal of Interactive Marketing, 18*(4), 7-21.

Kamins, M., Dreze, X., & Folkes, V. S. (2004). Effects of seller-supplied prices on buyers' product evaluations: Reference prices in an Internet auction context. *Journal of Consumer Research, 30*(4), 622-628.

Kauffman, R. J., & Walden, E. A. (2001). Economics and electronic commerce: A survey and directions for research. *International Journal of Electronic Commerce, 5*(4), 5.

Marcoux, A. (2003). Snipers, stalkers, and nibblers: Online auction business ethics. *Journal of Business Ethics, 46*(2), 163-173.

McDonald, C., & Slawson, V. (2002). Reputation in an Internet auction market. *Economic Inquiry, 40*(4), 633-650.

Ockenfels, A., & Roth, A. (2002). The timing of bids in Internet auctions: Market design, bidder behavior and artificial agents. *AI Magazine*, (Fall), 79-88.

Oh, W. (2002). C2C versus B2C: A comparison of the winner's curse in two types of electronic auctions. *International Journal of Electronic Commerce, 6*(4), 115-138.

Puto, C., Patton, W., & King, R. (1985). Risk handling strategies in industrial vendor selection decisions. *Journal of Marketing, 49*(1), 89-98.

Roth, A., & Ockenfels, A. (2002). Last-minute bidding and the rules for ending second-price auctions: Evidence from eBay and Amazon auctions on the Internet. *American Economic Review, 92*(4), 1093-1103.

Sandholm, T. (2000). Issues in computational vickrey auctions. *International Journal of Electronic Commerce*, *4*(3), 107.

Stafford, M. R., & Stern, B. (2002). Consumer bidding behavior on Internet auction sites. *International Journal of Electronic Commerce*, *7*(1), 135-150.

Standifird, S. S., Roelofs, M. R., & Durham, Y. (2004). The impact of eBay's buy-it-now function on bidder behavior. *International Journal of Electronic Commerce*, *9*(2), 167-176.

Ward, S. G., & Clark. J. M. (2002). Bidding behavior in on-line auctions: An examination of the eBay Pokemon card market. *International Journal of Electronic Commerce*, *6*(4), 139-155.

Wilcox, R. (2000). Experts and amateurs: The role of experience in Internet auctions. *Marketing Letters*, *11*(4), 363-374.

This work was previously published in the International Journal of E-Business Research, Vol. 4, Issue 1, edited by I. Lee, pp. 69-81, copyright 2008 by IGI Publishing (an imprint of IGI Global).

Chapter 11
Knowledge-Based Intermediaries

Levent V. Orman
Cornell University, USA

ABSTRACT

A new generation of intermediaries is predicted to flourish in the emerging electronic markets. They rely on new information technologies such as the semantic web, rule-based triggers, and knowledge-based constraint maintenance systems. These technologies do not automate or reduce intermediation, but inspire new types of intermediaries that rely on the technologies and complement them with human organizations. An inter-organizational architecture based on multiple levels of intermediation is described, and arguments are presented for its usefulness in emerging electronic markets.

INTERMEDIARIES

There have been many predictions and prescriptions for reduced intermediation in electronic markets. These predictions are based on the assumption of increasing reliance on information technologies to connect economic agents directly, in lieu of intermediaries (Bakos, 1998; Malone, 1989). These predictions, by and large, have not materialized, except for isolated instances of limited disintermediation in some information-intensive industries. In fact there is some evidence that there may be an increasing role for intermediaries in electronic markets, as compared with traditional markets (Andersen & Andersen, 2002; Sarkar et al, 1995). Intermediaries are third parties that facilitate economic transactions between the primary economic agents. As such, intermediation is an information-intensive activity. It requires considerable information about buyers and sellers to be collected, organized, processed, and distributed. The predictions for reduced intermediation are often based on the assumption that information technologies could completely automate intermediation, or so fundamentally simplify it that it can be absorbed into the operations of buyers and sellers. Information technologies are designed to reduce the cost of information-based activities, and hence they are poised to reduce the

DOI: 10.4018/978-1-60566-910-6.ch011

cost of intermediation, but not necessarily capable of automating them completely or even substantially. Since reducing the cost of a vital economic activity is likely to increase the utilization of that activity, it is reasonable to expect an increase, not a decrease, in intermediation in electronic markets, as compared with traditional markets. We will present four hypotheses in favor of substantially increased intermediation in electronic markets. Moreover, we will argue that the new intermediaries will need to be significantly different from the existing intermediaries, and they need to be explicitly designed to take advantage of the new technologies. We will take a design perspective and present a specific architecture for the new intermediaries.

1. **Information Technologies favor markets and outsourcing. Markets and outsourcing can benefit from intermediation**: Markets and hierarchies are often presented as two opposing organizational structures, and information technologies have been claimed to favor markets at the expense of hierarchies (Bichler, 2001; Malone, 1989). The argument is based on the inherent advantage of markets in creating specialization, and economies of scale, by outsourcing various activities to specialized businesses that perform them for all who need them. However, these advantages come at a cost of extensive communication and coordination required by markets. As information technologies reduce the cost of communication and coordination, increasing reliance on outsourcing and markets is expected, to take advantage of specialization and economies of scale. Yet, such outsourcing is not a trivial exercise, and it is not likely to be a simple shift from hierarchies to markets. It may require extensive intermediation to alleviate the control and coordination problems arising from extensive outsourcing and market orientation, since control and coordination are not mere communication problems, but require extensive information processing and analysis. Intermediation can reduce the cost of information processing and analysis through specialization (Sarkar et al, 1995).

2. **Markets and outsourcing lead to inter-organizational structures. Inter-organizational structures can benefit from intermediation:** There is considerable evidence that new information technologies led to increased outsourcing, but not necessarily a larger number of business partners, contradicting theoretical predictions of extensive outsourcing with more business partners through market transactions (Clemons, 1993). It is possible that this observation reflects merely an intermediate stage in the ongoing adaptation to new technologies, and fragmentation of outsourcing may yet to develop. Increased outsourcing creates a control and coordination problem with outside business partners, and the lack of new inter-organizational structures to deal with the increasing control problem might be limiting the fragmentation in the short term. New inter-organizational arrangements and further refinement of technologies would allow finer fragmentation of outsourced activities with many interchangeable parts, since that would create new efficiencies, if the control problems can be resolved effectively. These new organizational arrangements are likely to involve extensive intermediation, since that would allow indirect access to many business partners through intermediaries, and would provide the necessary control efficiently, without direct contact with each and every business partner (Waldfogel & Chen, 2006).

3. **Inter-organizational structures may require explicit design. Explicit design can benefit from intermediation:** Complex organizational structures do not always emerge spontaneously from evolution and incremental change encouraged by market forces, but they may need to be designed through a deliberate and prescriptive design process, involving radical redesign and extensive cooperation from many participants. Evolutionary and incremental changes do not always lead to optimum solutions, since market

forces emphasizing short term benefits may prevent radical redesign, and lead to long term problems. The future cannot always simply be predicted as a natural and evolving system, but it may need to be explicitly and collectively designed, but achieving such long-term planning from economic agents in a market economy is a fundamental challenge (Hevner et al, 2004). New technologies do not merely make existing organizational structures more or less efficient, favoring one over another, but they pose engineering and design problems. To create the best fit between technologies and organizations, both have to be redesigned, often leading to new technologies and new organizational structures. We will argue that such explicit design involving many economic agents may require extensive intermediation (Orman, 2008).

4. **Explicit design may require inter-organizational cooperation. Inter-organizational cooperation can benefit from intermediation:** Intermediaries can serve as agents of cooperation among their customers and members, in achieving collective and radical redesign of inter-organizational structures, by building communities of common interest, and attenuating the short term competition among their members that prevents long term planning. Some of this new intermediation would be mere expansion of the traditional roles intermediaries play (Andersen & Andersen, 2002). But we will argue that such inter-organizational cooperation may require new roles and activities for intermediaries, and even some new types of intermediaries (Zuboff, 2002).

Such extensive intermediation, necessary to manage and control a large number of business partners efficiently, is likely to change both markets and hierarchies, and create new structures that are neither pure markets nor pure hierarchies. Resulting structures may resemble inter-organizational networks, institutionalized by extensive intermediation (Thorelli, 1986). Highly intermediated markets create chains of interrelated markets which resemble inter-organizational hierarchies, similar to supply chains, due to long lasting and abstract transactions they encourage (Bichler, 2001; Sarkar et al, 1995). Highly intermediated hierarchies spread hierarchies over multiple organizations, and the resulting inter-organizational structures often resemble interconnected markets, due to the transaction orientation of the relationship (Andersen & Andersen, 2002; Orman, 2008). Moreover, both chains can have dynamically interchangeable components, converting hierarchies to networks of temporary connections (Hagel &Brown, 2002; Sawhney et al, 2004). In this environment, the relationship among the economic agents is neither as ephemeral as an ordinary market transaction, nor is it as permanent and all encompassing as an alliance or a merger. They are contractual transactions that involve continuing commitments, rather than individual products and services, as in supply chain relationships (Thorelli, 1986; McAfee 2005). Contractual transactions with a large number of interchangeable business partners are difficult to manage and control, and hence require a complex set of intermediaries. These intermediaries are starting to emerge, and likely to proliferate in highly developed electronic markets. We will refer to this new generation of intermediaries as "knowledge-based intermediaries", as opposed to "data-based intermediaries" currently dominating the electronic markets.

KNOWLEDGE-BASED INTERMEDIARIES

Intermediaries perform three information-based activities: posting, transforming, and matching. They post information regarding buyers and sellers in a central marketplace to reduce search costs; they transform and customize information and channel it to the right target to reduce processing costs; and they match buyers and sellers by evaluating and rating them with respect to each others' needs to reduce the analysis costs (Bakos, 1998; Sarkar et al, 1995). These activities are information intensive, and as such they are likely to benefit from information technologies, and they are likely to expand and change to take advantage of new information technologies. However, since there are many information technologies performing a variety of functions, and there are many types of intermediaries performing a variety of activities, not all intermediaries are likely to utilize the same technologies, in the same manner, to perform the same functions. It is possible to identify three types of intermediaries with respect to their use of information technologies: data-based intermediaries, trigger-based intermediaries, and constraint-based intermediaries. Data-based intermediaries are common in electronic markets. Trigger and constraint-based intermediaries are newly emerging, and we refer to them collectively as knowledge-based intermediaries. This classification of intermediaries is novel (Sarkar et al, 1995, Thorelli, 1986) and useful in understanding the information processing function of intermediaries, which leads to a design perspective to propose new intermediaries.

Data-based intermediaries collect and present data. Their primary focus is posting of data, although they do some transformation, and some very minimal matching. They are most abundant in current electronic markets in the form of electronic catalogs that collect and present data about products and services offered. Electronic catalogs have been the mainstay of electronic commerce and received significant attention as a new model for consumer-business interaction, resulting in a reduction of search costs for consumers, and increasing the reach of businesses into new markets (Sawhney et al, 2004). The technologies supporting electronic catalogs are the database technology for maintaining and searching them, and data warehousing and data integration technologies to create such catalogs by aggregating data from multiple businesses. Unfortunately, data integration technologies such as XML, ontologies, and semantic models often require application specific middleware (Erl, 2005). Consequently, data based intermediation rarely eliminates intermediaries completely, but relies on a complex human-machine organization that often requires specialized new intermediaries. Data-based intermediaries provide some information transformation by allowing the participants to set up filters to select relevant information, and also some minimal matching when they provide product evaluations and recommendations. The car sales site www.carpoint.com is a data-based intermediary. It collects data from car manufacturers and third parties, and posts it in an aggregate electronic catalog. The collection is sometimes automated but often requires application specific middleware; trust and reliability requires independence from auto manufacturers and encourages an independent intermediary; some transformation is provided by allowing customers to search and filter their choices; and minimal support for matching is provided in the form of recommendations and evaluations (Bichler, 2001).

Trigger-based intermediaries give a first glimpse into the new knowledge-based intermediaries. Their primary function is to transform data and customize it, although they sometimes do some analysis and matching also. They do not merely present information, but they push customized information to specific users when user-specific triggers are activated. Such conditional data delivery requires three components: extensive information about individual preferences and requirements, a dynamic sensor network to detect environmental conditions, and a triggering mechanism to match the detected conditions to an individual's specific requirements (Cook & Das, 2004). They can be viewed as powerful recommendation systems armed with detailed consumer data that are never available to traditional businesses, but only to intermediaries that serve as consumers' agents. Such conditional actions are the basis of knowledge-based systems when they are fully automated. However, business applications with complex and ambiguous triggers often require a human-machine organization (Cook &

Das, 2004). A second critical technology for the success of trigger-based intermediaries is sensor networks that sense the status of an organization (human or machine) and communicate it to a triggering system (Akyildiz, 2002). Creation and maintenance of such networks also require a human organization, even if the network itself is all automated. In fact, the sensor networks themselves may be human-machine systems when some observations require human senses and human judgment. These systems do minimal matching beyond linking observations to trigger conditions, and it is usually in the form of identifying inconsistencies among sensor observations, and resolving them by evaluating their credibility.

Early examples of trigger-based intermediaries are news delivery businesses such as www.worldflash.net and www.mercurynews.com that collect news from various news organizations and match them to triggers set up by individual readers and channel the relevant news items to those readers (Rust & Kannan, 2003). The triggers are often simple keywords and news class identifiers, but the intermediary also monitors what the reader has already read and uses it to modify the selection process. The monitoring requires an electronic sensor system, but in this case it is a rather simple collection of data markers called cookies on the reader's computer. Another early example is the maintenance unit of Caterpillar Corporation. They install wireless sensors on heavy equipment they sell, and then issue warnings about the maintenance status of the equipment. It is slightly more sophisticated in its use of sensors, but it is still limited to the narrow objectives of a single company with very limited information about the customer's status. Extension of such systems to a variety of consumer or industrial products would require large scale and diverse sensor networks and a variety of complex trigger systems. The complexity of sensor networks increase exponentially with scale, and decentralized peer-to-peer management of such networks is an ongoing research area [Akyildiz, 2002; Sheth et al, 2006). The car-sales intermediary example carpoint.com of Section 1 can be extended to become a trigger-based intermediary, by collecting extensive consumer and background information, using that information to trigger individualized warnings, and pushing triggered warnings to consumers' attention. Sensor networks can monitor the automobiles owned by specific consumers, and produce maintenance, repair, and replacement warnings. Another set of sensors can monitor the consumers' daily commute and traffic conditions, and issue route selection and travel delay warnings. A third set of sensors can monitor the new car models, new designs, and discounts, and issue personalized and just-in-time replacement and trade-in recommendations.

Constraint-based intermediaries are the new generation of knowledge-based intermediaries that are made possible by the new communication and information technologies. Inspired by knowledge-based systems that automatically maintain and enforce constraints, these new intermediaries engage in transactions on behalf of their customers, to maintain constraints and satisfy goals specified by their customers (Rust & Kannan, 2003). They don't simply issue warnings or push data as in trigger-based intermediaries, but they match buyers and sellers, and execute transactions on behalf of them on the basis of pre-specified goals and constraints. Such constraint maintenance problems are sometimes completely automated by knowledge-based and expert systems, but most business applications with complex constraints and goals require a human-machine organization including knowledge engineers and domain experts (Dechter, 2003). In fact, partial automation of such complex business problems led to the development of a new class of systems called "decision support systems". Decision support systems do not completely automate business processes, but need to be incorporated into the human organization (Orman, 2008a).

A simple early example of a constraint-based intermediary is www.priceline.com which receives

Table 1. Types of existing and proposed intermediaries, and their functions and information requirements.

Knowledge-Based Intermediaries			
	Data-Based	**Trigger-Based**	**Constraint-Based**
Function	Present Data, Respond to Queries, Facilitate Search	Push Customized Data, Make Recommendations, Observe events and Trigger Reports	Execute Transactions on Behalf of Customers, Maintain Optimum State, Observe Events and Take Necessary Actions
Information	Vendor and Product Information	Consumer Preferences and Constraints	Marketplace Information, Matching, Pricing, Constraints, and Optimization Algorithms

price constraints from buyers and sellers, matches them to each other, and executes appropriate transactions to satisfy their goals and constraints. This example is limited to price constraints and to specific preauthorized transactions. A more general business model involving a variety of consumer or industrial products would have a much larger attribute space and more complex goals and constraints to satisfy. The car-sales intermediary example carpoint.com of Section 1 can be extended to become a constraint-based intermediary by executing transactions on behalf of consumers on the basis of their preferences and constraints. The intermediary can schedule maintenance and repair appointments with appropriate mechanics. It can arrange replacement of cars at optimum times with the most appropriate replacement, since the relationship between the consumer and the intermediary is one of subscription to transportation services determined by preferences and constraints, rather than specific cars. The intermediary can also control the daily commute, but not by merely issuing warnings about traffic delays and routes as in trigger-based intermediaries, but by determining the optimum route and directing the driver, or by scheduling car pools, or by arranging limousine pick-up services, depending on the traffic conditions and scheduling requirements (Table 1).

FRAGMENTATION OF INTERMEDIARIES

Information technologies can have two opposing impacts on intermediaries. They may significantly simplify the intermediary function by partially or completely automating it, and hence allow the buyers and sellers to absorb that function, bypassing the intermediary. This outcome requires the buyers and sellers to expand their expertise into new areas to perform intermediary functions, and establish credibility with partners since the verification and guarantees provided by third parties would disappear. Alternatively, information technologies may reduce the cost of intermediation, and increase the reliance on intermediaries for an expanded range of functions. This outcome requires redefining and expanding the role of intermediaries, and creating new roles for them (Andersen &Andersen, 2002). We argue in this section that the latter is not only more likely from a predictive perspective, but also more desirable to actively pursue and promote from a design perspective. The framework and classification developed in Section 2 is used as a design tool for redefining and expanding the role of intermediaries, and creating new roles for them.

If intermediaries can discover new services enabled by technology, and integrate them with the old services, they will have an expanded role in commerce and increase their importance. If instead they remain in their traditional roles, and cannot find new useful functions to perform, they may be automated or simplified enough to be absorbed into buyers' or sellers' operations. To understand what new roles intermediaries can play, one needs to have an engineering and design framework to explore the functions of intermediaries and how new technologies create new opportunities for them within that framework. The framework described in Section 2 serves that purpose by classifying intermediaries and the roles they play. In this section we describe how new technologies impact each role, and what new opportunities and roles can be developed for each type of intermediary.

The success of a new design depends on three critical factors: the degree of automation made possible by technology, possible new roles for intermediaries created by technology, and the need for organizational independence in performing those intermediary functions. The first two of these factors are engineering and design issues, and the impact of information technologies on organizations often depends on the discovery of new designs and structures. The expected impact may not even materialize, if the new structures are not achievable from the current structures by following an incremental and organizationally feasible path, and a major radical overhaul of the organization is too risky to undertake. The path from the old structure to the new often requires discovering new roles and functions that benefit the old participants and encourages them to participate. A typical example is Barnes & Noble bookstores' initial difficulty to sell electronically to consumers, along with its physical stores, since that would cannibalize its own physical stores. Such channel conflict created by an existing structure often blocks the creation of a new structure, even if it is more efficient, until new roles and functions are invented to benefit the existing participants. In this case the new roles were physical stores serving as pick-up and delivery points for electronic purchases, and electronic kiosks in physical stores serving as customer service counters for non-inventoried items (Rust et al, 2004).

Data-based intermediaries are impacted by semantic web and ontology technologies, because they need to collect product data from a variety of vendors, and product evaluations and ratings from a variety of third parties (Alonso, 2003). As such, any technology that standardizes data representation and increases the semantic content of data models would enhance their ability to expand and provide an aggregate marketplace as agents of multiple vendors. However, data integration is not easily automatable because semantic heterogeneity continues to challenge researchers, and human control of aggregate electronic catalogs remains necessary for complete reliability. Moreover, even if this function is completely automated, it cannot be absorbed into the operations of any particular vendor. This function needs to be owned by an independent intermediary that serves a community of vendors, without a financial interest in any of them individually, to prevent any particular vendor owning the marketplace and gaining an unfair advantage over others. No vendor can have the trust and credibility to collect information about its competitors' products, and present it fairly in an aggregate marketplace. Finally, the cost of running a central marketplace requires specialized skills, and an independent intermediary can amortize those costs over multiple vendors. The semantic data modeling technologies can lower the cost of operating these intermediaries, making them even more desirable. Intermediaries are not inherently undesirable, but only because of the cost they add into a supply chain. As the cost of operating an intermediary drops, such intermediaries are likely to proliferate in electronic markets.

A typical example of an intermediary that emerged to represent a community of vendors is Orbitz which is an intermediary representing a large number of major airlines. It is in sharp contrast to an earlier attempt by American Airlines to absorb that function into its functions. That sole ownership of a marketplace could not be sustained in the long term. Another possible emerging example is home maintenance and security intermediaries that would collect and present data from all the local service businesses, and their third party evaluations and rankings, in addition to all the historical maintenance records of all the homes in the area irrespective of the vendor who provided that service, and also the municipal services provided in that locality, all from one central marketplace available to home owners to easily access, compare, and contrast.

Trigger-based intermediaries benefit from sensor networks that sense their environment, and trigger responses to it in the form of customized information pushed to customers as warnings and alerts. They cannot be easily automated, but require a human organization, because building and maintaining sensor networks is labor intensive, and requires ongoing maintenance of the hardware (Akyildiz, 2002). Sometimes even the network itself consists of human observers of the environment, since some observations require human capabilities and judgment. These intermediaries rely heavily on the data-based intermediaries for product information to properly interpret sensor data. Yet there are advantages to keeping them separate, because trigger-based intermediaries can benefit from being agents of buyers. Collecting data from the buyers' environment to detect the buyers' needs often requires extensive intrusion into buyers' operations with electronic and human sensors, and buyers' agents are more likely to be successful in gaining such access. Such data cannot always be turned over to data-based intermediaries, without creating a conflict of interest, because the data-based intermediaries can benefit from being sellers' agents, collecting and distributing product data. These functions cannot be easily absorbed into the buyers' operations either, since they require considerable specialized skills in building and maintaining sensor networks and knowledge based triggers. Moreover, some of these networks and triggers are common to many buyers, and the costs need to be amortized over a large number of buyers. Finally, such intermediaries have the potential of creating natural buyer communities, increasing their bargaining power, as in consumer cooperatives and group buying (Hagel & Brown, 2002). In this environment, one class of intermediaries represent communities of sellers; and another class of intermediaries represent communities of buyers; and business-consumer interaction takes on a completely new character, and starts to resemble the business-business interaction commonly observed in the modern supply chains (Hagel & Brown, 2002).

A typical example of a trigger-based intermediary representing a community of buyers is the Virtual Order system of Federal Express that monitors a supply chain for all members of the supply chain, and issues reports and alerts for all members involved. That is in sharp contrast to the supply chains of Dell or Wal-Mart where the dominant player owns the sensor network, and information generated about the status of the supply chain goes through the dominant player before it is transmitted to the other members, creating a myriad of distortions and delays in information delivery. There are many trigger-based intermediaries such as www.stockwatch.com that monitor the stock markets on behalf of their users and their specific portfolios, and issue alerts and warnings on the basis of predetermined user specific triggers.

Another possible advanced example is home maintenance and security intermediaries that install sensor networks at homes to detect safety, security, and maintenance problems, ranging from failures of the heating system to unauthorized intrusions to the home, and from energy efficiency problems to expiration dates of foods and medicines. Such sensor data trigger warnings and alerts to home owners or other interested parties. Clearly, the ownership of such monitoring devices should be independent of any individual service business, and the sensor networks should serve communities of home owners to share community based warnings, and to amortize the cost of network development over many customers.

Constraint-based intermediaries enforce their customers' goals and constraints by executing transactions on their behalf. The critical technologies are constraint-enforcing knowledge-based systems, but these systems are not likely to completely automate these activities (Dechter, 2003). Transaction selection to enforce user requirements and constraints requires considerable human judgment, and even capturing user requirements is an inexact science (Orman, 2008). These intermediaries require considerable information about the objectives and the constraints of both buyers and sellers, and they create novel marketplaces where transactions take place without an explicit participation by the buyers or the sellers. Such intermediaries cannot be the agents of the buyers or the sellers, since that would lead to an unfair advantage, but they need to be independent marketplaces to be entrusted with the details of member preferences. Members' constraints and goals reveal intimate details of their utility functions, and cannot be readily disclosed to competitors or even to business partners, but only to agents whose interests are correctly aligned, which often requires independent intermediaries. There are also advantages to keeping them separate from the data-based and trigger-based intermediaries, because those intermediaries can benefit even more from being the agents of sellers or buyers respectively. Moreover, the development of such knowledge-based decision support systems requires specialized skills, and need to be amortized over a large number of customers, which further encourages the reliance on independent intermediaries. The result is a significant fragmentation of intermediary services with three types of intermediaries.

In this environment of significant fragmentation, intermediaries need to rely on other intermediaries in a complex mosaic of interdependence (Hagel & Brown, 2002; Malone et al, 1989). That interdependence and fragmentation is likely to increase with the development of web services technologies, creating new intermediaries that are built on top of the existing intermediaries, to create new services. The new web services technologies are likely to encourage such fragmentation of intermediary services, because they make the services of one intermediary readily available to others to build on, by using standardized information exchange formats over the web such as web service description languages, web service directories, and ontologies, all based on the standard data model XML (Alonso, 2003; Waldfogel & Chen, 2006).

An example of a constraint-based intermediary is Li-Fung, a Chinese intermediary that coordinates a large number of member manufacturers to satisfy the needs of its customer firms which are often large retailers or down stream consumer product manufacturers. It relies on a variety of other intermediaries and an extensive rating and evaluation system that captures the capabilities, constraints, and limitations of its member manufacturers. It also relies on other intermediaries to monitor the needs of its customers through an electronic network of sensors, triggers, and alerts. It uses this knowledge to pick and choose the correct manufacturers, combine them dynamically with others, and link them dynamically for communication and coordination, to satisfy the needs and requirements of its customers. It provides all of these services hidden from the view of its customers, by acting as a constraint-based intermediary, and protecting its customers from the details of the supply chain, while satisfying the goals and constraints posed by them (Hagel & Brown, 2002). This behavior is similar to mutual funds that execute transactions on behalf of their customers to satisfy their pre-specified constraints, without consultation with them about each specific transaction. Another possible advanced example is the emerging home maintenance and security businesses that rely on electronic sensor and trigger systems, and execute transactions to remedy

Table 2. Types of knowledge-based intermediaries and their technical and organizational characteristics.

Knowledge-Based Intermediaries			
	Data-Based	**Trigger-Based**	**Constraint-Based**
Technology Used	Data Warehousing, Data Integration	Sensor Networks, Triggers, Rule-based systems	Constraint Enforcement, Decision Support, Optimization
Incentive Alignment	Seller alignment, for access to maximum product information	Buyer alignment, for access to consumer preferences	Independent alignment, for unbiased matching and transaction execution

problems. The transactions take advantage of extensive knowledge of the requirements and constraints of individual home owners, and also utilize other data-based intermediaries to access the schedules, capabilities, and offers of various service providers. The transactions range from scheduling service and repair visits to calling emergency and law enforcement agencies, from replenishment of grocery and medical supplies to changing energy and utility providers (Cook & Das, 2004).

IMPACT OF NEW INTERMEDIARIES

The extensive intermediation proposed in this article is likely to change the business-consumer interaction significantly. That interaction now focuses most attention on characterizing consumers by observing their past transactions, and statistically predicting their future transactions. When consumers do not participate in the marketplace directly, but collectively through a series of intermediaries, the dominant marketing paradigm may shift significantly. The current preoccupation with discovering consumer characteristics from their transactions would no longer be the dominant issue, since consumer characteristics would be readily available from the intermediaries that represent them, albeit collectively, since consumers would be incentivized to describe themselves to their agents in great detail. The new paradigm would shift the concern to discovering relevant transactions from consumer characteristics, with specialized intermediaries designing transactions for specific consumer groups. Such intermediaries would serve the dual purpose of channeling product information to the right consumer groups, and simultaneously protecting the consumers from irrelevant information. Such an efficient information exchange has the potential to benefit all, and create a significantly more efficient marketplace. In this marketplace, the information exchange between buyers and sellers would be significantly different, since they would not interact directly, but through a series of intermediaries. Customer relationship in this marketplace would resemble the business-business relationship of supply chain members, rather than the transaction data collection, statistical analysis, direct marketing, and mass media campaigns prevalent today.

The hierarchical relationship among the three types of intermediaries allows each to benefit from the services of those at the lower levels, but they remain separate both in terms of functionality and in terms of objectives. Data-based intermediaries represent sellers in the marketplace, but aggregation and comparison of data from a collection of sellers can benefit the buyers also. Existence of comparative data supplied by third party evaluations and ratings allows buyers to do objective product comparisons, and reduces the information asymmetry that ordinarily gives sellers an advantage, due to their superior knowledge of the products, as compared with the buyers. Sellers may also benefit from data-based intermediaries, because they provide easy access to customers in a central marketplace, and also they reduce the need for elaborate advertising campaigns to position products and to draw customers away from the competitors' products. Instead, competition is institutionalized and rationalized through comparative listings and objective ratings of data-based intermediaries. For example, Ebay's standard representation of products reduces the need for consumer-based advertising directed at Ebay customers. Carpoint's standard descriptions, side-by-side comparisons, ratings, and evaluations reduce the need for advertising campaigns directed at Carpoint's customers.

Trigger-based intermediaries represent buyer communities, but they can also benefit sellers by collecting sensor data from a collection of buyers and issuing warnings and alerts to all of them simultaneously. Such synchronization of data eliminates the possibility of arbitrage among buyers, since no buyer has an information advantage. No buyer would be able to undermine the price strategies of sellers by buying from sellers and selling to other buyers. But at the same time, the opportunities for price discrimination by sellers would be greatly reduced, because sellers do not have access to individual buyers, and cannot limit the availability of price and quality information to some buyers selectively. For example FedEx's Virtual Order system limits arbitrage and price discrimination within a supply chain since all members are notified of events and price changes simultaneously and completely. Membership in the Associated Press and other wire services prevents news organizations from competing on the basis of speed and timeliness of news, since they all receive the news at the same time on account of their membership in the wire services.

Constraint-based intermediaries are independent markets serving both buyers and sellers. As such, they reduce short-term hyper-competition, supply chain domination, and price wars among their members, and encourage a more cooperative relationship. They accomplish this by pushing their members into long term relationships with them, and by encouraging repetitive and long- lasting transactions. That makes it difficult for members to use short-term strategies to damage opponents, and then revert to a different long-term strategy. Constraint-based intermediaries are excellent tools for building communities naturally, without complex contracts and detailed negotiations. Consider the example of IT investments by businesses. IT investments are highly risky. Occasionally they confer overwhelming competitive advantage to an early adapter; but mostly they fail, or get imitated easily. Unfortunately, because of the outside chance of an overwhelming advantage to a competitor, all businesses are forced to invest heavily in all new technologies, leading to mounting costs, and even a reduction in overall productivity (Brynjolfsson, 2003). Outsourcing IT investments to a constraint-based intermediary, that follows a predetermined long-term investment and development strategy, for all of its member businesses, prevents an arms-race of technology investments that plagues many businesses. Another example is Li-Fung's member manufacturers, which are in a potentially long term relationship with it. This protects them from a devastating short-term attack

Table 3. Types of knowledge-based intermediaries and their inter-organizational unintended impact.

Knowledge-Based Intermediaries			
	Data-Based	**Trigger-Based**	**Constraint-Based**
Organizational Impact	Reduced mass advertising	Reduced arbitrage opportunity, Reduced price discrimination	Reduced price competition, Increased long-term cooperation

from other members, and allows them to build their capabilities within a long-term development strategy. Li-Fung has an incentive to protect its members from each other, to allow them plan for long term competitiveness which they may not be able to do alone because of insufficient resources or the intensity of short-term competition.

Knowledge-based intermediaries also have many non-profit and government applications. Data-based intermediaries can collect data about political candidates and present them in an easily searchable and comparative format to aid voters. League of Women Voters' web site is a typical example that contains considerable amount of data about the comparative positions of national political candidates. Trigger-based intermediaries can alert voters to specific issues and political events that are relevant to them as they develop. These alerts may contain specific passages from position papers of candidates, or portions of political speeches that relate to specific voters' concerns. Sierra Club, an environmental protection group, and moveon.org, a liberal political action group, perform such functions to organize their members to attend rallies, to contact their representatives, or simply to vote. Constraint-based intermediaries can take political action on behalf of their members on the basis of their expressed preferences, or sometimes on the basis of their implicit preferences inferred from their activities and memberships in various organizations. Such formulation and discovery of rational behavior on behalf of citizen or consumer groups is a new type of intermediary. Some simple traditional examples of constraint-based intermediaries are labor unions and elected representatives who take action and execute transactions on behalf of their members and constituents. However, widespread adoption of such intermediation, without creating any conflict of interest that currently plagues such institutions, would require considerable infrastructure, and significant changes in the design of markets and social institutions, as described in this article.

Such extensive intermediation also has implications in international relations. Outsourcing the development of weapons systems or monitoring of their development, as it is done now by United Nations, is an effective strategy to curb arms races. Extending such efforts to trade, environmental control, and labor policies is a promising area, if data collection and sensor networks can be deployed in large scale. Development of global intermediaries to implement or oversee environmental policies, international labor standards, or outsourcing and labor migration issues can benefit all parties by preventing market failures resulting from excessive short term competitive pressures.

VIRTUAL TRANSACTIONS

Virtual transactions are high-level transactions built by aggregating lower level transactions dynamically and at multiple levels. Corporations engage in virtual transactions routinely. A year-end dividend payout and a corporate merger are examples of high-level transactions that are aggregations of many individual dividend payouts, and many shareholder stock-swap transactions respectively. Consumers on the other hand are not always able to engage in virtual transactions. That greatly limits their economic efficiency and power. Knowledge-based intermediaries can aid in creating and executing virtual transactions. Transactions can be aggregated over three separate dimensions: over goods and services, over time, and over consumers. Bundling an automobile-insurance transaction with a home insurance transaction is an aggregation of services, and it can be accomplished by data-based intermediaries that link data about multiple products and services. Automatically updating and renewing an insurance policy every year is an example of aggregation over time, and it can be accomplished by trigger-based intermediaries that track and sense the external conditions and execute transactions when the conditions are satisfied. Bundling home insurance policies over many consumers such as a bulk insurance policy for all employees of a university is an example of aggregation over consumers, and it can be accomplished by constraint-based intermediaries that partition and aggregate consumers at multiple levels, and using multiple criteria. One can aggregate over multiple dimensions, at multiple levels, and sometimes conditionally, leading to complex multidimensional hierarchies. Consider the sale of a 10-year home-automobile insurance to all employees of a university who have not had a traffic accident during the past 5 years, and their home coverage is 80% or 90% of their home value depending on the neighborhood, and it is renewed every year at the level of inflation. A further aggregation of such transactions to all insurance needs of all academic employees over their life-time can be accomplished through multiple levels of conditional aggregation of various transaction types. Such high-level transactions can be described as abstract life-style choices, and then automatically translated into specific transactions for specific consumers by knowledge-based intermediaries [Orman 2008].

Such high-level aggregate transactions can be fulfilled by multiple levels of intermediaries over time. Various intermediaries commit to such high-level transactions at various levels of abstraction, and then arrange for lower-level intermediaries and vendors to fulfill individual transactions. Some vendors may not even exist at the time of the commitment, but they are recruited at-a-later time to fulfill the obligations. The specific products and services may not even exist at the time of the commitment, and such abstract commitments by a large number of consumers to future products and services may encourage experimentation with future business models.

Virtual transactions blur the distinction between shareholders and customers by allowing large-scale experimentation by consumers with new products and services, with the aid of multi-level knowledge-based intermediaries. Long-term aggregate transactions are commitments by large groups of consumers to future products and services, and as such they are similar to investments. The ability of multiple levels of intermediaries to support future products and services, long before they are developed, on the basis of high-level goals and targets, allows experimentation by individual vendors within the constraints of goals and targets set by various intermediaries operating at multiple levels of abstraction. Consider long-term home security services provided by constraint-based intermediaries, to a large community of homeowners, with a variety of dynamically changing services as the customers' needs change over time. Such long-term conditional commitments by large communities are large and long-term investments. They provide the necessary funds for intermediaries to continually improve their services by continually replacing some organizational components with new vendors, and by experimenting with new technologies and business models, without reselling, re-marketing, or seeking venture capital for every change [Orman, 2008].

CONCLUSION

A new generation of intermediaries has been introduced, and classified into three major types. They rely on new information technologies such as the semantic web, rule-based triggers, and knowledge-based constraint maintenance systems. A design science approach has been taken, and an inter-organizational architecture based on multiple levels of intermediation has been proposed. A variety of arguments have been presented for the usefulness of this architecture. All arguments have been posed as empirically testable hypotheses. Yet, as a design endeavor, the hypotheses all relate to future organizations that do not exist in great numbers, and empirical testing is suggested as future research, when these intermediaries are commercially developed, and come into wide-spread existence. Some early examples of knowledge-based initiatives are given throughout the article, and their advantages, strengths and weaknesses have been discussed. Hypothetical examples were added to the discussion to show how those early initiatives can be extended to become full-fledged knowledge-based intermediaries.

REFERENCES

Akyildiz, F. (2002). A Survey on Wireless Sensor Networks. *IEEE Communications*, *40*(8), 102–114. doi:10.1109/MCOM.2002.1024422

Alonso, G., et al. (2003). *Web Services*. Springer.

Anderson, P., & Anderson, E. (2002). The New E-commerce Intermediaries. *Sloan Management Review*, *43*(4), 53–62.

Bakos, Y. (1998). The Emerging Role of Electronic Markets on the Internet. *Communications of the ACM*, *41*(8), 35–42. doi:10.1145/280324.280330

Bichler, M. (2001). *The Future of E-Markets*. Cambridge University Press.

Brynjolfsson, E. (2003). IT Productivity Gap. *Optimize, 21*(7).

Clemons, E. K., Row, M. C., & Reddi, S. P. (1993). The Impact of Information Technology on the Organization of Economic Activity: The 'Move to the Middle' Hypothesis. *Journal of Management Information Systems*, *10*(2), 9–35.

Cook, D., & Das, S. (2004). *Smart Environments: Technology, Protocols, and Applications*. Wiley Pub.

Dechter, D. (2003). *Constraint Processing*. Morgan Kaufmann.

Erl, T. (2005). *Service Oriented Architecture (SOA): Concepts, Technology, and Architecture*. Prentice Hall.

Hagel, J., & Brown, J. S. (2002). *Out of the Box: Strategies for Achieving Profits Today & Growth Tomorrow Through Web Services*. HBS Press.

Hevner, A., March, S. T., Park, J., & Ram, S. (2004). Design Science Research in Information Systems. *MIS Quarterly*, *28*(1), 75–105.

Malone, T. W., Yates, J., & Benjamin, R. I. (1989). The Logic of Electronic Markets. *Harvard Business Review*, *67*(5), 166–170.

McAfee, A. (2005). Will Web Services Really Transform Collaboration? *Sloan Management Review*.

Orman, L. V. (2008). Service Semantics, Structure, and Design. *E-Service Journal*, *6*(2), 57–75. doi:10.2979/ESJ.2008.6.2.58

Orman, L. V. (2008a). Browsing Databases with Constraint Hierarchies. *Electronic Commerce Research*, *8*(1), 29–56. doi:10.1007/s10660-008-9016-y

Porter, M. (2001). Strategy and the Internet. *Harvard Business Review*, *79*(3), 63–78.

Rust, R. T., & Kannan, P. K. (2003). E-Service: A New Paradigm for Business in the Electronic Environment. *Communications of the ACM*, *46*(6), 36–42. doi:10.1145/777313.777336

Rust, R. T., Zeithaml, V. A., & Lemon, K. N. (2004). Customer-Centered Brand Management. *Harvard Business Review*, *82*(9), 110–118.

Sarkar, M. B., Butler, B., & Steinfield, C. (1995). Intermediaries and Cybermediaries: A Continuing Role for Mediating Players in the Electronic Marketplace. *Journal of Computer-Mediated Communication*, *1*(3).

Sawhney, M., Balasubramanian, S., & Krishnan, V. V. (2004). Creating Growth with Services. *Sloan Management Review*, *45*(2), 33–43.

Sheth, A., Verma, K., & Gomadam, K. (2006). Services Science: Semantics to Energize the Full Services Spectrum. *Communications of the ACM*, *49*(7). doi:10.1145/1139922.1139949

Thorelli, H. B. (1986). Networks: Between markets and hierarchies. *Strategic Management Journal*, *7*, 37–51. doi:10.1002/smj.4250070105

Waldfogel, J., & Chen, L. (2006). Does Information Undermine Brand? Information Intermediary Use and Preference for Branded Web Retailers. *The Journal of Industrial Economics*, *54*(4), 425–449. doi:10.1111/j.1467-6451.2006.00295.x

Zuboff, S. (2002). *The Support Economy: Why Corporations are failing Individuals and the Next Episode of Capitalism*. Viking Books.

Chapter 12
Strategy to Regulate Online Knowledge Market:
An Analytical Approach to Pricing

Zuopeng (Justin) Zhang
State University of New York at Plattsburgh, USA

Sajjad M. Jasimuddin
Aberystwyth University, UK

ABSTRACT

This chapter studies different levels of pricing strategies for an online knowledge market, where consumers ask and experts answer questions to make knowledge transactions. Consumers optimally price their questions to obtain answers and a firm manages the online knowledge market by determining the optimal price allocation to experts. This research identifies two types of consumers, spin-off and mainstream, depending on whether additional utilities will be derived from the market. In addition, we investigate how the firm can use minimal and maximal posting prices to regulate the knowledge market.

INTRODUCTION

Internet technologies have revolutionized the ways firms do business and the shopping behaviors of consumers. Firms launch electronic storefronts to advertise their products and attract consumers to shop online. The proliferation of electronic intermediaries make online shopping significantly easy because online shopping can save time and transportation costs as well as provide customers additional benefits such as product reviews, comparison of similar products, best price search, and other advantages which are difficult to obtain when they shop in local stores.

Recent years have also witnessed the steady growth of online knowledge market specializing in various domain knowledge and pricing systems. There are various online knowledge markets available in the economy. For example, Intota Expert Knowledge Services (intota.com) specialize in science and engineering, materials science, industry and technology, and business question answering. Kasamba (kasamba.com) has experts in a field whom customers contact directly with their questions and a bid price. Allexperts (allexperts.com) offers confidential services and uses direct email to experts for customers' questions. SwapSmarts (swapsmarts.com) allows users to choose prices

DOI: 10.4018/978-1-60566-910-6.ch012

for their posted questions. Uclue (uclue.com), following the prior business model of Google Answers, is the online knowledge market place where potential buyers and sellers can meet to transact electronically. In contrast to the traditional transactions of physical goods, buyers set their prices for the "goods", the knowledge, they want to purchase and sellers choose from available offers from buyers to make transactions. At Uclue, customers post their questions and set a price between $10 and $400. Experts hired by Uclue browse all the posted questions and decide whether or not to answer the questions based on their own valuations. A question can only be answered by one expert and once the answer is complete, 75% of the price for the question will go to the expert and the other 25% will remain with Uclue for its maintenance fee. These business practices provide excellent examples for research on online knowledge markets.

Given the growing prevalence of online knowledge markets, it is naturally interesting to understand the working mechanisms supporting online knowledge markets. This chapter investigates different levels of pricing strategies for an online knowledge market. First, we analyze a consumer's pricing strategy and identify two types of consumers on the knowledge market: *spin-off* and *mainstream* consumers, based on whether additional utilities are derived from the knowledge market. Second, we study the reasons of specifying minimal and maximal posting prices for an online knowledge market; the firm may be able to eliminate some spin-off consumers by designating a minimal posting price and increase its profit by mandating a maximal posting price. Third, we explore the optimal pricing strategy for experts, i.e., the proportion allocated to experts, by comparing different effects on market structure, transaction price, and the probability of questions being answered when the allocation proportion changes.

The chapter proceeds as follows. Next section reviews relevant literature from the perspectives on knowledge market. The third section outlines an analytical model of online knowledge markets. The fourth section details our analysis and discussion and the last section concludes the chapter.

PRIOR LITERATURE

In this section, we review prior literature related with our research. Beginning with the literature review on knowledge market in organizations, we then continue with the relevant literature on electronic market and online knowledge market. Finally, we differentiate our study with prior research.

The principle of knowledge market has been recently applied to facilitate knowledge transfer within organizations, which has gained growing attentions from researchers. Davenport and Prusak (1998) illustrate the concept of internal knowledge market within organizations and propose to employ the necessary IT support and the indispensable incentives to build an effective internal knowledge market for knowledge transfer. Following the initial idea of knowledge market within an organization, Ba, Stallaert, and Whinston (2001) demonstrate that knowledge components can be optimally traded with a Grove-Clarke like mechanism within different bundles in an internal organization market so that a firm can optimally choose the knowledge bundles for investment. Desouza, Awazu, Yamakawa, and Umezawa (2005) develop a mathematical analytics to show the viability of the market mechanism for knowledge management in organizations. Mueller, Spiliopoulou, and Lenz (2002) formally consider the electronic marketplace as an approach to sharing knowledge assets and investigate the characteristics of knowledge as tradable goods on the e-marketplace within two types of frameworks: the pricing system and the quality evaluation method. Drawing from the lessons from mini-cases, Desouza and Awazu (2003) define the necessary components of an internal knowledge market and outline several important caveats in association with economics literature when devising the market, including market of lemons, chicken-and-egg predicament, black markets, and advertising strategies.

Several researchers have investigated online marketplace. Most notably, Bakos (1997) regards an online marketplace as a special type of electronic marketplace and proposes that electronic marketplaces reduce inefficiencies by lowering buyers' cost to acquire information about sellers' prices and product offerings. Bakos (1998) also adds that electronic marketplaces serve the role of matching buyers and sellers and facilitating transactions, where increasing differentiations and lower cost of product information can be observed. Nevertheless, there lacks research on online knowledge markets, especially how they function and what types of consumers they deal with.

Some studies have recently attempted to study the impact and implications of an online knowledge market from the inspirations of Google Answers. In a study conducted by Cornell University Library (Kenney, McGovern, Martinez, and Heidig, 2003), its digital reference services are compared with those of Google Answers, which provides an opportunity for librarians to borrow valuable insights from Google's approach to service development and delivery. Rafaeli, Raban, and Ravid (2007) investigate the antecedents, inhibitors, and catalysts of providing information and knowledge by studying the behavior of about 500 'Experts' in the Google Answers (GA) online service and over 70,000 questions posed on this system over a 29 month period. They verify some of the hybrid theories of information provision behavior mostly in laboratory settings and conclude that the participation of experts in GA is related with a mixture of economic and social incentives. Cahill (2007) compares three models of online information delivery: the former Google Answers service, library-run Virtual Reference Services, and global knowledge forums, where individuals gather voluntarily to share their opinions, search skills, and knowledge with others. From the behavioral perspective, Edelman (2004) analyzes all auctions since the inception of the Google Answers service. According to his findings, there exist certain notable trends in experts' behaviors. More experienced experts provide answers with more added value asked by the requestors, receiving higher rankings as a result. An expert's rate of earnings increases in his(her) experience. In addition, an expert who only answers questions in particular categories will provide answers of higher quality but earn less.

However, these studies have not investigated the working mechanism of an online knowledge market, i.e., the pricing strategies of consumers and the firm who maintains the online knowledge market. Our research addresses this gap by specifically investigating the model applied by Uclue and Google Answers and makes contributions to the literature by identifying two types of consumers on the knowledge market. Our assertion is different from the prior literature on horizontally differentiated markets (Anderson & Palma, 1992; Bockem, 1994; Cohen & Whang, 1997; Lancaster, 1990), where consumers are assumed to have different preferences on products with respect to quality, color, or other characteristics. Our research does not make assumptions on consumers' preferences and they are only different with respect to their searching skills. We find that even the consumers with the same searching skills can belong to different type, either the spin-off or the mainstream type, depending on whether they are utilizing the knowledge market seriously to derive additional utilities. Our analytical model and its detailed analysis and discussion start from the next section.

THE MODEL OF ONLINE KNOWLEDGE MARKET

This section outlines our analytical model and shows some preliminary results for further analysis. We first describe the model with some general assumptions and then formulate the decision problem for each participant on the knowledge market.

We consider a firm that hires knowledge experts to answer questions posted by customers on the online knowledge market. The competition for people to be recruited as an expert by the firm is fierce, so the firm can always find the most intelligent and capable individuals to work as experts for the knowledge market. Customers visit the knowledge market, post their questions, and set the prices p for them. Since customers do not know who will eventually answer their questions before setting the prices, they may price their questions as suggested by the firm's guidelines. For instance, Uclue posts the pricing guidelines on its website as follows:

That depends on what you're expecting from the Uclue researchers. A question that can be answered quickly, with little written content (e.g. What are a few good sites for information on global climate change?) can usually be handled for $10. But a more involved question, requiring numerous links and researcher-written content (e.g. Please explain the scientific and policy debates surrounding global warming, showing how these have changed over the past few decades?) can involve many hours of a researcher's time, and should be priced accordingly. A good rule of thumb is that the researcher time needed to answer a question should be priced from $25 to $50 per hour.

When a question is posted on the knowledge market, all the experts will have the equal opportunity to answer it, depending on who discovers it first, because of the locking mechanism employed by the firm. In other words, an open question can be locked by an expert for a certain period of time so that only one expert is allowed to work on it at one time. Consequently, either the question is answered and not available any more or it becomes open again to all experts if the expert who locked it was not able to answer it. Therefore, experts try to catch questions within their knowledge domains as quickly as they can from the knowledge market. However, an expert cannot negotiate price with customers, but will inspect the difficulty of each question and the price tag attached to it to decide whether or not to answer the question. A question m has its type q_m of difficulty which is distributed with the probability function $H(q_m)$ between 0 and 1. A more difficult question (higher q_m) requires an expert to have a higher knowledge level or exert more effort to answer it.

Therefore, an expert tries to maximize his(her) total surplus by determining whether or not to answer the questions he(she) observes from the market. For a specific question m priced at p, if he(she) answers it, an expert will get a net payoff as

$$\pi_r = \alpha p - \theta(p, k_i) \cdot q_m, \tag{1}$$

where α is the proportion allocated by the firm for answering each question with price p and $\theta(p,k_i)$ is the disutility an expert with knowledge level k_i will incur by answering this question, including the effort cost and the risk of getting a bad evaluation. The knowledge level k_i is not observable, but can be inferred from a known distribution $G(k_i)$. We assume that $\theta(p,k_i)$ is an increasing function in p and a decreasing function in k_i with $\theta(0,k_i) = 0$. It is also assumed that $\partial^2\theta/\partial k_i^2 \geq 0$ and $\partial^2\theta / \partial p^2 \geq 0$. Consequently, the following lemma can be obtained.

Lemma 1. *Only those experts with knowledge levels $k_i \geq \hat{k}_i$ will answer the question m with a price p, where $\hat{k}_i$ is the threshold knowledge level such that*

Table 1. Summary of notation

α	proportion of price allocated to experts
$c(\cdot)$	cost of finding the answer with other methods
k_i	expert *i*'s knowledge level
$F(s_j)$	distribution of consumer skill level
$G(k_i)$	distribution of knowledge level
$H(q_m)$	distribution of question difficulty
p	price of a question set by knowledge consumers
$\overline{p}$	minimal posting price on the knowledge market
$\hat{p}$	maximal posting price on the knowledge market
$\rho(\cdot)$	probability of yielding an additional utility
q_m	difficulty of a question *m*
s_j	level of search skills for consumer *j*
$\theta(\cdot)$	disutility of an expert answering a question

$$\alpha p = \theta(p, \hat{k}_i) \cdot q_m. \qquad (2)$$

Proof. See Appendix A. ■

This lemma identifies the threshold knowledge level $\hat{k}_i$; for those experts with a knowledge level higher than or equal to $\hat{k}_i$, they will prefer to answer the question *m* with a price *p* since they will get a non-negative surplus, whereas for those lower than $\hat{k}_i$, they will not answer the question. From another perspective, if a question *m* is more difficult (q_m is larger), it will require an expert to be more knowledgeable to be able to answer it because the threshold knowledge level $\hat{k}_i$ increases in q_m.

In addition, the above lemma assumes that an expert's disutility in answering a question is positively related to its price, which stems from the reputation(rating) system adopted by the firm for the knowledge market. In general, questions with a higher price are more vulnerable to getting a lower rate from their originators for the quality of the answer because those questions are generally more difficult and customers are less easily pleased for a larger amount of payment. In this sense, although questions with high price tags seem more profitable, they are more susceptible to bad evaluations; therefore, experts from a long term perspective may not be willing to risk their reputations to answer those expensive questions. In contrast, questions with low price tags, although they are less profitable, may be more popular to experts because they may help build the long-term reputations on the market.

Following the above lemma, a question with a price *p* to be answered is $\int_{\hat{k}_i}^{1} g(k_i)dk_i$, where $g(k_i)$ is the probability density function of experts' knowledge levels and we assume that k_i is uniformly distributed in [0,1]. From the above discussion, it can be observed that the probability of a question to be answered decreases with its price, whereas the quality of the answer increases in its price because it will be answered by an expert with a higher knowledge level. If the question *m* is not answered, a customer *j* will incur a cost $c(s_j, q_m)$ for exerting effort to get the answer by himself(herself) or from other means in order to enjoy the utility *u*, where s_j describes the ability of the customer *j* in finding the answer without the help of knowledge market. For instance, customers may use online search engines to search for the relevant documents by themselves. The amount of time they spend varies with their skills in using the search engine. Typically, a more skillful consumer (a higher s_j) incurs less cost $c(s_j, q_m)$ for finding the answer to the same question. If his(her) question m is finally answered, the customer who sets the price p for his(her) question m may get additional utility $\Delta u(q_m)$ with the probability $\rho(p)$, which increases in the price. The probability $\rho(p)$ of obtaining additional utility being an increasing function of the price can be justified from the following two aspects. First, questions with higher prices will be answered by experts with higher knowledge levels; therefore, consumers may get better answers if they set higher prices for their questions. Second, as shown later, a question with a higher price implies that its originator is more knowledgeable or values his(her) question more; hence, experts will try to provide a better answer in order to be better evaluated.

A consumer's level of searching skills s_j is unknown to experts and the firm and can only be perceived from a probability distribution function $f(s_j)$. Therefore, a customer with the skill level s_j maximizes his(her) expected surplus,

$$\pi_c = [u + \rho(p) \cdot \Delta u(q_m) - p] \cdot$$

$$\int_{\hat{k}_i}^{1} g(k_i)dk_i + [u - c(s_j, q_m)] \cdot \left[1 - \int_{\hat{k}_i}^{1} g(k_i)dk_i\right], \quad (3)$$

by finding a best price p for his(her) question on the knowledge market. Given the questions posted on the knowledge market, the firm maximizes its expected payoff from these questions,

$$\pi = E_{q_m} E_{s_j} \left[(1-\alpha) \cdot p \cdot \int_{\hat{k}_i}^{1} g(k_i)dk_i\right], \quad (4)$$

by specifying the appropriate allocation proportion α. The allocation proportion α, the actual proportion of payment to experts for answering each question, can be regarded as the incentive to induce experts to answer questions because they get more out of answering each question for a larger α. In addition, it can be inferred from Lemma 1 that a larger α increases the success rate of knowledge "transaction" (questions being answered) since the threshold knowledge level $\hat{k}_i$ decreases in α. Intuitively, if an expert can get more payoff from answering each question(or if the firm assigns a higher α), he(she) will be willing to answer those questions with lower prices because he(she) will be able to extract positive surpluses from these questions with a higher allocation proportion α. Therefore, each expert will be willing to answer more questions with a larger allocation proportion α, which implies that a question will become lucrative to more experts, increasing its probability of being answered. On the negative side, increasing the allocation proportion α may jeopardize the quality of answers to each question and decrease the firm's net profit. Hence, the firm has to balance these tradeoffs so as to choose a best allocation proportion to experts.

ANALYSIS AND DISCUSSION

This section details our analysis and discussion on various pricing strategies of an online knowledge market. Beginning with the analysis of individual consumers' best pricing strategy, we identify two types of consumers on the knowledge market: *spin-off information consumers* and *mainstream knowledge consumers*, with regard to whether additional utilities can be derived from the knowledge market. Based upon this differentiation, we then discuss the firm's purpose of specifying minimal and maximal posting prices and finally explore the firm's optimal allocation proportion to experts.

Consumers' Pricing Strategy

A knowledge consumer's best pricing strategy is investigated in this subsection, leading to the differentiation of two types of consumers on the knowledge market.

First of all, a knowledge consumer will always set his(her) best price for a posted question on the market. Assuming a uniform distribution for $G(k_i)$, we can rewrite a consumer j's total surplus from Equation (3) as

$$\pi_c = (u - c(s_j, q_m)) + [\rho(p)\Delta u + c(s_j, q_m) - p] \cdot (1 - \hat{k}_i),$$

where the second term is contingent on $\hat{k}_i$, the lowest knowledge level required to answer the question, which fluctuates with the customer's price p according to Lemma 1. Hence, the customer can price his(her) question according to the following lemma.

Lemma 2. *A knowledge consumer will set his(her) question with the optimal price p^* where*

$$\frac{\rho'(p^*)\Delta u - 1}{\rho(p^*)\Delta u + c(s_j, q_m) - p^*} = \frac{\partial \hat{k}_i / \partial p}{1 - \hat{k}_i} \quad (5)$$

and $\rho(p^*)\Delta u \geq p^* - c(s_j, q_m)$.

Proof. See Appendix B. ■

Lemma 2 suggests that there exists a best price p^* for a customer to post a specific question to maximize his(her) surplus because the LHS of Equation (5) decreases and the RHS increases in p. Therefore, an intersection always exists between the curves of LHS and RHS with respect to the price p. For a consumer with a better skill level (larger s_j) or a larger additional utility, the LHS increases, resulting in a larger p^*. However, the condition, $\rho(p^*)\Delta u \geq p^* - c(s_j, q_m)$, has to be satisfied, which implies that a customer should always offer a price p^* less than his(her) own cost $c(s_j, q_m)$ of finding the answer unless the expected additional utility $\rho(p^*)\Delta u$ he(she) may derive from the knowledge market can be justified by offering the price.

A knowledge consumer's pricing strategy critically depends on the additional utility Δu he(she) may obtain from the knowledge market. It can be inferred from Equation (5) that under certain conditions, a knowledge consumer will be willing to pay a higher price for a question if the answer can bring more additional utility Δu, which is stated in the following proposition.

Proposition 1. *A consumer will set a higher price p^* for his(her) question if he(she) can derive more additional utility Δu from it and*

$$\frac{\rho'(p^*)}{\rho(p^*)} > \frac{1}{p^* - c(s_j, q_m)}. \quad (6)$$

Proof. See Appendix C. ■

The condition in Proposition 1 parallels that in Lemma 2, implying again that the cost $c(s_j, q_m)$ serves as a critical point for a consumer to decide whether or not to increase his(her) price when he(she) may derive more additional utility from the answer of the question. The condition suggests that when p is less than $c(s_j, q_m)$, the condition always holds, whereas when p is greater than $c(s_j, q_m)$, the condition may not hold. When the additional utility keeps increasing, the optimal price increases as well until the critical point $c(s_j, q_m)$, beyond which point the consumer has to consider whether keeping increasing the optimal price will increase the probability of obtaining the additional utility so that the increased cost of payment can be compensated.

In summary, when a consumer seriously seeks answers from the knowledge market to derive additional utilities, he(she) may find a best price to post his(her) question on the market under certain general conditions. However, when the additional utility Δu is very small, especially when $\Delta u \to 0$, a consumer's pricing strategy will change dramatically, which is presented in the next proposition.

Proposition 2. *When $\Delta u \to 0$*, or no additional utility is derived from the answer of his(her) question, a consumer will set the price at $\bar{p}$ for his(her) question, where $\bar{p}$ is the minimal price required to post a question on the market.

Proof. See Appendix D. ■

The phenomena described in Proposition 2 can be truly observed from the knowledge market previously maintained by Google Answers; about 15% of questions were priced at the minimal posting price $2. In such case, the consumers do not care whether they will be able to derive any additional utilities from the market, but regard the market as an alternative to find answers by themselves. Therefore, the prices set by these consumers do not properly reflect the value of their questions. Nevertheless, for those consumers who utilize the knowledge market to derive additional utilities, their prices will be set according to Equation (5); the more utility they expect to get, the higher prices they are willing to pay. Following this rational, the consumers on the knowledge market can be differentiated into the following two categories.

- **spin-off information consumer**: those who are only interested in using the knowledge market as an alternative to finding information without value added to their utilities.
- **mainstream knowledge consumer**: those who utilize the knowledge market to acquire knowledge to derive additional utilities.

As we have assumed, the additional utility is related with the difficulty q_m of the question, so a more difficult question will give a consumer more additional utility, which, however, is only applicable to mainstream knowledge consumers, but not to spin-off consumers because there is no additional utility attached to their questions no matter how difficult they are. In fact, the highest price that a spin-off consumer with the skill level s_j is willing to pay for the answer of a question m is just $c(s_j, q_m)$, whereas a mainstream consumer with the same skill level will be willing to pay an amount higher than $c(s_j, q_m)$ as long as he(she) may derive additional utility from the answer of his(her) question.

The existence of the spin-off consumers has some negative effects on the knowledge market because prices do not reflect their true values as they just post their questions by offering a very low price. As a former expert of Google Answers, West (2002) made the similar observations as follows. "Of the questions I picked, some fell in my areas of expertise — technical support issues, historical facts, etc. — and some were just plain odd. One person offered a few bucks for someone to tell him a joke he hadn't heard before. Another wanted psychic advice, or barring that, a humorous reply. People used the service as an impromptu temp agency, offering a few dollars for someone to test drive a Web site or to make business appointments for them. While the service was intended to offer answers to factual questions, people tried to push the envelope any way they could."

Although the group of spin-off consumers seems to generate some undesirable outcomes to the knowledge market, it may also benefit the firm to retain some of them for several reasons. First of all, spin-off consumers are usually those who keep abreast of the state-of-art technologies and new phenomena. The "word of mouth" effect through spin-off consumers can bring more customers to the market. Second, network effects require the firm to attract more questions to the knowledge market so that more consumers may be enticed to participate in the "asking-and-answering" process. Third, maintaining a certain volume of questions on the knowledge market may alleviate the malicious competition among experts. Therefore, it is also necessary to welcome some spin-off consumers on the knowledge market.

Minimal and Maximal Posting Prices

In this subsection, we analyze the purpose of the firm specifying the minimal and maximal posting prices on the knowledge market.

First of all, designating a minimal posting price has two effects on the knowledge market. Next proposition illustrates the effect of a minimal posting price on the spin-off consumers of the knowledge market.

Proposition 3. *A spin-off consumer j will not post his(her) question m on the knowledge market if the minimal posting price* $\overline{p}$ *specified by the firm is greater than* $c(s_j, q_m)$.

Proof. See Appendix E. ■

Proposition 3 implies that the firm can specify the minimal posting price $\overline{p}$ on the knowledge market to reduce the proportion of spin-off consumers. Given the minimal posting price $\overline{p}$ specified by the firm, it can be inferred that spin-off consumers with the skill level $s_j \geq \overline{s}_j$ will not be served on the knowledge market, where

$$c(\overline{s}_j, q_m) = \overline{p},$$

and $\overline{s}_j(\overline{p})$ is the threshold skill level for the minimal posting price. The minimal posting price $\overline{p}$ determines the proportion of spin-off consumers to be served on the knowledge market. When the firm increases the minimal posting price $\overline{p}$, the threshold skill level $\overline{s}_j(\overline{p})$ decreases, so more spin-off consumers will be eliminated from the knowledge market.

In addition to its effect on the spin-off consumers, the minimal posting price also affects the mainstream consumers, but in a different way. Next proposition summarizes this effect.

Proposition 4. *All the mainstream consumers will be served by the knowledge market no matter what a minimal posting price is specified. In addition, for a minimal posting price* $\overline{p}$, *those mainstream consumers with the skill level* $s_j < s_j^l(\overline{p})$ *will be mandated to price their questions at* $\overline{p}$, *where the skill level* $s_j^l(\overline{p})$ *of mainstream consumers is determined by*

$$\frac{\rho'(\overline{p})\Delta u - 1}{\rho(\overline{p})\Delta u + c(s_j^l(\overline{p}), q_m) - \overline{p}} = \frac{\partial \hat{k}_i / \partial p}{1 - \hat{k}_i}.$$

Proof. See Appendix F. ■

Proposition 3 and 4 suggest that less skilled consumers, either spin-off or mainstream types, are generally less sensitive to the minimal posting price specified by the firm for the knowledge market. For spin-off consumers, the maximal price they are willing to pay to obtain a positive surplus from the market is $c(s_j, q_m)$, which decreases in the consumer's skill level. For mainstream consumers, their maximal price for a positive surplus is determined from $\tilde{p} = \rho(\tilde{p})au + c(s_j, q_m)$, where $\tilde{p}$ also decreases in the skill level s_j. Therefore, a less skilled consumer for the knowledge market always has a larger support of price than a more skilled one. Intuitively, this result directly relates to our assumption that a less skilled consumer incurs a higher cost for finding an answer to the question by himself(herself) than a more skilled one. Therefore, a less skilled consumer is willing to pay more for answers to his(her) question.

In summary, by specifying a minimal posting price $\overline{p}$ for the knowledge market, the firm can eliminate some spin-off consumers and mandate some mainstream consumers to increase their posting prices to $\overline{p}$. The reduced proportion of spin-off consumers on the knowledge market is thus $\overline{s}_j(\overline{p}) \cdot \gamma$, where γ is the originally estimated proportion of spin-off consumers.

Having described the effects of the minimal posting price, we next turn to analyze the influence of a maximal posting price on the knowledge market. As we have illustrated, the existence of spin-off consumers is due to the fact that some consumers do not derive additional utilities from the market ($\Delta u = 0$), so that the term $\rho(p)\Delta u$ is reduced into zero and a consumer's additional surplus from the market ($[c(s_j, q_m) - p](1 - \hat{k}_i)$) becomes a decreasing function of price p. In addition to the effect of the additional utilities Δu, the probability function $\rho(p)$ also has the similar effect when $\rho(p) \rightarrow 1$. When a consumer realizes that the probability function has such property that when $p \geq \hat{p}, \rho(p) = 1$ and $\rho'(p) = 0$, he(she) will price his(her) question at $\hat{p}$ instead of p^* when $p^* > \hat{p}$. Suppose a consumer is not aware of this property of the probability function, he(she) will still set a very high price p^* for his(her) question. The consumer's total surplus in this situation will be

$$\pi_c = (u - c(s_j, q_m)) + [\rho(p^*)\Delta u + c(s_j, q_m) - p](1 - \hat{k}_i(p^*))$$
$$\approx (u - c(s_j, q_m)) + [\Delta u + c(s_j, q_m) - p](1 - \hat{k}_i(p^*)).$$

Since we know that $\rho(p) \approx 1$ when $p \in [\hat{p}, p^*)$, then the consumer will be better off by setting his(her) price at $\hat{p}$, which barely reduce his(her) utility but increase the probability for his(her) question to be answered.

Therefore, by announcing a maximal posting price, the firm implicitly informs consumers of the boundary price $\hat{p}$ so that they will be better off. In addition, the firm will also be better off, which is shown in the following proposition.

Proposition 5. *The firm is better off by specifying a maximal posting price* $\hat{p}$ *such that when all* $p^* > \hat{p}$, $\hat{k}_i(p^*) \to 1$.

Proof. See Appendix G. ■

In summary, the specified maximal posting price not only may increase a mainstream consumer's surplus, but also makes the firm better off.

Firm's Optimal Decisions

The firm should determine the appropriate allocation proportion to experts, based on the specified minimal and maximal posting prices on the knowledge market in order to maximize its total expect profit. Therefore, the firm's decision problem can be formulated as

$$\max_{\alpha} E_{q_m}[\pi_1 + \pi_2 + \pi_3 + \pi_4], \tag{7}$$

where

$$\pi_1 = \gamma \int_0^{\overline{s}_j(\overline{p})} \left[(1-\alpha)\overline{p} \int_{\hat{k}_i(\overline{p})}^1 g(k_i)dk_i \right] f(s_j)ds_j,$$

$$\pi_2 = (1-\gamma) \int_0^{s_j^l(\overline{p})} \left[(1-\alpha)\overline{p} \int_{\hat{k}_i(\overline{p})}^1 g(k_i)dk_i \right] f(s_j)ds_j,$$

$$\pi_3 = (1-\gamma) \int_{s_j^l(\overline{p})}^{\hat{s}_j(\hat{p})} \left[(1-\alpha)p^* \int_{\hat{k}_i(p^*)}^1 g(k_i)dk_i \right] f(s_j)ds_j,$$

$$\pi_4 = (1-\gamma) \int_{\hat{s}_j(\hat{p})}^1 \left[(1-\alpha)\hat{p} \int_{\hat{k}_i(p)}^1 g(k_i)dk_i \right] f(s_j)ds_j.$$

As shown in Figure 1, the firm's total profit consists of four components. The first component π_1 is the profit from the remaining spin-off consumers by specifying the minimal posting price $\overline{p}$, in which γ is the initial proportion of spin-off consumers on the knowledge market. The second is the profit from mainstream consumers with the minimal posting price, the third is from mainstream consumers who set their optimal prices in between the minimal and maximal posting prices, and the fourth is again from mainstream consumers who are forced to have their questions priced at the maximal level $\hat{p}$.

According to our previous analysis, when the firm increases the minimal posting price $\overline{p}$ and decreases the maximal posting price $\hat{p}$, $\overline{s}_j(\overline{p})$ decreases, $s_j^l(\overline{p})$ increases, and $\hat{s}_j(\hat{p})$ decreases. Therefore, less consumers will be able to price their questions at their optimal ones p^* and more mainstream consumers will price their questions at either the minimal or maximal price while more spin-off consumers will be eliminated from the market. Therefore, the firm may regulate the knowledge market by designating different minimal and maximal posting prices.

Due to the complex nature of the firm's maximization problem, the closed-form solution of α is not available. However, Table 2 demonstrates how the profit structure of each component will change when the allocation proportion α to experts increases, which gives the intuition for the firm to practically assign the allocation proportion α. Essentially, the firm has to balance the tradeoff between the gross revenue of each question and the payment to experts; by increasing the allocation

Figure 1. The firm's profit: four components

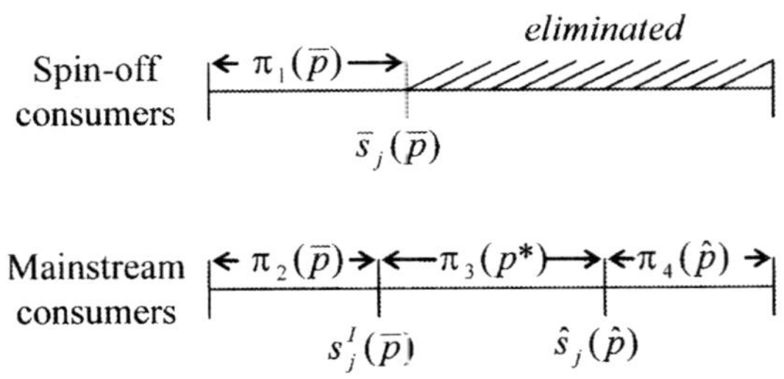

to experts, the firm may increase the total gross revenue from each question but have to allocate more to experts.

CONCLUSION

The development of web and information technologies provides real-time access for people to seek and acquire information and knowledge from online sources. Based on the best practice from Uclue and the prior services provided by Google Answers, we present and study an analytical model of online knowledge market, where knowledge experts and consumers trade knowledge.

We analyze the pricing strategies from both consumers' and a firm's perspectives. First, we study a consumer's optimal strategy to price his(her) question. Contingent on whether the knowledge market is used to derive additional utilities, we differentiate two types of consumers on the market: spin-off and mainstream customers. Second, we investigate the purpose of specifying the minimal and maximal posting prices on the market. By specifying different minimal and maximal posting prices, the firm can effectively moderate the structure of the online knowledge market. Finally, we explore the firm's optimal allocation proportion to experts. The firm has to balance the tradeoff between its payment to experts and its increased benefit in order to choose the best allocation to experts.

This study serves as our initial attempt to fully understand the pricing mechanism on the knowledge market. Future research may empirically verify the results in this study, incorporate network effect in the model, and explore the variations of pricing strategies on the knowledge market.

REFERENCES

Anderson, S., & Palma, A. (1992). The Logit as a Model of Product Differentiation. *Oxford Economic Papers*, *44*, 51–67.

Table 2. Changes of profit components with respect to allocation proportion α

Component	Organizational proportion	Selling price	Market share	Probability of questions being answered
π_1	decrease	same	Same	increase
π_2	decrease	same	Decrease	increase
π_3	decrease	increase	Increase	increase
π_4	decrease	same	Increase	increase

Ba, S., Stallaert, J., & Whinston, A. B. (2001). Optimal Investment in Knowledge within a Firm using a Market Mechanism . *Management Science, 47*(9), 225–239. doi:10.1287/mnsc.47.9.1203.9781

Bakos, Y. (1997). Reducing Buyer Search Costs: Implications for Electronic Marketplaces. *Management Science, 43*(12), 1676–1692. doi:10.1287/mnsc.43.12.1676

Bakos, Y. (1998). The Emerging Role of Electronic Marketplaces on the Internet. *Communications of the ACM, 41*(8), 35–42. doi:10.1145/280324.280330

Bockem, S. (1994). A Generalized Model of Horizontal Product Differentiation. *The Journal of Industrial Economics, 42*, 287–298. doi:10.2307/2950571

Cahill, K. (2007). Worth the Price? Virtual Reference, Global Knowledge Forums, and the Demise of Google Answers. *Journal of Library Administration, 46*(3/4), 73–86.

Cohen, M., & Whang, S. (1997). Competing in Product and Service: A Product Life-Cycle Model. *Management Science, 43*, 535–545. doi:10.1287/mnsc.43.4.535

Davenport, H. T., & Prusak, L. (1998). *Working Knowledge: how organizations manage what they know*. Boston, MA: Harvard Business School Press.

Desouza, K. C., & Awazu, Y. (2003). Constructing internal knowledge markets: Considerations from mini cases. *International Journal of Information Management, 23*, 345–353. doi:10.1016/S0268-4012(03)00056-2

Desuoza, K., Awazu, Y., Yamakawa, S., & Umezawa, M. (2005). Facilitating knowledge market mechanism. *Knowledge and Process Management, 12*(2), 99–107. doi:10.1002/kpm.226

Edelman, B. (2004). *Earnings and Ratings at Google Answers* (Working draft). Harvard Law School, Harvard University.

Kenney, A. R., McGovern, N. Y., Martinez, I. T., & Heidig, L. J. (2003). Google Meets eBay-What Academic Librarians Can Learn from Alternative Information Providers. *D-Lib Magazine, 9*(6). doi:10.1045/june2003-kenney

Lancaster, K. (1990). The Economics of Product Variety: A Survey. *Marketing Science, 9*, 189–206. doi:10.1287/mksc.9.3.189

MÄuller. R. M., Spiliopoulou, M., & Lenz, H.-J. (2002). Electronic Marketplaces of Knowledge: Characteristics and Sharing of Knowledge Assets. In *Proceedings of the International Conference on Advances in Infrastructure for e-Business, e-Education, e-Science, and e-Medicine on the Internet (SSGRR 2002w)*.

Rafaeli, S., Raban, D. R., & Ravid, G. (2007). How social motivation enhances economic activity and incentives in the Google Answers knowledge sharing market. *International Journal of Knowledge and Learning, 3*(1), 1–11. doi:10.1504/IJKL.2007.012598

West, J. (2002). Information for Sale: My Experience with Google Answers. *Searcher - The Magazine for Database Professionals, 10*(9), 14.

APPENDIX

A. Proof of Lemma 1

Proof. An expert will only answer a question when his(her) total payoff from this question is positive. For a given question m with price p, only those experts with $k_i \geq \hat{k}_i$ will have positive payoffs, where $\alpha p = \theta(p, \hat{k}_i) \cdot q_m$.

B. Proof of Lemma 2

Proof. The first order condition of π_c with respect to p is

$$(1-\hat{k}_i)(\rho'(p)\Delta u - 1) - (\rho(p)\Delta u + c(s_j, q_m) - p)\frac{\partial \hat{k}_i}{\partial p} = 0$$

and the second order derivative is

$$\frac{\partial^2 \pi_c}{\partial p^2} = (1-\hat{k}_i)\rho''(p)\Delta u - (\rho(p)\Delta u + c(s_j, q_m) - p)\frac{\partial^2 \hat{k}_i}{\partial p^2} < 0$$

if $\rho(p^*)\Delta u \geq p^* - c(s_j, q_m)$, because $\partial^2 \hat{k}_i / \partial p^2 > 0$, which can be derived from Lemma 1 and our assumptions about an expert's disutility that $\partial^2 \theta / l k_i^2 \geq 0$ and $\partial^2 \theta / \partial p^2 \geq 0$.

C. Proof of Proposition 1

Proof. The derivative of RHS of Equation (5) with respect to Δu is zero. Hence, the optimal price p^* will increase with the additional utility Δu if $\rho'(p^*)(c(s_j, q_m) - p^*) + \rho(p^*)$, the derivative of LHS of Equation (5) with respect to Δu is greater than zero, which is true if the above condition holds.

D. Proof of Proposition 2

Proof. When $\Delta u \to 0$, a knowledge consumer's expected payoff is

$$\pi_c = (u - c(s_j, q_m)) + [c(s_j, q_m) - p](1-\hat{k}_i),$$

where the second term, the benefit derived from the knowledge market, decreases in the price p. Therefore, the optimal price is achieved as the boundary solution, i.e., the consumer will set the price at $\overline{p}$, the minimal price to post a question on the market.

E. Proof of Proposition 3

Proof. If the firm specifies a minimal posting price as $\overline{p}$, then a spin-off consumer's expected payoff by posting a question m is

$$\pi_c = (u - c(s_j, q_m)) + [c(s_j, q_m) - p](1 - \hat{k}_i),$$

whose second term is positive and decreasing in p when $p < c(s_j, q_m)$ and negative when $p > c(s_j, q_m)$. Therefore, the consumer will post a question if the minimal posting price $\overline{p} \le c(s_j, q_m)$. Otherwise, he(she) is better off by not posting the question on the knowledge market.

F. Proof of Proposition 4

Proof. For questions with the difficulty type as q_m, mainstream consumers with the skill level $s_j < s_j^I(\overline{p})$ will price their questions at $\overline{p}$, where the threshold skill level $s_j^I(\overline{p})$ of consumers is from

$$\frac{\rho'(\overline{p})\Delta u - 1}{\rho(\overline{p})\Delta u + c(s_j^I(\overline{p}), q_m) - \overline{p}} = \frac{\partial \hat{k}_i / \partial p}{1 - \hat{k}_i}.$$

However, a sufficient condition for all the consumers with the skill level $s_j \in (0, s_j^I(\overline{p})]$ to remain served by the market is $s_j^P(\overline{p}) \le s_j^I(\overline{p})$, where the threshold skill level $s_j^P(\overline{p})$ is determined from

$$\overline{p} = \rho(\overline{p})\Delta u + c(s_j^P(\overline{p}), q_m).$$

This is to ensure that all the mainstream consumers with the skill level $s_j \in (0, s_j(\overline{p})]$ have positive surpluses by posting their questions with the minimal price $\overline{p}$ on the market. By comparing the two threshold skill levels, it is not difficult to see that consumers who are mandated to price their questions at the minimal posting price always have positive surpluses. First, it can be seen that at the minimal posting price, a customer whose skill level is $s_j^I(\overline{p})$ always has a positive surplus because the maximal price for the skill level $s_j^I(\overline{p})$ to have a positive surplus is greater than $\overline{p}$. Second, for a consumer with the skill level $s_j < s_j^I(\overline{p})$, the maximal prices for him(her) to have a positive surplus is greater than that for the skill level $s_j^I(\overline{p})$, which is certainly greater than $\overline{p}$.

G. Proof of Proposition 5

Proof. The firm's net benefit for a posted question m is

$$\pi^m(p^*) = (1 - \alpha) \cdot p^*(1 - \hat{k}_i(p^*)).$$

Instead, if the firm sets a maximal posting price as $\hat{p}$ that is greater than p^*, then the firm's net benefit for this question m will be

$$\pi^m(\hat{p}) = (1 - \alpha) \cdot \hat{p} \cdot (1 - \hat{k}_i(\hat{p})).$$

It can be observed that $\pi^m(p^*) \to 0$ when $\hat{k}_i(p^*) \to 1$. Therefore, the firm will be better off by specifying a maximal posting price $\hat{p}$ such that $\pi^m(\hat{p}) > \pi^m(p^*)$.

Chapter 13
Product Choice Strategy for Online Retailers

Ruiliang Yan
Virginia State University, USA

Amit Bhatnagar
University of Wisconsin-Milwaukee, USA

ABSTRACT

An important strategic issue for managers planning to set up online stores is the choice of product categories to retail. While the "right" product category would depend on a number of factors, here we focus on the following two factors: compatibility of the product with the online channel, and the competition between the traditional brick and mortar channel and the online channel. This is to acknowledge two well-known facts: Certain products are more suitable for selling through the Web than through other channels; and an online retailer competes with not only other online retailers, but also traditional brick and mortar retailers. To determine the right product category, we develop a game theoretical model that allows for competition between the retailers. We study both Stackelberg and Bertrand competition models, as these two models capture the essence of different types of competition on the Web. Based on our results, we propose that, under all types of competition, the optimal product is one that is only moderately compatible with the Internet.

INTRODUCTION

There are a number of statistical reports, from the industry and the government, which show that commerce on the Internet is growing at a healthy rate. For instance, according to "The 2006 State of Retailing Online", the ninth annual report published by Shop.org and conducted by Forrester Research, online retail sales are expected to hit $211.4 billion in 2006, a 20% gain over revenues of $176.4 billion in 2005. Again according to the Department of Commerce, quarterly e-commerce sales in the third quarter of 2006 increased 20.4% from the third quarter of 2005. This growth in commerce on the Internet has attracted the attention of business managers, who now accept the Internet as a viable channel to distribute products to consumers.

Managers who are planning to set up online stores need to make a number of strategic decisions, one of which is simply what product to sell. This can be a critical decision, because some products have a greater likelihood of succeeding on the Web. For instance, according to "The 2006 State of Retailing Online" report, the top-selling categories on the Internet are computer hardware and software ($16.8 billion); autos and auto parts ($15.9 billion); and apparel, accessories, and footwear ($13.8 billion). Cosmetics and fragrances ($800 million) and pet supplies ($500 million) are expected to experience over 30% growth in 2006, more than any other categories. A business manager who studies these numbers may be tempted to choose a category, such as computer hardware, software, or auto parts, which sells in large quantities on the Web. But is this the right strategy?

An important question here would be, why do some categories do very well on the Web and some others do not. The reason is that some products have characteristics that have synergies with the characteristics of the Internet, making it advantageous for consumers to buy these products on the Internet. For instance, software does so well because the Internet allows consumers to download software from the Internet onto their computers. The digital nature of software is very compatible with the digital nature of the Internet. To capture this synergy issue, we introduce a new index, *Web-product compatibility,* which measures the extent of synergy between the characteristics of a product and the Internet. Its value varies from zero to one, where zero signifies no compatibility and one stands for complete compatibility. In a similar vein, Balasubramanian (1998) assumes that the fit with the direct channel varies across product categories. While business managers would be tempted to pick a product that has a *Web-product compatibility* of one, it is not clear if this is the optimal strategy. This is the question that we attempt to answer in this article.

There would always be some products (e.g., music, airline tickets, etc.) where the online channel enjoys overwhelming advantage over the traditional channel, that is, Web-product compatibility is greater than one. Here the decision, whether to choose the category or not, is straightforward. Since the online stores would dominate the traditional brick and mortar stores and, over a period of time, the traditional stores would disappear and only the online stores would remain, the recommendation would be a clear "yes". For illustration, notice the demise of traditional music stores and travel agents. The question is much more complicated for those products where online stores have certain advantages and disadvantages as compared to brick and mortar stores. In such a scenario, neither of the two types of stores enjoys any clear advantage over the other. In this article, we consider only those products for which Web-product compatibility is less than one.

We develop a game theoretical model to study how *Web-product compatibility* impacts the profits of an online retailer that is in competition with a brick and mortar retailer. We consider two competitive market settings, the Stackelberg and Bertrand competition models. Based on the optimal profits, we argue that under all kinds of competitive settings, the optimum product to retail would be the one that has only moderate *Web-product compatibility.* This is a counterintuitive finding and the main contribution of this study.

The rest of our article is organized as follows. The second section provides a summary of the relevant literature. The third section presents our modeling framework. We assume that both the traditional and the brick and mortar retailer are horizontally integrated, and then we determine the optimal pricing policies and joint profits. The fourth section introduces Web-product compatibility, and studies its impact on the profits of online and traditional retailers under different types of competitive settings, the Stackelberg and Bertrand competition models. In the fifth section, we illustrate the influence of Web-product compatibility on the profits of online and traditional retailers by means of a computational study. The sixth section has the concluding remarks and managerial implications.

LITERATURE REVIEW

In this section, we review the relevant marketing literature to position our article. First, we review some of the literature that examines issues which arise when manufacturers sell through multiple channels. In particular, we focus on the literature that explicitly accounts for the presence of the Internet channel. Next, we review some of the literature that shows that, from a consumer's perspective, purchasing on the Internet is not the same as purchasing from a traditional retail channel. Therefore, the utility that a consumer derives from purchasing some specific product on the Internet is less than that derived from purchasing the same product from a traditional brick and mortar store.

Multi-Channel Competition

There is a long marketing tradition of studying issues that arise from selling across multiple channels. With the emergence of the Internet as a viable channel of distribution, study of multi-channel competition has acquired an additional importance. Researchers have examined a variety of issues. Balasubramanian (1998) modeled competition in a multiple-channel environment from a strategic perspective. The level of information disseminated by the direct marketer is shown to have strategic implications, and the author showed that level of market coverage can be used as a mechanism to control competition. Rhee and Park (1999) studied the multiple-channel design problem when there are two distinct consumer segments: a price-sensitive segment and a service-sensitive segment. They showed that multiple channels are optimal when the segments are similar in the values that they assign to the retail services. Levary and Mathieu (2000) studied the profits of physical retail stores, online stores, and hybrid retails. They argued that, in the future, hybrid retails would have the maximum optimal profits. Geyskens, Gielens, and Dekimpe (2002) found that powerful firms with a few direct channels achieve better financial performance than less powerful firms with broader direct-market offerings. King, Sen, and Xia (2004) used a game-theoretic approach to study the impact of Web-based e-commerce on a retailer's distribution channel strategy. They showed that the multi-channel strategy followed by retail firms is an equilibrium outcome of the game resulting from competitive pressure by other retailers.

Some of the studies specifically focus on the price competition between traditional retailers and brick and mortar retailers. For instance, Brynjolfsson and Smith (2000) showed through an empirical study that the prices are 9-16% lower on the Internet than in conventional outlets. Smith, Bailey, and Brynjolfsson (2000) did an empirical study to show that online prices for digital products are lower than traditional brick and mortar prices. Yao and Liu (2005), Cattani, Gilland, Heese, and Swaminathan (2006), and Chiang, Chhajed, and Hess (2003) use a game-theoretic model to study the price competition between a manufacturer's direct channel and its traditional channel partner. Chiang, Chhajed, and Hess (2003) argue that the vertically-integrated direct channel allows a manufacturer to constrain the partner-retailer's pricing behavior. However, this may not always be detrimental to the retailer because it may be accompanied by wholesale price reduction. Our article also follows this approach, where we model the price competition between the online retailer and the traditional brick and mortar retailer.

Web-Product Compatibility

Kwak, Fox, and Zinkhan (2002) did an empirical study to examine online purchasing across seven categories: books, information, or magazines; communications services (e.g., Internet phone services); computer-related products and services; education; electronics; entertainment; Internet-related products and services; music and videos; and travel and vacations (e.g., airplane tickets). They found that computer-related products and services, and books, information, and magazines were the most frequently-purchased products, whereas electronics and entertainment were the least-frequently purchased. Liang and Huang (1998) were probably the first to provide a rationale for why some products are more suitable for marketing on the Web than others. Their main thesis is that consumers would go to that channel where the transaction cost is the lowest. Bhatnagar, Misra, and Rao (2000) identified some of the product characteristics that decrease the value of a product when it is purchased on the Web as compared to from a traditional retail channel. Consumers perceive a high degree of risk when a product is technologically complex, satisfies ego-related needs, has a high price, and when "touch and feel" is important. The increase in risk lowers the consumer value. According to Lal and Sarvary (1999), consumers need to gather information about two types of product attributes: digital attributes (which can be communicated on the Web at very low cost) and non-digital attributes (for which physical inspection of the product is necessary). Evidently, consumers would be less willing to buy products with a high proportion of non-digital attributes. It is also difficult to return products on the Web. According to Wood (2001), consumer purchase on the Web is more risky because of the lack of experiential information about product return policies. According to Korgaonkar, Silverblatt, and Girard (2006), credence products are less likely to sell on the Web as compared to search or experience products. All of these factors reduce consumers' utility for products bought online. Therefore, products that are most likely to sell on the Web have Web-product compatibility close to one, and those that are less likely to sell have Web-product compatibility close to zero.

MODEL FRAMEWORK

In this section, we lay out our basic market structure.

The Online and Traditional Retailers Demand Functions

We consider a market setting where online retailers and traditional brick and mortar retailers sell the same product, and compete with each other. In this market, customers can purchase the product from either the online retailer or the traditional retailer. Let the value obtained by consumers from purchasing a product from the traditional retailer be v. We assume consumers are heterogeneous in the value that they obtain, and for analytic simplicity, we assume that v is uniformly distributed within the consumer population from 0 to 1. If the price of the product in a traditional store is p_2, then the consumer surplus in the traditional channel would be $v - p_2$. All consumers with positive consumer surplus (i.e., v is greater than p_2) will buy this product at the traditional channel. The marginal consumer whose valuation v^r equals p_2 is indifferent between buying from the traditional retailer and not at all.

The value obtained by consumers when the same product is purchased online would be less than v. We capture the decrease in value by the parameter θ, which stands for Web-product compatibility. Therefore, the value of the product when purchased on the Web is θv. If the price of the product at an online store is p_1, the resulting consumer surplus would be $\theta v - p_1$. All consumers whose consumer surplus at the online retailer is positive (i.e., $\theta v - p_1 \geq 0$) would consider buying

Figure 1. Schematic graph of the market structure

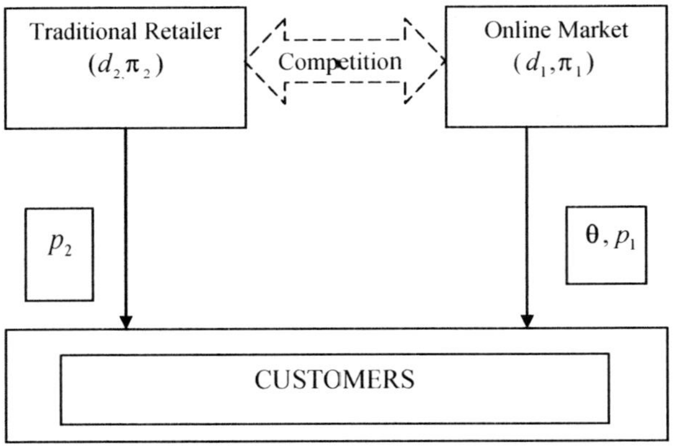

The symbols are as follows,
π_1: *Online Retailer's profits*
π_2: *Traditional Retailer's profits*
d_1: *Demand in the online retail store*
d_2: *Demand in the traditional retail store*
p_1: *Price in the online retail store*
p_2: *Price in the traditional retail store*
θ: *Web-product compatibility*

from the online retailer. The marginal consumer whose value v^d equals p_1/θ is indifferent to buying from the online channel, or not at all.

Since consumers can buy from either retailer, they would prefer the retailer where they derive more surplus (i.e., the online or the traditional retailer: $\theta v - p_1$ versus $v - p_2$). If $v - p_2 \geq \theta v - p_1$, then the traditional retailer would be weakly preferred to the online retailer. The marginal consumer would be one who is indifferent between the two retailers and whose value v^{dr} equals $(p_2 - p_1)/1 - \theta$. Consumers, whose value exceeds this, would prefer the traditional retailer. It can be shown that when $v^d < v^r$, then $v^d < v^r < v^{dr}$, and when $v^d > v^r$, then $v^d > v^r > v^{dr}$. In the former case, all consumers with value in the interval $[v^d, v^{dr}]$ prefer to buy from the online retailer, and all those in the interval $[v^{dr}, 1]$ prefer to buy from the traditional retailer. Consumers whose value is in the interval $[0, v^d]$ will not buy the product from either retailer. Chiang, Chhajed, and Hess (2003) have used a similar market structure, and they show that demand faced by the two retailers would be,

$$d_1 = \begin{cases} \dfrac{\theta p_2 - p_1}{\theta(1-\theta)} & p_2 \geq \dfrac{p_1}{\theta} \\ 0 & otherwise \end{cases} \tag{1}$$

$$d_2 = \begin{cases} 1 - \dfrac{p_2 - p_1}{1-\theta} & p_2 \geq \dfrac{p_1}{\theta} \\ 1 - p_2 & otherwise \end{cases} \tag{2}$$

where, d_1 is the demand at the online retailer and d_2 is the demand at the traditional retailer.

Profit Functions of Online and Traditional Retailers

For the sake of simplicity, we assume that the marginal costs of production and retailing are identical for the two retailers, and are zero. Therefore, in a given period, the traditional retailer's profit would be,

$$\pi_2 = p_2 d_2. \tag{3}$$

An online retailer's profit would be,

$$\pi_2 = p_1 d_1. \quad (4)$$

If the two retailers are horizontally integrated, the total profit would be,

$$\pi_I = p_1 d_1 + p_2 d_2. \quad (5)$$

If the same company owns the two channels, there would be no competition between them. The decision-making is centralized at the corporate headquarters. Therefore, the firm would seek to set prices that maximize the total profits of the whole supply chain, which can be expressed as the following maximization problem,

$$Max\ \pi_I = \pi_1 + \pi_2 = p_1 d_1 + p_2 d_2.$$

Maximizing this equation with respect to p_1 and p_2, we obtain

$$p_1^I = \frac{\theta}{2}$$

and

$$p_2^I = \frac{1}{2}.$$

Substituting these values in Equations 1, 2, 3, 4, and 5, we get,

$$d_1^I = 0,\ d_2^I = \frac{1}{2},\ \pi_1^I = 0,\ \pi_2^I = \frac{1}{4},\ \pi_I = \frac{1}{4}. \quad (6)$$

For the proof, please see Appendix A.

When a firm uses a dual marketing channel strategy, the optimal demand for online retailer is $d_1^I = 0$ and the total profit is equal to the profit when the product is sold only through a traditional retailing channel ($\pi_I = \pi_2$). The rational for this is: Since the value that consumers derive from the online channel is less than what they derive from the traditional channel, it is better to sell through the traditional medium.

THE STACKELBERG AND BERTRAND COMPETITION

When the two channels are not horizontally integrated, they would compete to maximize their market share. There can be different types of competition between the retailers. In the marketing literature, two kinds of competitive games are generally considered, Stackelberg and Bertrand models. We consider both these types of competition, in order to be as inclusive as possible. In what follows, we first consider the Stackelberg game, followed by the Bertrand game.

Stackelberg Competition

First, we consider the Stackelberg competition, where one of the two retailers (market leader) moves first and sets its price, independent of any competing retailer. The second retailer acts as the market follower and sets its price, taking into account the price set by the market leader. Examples of Stackelberg competition in the real world abound. For example, Amazon.com works as a pure online retailer, and Cody's Books works as a traditional retailer; they both sell the same product, books. The competition between them would be like a Stackelberg competition, where Amazon sets the price first, and then Cody's Books sets its price, taking into account the Amazon prices. Another example would be an online travel site like Expedia or Orbitz competing with a traditional neighborhood travel agent. Expedia would determine its own price, and then the traditional travel agent would determine its price, keeping account of the Expedia prices.

We consider two scenarios: first, the online retailer as the market leader, and then the traditional retailer as the market leader.

In the first scenario, we assume that the online retailer (the leader) announces its price p_1 to maximize its profit. In response to p_1, the traditional retailer (the follower) updates its retail price p_2 to maximize its retail profit. Given the above

Table 1. Market strategies of online (market leader) and traditional retailer (market follower) under Stackelberg Equilibrium

		θ's impact
Price		
Online Retailer, p_1	$p_1^S = \frac{\theta(1-\theta)}{2(2-\theta)}$	$\frac{\partial(\partial p_1^S)}{\partial\theta^2} = -\frac{2}{2(2-\theta)^3} < 0$
Traditional Retailer, p_2	$p_2^S = \frac{(1-\theta)(4-\theta)}{4(2-\theta)}$	$\frac{\partial p_2^S}{\partial\theta} = -\frac{6-4\theta+\theta^2}{4(2-\theta)^2} < 0$
Demand		
Online Retailer, d_1	$d_1^S = \frac{1}{4}$	$\frac{\partial d_1^S}{\partial\theta} = 0$
Traditional Retailer, d_2	$d_2^S = \frac{4-\theta}{4(2-\theta)}$	$\frac{\partial d_2^S}{\partial\theta} = \frac{1}{2(2-\theta)^2} > 0$
Profit		
Online Retailer, π_1	$\pi_1^S = \frac{\theta(1-\theta)}{8(2-\theta)}$	$\frac{\partial(\partial\pi_1^S)}{\partial\theta^2} = -\frac{1}{2(2-\theta)^3} < 0$
Traditional Retailer, π_2	$\pi_2^S = \frac{(4-\theta)^2(1-\theta)}{16(2-\theta)^2}$	$\frac{\partial\pi_2^S}{\partial\theta} = -\frac{16-12\theta+6\theta^2-\theta^3}{16(2-\theta)^3} < 0$

For the proof, please see Appendix B

structure, we can obtain the online and traditional retailers' prices, demands, and profits.

In order to examine the effect of Web-product compatibility on the profits of online and traditional retailers, we differentiate their profits with respect to θ. Based on this, we obtain proposition 1.

Proposition 1a: *The profit of the online retailer first increases and then decreases in Web-product compatibility, and*
Proposition 1b: *The profit of the traditional retailer decreases monotonically in Web-product compatibility.*

For the proof, please see Appendix C.

Proposition 1 shows that the profit of the online retailer has a concave relationship with Web-product compatibility. However, the profit of the traditional retailer decreases as the compatibility of a product to the Web increases. The rational is that as compatibility of a product to the Web increases, (before the Web-product compatibility reaches a threshold value θ^*, i.e., $\theta < \theta^*$), the online retailer increases its price; since its demand is constant, an increase in price leads to an increase in its profit. However, when the Web-product compatibility is larger than the threshold value, $\theta \geq \theta^*$, it leads to a very serious and intense competition with the traditional retailer. Since now the online channel becomes a serious alternative to the traditional retailer, the traditional retailer will have to dramatically cut its retail price, and the online retailer will have to respond by cutting its price, too.

Table 2. Market strategies of traditional (market leader) and online retailer (market follower) under Stackelberg equilibrium

		θ's impact
Price		
Online Retailer , p_1	$p_1^S = \frac{\theta}{2}(1 - \frac{1}{2-\theta})$	$\frac{\partial(\partial p_1^S)}{\partial\theta^2} = -\frac{2}{(2-\theta)^3} < 0$
Traditional Retailer, p_2	$p_2^S = 1 - \frac{1}{2-\theta}$	$\frac{\partial p_2^S}{\partial\theta} = -\frac{1}{(2-\theta)^2} < 0$
Demand		
Online Retailer, d_1	$d_1^S = \frac{1}{4-2\theta}$	$\frac{\partial d_1^S}{\partial\theta} = \frac{1}{2(2-\theta)^2} > 0$
Traditional Retailer, d_2	$d_2^S = \frac{1}{2}$	$\frac{\partial d_2^S}{\partial\theta} = 0$
Profit		
Online Retailer, π_1	$\pi_1^S = \frac{\theta(1-\theta)}{4(2-\theta)^2}$	$\frac{\partial(\partial\pi_1^S)}{\partial\theta^2} = -\frac{3\theta}{2(2-\theta)^4} < 0$
Traditional Retailer, π_2	$\pi_2^S = \frac{1-\theta}{2(2-\theta)}$	$\frac{\partial\pi_2^S}{\partial\theta} = -\frac{1}{2(2-\theta)^2} < 0$

For the proof, please see Appendix B

By comparing the effect of Web-product compatibility on the profits of online and traditional retailers, we obtain Proposition 2.

Proposition 2: *The Web-product compatibility has a larger marginal impact on the profit of the traditional retailer than on the profit of the online retailer.*

For the proof, please see Appendix D.

Proposition 2 shows that in a competitive market, Web-product compatibility plays a more important role in the traditional retailer' profit than in the online retailer' profit. That means a small increase in online business will have a larger impact on the traditional retailer's business.

In the second scenario, we assume that the traditional retailer (the leader) first announces a retail price p_2 to maximize its profit. In response to p_2, the online retailer (the follower) updates its price p_1 to maximize its profit. Given the above structure, we obtain the online and retail prices, demands, and profits as in Table 2.

When the traditional retailer is the leader and the online retailer is the follower, as Web-product compatibility increases, the profit of the online retailer first increases and then decreases, and the profit of the traditional retailer decreases continually. As it turns out, these results are similar to Proposition 1, and the reason is the same. We found that, in the first case, the Web-product fit has a larger impact on the profits of the traditional retailer than on the profits of the online retailer. We also found that this result holds for the second case as well.

The Bertrand Competition Model

Under the Bertrand competition, the online and the traditional retailer have equal power in the

Table 3. Market strategies of online and traditional retailer under Bertrand equilibrium

		θ's impact
Price		
Online Retailer, p_1	$p_1^B = \frac{\theta(1-\theta)}{4-\theta}$	$\frac{\partial(\partial p_1^B)}{\partial\theta^2} = -\frac{24}{(4-\theta)^3} < 0$
Traditional Retailer, p_2	$p_2^B = \frac{2(1-\theta)}{4-\theta}$	$\frac{\partial p_2^B}{\partial\theta} = -\frac{6}{(4-\theta)^2} < 0$
Demand		
Online Retailer, d_1	$d_1^B = \frac{1}{4-\theta}$	$\frac{\partial d_1^B}{\partial\theta} = \frac{1}{(4-\theta)^2} > 0$
Traditional Retailer, d_2	$d_2^B = \frac{2}{4-\theta}$	$\frac{\partial d_2^B}{\partial\theta} = \frac{2}{(4-\theta)^2} > 0$
Profit		
Online Retailer, π_1	$\pi_1^B = \frac{\theta(1-\theta)}{(4-\theta)^2}$	$\frac{\partial(\partial\pi_1^B)}{\partial\theta^2} = -\frac{2(8+7\theta)}{(4-\theta)^4} < 0$
Traditional Retailer, π_2	$\pi_2^B = \frac{4(1-\theta)}{(4-\theta)^2}$	$\frac{\partial\pi_2^B}{\partial\theta} = -\frac{4(2+\theta)}{(4-\theta)^3} < 0$

For the proof, please see Appendix E

market. The online retailer determines its price p_1 to maximize its profit π_1. Similarly, the traditional retailer determines its retail price p_2 to maximize its profit π_2. There is no price leader in this market, and both market members make price decisions independently of each other. Given the above structure, we obtain the online and traditional retail prices, demand, and profit as follows.

We saw in proposition 1 that under the Stackelberg competition, as Web-product compatibility increases, the profit of the online retailer has a concave shape, and the profit of the traditional retailer decreases monotonically. As it turns out, these results hold for the Bertrand competition as well, and the reason is the same. We found that, in the Stackelberg case, the Web-product fit has a larger impact on the profits of the traditional retailer than on the profits of the online retailer. We also found that these results hold for the Bertrand competition as well. Next, we compare the competitive pricing strategies under different market structures and identify the market structure that the online retailer and the traditional retailer should employ respectively in order to obtain better profits.

By comparing the pricing strategies between the Bertrand and the Stackelberg equilibriums, we obtain proposition 3.

Proposition 3: *The prices of both the traditional and the online retailer under Stackelberg competition equilibrium are higher than under Bertrand competition equilibrium. This means,* $p_1^S > p_1^B$ *and* $p_2^S > p_2^B$.

For the proof, please see Appendix F.

Proposition 3 shows that the prices of both the online and the traditional retailer under the Stackelberg competition structure are always higher than under the Bertrand competition structure.

The bottom line, however, is the profit, not the prices. Therefore, it is critical to find under which marketing structure the online retailer and the traditional retailer can derive more profits. This can be determined by comparing the profits between the Stackelberg and the Bertrand competition models. Accordingly, we obtain proposition 4,

Proposition 4: *The profits of both the traditional and the online retailer are higher under the Stackelberg competition equilibrium than under the Bertrand competition equilibrium. That is,* $\pi_1^S > \pi_1^B$ *and* $\pi_2^S > \pi_2^B$.

For the proof, please see Appendix G.

Proposition 4 indicates that both the online and the traditional retailers would always prefer the Stackelberg competitive setting.

To ascertain optimal marketing strategies for the retailers, we study the above two propositions in conjunction. When $0 < \theta < 1$ and the two retailers sell the same product competitively, Web-product compatibility plays a strategic impact on their profits. The market competition will be more intense at higher levels of Web-product compatibility, and this would lead to a "price war" between the two types of retailers. Therefore, the profit of the online retailer is related concavely with Web-product compatibility, and the profit of the traditional retailer is decreasing monotonically in Web-product compatibility. However, both the online retailer and the traditional retailer can adopt a leader-follower strategy to improve their profits.

COMPUTATIONAL STUDY

In this section, we do some numerical analysis to illustrate our findings. We focus on situations where both the online and the traditional retailer profits are influenced by Web-product compatibility. We examine the impact of θ on each retailer's profit when θ varies from 0 to 0.99. θ is Web-product compatibility of products such as software, hardware, shoes, clothes, computers, and so forth, and therefore all of its values are feasible in a business market. We varied the value of θ from 0 to 0.99, and substituted these values into the two retailers' profit functions under the two different competitive structures, the Stackelberg and the Bertrand competition models.

Figure 2 shows that, under the Stackelberg competition, the profits of both the online and the traditional retailers will be impacted by the Web-product fit. The profit of the online retailer is concavely related to the value of θ, and the profit of the traditional retailer is continuously decreasing with the value of θ. Also, the value of θ has a larger impact on the traditional retailer' profit than on the online retailer' profit in the Stackelberg competition.

We further investigated the effect of θ on the two retailers' profits in the Bertrand case. As it turns out, the above results hold for the Bertrand case as well (please see Figure 3). The profit of the online retailer is concavely related to the value of θ, and the profit of the traditional retailer decreases monotonically with θ. Also, the value of θ has a larger impact on the traditional retailer' profit than on the online retailer' profit under the Bertrand competition. In Figure 4 and 5, we compare the profits of the two retailers under different competitive scenarios. It is clear that under the Stackelberg equilibrium, both the retailers would make a higher profit.

CONCLUSION AND MANAGERIAL IMPLICATIONS

The contributions of this study are both theoretical and substantive in nature. In this article, we study the strategic role of Web-product compatibility in online retailer-traditional retailer competition. We derive strategies for the two types of retailers

Figure 2. The effect of θ *on the retailer's profits in the Stackelberg competition*

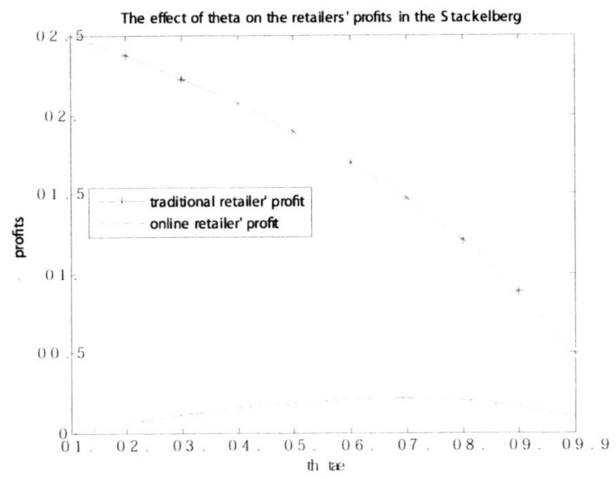

Figure 3. The effect of θ *on the retailer's profits in the Bertrand competition*

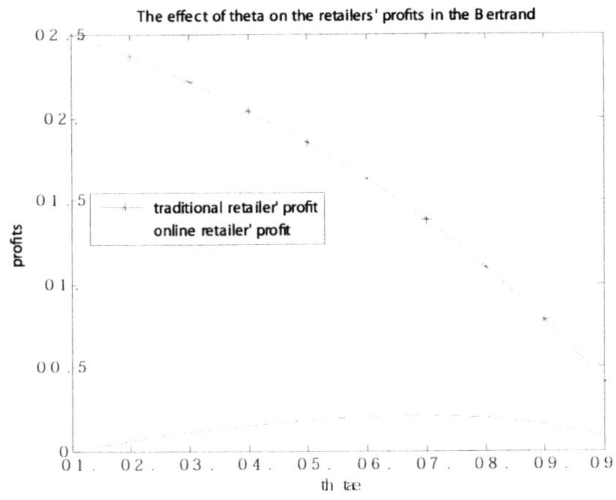

Figure 4. The effect of θ *on the traditional retailer's profits in the Stackelberg and Bertrand competition*

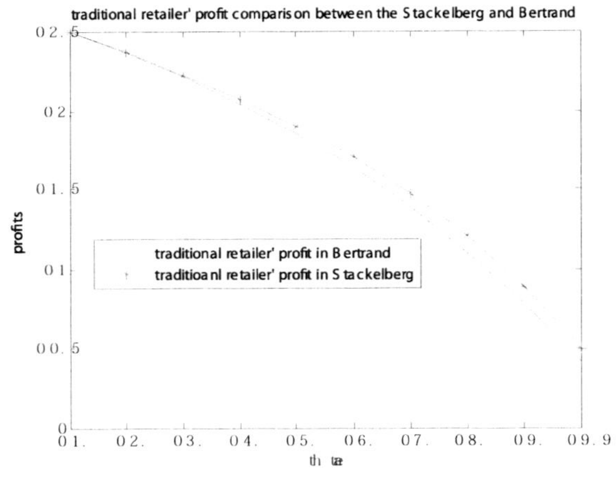

Figure 5. The effect of θ *on the online retailer's profits in the Stackelberg and Bertrand competition*

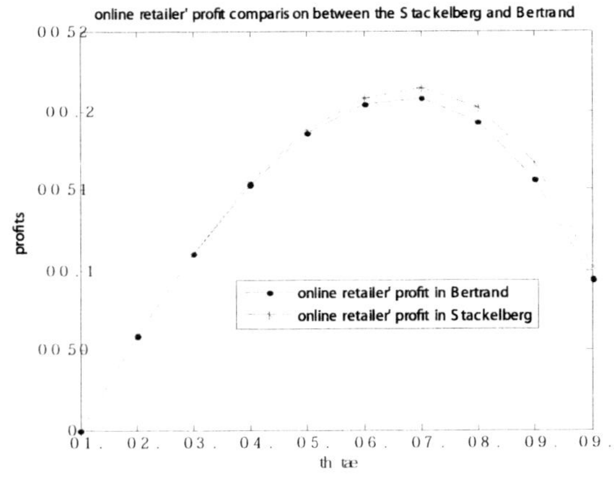

under Bertrand and Stackelberg competition models. Our results indicate that when Web-product compatibility is greater than zero and less than 1($0 < \theta < 1$), the profits of the online retailers are related concavely to Web-product compatibility, but the profits of the traditional retailer decrease monotonically with Web-product compatibility. Also, Web-product compatibility has a larger impact on the profits of the traditional retailer. Both the online retailer and the traditional retailer can adopt a leader-follower strategy to increase their profits.

This research has serious implications for product-choice decisions of online retailers. An important topic of industry symposium, consultant talks, Internet marketing books, and so forth, is, what are good products to retail on the Web. Most of the industry practitioners and consultants tend to rate the products based on the benefits that would accrue to consumers when they buy these products online. The argument seems to be that consumers would prefer buying those products online where they would have lots of advantages as compared to the traditional channel. However, if products are very compatible with the Internet, they do very well, but this comes at the expense of traditional retailers. Generally, this can trigger an intense and serious competition in the industry leading to "price war", somewhat similar to the price wars seen in the airline industry. Since traditional retailers have deeper pockets, they are likely to prevail in these kinds of war, and therefore online retailers should avoid getting into a price war. They can do this by retailing those product categories that are only moderately suited to the Internet. In these products, consumers enjoy some advantages when they shop for these products online, but they also enjoy some disadvantages vis-à-vis traditional retailers. This leads to segmentation of the market, with the segment that values the advantages of the Internet preferring the online channel, and the segment that values the advantages of the traditional channel favoring the traditional channel. This kind of market segmentation avoids a head-on competition and debilitating price wars. This finding is our main contribution to this literature.

The research contained in this study may be extended in a number of different ways. In this article, we assumed that consumers have perfect information. However, information with the consumers could be incomplete, and we can explore the competitive equilibrium under incomplete information settings. We can also extend our study by including more than one traditional retailer and one online retailer. Since the subject of market competition and coordination is of interest to researchers and practitioners, further research can be carried out on the methods and plans that can be used by an online retailer to coordinate with traditional retailers, such as cooperative advertising, information sharing, and so forth.

REFERENCES

Baker, W., Marn, M., & Zawada, C. (2001). Price smarter on the net. *Harvard Business Review, 79*, 122-127.

Balasubramanian, S. (1998). Mail versus mall: A strategic analysis of competition between direct marketers and conventional retailers. *Marketing Science, 17*(3), 181–195.

Bhatnagar, A., Misra, S., & Rao, H. R. (2000). On risk, convenience, and Internet shopping. *Communications of the ACM, 43*(11), 98-105.

Brynjolfsson, E., & Smith, M. D. (2000). Frictionless commerce? A comparison of Internet and conventional retailers. *Management Science, 46*(4), 563-585.

Cassar, K., Swerdlow, F. S., Kim, H., & Johnson, M. (2000). Local commerce: Internet threat mandates brick-and-mortar reconstruction. *Jupiter Vision Report,* (11), January 3.

Cattani, K. D., Gilland, W. G., & Swaminathan, J. M. (2006). Boiling frogs: Pricing strategies for a manufacturer adding a direct channel that competes with the traditional channel. *Production and Operations Management, 15*(1), 40-57.

Chiang, W. Y., Chhajed, D., & Hess, J. D. (2003). Direct marketing, indirect profits: A strategic analysis of dual-channel supply chain design. *Management Science, 49*(1), 1-20.

Geyskens, I., Gielens, K., & Dekimpe, M. G. (2002). The market valuation of Internet channel additions. *Journal of Marketing, 66*, 102-119.

Johannes, L. (2000). Competing online, drugstore chains virtually undersell themselves --- their Web sites charge less than their own stores, with some strings attached. *Wall Street Journal,* (January 10, 2000) B1.

King, R. C., Sen, R., & Xia, M. (2004). Impact of Web-based e-commerce on channel strategy in retailing, *International Journal of Electronic Commerce, 8*(3), 103-130.

Korgaonkar, P., Silverblatt, R., & Girard, T. (2006). Online retailing, product classifications, and consumer preferences. *Internet Research, 16*(3), 267-287.

Kwak, H., Fox, R. J., & Zinkhan, G. M. (2002). What products can be successfully promoted and sold via the Internet? *Journal of Advertising Research, 2*, 23-38.

Lal, R., & Sarvary, M. (1999). When and how is the Internet likely to decrease price competition? *Marketing Science, 18*(4), 485-503.

Levary, R., & Mathieu, R. G. (2000). Hybrid retail: Integrating e-commerce and physical stores. *Industrial Management, 42*(5), 6-13.

Liang, T., & Huang, J. (1998). An empirical study on consumer acceptance of products in electronic markets: A transaction cost model. *Decision Support Systems, 24*, 29-43.

McIntyre, K., & Perlman, E. (2000). Nike - Channel conflict. Stanford GSB Case, EC-9B.

Rhee, B., & Park, S. (1999). *Online store as a new direct channel and emerging hybrid channel system* (Working Paper). Hong Kong University of Science and Technology, Clear Water Bay, Kowloon, Hong Kong.

Rovenpor, J. (2003). Explaining the e-commerce shakeout: Why did so many Internet-based businesses fail? *E-Service Journal, 3*(1), 53-76.

Smith, M. D., Bailey, J., & Brynjolfsson, E. (2000). Understanding digital markets: Review and assessment. In E. Brynjolfsson & B Kahin (Eds.), *Understanding digital economy.* Cambridge, MA: MIT Press.

Wood, S. T. (2001). Remote purchase environments: The influence of return policy leniency on two-state decision processes. *Journal of Marketing Research, 38*(2), 156-169.

Yao, D. Q., & Liu, J. (2005). Competitive pricing of mixed retail and e-tail distribution channels. *Omega, 33*(25), 235-247.

from free to thousands of dollars. Meanwhile, an e-vendor will be examined by the seal issuer for conformity to the seal issuer's standards and principles. A qualified e-vendor earns the right to display these seals on its Web site. When a seal is clicked, a detailed disclosure of the principles ensured by the seal issuer is displayed to online shoppers.

By voluntarily placing itself under the scrutiny of a third party, a small, less-established e-vendor communicates to online shoppers that it runs a trustworthy business. If trust can be promoted through display of such Web seals, a small e-vendor will have a fair chance to compete with well-established names in e-markets.

The academic community has started to explore the impact of Web seals in promoting consumers' trust and increasing Web sales (Hu, Lin, & Zhang, 2002; Kovar, Burke, & Kovar, 2000; Noteberg, Christaanse, & Wallage, 2003). However, more research in this area is needed. Prior studies on Web seals were conducted from an empirical perspective. The literature lacks a theoretical analysis of how Web seals as trust-enhancing tools affect consumers' shopping decisions. The current research aims to fill this gap.

The first objective of this article is to explore the conditions under which small, less-established e-vendors can persuade online shoppers to purchase from them rather than from a highly reputable, well-known e-vendor. We establish a decision-making model for risk-neutral and rational online shoppers and derive a set of shoppers' decision-making criteria about whether and when to trade with a less-established e-vendor than with a well-established one. Furthermore, we analyze the effect of a Web seal in helping an e-vendor win consumer trust. Based on this analysis, we demonstrate the conditions under which the price effect, the seal effect, and the reputation effect play major roles in a shopper's decision about vendor selection.

The second objective of the article is to present evidence on how much social welfare a Web seal can produce. A social welfare analysis is conducted to reveal the positive social impact associated with the participation of less-established e-vendors and the introduction of Web seals in e-markets. The results of this article provide insights for small and less-established e-vendors on gaining competitive advantages and prospering in e-markets; the results will also help social planners advance e-commerce to its true potential.

The rest of the article is organized as follows. We first provide a literature review and background on third-party Web seals. Then, we develop a decision-making model for an online shopper. The model indicates that a shopper's decision is subject to a set of criteria, including price advantage of a less-established e-vendor over a well-established e-vendor, a shopper's reservation value, proportion of honest to strategic e-vendors in e-markets, the degree of cheating from a strategic trader, and the "safeguard" effect of a Web seal in preventing a strategic e-vendor from cheating. We study how seals used as trust-enhancing tools affect a shopper's decision and thus change the structure of e-markets. We later analyze how the participation of small, less-established e-vendors and the adoption of Web seals in e-markets increase social welfare. We conclude with an analysis of our research results and offer suggestions for future research.

LITERATURE REVIEW

As the Internet's popularity as a distribution medium increases, the issue of Web trust has attracted the attention of many people. Trust is widely regarded as critical to consumer's adoption of e-commerce (e.g., Gefen, 2002; Gefen, Karahanna, & Straub, 2003; Jarvenpaa, Tractinsky, & Vitale, 2000; McKnight, Choudhury, & Kacmar, 2002; Stewart, 2003; Yoon, 2002). McKnight et al. (2002) define trust as a multidimensional construct with two interrelated components—trusting beliefs and trusting intentions. They find that trusting

APPENDIX A

$$d_2 = 1 - \frac{p_2 - p_1}{1-\theta},\ p_2 \geq \frac{p_1}{\theta} \tag{A1}$$

$$d_1 = \frac{\theta p_2 - p_1}{\theta(1-\theta)},\ p_2 \geq \frac{p_1}{\theta} \tag{A2}$$

The profit for the manufacturer is as follows:

$$\pi_1 = p_1 d_1 \tag{A3}$$

Similarly for the traditional retailer, the profit is as follows:

$$\pi_2 = p_2 d_2 \tag{A4}$$

Therefore, the total profit of a vertically-integrated firm can be written as follows:

$$\pi_I = p_1 d_1 + p_2 d_2 \tag{A5}$$

By differentiating π_I on p_1 and π_T on p_2, respectively, and letting $(\partial\pi_1/\partial p_1) = 0$ and $(\partial\pi_1/\partial p_2) = 0$, we obtain:

$$p_1^I = \frac{\theta}{2} \tag{A6}$$

$$p_2^I = \frac{1}{2} \tag{A7}$$

Further, substituting (A6) and (A7) into demand and profit functions, we obtain the demands and profits as follows.

$$d_1^V = 0 \tag{A8}$$

$$d_2^V = \frac{1}{2} \tag{A9}$$

$$\pi_v = \frac{1}{4} \tag{A10}$$

APPENDIX B

First case: The online retailer announces its price first, and then the traditional retailer announces its retail price.

Under the Stackelberg competition, taking derivative of (A4) with respect to p_2 and letting $(\partial\pi_2/\partial p_2) = 0$ yields,

$$p_1^S = \frac{\theta(1-\theta)}{2(2-\theta)} \tag{A11}$$

Substituting (A11) into (A3), and then by differentiating π_1 with respect to p_1 and letting $(\partial\pi_1/\partial p_1) = 0$, we obtain:

$$p_2^S = \frac{(1-\theta)(4-\theta)}{4(2-\theta)} \tag{A12}$$

Substituting (A12) into (A11) and demand function, we obtain:

$$p_2^S = \frac{(1-\theta)(4-\theta)}{4(2-\theta)} \tag{A13}$$

$$d_1^S = \frac{1}{4} \tag{A14}$$

Substituting the above values into all of the demand and profits functions, we obtain,

$$d_2^S = \frac{4-\theta}{4(2-\theta)} \tag{A15}$$

$$\pi_1^S = \frac{\theta(1-\theta)}{8(2-\theta)} \tag{A16}$$

$$\pi_2^S = \frac{(1-\theta)(4-\theta)^2}{16(2-\theta)^2} \tag{A17}$$

Second case: The traditional retailer announces its price first, and then the online retailer announces its price later.

Under the Stackelberg competition, taking derivative of (A3) with respect to p_1 and letting $(\partial\pi_1/\partial p_1) = 0$ yields,

$$p_1^S = \frac{p_2\theta}{2} \tag{A18}$$

Substituting (A18) into (A4), and then by differentiating π_2 with respect to p_2 and letting $(\partial\pi_2/\partial p_2) = 0$, we obtain:

$$p_2^S = 1 - \frac{1}{2-\theta} \tag{A19}$$

Substituting (A19) into (A18) and demand function, we obtain:

$$p_1^S = \frac{1}{2}\theta(1 - \frac{1}{2-\theta}) \tag{A20}$$

$$d_2^S = \frac{1}{2} \tag{A21}$$

Substituting the above values into all of the demand and profits functions, we obtain,

$$d_1^S = \frac{1}{2(2-\theta)} \tag{A22}$$

$$\pi_1^S = \frac{\theta(1-\theta)}{4(2-\theta)^2} \quad \text{(A23)}$$

$$\pi_2^S = \frac{(1-\theta)}{2(2-\theta)} \quad \text{(A24)}$$

APPENDIX C

First case: The online retailer announces its price first, and then the traditional retailer announces its retail price later.
By differentiating π_1^S and π_2^S (from the Stackelberg equilibrium) each on θ, we obtain:

$$\frac{\partial(\partial\pi_1^S)}{\partial\theta^2} = -\frac{1}{2(2-\theta)^3} < 0 \quad \text{(A25)}$$

$$\frac{\partial\pi_2^S}{\partial\theta} = -\frac{16-12\theta+6\theta^2-\theta^3}{16(2-\theta)^3} < 0 \quad \text{(A26)}$$

Second case: The traditional retailer announces its price first, and then the online retailer announces its price.
By differentiating π_1^S and π_2^S (from the Stackelberg equilibrium) each on θ, we obtain:

$$\frac{\partial(\partial\pi_1^S)}{\partial\theta^2} = -\frac{3\theta}{2(2-\theta)^4} < 0 \quad \text{(A27)}$$

$$\frac{\partial\pi_2^S}{\partial\theta} = -\frac{1}{2(2-\theta)^2} < 0 \quad \text{(A28)}$$

Thus, propsition 1 is proved.

APPENDIX D

First case: The online retailer announces its price first, and then the traditional retailer announces its retail price.

By differentiating π_2^S and π_1^S on θ, we obtain:

$$\frac{\partial\pi_2^S}{\partial\theta} = -\frac{16-12\theta+6\theta^2-\theta^3}{16(2-\theta)^3}$$

and

$$\frac{\partial\pi_1^S}{\partial\theta} = \frac{2-4\theta+\theta^2}{8(2-\theta)^2};$$

Next, it is easy to prove that

$$\frac{\partial \pi_1^S}{\partial \theta} - \frac{\partial \pi_2^S}{\partial \theta} = \frac{16 - 12\theta + 6\theta^2 - \theta^3}{16(2-\theta)^3} + \frac{2 - 4\theta + \theta^2}{8(2-\theta)^2} > 0;$$

Second case: The traditional retailer announces its price first, and then the online retailer announces its price.

By differentiating π_2^S and π_1^S on θ, and let

$$\frac{\partial \pi_1^S}{\partial \theta} - \frac{\partial \pi_2^S}{\partial \theta},$$

we obtain:

$$\frac{\partial \pi_1^S}{\partial \theta} - \frac{\partial \pi_2^S}{\partial \theta} = \frac{6 - 5\theta}{4(2-\theta)^3} > 0$$

Thus, proposition 2 is proved.

APPENDIX E

Under the Bertrand competition, we differentiate π_1 on p_1 and π_2 on p_2 and let $(\partial\pi_1/\partial p_1) = 0$ and $(\partial\pi_2/\partial p_2) = 0$. Then, by substituting p_1 intop_2, we obtain:

$$p_1^B = \frac{\theta(1-\theta)}{4-\theta} \tag{A29}$$

Similarly, we obtain:

$$p_1^B = \frac{\theta(1-\theta)}{4-\theta} \tag{A30}$$

$$d_1^B = \frac{1}{4-\theta} \tag{A31}$$

$$d_2^B = \frac{2}{4-\theta} \tag{A32}$$

$$\pi_1^B = \frac{\theta(1-\theta)}{(4-\theta)^2} \tag{A33}$$

$$\pi_2^B = \frac{4(1-\theta)}{(4-\theta)^2} \tag{A34}$$

APPENDIX F

First case: The online retailer announces its price first, and then the traditional retailer announces its retail price.

Let $p_1^S - p_1^B$, so

$$p_1^S - p_1^B = \frac{\theta^2(1-\theta)}{2(4-\theta)(2-\theta)} > 0$$

Similarly, we have

$$p_2^S - p_2^B = \frac{\theta^2(1-\theta)}{4(4-\theta)(2-\theta)} > 0;$$

Second case: The traditional retailer announces its price first, and then the online retailer announces its price.

Let $p_1^S - p_1^B$, so

$$p_1^S - p_1^B = \frac{\theta^2(1-\theta)}{2(4-\theta)(2-\theta)} > 0$$

Similarly, we have

$$p_2^S - p_2^B = \frac{\theta(1-\theta)}{(4-\theta)(2-\theta)} > 0;$$

Therefore, we obtain: $p_1^S > p_1^B$ and $p_2^S > p_2^B$, thus proposition 3 is proved.

APPENDIX G

First case: The online retailer announces its price first, and then the traditional retailer announces its retail price.

Let $\pi_1^S - \pi_1^B$, so

$$\pi_1^S - \pi_1^B = \frac{\theta^3(1-\theta)}{8(4-\theta)^2(1-\theta)} > 0$$

Similarly, we have

$$\pi_2^S - \pi_2^B = \frac{\theta^2(1-\theta)}{4(4-\theta)(2-\theta)}\left(\frac{4-\theta}{4(2-\theta)} + \frac{2}{2-\theta}\right) > 0;$$

Second case: The traditional retailer announces its price first, and then the online retailer announces its price.

Let $\pi_1^S - \pi_1^B$, so

$$\pi_1^S - \pi_1^B = \frac{\theta^2(1-\theta)(8-3\theta)}{4(4-\theta)^2(2-\theta)^2} > 0$$

Similarly, we have

$$\pi_2^S - \pi_2^B = \frac{\theta^2(1-\theta)}{2(4-\theta)^2(2-\theta)} > 0$$

Therefore, we obtain: $p_1^S > p_1^B$ and $p_2^S > p_2^B$, thus proposition 4 is proved.

This work was previously published in the International Journal of E-Business Research, Vol. 4, Issue 1, edited by I. Lee, pp. 22-39, copyright 2008 by IGI Publishing (an imprint of IGI Global).

Chapter 14
Can Web Seals Work Wonders for Small E-Vendors in the Online Trading Environment?
A Theoretical Approach

Xiaorui Hu
Saint Louis University, USA

Yuhong Wu
William Paterson University, USA

ABSTRACT

Trust is a major issue in e-markets. It is an even more prominent issue when online shoppers trade with small, less-established e-vendors. Empirical studies on Web seals show that small e-vendors could promote consumers' trust and increase Web sales by displaying Web seals of approval. This article takes a theoretical approach to examine online trading when seals are used in e-markets. We establish an online shopper's decision-making model to reveal the online shopper's decision-making criteria. Criteria include when to trade with a well-established e-vendor and when to trade with a small, less-established e-vendor, with or without a Web seal. Based on our analysis of the research results, we reveal the price effect, the seal effect, the reputation effect, and their impact on a shopper's decision-making process. Meanwhile, a social welfare analysis is conducted to further demonstrate the positive impact of Web seals on small, less-established e-vendors.

INTRODUCTION

The Internet and World Wide Web have emerged as powerful media for communication and merchandise distribution. Cyberspace retailers are prospering and are now a potential threat to traditional retailers (Sanderson, 2000). Retail e-commerce sales have experienced fast and stable growth; these sales reached $110 billion in 2006, up 30% from 2005 and almost 100% from 2003.

Meanwhile, e-commerce as a percentage of total sales has also steadily increased from 0.9% in 2000 to 2.8% in 2006 (U.S. Census Bureau, 2007).

Rapid growth in electronic commerce is largely due to its inherent advantages as a medium. Internet shopping has the merits of convenience (such as 24-hour availability, no travel cost, and an easy global reach), the ability to quickly and cost-effectively search product and service information, and the ability to obtain competitive prices through comparison among various e-retailers (Kau, Tang, & Ghose, 2003; Peterson, Balasubramanian, & Bronnenberg, 1997; Wolfinbarger & Gilly, 2001).

Although Internet sales are booming, many people still perceive the risks associated with online trading to be high. Consumers' inability to inspect online products and merchants results in uncertainty about product quality and distrust of e-vendors (Bhatnagar, Mishra, & Rao, 2000; Van den Poel & Leunis, 1999). In addition, consumers are concerned about privacy and security issues associated with online shopping, because their personal identification information and other important data (such as credit card numbers) could be inappropriately handled or even misused (Bhimani, 1996; Ford & Baum, 1997; Griffin, Ladd, & Whitehead, 1997; Miyazaki & Fernandez, 2001). Some recent statistics reveal how serious these issues have become. According to Internet Fraud Watch (2005) the National Consumers League (NCL) received 12,315 complaints in 2005, compared to 10,794 in 2004. In 2005, the average loss per complaint was $1,917, and double that in 2004 ($894 average loss per complaint). Meanwhile, the Internet Crime Complaint Center (IC3, 2007) received its millionth complaint on June 11, 2007. Since beginning operation in May 2000, IC3 has referred 461,096 criminal complaints to federal, state, and local law enforcement agencies around the country for further investigation. The total dollar loss from all these referred cases is estimated to be $647.1 million (see http://www.ic3.gov).

Consumers enjoy the convenience, price advantage, and other benefits of online shopping, but they want to be protected from Internet fraud and other malicious activities. Online shoppers are advised to do business only with those with whom they have had favorable experiences. When consumers become interested in trading with less-established e-vendors, they are advised to thoroughly research these e-vendors. Recommendations include checking with state or local consumer protection agencies or the Better Business Bureau and reviewing other customers' feedback about a specific vendor (Internet Fraud Watch, 2007; U.S. Department of Justice, 2007). Given that most consumers are reluctant to spend the time and effort to perform thorough background checks on small online businesses, risk-averse consumers generally will conduct business only with well-established e-vendors (Lasica, 1999). This implies that companies with an established reputation either offline (e.g., Wal-Mart, Sears) or online (e.g., eBay.com, Amazon.com) enjoy a competitive advantage in e-markets.

One might wonder whether these small, less-established e-vendors could find methods to attract online shoppers and eventually prosper in e-markets. As a matter of fact, it is the active participation of small entrepreneurs in e-markets that fosters competition, which benefits consumers with lower prices, more choices, and better services.

Recently, the use of third-party Web seals (also called Internet seals of approval) as trust-enhancing mechanisms has attracted attention from both practitioners and academic researchers. These seals attempt to address consumers' various concerns—such as information privacy, transaction security, and complete/accurate transactions—about online shopping (Cook & Luo, 2003; Kimery & McCord, 2006). A few Internet seals are becoming well known in e-markets, such as Trust*e*, VeriSign, WebTrust, Good Housekeeping, and BBB*OnLine*. The seal issuer charges e-vendors for enrollment, with fees ranging

intentions and trusting beliefs significantly affect consumers' behavioral intentions. Jarvenpaa et al. (2000) find that trust affects consumer's attitude and willingness to purchase from an online store. Gefen (2002), Gefen et al. (2003), and Stewart (2003) also suggest that consumers' trust toward an e-vendor plays an important role in determining their online purchasing intentions. Yoon (2002) presents a model for the antecedents and consequences of consumer trust in online purchase decision making. The research results confirm that Web site trust, along with Web-site satisfaction and Web-site awareness, influences consumers' online purchase intentions.

Various studies also explore the antecedents to online trust and provide recommendations to e-vendors on how to promote online trust. Jarvenpaa et al. (2000) find that size and reputation of an Internet-based store significantly impact consumer's perceived trust in the store. Yoon (2002) proposes that Web-site trust was impacted by site properties, including image-related variables such as company awareness and company reputation. In addition, Web site trust is also driven by such personal traits as familiarity with and prior satisfaction with e-commerce. Metzger (2006) also suggests that vendor reputation is important in influencing consumers' trust. Lee and Turban (2001) propose that consumer trust in Internet shopping is influenced by such factors as trustworthiness of the Internet merchant, trustworthiness of the Internet as a shopping medium, and infrastructural/contextual factors (e.g., security, third-party certification). Urban, Sultan, and Qualls (2000) believe that to build Web-site trust, e-vendors should maximize such trust-building cues as third-party seals of approval, security technology, and/or security/privacy policies, provide unbiased and complete information, use virtual advisor technology, and fulfill expectations.

Third-party seals of approval have a fairly long history. Such approval certifies that the product or service bearing the seal meets the requirement of the seal-granting organization, and use of a seal aims to give consumers some assurance from a third party as to the quality and/or other important features of a product. For example, the Good Housekeeping seal assures buyers that if the product bearing the seal proves to be defective within two years of purchase, Good Housekeeping will replace the product or refund the purchase price. Institutions that grant seals and certifications can be independent testing companies, professional organizations, and/or government agencies. Seals of approval work to raise consumers' confidence in a particular product and are believed to significantly influence consumers' choice behavior (Parkinson, 2002).

A third-party seal is a cue that signals information to consumers (Wang, Beatty, & Foxx, 2004). In e-markets, because the true type (honest or strategic) of the e-vendor is unknown to the consumer, the e-vendor can use a seal to signal its honesty. Such a signaling process works through the seal issuer. The seal issuer, which has more information about the e-vendor than shoppers do, extends its own reputation to the e-vendor and takes on potential risk if the e-vendor is dishonest. The seal issuer generally will examine the seller and/or the product that bears the seal. In return, the seller is charged a fee. Shoppers who have little information about the seller tend to believe that a seller/product that bears a seal is trustworthy.

For example, WebTrust is an assurance service jointly developed by the American Institute of Certified Public Accountants (AICPA) and the Canadian Institute of Chartered Accountants (CICA). Any site displaying a WebTrust seal is guaranteed to have (1) been examined by a trained and licensed public accounting firm; (2) disclosed its business practices; (3) been audited to prove the site actually follows those practices; (4) met international Trust Services Standards; and (5) an audit report, which is based on one or more Trust Services Principles, linked to the seal (see http://www.cpaWebtrust.org/seal_info.

htm). An e-vendor displaying a WebTrust seal thus signals to consumers that it pursues a high standard of business practices. Indeed, one may argue that because there is little regulation in e-markets, the potential need for Web seals is huge. Signaling can be powerful, but it is also important that the targeted market and consumers adequately understand and trust the signals for them to play their role in assuring consumer trust (Kimery & McCord, 2006; Odom, Kumar, & Saunders, 2002).

Some academic research, although still limited, specifically focuses on third-party-assured Web seals and how such seals help e-vendors build consumers' trust toward the site and attract consumers. Kovar et al. (2000) find that consumers' intent to purchase online is positively related to (1) their degree of attention to the WebTrust seal, (2) their exposure to WebTrust advertising, and (3) their knowledge of CPAs. Hu et al. (2003) find that some Web seals, such as BBB*OnLine* and AOL Certified Merchant Guarantee seals, do promote consumers' willingness to buy from an e-vendor that displays such seal(s). Noteberg et al. (2003) discover that third-party assurance significantly increases consumers' likelihood of purchasing from an e-vendor and also reduces consumers' concerns about privacy and transaction integrity, concerns that might inhibit purchases. Odom et al. (2002) also find that Web seals influence consumers' online purchasing decisions. They suggest a relationship between consumers' ability to recognize a brand of Web seal and the seal's ability to influence consumers' online purchasing decisions. Nikitkov (2006) examines the use of seals on the eBay auction site, where buyers act with their own money and also have time to study the site. The results confirm that the presence of a seal on the seller's Web page does impact actual consumer purchasing behavior in both auction and posted-price contexts. Wakefield and Whitten (2006) further the study on Web seals by examining the role played by third-party organization (TPO) credibility on Internet users' attitude toward Web sites. Their finding suggests that TPO credibility reduces consumers' perception of purchasing risk and increases their trusting attitude toward an e-vendor.

The literature enriches our understanding about the important role of online trust and empirically confirms that third-party Web seals help promote consumers' trust and increase their purchase intention from sites that display such seals. However, there is no comprehensive theoretical analysis of how the use of Web seals changes the structure of e-markets and how their use increases the competitiveness of the e-markets by altering the decisions and strategies of both online shoppers and e-vendors. The current study attempts to fill this research gap. The results of this study show how small, less-established e-vendors, by promoting trust through the adoption of Web seals, can compete with well-established e-vendors and gain market share from peers that do not deploy a trust-promoting seal. Therefore, the research shows that competitive markets can be established; these markets not only grant online shoppers more consumer surplus, but also promote a more secure online shopping environment, which benefits society as a whole.

ONLINE SHOPPERS' DECISION-MAKING MODEL

Basic Model Set-Up

In this model, we assume two types of e-vendors in e-markets: the high reputation type (HH-type) and the low reputation type (L-type). HH-type e-vendors are those with well-established reputations that trade honestly and that are trusted by online shoppers. Examples of HH-type e-vendors include Walmart.com, Dell.com, and Amazon.com. L-type e-vendors are those that have less-established reputations. Examples include new and small market entrants. Among L-type e-vendors, we assume two sub-types: honest (denoted as LH) and strategic (denoted as LS). LH-type e-vendors are honest, have higher moral standards, and are committed to fair business practices. They do not intentionally deceive their customers for illegal profits (in this article, we ignore the case of misrepresentation, i.e., unintentional mishandling from HH- or LH-type e-vendors). LS-type e-vendors intend to deceive online customers in order to reap higher illegal profits. However, LS-type e-vendors might not cheat on every trade if certain mechanisms are in place to restrain them from cheating.

We make several assumptions before presenting the decision-making model:

1. Shoppers are rational and risk neutral. They set their goals to maximize the expected trading surplus/utility. We further assume that utility is linear.
2. The identities of HH-type e-vendors are common knowledge among online shoppers. However, among low reputation e-vendors, shoppers cannot differentiate between LH-type and LS-type. Only L-type e-vendor knows its true type.
3. Due to the differences in reputation, both LH- and LS-type e-vendors offer lower prices than HH-type e-vendors. This price advantage serves as an incentive for online shoppers to do business with the low reputation e-vendors.
4. Both LH- and LS-type e-vendors offer the same price. Although some LS-type e-vendors might use significantly lower prices to attract shoppers, we ignore that case in this article.
5. Due to various factors such as price advantage, convenience, and information availability, shopping at a physical store is assumed to be less desirable in this model. Therefore, we ignore the option of shopping at a local brick-and-mortar store. Readers can assume that the utility of shopping at a local retail store is lower than the expected utility a consumer can achieve online.
6. Shoppers can purchase a product from either a high reputation or a low reputation e-vendor. That is, product availability is not an issue and cannot be used as a competitive advantage.

Part of the decision-making process for shoppers is evaluating the expected payoff between taking the price advantage and bearing potentially higher risk (i.e., shopping from L-type e-vendors) and taking the reputation advantage and paying a higher price (i.e., shopping from HH-type e-vendors).

We use HH-type e-vendors as a benchmark. The price an HH-type e-vendor offers is denoted as P_H, and the price an L-type e-vendor offers is denoted as P_L. We assume $P_H > P_L$, although in special cases, HH-type e-vendors might of-

Figure 1. The decision tree for an online shopper (before seal is introduced)

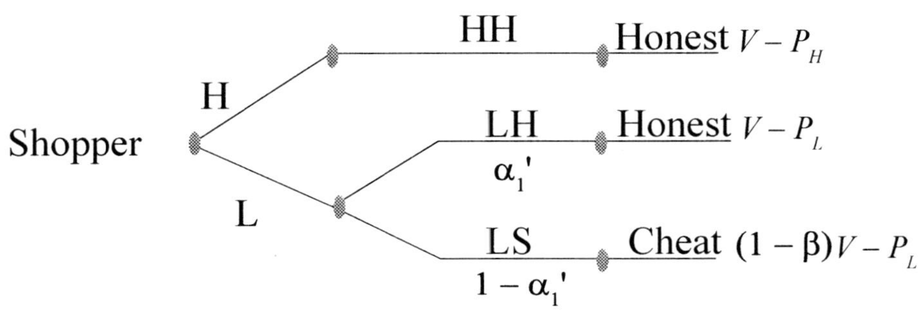

fer cheaper prices than L-type e-vendors. The reservation value of a shopper for the underlying merchandise is denoted as V. The risks associated with HH- and LH-type e-vendors are normalized as zero due to their honesty. However, there are risks associated with trading with LS-type e-vendors. The decision tree of a shopper is shown in Figure 1.

Meanwhile, we assume in e-markets among the low reputation e-vendors, α portion of them are of the honest type (LH), whereas 1-α portion of them are of the strategic type (LS). The parameter α is within the range of [0, 1], and its value is determined by nature. In addition, we assume that shoppers, although not knowing the true type of each individual L-type e-vendor, have a perception of the approximate value of α. That perception can be formed from watching news, reading general reports about Internet safety, sharing thoughts in different forums, and similar methods.

Decision Criteria

Scenario I: Before a Seal Is Introduced to E-Markets

First, we discuss the scenario in which no seal exists in e-markets. The expected utility for a shopper under the benchmark case (i.e., trading with a high-reputation e-vendor) is the difference between the shopper's valuation of the underlying product or service and the price a shopper pays. It is represented as follows: $EU(HH) = V - P_H$.

When trading with an L-type e-vendor, a shopper cannot differentiate between LH-type and LS-type. If a shopper meets an LH-type e-vendor, the expected utility for the shopper is represented as follows: $EU(HH) = V - P_L$, as the shopper gets reservation value of the merchandise and pays the low price of P_L.

However, if a shopper encounters an LS-type e-vendor, he or she might be deceived and receive only partial reservation value, $(1 - \beta)V$, where $\beta \in [0, 1]$. We call β the cheating portion. Thus, the expected utility for the shopper is represented as follows: $EU(LS) = (1 - \beta)V\text{-}P_L$. The worst case is that the shopper receives nothing from the e-vendor (i.e., when β equals one) and realizes an expected utility of $- P_L$.

Meanwhile, in e-markets, though the real portion α cannot be known to shoppers, they can have an expectation of α. In this article, we denote the shoppers' perception of the value α as α',which can be viewed as the shoppers' perceived risk factor (Cunningham, 1967), and we know that shoppers make decisions based on their perception of risk.

We first explore the conditions under which a shopper is willing to purchase from a low reputation e-vendor rather than from a high reputation e-vendor. As an L-type e-vendor can only lower its price to compete with HH-type e-vendors, we focus on finding the threshold for that price.

Proposition 1: *When*

$$\alpha' > 1 - \frac{P_H - P_L}{\beta V},$$

shoppers will purchase from an L-type e-vendor rather than from an HH-type e-vendor.

Proof: *We compare the expected payoff of trading with L-type e-vendors, EU(L), and the expected payoff of the benchmark case, EU(HH),*

$$\begin{aligned} EU(L) - EU(HH) &= \\ &\alpha'(V - P_L) + (1 - \alpha')[(1\text{-}\beta)V\text{-}P_L] - (V - P_H) = \\ &(P_H - P_L) - (1 - \alpha')\beta V \end{aligned}$$

When the above condition is greater than zero, we have the following condition:

$$(1 - \alpha') < \frac{(P_H - P_L)}{\beta V}.$$

Several observations are found in this proposition. First, the relative price advantage

$$\frac{(P_H - P_L)}{V}$$

is the key factor that draws shoppers to L-type e-vendors. If an L-type e-vendor can offer a price lower than $P_H - (1 - \alpha')\beta V$, it will win shoppers from the HH-type e-vendors. Second, the higher the proportion of honest e-vendors α' among low reputation e-vendors, the more comfortable a shopper feels trading with an L-type e-vendor; thus, the higher the threshold price P_L can be. Third, the lower the expected cheating portion β is, the more likely a shopper will engage in a trade with an L-type e-vendor.

Scenario II: When a Seal is Introduced to E-Markets

Before the introduction of Web assurance seals in e-markets, less-established e-vendors could mainly win customers by reducing their product prices. The price factor was the only thing they could manipulate. Meanwhile, they were subject to the uncontrollable factors of the general perception of e-markets, such as the shoppers' perception of α' and β.

When a third party aims to ensure the trustworthiness of an e-vendor and regulates the e-vendor to abide by fair business practices, the decision-making process for online shoppers changes. Studies have shown that when shoppers are shopping online, they tend to pay attention to the various assurance seals the vendor presents on the Web site. In this study, we focus on this type of shopper, who voluntarily seeks assurance on the e-vendor's Web site and responds to that assurance if it is presented.

E-vendors who earn the right to display a well-trusted seal voluntarily agree to (1) abide by certain rules and (2) put themselves under the scrutiny of an independent third party. Therefore, such e-vendors are signaling to online shoppers that they are serious and honest merchants. As the true type of an e-vendor (honest or strategic) is unknown to shoppers, signaling can be a powerful tool for an LH-type e-vendor to convey its true type to shoppers. Previous research has proven this to be true (Kimery & McCord, 2006; Odom et al., 2002).

In this model, we assume that HH-type e-vendors are not interested in any seal. This assumption is reasonable. HH-type e-vendors have a high reputation and are believed by consumers to be trustworthy. When HH-type e-vendors do want to address consumers' concerns over an online transaction, they can post their policy on their Web sites (i.e., provide self-claimed assurance) rather than resort to a third party. For example, Walmart.com has assurance on information privacy and transaction security. Wal-Mart guarantees that it does not sell customers' personal information to any third parties under any circumstances. It also assures customers that it uses secure sockets layer (SSL) technology to encrypt and encode sensitive information before information is sent over the Internet (www.walmart.com). Wal-Mart already has a high enough reputation that its guarantees/claims are well trusted by online shoppers. Noteberg et al. (2003) empirically confirm that for unknown e-vendors, consumers' purchase intention increases significantly when there is third-party assurance (as opposed to self-claimed assurance), whereas for known e-vendors, whether the assurance is self-claimed or from a third party makes no significant difference to consumers.

For low reputation e-vendors, as they are less established and generally unknown to online shoppers, Web seals can provide an opportunity to establish their online reputations. Both LH- and LS-type e-vendors may consider adopting a seal. LH-type e-vendors may consider employing a seal to indicate its true type. However, some LS-type e-vendors may adopt a seal for other reasons. Motivations for seal adoption are complicated for LS-type e-vendors. Some are aiming to improve their reputation and obtain normal profits for a while; some are trying to establish reputations for future cheating actions; and others might intend to effectively deceive the shoppers and gain illegal profits for the short run. Although the various motivations of seal adoption from LS-type e-vendors are beyond the scope of the article, we can model the general actions derived from these motivations in a decision tree. The decision tree for a shopper is shown in Figure 2.

We assume that when an LS-type e-vendor employs no seal (i.e., it is not under the scrutiny of a third party), it will cheat. Then a shopper trading with this LS-type e-vendor might receive only partial reservation value, $1 - \beta V$, and pay the price P_L. Thus, the shopper's expected utility is $(1 - \beta)V - P_L$, the same as that in Scenario I. When LS-type e-vendors employ a seal, we assume among them θ portion will act honestly, and the other $1 - \theta$ portion will cheat, where θ is within the range of [0, 1]. There are two main reasons for LS-type e-vendors to act honestly when they adopt a seal. One is that the scrutiny of the seal issuer induces them to act honestly, and the other is that they want to use the seal as a trust-promoting tool to earn solid reputations in e-markets. When an LS-type e-vendor acts honestly, a shopper's payoff is $V - P_L$; otherwise, the payoff is $(1 - \beta)V - P_L$.

We assume that in e-markets, the δ portion of the low reputation e-vendors employs a seal, while the other $1 - \delta$ portion does not. Consumers can observe whether an e-vendor adopts a seal or not when visiting its Web site, although consumers do not know whether the specific e-vendor they are trading with is honest or strategic. An LS-type e-vendor might cheat even with a seal on its Web site, while an LH-type e-vendor will not cheat even without a seal on its Web site.

A shopper must make several decisions: (1) whether he or she should trade with a low reputation e-vendor and (2) when he or she trades with a low reputation vendor, whether he or she should trade with one that adopts a seal or one that does not.

Because seal adoption can be observed by online shoppers before they make their purchase decisions, we analyze the expected utility of online shoppers under two distinct cases: trading with an L-type e-vendor with a seal on its Web site and trading with one that does not display a seal.

The expected utility of trading with an L-type e-vendor with a seal is denoted as $EU(L_{Seal})$ and with an L-type e-vendor without a seal as $EU(L_{NoSeal})$. The payoffs are shown below:

Figure 2. Decision tree for an online shopper (with a seal case)

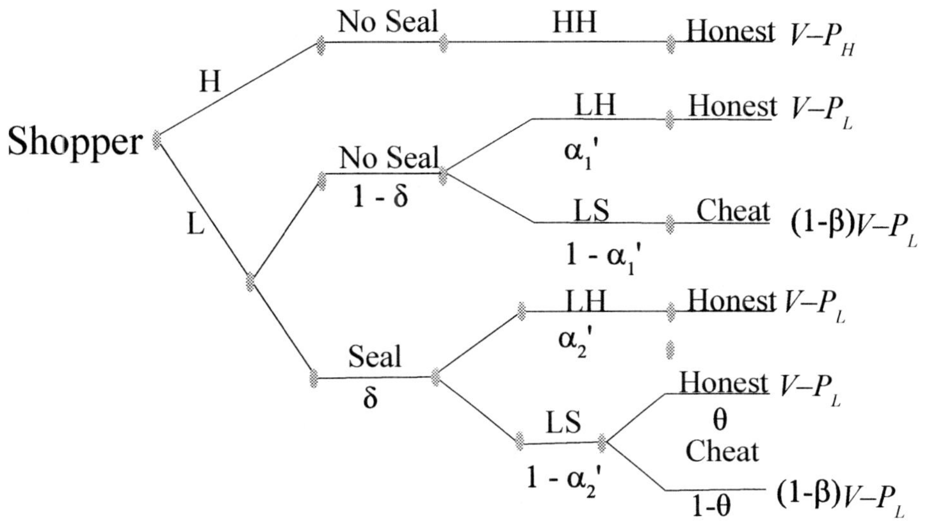

$$EU(L_{NoSeal}) = \alpha_1'(V - P_L) + (1 - \alpha_1')[(1-\beta)V - P_L]$$

$$EU(L_{Seal}) = \alpha_2'(V - P_L) + (1 - \alpha_2')\{\theta(V - P_L) + (1 - \theta)[(1-\beta)V - P_L]\}$$

In the following sections of the article, we propose to use the same parameter α' for both α_1' and α_2'. Readers might argue that α_1' and α_2' are different because they belong to two distinct groups: seal-adopting group and no-seal group. However, we have little information to project whether the percentage of the moral trader α_1' should be higher or lower in the seal-adopting group than in the no-seal group. Therefore, we use the same parameter to derive possible intuitions for the model. Meanwhile, we encourage readers to apply their own perception of α_1' and α_2' to evaluate their utility and to guide their decision making.

From Figure 2, we can see that, on average, shoppers can gain higher expected payoff when trading with an L-type e-vendor with a seal than one without a seal. When an L-type e-vendor adopts a seal, the likelihood of it acting honestly increases. If we compare the expected payoffs of $EU(L_{Seal})$ and $EU(L_{NoSeal})$, we find that $EU(L_{Seal})$ is always higher than or equal to $EU(L_{NoSeal})$ (i.e., $EU(L_{Seal}) - EU(L_{NoSeal}) = (1 - \alpha')\beta\theta V \geq 0$).

The above expression is equal to zero only when α' equals 1, or when θ or β equals 0. When α' equals one, or β equals zero, all L-type e-vendors are of the honest type. E-markets are only filled with LH-type e-vendors; thus, a seal's effect for promoting trust diminishes. When θ equals 0, LS-type e-vendors will always cheat, whether they employ a seal or not. In such cases, a seal has absolutely no power to confine (i.e., positively impact) a strategic e-vendor's cheating behavior. Thus, the trustworthiness of a seal is significantly reduced. Consequently, no rational shopper would pay attention to a Web seal.

Those are the two extreme cases. In reality, quite neither case is quite possible. In all other cases, a seal has effect and the expected payoff of trading with an e-vendor adopting a seal is always higher than the expected payoff of trading with an e-vendor without a seal. This is useful information for LH-type e-vendors who are attempting to adopt a seal. Empirical analyses also confirm this theoretical result (e.g., Kovar et al., 2000; Noteberg et al., 2003).

Our key questions then are with the help of a trusted third party, whether a low reputation e-vendor can win business over HH-type e-vendor if it offers a lower price and employs a Web seal, and whether this price threshold can be somewhat higher than that in Scenario I.

Proposition 2: *When the following condition holds, shoppers will purchase from a low reputation e-vendor with a seal rather than from a high reputation e-vendor.*

$$\theta > 1 - \frac{P_H - P_L}{\beta V(1-\alpha')} \quad \text{or} \quad \alpha' > 1 - \frac{P_H - P_L}{\beta V(1-\theta)}.$$

Proof: We compare the expected payoff of trading with an L-type e-vendor with a seal, $EU(L_{Seal})$, and the expected payoff of the benchmark case $EU(HH)$.

$$EU(L_{Seal}) - EU(HH) = \alpha'(V - P_L) + (1 - \alpha')\{\theta(V - P_L) + (1 - \theta)[(1 - \beta)V - P_L]\} - (V - P_H) = (P_H - P_L) - (1 - \alpha')(1 - \beta)(1 - \theta)V$$

When the above condition is greater than zero, we find the following condition:

$$(1 - \alpha')\beta(1 - \theta) < \frac{(P_H - P_L)}{V}.$$

All the parameters in the equation contribute to the balance of this condition and jointly impact a shopper's decision. Several observations are found. First, when all other parameters are fixed, as the relative price advantage between the high and low reputation type

$$\frac{(P_H - P_L)}{V}$$

is enlarged, the above condition will be easily satisfied, and shoppers will be drawn to the low reputation e-vendor with a seal and enjoy the price advantage. Second, compared to the price threshold in Proposition 1, we find that the required price for an L-type e-vendor to win shoppers over in Scenario II is indeed higher than that in Scenario I. This demonstrates that with the help of Web seals, an L-type e-vendors can compete more efficiently with HH-type e-vendors, and earn a higher profit margin. Finally, as any of the parameters α', $(1 - \beta)$, and θ increase, shoppers are more likely to trade with a low reputation e-vendor. When α' increases, it indicates that in e-markets more low reputation e-vendors are of the honest type. Therefore, shoppers' perceived risks for e-markets are reduced, and they are more willing to trade with an L-type e-vendor with a seal. Similarly, θ indicates the percentage of the strategic type e-vendors that trade honestly when a seal is adopted. Therefore, when θ increases, a shopper's perceived risk decreases, and his or her willingness to purchase from an L-type e-vendor increases. In the same fashion, β reflects potential loss per trade; when the potential loss is reduced, a shopper's gain per trade increases, and his or her interest in trading with an L-type e-vendor increases.

We further explore the question of whether a seal can create a price premium for an e-vendor.

Proposition 3: *A seal can create a price premium for an L-type e-vendor with a seal, and the price premium can be up to* $(1 - \alpha')\beta\theta V$.

Proof: *We set the expected payoffs of* $EU(L_{Seal})$ *and* $EU(L_{NoSeal})$ *to be equal and found how much more a vendor with a seal can charge online shoppers without losing them to lower price competitors without a seal. We denoted the price premium as K.*

$$EU(L_{Seal}) - EU(L_{NoSeal}) = 0$$

$$\alpha'(V - P_L - K) + (1 - \alpha')\{\theta(V - P_L - K) + (1 - \theta)[(1 - \beta) V - P_L - K]\} = \alpha'(V - P_L) + (1 - \alpha')[(1 - \beta) V - P_L]$$

Thus, $K = (1 - \alpha')\beta\theta V$ *holds.*

Therefore, as long as the price premium is between the range of $[0, (1 - \alpha')\beta\theta V]$, shoppers are more likely to purchase from an e-vendor with a seal than one without a seal. The seal premium can help the seal adopter offset some of the cost associated with the seal and gain competitive advantage. The issues of how a vendor decides the level of premium are reserved for future research. In the following discussion, we apply the same low price rule to L-type e-vendors both with and without a seal.

Price Effect, Seal Effect, and Reputation Effect

To demonstrate the dynamics of these propositions and shoppers' decision-making criteria, we draw Figures 3, 4, and 5 to show the range of various parameters and the criteria for a shopper's decision making.

In Figure 3, Range I indicates that when a shopper perceives that the proportion of honest e-vendors (α') among the low reputation types is relatively low, which is within the range of

$$[0, 1 - \frac{(P_H - P_L)}{\beta V(1 - \theta)}],$$

the shopper will surely (100% likelihood) purchase from a high reputation e-vendor even given that the low reputation e-vendor provides a price advantage and may display a seal on its Web site. We call Range I the reputation effect range. When the e-markets are perceived to be risky, and the price advantages are not significant enough, the reputation effect prevails.

However, a shopper's decision starts to change when the perceived honesty level among low reputation e-vendors is enhanced. Specifically,

Figure 3. Shopper's decision making w.r.t α'

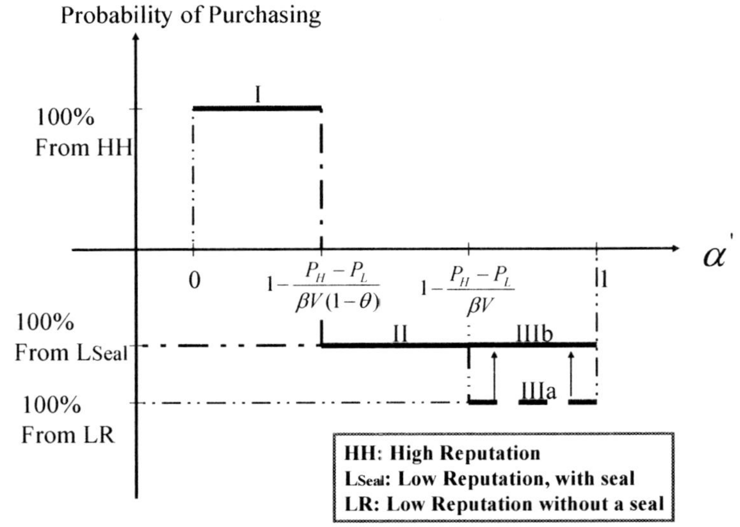

this is the range where the perceived α' is within the range of

$$[1-\frac{(P_H-P_L)}{\beta V(1-\theta)}, 1-\frac{(P_H-P_L)}{\beta V}].$$

Now, shoppers are drawn to the low reputation e-vendors that display a seal on their Web site (Range II) and will purchase from them rather than from an HH-type e-vendor. We call Range II the seal effect range. A seal signals to the shoppers that the e-vendor might be honest; as a seal does induce an extra θ portion of LS-type e-vendor to trade honestly, the perceived risk decreases in doing business with such a seal-bearing vendor. Online shoppers therefore are attracted to purchase from a low reputation vendor with a seal.

Range III implies that when the expected percentage of LH-type e-vendors is higher than

$$1-\frac{(P_H-P_L)}{\beta V},$$

the shoppers would even prefer to purchase from low reputation e-vendors that adopt no seal than from the HH-type e-vendors. Range III combines the effects of price advantage and a safer online environment, which draw shoppers to L-type e-vendors. We call this range the combined effect range.

However, prior discussion indicates that, on average, shoppers are better off purchasing from an L-type e-vendor with a seal than from an L-type e-vendor without a seal. Therefore, if a shopper can choose between L-type e-vendors with or without a seal, they will definitely choose to trade with one that adopts a seal. Then the range IIIa no longer exists; only Range IIIb sustains (i.e., only L-type e-vendors that adopt a seal can establish a trade).

Analysis w.r.t. ε

Figure 4 presents the relationship among the parameters from another perspective. Most of the intuitions are the same as those demonstrated in Figure 3. However, a clearer view of how the relative price advantage affects shoppers' purchasing decisions is shown in Figure 4. We define the relative price advantage as ε, where

$$\varepsilon = \frac{(P_H-P_L)}{V}.$$

Figure 4. Shoppers' decision making w.r.t ε'

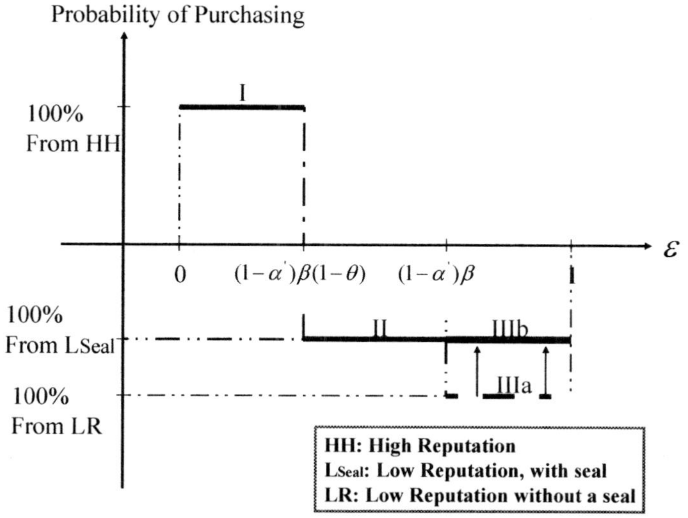

When ε is relatively low, a shopper will purchase from HH-type e-vendors. As the relative price advantage is enhanced, the low reputation e-vendors with a seal start to gain customers. Similar to the discussion above, the sustainable equilibrium in Range III is IIIb.

Analysis w.r.t. *V*

Similarly, we draw the above relationships in terms of consumer's reservation value in Figure 5. The higher the reservation value, the lower the relative price advantage is. Therefore, when a shopper has a reservation value higher than

$$\frac{(P_H - P_L)}{(1-\alpha')\,\beta(1-\theta)},$$

he or she would like to pay the higher price and avoid any potential risk (i.e., a shopper would purchase from HH-type e-vendors). When a shopper's reservation value is between

$$\frac{(P_H - P_L)}{(1-\alpha')\,\beta(1-\theta)}$$

and

$$\frac{(P_H - P_L)}{(1-\alpha')\,\beta},$$

the seal's effect emerges, and the shopper will take the price advantage and purchase from an L-type e-vendor with a seal. When a shopper's reservation value is lower than

$$\frac{(P_H - P_L)}{(1-\alpha')\,\beta},$$

he or she would rather purchase from an L-type e-vendor without any seal than from an HH-type e-vendor. As discussed earlier, this range will not hold. Meanwhile, two reservation values of the cutoff points,

$$\frac{P_L}{\{\alpha' + (1-\alpha')[\theta + (1-\theta)(1-\beta)]\}}$$

and

$$\frac{P_L}{[1-(1-\alpha')\beta]}'$$

Figure 5. Shoppers' Decision Making w.r.t V

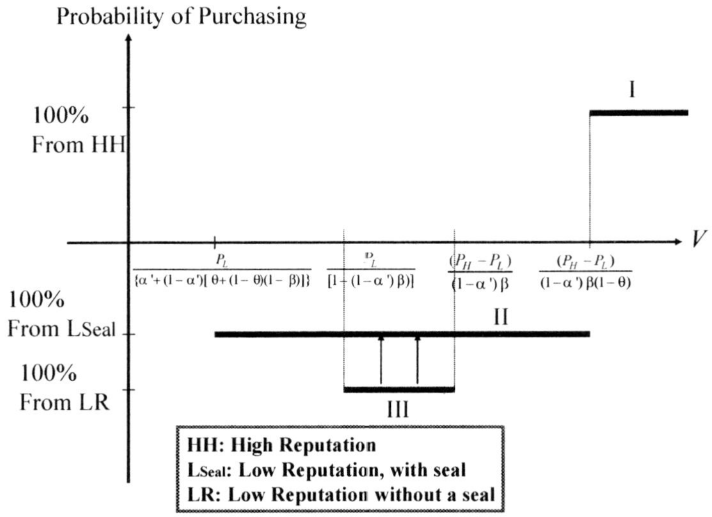

are derived from setting the expected utilities of $EU(L_{Seal})$ and $EU(L_{No\ Seal})$ equal to zero.

We assume that the reasonable orders of these four cutoff points are as presented in Figure 5. However, two of the cutoff points,

$$\frac{P_L}{[1-(1-\alpha')\beta]}$$

and

$$\frac{(P_H-P_L)}{(1-\alpha')\beta},$$

might switch places. If that is the case, Range III will be eliminated.

SOCIAL WELFARE ANALYSIS

From the afore-mentioned discussion, we see that the use of Web seals in e-markets encourages online shoppers to trade with lower priced, less-established e-vendors, and the reputation, price, and seal effects play important roles under different market conditions. In this section, we analyze the contribution of the third-party Web seals to social welfare. As Web seals help motivate some shoppers to purchase from less-established e-vendors, a lower settlement price is achieved for these shoppers, and social welfare might be enhanced as a result. We discuss three cases to clearly demonstrate how social welfare is enhanced by the participation of low reputation e-vendors and through the use of Web seals in e-markets.

Case I: Only HH-Type E-Vendors in E-Markets

When only well-established e-vendors are in e-markets, consumers can only purchase from them, with the price P_H. We assume a downward-sloping linear demand to simplify the chart formation. HH-type e-vendors will charge P_H and sell q_H for the underlying product, with a marginal cost of C. Online shoppers who have reservation value higher than P_H will purchase the product at that price.

Therefore, consumer surplus is the triangle area above the price line P_H and under the demand curve, which includes the areas from 1 to 6. The vendor's profit is the shaded area A in Figure 6.

Figure 6. Social Welfare with only HH-type e-Vendors in e-markets

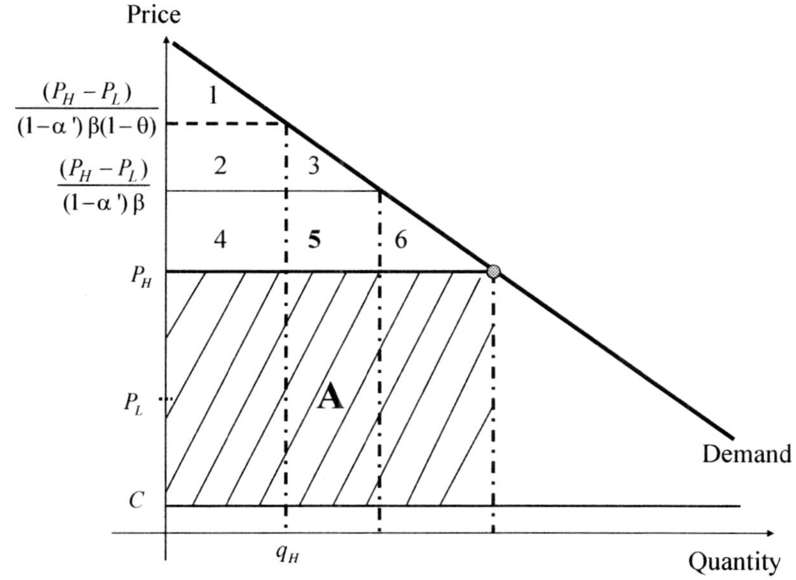

Case II: Less-Established E-Vendors Enter E-Markets (No Seal is Introduced)

When small, less-established e-vendors enter e-markets, consumers can either purchase from HH-type e-venders or from L-type e-vendors. As an L-type e-vendor charges a lower price, some shoppers might take the price advantage and enjoy more consumer surplus. According to a shopper's decision-making criteria, shoppers who have reservation values higher than

$$\frac{(P_H - P_L)}{(1-\alpha')\beta}$$

will still purchase from HH-type e-vendors, paying the price of P_H; whereas others with reservation values from

$$\frac{P_L}{[1-(1-\alpha')\beta]}$$

up to

$$\frac{(P_H - P_L)}{(1-\alpha')\beta}$$

will purchase from low reputation e-vendors—these shoppers will take the price advantage of paying P_L but bear the potential risk.

If we momentarily set aside the potential risk and assume that all trades are carried out honestly, the social welfare chart changes to Figure 7.

Compared with Figure 6, the consumer surplus increases, the low reputation e-vendor's profit increases, and the high reputation e-vendor's profit declines. Meanwhile, more shoppers make the purchases and enjoy the products. The sales rise from the original q_H to q_L . Social welfare increases. The total increased social welfare is shown in areas 9, 12, 16, and the bubble area under 16. However, due to the potential risks associated with trading with low reputation e-vendors, some of the trades might not be completed successfully. From a social planner's point of view, the money moved from consumers' pockets into those of the e-vendors' will not affect social welfare. However, one part of the social welfare still can be affected,

Figure 7. Social Welfare with L-type e-Vendors in e-markets (before seal is introduced)

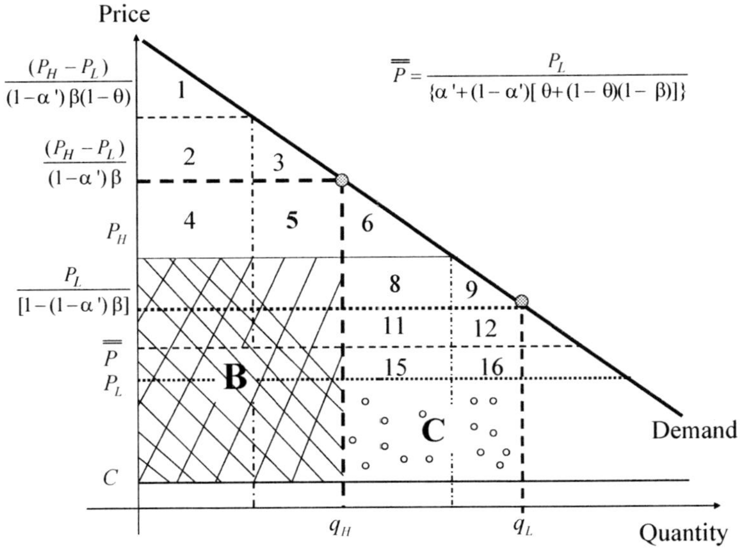

which is the loss of consumer surplus. Thus, the increases in social welfare in Figure 7 should be partially discounted.

In addition, whether the increase in social welfare is positive or negative depends on the market conditions (in other words, on the parameters in e-markets). With a demand like what we project and a relatively small portion of strategic e-vendors in e-markets, social welfare is more likely enhanced by the intensification of competition due to the participation of small, less-established e-vendors. Meanwhile, even though we use the same price for the HH-type e-vendors between Case I and Case II, due to the competition brought in by low reputation e-vendors, the price charged by HH-type e-vendors is more likely to be lowered in Case II than in Case I, yielding more consumer surplus to the shoppers.

Case III: Seal is Used by Less Established E-Vendors in E-Markets

When seals are introduced to e-markets, shoppers with reservation values higher than

$$\left(\frac{(P_H - P_L)}{(1-\alpha')\beta(1-\theta)}\right)$$

will still purchase from an HH-type vendor and pay the price of P_H. Shoppers with lower reservation values will purchase from a low reputation e-vendor with a seal and pay the price of P_L. According to the assumed orders of the four cutoff points shown in Figure 5, we can draw the new social welfare in Figure 8.

Figure 8 indicates that only those shoppers who have high reservation values will purchase from HH-type e-vendors, and the quantity demanded is q_H. For those shoppers, the consumer surpluses are the areas 1, 2, and 4. The e-vendor's profit is the area D. Other shoppers purchase from the low reputation e-vendors with a seal, pay P_L, and the consumer surpluses are the areas 3 and 5 through 17. The e-vendor's profit is the bubble area E.

Comparing Figures 8 and 7, consumer surplus is definitely enhanced; the vendor profit is reallocated, shifting from HH-type e-vendors to L-type e-vendors. The social welfare has been increased due to the introduction of seals.

Meanwhile, uncertainty about low reputation e-vendors still exists. The increased social welfare is also subject to some degree of discount. However, as all of the trades are conducted with either HH-type e-vendors or L-type e-vendors that

Figure 8. Social welfare when seal is employed by L-type e-Vendors

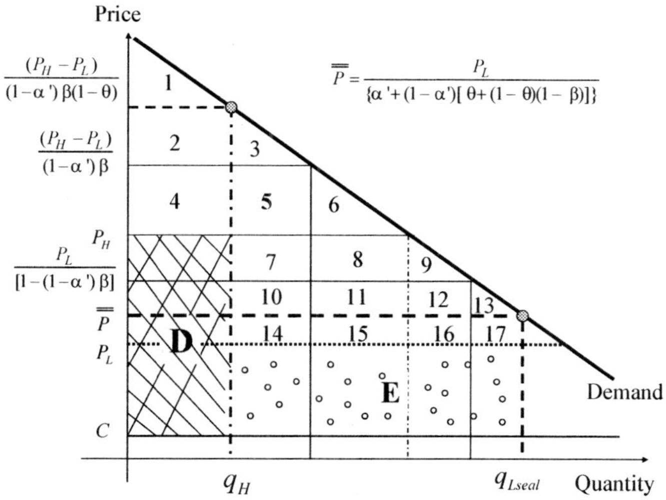

adopt a seal, the potential risks are relatively low. Therefore, we believe that the social welfare is enhanced significantly by the services provided by third-party Web seals.

CONCLUSION AND FUTURE RESEARCH

Due to the nature of e-markets, it is difficult for consumers to assess the trustworthiness of an e-vendor. The issue is more prominent when an e-vendor is small and less-established. The display of a third-party-assured Web seal has been adopted by small e-vendors as a trust-building practice. Though seals of approval have long existed offline, the use of such seals on the Web is a recent phenomenon. The literature needs a theoretical analysis to examine how online trading is impacted when seals of approval are used in e-markets.

In this article, we establish a decision-making model for online shoppers which examines when they would choose to purchase from a small, less-established e-vendor rather than from a well-established one. This model indicates that when the same product can be purchased from either a less-established e-vendor or a well-established e-vendor, relative price advantage is a key factor in attracting consumers to the less-established e-vendors. Small e-vendors in e-markets compete with big names through price. When third-party Web seals are introduced to e-markets, the competition structure starts to change. Web seals, due to their scrutiny functions and the reputation of the seal issuers, can signal online shoppers of the trustworthiness of the e-vendor, thus enhancing both the online shoppers' trust of the e-vendor and the likelihood of purchasing from it. Consequently, a small e-vendor with a seal enjoys a competitive advantage over other small e-vendors that have no seal. In addition, small e-vendors with seals are able to compete more efficiently with well-established e-vendors. Besides price advantage, the effect of assurance seals in preventing e-vendors from behaving strategically also plays an important role.

Based on the analytical results, this research presents how and when the price effect, the seal effect, and the reputation effect play a major role in an online shopper's decision making. When the online environment is perceived to be risky, reputation effect plays a major role, and shoppers purchase from well-established e-vendors only. As the perceived proportion of honest e-vendors among less-established ones increases and the use of Web seals converts a great number of strategic e-vendors into honest businesses, the seal effect starts to play a major role. Small e-vendors with seals start to win shoppers from the well-established e-vendors. In fact, when the relative price advantage is high and when online trading is perceived to be less risky, shoppers would rather purchase from a less-established e-vendor with no seal than from a well-established e-vendor.

Throughout the analysis, price effect is a key factor that attracts consumers to small e-vendors. From a social planner's point of view, participation of small e-vendors and the use of Web seals directly or indirectly fosters competition in e-markets, which, in turn, enhances social welfare. As a result, shoppers enjoy lower prices, low reputation e-vendors obtain a better chance to compete in e-markets, and the overall social welfare is enhanced as e-markets become more competitive and secure.

However, the current research has its limitations as well. We applied a representative model in the shopper's decision-making process. We assumed all shoppers are the same and we intentionally ignored the possible distribution of the reservation values for the underlying merchandise of all shoppers. In reality, shoppers might possess differentiated reservation values. Thus, their purchase decisions might not be exactly the same as presented in Figures 3 through 5. Meanwhile, shoppers reflect a different risk aversion, which will induce differentiated decision making when facing the same market situation. In addition, the model is presented in an abstract form. If we gather empirical data to support the model setting, it will benefit online shoppers, small e-vendors, and social planners in understanding the model.

Future research can be conducted to enrich the current model and research stream. Three streams are identified. First, we can allow consumers to have different risk perceptions toward online shopping. Based on their risk tolerance, consumers can be risk-seeking, risk-neutral, or risk-averse. A consumer's risk perception could affect his or her purchasing decision, and this factor could be incorporated into the decision-making model. Meanwhile, the price charged for a specific product by a less-established e-vendor with a seal or one with no seal can be projected as being different. A seal may earn a premium and an e-vendor with a seal might charge a higher price than one without a seal. Future research can study how the price premium changes the decision-making criteria of online shoppers and the competition structure of the e-markets. Finally, e-vendors themselves are active decision makers as well. Their decision with regard to seal adoption is based on their analyses of the benefits provided vs. the expenses incurred by the seal, as well as the overall condition of the e-markets. Exploration of a decision-making model from an e-vendor's perspective could also be a future research direction.

For practitioners and academic researchers to gain a fuller understanding of the role of Web seals in online trading, more research needs to be done in this area. As discussed, the extent to which Web seals impact online shopper's decision making is still a debated topic, but past research (Kovar et al., 2000; Odom et al., 2002) agrees that the effectiveness of a Web seal in affecting a consumer's purchasing decision is related to the consumer's knowledge and familiarity with the seal and the seal issuer. Thus, an important task for seal-issuing organizations and small e-vendors that adopt those seals is to educate online shoppers about the benefits of the seals. In the meantime, selecting a well-known Web seal can also help small e-vendors win more consumers. As the use of Web seals gains in popularity, better understanding of how e-vendors can take full advantage of these benefits is compelling enough to warrant more research.

REFERENCES

Bhatnagar, A., Misra, S., & Rao, H. R. (2000). On risk, convenience, and Internet shopping behavior. Communications of the ACM, 43(11), 98-105.

Bhimani, A. (1996). Securing the commercial Internet. *Communications of the ACM, 39*(6), 29-31.

Cook, D. P., & Luo, W. (2003). The role of third-party seals in building trust online. *E-Service Journal, 2*(3), 71-84.

Cunningham, S. M. (1967). The major dimensions of perceived risk. In *Risk taking and information handling in consumer behavior* (pp. 82-108). Boston: Harvard University Press.

Ford, W., & Baum, M. S. (1997). *Secure electronic commerce.* Upper Saddle River, NJ: Prentice Hall.

Gefen, D. (2002). Customer loyalty in e-commerce. *Journal of the Association for Information Systems, 3*(1), 27-51.

Gefen, D., Karahanna, E., & Straub, D. W. (2003). Trust and TAM in online shopping: An integrated model. *MIS Quarterly, 27*(1), 51-90.

Griffin, K., Ladd, P., & Whitehead, R. (1997). Internet commerce: Security is still a concern. *The Review of Accounting Information Systems, 2*(1), 67-71.

Hu, X., Lin, Z., & Zhang, H. (2003). Trust promoting seals in electronic markets: An exploratory study of their effectiveness for online sales promotion. *Journal of Promotion Management, 9*(1/2), 163-180.

Internet Crime Complaint Center. (2007). *The Internet crime complaint center hits 1 million!* Retrieved February 13, 2008, from http://www.ic3.gov/media/2007/070613.htm

Internet Fraud Watch. (2005). *Internet fraud statistics report.* Retrieved September 24, 2007, from http://www.fraud.org/internet/intstat.htm

Internet Fraud Watch. (2007). *Internet fraud tips.* Retrieved February 13, 2008, from http://www.fraud.org/tips/internet/general.htm

Jarvenpaa, S., Tractinsky, L. N., & Vitale, M. (2000). Consumer trust in an Internet store. *Information Technology and Management, 1*(1/2), 45-71.

Kau, A. K., Tang, Y. E., & Ghose, S. (2003). Typology of online shoppers. *The Journal of Consumer Marketing, 20*(2/3), 139-156.

Kimery, K. M., & McCord, M. (2006). Signals of trustworthiness in e-commerce: Consumer understanding of third-party assurance seals. *Journal of Electronic Commerce in Organizations, 4*(4), 52-74.

Kovar, S. E., Burke, K. G., & Kovar, B. R. (2000). Consumer responses to the CPA WEBTRUST assurance. *Journal of Information Systems, 14*(1), 17-35.

Lasica, J. D. (1999). *The confidence game.* Retrieved September 24, 2007, from http://www.jdlasica.com/idrive/TIS/confidence.html

Lee, M. K. O., & Turban, E. (2001). A trust model for consumer Internet shopping. *International Journal of Electronic Commerce, 6*(1), 75-91.

McKnight, D. H., Choudhury, V., & Kacmar, C. (2002). The impact of initial consumer trust on intentions to transact with a Web site: A trust building model. *Journal of Strategic Information Systems, 11,* 297-323.

Metzger, M. J. (2006). Effects of site, vendor, and consumer characteristics on Web site trust and disclosure. *Communication Research, 33*(3), 155-179.

Miyazaki, A. D., & Fernandez, A. (2001). Consumer perceptions of privacy and security risks for online shopping. *The Journal of Consumer Affairs, 35*(1), 27-44.

Nikitkov, A. (2006). Information assurance seals: How they impact consumer purchasing behavior. *Journal of Information System. 20*(1), 1-17.

Noteberg, A., Christiaanse, E., & Wallage, P. (2003). Consumer Trust in Electronic Channels. *E- Service Journal, 2*(2), 46-67.

Odom, M. D., Kumar, A., & Saunders, L. (2002). Web assurance seals: How and why they influence consumers' decisions. *Journal of Information Systems, 16*(2), 231-250.

Parkinson, T. L. (2002). The role of seals and certifications of approval in consumer decision-making. *Journal of Consumer Affairs, 9,* 1-14.

Peterson, R. A., Balasubramanian, S., & Bronnenberg, B. J. (1997). Exploring the implications of the Internet for consumer marketing. *Journal of the Academy of Marketing Science, 25*(4), 329-346.

Sanderson, B. (2000). Cyberspace retailing a threat to traditionalists. *Retail World, 53*(14), 6-7.

Stewart, K. J. (2003). Trust transfer on the World Wide Web. *Organization Science, 14*(1), 5-17.

Urban, G. L., Sultan, F., & Qualls W. J. (2000). Placing trust at the center of your Internet strategy. *Sloan Management Review, 42*(1), 39-48.

U.S. Census Bureau. (2007). *Quarterly retail e-commerce sales.* Retrieved February 13, 2008, from http://www.census.gov/mrts/www/ecomm.html

U.S. Department of Justice. (2007). *Internet and telemarketing fraud.* Retrieved February 13, 2008, from http://www.usdoj.gov/criminal/fraud/internet/#howtodeal

Van den Poel, D., & Leunis, J. (1999). Consumer acceptance of the Internet as a channel of distribution. *Journal of Business Research, 45,* 249-256.

Wakefield, R. L., & Whitten D. (2006). Examining user perceptions of third-party organization credibility and trust in an e-retailer. *Journal of Organizational and End User Computing, 18*(2), 1-19.

Wang, S., Beatty, S. E., & Foxx, W. (2004). Signaling the trustworthiness of small online retailers. *Journal of Interactive Marketing, 18*(1), 53-69.

Wolfingarger, M., & Gilly, M. C. (2001). Shopping online for freedom, control, and fun. *California Management Review, 43*(2), 34-55.

Yoon, S. J. (2002). The antecedents and consequences of trust in online-purchase decisions. *Journal of Interactive Marketing, 16*(2), 47-63.

This work was previously published in the International Journa! of E-Business Research, Vol. 4, Issue 3, edited by I. Lee, pp. 20-39, copyright 2008 by IGI Publishing (an imprint of IGI Global).

Chapter 15
Analysis of the Relationship Existing between Business Commercial Information Technologies[1]

Blanca Hernández
University of Zaragoza, Spain

Julio Jiménez
University of Zaragoza, Spain

M.José Martín
University of Zaragoza, Spain

ABSTRACT

The objective of this work is to analyse the importance of firms' previous experience with different information technologies (Internet, EDI) in their implementation of e-CRM and B2B e-commerce. Moreover, we also study the role of e-CRM in B2B development. With this objective, we have analyzed 109 firms belonging to the IT sector. The results show that experience with IT such as EDI or the Internet has a direct influence on the use of e-CRM. There is also a direct and positive transmission of knowledge from e-CRM to B2B e-commerce, even though they have not yet been adopted intensively by firms. Firms need to be aware of the interrelations that exist between the different information technologies. The experience accumulated from using an IT can be considered an important aspect of organisational knowledge, which allows firms to obtain a number of benefits as a result of applying other IT that are complementary.

INTRODUCTION

At the beginning of the 21st century, firms still find it difficult to recognise the strategic value of adopting information technologies (IT) (Von Krogh *et al.*, 2000; Nonaka *et al.*, 2000). While some research considers that IT do not positively impact on firms' productivity (Carr, 2003), another states that IT are an inexhaustible source of business opportunities for the modern firm (Rayport & Sviokla, 1994). Therefore, companies could apply IT to generate

DOI: 10.4018/978-1-60566-910-6.ch015

important competitive advantages and to differentiate themselves (Grant, 1996; Teece, 2000).

Of all the IT that have emerged in recent years, the Internet has been one of those that has generated most interest. So too, clearly, has electronic commerce (or e-commerce), which derives from the Internet, and redefines some of the variables and elements of traditional exchanges (Webb, 2002). B2B e-commerce enables a business to interact with another business electronically, in particular via the web (Zeng *et al.*, 2003). There are many benefits of B2B e-commerce. We highlight the following: increasing productivity, reducing potential staff overhead, and clearing audit training (Yang & Papazoglou, 2000). In spite of these benefits, firms have not adopted B2B e-commerce unanimously and homogenously. Observing the different levels of acceptance, some studies have suggested the need to know which factors influence on a firm's technological behaviour. One factor that appears to be of considerable importance in the implementation process is IT experience, i.e. previous knowledge obtained by the firm from other IT related to the tool being analysed.

How do existing IT affect the adoption of new IT and enable transactions between businesses? This question has been an object for concern of firms for decades and continues to increase as new IT for commercial management appear (Chatterjee & Ravichandran, 2004). The objective of the current study is to analyse the importance of the firm's previous experience with IT in its acceptance of e-CRM and in the development of B2B e-commerce. With this objective, we look into the question of whether previous use of IT is a requisite for its implementation (such as the Internet) or related to commercial management (such as e-CRM and Electronic Data Interchange, EDI) and if it generates a greater affinity and an appropriate internal structure for the successful implementation of B2B e-commerce, as has been suggested by Riggins and Rhee (1999), Watson and McKeown (1999) and Ngai and Wat (2002). If these propositions are fulfilled, technological experience could be considered as being a key factor for organisational knowledge, which would allow firms to obtain sustainable competitive advantages as a result of applying previous IT.

EDI is defined as cooperative interorganizational systems that allow trading partners to exchange structured business information electronically between separate computer applications (Swatman & Swatman, 1992; Asher, 2007). With regards to CRM, there is no universally accepted definition of this term, so this work follows that given by Kincaid (2003): "CRM is the strategic use of information, processes, technology and people to manage the customer's relationship with your company (Marketing, Sales, Services and Support) across the whole customer life cycle" (p. 41). Consequently, e-CRM is CRM software on the Internet, or Internet-based CRM. Finally, by B2B we consider interfirm e-commerce in its more restrictive meaning, i.e., conducting commercial transactions through the World Wide Web, not including advertising on the Internet or simple e-mail.

In the following section we review the literature on the role of acquired experience and compatibility in the acceptance of other IT. Moreover, we look further into the characteristics of the main IT considered in the study: EDI, Internet, e-CRM and B2B e-commerce. In the following sections, we detail the empirical analyses carried out, the results obtained and the conclusions that we draw from them.

TECHNOLOGICAL EXPERIENCE

The implementation of the Internet as a distribution and supply channel has been a key change in the evolution of many firms in recent years. Internet and other online IT are considered as a major accelerator in realizing closer forms of collaboration between business partners (Legner, 2008), so they create enormous opportunities for improving the scope and strength of B2B customer relationships

(Merrilees & Fenech, 2007). However, despite the fact that adopting new technological systems can be regarded as an opportunity for modern firms, not all of them have chosen to adopt IT for carrying out their operations. Indeed, only a few of them have applied online IT related to their commercial management such as EDI, e-CRM software and B2B e-commerce (Asher, 2007).

In this context, some researchers have investigated the various factors determining users' intentions and/or level of acceptance of IT, using empirically validated models for that purpose (Davis *et al.*, 1989; DeLone & McLean, 1992; Rogers, 1995). One of these factors, labelled "perceived compatibility", reflects the degree to which individuals regard using an innovation to be consistent with their values, socio-cultural beliefs and past and present experiences (Rogers, 1983, 1995). This author argues that certain innovations have close links to each other, so the existence of previous experience with some of them is strongly correlated with the subsequent adoption of related innovations. On the basis of this idea, the concept of "technology clustering" is introduced and subsequently used by various authors (Leung, 2001; Eastin, 2002; Tsai, 2005). A technology cluster is defined as the set of elements the user perceives to be interrelated and determinants of the level of adoption of subsequent technologies.

The knowledge the firm acquires previously by using other IT should also be considered when explaining its subsequent technological behaviour (Bennett *et al.*, 2005), since it will increase the user's motivation and mitigate the resistance to using new IT (Dewar & Dutton, 1986; Kearns & Sabhewarl, 2007). Moreover, this knowledge modifies the organisational culture, and conditions the level of future development (Dishaw & Strong, 1999). Acquired capabilities allow the firm to adapt its activities to the new opportunities being opened up by new IT, and provide a fundamental support for subsequent IT diffusion (Ettlie, 1990; Tiessen *et al.*, 2001; Mirchandani & Motwani, 2001).

Park *et al.* (2004) demonstrate that a firm's orientation towards new technology and information sharing as well as its trust based working culture increase the acceptance of new IT. Hence, in the context of utilizing IT, technological orientation should help to promote an environment conducive to learning and incorporating the new IT into standard operational use (Kim *et al.*, 2009). More specifically, Premkumar and Roberts (1999) analyse the intensity with which firms use several information technologies (EDI, online access to data, email, or Internet), and establish a series of explanatory factors of this variable. They find that the degree of compatibility and experience differs between adopters and non-adopters, considering both variables as determinants of the firm's behaviour. Experience is an important factor in learning, since firms tend to use and build their knowledge over the foundation of the already-existing information (Cohen & Levinthal, 1990). As Power (2005) asserts, a firm's technological knowledge determines its capability to effectively implement new IT.

Some researchers have analysed the influence of past experience in IT related to a firm's commercial management, such as EDI and B2B e-commerce. Implementing B2B requires certain technological knowledge, so that the previous orientation of some firms towards EDI-related systems facilitates its development, due in large part to the generation of affinity and the simplification of the learning process (Angeles *et al.*, 2001; Hsieh & Lin, 2004). Jiménez and Polo (1998) empirically demonstrate -for a particular country and sector- that in the early years of EDI, a fundamental factor for EDI diffusion was the level of technological experience existing in the company, a result of the implementation of some antecedent information tools, such as fax or videotext. In this way, the so-called "technological sophistication" positively influences the perception of the ease of use of other IT and generates an organisational culture eager to accept them (Jiménez & Polo, 1998).

More recently, Kaefer and Bendoly (2004) investigate differences in the commercial development of firms on the basis of different levels of knowledge about EDI. These authors demonstrate that accepting EDI is a key factor for explaining the level of progress, so that firms that are more experienced in using this IT also manifest a greater flexibility in applying B2B in the future and find it easier to use. Similarly, Patterson *et al.* (1997) and Härtel (1999) consider compatibility derived from experience to be associated with the efficiency the firm has achieved in using previous IT, and with the way this experience affects the development of B2B e-commerce.

We can therefore state that experience transforms the firm's responses and reactions, and increases the probability it will adopt other IT linked to the earlier ones (Ward & Lee, 2000; Dahlen, 2002).

EDI, INTERNET, E-CRM AND B2B E-COMMERCE

Our study begins with the idea that acceptance of previous IT is a key aspect in the application of other technologies, given that the firm's perceptions and use of previous IT (technology clustering) make up technological knowledge that helps to determine the extent to which B2B is developed. We include those IT that the literature regards as being closely linked to commercial management, and whose previous knowledge generates experience in the firm's use of technology: EDI, Internet and e-CRM (Truman, 2000; Subramani & Walden, 2001; Lee *et al.*, 2003a).

Firstly, we considered EDI, because of the fact that the literature regards this technology as a precursor to current e-commerce (Lee *et al.*, 2003a; Teo & Ranganathan, 2004; Kaefer & Bendoly, 2004). Among the benefits of EDI are cost reduction, decrease in the amount of paperwork, faster turnaround, improved customer relationship and customer service, and potential strategic advantage (Subramani, 2004; Jai-Yeol *et al.*, 2005). Likewise, EDI has been adopted by many businesses worldwide as a vehicle for eliminating the paper-work associated with business transactions, thereby eliminating errors caused by entering data manually (Hsieh & Lin, 2004).

EDI can be considered as being the grandfather of B2B e-commerce, since it has the ability of expressing data in a simple format and sending it to someone else (Hsieh & Lin, 2004). Furthermore, researchers have shown that knowledge of other IT applications facilitates the assimilation of EDI (O'Callaghan *et al.*, 1992; Emmelhainz, 1993; Jiménez & Polo, 1998).The link between this IT and B2B e-commerce leads us to suggest that this phenomenon would be repeated between them. Although the adoption of EDI may involve factors different from other IT related to a firm's commercial management, they are very similar in nature (Cho, 2006). Therefore, implementing EDI favours the subsequent acceptance of IT such as the Internet and e-CRM, and speeds up the development of the B2B e-commerce. Following Kaefer and Bendoly (2004), B2B e-commerce rectifies EDI weaknesses, so it can be regarded as a direct substitute of it. Consequently, our proposals are:

H1: The firm's experience in using EDI has a positive influence on the implementation of the Internet.

H2: The firm's experience in using EDI has a positive influence on the implementation of e-CRM.

H3: The firm's experience in using EDI has a positive influence on the development of B2B e-commerce.

After considering the effect of EDI, we introduced the knowledge derived from using the Internet, given that this is a fundamental tool for the evolution of the firm's exchanges from traditional environments towards new technological support of sales management (e-CRM) and sales channels (B2B e-commerce) (Kafer & Bendoly, 2004; Soliman & Janz, 2004).

On the one hand, the e-CRM information system has recently evolved further as a result of the emergence of IT such as the Internet and web technologies, which allow it to integrate and simplify all customer-related processes (Plakoyiannaki & Tzokas, 2002; Tung, 2007). Therefore, the experience of the firm in relation to the Internet positively influences on the development of e-CRM. On the other hand, firms should have the ability to capture, integrate and distribute data obtained from the Internet. The correct management of the knowledge derived from the Internet is a key requisite to obtaining the highest returns from transactions conducted through B2B e-commerce.

H4: The firm's experience in using the Internet has a positive influence on the implementation of e-CRM.

H5: The firm's experience in using the Internet has a positive influence on the development of B2B e-commerce.

Finally, we included the use of e-CRM as an antecedent to B2B development. There are several points that link e-CRM and B2B (their principal similarities and differences can be seen in Table 1). Firstly, e-CRM software is a relatively novel IT, like B2B e-commerce, so both of them have gone through a similar process of evolution (Kincaid, 2003; Ngai, 2005). Moreover, sales are one of the three principal functional areas that CRM comprises (marketing, sales and services, and support), so it represents another important link between both technologies (West, 2001; Xu *et al.*, 2002; Kincaid, 2003). Firms capture and analyse customer information through e-CRM; later, this information is used for their strategic marketing plans (Ngai, 2005). Thus, e-CRM supports and improves B2B marketing decision-making (Noori & Salimi, 2005).

Regular use of e-CRM allows firms to gain good experience in online commercial systems, so using e-CRM increases the subsequent use of the Internet to do business with other firms: i.e., B2B e-commerce (Bauer *et al.*, 2002; Kincaid, 2003; Teo & Ranganathan, 2004; Noori & Salimi, 2005). A greater experience with technologies such as e-CRM will raise the use of B2B e-commerce, which eventually translates into a higher proportion of business with other firms via the Internet. Consequently, we defend the positive influence of e-CRM experience on B2B development, as has been demonstrated by Bauer *et al.* (2002), Kincaid (2003), Teo and Ranganathan (2004) and Noori and Salimi (2005):

H6: The firm's experience in using e-CRM has a positive influence on the development of B2B e-commerce.

On the basis of these hypotheses, we formulate the model that is shown in Figure 1.

METHODOLOGY

We used surveys to collect data, sending questionnaires by post and e-mail to a group of Spanish firms (449 firms) belonging to a sector closely linked to the systems under analysis, i.e., new technology or IT firms (SIC code 737), which have become increasingly important in the economy in recent years. Given the nature of their business, these firms are in direct contact with IT and possess considerable experience or know-how in this area. An ad-hoc analysis on how firms which produce IT as an *output* also use it as an *input* in their productive process is essential to studying the perceptions affecting IT acceptance and the synergies derived from the previous use of other systems.

After the initial refinement process, we obtained a final sample of 109 valid cases. As in

Table 1. Similarities and differences between e-CRM and B2B e-commerce

Similarities	
Application anytime and anywhere	Since both IT are online operations, they are available to other companies online anytime and anywhere
Broadening and deepening hub services	Necessary for the success of both IT.
Customer research	Need for comprehensive customer research regarding what other companies need and who the target customers are. To conduct customer research, both need strong technology supports such as data warehousing and database management.
Platform of operating system support	Need for similar platforms to operate
Data-mining capability	Need for data-mining capability and systems tend to overlap in certain functions
Flexibility	Need for flexibility to meet unpredictable growth and demand
Involvement in business transactions	Involved in every process in business transactions including marketing, sales, billing and shipping, and customer services
One-to-one marketing	Need for one-to-one marketing to customize their products and services for customers
Differences	
Dependence	CRM is more involved in intranet and has more emphasis on business processes
Automation	CRM has its own features of marketing automation and sales automation, while B2B does not necessarily need these features
Users-customers	Much harder for B2B users than for CRM users to switch to different products and service providers.
Customers	B2B can generate a critical mass of customers, while CRM may or may not create this, depending on the size of customers it has
Parties involved	B2B customers in the industry tend to be generalized while CRM has specific customers and needs vendor suppliers.
Features	CRM is a concrete package while B2B is a collection of technologies

Adapted from Zeng *et al.* (2003)

Figure 1. Study Model

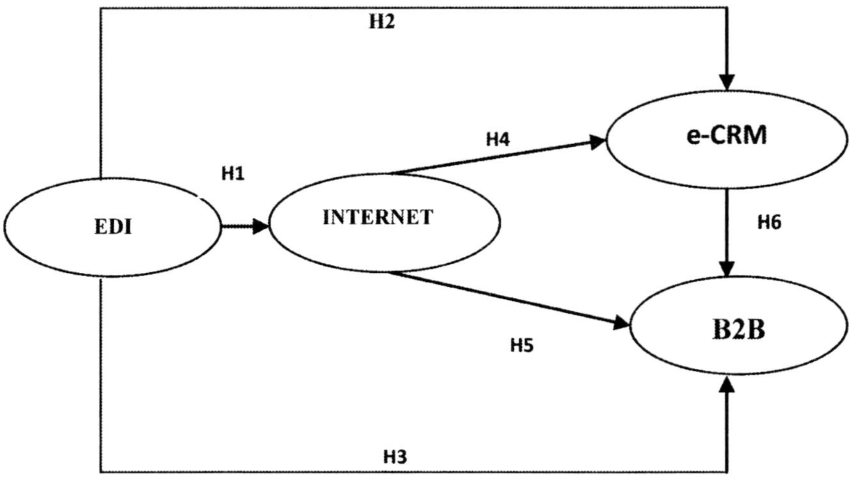

previous research (Riemenschneider *et al.*, 2003; Grandon & Pearson, 2004; Carayannis & Turner, 2006), the information refers to the firm as a whole and not to each of its employees, so the questionnaires were directed to the managers who take decisions regarding IT, who were requested to reply on behalf of the company.

Based on a review of related literature (e.g. Gefen and Straub, 1997; Igbaria *et al.*, 1997; Lee *et al.*, 2003b; Shih, 2004; Shang *et al.*, 2005), the items designed for measuring Internet, EDI and e-CRM experience used a 7-point scale (1= *strongly disagree*, 7= *strongly agree*), while for B2B e-commerce we used the percentage of business with other companies via the Internet. As this is directly observable, the factor was made up using this single indicator (see Appendix).

With regard to the indicators for each IT, the questionnaire included questions about the firm's intensity of use and about these technologies' ease of use and usefulness. These latter indicators were included in the scale in light of empirical evidence that perceptions about a particular IT serve as indicators to measure the level of objective development and acceptance of IT (Davis, 1989; Davis *et al.*, 1989; Venkatesh and Davis, 2000; Amoako-Gyampah & Salam, 2004). In this way, positive perceptions about a specific IT (Internet, EDI, e-CRM) incentivise the user to apply the IT more, at the same time as they generate a predisposition to accept other, more complex tools that are related to IT. Thus, the experience with each technology is represented through a technology acceptance model (TAM), which reflects the affinity and knowledge generated in the firm as a result of its application.

In the following sections we describe the validation process carried out by exploratory and confirmatory factor analyses, which appropriately refined the indicators involved in the study. We show the structural model and the causal relations postulated between the constructs, thereby testing the hypotheses formulated.

FACTOR ANALYSES

An exploratory analysis was carried out, with the aim of approximating the underlying structure of the model and refining the constructs that represent the variables included in the questionnaire.

The refining process consists of eliminating those indicators with a low item-total correlation (Nurosis, 1993), or those whose elimination raises the value of Cronbach's alpha (Nunnally, 1978). The item-total correlations easily exceeded the minimum acceptable value of 0.3 in all cases. Cronbach's alpha achieved values of 0.811 for Internet, 0.863 for EDI and 0.841 for e-CRM.

The following step was to study the unidimensionality of the proposed factors by the exploratory factor analyses, using the "principal axis factoring" method and varimax rotation, as the literature recommends (Kaiser, 1970; McDonald, 1981; Hair *et al.*, 1999). In all cases one factor was extracted through the eigenvalue criterion; the factor loadings exceeded the minimum threshold of 0.5 (Hair *et al.*, 1999), and the variances explained by the factors were as follows: 71.66% for Internet, 69.63% for EDI and 65.50% for e-CRM. Our results presented satisfactory and stable values as they have been tested through rotation in all existing methods (Oblimin, Quartimax, Equamax and Promax).

The second phase of the study was to carry out a confirmatory factor analysis jointly for all constructs making up the model. In this way, we continued with the process of refinement and validation of the scales and evaluated the reliability and validity of the proposed dimensions, applying structural equation modelling (SEM) techniques using the statistics package EQS 5.7b. The estimation method used was robust maximum likelihood, since our data do not comply with the normality assumption (Bentler, 1995).

The scale refinement process continued with the three criteria proposed by Jöreskog and Sörbom (1993): weak convergence, strong convergence and explanatory coefficient. The weak convergence

Table 2. Confirmatory factor analysis: validity, reliability and adjustment fit measures

	ITEM	P	AVE	R^2	Lambda	F	Interval
Internet experience (F1)	INT_1	0.904	0.763	0.95	0.975	F1- F2	(0.033- 0.233)
	INT_2			0.88	0.938	F1- F3	(-0.016- 0.408)
	INT_3			0.458	0.677	F1- F4	(-0.185- 0.259)
EDI experience (F2)	EDI_1	0.872	0.694	0.71	0.843	F2- F3	(0.132- 0.44)
	EDI_2			0.659	0.812	F2- F4	(-0.124- 0.372)
	EDI_3			0.712	0.844	F3- F4	(0.037- 0.373)
e-CRM experience (F3)	CRM_1	0.849	0.654	0.801	0.895		
	CRM_2			0.62	0.788		
	CRM_3			0.54	0.735		
B2B (F4)	B2B	0.85	0.85	0.85	0.922		

Absolute Fit	**Incremental Fit**	**Parsimony Fit**
GFI= 0.946	NFI = 0.946	χ^2/ d.f.= 1.10
MFI = 0.987	NNFI = 0.993	
RMSR = 0.033	CFI= 0.995	
RMSEA = 0.031	IFI= 0.995	

criterion, according to Steenkamp and Van Trijp (1991), analyses the significance of the factorial regression coefficients between the indicators and their latent variable. The non-significant indicators should be eliminated (student's t >2.58; p= 0.01). The strong convergence criterion implies eliminating non-substantial indicators, i.e. those whose standardised coefficient is less than 0.5 (Hildebrandt, 1987). The explanatory coefficient of the indicator should exceed 0.3. All indicators achieved acceptable values in these three criteria (Table 2), so we went on to analyse the validity and reliability of the factors making up the model (Churchill, 1979; Gerbing & Anderson, 1988).

As we explained above, the reliability was initially tested using Cronbach's alpha. But due to the exploratory nature of the initial analysis, this property was also verified using the composite reliability index of the construct (ρ) (Jöreskog, 1971) and the average variance extracted (AVE) (Fornell & Larcker, 1981) (Table 2).

With regard to the validity of the scales as instruments to measure the concepts that they are intended to measure, content validity and construct validity were calculated. The former is guaranteed by the review of the literature that we have carried out.

Construct validity is made up of two fundamental types of validity: convergent and discriminant. With regard to convergence, in Table 2 we can see that the factor loadings exceed 0.5 and are statistically significant at the 99% level, so the measurement scales satisfy this property (Bagozzi & Yi, 1988; Anderson & Gerbing, 1988; Steenkamp & Van Trijp, 1991). Discriminant validity implies that the factor being analysed should be significantly distant from other constructs with which it is not theoretically related (Lehmann *et al.*, 1999). This was confirmed through three different criteria:

1. The correlations between the different factors making up the model do not exceed 0.8, as this would indicate a low discrimination between them (Bagozzi, 1994).
2. We estimated confidence intervals of the correlations between the different latent variables, ensuring that none of them

Table 3. Goodness-of-fit indices of structural analysis

Absolute Fit		Incremental Fit		Parsimony Fit	
Indicator	Value	**Indicator**	Value	**Indicator**	Value
GFI	0.946	**NFI**	0.946	χ^2/ d.f.	1.10
MFI	0.987	**NNFI**	0.993		
RMSR	0.033	**CFI**	0.995		
RMSEA	0.031	**IFI**	0.995		
P X^2	0.559				

contained the value 1 (Bagozzi & Yi, 1988) (Table 2).

3. We confirmed that the variance extracted from each factor was greater than the variance shared with the remaining constructs (Wang *et al.*, 2003; Dholakia, Bagozzi & Klein, 2004).

Moreover, the model presents an acceptable goodness of fit, since all of the absolute, incremental and parsimony fit indices achieve optimal values (Hair *et al.*, 1999) (Table 2).

RESULTS

Having refined the scales and validated the measurement model, we went on to test the proposed hypotheses through a structural relations model. We also confirmed that the goodness of fit exceeds the theoretically recommended limits (Table 3).

The results obtained (Table 4 and Figure 2) show that implementing an EDI system in the firm generates knowledge that positively influences the acceptance of other information systems, such as Internet (γ= 0.131), or e-CRM. The last relation is both direct (γ= 0.265), and indirect via the experience acquired from the Internet, with global weight amounting to 0.286. Thus, H1 and H2 are satisfied.

The knowledge generated from using EDI and Internet does not favour the implementation of B2B, so H3 and H5 are rejected. The relation between Internet and implementation of e-CRM is significant, so H4 is fulfilled. Finally, the relation between e-CRM and B2B e-commerce has achieved a statistically significant weight of 0.19, so H6 is satisfied.

The objective of this study has been to determine how

Thus, knowledge of those systems that are fundamental for the execution and development

Table 4. Results of the structural model

Hypotheses		
H1	**EDI→ INT**	**γ= 0.131***
H2	**EDI→ e-CRM**	**γ= 0.265****
H3	**EDI→ B2B**	γ= 0.07
H4	**INT→ e-CRM**	**ß= 0.161***
H5	**INT→ B2B**	**ß**= 0.01
H6	**e-CRM→ B2B**	**ß= 0.19***

** Significant at 99% level; * Significant at 95% level.

Figure 2. Results of causal model (standardized coefficients and t-statistics)

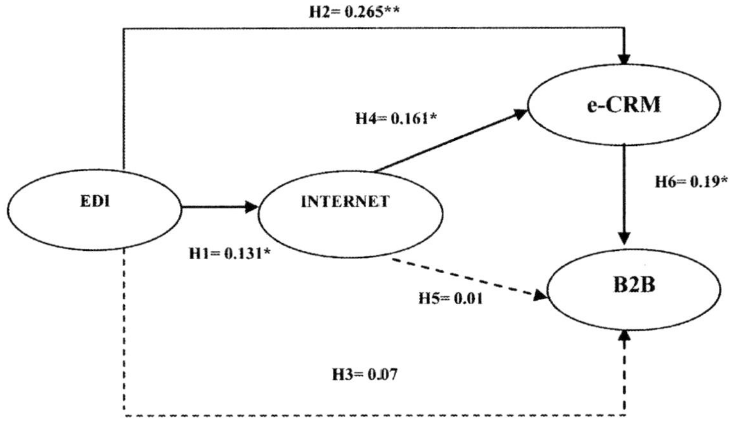

** Significant at 99% level; * Significant at 95% level.

of e-CRM, such as Internet or EDI, does not directly affect B2B e-commerce. Nevertheless, both EDI and Internet influence B2B indirectly through e-CRM. We can therefore see that the firm's technological evolution depends on the knowledge it acquires from using previous IT, and on the general technological culture created. This reduces the perception of risk associated with adopting and using innovations.

DISCUSSION OF RESULTS

The experience the firm gains from its past IT use affects its subsequent acceptance of e-CRM and B2B. For this purpose, we have analysed the influence of other IT that might have generated some knowledge: the Internet and EDI. Likewise, we have also analysed the role of e-CRM in B2B acceptance.

The analyses show that EDI directly influences the firm's use of Internet and e-CRM. This result is coherent with the research conducted by Hsieh and Lin (2004). They assert it is expected that there was a rapid move toward extending the use of the Internet as an alternative method of delivering EDI, since this could allow more widely distributed participation. Moreover, the Internet positively modifies implementation of e-CRM. With regard to the importance of each tool, we find that EDI has the greatest weight in the development of e-CRM. This is due to the fact that EDI was the first IT that allowed firms to establish e-commerce relationships. Likewise, the experience and knowledge acquired previously with EDI and the Internet influences the level of development of other subsequent IT, thereby facilitating the adoption of the new channel and providing the firm with synergies as a result of the aspects that they all have in common.

Although the direct effects of the Internet and EDI on B2B e-commerce are not significant, these effects take place indirectly through e-CRM. This finding may be due to the economic activity of the firms in our study (IT sector), since their knowledge of different technologies is higher than in other sectors. Their knowledge facilitates the acceptance of new IT, e.g. e-CRM, which have not yet been intensively applied by other firms. This is the reason why the experience previously acquired (Internet and EDI) has already been assumed by these firms, and the influence of these tools on B2B e-commerce is indirect.

We have also demonstrated that e-CRM has a significant influence on B2B e-commerce. Nevertheless, such effect is not as high as could be expected. This is probably due to the fact that both IT are still in an introduction phase, which is why the relationship existing between them has not reached the intensity that it will achieve in the near future. Along this line, Kim *et al.* (2009) obtained similar results.

In summary, we emphasize that the results obtained validate the relationship existing between the IT that has been already used in the commercial management and others that are appearing in the market. Thus, the investment in a certain technology means the firm will achieve some experience which will progressively redeem itself during the start-up and use of the new IT. These results are coherent with previous research (Gatignon & Xuereb, 1997; Kim *et al.*, 2009), which showed that technologically-oriented firms are more willing to acquire new sophisticated IT, and that they are also motivated to use their knowledge to build new solutions. This technological orientation refers to both the ability and the incentive to apply new IT (Zhang & Tansuhaj, 2007).

IMPLICATIONS

Our findings offer important strategic implications for firms and their process of technology acceptance, arguing that a high level of use of a certain IT is conducive to creating benefits for the user firm in terms of knowledge and ability.

Firms should be aware of the interrelations that exist between the different information technologies. Investing in a specific tool can help them to accept subsequent IT, as well as to exploit them more effectively. Likewise, the knowledge accumulated from using a technological innovation can be considered as an important aspect of organisational knowledge, which allows firms not only to obtain sustainable competitive advantages, but also to gain a number of benefits as a result of applying systems that are complementary. These benefits incentivise firms to acquire knowledge that they can exploit in numerous future activities and that will improve their efficiency. In this way, this learning process reduces the perception of risk associated with implementing a new system, creating a global compatibility that significantly affects the level of future development.

The diffusion of different IT in the firm requires previous technological experience with which to sustain the organisation's capacity to transform and use the information acquired previously in IT-related knowledge. This knowledge (EDI, Internet and e-CRM) minimises, for example, the impact of implementing the new electronic market (B2B), at the same time as it facilitates its subsequent development in the firm.

In order to encourage e-CRM-B2B integration, firms need to be aware of the advantages that could be derived from a joint application. Among these, we should mention the possibility of reducing the cost of quality management, carrying out more efficient marketing campaigns and taking advantage of new opportunities for competing in the information space (Zeng *et al.*, 2003). Each IT mutually complements with those capabilities that it does not have in its functions and makes both B2B and e-CRM systems more powerful and sufficient to provide better products and services to customers.

Firms need to invest significant financial resources in technology, in an attempt to achieve a general level of knowledge for the whole company that will help them to differentiate themselves from their rivals. We have therefore shown in this research that previous technologies generate a key compatibility that firms can exploit. Our general recommendation would then be for firms to introduce a corporate culture based on IT, trying not to get bogged down in well-known, established systems, and to continually improve the computerisation of their traditional functions. Moreover, we suggest that companies invest in training and information for their employees regarding these new IT, irrespective of their position in the firm. Training may make the technology easier to use and more familiar. Likewise, it is important that all staff, from the top manager down to the lowest employee, be aware of the benefits inherent to the technology.

With regards to principal limitations and future lines of research, we would first note that the analysis carried out here is based on cross-sectional data. We have therefore been able to determine the relevance of experience in firms' technological development, but we cannot say anything about how the Internet and EDI have evolved over time. As a future line of research, we propose to examine the importance of experience through a longitudinal study that analyses the variations occurring in each tool that influences e-CRM and B2B e-commerce acceptance.

Another possible extension would be to test the model including several control variables such as size and industry. With this objective in mind, we would test our hypotheses on a sample of firms belonging to different economic sectors whose productive activities are not related to IT. Moreover, we could test the hypotheses considering firms grouped according to their size. These analyses would determine whether or not there are any significant differences between sectors and sizes of firms, and evaluate the relevance of technological experience for IT adoption and implementation in every company.

REFERENCES

Amoako-Gyampah, K., & Salam, A. F. (2004). An extension of the technology acceptance model in an ERP implementation environment. *Information & Management*, *41*, 731–745. doi:10.1016/j.im.2003.08.010

Anderson, J. C., & Gerbing, D. W. (1988). Structural equation modelling in practice: A review and recommend two-step approach. *Psychological Bulletin*, *103*(3), 411–423. doi:10.1037/0033-2909.103.3.411

Angeles, R., Corritore, C., Choton, S., & Nath, R. (2001). Success factors for domestic and international electronic data interchange (EDI) implementation for US firms. *Information & Management*, *21*, 329–347. doi:10.1016/S0268-4012(01)00028-7

Asher, A. (2007). Developing a B2B e-commerce implementation framework: a study of EDI implementation for procurement. *Information Systems Management*, *24*(4), 373–390. doi:10.1080/10580530701586144

Bagozzi, R. (1994). *Structural equation model in marketing research.* Basic Principles, Principles of Marketing Research. Oxford: Blackwell Publishers.

Bagozzi, R., & Yi, Y. (1988). On the evaluation of structural equation models. *Academy of Marketing Science*, *16*(1), 74–94. doi:10.1007/BF02723327

Bauer, H. H., Grether, M., & Leach, M. (2002). Building Customers Relations over the Internet. *Industrial Marketing Management*, *31*(2), 55–63. doi:10.1016/S0019-8501(01)00186-9

Bennett, R., Härtel, C., & McColl-Kennedy, J. R. (2005). Experience as a moderator of involvement and satisfaction on brand loyalty in a business-to-business setting 02-314R. *Industrial Marketing Management*, *34*, 97–107. doi:10.1016/j.indmarman.2004.08.003

Bentler, P. M. (1995). *EQS structural equations program manual.* CA: Multivariate Software, Inc.

Carayannis, E. G., & Turner, E. (2006). Innovation diffusion and technology acceptance: The case of PKI technology. *Technovation*, *26*(7), 847–855. doi:10.1016/j.technovation.2005.06.013

Carr, N. (2003). IT doesn't matter. *Harvard Business Review*, (May): 5–12.

Chatterjee, D. & Ravichandran, T. (2004). Beyond exchange models: understanding the structure of B2B information systems. *Information Systems and e-Business Management, 2*, 169-186.

Cho, V. (2006). Factors in the adoption of third-party B2B portals in the textile industry. *Journal of Computer Information Systems*, *46*(3), 18–31.

Churchill, J. R. (1979). A paradigm for developing better measures of marketing construct. *JMR, Journal of Marketing Research*, (February): 64–73. doi:10.2307/3150876

Cohen, W. M., & Levinthal, D. A. (1990). Absorptive capacity: A new perspective on learning and innovation. *Administrative Science Quarterly*, *35*(1), 1–14. doi:10.2307/2393553

Dahlen, M. (2002). Learning the web: Internet user experience and response to web marketing in Sweden. *Journal of Interactive Advertising*, *3*(1).

Davis, F., Baggozi, R., & Warshaw, P. (1989). User acceptance of computer technology: A comparison of two theoretical models. *Management Science*, *35*(8), 982–1002. doi:10.1287/mnsc.35.8.982

Davis, F. D. (1989). Perceived usefulness, perceived ease of use and user acceptance of information technology? *MIS Quarterly*, *13*(3), 319–339. doi:10.2307/249008

DeLone, W. H., & McLean, E. R. (1992). Information systems success: the quest for the dependent variable. *Information Systems Research*, *3*, 60–95. doi:10.1287/isre.3.1.60

Dewar, R. D., & Dutton, J. E. (1986). The adoption of radical and incremental innovation: An empirical analysis. *Management Science*, *32*, 1422–1433. doi:10.1287/mnsc.32.11.1422

Dholakia, U., Bagozzi, R., & Klein, L. (2004). A social influence model of consumer participation in network and small-group-based virtual communities. *International Journal of Research in Marketing*, *21*, 241–263. doi:10.1016/j.ijresmar.2003.12.004

Dishaw, M., & Strong, D. (1999). Extending the technology acceptance model with task-technology fit constructs. *Information & Management*, *36*(1), 9–21. doi:10.1016/S0378-7206(98)00101-3

Eastin, M. S. (2002). Diffusion of e-commerce: An analysis of the adoption of four e-commerce activities. *Telematics and Informatics*, *19*, 251–267. doi:10.1016/S0736-5853(01)00005-3

Emmelhainz, M. A. (1993). *EDI. A total management guide* (2nd ed.). Van Nostrand Reinhold, New York.

Ettlie, J. E. (1990). What makes a manufacturing firm innovative? *Academy of Management Executive*, *4*(4), 7–20.

Fornell, C., & Larcker, D. (1981). Structural equation models with unobserved variables and measurement error. *JMR, Journal of Marketing Research*, *36*(3), 39–50. doi:10.2307/3151312

Gatignon, H., & Xuereb, J. (1997). Strategic orientation of the firm and new product performance. *Journal of Business Research*, *41*, 161–174.

Gefen, D., & Straub, D. W. (1997). Gender differences in the perception and use of e-mail: an extension to the technology acceptance model. *MIS Quarterly*, *21*(4), 389–400. doi:10.2307/249720

Gerbing, D. W., & Anderson, J. C. (1988). An updated paradigm for scale development incorporating unidimensionality and its assessment. *JMR, Journal of Marketing Research*, *15*, 186–192. doi:10.2307/3172650

Grandon, E., & Pearson, J. M. (2004). Electronic commerce adoption: An empirical study of small and medium US businesses. *Information & Management, 42*, 197–216.

Grant, R. M. (1996). Towards a knowledge-based theory of the firm. *Strategic Management Journal, 17*, 109–122. doi:10.1002/(SICI)1097-0266(199602)17:2<109::AID-SMJ796>3.0.CO;2-P

Hair, J. F., Anderson, R. E., Tatham, R. L., & Black, W. C. (1999). *Multivariate data analysis*. NJ: Prentice Hall.

Härtel, C. E. (1999). Development and test of a two-stage model of performance appraisal. *Asia Pacific Journal of Human Resources, 37*(2), 76–91. doi:10.1177/103841119903700207

Hildebrandt, L. (1987). Consumer retail satisfaction in rural areas: A reanalysis of survey data. *Journal of Economic Psychology, 8*, 19–42. doi:10.1016/0167-4870(87)90004-3

Hsieh, C., & Lin, B. (2004). Impact of standarization on EDI in B2B development. *Industrial Management & Data Systems, 104*(1/2), 68–77. doi:10.1108/02635570410514106

Igbaria, M., Zinatelli, N., Cragg, P., & Cavaye, A. L. M. (1997). Personal computing acceptance factors in small firms: A structural equation model. *MIS Quarterly, 21*(3), 279–302. doi:10.2307/249498

Jai-Yeol, S., Narasimhan, S., & Riggins, F. J. (2005). Effects of relational factors and channel climate on EDI usage in the customer-supplier relationship. *Journal of Management Information Systems, 22*(1), 321.

Jiménez, J., & Polo, Y. (1998). International diffusion of a new tool: The case of electronic data interchange in the retailing sector. *Research Policy, 26*(7-8), 811–827. doi:10.1016/S0048-7333(97)00045-0

Jöreskog, K. (1971). Statistical analysis of sets of cogeneric tests. *Psychometrika, 36*, 109–133. doi:10.1007/BF02291393

Jöreskog, K., & Sörbom, D. (1993). LISREL 8 structural equation modelling with the simplis comand language. *Scientific software International*. Chicago-Illinois.

Kaefer, F., & Bendoly, E. (2004). Measuring the impact of organizational constraint on the success of business e-commerce efforts: A transactional focus. *Information & Management, 41*, 529–541. doi:10.1016/S0378-7206(03)00088-0

Kaiser, H. F. (1970). A second-generation little jiffy. *Psychometrika, 35*, 401–415. doi:10.1007/BF02291817

Kearns, G. S., & Sabhewarl, R. (2007). Antecedents and consequences of information systems planning integration. *IEEE Transactions on Engineering Management, 54*(4), 628–643. doi:10.1109/TEM.2007.906848

Kim, N., Pae, J. H., Han, J. K., & Srivastava, R. K. (in press). Utilization of business technologies: Managing relationship-based benefits for buying and supplying firms. *Industrial Marketing Management*.

Kim, N., & Srivastava, R. K. (1998). Managing intraorganizational diffusion of technological innovations. *Industrial Marketing Management, 27*, 229–246. doi:10.1016/S0019-8501(97)00059-X

Kincaid, J. W. (2003). *Customer Relationship Management: Getting it Right!* Upper Saddle River, NJ: Prentice Hall PTR.

Lee, S. C., Pak, B. Y., & Lee, H. G. (2003a). Business value of B2B electronic commerce: The critical role of inter-firm collaboration. *Electronic Commerce Research and Applications, 2*, 350–361. doi:10.1016/S1567-4223(03)00003-6

Lee, Y., Kozar, K. A., & Larsen, K. R. T. (2003b). The technology acceptance model: past, present, and future. *Communications of the Association for Information Systems*, *12*, 752–780.

Legner, C. (2008). The evolution of B2B e-services from first generation e-commerce solutions to multichannel architectures. *Journal of Electronic Commerce in Organizations*, *6*(2), 58–77.

Lehmann, D. R., Grupta, S., & Steckel, J. H. (1999). *Marketing Research*. New York: Addison-Wesley.

Leung, L. (2001). College student motives for chatting on ICQ. *New Media & Society*, *3*(4), 483–500. doi:10.1177/14614440122226209

McDonald, R. (1981). The dimensionality of test and items. *The British Journal of Mathematical and Statistical Psychology*, *34*, 110–117.

Merrilees, B., & Fenech, T. (2007). From catalog to Web: B2B multichannel marketing strategy. *Industrial Marketing Management*, *37*(1), 44–49. doi:10.1016/j.indmarman.2006.06.014

Mirchandani, A. A., & Motwani, J. (2001). Understanding small business electronic commerce adoption: An empirical analysis. *Journal of Computer Information Systems*, *41*(3), 70–73.

Ngai, E. W. T. (2005). Customer Relationship Management Research, (1999-2002) an Academic Literature Review and Classification. *Marketing Intelligence & Management*, *23*(6), 582–605. doi:10.1108/02634500510624147

Ngai, E. W. T., & Wat, F. K. (2002). A literature review and classification of electronic commerce research. *Information & Management*, *39*, 415–429. doi:10.1016/S0378-7206(01)00107-0

Nonaka, I., Toyama, R., & Nagata, A. (2000). A firm as a knowledge-creating entity: A new perspective on the theory of the firm. *Industrial and Corporate Change*, *9*(1), 1–20. doi:10.1093/icc/9.1.1

Noori, B., & Salimi, M. H. (2005). A decision-support system for business-to-business marketing . *Journal of Business and Industrial Marketing*, *20*(4-5), 226–236. doi:10.1108/08858620510603909

Nunnally, J. (1978). *Psychometric Theory* (2nd ed.). New York: McGraw-Hill.

Nurosis, M. (1993). *Estatistical Data Analysis*. SPSS Inc.

O'Callaghan, R., Kaufmann, P. J., & Konsynski, B. R. (1992). Adoption correlates and share effects of electronic data interchange systems in marketing channels. *Journal of Marketing*, *56*, 45–56. doi:10.2307/1252041

Park, H., Ribiere, V., & Schulte, W. D. (2004). Critical attributes of organizational culture that promote knowledge management technology implementation success. *Journal of Knowledge Management*, *8*(3), 106–117. doi:10.1108/13673270410541079

Patterson, P. G., Johnson, L. W., & Spreng, R. A. (1997). Modelling the determinants of customer satisfaction for business-to-business profesional service. *Journal of the Academy of Marketing Science*, *25*(1), 4–17. doi:10.1007/BF02894505

Plakoyiannaki, E., & Tzokas, N. (2002). Customer relationship management: a capabilities portafolio perspective. *Journal of Database Marketing*, *9*(3), 228–237. doi:10.1057/palgrave.jdm.3240004

Power, D. (2005). Implementation and use of B2B-enabling technologies: five manufacturing cases. *Journal of Manufacturing Technology Management*, *16*(5/6), 554–572. doi:10.1108/17410380510600527

Premkumar, G., & Roberts, M. (1999). Adoption of new information technologies in rural small business. *OMEGA-The International Journal of Management Science*, *27*, 467–484. doi:10.1016/S0305-0483(98)00071-1

Rayport, J., & Sviokla, J. (1994). *From market-place to market space.* Ed. Materials, Harvard University.

Riemenschneider, C., Harrison, D. A., & Myky-tyn, P. P. J. (2003). Understanding ICT adoption decisions in small business: Integrating current theories. *Information & Management*, *40*(4), 269–285. doi:10.1016/S0378-7206(02)00010-1

Riggins, F. J., & Rhee, H. S. (1999). Developing the learning network using extranets. *International Journal of Electronic Commerce*, *4*, 65–84.

Rogers, E. M. (1983). *Diffusion of Innovations*. New York: The Free Press.

Rogers, E. M. (1995). *Diffusion of Innovations* (4th ed.). New York: The Free Press.

Shang, R., Chen, Y., & Shen, L. (2005). Extrinsic versus intrinsic motivations for consumers to shop on-line. *Information & Management*, *42*(3), 401–413. doi:10.1016/j.im.2004.01.009

Shih, H. (2004). An empirical study on predicting user acceptance of e-shopping on the web. *Information & Management*, *41*, 351–368. doi:10.1016/S0378-7206(03)00079-X

Soliman, K., & Janz, B. (2004). An exploratory study to identify the critical factors affecting the adoption of internet-based interorganizational information systems. *Information & Management*, *41*, 697–706. doi:10.1016/j.im.2003.06.001

Steenkamp, J. P., & Van Trijp, H. C. M. (1991). The use of Lisrel in validating marketing constructs. *International Journal of Research in Marketing*, *8*, 283–299. doi:10.1016/0167-8116(91)90027-5

Subramani, M. (2004). How do suppliers benedif from information technology use in supply chain relationships? *MIS Quarterly*, *28*(1), 45–73.

Subramani, M., & Walden, E. (2001). The impact of e-commerce announcements on the market value of firms. *Information Systems Research*, *12*(2), 135–154. doi:10.1287/isre.12.2.135.9698

Swatman, P. M. C., & Swatman, P. A. (1992). EDI system integration: a definition and literature survey. *The Information Society*, *8*, 169–205.

Teece, D. J. (2000), *Managing intellectual capital.* UK: Oxford University Press.

Teo, T., & Ranganathan, C. (2004). Adopters and non-adopters of business-to-business electronic commerce in Singapore. *Information & Management*, *42*, 89–102.

Tiessen, J., Wright, R., & Turner, I. (2001). A model of e-commerce use by internationalizing SMEs. *Journal of International Management*, *7*, 211–233. doi:10.1016/S1075-4253(01)00045-X

Truman, G. (2000). Integration in electronic exchange environments. *Journal of Management Information Systems*, *17*(1), 209–244.

Tsai, D. H. A. (2005). Knowledge spillovers and high-technology clustering: evidence from Taiwan's Hsinchu Science-Based Industrial Park. *Contemporary Economic Policy. Huntington Beach*, *23*(1), 116–139.

Tung, F.-C. (2007). Using the e-CRM information system in the high-tech industry: predicting salesperson intentions. *Journal of American Academy of Business*, *11*(2), 131–136.

Venkatesh, V., & Davis, F. D. (2000). A theoretical extension of the technology acceptance model: Four longitudinal field studies. *Management Science*, *46*(2), 186–204. doi:10.1287/mnsc.46.2.186.11926

Von Krogh, G., Ichijo, K., & Nonaka, I. (2000). *Enabling knowledge creation: How to unlock the mystery of tacit knowledge and release the power of innovation*. New York: Oxford University Press.

Wang, Y. S., Wang, Y., Lin, H. H., & Tang, T. I. (2003). Determinants of user acceptance of Internet banking: An empirical study. *International Journal of Service Industry Management*, *14*(5), 501–519. doi:10.1108/09564230310500192

Ward, M., & Lee, M. (2000). Internet shopping, consumer search and product branding. *Journal of Product and Brand Management*, *9*(1), 6–20. doi:10.1108/10610420010316302

Watson, R., & McKeown, P. (1999). Manheim auctions: transforming interorganizational relationships with an extranet. *International Journal of Electronic Commerce*, *3*(4), 29–46.

Webb, K. (2002). Managing channels of distribution in the age of electronic commerce. *Industrial Marketing Management*, *31*, 95–102. doi:10.1016/S0019-8501(01)00181-X

West, J. (2001). Customer relationship management and you. *IIE Solutions*, *33*(4), 34–37.

Xu, Y., Yen, D. C., Lin, B., & Chou, D. C. (2002). Adopting Customer Relationship Management Technology. *Industrial Management & Data Systems*, *102*(8/9), 442–452. doi:10.1108/02635570210445871

Yang, J., & Papazoglu, M. P. (2000). Interoperation support for electronic business. *Communications of the ACM*, *43*(6), 39–47. doi:10.1145/336460.336473

Zeng, Y. E., Wen, H. J., & Yen, D. C. (2003). Customer-relationship management (CRM) in business to business e-commerce. *Information Management & Computer Security*, *11*(1), 39–44. doi:10.1108/09685220310463722

Zhang, M., & Tansuhaj, P. (2007). Organizational culture, information technology capability, and performance: The case of born global firm. *The Multinational Business Review*, *15*(3), 43–77.

ENDNOTE

[1] The authors wish to express their gratitude for the financial support received from the Spanish Government CICYT (ECO 2008-04704), the Aragón Regional Government (Generés S-09; DGA 138/08) and Catedra Telefonica of the University of Zaragoza (267-184).

APPENDIX: MEASUREMENT SCALE

Table 5.

	ITEM	ANALYSIS*
INTERNET		
Internet is intensively applied in the performance of my organization's business activity	INT_1	Accepted
Internet is useful in the performance of my organization's business activity	INT_2	Accepted
In general, Internet is easy to use in the performance of my organization's business activity	INT_3	Accepted
EDI		
EDI is intensively applied in the performance of my organization's business activity	EDI_1	Accepted
EDI is useful in the performance of my organization's business activity	EDI_2	Accepted
In general, EDI is easy to use in the performance of my organization's business activity	EDI_3	Accepted
e-CRM		
e-CRM software is intensively applied in the performance of my organization's business activity	e-CRM_1	Accepted
e-CRM software is useful in the performance of my organization's business activity	e-CRM_2	Accepted
In general, e-CRM software is easy to use in the performance of my organization's business activity	e-CRM_3	Accepted
B2B e-commerce		
Approximately, what proportion of your business do you do with other companies via Internet (B2B)?		

Chapter 16
Building Business Value in E-Commerce Enabled Organizations:
An Empirical Study

M. Adam Mahmood
University of Texas at El Paso, USA

Leopoldo Gemoets
University of Texas at El Paso, USA

Laura Lunstrum Hall
University of Texas at El Paso, USA

Francisco J. López
Macon State College, USA

ABSTRACT

This research attempts to identify critical e-commerce success factors essential for building business value within e-commerce enabled organizations. It is important to identify the critical success factors that organizations must pursue in order to facilitate a successful transformation from traditional brick-and-mortar organizations to click-and-brick business models. Diffusion theory is used to demonstrate how these success factors create business value within an organization. The research model is fully grounded in information technology business value and productivity literature (e.g., Kauffman & Kriebel (1988), Mahmood, Gemoets, Hall, & Lopez (2008) Mahmood & Mann (1993), and Zhu (2004)). The manuscript utilizes an existing sample set consisting of a population of more than 550 company executives who are successfully implementing e-commerce strategies. The research examines constructs found in the literature and focuses on two importance dimensions of creating business value through e-commerce strategies: IT alignment to organizational strategies (ITOrS) and the quality and effectiveness of existing online systems (OnSQE). Critical success factors for e-commerce business success were found to include ITOrS (IT alignment to organizational strategies), IOrSA (Quality and effectiveness of online systems,

DOI: 10.4018/978-1-60566-910-6.ch016

OnSE (Online systems efficiency), and OnSQE (Online systems quality and effectiveness). The research produces empirical evidence that strategic decision making concerning implementation of e-commerce technologies and alignment with top management strategic planning is critical to the success of creating business value for e-commerce enabled organizations. The manuscript concludes with limitations of the research and implications for future research studies.

INTRODUCTION

In today's economic environment it is vital that organizations invest in resources that will build value throughout the entire organization. Implementation of e-commerce strategies have become a popular way of increasing business value. Social networking through popular mediums such as Facebook and MySpace, advanced mobile communication devices, and widespread accessibility has redefined e-commerce and has resulted in an explosion of increased opportunities for e-commerce enabled organizations.

Many traditional brick-and-mortar companies have invested and continue to invest heavily in e-commerce technologies due to a huge increase in online opportunities. e-Marketer (June, 2008) reports:

In 2007, 133.1 million individuals, nearly four-fifths of US Internet users, shopped online. By 2012, the total will be closer to 158.2 million, or 82.5% of Internet users. From 2007 to 2012, the number of new online shoppers in the US is expected to grow at a 3.5% average annual rate. Also in 2007, 110.7 million individuals, nearly two-thirds of US Internet users, made at least one online purchase. By 2012 the number of online buyers is expected to be 141 million, or 73.5% of Internet users. From 2007 to 2012, the number of new online buyers in the US will grow at a 5% average annual rate. (p.1)

Forrester Research (2007) estimates that online sales will reach $204 billion this year and $335 billion by 2012. E-commerce currently accounts for 6 percent of all retail sales in the United States. Although forecasts for retailers are currently dismal it is believed that e-commerce retailers will fare better than their brick-and-mortar counterparts (e-Marketer, 2008).

The information technology (IT) productivity and business value literature report performance and productivity gains for organizations that evolve to click-and-brick structures. There are many well known examples of click-and-brick organizations including Best Buy, Wal-Mart, Target, Walgreens, Sears, Cisco Systems, Dell Computers, and Boeing Corporation. These are excellent examples of large click-and-brick organizations that have achieved a significant economic benefit by using e-commerce technologies. Cisco, the single largest user of e-commerce in the world, attributes 90% of its 2000 sales to online sales. Also, 82% of its customer inquiries are handled online (McIlvaine, 2000) and 83% of questions concerning support are answered through. Cisco's web based self service tools ("Customer Care," 2001) handle 82% of customer inquiries and 83% of technical support requests online. Dell, ranked 49th in the Top 50 Internet properties logged 9.2 million first time visitors with sales of more than $1 million dollars in PC sales online, everyday. Dell Computer reported over 250% return on invested capital from its logistics and order fulfillment systems (Dell.com, November 2000).

The importance of the present research stems from the fact that there is very little empirical evidence in the IT productivity and business value literature regarding the critical success factors from e-commerce business initiatives, especially for large click-and-brick companies (Brynjolfsson and Kahim, 2000).

The fundamental objective of the present research is to define the success indicators of business value in e-commerce technology enabled organizations. The present research investigates firms that are successful in using e-commerce technology and determines the critical success factors responsible for building business value within the organizations.

THEORETICAL FOUNDATIONS

Diffusion theory studies the adoption of innovations through (or as a function of) time and identifies and understands the factors that influence adoption behavior. Zhu (2004) assessed e-commerce payoffs indirectly via an interaction effect with IT infrastructure. He found a positive interaction effect between IT infrastructure and e-commerce capability. He also found that this relationship positively contributes to firm performance in terms of sales per employee, inventory turnover, and cost reduction. Clearly Zhu's study did not look at the stand-alone impact of e-commerce technologies on a firm's performance. In addition, Zhu used the resource-based theory to ground his research whereas we used Rogers' (1983) diffusion theory to provide a foundation for our research. While it is true that price (reduction) is an important factor, our contention is that diffusion may be strong even if prices go up (so long as the benefit more than offsets the cost). Therefore, we propose the incorporation of the cost-benefit relation in diffusion models. A complete and thorough analysis would study these models using longitudinal data. Unfortunately, such data are not yet available. Thus this research requires two steps. The first consists of identifying the factors that result in e-commerce business success while the second step incorporates those factors into diffusion models. We are unable to validate these models in the present research for the lack of longitudinal data. In the present research, we use cross-sectional data to analyze the relationship among the following factors that were selected based on a thorough literature review: a. Inter-organizational systems availability (IOrSA): extent to which e-commerce has helped in integrating the different systems and made workflow processes easier; b. Online systems efficiency (OnSE): availability of uniform operating and highly automated mechanisms using e-commerce technologies; c. IT Alignment to organizational strategies (ITOrS): an indication of how much support for Internet-enabled IT there is in the organization; d. online systems quality and effectiveness (OnSQE): measured in terms of e-commerce site design and availability; and, e. e-commerce business success (ECBS). Since the latter clearly indicates "benefits," we propose (based on our findings in the present research) incorporating IOrSA, OnSE, ITOrS, and OnSQE in e-commerce diffusion models as indicators of the cost-benefit relationship. Later on, when historical data become available, we will build on the ideas of Gurbaxani and Mendelson (2001) and develop and test (i.e., curve-fit) the corresponding models. Even though it is not yet possible to conduct this research because of lack of data, this idea is an important contribution of the present research.

LITERATURE REVIEW

As stated earlier, there is little or no empirical research in the area of e-commerce business value but some important related concepts that have been identified include business value; impact of e-commerce; and success and failure of e-commerce businesses. It is, however, possible to draw useful insights from the IT business value and other related literature. There are a number of studies on factors contributing to the success or failure of an IT system that can be compared to e-commerce success and failure. Thus, the IT business value literature provides background information and concrete theoretical support to ground our e-commerce success model. Next, a review of IT business value literature is provided below that suggests incorporation of five constructs in our analysis.

IT Alignment to Organizational Strategies (ITOrS)

This construct measures the degree to which the strategies, goals, and objectives of a firm's information technology are aligned to its parent organization's strategies, goals, and objectives. When this occurs, top management support for IT initiatives appears to be stronger which, as suggested in previous research, seems to be a critical success factor for any IT project. This type of support takes various forms including appointing an executive level manager as the Chief Information/Technology Officer and allowing IT to influence, through the adoption of email and other internet technologies, the way the company conducts business. Thus, indicators of alignment include the presence of an IT manager with executive authority; the level of support given by executives to e-commerce and other IT-related initiatives; and whether the organization has a learning and adaptive culture that allows innovations to take root in its functional environment.

Segars and Grover (1998) evaluate the strategic impact of IT. They suggest that paying attention to aligning IT strategy with business strategy, understanding the systems processes, and the support of management and end-user groups are crucial to IT planning success. Barua, Kriebel, and Mukhopadhyay (1991) also analyze the strategic use and impact of IT implementation. Reich and Benbasat (2000) found that communication between IT and business executives, the level of connection between IT and business planning processes, and the extent of shared domain knowledge leads to better alignment of IT and business strategies in the short and long terms. Feeny and Ives (1990) also focus on sustained strategic advantage generated through IT applications. Teo and Ang (1999) identify top management commitment to the strategic use of IT, IS management business knowledge, and top management confidence in the IS department as critical success factors for an organization. Teo and King (1996, 1997) studied potential inhibitors and facilitators of development of successful IT applications with strategic value. They found that integration of business planning and IT planning results in IT being able to support business strategies more effectively. Kao and Decou (2003) focused on the importance of a strategy-based e-commerce planning model which included seven dimensions with strategy at the core. Clemons and Wang (2000) also provided strategies for electronic commerce initiatives. Shin (2001) also suggested the use of e-commerce for gaining organizational competitive advantage. Additionally, it is obvious that some companies have an explicit way of aligning IT to their organizational strategy. Wal-Mart, for example, has separate e-commerce headquarters in California to fulfill the company's e-commerce business needs.

The following research studies also identify other factors related to the success or value of IT. Teo and King (1997) address the importance of alignment of IT planning with business planning and determine that the business knowledge of the IT executive is the most significant factor in influencing the integration process. Dos Santos and Sussman (2000) indicate that IT investments fail to have an impact on some firms' value because the companies fail to prepare or respond well to the structural changes of the firm caused by IT.

Inter-Organizational Systems Availability (IOrSA)

IOrSA refers to the extent to which e-commerce has helped in integrating the different systems and made workflow processes easier between different organizations, thus creating e-commerce business value. E-commerce technologies such as extranets allow businesses to connect their suppliers, customers, and other business partners that result in competitive advantages. E-commerce adds value to a firm by introducing a new channel for buying, selling, and providing information to appropriate stake holders. Some of the readily apparent benefits are reduction in employee costs and communication costs. Hidden benefits include better relationships with upstream and downstream business partners because of a uniform and secure communication system.

Dai and Kauffman (2002) point out that firms can conduct successful B2B transactions by creating inter-organizational systems. Barua, Konawa, Whinstone, and Yin (2001) indicate that the readiness of business partners to implement e-commerce technologies is critical to achieving business excellence and online systems efficiency. Bakes (1991) uses economic theory to understand why electronic marketplaces work and when they are of strategic value. Electronic marketplaces, according to Bakos (1991), improve inter-organizational coordination and reduce search costs. Bakos (1997) also points out that commodity and differentiated markets respond differently to the integration enabled by electronic marketplaces. Reduction in search costs, for example, occurs differently in these markets. In a subsequent paper, Bakos (1998) reinforces the points made in the first two by explaining that electronic marketplaces create more efficient and friction-free markets. Steinfield, Markus, and Wigand (2005) used the industry level of analysis to explore inter-organizational systems in the home mortgage industry. Kurnia and Johnston (2005) developed a case study of category management adoption in Australia to model inter-organizational system adoption. According to Benjamin and Scott (1988), one of the reasons for IT success is that it enables, through online databases and telecommunication networks, new forms of integration that result in better cost performance and increased data integration. Kickul and Gundry (2001) argue that maintaining better relationships and integrating with suppliers is a crucial element of the company sustenance process. Johnston and Vitale (1988) suggested that, if carefully identified and used, inter-organizational systems give significant competitive advantages to organizations.

Online Systems Efficiency (OnSE)

Automated, uniform operating mechanisms through e-commerce technologies that function in tandem with existing mechanisms normally result in better online systems efficiency and cost savings. Also, online customer service in terms of FAQ's, chat rooms, and a link to call centers indicate high levels of online systems efficiency.

Banker, Kauffman, and Morey (1990) distinguish between impacts of IT investment on competitive efficiency and on online systems efficiency. Mukhopadhyay and Kekre (1995) study an electronic data interchange system at Chrysler over a period of 10 years and observe that it has caused massive cost savings and has imparted system-wide discipline and integrative value to the company. In 2001, Molla and Licker (2001) extended the IS success model to E – commerce including the construct of system efficiency. Shao and Lin (2001, 2002) define technical efficiency as actual output versus expected output and find a positive correlation between investment in IT and technical efficiency improvement. Stratopoulos and Dehning (2000) show that firms making successful investments in IT are more successful at solving the productivity paradox than those that make failed or abortive investments in IT. Hitt and Brynjolfsson (1997) analyze the effects of IT on the internal firm organization and find that IT is associated with decentralization of authority, increased knowledge work, and decreased observability. Brynjolfsson and Hitt (1998) find that productivity payoffs from computerization are not automatic, but part of a series of productivity changes that eventually make financial sense. Barua, Konana, Whinston, and Yin (2001) show that the readiness of business partners to implement e-commerce technologies is critical to achieve business excellence and online systems efficiency. Lee and Barua (1999) find positive correlation between IT inputs and productivity. Tsay and Agrawal (2004) identified various sources of inefficiency in the effect of channel conflict on system efficiency including pricing changes, demand functions, and structural properties.

Online System Quality and Effectiveness (OnSQE)

OnSQE has to do with aspects related to online presence effectiveness through the e-commerce site design and availability. Previous research has analyzed factors that make a website successful, which is associated to overall e-business success. E-commerce, via websites, has dramatically improved the interaction of companies with customers. The ability to offer products and services to customers worldwide on a 24/7 basis is a value-adding attribute of an e-business. Also, the website is a means that firms can use to influence customer's perceptions of its business. Customers perceive website security and access time of the site as critical. This is a reason why firms try to improve users' perceptions of security by making their website secure and aligning with recognizable internet security agencies and protocols. Other important factors include ease of use, quality of design, and value for the customer.

Moon and Kim (2001) argue that the acceptance of the World Wide Web (WWW) is similar to the "perceived ease of use" (PEOU) component of the Technology Acceptance Model. Acceptance of a new technology is likely to vary depending on the kind of technology, the target users, and the context. Moon and Kim (2001) measure online presence effectiveness through the design and availability aspects of the e-commerce site. Gefen and Straub (2000) also argue that the PEOU plays an important role in the actual use and success of systems. A survey by Liu, Arnett, and Litecky (2000) indicates that the attractiveness, quality of design, and information available on the e-commerce site are the most important factors influencing the purchase decision of a customer. Zhang, Von Dran, Small, and Barcellos (1999, 2000) use Herzberg's hygiene-motivation theory to study how the general 'hygiene' or perceptual quality of a website affects users' motivation to use the website. Applying the Kano model of quality to the design of web sites, Zhang and Von Dran (2001) study features that generate 'delight', 'motivation,' and 'loyalty' of website users. Lederer, Maupin, Sena, and Zhuang (2000) point out that the "ease of use" quality is very important for successful web sites. Keeney (1999) stresses the importance of identifying value propositions and developing an e-commerce value model for the customer. Cao, Zhang, and Seydel (2005) conducted an empirical study on web-site quality in a B2C environment and identified, using the IS success model, what constitutes the web site quality and effectiveness. Pather, Erwin, and Remenyi (2003) attempted to measure the effectiveness of e-commerce using a conceptual model. Rose and Straub (1999) identify excessive download time as one of the most serious technological impediments to e-commerce.

E-Commerce Business Success (ECBS)

Performance, productivity, and perception are some factors that can be used to measure ECBS. Performance has been measured in the IT business value literature, in terms of financial ratios: return on investment (ROI), return on equity (ROE), return on sales (ROS), growth in revenue (GINR), and net income over invested capital (NIC). Cron and Sobol (1983) and Dos Santos, Peffers, and Mauer (1993) employed ROI. Barua, Kriebel, and Mukhopadhyay (1995); Hitt and Brynjolfsson (1994); and Strassman (1990) used NIC. Hitt and Brynjolfsson (1994) and Mahmood and Mann (1993) applied ROS. Woo and Willard (1983) used GINR. Dehning and Richardson (2002) analyze the impact of IT through ROI, ROE, and ROS that can capture system success.

Likewise, productivity has been assessed in terms of two ratios: sales to total assets (STA) and sales by employee (SE). Brynjolfsson and Hitt (1993) used a measure similar to STA, total sales. Strassman (1990) employed SE. Perception has been measured through company's image, customer satisfaction, product service innovation, and the number of return customers. The first three create customer loyalty that results in return purchases. Loyalty is one of the most significant contributors to business profitability (Turban, King, Lee, Warkentin, and Chung (2002)). It can also reduce costs in the sense that it costs five to eight times more to acquire a new customer than to keep an existing one. A comprehensive review of the use of the Internet to foster customer loyalty is provided by Reichheld (2001).

In spite of the fact that initial costs are high, continuous support for e-commerce strategy is essential for the success of e-commerce initiatives. Zott, Amit, and Donlevy (2000) study the most successful strategies among European firms to create e-business value. Peteraf (2000) looks at strategies that give competitive advantage to the firm from a resource-based perspective of the firm. Zhu (2004) focused on the firm level to develop a research framework which described the relationships between IT infrastructure and e-commerce capability. Barua *et al.* (2001) observe that e-business affects large and small companies differently. Smaller companies experience a quick impact because of immediate expanded geographic reach. Larger companies face more complexity and need to pay considerable attention to the drivers and need to establish an appropriate infrastructure before acceptable payback is received. Subramani and Walden (2001) analyze the impact of e-commerce investment on the value perceived by investors. They found a positive correlation between e-commerce initiatives announcements and higher value perceived by investors.

Dekleva (2000) identifies four environmental variables that affect e-businesses: building trust so consumers engage in e-commerce, establishing a legal framework for e-commerce operations, enhancing information systems infrastructure by improving technical resources, and maximizing the benefits provided by such systems via increased integration across systems. Amit and Zott (2001) identified four value creating components in e-commerce companies: efficiency, complimentarity, lock-in, and novelty. Lee and Clark (1997) analyze factors contributing to the successful implementation of electronic market systems.

There are other factors that affect ECBS. These are similar to what occurs with IT investment (Mahmood *et al.*, 2000, 2001) as such some of the relevant IT-related literature is also covered in this section. Melville, Kraemer, and Gurbaxani (2004) used an integrative model of IT business value to describe the relationship between information technology and organizational performance. Chan (2000) points out the need to take 'soft' factors into account when measuring the value of IT. Chircu and Kauffman (2000) consider hard (e.g., better financial performance and increase in sales) and soft (e.g., better market position and better supplier relationship) IT benefits.

DeLone and McLean (1992) review existing research (180 articles) on MIS success and identify six IS success dimensions: system quality, information quality, use, user satisfaction, individual impact, and organizational impact. Davern and Kauffman (2000) differentiate the value realizable from IT into: a) potential value, in the areas of the organization that would have an impact that current systems fail to provide, and b) realizable value, which is the value that can be derived from the system considering the assets that exist in the firm. Kim and Peterson (2001) identify five factors that contribute to IS success from a developer's perspective: management and user input, project management, characteristics of the project leader, methodology, and characteristics of the team members. Teo and Ang (2000) analyze how IT planning leads to IT success. Teo and Ang

Figure 1. E-Commerce Business Value Model

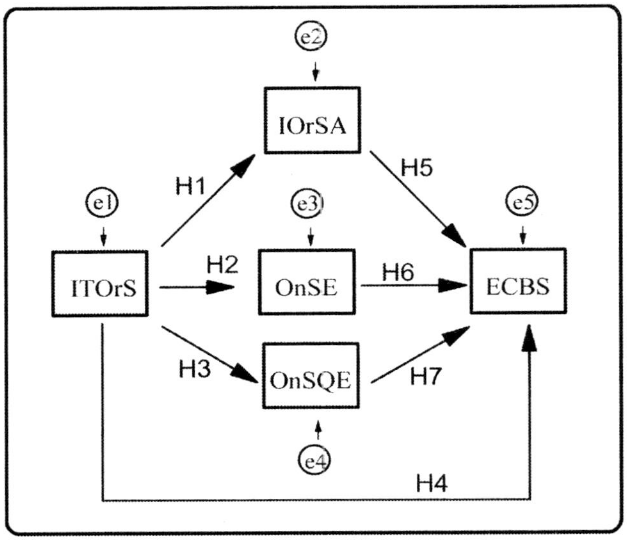

ITOrS - IT Alignment to Organizational Strategies
IOrSA - Inter-Organizational Systems Availability
OnSE - Online Systems Efficiency
OnSQE - Online System Quality and Effectiveness
ECBS - Electronic Commerce Business Success

(2001) also study major existing problems during the IT planning process. Dos Santos, Peffers, and Mauer (1993) and Im, Dow, and Grover (2001) discuss the effects of IT investments on a firm's value. Josefek and Kauffman (1997) analyze the potential profitability or success of innovative information systems.

As explained earlier, this work is the first of two research steps in the analysis and development of theory on e-commerce technology enabled business value. The fundamental objective and contribution of the current (first step) exploratory research is to formulate and test a model for value creation in e-businesses. Thus, the model can function as a reference framework for strategic managers by offering guidelines for e-business initiatives. Also, the constructs developed in the model can serve as a foundation for further investigation of different e-commerce drivers and their relationships with business value measures. There may be additional environmental factors associated with successful e-commerce implementations. A further goal of this article is to establish theoretically grounded constructs, factors, and ideas that can be used for further research in this field and to enhance the present work. The importance of the research study stems from the fact that it is the first empirical study that directly addresses the business value of e-commerce technologies enabled business initiatives.

MODEL AND HYPOTHESES

We used the five constructs that we identified in the IT business value/success literature to build a model that captures the drivers of e-business success and relationships among these drivers. The model appears in Figure 1.

The model suggests the way IT alignment to organizational strategies (ITOrS), inter-organizational systems availability (IOrSA), online system efficiency (OnSE), and online system quality and effectiveness (OnSQE) interact with each other and together affect e-commerce business success (ECBS). The model also helps one understand the tasks a brick-and-mortar company has to complete as well as the processes that it needs to undertake for it to be successful in an e-commerce initiative.

Hypotheses

Hypothesis 1: This proposition examines the relationship between the degrees of IT alignment to organizational strategies (ITOrS) and Inter-organizational systems availability (IOrSA). ITOrS, as stated earlier, measures the degree to which the strategies, goals, and objectives of a firm's information technology are aligned to its parent organization's strategies, goals, and objectives. When this occurs, top management support for IT initiatives appears to be stronger, which appears to be a critical success factor for any IT project. Indicators of alignment include the presence of an IT manager with executive authority, the level of support given by executives to e-commerce initiatives, and whether the organization has a learning and adaptive culture that allows innovations to take place in its functional environment. What influence might that have on the inter-organizational system availability? If indicators of alignment are present in the firm it should have a positive impact on the extent to which these indicators has helped in integrating the different systems and made workflow processes easier between different organizations. Barua, Konana, Whinston, and Yin (2001) indicate that the readiness of business partners to implement e-commerce technologies is critical to achieving business excellence and online systems efficiency. Thus the degree of IT alignment to organizational strategy will have a positive influence on workflow processes and integration of systems, creating e-commerce business value. Segars and Grover (1998) evaluate the strategic impact of IT. They suggest that paying attention to aligning IT strategy with business strategy, understanding the systems processes, and the support of management and end-user groups are crucial to IT planning success. These lead to the following hypothesis

H1: There is a positive correlation between high levels of IT alignment to organizational strategies *(ITOrS)* and high levels of inter-organizational systems availability *(IOrSA)*.

Hypothesis 2: This proposition considers the relationship between the level of IT alignment to organizational strategies (ITOrS) and the likelihood of obtaining online system efficiency (OnSE). While ITOrS is measured in terms of top management support, existence of a top IT executive, and other similar factors. OnSE is measured by way of impact on productivity and efficiency measures. Online systems with impact on productivity and efficiency will garner more management support which in turns will ensure that these systems are more aligned toward organizational goals and strategies. Huselid and Becker (1997) propose the existence of synergies between implementing efficient systems and their strategic alignment. Sanders and Premus (2002) classify firms into high-, medium-, and low-level users of IT depending on their level of IT sophistication. They cite efficiencies gained via IT in terms of automation as an enabler for managers to focus on strategic issues and competencies. Lederer, Mirchandani, and Sims (2001) investigate whether strategic advantage can be derived from the World Wide Web. They propose that the efficiencies created by web-enabled systems affect strategic alignment positively by improving customer relations. Banker, Kauffman, and Morey (1990) distinguish between impacts of IT investment on competitive efficiency and on online systems efficiency. Mukhopadhyay and Kekre (1995) study an electronic data interchange system at Chrysler over a period of 10 years and observe that it has caused massive cost savings and has imparted system-wide discipline and integrative value to the company. Shao and Lin (2001, 2002) define technical efficiency as actual output versus input. This leads to the following hypothesis:

H2: There is a positive correlation between businesses with higher levels of IT alignment with organizational strategies (ITOrS) and a high level of online systems efficiency (OnSE).

Hypothesis 3: This hypothesis examines the relationship between ITOrS and online systems quality and effectiveness (OnSQE). The more the company's IT is aligned to the organization's strategies the more successful it will be at obtaining management support for creating higher quality and effective online presence. Also, since the online interface presents the so-called gateway to the company, top management is eager to portray an elegant interface reflecting a better image of the company. Kowtha and Choon (2001) mention how the sophistication of a firm's website reflects the strategic priorities of the firm. They suggest that critical competencies in e-commerce have little to do with technology and more with managerial domain and priorities. They furthermore suggest that strategic commitment has substantive and high significant effects on website development. Peak and Guynes (2003, 2003-1) point out that aligning IT with organizational strategies improves information quality which results in improved quality of products and services. These lead to the following hypothesis

H3: There is a positive correlation between businesses with higher level of IT alignment to organizational strategies (ITOrS) and online system quality and effectiveness (OnSQE)

Hypothesis 4: This hypothesis examines the critical relationship between ITOrS and e-commerce business success (ECBS). Some studies mention top management commitment and involvement in IT projects as singularly important factors for success of IT systems. In fact, lack of their commitment and involvement is pointed out to be among the top ten problems that lead to failure of IT projects (Johnson, Boucher, Conners, and Robinson, 2001). Pollalis (2003) confirms that alignment of IT to organizational strategies leads to better performance of banks. Sun and Hong (2002) find wide support in the literature for a significant direct effect of strategic alignment on business performance in the field of manufacturing strategy. Burn and Szeto (1999) find that it is critically important to align business and IT planning. They point out that business success depends on the linkage of business strategy; IT strategy; organizational infrastructure and processes; and IT infrastructure and processes. They also argue that the role of IT management is to lead and maintain a close alignment between the IT function and business strategy. Brown (2003) suggests that developing appropriate strategic performance metrics for the IT human resource function creates a closer IT alignment with strategic objectives. These lead to the following hypothesis.

H4: There is a positive correlation between businesses with higher level of IT alignment to organizational strategies (ITOrS)and e-commerce business success (ECBS).

Hypothesis 5: This hypothesis examines the critical relationship between Inter-organizational systems availability (IOrSA) and E-commerce business success (ECBS). Dai and Kauffman (2002) point out that firms can conduct successful B2B transactions by creating inter-organizational systems. Barua, Konana, Whinston and Yin (2001) indicate that the readiness of business partners to implement e-commerce technologies is critical to achieving business excellence and online systems efficiency. According to Bakos (1991) electronic marketplaces improve inter-organizational coordination and reduce search costs. Bakos (1998) further explains that electronic marketplaces create more efficient, friction-free markets. According to Benjamin and Scott (1988) one of the reasons for IT success is that it enables, through online databases and telecommunication networks, new forms of integration that result in better cost performance and increased data integration. Kickul and Gundry (2001) argue that maintaining better relationships and integrating with suppliers is a crucial element of the company sustenance process. Johnston and Vitale (1988) defend that if carefully identified and used, inter-organizational systems give significant competitive advantages to organizations. Barua, Konana, Whinston, and Yin (2001) show that the readiness of business partners to implement e-commerce technologies is critical to achieve business excellence and online systems efficiency. Epstein (2000) says that integration of processes and systems directly translates into business success. Huang, Chen, and Frolick (2002) argue that introducing the web as a way of integrating data leads to better business value. Epstein (2000) says that integration of processes and systems directly translates into business success. Huang, Chen, and Frolick (2002) argue that introducing the web as a way of integrating data leads to better business value. Dai and Kauffman (2002) point out that firms can conduct successful B2B transactions by creating inter-organizational systems. At least one empirical study mention efficiency benefits from the implementation of integrated inter-organizational systems. Mebane Packaging, for example, achieved better efficiency, and the chain of supermarkets Somerfield Stores improved its business efficiencies with its systems integration project (Thomas, 2003). These lead to the following hypothesis:

H5: There is a positive correlation between businesses with higher level of Inter-organizational systems availability (IOrSA) and e-commerce business success (ECBS).

Hypothesis 6: Firms with efficient e-commerce systems should be at a relatively better position to achieve better business success. Lederer, Mirchandani, and Sims (2001) point out that strategic advantage from the World Wide Web is created through increased efficiency of processes. Mukhopadhyay and Kekre (1995) study an electronic data interchange system at Chrysler over a period of 10 years and observe that it has caused massive cost savings and has imparted system-wide discipline and integrative value to the company. Stratopoulos and Dehning (2000) show that firms making successful investments in IT are more successful at solving the productivity paradox than those that make failed or abortive investments in IT. Hitt and Brynjolfsson (1997) analyze the effects of IT on the internal firm organization and find that IT is associated with decentralization of authority, increased knowledge work, and decreased observability. These lead to the following hypothesis.

H6: There is a positive correlation between businesses with higher online systems efficiency (OnSE) and e-commerce business value (ECBS).

Hypothesis 7: This hypothesis purports that the quality and effectiveness of the website (online presence) drives e-commerce success. Zhu and Kraemer (2002) point out that better e-commerce capability serves to improve the effectiveness of investments on e-commerce initiatives through firm performance. Chaudhury, Mallick, and Rao (2001) mention that enhancement of website quality can improve business value. Huang, Chen, and Frolick (2002) say that evaluating web data quality is important to effectively determining the business value of online data. A survey by Liu, Arnett, and Litecky (2000) indicates that the attractiveness, quality of design, and information available on the e-commerce site are the most important factors influencing the purchase decision of a customer. Lederer, Maupin, Sena, and Zhuang (2000) point out that the "ease of use" quality is very important for successful web sites. Rose and Straub (1999) identify excessive download time as one of the most serious technological impediments to e-commerce. These lead to the following hypothesis.

H7: There is a positive correlation between businesses that have a higher level of online system quality and effectiveness (OnSQE) and e-commerce business success.

Data Collection

Based on the existing research literature, we designed a questionnaire (Appendix) to gather information on e-business success. It contains 31 items. The respondents were asked to answer each question on a seven-point Likert scale with values ranging from 1 (strongly disagree) to 7 (strongly agree). Each construct was measured using a series of items. ITOrS, for example, comprises five items: *i*) alignment of IT strategies with top management strategies, *ii*) whether IT is considered a part of the long term strategies, *iii*) whether IT executives have decision making roles, *iv*) IT structure features, and *v*) overall organizational learning environment. IOrSA was measured in terms of whether: *i*) the firm has an Internet-enabled system to share information among upstream and downstream entities, *ii*) there is an effective automated order changing system, *iii*) data is automatically transmitted and processed, *iv*) it is possible to track inventory and purchasing continuously, *v*) there exists an online procurement system, and *vi*) internet expertise is an important selection criteria for suppliers/vendors. OnSE contains five items: *i*) degree of online business transactions, *ii*) availability of online customer service, *iii*) availability of a highly automated order tracking system, *iv*) possibility of resolving customer requests online, and *v*) availability of a system to monitor orders continuously. OnSQE is based on five e-commerce website features: *i*) security, *ii*) attractiveness, *iii*) navigability, *iv*) flexibility, and *v*) continuous availability of the website. Finally, ECBS is measured in terms of *i*) return on investment; *ii*) return on sales; *iii*) growth in revenue; *iv*) net income over invested capital; *v*) sales over total assets; *vi*) sales by employee; *vii*) company image; *viii*) customer satisfaction; *ix*) product /service innovation; and *x*) return of customers.

The data used to test the hypothesized model comes from surveying 550 companies which have been identified as premier companies that have successfully implemented e-commerce technologies or strategies and have been able to create positive value either operationally or financially. These firms were listed in *Information week 500* and *Internet week 100*, two premier computer magazines that survey innovative and successful uses of IT for competitive advantage. A copy of the instrument was mailed to IT executives of these companies. The respondents were allowed to reply via postal mail or answer the questionnaire online. In order to increase the response rate, we sent the instrument to local representatives (managers of the stores) of national chains (e.g., Target and Dillards). Also, we surveyed a group of Boeing executives attending an MBA-level e-commerce class. Individual identity and responses were masked to ensure confidentiality.

We received 43 responses, seven of which were incomplete. Incomplete responses were not used for analyzing data. Forty one companies had more than 1 million in revenues. The other two companies did not provide revenue information. About 42% of the respondents were executive technology officers; 32% middle managers; and 26% store managers or associates. About 2% of the responses were from companies that engage in all three modes of business: Business to Business (B2B), Business to Consumer (B2C), and Consumer to Consumer (C2C). Overall, there were 60% in B2B, 72% in B2C, and 2% in C2C. About 44% were involved in both B2B and B2C. Around 14% were exclusively B2B and 23% exclusively B2C. The average number of employees of these firms was 98,000. Surveying took about 6 months during the summer and fall of 2002.

RESULTS

Analysis

In this section we report statistical results with respect to constructs validations; model fit; and hypotheses and paths analysis. The Partial Least Square (PLS) approach is typically suited for problems dealing with a small sample size and is often used for exploratory model testing and validation. We have, however, used an AMOS/LISREL approach to analyze our structural equation model because it also yields similar results for a small sample size while offering more statistics (AMOS (Arbuckle, J.L., 1989). In the past, the AMOS/LISREL approach has been used for analyzing small data sets. Wold (1989), for example, used this approach to analyze a model based on a data set consisting of 10 cases and 27 variables.

Unidimensionality

To verify that all items loaded well in their assigned constructs, we used factor analysis to conduct unidimensionality tests with a reference norm of 0.40 as suggested by Mahmood and Sniezek (1989). Table 1 shows results of this test. With the exception of item #2, all remaining five items in construct IOrSA load satisfactorily (> 0.40). All items in OnSE load well (> 0.46). All 5 items in ITOrS load strongly (> 0.82) with the exception of item #3 which loads at 0.59. All five items in OnSQE load well (> 0.58). The 10 items or questions in ECBS, the terminal construct of the model, load well (> 0.59). The average variance extracted (from communalities from factor analysis) is 0.54, 0.42, 0.62, 0.37, and 0.57 for IOrSA, OnSE, ITOrS, OnSQE, and ECBS, respectively.

Reliability

Cronbach's coefficient alpha, one of the most widely used reliability tests, was carried out to ensure that the items for each factor were internally related in the manner expected. Cronbach's alpha is based on "internal consistency" of a test: the degree to which variables in the measurement set are homogeneous. We used the Cronbach alpha to measure convergent validity as a reliability test for the five constructs in the model. ITOrS, IOrSA, OnSE, OnSQE, and ECBS have reliability scores of 0.90, 0.76, 0.67, 0.81, and 0.93, respectively. Nunnally (1978) suggested a Cronbach alpha threshold level of 0.60 for exploratory research. All five constructs met this threshold level. More recently, Hair et al. (1998) suggested the threshold

Table 1. Scale Development

Construct	Number of Items or questions	Mean	Standard Deviation	Cronbach Alpha	Factor Loadings	Variance Extracted
IOrSA	6	5.00	0.90	0.76	0.501, 0.148, 0.614, 0.763, 0.683, 0.648	0.54
OnSE	5	4.98	0.92	0.67	0.584, 0.701, 0.545, 0.613, 0.459	0.42
ITOrS	5	5.55	1.01	0.90	0.861, 0.881, 0.592, 0.837, 0.884	0.62
OnSQE	5	5.24	1.00	0.81	0.748, 0.752, 0.890, 0.583, 0.677	0.37
ECBS	10	5.29	0.98	0.93	0.767, 0.796, 0.807, 0.792, 0.751, 0.590, 0.794, 0.833, 0.880, 0.585	0.57

Table 2. Correlations among constructs

Construct	IOrSA	OnSE	ITOrS	OnSQE	ECBS
IOrSA	1.000				
OnSE	0.520	1.000			
ITOrS	0.641	0.464	1.000		
OnSQE	0.373	0.471	0.403	1.000	
ECBS	0.498	0.467	0.609	0.558	1.000

values of 0.60 for exploratory research and 0.80 for confirmatory research. Since the present research is an exploratory study, the aforementioned constructs are also reliable using the guidelines suggested by Hair et al. (1998).

Validity

Table 2 presents the correlations between constructs. Demonstrating divergent or discriminant validity requires that the constructs do not correlate highly with each other (Campbell and Fiske (1959)). As can be seen, almost all the correlations are below 0.50 which indicates that these constructs are valid independently. Note that IOrSA-OnSE, IOrSA-ITOrS, ITOrS-ECBS and OnSQE-ECBS involve some measure of overlap. It is not surprising that all constructs are correlated to some extent as the various processes contributing to the success of an e-business can be expected to interweave and overlap among themselves.

Since the aforementioned analysis does not provide unambiguous results on discriminant validity, we have decided to use a more rigorous factor-based procedure known as the average variance extracted (AVE) method proposed by Fornell and Larcker (1981) (see Tables 3 and 4). The formula for calculating the AVE is provided at the bottom of Table 3. Using this method, one may conclude that constructs are different if the AVE for a given set of constructs is greater than their shared variance. Table 4 provides a matrix of squared covariance of each construct with each other construct. The diagonal elements are replaced with the AVE for the column construct. If there is a discriminant validity among the constructs, then the diagonal element for a given construct (column) should be larger than any of the squared covariances in the column or row in which it is found. Using this benchmark, there is unequivocal evidence that all constructs included in the present study pass the test of discriminant validity.

Results

AMOS results showed that the paths ITOrS →IOrSA, ITOrS → OnSE, ITOrS → OnSQE, and ITOrS → ECBS, and OnSQE → ECBS are significant, with coefficients 0.64 (significant at the .01 level), 0.46 (significant at the .05 level), 0.40 (significant at the .05 level), 0.39 (significant at the .05 level), and 0.34 (significant at the .05 level), respectively. Paths IOrSA → ECBS and OnSE → ECBS, with regression weights of 0.09 and 0.09, respectively, were not significant. The model itself is significant at 0.034 with a chi-square value of 8.681 and 3 df (the chi-square value for SEM type of analyses should not be significant if there is a good model fit to the data) (see Figure 2). The chi-sq is, however, too sensitive to sample size (Bentler and Bonnett (1980)). Instead, for the present research, the chi- sq/df ratio is being used to test the model fit (see Table 5). This ratio is less than 5, the threshold suggested by Hayduck (1987) to move forward with further analyses.

Table 3. Average Variance Extracted (AVE)

Construct	S1	S2	AVE
IOrSA	2.12	3.88	0.35
OnSE	1.72	3.28	0.34
ITOrS	3.35	1.65	0.67
OnSQE	2.72	2.29	0.54
ECBS	5.85	4.15	0.59

S1 = Sum of squared factor loadings of the indicator variables on the factor representing the construct
S2 = Quantity (1 – the squared loading) summed for all indicators
AVE = S1/(S1 + S2)

Table 4. Squared covariance and AVE

	ITOrS	IOrSA	OnSE	OnSQE	ECBS
ITOrS	0.35	0.331	0.198	0.181	0.365
IOrSA	0.331	0.34	0.186	0.116	0.181
OnSE	0.198	0.186	0.67	0.212	0.182
OnSQE	0.181	0.116	0.212	0.54	0.318
ECBS	0.365	0.181	0.182	0.318	0.59

Table 5. Model Fit Statistics

Model	Chi sq/ df	CFI	GFI	IFI
M1: Hypothesized Model	2.89	0.90	0.91	0.91
M2: Independent Model	6.61	0.00	0.49	0.00

Given that the sample size for the present research is small, the comparative fit index (CFI) is first used to test fit between the model and the data. The CFI value for the proposed model is .90 which satisfies the standard suggested by Hu and Bentler (1999). To further test the model-to-data fitness, the goodness-of-fit index (GFI) is used. The value for GFI for the model is 0.91 which surpasses a conservative value on .90 recommended by Kline (1998). The final index reported in the present analysis is the incremental fit index (IFI). The IFI value for the present model is 0.91. A value above .90 is an acceptable fit. Aforementioned results related to the proposed model in Figure 2 reveal a good fit of the model to the data.

DISCUSSION

Our results confirm H1, that IT alignment to Organizational Strategies (ITOrS) and Inter-Organizational System Availability (IOrSA) are critical success factors for building business value in e-commerce enabled organizations. The results show that evidence of strong decision making power of a chief information officer, strong structure for technology and planning, strong alignment of short term and long term IT strategy with top management strategy, and a positive technology learning environment results in more sophisticated systems integration within the organization. These results are supported by

Figure 2. Results of the Hypothesized Model

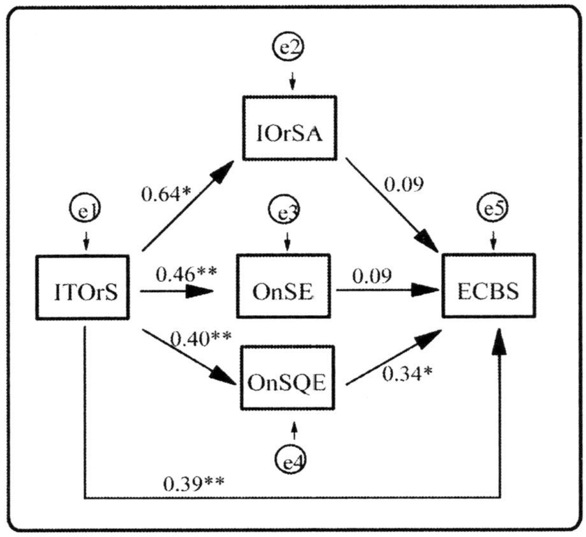

* Significant at the .01 level
** Significant at the .05 level.

ITOrS - IT Alignment to Organizational Strategies
IOrSA - Inter-Organizational Systems Availability
OnSE - Online Systems Efficiency
OnSQE - Online System Quality and Effectiveness
ECBS - Electronic Commerce Business Success

Barua, Konana, Whinston, and Yin (2001) when they indicated that the readiness of business partners to implement e-commerce technologies is critical to achieving business excellence and online systems efficiency. Our results are also supported by Segars and Grover (1998) when they suggested that paying attention to aligning IT strategy with business strategy, understanding the systems processes, and the support of management and end-user groups are crucial to IT planning success.

In the same way, strong ITOrs result in a high quantity or percentage of online business and highly automated systems. Our results confirm that higher ITOrS is a critical success factor for achieving online systems efficiency (OnSE) (H2). Our results, as such, support Huselid and Becker's (1997) suggestion about the existence of synergies between implementing efficient systems and their strategic alignment. Results of the present research study also support Lederer, Mirchandani, and Sims' (2001) assertion that the efficiencies created by web-enabled systems affect strategic alignment positively by improving customer relations. Our results are also in line with Mukhopadhyay and Kekre's (1995) findings on EDI at Chrysler that suggest that the EDI system has caused massive cost savings and has imparted system-wide discipline and generated integrative value for the company.

Our results confirm that ITOrS is a critical success factor for Online System Quality and Effectiveness (OnSQE) (H3). In other words, the more a company's IT is aligned to its parent organization's strategies the more successful it will be in obtaining management support for creating higher quality and effective online presence. These results are in line with what is suggested in the literature. Peak and Guynes (2003, 2003-1), for example, point out that aligning IT with organizational strategies improves information quality which resulted in improved quality of products and services. Kowtha and Choon (2001) mention how the sophistication of a firm's website reflects the strategic priorities of the firm. They further suggest that strategic commitment has substantive and high significant effects on website development.

The results of the present research support the critical relationship between IT alignment to organizational strategies and e-commerce business success (ECBS) (H4) by finding that this relationship is significant and positive. This is also in line with what is suggested in the literature. Pollalis (2003), for example, confirms that alignment of IT to organizational strategies leads to better performance of banks. Sun and Hong (2002) find support in the literature for a significant and direct effect of strategic alignment on business performance in the field of manufacturing strategy. Burn and Szeto (1999) find that this alignment is critically important for business success.

Our results also support the critical relationship between quality and effectiveness of online systems (OnSQE) and e-commerce business success (H7). We are, therefore, able to agree with Zhu and Kraemer (2002) when they pointed out that better e-commerce capability serves to improve the effectiveness of investments on e-commerce initiatives through firm performance. We are also in a position to agree with Chaudhury, Mallick, and Rao (2001) when they mentioned that enhancement of website quality can improve business value. The same goes for the study by Huang, Chen, and Frolick (2002) when they stated that evaluating web data quality is important to effectively determining the business value of online data.

Our results do not support the relationship between inter-organizational system availability (IOrSA) and ECBS (H5) as critical success factors. We are, therefore, unable to support Bakos' (1991) assertion that electronic marketplaces create inter-organizational coordination and reduce search costs and increase economic efficiencies. We are also unable to concur with Kickul and Gundry's (2001) argument that maintaining better relationships and integrating with suppliers is a crucial element of the company sustenance process. We are also unable to agree with Johnston and Vitale's (1988) contention that, if carefully identified and used, inter-organizational systems give significant competitive advantages to organizations.

Our results also do not support the path between online system efficiency and e-commerce business success (H6). It is not known why this path is not significant. One would normally expect for firms with efficient e-commerce systems to be at a relatively better position to achieve better business success. We are, therefore, unable to support Hitt and Brynjolfsson's (1997) findings with regard to IT's association with decentralization of authority, increased knowledge work, and decreased observability. We are also unable to agree with Stratopoulos and Dehning (2000) when they stated that firms making successful investments in IT are more successful at solving the productivity paradox than those that make failed or abortive investments in IT.

CONCLUSION

In summary, our results clearly identify IT alignment to organizational strategies (ITOrS) and online system quality and effectiveness (OnSQE) as critical success factors toward achieving e-business success (ECBS). Our results also suggest that ITOrS impacts inter-organizational system availability (IOrSA), online system efficiency (OnSE), quality and effectiveness of online systems (OnSQE), and e-commerce business success (ECBS). Our results also suggest that OnSQE is a critical success factor towards ECBS.

Our results showed that IOrSA and OnSE played no significant role towards e-business success. It is obvious, at least according to the present research, the quality and effectiveness of these systems are much more important than their availability and efficiency. These relationships need to be investigated further in future research studies.

Implications of the Research

The present research has revealed critical success factors for building business value in e-commerce enabled organizations. The research reveals to practioners that it is crucial to develop strong IT planning strategies for the implementation of e-commerce technologies within organizations. Organizations must understand that the increase in business value of the organization depends on the quality and effectiveness of the systems design.

Limitations and Future Research

Although critical success factors for building business value in e-commerce enabled organizations have been identified, further research should include a study determining the strength of the relationships between the constructs as well as their validity across varied e-commerce sectors. The scope of the study should also be expanded in order to include a larger sample size.

ACKNOWLEDGMENT

Special acknowledgement to Ritesh Mariadas, University of Texas at El Paso

REFERENCES

Amit, R., & Zott, C. (2001). Value creation in e-business. *Strategic Management Journal*, *22*, 493–520. doi:10.1002/smj.187

Arbuckle, J. L. (1989). AMOS: Analysis of moment structures. *The American Statistician*, *43*, 66–67. doi:10.2307/2685178

Bakos, J. Y. (1991). A strategic analysis of electronic marketplaces. *MIS Quarterly*, *15*(3), 295–310. doi:10.2307/249641

Bakos, J. Y. (1991). Information Links and Electronic Marketplaces: Implications of Inter-organizational Information Systems in Vertical Markets. *Journal of Management Information Systems*, *8*(2), 1991.

Bakos, J. Y. (1997). Reducing buyer search costs: Implications for electronic marketplaces. *Management Science*, *43*(12), 1676–1692. doi:10.1287/mnsc.43.12.1676

Bakos, J. Y. (1998). The Emerging Role of Electronic Marketplaces on the Internet. *Communications of the ACM*, *41*(8), 35–42. doi:10.1145/280324.280330

Banker, R. D., Kauffman, R. J., & Morey, R. C. (1990). Measuring gains in operational efficiency from information technology: A study of the positran deployment at Hardee's Inc. *Journal of Management Information Systems*, *7*(2), 29–54.

Barua, A., Konana, P., Whinston, A. B., & Yin, F. (2001). Driving e-business excellence. *MIT Sloan Management Review*, *43*(1), 36–44.

Barua, A., Konana, P., Whinston, A. B., & Yin, F. (2001a). Measures for e-business value assessment. *IT Pro*, 35-39.

Barua, A., Kriebel, C. H., & Mukhopadhyay, T. (1991). An economic analysis of strategic information technology investments. *MIS Quarterly*, *15*(3), 313–331. doi:10.2307/249643

Barua, A., Kriebel, C. H., & Mukhopadhyay, T. (1995). Information technologies and business value: An analytic and empirical investigation. *Information Systems Research*, *6*(1), 3–23. doi:10.1287/isre.6.1.3

Benjamin, R. I., & Scott, M. S. S. (1988). Information technology, integration, and organizational change. *Interfaces*, *18*(3), 86–98. doi:10.1287/inte.18.3.86

Bentler, P. M., & Bonett, D. G. (1980). Significance tests and goodness of fit in the analysis of covariance structures. *Psychological Bulletin*, *88*(3), 588–606. doi:10.1037/0033-2909.88.3.588

Brown, C. V. (2003). Performance metrics for IT human resource alignment. *Information Systems Management*, *20*(4), 36–42. doi:10.1201/1078/43647.20.4.20030901/77291.6

Brynjolfsson, E., & Hitt, L. (1993). Is information technology spending productive? New evidence and new results. In *Proceedings of the 14th Annual International Conference on Information Systems.*

Brynjolfsson, E., & Hitt, L. (1998). Beyond the productivity paradox. *Communications of the ACM*, *41*(8), 41–55. doi:10.1145/280324.280332

Brynjolfsson, E., & Kahim, B. (Eds.). (2000). *Understanding the Digital Economy*. Cambridge: MIT Press.

Burn, J. M., & Szeto, C. (1999). A comparison on the views of business and IT management on success factors for strategic alignment. *Information & Management*, *37*(4), 197–216. doi:10.1016/S0378-7206(99)00048-8

Campbell, & Fiske (1959). Convergent and discriminant validation by the multitrait-multimethod matrix. *Psychological Bulletin*, *56*(2), 81-105.

Cao, M., Zhang, Q., & Seydel, J. (2005). B2C e-commerce web site quality: an empirical examination. *Industrial Management & Data Systems*, *105*(5), 645–661. doi:10.1108/02635570510600000

Chan, Y. E. (2000). IT value: The great divide between qualitative and quantitative and individual and organizational measures. *Journal of Management Information Systems*, *16*(4), 225–261.

Chaudhury, A., Mallick, D. N., & Rao, H. R. (2001). Web channels in e-commerce. *Communications of the ACM*, *44*(1), 99–104. doi:10.1145/357489.357515

Chircu, A. M., & Kauffman, R. J. (2000). Limits to value in electronic commerce-related IT investments. *Journal of Management Information Systems*, *17*(2), 59–80.

Clemons, E., & Wang, Y. (2000). Special Issue: Technology Strategy for Electronic Marketplaces. *Journal of Management Information Systems*, *17*(2), 5–7.

Cron, W., & Sobol, M. (1983). The relationship between computerisation and performance: a strategy for maximising the economic benefits of computerisation. *Information & Management*, *6*, 171–181. doi:10.1016/0378-7206(83)90034-4

Customer Care - Best Practices @ Cisco. (2001). Retrieved 2001, from http://www.cisco.com/warp/public/779/ibs/solutions/customer/practices/

Dai, Q., & Kauffman, R. J. (2002). Business models for internet-based B2B electronic markets. *International Journal of Electronic Commerce*, *6*(4), 41–72.

Dai, Q., & Kauffman, R. J. (2002). B2B E-Commerce Revisited: Leading Perspectives on the Key Issues and Research Directions. *Electronic Markets*, *12*(2).

Davern, M. J., & Kauffman, R. J. (2000). Discovering potential and realizing value from information technology investments. *Journal of Management Information Systems*, *16*(4), 121–143.

Dehning, B., & Richardson, V. J. (2002). Returns on investments in information technology: A research synthesis. *Journal of Information Systems*, *16*(1), 7–30. doi:10.2308/jis.2002.16.1.7

Dekleva, S. (2000). Electronic commerce: A half-empty glass. *Communications of the Association for Information Systems*, *3*(18).

DeLone, W. H., & McLean, E. R. (1992). Information systems success: The quest for the dependent variable. *Information Systems Research, 3*(1), 60–95. doi:10.1287/isre.3.1.60

Dos Santos, B. L., Peffers, K., & Mauer, D. C. (1993). The impact of information technology investment announcements on the market value of the firm. *Information Systems Research, 4*(1), 1–23. doi:10.1287/isre.4.1.1

Dos Santos, B. L., & Sussman, L. (2000). Improving the return on IT investment: the productivity paradox. *International Journal of Information Management, 20*, 429–440. doi:10.1016/S0268-4012(00)00037-2

e-Marketer (June, 2008). *June 2008: U.S. Retail E-Commerce*. Retrieved 2008 from http://www.iab.net/insights_research/530422/1675/334589.

Epstein, M. J. (2000). Organizing your business for the internet evolution. *Strategic Finance, 82*(1), 56–60.

Feeny, D. F., & Ives, B. (1990). In search of sustainability: Reaping long-term advantage from investments in information technology. *Journal of Management Information Systems, 7*(1), 27–46.

Fornell, C., & Larker, D. F. (1981). Evaluating structural equation models with unobservable variables and measurement error. *JMR, Journal of Marketing Research, 18*(1), 39–50. doi:10.2307/3151312

Gefen, D., & Straub, D. W. (2000). The relative importance of perceived ease of use in IS adoption: A study of e-commerce adoption. *Journal of the Association for Information Systems, 1*(8).

Gurbaxani, V., & Mendelson, H. (2001). An Integrative Model of Information Systems Spending Growth. *Information Systems Research, 1*(1), 23–46. doi:10.1287/isre.1.1.23

Hair, J. F., Anderson, R. E., Tatham, R. L. T., & Black, W. C. (1998). *Multivariate data analysis*. Upper Saddle River, NJ: Prentice-Hall Inc.

Hitt, L. M., & Brynjolfsson, E. (1994). The Three Faces of IT Value: Theory and Evidence. In *Proceedings of the International Conference on Information Systems* (pp. 263-277).

Hitt, L. M., & Brynjolfsson, E. (1997). Information technology and internal firm organization: An exploratory analysis. *Journal of Management Information Systems, 14*(2), 81–101.

Huang, Z., Chen, L., & Frolick, M. N. (2002). Integrating web-based data into a data warehouse. *Information Systems Management, 19*(1), 23–34. doi:10.1201/1078/43199.19.1.20020101/31473.4

Huselid, M. A., & Becker, B. E. (1997). The impact high performance work systems, implementation effectiveness, and alignment with strategy on shareholder wealth. *Academy of Management Proceedings* (pp. 144-148).

Im, K. S., Dow, K. E., & Grover, V. (2001). Research report: A reexamination of IT investment and the market value of the firm - An event study methodology. *Information Systems Research, 12*(1), 103–117. doi:10.1287/isre.12.1.103.9718

Johnson, C. A., et al. (2003). *US eCommerce Overview: 2003 - 2008*. Cambridge, MA: Forrester Research, Inc.

Johnson, J., Boucher, K., Connors, K., & Robinson, J. (2001). Collaboration: Development and Management Collaborating of Project Success. *Software Magazine*. Retrieved from http://www.softwaremag.com/archive/2001feb/CollaborativeMgt.html

Johnson, J., Boucher, K. D., Conners, K., & Robinson, J. (2001). The Criteria for Success. *Software Magazine, 21*(1), s3–s11.

Johnston, H. R., & Vitale, M. R. (1988). Creating competitive advantage with interorganizational information systems. *MIS Quarterly*, 152–165.

Josefek, R. A. Jr, & Kauffman, R. J. (1997). Dark pockets and decision support: The information technology value cycle in efficient markets. *Electronic Markets*, *7*(3), 36–42. doi:10.1080/10196789700000036

Kao, D., & Decou, J. (2003). A strategy-based model for e-commerce planning. *Industrial Management & Data Systems*, *103*(4), 238–252. doi:10.1108/02635570310470638

Kauffman, R. J., & Kriebel, C. H. (1988). Modeling and measuring the business value of information technologies. In P.A. Strassman, P. Berger, E.B. Swanson, C.H. Kriebel and R.J. Kauffman (Eds.), *Measuring the Business value of Information Technologies*. Washington, D.C.: ICIT Press.

Keeney, R. L. (1999). The value of internet commerce to the customer. *Management Science*, *45*(4), 533–542. doi:10.1287/mnsc.45.4.533

Kickul, J., & Gundry, L. K. (2001). Breaking through boundaries for organizational innovation: New managerial roles and practices in e-commerce firms. *Journal of Management*, *27*, 347–361. doi:10.1016/S0149-2063(01)00095-2

Kim, C. S., & Peterson, D. K. (2001). Developers' perceptions of information systems success factors. *Journal of Computer Information Systems*, *41*(2), 29–36.

Kline, R. B. (1998). *Principles and Practice of Structural Equation Modeling*. New York: Guilford Press.

Kowtha, N. R., & Choon, T. W. I. (2001). Determinants of website development: a study of electronic commerce in Singapore. *Information & Management*, *39*(3), 227–242. doi:10.1016/S0378-7206(01)00092-1

Kraemer, K. L., & Dedrick, J. (2002). Strategic use of the Internet and e-commerce: Cisco Systems. *The Journal of Strategic Information Systems*, *11*, 5–29. doi:10.1016/S0963-8687(01)00056-7

Kurnia, S., & Johnston, R. (2005). The Dynamic Interactional Model of Inter-Organizational System Adoption: The Case of Category Management Adoption in Australia, *Asia Pacific . Management Review*, *10*(1).

Lederer, A. L., Maupin, D. J., Sena, M. P., & Zhuang, Y. (2000). The technology acceptance model and the world wide web. *Decision Support Systems*, *28*, 269–282. doi:10.1016/S0167-9236(00)00076-2

Lederer, A. L., Mirchandani, D. A., & Sims, K. (2001). The Search for Strategic Advantage from the World Wide Web. *International Journal of Electronic Commerce*, *5*(4), 117–133.

Lee, B., & Barua, A. (1999). An integrated assessment of productivity and efficiency impacts of information technology investments: Old data, new analysis and evidence. *Journal of Productivity Analysis*, *12*, 21–43. doi:10.1023/A:1007898906629

Lee, H. G., & Clark, T. H. (1997). Market process reengineering through electronic market systems: Opportunities and challenges. *Journal of Management Information Systems*, *13*(3), 113–136.

Liu, C., Arnett, K. P., & Litecky, C. (2000). Design quality of websites for electronic commerce: fortune 1000 webmasters' evaluations. *Electronic Markets*, *10*(2), 120–129. doi:10.1080/10196780050138173

Mahajan, & Peterson, R. A. (1985). *Models for Innovation Diffusion*. SAGE Publications, Inc.

Mahmood, M. A., Gemoets, L., Lopez, F., & Hall, L. (2008). Measuring E-Commerce Technology Enabled Business value: An Exploratory Research. *International Journal of E-Business Research*, *4*(2), 48–68.

Mahmood, M. A., Hall, L., & Swanberg, D. (2001). Factors affecting information technology usage: A meta-analysis of the empirical literature. *Journal of Organizational Computing and Electronic Commerce*, *11*(2), 107–130. doi:10.1207/S15327744JOCE1102_02

Mahmood, M. A., & Mann, G. J. (1993). Measuring the organizational impact of information technology investment: An exploratory study. *Journal of Management Information Systems*, *10*(1), 97–122.

Mahmood, M. A., & Mann, G. J. (2000). Impacts of information technology investment on organizational performance. *Journal of Management Information Systems*, *17*(1), 3–10.

Mahmood, M. A., & Sniezek, J. A. (1989). Defining Decision Support Systems: An Empirical End-User Assessment. *INFOR -- Canadian Journal of Operational Research and Information Processing*, 27, 253-271.

McIlvaine, B. (2000). John Chambers. *Electronic Buyers'. News*, *1242*, 42.

Melville, N., Kraemer, K., & Gurbaxani, V. (2004). Information Technology and Organizational Performance: An Integrative Model of IT Business value. *MIS Quarterly*, *23*(2), 165–173.

Molla, A., & Licker, P. S. (2001). E-Commerce Systems Success: An Attempt to Extend and Respecify the Delone and Maclean Model of IS Success. *Journal of ECommerce Research*, *2*(4).

Moon, J. W., & Kim, Y. G. (2001). Extending the TAM for a world-wide-web context. *Information & Management*, *38*, 217–230. doi:10.1016/S0378-7206(00)00061-6

Mukhopadhyay, T., & Kekre, S. (1995). Business value of information technology: A study of electronic data interchange. *MIS Quarterly*, *19*(2), 137–156. doi:10.2307/249685

Nunnally, J. C. (1978). Psychometric theory (2nd Ed.). New York: McGraw-Hill.

Pather, S., Erwin, G., & Remenyi, D. (2003). Measuring e-Commerce effectiveness: a conceptual model. In J. Eloff, A. Engelbrecht, P. Kotzé, & M. Eloff (Eds.), *Proceedings of the 2003 Annual Research Conference of the South African institute of Computer Scientists and information Technologists on Enablement Through Technology* (September 17-19, 2003) (pp. 143-152). ACM Press.

Peak, D., & Guynes, C. S. (2003). The IT alignment planning process. *Journal of Computer Information Systems*, *44*(1), 9–15.

Peak, D., & Guynes, C. S. (2003). Improving information quality through IT alignment planning: a case study. *Information Systems Management*, *20*(4), 22–29. doi:10.1201/1078/43647.20.4.20030901/77289.4

Peteraf, M. A. (2000). The cornerstones of competitive advantage: A resource-based view. *Strategic Management Journal*, *14*(3), 179–191. doi:10.1002/smj.4250140303

Pollalis, Y. A. (2003). Patterns of co-alignment in information-intensive organizations: business performance through integration strategies. *International Journal of Information Management*, *23*(6), 469–492. doi:10.1016/S0268-4012(03)00063-X

Reich, B. H., & Benbasat, I. (2000). Factors that influence the social dimensions of alignment between business and information technology objectives. *MIS Quarterly*, *24*(1), 81–113. doi:10.2307/3250980

Reichheld, F. F. (2001). Lead for Loyalty. *Harvard Business Review*, *79*(7), 76–84.

Riggins, F. J., & Mukhopadhyay, T. (1994). Interdependent benefits from interorganizational systems: Opportunities for. *Journal of Management Information Systems, 11*(2), 37–57.

Rogers, E. M. (1983). *Diffusion of innovations*, New York: The Free Press.

Rose, G. M., & Straub, D. W. (1999). The Effect of Download Time on e-Commerce: The Download Time Brand Impact Model. In *Proceedings of the Ninth Workshop on Information Technologies and Systems* (pp. 167-172).

Sanders, N. R., & Premus, R. (2002). IT applications in supply chain organizations: a link between competitive priorities and organizational benefits. *Journal of Business Logistics, 23*(1), 65–83.

Segars, A. H., & Grover, V. (1998). Strategic information systems planning success: An investigation of the construct and its measurement. *MIS Quarterly*, (June): 139–163. doi:10.2307/249393

Shao, B. B. M., & Lin, W. T. (2001). Measuring the value of information technology in technical efficiency with stochastic production frontiers. *Information and Software Technology, 43*, 447–456. doi:doi:10.1016/S0950-5849(01)00150-1

Shao, B. B. M., & Lin, W. T. (2002). Technical efficiency analysis of information technology investments: a two-stage empirical investigation. Information & Management, 39, 391–401. doi:10.1016/S0378-7206(01)00105-7

Shin, N. (2001). Strategies for competitive advantage in electronic commerce. *Journal of Electronic Commerce Research, 2*(4), 34–41.

Siau, K. (2003). Interorganizational systems and competitive advantages - lessons from history. *Journal of Computer Information Systems, 44*(1), 33–39.

Steinfield, C., Marcus, M. L., & Wigand, R. (2005). Exploring Interorganizational Systems at the Industry Level of Analysis: Evidence from the U.S. Home Mortgage Industry. *Journal of Information Technology, 20*(4). doi:10.1057/palgrave.jit.2000051

Strassman, P. (1990). *The Business value of Computers*. The Economic Press.

Stratopoulos, T., & Dehning, B. (2000). Does successful investment in information technology solve the productivity paradox? *Information & Management, 38*, 103–117. doi:10.1016/S0378-7206(00)00058-6

Subramani, M., & Walden, E. (2001). The impact of e-commerce announcements on the market value of firms. *Information Systems Research, 12*(2), 135–154. doi:10.1287/isre.12.2.135.9698

Sun, H., & Hong, C. (2002). The alignment between manufacturing and business strategies: its influence on business performance. *Technovation, 22*(11), 699–705. doi:10.1016/S0166-4972(01)00066-9

Teo, T. S. H., & Ang, J. S. K. (1999). Critical success factors in the alignment of IS plans with business plans. *International Journal of Information Management, 19*, 173–185. doi:10.1016/S0268-4012(99)00007-9

Teo, T. S. H., & Ang, J. S. K. (2000). How useful are strategic plans for information systems? *Behaviour & Information Technology, 19*(4), 275–282. doi:10.1080/014492900500086381

Teo, T. S. H., & Ang, J. S. K. (2001). An examination of major IS planning problems. *International Journal of Information Management, 21*, 457–470. doi:10.1016/S0268-4012(01)00036-6

Teo, T. S. H., & King, W. R. (1996). Assessing the impact of integrating business planning and IS planning. *Information and Management*, 30, 309-321.

Teo, T. S. H., & King, W. R. (1997). Integration between business planning and information systems planning: An evolutionary-contingency perspective. *Journal of Management Information Systems*, *14*(1), 185–214.

Thomas, D. (2003). Supermarket boosts efficiency tenfold by integrating systems. *Computer Weekly*, 2/20/2003, 12.

Tsay, A., & Agrawal, N. (2004). Channel Conflict and Coordination in the E-Commerce Age. *Production and Operations Management*, *13*(1), 93–110.

Turban, E., King, D., Lee, J., Warkentin, M., & Chung, H. M. (2002). *Electronic Commerce 2002: A Managerial Perspective*. Upper Saddle River, NJ: Pearson Education.

Wold, H. (1989). Introduction to the Second Generation of Multivariate Analysis. In H. Wold (Ed.), *Theoretical Empiricism* (pp. vii-xl). New York: Paragon House.

Woo, C. Y., & Willard, G. E. (1983). *Performance representation in business policy research: discussion and recommendation*. Paper presented at the 23rd Annual National Meetings of the Academy of Management, Dallas.

Zhang, P., & Von Dran, G. M. (2001). Expectations and rankings of website quality features: Results of two studies on user perceptions. In *Proceedings of the 34th Hawaii International Conference on System Sciences,* HICSS-34(7), 7019.

Zhang, P., Von Dran, G. M., Small, R. V., & Barcellos, S. (1999). Websites that Satisfy Users: A Theoretical Framework for Web User Interface Design and Evaluation. In *Proceedings of the 32nd Hawaii International Conference on System Sciences,* HICSS-32(2), 2016.

Zhang, P., Von Dran, G. M., Small, R. V., & Barcellos, S. (2000). A Two-Factor Theory for Website Design. In *Proceedings of the 33rd Hawaii International Conference on System Sciences*, HICSS-33(6), 6026.

Zhu, K. (2004). The Complementarity of Information Technology Infrastructure and E-Commerce Capability: A Resource Based Assessment of their Business value. *Journal of Management Information Systems*, *21*(1), 187–202.

Zhu, K., & Kraemer, K. L. (2002). e-Commerce metrics for net-enhanced organizations: Assessing the value of e-Commerce to firm performance in the manufacturing sector. *Information Systems Research*, *13*(3), 275–295. doi:10.1287/isre.13.3.275.82

Zott, C., Amit, R., & Donlevy, J. (2000). Strategies for value creation in e-commerce: Best practice in europe. *European Management Journal*, *18*(5), 463–475. doi:10.1016/S0263-2373(00)00036-0

APPENDIX

Success Factors of Internet Enabled Business Questionnaire

(On a Likert Scale with values ranging from 1, strongly disagree, to 7, strongly agree)

- IT Alignment to Organizational Strategies (ITOrS)
  1) There is an alignment of information technology (IT) strategy and top management strategy
  2) There is agreement within the company that information technology is part of long term strategy
  3) The Chief Information Officer has significant decision-making power
  4) There is a strong structure within the company for information technology planning and implementation
  5) There is a positive environment for organizational learning associated with the use of new information technology
- Inter-Organizational Systems Availability (IOrSA)
  6) An Internet-enabled uniform system of information sharing is available
  7) An automatic change order system is available
  8) The system permits highly automated transmitting and processing of data
  9) Inventory and Purchase tracking systems are continuously monitored
  10) The online procurement system is satisfactory
  11) Internet expertise is a selection criteria for suppliers/vendors
- Online Systems Quality and Effectiveness (OnSQE)
  12) The website used for purchasing and customer relations is highly secure
  13) The website used for purchasing and customer relations is visually attractive
  14) The website used for purchasing and customer relations is easily navigable
  15) The website used for purchasing and customer relations offers personalized logons
  16) The website used for purchasing and customer relations is consistently accessible without experiencing loading delays
- Online Systems Efficiency (OnSE)
  17) There is a high quantity or percentage of online business
  18) Online customer service is available
  19) Customers requests are resolved online
  20) Continuous monitoring of orders is available
  21) A highly automated order tracking system is available
- E-commerce business success (ECBS)
  22) Return on Investment has increased
  23) Return on Sales has increased
  24) Growth in Net Revenue has increased
  25) Net Income over Invested Capital has increased
  26) Sales over total assets has increased
  27) Return on Sales per employee has increased

28) The organization has a positive company image
29) Customer satisfaction is high
30) The organization engages in product/service innovation
31) There are a large number of return customers

Organizational Characteristics
Company Name
Location
Primary products
Circle all that applies. Does your business do…?
Business to Business Business to Consumer Consumer to Consumer
Annual sales (US $ billions) <1 | 2 - 50 | 51 - 100 | 101 - 150 | 151 – 200
Total number of employees
Company's primary customers
Personal Characteristics (This information is entirely voluntary, all information will be kept confidential)
Name
Position/Title
Phone
Email
I would like a copy of the results emailed to me at _______________ (e-mail address)

Chapter 17
Small Business Performance Impacts of Information Systems Strategic Orientation

R. Rajendran
Sri Ramakrishna Institute of Technology, India

K. Vivekanandan
Bharathiar University, India

ABSTRACT

Businesses invest in developing information systems resources to gain competitive advantages. Literature has demonstrated the requirement of strategic alignment in converting these competitive advantages into sustained superior business performance. The knowledge of information systems strategic orientation and its impact on business performance will enable these businesses to fine tune their strategic information systems applications portfolio in achieving required strategic alignment. This chapter describes a research study that focuses on the information systems strategic orientation of small businesses and investigates its relationship with their perceived business performance. The organizational impact of adoption of the initial stages of electronic business development is also examined. The data were collected from small businesses on nine strategy areas, through mail survey. The result reveals three multifaceted dimensions of information systems strategic orientation. These dimensions of strategic orientation have significantly influenced their business performance. This phenomenon is explained with a model named Linear Strategic Alignment Model. For the adopters of Web presence, all these three dimensions remain significant in explaining their business performance.

INTRODUCTION

Small businesses are an important and integral part of every nation's economy and their contributions are significant in the present business environment of globalisation and digitization. In response to changes in their environment, these small businesses are investing in Information Technologies at an increased rate to develop information systems to support their business strategy to gain best possible returns (Hoving, 2007). The small businesses use the Internet and establish Web presence as a complement to traditional way of competing. Weill (1990) found that investment in strategic information

DOI: 10.4018/978-1-60566-910-6.ch017

systems, rather operational information systems, was risky but with a potential for high return in the long run. The Internet architecture has turned information systems into a far more powerful tool for strategy (Porter, 2001).

The translation of information systems investment into the attainment of competitive advantage and increased business performance (Dhillon, 2008) are the focus of the attention of these small businesses. The knowledge about the extent to, and manner in which information systems complement company strategy (Furrer, Thomas & Goussevskaia, 2008) will help small firms to prioritize relative information systems investments. This enables small businesses to adjust portfolios of strategic information systems (Weill & Aral, 2006) so that they could provide more business support that leads to superior business performance.

The present chapter describes a research study that examines the information systems strategic orientation in small businesses and presents its impact on business performance. To present the consequences of adoption of Web presence, one of the earlier stages of electronic business development (Figure 1), the impact of Web site ownership on the degree and the direction of this relationship are also investigated.

The subsequent sections present the review of literature on strategic management of information systems in small businesses and describe the research design used by the study and it is followed by the presentation of results. Then the research findings and their implications are discussed. The chapter concludes with the summary of the study and its contributions.

LITERATURE REVIEW

Businesses allocate resources to develop information systems because it is believed that these

Figure 1. E-Business development. (Source: E-Commerce and Development Report 2004, United Nations, Geneva, 2004, p 53)

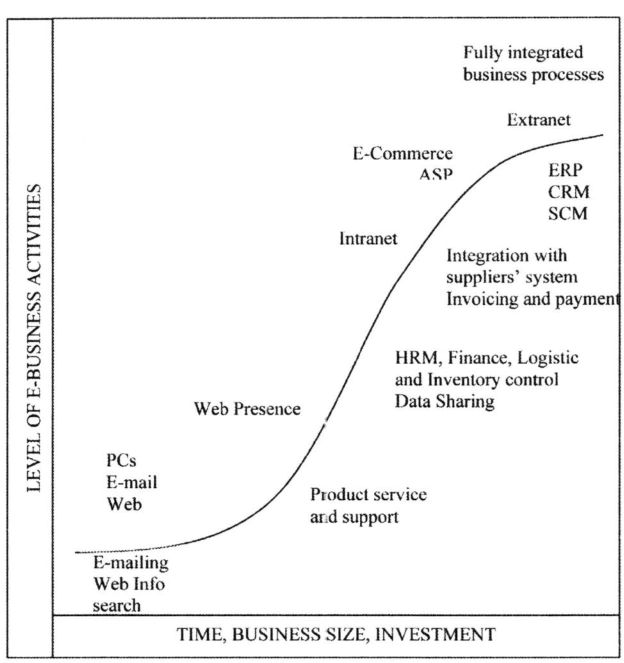

investments provide them with competitive advantages and economic returns. While small businesses have been traditionally seen reluctant to develop information systems strategy (Hagmann & McCahon, 1993; Mehrtens, Cragg, & Mills, 2001), evidence over the past decade shows an increase in strategic use of information systems in small businesses (Naylor & Williams, 1994; Poon, 2000).

The information systems have evolved from its traditional orientation administrative support toward a more strategic role within an organization (Henderson & Venketraman, 1993). Blili and Raymond (1993) emphasize the need for information systems strategic planning to create information systems based strategic advantages. Levy and Powell (2000) propose an approach (Figure 2) to information systems strategy development for small businesses.

For small businesses, the strategy execution perspective (Figure 3) proposed by Henderson and Venkatraman (1993) is more appropriate. This perspective is anchored on the notion that a business strategy is the driver of both organiza-

Figure 2. Information systems strategy approach for SMEs (Levy and Powell, 2000).

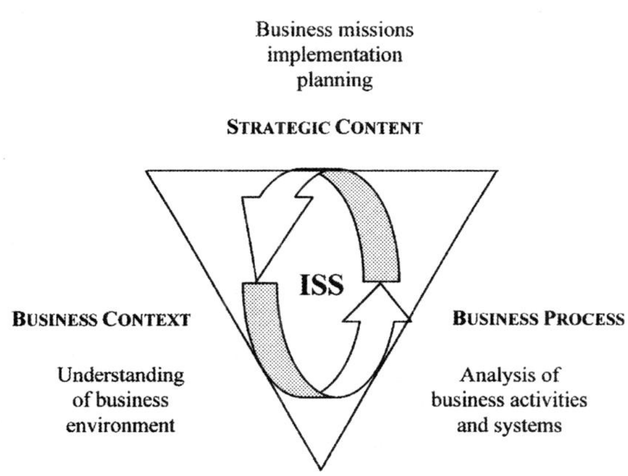

Figure 3. Strategy alignment model with strategy execution alignment perspective (Henderson and Venkatraman, 1993).

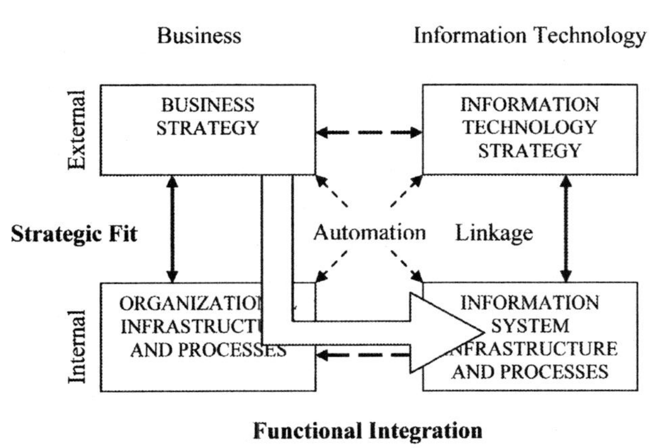

tional design choice and the design of information systems infrastructure (Hevner, March, & Park, 2004).

They argue that the top management should play the role of strategy formulator to articulate the logic and choices pertaining to business strategy, whereas the role of the information systems manager should be that of a strategy implementer, who efficiently and effectively designs and implements the required information infrastructure and processes that support the chosen business strategy.

The resulted information systems strategy with implemented information systems infrastructure and processes, constitute the information systems resource for small business. The organizational performance impact of the information systems resource is commonly referred to as IT Business Value (Melville, Kraemer, & Gurbaxani, 2004). The process of IT business value generation is shown in Figure 4.

Wade and Hulland (2004) describe information systems resources using six resource-attributes viz., value, rarity, appropriability, imitability, substitutability, and mobility based on the finding of prior information systems research. In resource-based view (Figure 5), these information systems resource attributes will enable a firm to achieve competitive advantages over others and lead to superior long-term performance (Rivard, Raymond & Verreault, 2006).

However, resources rarely act alone in creating and sustaining competitive advantage. The information systems resources normally act in conjunction with other firm resources to provide strategic benefits (Ravichandran & Lertwongsatien, 2002). Benjamin and Levinson (1993) conclude that performance depends on how information systems resource is integrated with organizational, technical, and business resources. Chan, Huff, Barclay, and Copeland (1997) argue that the impact of information systems on business performance is intermediated by the alignment between information systems strategy and business strategy.

Luftman, Lewis, and Oldach (1993) recognize that for companies to succeed the alignment of information systems strategy with the business strategy is a necessity. Alignment (Chan & Reich, 2007) expresses an idea that the objective of design for an organizational structure or its information systems must match its context in order to be effective (Iivari, 1992; Gregor, Hart & Martin, 2007). Strategic orientation expresses this context and its relationship with business performance, will set the direction to the measurement of strategic alignment. The moderation model of strategic alignment suggests that the strategic orientation of a business determines the relative importance of the alignment dimensions. Strategic orientation of information systems indicates the degree

Figure 4. IS business value model (Melville et al., 2004).

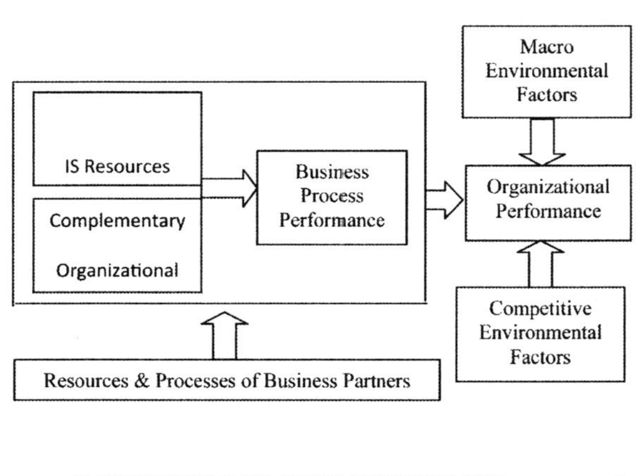

of information systems support for each strategic alignment dimension.

The strategic orientation of the existing portfolio of information systems applications, representing the general pattern of realized information systems strategy provides valuable predictive information regarding perceived business performance. Venkatraman (1989) identifies key traits of business strategic orientation based on the theoretical perspective and specifies the following six characteristics as dimensions a priori in developing valid measurement for strategic orientation of business enterprises (STROBE): Aggressiveness, Analysis, Defensiveness, Futurity, Proactiveness and Riskiness. Chan et al. (1997) use these dimensions to hypothesize the structure of information systems strategic orientation. But the empirical findings of their study suggest a parsimonious taxonomy of three generic realized information systems strategies. The three core dimensions of information systems strategic orientation that emerged are information systems support for anticipation, information systems support for analysis and information systems support for action. But the emergence of these dimensions is ignored in the further analyses of their study. Their further study (Chan, Huff & Copeland, 1998) reveals four dimensions: Anticipation, Armour, analysis, and Action.

However, they conclude that the concept of information systems strategic orientation is somewhat novel and is an area ripe for future information systems strategy research. Thus, a relevant question for small business information systems strategy research study could be to derive information systems strategic orientation and examine the efficacy of the dimensions of orientation to generate business value within the conditions set by the information systems resources (Figure 6). The approach represented in the conceptual model is to empirically derive dimensions of information systems strategic orientation, a posteriori.

Figure 5. Resource-based view of IS (Wade and Hulland, 2004).

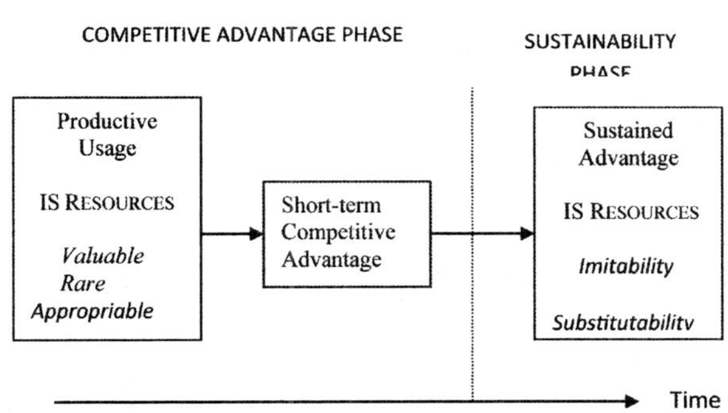

Figure 6. Conceptual model.

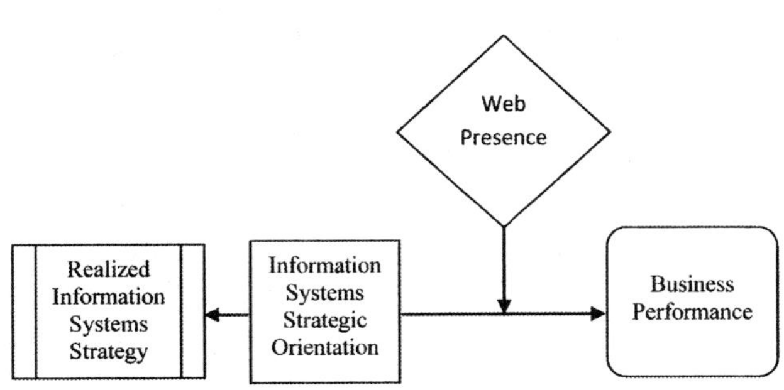

The recent streams of studies on net-enhanced large business organizations suggest that E - Business initiatives tend to make information systems resources more valuable. Zhu and Kraemer (2002) emphasize that Web presence promotes the business value generation capability of information systems resources. To validate this finding in the small business context, the role of the Web presence in determining the relationship between information systems strategic orientation and their business performance is also investigated.

DESIGN OF THE RESEARCH STUDY

Research Instrument

The studies of business strategy in small businesses provide evidence that small businesses have to adopt numerous strategies. Tallon (2007) recommends multi-focused business strategies for higher business payoff. Storey (1994) identifies strategy as one of the three main components that contribute towards growth among small businesses. Sougata (2004) argues that the environment played a significant role in shaping business strategy during reforms. These studies have drawn on typologies based on large firms viz. Ansoff (1965)'s matrix of strategies (Barkham, Gudgin, Hart, & Hanvey., 1996; Hewitt-Dundas & Roper, 1999) and Porter (1980)'s generic strategies (Namiki, 1988; Reid, 1993; Kakati & Dhar, 2002). Julien, Joyal, Deshaies, and Ramangalahy (1997) find that exporters compete on price, technical superiority, product quality, and customer service. Gunasekaran, Okko, Martikainen, and Yli-Olli (1996) identify productivity and quality improvement strategies based on cost control, improving quality, new product, lower price, fast delivery, and increased market share, for small and medium enterprises in the manufacturing sector.

These studies have produced different typologies and have failed to provide a consensus model of strategy for small businesses (Southern & Tilley, 2000). As the approach to strategy formation in small business is informal, inexplicit, intuitive, and incremental (Mintzberg, 1988), the explicit identification of strategy is found to be more difficult (Lefebvre, Langley, Harvey, & Lefebvre, 1992). Cragg, King, and Hussin (2002) extracted key factors that contributed towards small business competitiveness from these studies. Pretesting with practicing owner/managers of small businesses, they refined this list of business strategy items. For the present study, we consider these nine strategies (Table 1) as business strategies of small businesses.

As explained in the earlier paragraphs, the planning and development of strategy in small businesses are embryonic and informal. To capture the actual and realized deployment of information systems applications, we designed the question in the instrument for information systems strategies, around the same nine business strategies shown in Table 1 (Chan et al., 1997; Cragg et al., 2002). For each business strategy item (question), a parallel information systems strategy item was created to assess the extent to which the information systems support that particular aspect of business strategy. We used a Five point Likert scale for measurement.

From business perspective, business performance is a complex and multifaceted concept (Venkatraman & Ramanujam, 1986). Strategic management research literature proposes a subjective approach to measure business performance and it is appropriate in a small business context where financial data are often unavailable or unreliable (Dess, 1984; Sapienza, Smith, & Gannon, 1988). Khandwalla (1977) developed a four items (long term profitability, sales growth, availability of financial resources, and image & client loyalty) instrument to measure business performance based on the owner/manager's subjective assessment of the company's performance relative to its competitors. Raymond, Pare, and Bergeron (1995) and Cragg et al. (2002) validated this business performance instrument in the small business context and

Table 1. Business strategies of small businesses

Sl. No.	Business Strategy
1	Pricing Strategy
2	New Market Strategy
3	New Product Strategy
4	Quality Service Strategy
5	Quality Product Strategy
6	Intensive Marketing Strategy
7	Process Efficiency Strategy
8	Product Differentiation Strategy
9	Product Diversification Strategy

we deemed it appropriate for the present study. We confirmed the suitability and face validity of the instrument along with business and information systems strategies during the pre-testing stage of the questionnaire development.

The status and the usage of information systems infrastructure of small manufacturing knitwear exporters with a Web site differ significantly from that of exporters not having a Web site (Vivekanandan & Rajendran, 2005). To examine the contingency effect of Web presence on the linkage between information systems strategic orientation and business performance, we made necessary provisions in the questionnaire to collect details about their Web presence.

RESEARCH METHOD

We conducted a mail questionnaire survey among the small businesses of Tirupur, India. This cluster of small manufacturing businesses is well known for its excellent export performance and its participation in the global apparel supply chain as a quality supplier (Vivekanandan & Rajendran, 2006). We identified one thousand and one hundred Knitwear apparel manufacturing exporters. We selected the manufacturing sector as they could provide a range of levels of information systems sophistication (Cragg & King, 1993; Rajendran, 1999). We planned to send each questionnaire with a prepaid business reply envelope and a letter explaining the purpose of the study.

We pretested the questionnaire with two professionals associated with small businesses and further with the owner/managers of five leading exporting organizations. Based on their comments we suitably modified the questionnaire. Further, we conducted a pilot test among a randomly selected one hundred and fifty exporters. It resulted in minor modifications in the questionnaire. Thus we refined the questionnaire at three stages (Dillman, 1978).

We sent the refined questionnaires to other nine hundred and fifty exporters and we received back in total one hundred and twenty nine filled in useable questionnaires. To assess the non-response bias, we compared the first thirty and last thirty responses on the nine information systems strategy items (Armstrong & Overton, 1982). The Mann Whitney test revealed that the differences are not significant except for the new product strategy and so concluded that the non-response bias is not a significant factor that could affect the results of the data analysis.

RESULTS OF DATA ANALYSIS

The table 2 presents the profile of the respondents. The table 3 and table 4 show their mean score with standard deviation for their business performance and information systems strategies.

Factor analysis is a multivariate interdependency technique used for data reduction and structure simplification (Hair, Anderson, Tatham, & Black, 1998). We conducted the Bartlett test of sphericity to assess the appropriateness of factor analysis and it was found satisfactory (Sig. 0.000). As the increasing the sample size causes the Bartlett test to become more sensitive, we further used the measure of sampling adequacy (MSA) to reassess the appropriateness of factor analysis. The MSA index of 0.845 reveals the meritorious nature of the data for factor analysis.

Under factor analysis procedure, we used the principal component analysis to extract minimum number of factors that explain maximum percentage of variation. We applied the Varimax rotation with Kaiser Normalization to simplify the revealed structure.

The three factors extracted explain 76% of variation and we consider these factors as the dimensions of information systems strategic orientation. We label these dimensions of information systems strategic orientation as 1. Cost-Quality Leadership, 2. Product Development and 3. Market Development. The rotated component matrix of the simplified structure (Table 5) reveals the convergent and discriminant validity of the above three constructs. The reliability coefficient, Cronbach's alpha being the most widely used measure (Peter, 1981) is used to assess the internal consistency of these constructs. The values of Cronbach's alpha are 0.83, 0.88 and 0.78 and these are well above the lower limit of 0.70 (Robinson, Shaver, & Wrightsman, 1991; Nunnally, 1978). The factor

Table 2. Profile of the respondents

Description	Range	Frequency	Percent
Company Age	Up to 10 yrs	34	26.4
	10 to 20 yrs	70	54.3
	Above 20 yrs	25	19.4
Ownership Status	Proprietorship	38	29.5
	Partnership	73	56.6
	Private limited	17	13.2
	Public limited	1	00.8
Growth Stage	Conceptual	2	01.6
	Survival	15	11.6
	Stabilization	30	23.3
	Growth Orientation	48	37.2
	Rapid Growth	20	15.5
	Resource Maturity	14	10.9
Internet Experience	More than 3 yrs	107	82.9
	2 – 3 years	12	09.3
	1 – 2 years	7	05.4
	Less than one year	2	01.6
	Not applicable	1	00.8
Web Presence	Ownership	79	61.2

scores are computed based on the factor loadings of all variables on each factor, to replace the original scores of nine information systems strategies.

The Business performance is the dependent variable in this study. We conducted the factor analysis to convert the multiple measure of business performance into a single composite measure. However, the Principal Component Analysis resulted in a single factor that explained 56% variance with construct validity of 0.73 (Cronbach's alpha). The factor score is generated as a composite measure of business performance.

As the primary object of this study is to explore and explain the relationship between the dimensions of information systems strategic orientation and business performance, we used the multiple regression analysis. The multiple regression analysis is a multivariate dependency technique used to analyze the relationship between a single dependent (criterion) variable and several independent (predictor) variables.

The regression equation generated is a linear combination of the independent variables that best explains and predicts the dependent variable. It is the regression variate that is formed by a set of weighted independent variables. The weights (regression coefficient) represent the relative contribution of the independent variables to the overall prediction and facilitate interpretation on the influence of each variable in making the prediction. The correlations among the independent variables are also referred to as multicollinearity. Multicollinearity reduces the variables' predictive power and complicates the interpretation process (Hair et al., 1998).

We conducted the multiple regression analysis with the business performance as dependent variable and the dimensions of information systems

Table 3. Business performance score

Performance Criteria	Mean	Std Deviation
Public image and client loyalty	4.15	0.77
Sales Growth	3.87	0.72
Financial Resources	3.69	0.75
Long term profitability	3.68	0.84

Scale: 1- Strongly Disagree, 5 – Strongly Agree

Table 4. Information systems strategy score

Information Systems Strategy	Mean	Std Deviation
Quality Service	3.93	0.97
Process Efficiency Improvement	3.84	0.95
Cost reduction	3.83	0.95
New Market Expansion	3.71	0.85
Quality Product	3.67	0.90
Intensive Marketing	3.47	0.80
Product Differentiation	3.29	0.84
New Product	3.22	0.93
Wide Product Range	3.19	0.85

Scale: 1- Strongly Disagree, 5 – Strongly Agree

Table 5. Simplified structure of information systems strategic orientation (rotated component matrix)

Information Systems Strategy	Component		
	Factor 1	Factor 2	Factor 3
Process efficiency improvement	**.79**	.26	.22
Cost reduction	**.79**	.20	.02
Quality service	**.78**	.17	.34
Quality product	**.66**	.35	.18
Wider product range	.21	**.89**	.08
New products	.27	**.84**	.24
Product differentiation	.31	**.78**	.20
New market expansion	.14	.12	**.90**
Intensive marketing	.29	.27	**.80**

Extraction method: Principal component analysis
Rotation method: Varimax with Kaiser Normalization

strategic orientation as independent variables. To ensure the minimization of impact of multicollinearity, we used the factor scores generated in the factor analysis with Varimax as an orthogonal rotation method, in the regression procedure.

The coefficient of determination (R^2) for the regression model generated is 0.252 with adjusted R^2 equal to 0.234. The model is statistically significant (ANOVA – Sig. 0.000). All the three dimensions of information systems strategic orientation significantly influence the business performance. The details of regression coefficient and their significance are shown in Table 6.

To assess the influence of Web presence, we conducted the regression analyses independently for exporters having Web site and for others. Table 7 presents the results.

DIMENSIONS OF INFORMATION SYSTEMS STRATEGIC ORIENTATION

The results of the preliminary analysis of the data show that 74% of the respondent businesses are more than 10 years old and 89% have reached the business growth stage of stabilization and beyond. As expected, the proprietorship and partnership is predominant (eighty six percent). All the exporters have Internet connectivity and two third of

Table 6. Results of regression analysis coefficients and its significance

Predictors	Unstandardized Coefficients		Standardized Coefficients	t	Sig.
	B	Std. Error	Beta		
(Constant)	.000	.077		.000	1.000
Cost-Quality Leadership	.355	.077	.355	4.586	**.000**
Product Development	.261	.077	.261	3.374	**.001**
Market Development	.242	.077	.242	3.124	**.002**

Dependent variable: Overall business performance

Table 7. Results of regression analysis –Web site owners and non-owners

Model Component *Independent Variables*	Exporters with Web site	Exporters without Web site
Cost-Quality Leadership	0.404	0.179
Regression Coefficient	0.403	0.176
Beta Coefficient	4.022	1.244
t value	**0.000**	0.220
Sig.		
Product Development	0.332	0.117
Regression Coefficient	0.351	0.111
Beta Coefficient	3.593	0.845
t value	**0.001**	0.402
Sig.		
Market Development	0.233	0.333
Regression Coefficient	0.237	0.341
Beta Coefficient	2.368	2.418 **0.020**
t value	**0.020**	
Sig.		
Model Fit	0.285	0.206
R^2	0.257	0.154
Adjusted R^2	**0.000**	**0.013**
ANOVA Sig.		

them have Web presence. Eighty three percent of the respondent has more than three years of experience in using Internet. This indicates their high receptivity to the adoption of initial stages of electronic business practices.

The quality service strategy receives the highest mean score and it is followed by process efficiency improvement and cost reduction strategies. The mean score of all the other strategies are also above 3.00 and the overall mean score is 3.57. The exporters perceive that the information systems in general support their business strategies. Thus, these small businesses are upgrading their ways of competing that can lead to successful national economic development (Porter, 2004). These small businesses assess their competitive positions in all four criteria of business performance as strong.

The factor analysis reveals the three dimensions of information systems strategic orientation. The strategies of process efficiency improvement, cost reduction, quality service, and quality product emerge as indicators of the information systems strategic orientation dimension, Cost-Quality Leadership. The other two dimensions of information systems strategic orientation are Product Development (wider product range, new product, and product differentiation strategies) and Market Development (new market expansion and intensive marketing strategies). These are in line with Ansoff's matrix of strategy and Porter's generic strategies.

The defensiveness is the predominant characteristic of the Cost-Quality Leadership dimension; The Proactiveness is the predominant characteristic of the Product Development dimension and Aggressiveness is the predominant characteristic of Market Development dimension (Venketraman, 1989). All these characteristics collectively describe the information systems strategic orientation.

The extraction of a single factor from multiple measures of business performance indicates that the four different criteria reflect the overall business performance. The factor score is an indicator of their business performance and is used as dependent variable in the regression analyses. Multiple regression analysis provides insight into the relative importance of each dimension of information systems strategic orientation in the prediction of business performance. The order of importance is cost-quality leadership, product development and then market development. The strategic orientation, like strategic alignment, is a process and not an event. The information systems strategic orientation is strategy specific and industry oriented. However, the strategic orientation analysis has set the direction to the measurement of alignment and its linkage with business performance.

The Linear Strategic Alignment Model (Figure 7) represents the above phenomenon. The business strategy drives the information strategy. The small business performance impact of realized information systems is intermediated by its strategic orientation. The dimensions of strategic orientation determine the degree of this performance impact.

Consistent with the finding of Burn and Szeto (2000), the model reveals the technology transformation perspective for information systems strategic planning. Burn and Szeto find that the perspectives of IT managers and business managers are remarkably consistent with both groups selecting technology transformation as their overall perspective and business strategy as the driver.

To sustain the strategic advantage and thereby to maintain the superior business performance in the dynamic business environment, the 'business value chain' is to be agile and flexible. The findings of the recent studies (Sledgianowski & Luftman, 2005; Strnadl, 2006; Huang & Hu, 2007) will be useful in this direction. However, the Linear Strategic Alignment Model is to be validated in the different contexts.

The knowledge about the predictive value of the information systems strategic orientation is highly useful in understanding the business value generating process of information systems resource deployment in a given business setting. This facilitates the fine-tuning of the information systems investment and adjusting the portfolio of information systems applications by knowing the efficacy of a particular information systems strategy to attain certain ends within a particular setting. And strategic orientation as a process does not normally lead to competitive convergence (Porter, 2001).

RESEARCH AND MANAGERIAL IMPLICATIONS

Implication for Strategy Research

The major contribution of the present study is the revelation of three core multifaceted dimensions of information systems strategic orientation in small business context. This emphasizes that small businesses explicitly indulge in information systems strategic planning for business performance management (Frolick & Ariyachandra, 2006). Future research could focus on small business information systems strategic planning and investigate their strategy making process (Miller, 1987)

As the newer strategic management research paradigm explicitly separate goals from strategy, the information systems strategy could also be

Figure 7. Linear strategic alignment model.

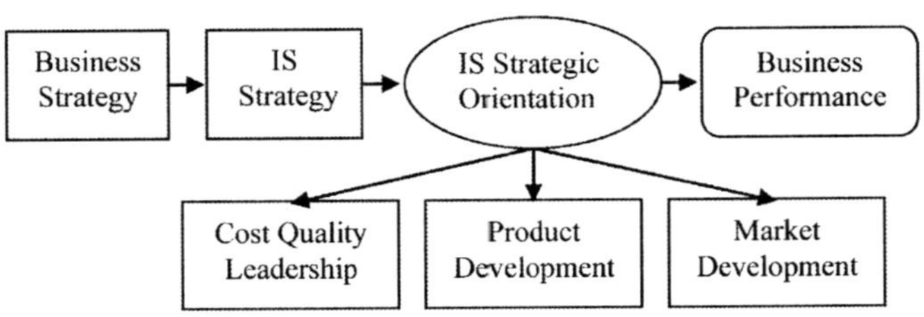

viewed as means to attain certain ends within a particular setting. Empirically deriving dimensions of strategic orientation a posteriori has certain limitations (Venkatraman, 1989). A valid operational measure could be developed specifying the identified dimensions a priori.

In net-enabled organizations (Straub, Hoffman, Weber, & Steinfield, 2002), strategy is fast becoming a dynamic process of recreating and executing innovative options to gain and sustain competitive advantages (Teece, Pisano, & Shuen, 1997). The insight gained through the present study into the relationship among core dimensions of information systems strategic orientation in their prediction of business performance could be used in assessing and choosing emerging and enabling information technologies (ET). Selecting ET is the first stage in Net-Enabled Business Innovation Cycle (Wheeler, 2002) that asserts that choosing IT proceeds rather than aligns with business strategy in developing dynamic capabilities (Eisenhardt & Martin, 2000) to turn timely net-enabled business innovations into customer value (Chen, Chen, & Wu., 2005).

Future research studies could investigate whether the contingent effect of the Web presence on the relationship between information systems strategic orientation and business performance is a direct one or intermediated by any other factor. The capabilities of the Web site could also be examined in detail to ascertain its role (Whinston & Geng, 2004) in determining the degree and character of association between information systems strategic orientation and business performance.

Implications for E-Business Development

The Web presence strengthens the relationship between information systems strategic orientation and business performance as a 'promoting' variable. This emphasizes the strategic benefits of adoption of Web presence, one of the initial stages of electronic business development. The market development dimension of information systems strategic orientation is equally significant for exporters who have not yet adopted Web presence (Table 7). Even though their regression model explains only 15% of the variation in their business performance, the model remains significant.

It seems that their participation in the global production networks, and the extent of trade liberalization forced these exporters to adopt the first stage of electronic business development viz. e-mailing and Web info search as a means of expanding their market. The near universal desire of business to gain advantages over their competitors, in addition to extend their markets, reach new markets, and protect existing markets, is perhaps the most significant force (Gibbs, Kraemer, & Dedrick, 2003), driving these exporters to move to the next stage of electronic business development viz. Web presence. It appears that Web presence creates information visibility (Straub et. al., 2002) forcing the small businesses to improve their internal processes and strategic positioning that in turn lead to superior business performance.

As the strategic planning in small businesses being incremental in nature (Mintzberg, 1988), the demonstration of the beneficial results for adopters will enable the small businesses to move forward in the electronic business development. Rogers (1983) argues that change agents should recognize their responsibility for the consequences of the innovation they advocate. Thus, the results of the present study have practical implications to government and non-governmental organizations that promote the diffusion of electronic business adoption in small businesses.

CONCLUSION

The small businesses are investing in information and communication technologies to develop information systems applications to support their business strategy and thereby establish a competitive advantage based on the distinctive capability created in their markets. However, these small businesses struggle to achieve business benefits from their information systems investments and in particular to obtain a sustained competitive advantage and superior business performance. To explore the relationship between the strategic orientation of these information systems and business performance, a study was designed. The mail survey was conducted among nine hundred and fifty small businesses manufacturing and exporting knitwear apparels.

The results reveal the three general patterns of their realized information systems strategies viz. cost-quality leadership, product development, and market development. These dimensions of information systems strategic orientation have strong positive impact on their business performance. The Linear Strategic Alignment Model represents these impacts of strategic orientation of their information systems strategy that are driven by their business strategy. The consequences of their adoption of Web presence promote the degree of the linkage between information systems orientation and their business performance. This chapter demonstrates the business value of information systems investment and adoption of initial stages of electronic business development. The small businesses will find the Linear Strategic Alignment Model highly helpful in understanding the 'business value chain' of their business-IS strategies.

REFERENCES

Ansoff, H. I. (1965). *Corporate Strategy*. New York: McGraw Hill.

Armstrong, J. B., & Overton, T. A. (1982). Estimating non-response bias in mail survey. In A. K. Jain, C. Pinson, B. T. Ratchford (Eds.), *Marketing Research: Application and Problems*. Chinchester: Wiley.

Barkham, R., Gudgin, G., Hart, M., & Hanvey, E. (1996). *The determinants of small firm growth: An inter-regional study in the United Kingdom, 1986-1990*. London: Jessica Kingsley.

Benjamin, R. I., & Levinson, E. (1993). A framework for managing IT enabled change. *Sloan Management Review*, *1993*(Summer), 22–33.

Blili, S., & Raymond, L. (1993). Information technology – threats and opportunities for small and medium sized enterprises. *International Journal of Information Management*, *13*, 439–448. doi:10.1016/0268-4012(93)90060-H

Burn, J. M., & Szeto, C. (2000). A comparison of the views of business and IT management on success factors for strategic alignment. *Information & Management*, *37*(4), 197–216. doi:10.1016/S0378-7206(99)00048-8

Chan, E. Y., Huff, L. S., Barclay, W. D., & Copeland, G. D. (1997). Business strategic orientation, information systems strategic orientation, and strategic alignment. *Information Systems Research*, *8*(2), 125–150. doi:10.1287/isre.8.2.125

Chan, E. Y., Huff, S. L., & Copeland, G. D. (1998). Assessing realized information systems strategy. *Strategic Information Systems*, *6*(4), 273–298. doi:10.1016/S0963-8687(97)00005-X

Chan, E. Y., & Reich, H. B. (2007). IT alignment: What have we learned? *Journal of Information Technology*, *22*, 297–315. doi:10.1057/palgrave.jit.2000109

Chen, Y., Chen, G., & Wu, S. (2005). Issues and opportunities in E-Business research: A Simonian perspective. *International Journal of E-Business Research*, *1*(1), 37–56.

Cragg, P., King, M., & Hussin, H. (2002). IT alignment and firm performance in small manufacturing firms. *The Journal of Strategic Information Systems*, *11*(1), 109–132. doi:10.1016/S0963-8687(02)00007-0

Cragg, P. B., & King, M. (1993). Small firm computing – motivators and inhibitors. *MIS Quarterly*, *17*, 47–60. doi:10.2307/249509

Croteau, A., & Bergeron, F. (2001). An information technology trilogy: Business strategy, technological deployment and organization performance. *The Journal of Strategic Information Systems*, *10*(2), 77–99. doi:10.1016/S0963-8687(01)00044-0

Dess, G. G., & Robinson, R. B. (1984). Measuring organizational performance in the absence of objective measure: The case of the privately-held firm and conglomerate business unit. *Strategic Management Journal*, *5*, 265–273. doi:10.1002/smj.4250050306

Dhillon, G. (2008). Organizational competence for harnessing IT: A case study. *Information & Management*, *45*, 297–303. doi:10.1016/j.im.2008.01.008

Dillman, D. A. (1978). *Mail and Telephonic Surveys: The Total Design Method*. New York: Wiley.

Eisenhardt, K. M., & Martin, J. A. (2000). Dynamic capabilities: what are they? *Strategic Management Journal*, *21*(10), 1105–1121. doi:10.1002/1097-0266(200010/11)21:10/11<1105::AID-SMJ133>3.0.CO;2-E

Frolick, N. M., & Ariyachandra, R. T. (2006). Business performance management: one truth. *Information Systems Management*, *2006*(Winter), 41–48. doi:10.1201/1078.10580530/45769.23.1.20061201/91771.5

Furrer, O., Thomas, H., & Goussevskaia, A. (2008). The structure and evolution of the strategic management field: A context analysis of 26 years of strategic management research. *International Journal of Management Reviews*, *10*(1), 1–23. doi:10.1111/j.1468-2370.2007.00217.x

Gibbs, J., Kraemer, K. L., & Dedrick, J. (2003). Environment and policy factors shaping global E-Commerce diffusion: A cross-country comparison. *The Information Society*, *19*, 5–18. doi:10.1080/01972240309472

Gregor, S., Hart, D., & Martin, N. (2007). Enterprise architecture: Enablers of business strategy and IS/IT alignment in government. *Information Technology & People*, *20*(2), 2007. doi:10.1108/09593840710758031

Gunasekaran, A., Okko, P., Martikainen, T., & Yli-Olli, P. (1996). Improving productivity and quality in small and medium enterprises: Cases and analysis. *International Small Business Journal*, *15*(1), 59–72. doi:10.1177/0266242696151004

Hagmann, C., & McCahon, C. (1993). Strategic information systems and competitiveness. *Information & Management*, *25*, 183–192. doi:10.1016/0378-7206(93)90067-4

Hair, J. F., Anderson, R. E., Tatham, R. L., & Black, W. C. (1998). *Multivariate Data Analysis, Fifth edition*. Delhi: Pearson Education.

Henderson, J. C., & Venkatraman, N. (1993). Strategic alignment: Leveraging information technology for transforming organizations. *IBM Systems Journal*, *32*(1), 4–16.

Hevner, R. A., March, T. S., & Park, J. (2000). Design science in information systems research. *MIS Quarterly*, *28*(1), 75–105.

Hewitt-Dundas, N., & Roper, S. (1999). Overcoming customer dependency: A case study of the strategies of small consumer food producers in Northern Ireland. *International Small Business Journal*, *17*(4), 49–65. doi:10.1177/0266242699174003

Hoving, R. (▪▪▪). (2207). Information technology leadership challenges – Past, present and future. *Information Systems Management*, *24*, 147–153. doi:10.1080/10580530701221049

Huang, C. D., & Hu, Q. (2007). Achieving IT – business strategic alignment via enterprise-wide implementation of balanced scorecards. *Information Systems Management, 24*, 173–184. doi:10.1080/10580530701239314

Iivari, J. (1992). The organizational fit of information systems. *Journal of Information Systems, 2*, 3–29. doi:10.1111/j.1365-2575.1992.tb00064.x

Julien, P., Joyal, A., Deshaies, L., & Ramangalahy, C. (1997). A typology of strategic behavior between small and medium sized exporting businesses: A case study. *International Small Business Journal, 15*(2), 33–50. doi:10.1177/0266242697152002

Kakati, P. R., & Dhar, U. R. (2002). Competitive strategies and new venture performance. *Vikalpa, 27*(3), 13–26.

Khandwalla, P. N. (1977). *The Design of Organization*. New York: Harcourt Brace Jovanovich.

Lefebvre, L. A., Langley, A., Harvey, J., & Lefebvre, E. (1992). Exploring the strategy-technology connection in small manufacturing firms. *Production and Operations Management, 1*(3), 1–17.

Levy, M., & Powell, P. (2000). Information systems strategy for small and medium sized enterprises: An organizational perspective. *The Journal of Strategic Information Systems, 9*(1), 63–84. doi:10.1016/S0963-8687(00)00028-7

Luftman, J. N., Lewis, P. R., & Oldach, S. H. (1993). Transforming the enterprise: The alignment of business and IT strategies. *IBM Systems Journal, 32*(1), 198–221.

Mehrtens, J., Cragg, P. B., & Mills, A. M. (2001). A model of Internet adoption by SMEs. *Information & Management, 26*, 1–12.

Melville, N., Kraemer, K., & Gurbaxani, V. (2004). Review: information technology and organizational performance: An integrative model of IT business value. *MIS Quarterly, 28*(2), 283–322.

Miller, D. (1987). Strategy making and structure: Analysis and implications for performance. *Academy of Management Journal, 30*(1), 7–32. doi:10.2307/255893

Mintzberg, H. (1988). Generic strategies: Toward a comprehensive framework. *Advances in Strategic Management, 5*, 1–67.

Namiki, N. (1988). Export Strategy for small business. *Journal of Small Business Management, 26*(2), 33–37.

Naylor, J. B., & Williams, J. (1994). The successful use of IT in SMEs on Merseyside. *European Journal of Information Systems, 3*(1), 48–56. doi:10.1057/ejis.1994.6

Nunnally, J. (1978). *Psychometric Theory*. New York: McGraw-Hill.

Peter, J. P. (1981). Construct Validity: A review of basic issues and marketing practices. *JMR, Journal of Marketing Research, 18*(5), 133–145. doi:10.2307/3150948

Poon, S. (2000). Business environment and Internet commerce benefit – a small business perspective. *European Journal of Information Systems, 9*(2), 72–81. doi:10.1057/palgrave/ejis/3000361

Porter, M. E. (1980). *Competitive strategy – technologies for analyzing industries and competitors*. New York: Free Press.

Porter, M. E. (2001). Strategy and the Internet. *Harvard Business Review, 2001*(March), 63–78.

Porter, M. E. (2004). Building the microeconomic foundations of prosperity: Findings from the business competitiveness index. In *The Global Competitiveness Report 2003-2004*. World Economic Forum, Rajendran, R. (1999). *A study on computer based information systems of small-scale industries in Coimbatore, India*. Unpublished master thesis, Bharathiar University, Coimbatore.

Ravichandran, T., & Lertwongsatien, C. (2002). Impact of information systems resources and capabilities on firm performance: A resource-based perspective. In *Proceedings of 23rd International Conference on Information System*, 577-582.

Raymond, L., Pare, G., & Bergeron, F. (1995). Matching information technology and organizational structure: An empirical study with implication for performance. *European Journal of Information Systems*, *4*(1), 3–16. doi:10.1057/ejis.1995.2

Reid, G. C. (1993). *Small business enterprise: An economic analysis*. London: Routledge.

Rivard, S., Raymond, L., & Verreault, D. (2006). Resource-based view and competitive strategy; An integrated model of the contribution of information technology to firm performance. *The Journal of Strategic Information Systems*, *156*, 29–50. doi:10.1016/j.jsis.2005.06.003

Robinson, J. P., Shaver, P. R., & Wrightsman, L. S. (1991). *Criteria for scale selection.* In *Measures of Personality and Social Psychological Attitudes*. San Diego: Academic Press.

Rogers, M. E. (1983). *Diffusion of innovations* (3rd ed.). New York: Free Press.

Sapienza, H. J., Smith, K. G., & Gannon, M. J. (1988). Using subjective evaluation of organizational performance in small business research. *Entrepreneurship Theory and Practice*, *1988*(Winter), 45–53.

Sledgianowski, D., & Luftman, J. (2005). IT-business strategic alignment maturity: A case study. *Journal of Cases on Information Technology*, *7*(2), 102–120.

Sougata, R. (2004). Environment-strategy-performance linkages: A study of Indian firms during economic liberalization. *Vikalpa*, *29*(2), 9–23.

Southern, A., & Tilley, F. (2000). Small firms and information and communication technologies: Toward a typology of ICT usage. *New Technology, Work and Employment*, *15*(2), 115–123. doi:10.1111/1468-005X.00070

Storey, D. J. (1994). *Understanding the small business sector*. London: Routledge.

Straub, D. E., Hoffman, D. L., Weber, B. W., & Steinfield, C. (2002). Measuring e-Commerce in net-enabled organizations: An introduction to the special issue. *Information Systems Research*, *13*(2), 115–124. doi:10.1287/isre.13.2.115.86

Strnadl, F. C. (2006). Aligning business and IT: The process driven architecture model. *Information Systems Management*, *2006*(Fall), 67–77. doi:10.1201/1078.10580530/46352.23.4.20060901/95115.9

Teece, D. J., Pisano, G., & Shuen, A. (1997). Dynamic capabilities and strategic management. *Strategic Management Journal*, *18*(7), 509–533. doi:10.1002/(SICI)1097-0266(199708)18:7<509::AID-SMJ882>3.0.CO;2-Z

Venkatraman, N. (1989). Strategic orientation of business enterprises: the construct, dimensionality and measurement. *Management Science*, *35*(8), 942–962. doi:10.1287/mnsc.35.8.942

Venkatraman, N., & Ramanujam, V. (1986). Measurement of business performance in strategy research: A comparison of approaches. *Academy of Management Review*, *11*, 801–814. doi:10.2307/258398

Vivekanandan, K., & Rajendran, R. (2005). Adoption of electronic commerce: An exploratory study of Tirupur knitwear apparel exporters. *Organisational Management*, *21*(3), 5–11.

Vivekanandan, K., & Rajendran, R. (2006). Export marketing and the World Wide Web: perceptions of export barriers among Tirupur knitwear apparel exporters - An empirical analysis. *Journal of Electronic Commerce Research, 7*(1), 27–40.

Wade, M., & Hulland, J. (2004). Review: The resource-based view and information systems research: Review, extension and suggestions for future research. *MIS Quarterly, 28*(1), 107–142.

Weill, P. (1990). Strategic investment in information technology - an empirical study. *Information Age, 12*(3), 141–147.

Wheeler, C. B. (2002). NEBIC: A dynamic capabilities theory for assessing net-enablement. *Information Systems Research, 13*(2), 125–146. doi:10.1287/isre.13.2.125.89

Whinston, B. A., & Geng, X. (2004). Operationalising the essential role of the information technology artifact in information systems research: Gray area, pitfalls, and the importance of strategic ambiguity. *MIS Quarterly, 28*(2), 149–159.

Zhu, K., & Kraemer, L. K. (2002). E-Commerce metrics for net-enhanced organizations: Assessing the value of e-commerce to firm performance in the manufacturing sector. *Information Systems Research, 13*(3), 275–295. doi:10.1287/isre.13.3.275.82

Chapter 18
E-Business and Nigerian Financial Firms Development:
A Review of Key Determinants

Uchenna Cyril Eze
Multimedia University, Malaysia

ABSTRACT

This research discusses Nigerian financial firms' perspectives on key determinants of e-business deployment. It explores possible differences that exist among financial firms using in-house e-business capabilities and those that outsource their e-business capabilities. This chapter contributes to the few pieces of literature on e-business experiences among firms operating in Africa, particularly Nigeria. The Technology-Organization-Environment (TOE) model underpins the conceptual framework for this chapter. The independent variables are the firm, technological and environmental factors while e-business use constitutes the dependent variable. The findings reveal that all the factors were significant, but that environmental factors were key determinants of e-business use among the firms. In addition, this study reveals practitioners' interests in Nigerian government agencies to maintain and enhance the existing e-business legal, regulatory and security frameworks in the country. By extension, this could enable greater e-business use in firms, which could improve the overall economy.

INTRODUCTION

The increasing globalization, enhanced by advances in information system (IS), has spurred national and regional markets integration, international production, distribution, marketing, and consumption. eGlobal report reveals that as the world Internet access grows, the number of active (those who spend at least one hour per week online) users rose to about 640.2 million in 2004 (eGlobal Report, 2005). Similarly, the eGlobal report reveals various forecasts of total world e-commerce revenues by year 2004 ranging from a conservative value of $963 billion to $4 trillion. International Data Corporation (IDC) projects e-commerce revenue by 2004 at $2.8 trillion while Goldman Sachs & Co. estimates $3.48 trillion (IDC, 2002). The report also indicates that 2003-2005 growth of e-commerce will be driven by

DOI: 10.4018/978-1-60566-910-6.ch018

the B2B segment not business-to-consumer (B2C) transactions. According to United State Census Bureau 2009, total e-commerce sales for United States for 2008 were estimated at $133.6 billion, an increase of 4.6 percent from 2007. Total retail sales in 2008 decreased 0.6 percent from 2007. E-commerce sales in 2007 and 2008 accounted for 3.2% and 3.3% of total sales, respectively. Based on the foregoing, this paper examines electronic business use in Nigerian financial firms using the technology-organizational-environment framework (TOE).

Electronic business (e-business) is often regarded as generally focused on e-commerce, however, the true definition is much broader. The Aberdeen Consulting Group defines e-business as the automation of the entire spectrum of interactions between enterprises and their distributed employees, trading partners, suppliers and customers (Intel, 2003). E-business encompasses the application of electronic systems to transform functional processes (Xu, Rohatgi, & Yangxing, 2007). Both definitions include a broad range of business processes such as multi-entity product design collaboration, electronic product marketing and information sharing, e-commerce sales of product to consumers or between firms or governments, internal business process re-engineering, multi-entity supply chain collaboration and customer relationship management. The operational definition of e-business in this paper includes all business transactions firms conduct using open standard [e.g., the Internet] and/or closed standard networks [e.g., Electronic Data Interchange (EDI)].

Revenue generated from e-business support and related services grow at a rapid rate in Nigeria, despite the insecurity in online economic transactions. With the liberalization of the telecommunication sector and the introduction of GSM services, experts predict a boom for e-business activities across Nigeria including Lagos. Lagos is the commercial and economic hub of Nigeria attracting key investments in financial activities and information and communication technology (ICT). As the fastest growing industry in Nigeria, the ICT industry is projected to be the next foreign direct investment driver in the next decades. E-business value was $18.2billion in 2007 up 22.9% from USD14.8billion in 2006. E-business value for 2006 and 2007 were $16.5billion and 17.7billion, respectively (Ujah, 2008). The financial services industry (FSI) is a significant source of development and growth in Lagos. The Lagos corporate sector, especially the financial and oil industries are all expected to experience increased growth in their future EB activities.

Notwithstanding enormous corporate interest and many success stories on e-business, there are few contextual studies on factors that determine its deployment in Africa. In addition, prior studies on e-business focused primarily on the statistics and growth patterns of e-business in terms of usage across industries and countries located in the European Union, United States of America and Asia (Eze and Gilbert, 2004; Goren and Turban, 1996). Similarly, Nigeria, like most African countries, provides little empirical studies that explain the dynamics of e-business deployment in firms (Eze, 2006; Eze and Gilbert, 2008). The goal of this paper, however, is to contribute empirical material to the scanty literature on e-business development in Africa by examining factors that determine e-business use among Nigerian financial firms. This paper will also examine the dynamics of outsourced and in-house e-business applications among Nigerian financial firms. These may provide some bases for additional future researches that might address specific issues in e-business developments among firms from a comparative perspective.

THE CONTEXT FOR E-BUSINESS AND THE FINANCIAL INDUSTRY

Nigeria is located at the south-western coast of West Africa, consists of 36 states, and covers an area of about 923,768 sq km. Nigeria shares borders with Benin, Cameroon, Chad and Niger. Its population, estimated at 130 million (comprising about 250 recognized groups, many divided into subgroups of considerable social, economic and political significance), is sparsely and densely distributed in rural areas and cities (*land*: 910,768 sq km, *water*: 13,000 sq km). Nigeria is slightly more than twice the size of California which makes it extremely difficult to provide universal communications access to its entire population. EB activities between government agencies and private entities have not taken strong roots except in the case of government agencies such as the Central Bank of Nigeria and the private banks including oil companies. There is the potential for growth in this area and evidence suggests that more sectors of the economy are following the examples set by the banking and other financial services firms in deploying e-business innovation.

As noted earlier, Lagos, the specific site for this research, is the industrial and commercial capital of Nigeria, regionally known as the growth pole and center for local/international financial institutions and IT headquarters (Ujah, 2004). Lagos has a population of 13.6 million and is the most important city in Nigeria because of its strategic position in the country's development process. e-business development in Lagos is driven mainly by the private sector. Firms, particularly banks, oil, and telecommunication players are very active in e-business operations supported by quasi-governmental agencies such as the National Communication Commission and the Nigeria Internet Group. The government of Nigeria is gradually developing a solid e-business infrastructure to enhance operational activities among business partners. Over 75% of manufacturing and trading firms in Lagos use Electronic Data Interchange (EDI) as their key operating platform. With the rapid diffusion of the Internet, greater numbers of firms in Nigeria are moving their operations from the proprietary networks to the World Wide Web.

The nature of IS in Nigerian FSI is complex and diverse. On the front end, firms use IS to execute and record customer transactions, whether they are handled in person, by phone, by electronic funds transfer, or on the Internet. On the back end, funds are transferred among firms and individuals via electronic transfer system, such as Swift and giro, which handle hundreds of trillions of dollars in transactions yearly. Because of the less sophisticated Internet systems in Nigeria, most financial firms use EDI and EDI over the Internet as their main platform for e-business activities.

The FSI already has benefited from network-enhanced connectivity and interactivity for more efficient information exchange (Fan et al., 2000) and has experienced improvement in operating efficiencies and reduction in operating costs. Clemons and Hitt (2001) demonstrated significant cost savings, owing to electronic business, in product origination processes and operational activities, such as billing and document processing, for a variety of sectors ranging from brokerage, mortgage, and insurance to credit cards. Deploying e-business technologies and practices, however, has been gradual, limited by security's network structure and international legal, cultural and tax concerns (Olazabal, 2002). A case in point is Nigeria, where online security related issues are not only a constant challenge but also a growing risk to the growth and advancement of the financial industry.

E-business systems still have the potential to add significant value to all key functional areas in Nigerian financial firms, especially the potential for integrated Web-based applications to improve customer services. Loan applications and insurance forms can be filed out, stock trades initiated, bills, and fund transferred on-line. Customers can now access research tools such as mortgage calculators or retirement planning applications, credit status, and account information online. Applications based on open Internet standard can enable data sharing across firms in this industry. Internally, e-business applications can also improve integration of various proprietary systems. The challenge, however, is for the Nigerian government to strengthen its social, economic, technical and legal infrastructure to enable accelerated growth of e-business, not only in the FSI but also in key sectors of the economy.

CONCEPTUAL FRAMEWORK

The TOE framework, developed by Tornatzky and Fleischer (1990), identifies three aspects of an enterprise that influence the process by which the enterprise implements an innovation. These aspects are technology, organization and environment. The technology factors describe the external and internal technology relevant to the firm, which includes existing technologies in the firm and those technologies available in the market. Organization factors define the size, scope and the management structure: the quality of human resources, risk taking ability, management policies and support and available slack resources. Environment factors define conditions in which the firm conducts its business, which includes competitors, customers, suppliers, and the government. These three factors influence innovation deployment process and consequently the performance of the firm. Although a firm has its contextual strategic approaches to address its environment, identifying the prevalent factors, which firms consider when making investment decisions on an innovation, is very important (Zhu et al., 2004, Teo and Pian, 2003; Zhu et al., 2002; Kendall et al, 2001). The context and TOE factors firms consider when deploying innovations vary depending on their specific needs, internal and external constraints, and the objective of the respective firms.

Applying TOE framework in IS research underscores the significance of matching information-processing requirements, often determined by environmental and firm factors, with information processing capabilities of the firm. Researchers apply this framework extensively in various areas like economics, finance, and planning (Eccles and Crane, 1988), theorizing new emerging network firms and innovation deployment. Grover (1993), for example, derived empirically a model of five factors (firm policy, technological, environmental, inter-firm and support factors), which influence a firm's decision to deploy customer-based inter-organizational systems. Weill and Olson (1989), on the other hand, identified seven useful factors (environment, technology, task, individual, strategy, structure, and size) in their meta-analysis of contingency research in business information systems. In addition, King and Sabherwal (1992) studied variables in the environmental, firm, and the IS contexts. Several other researchers including Rogers (1995) identified few factors that influence the adoption of innovation including. Although, some factors identified within the TOE framework may vary across studies and research samples, the framework has consistent empirical support. Building on the empirical evidence and literature review, TOE framework is considered appropriate for this study, and this paper extends the theoretical framework to e-business practices in a region less researched, Nigeria.

The review in this research indicates that existing literature focused mainly on technology adoption (Zhu et al. 2004). Only a few researches have directly examined how TOE factors affect the ability of firms to derive value from e-business innovation (Zhu et al., 2002, 2004; Ramamurthy et al., 1999; Iacovou et al., 1995). Zhu et al (2004) argue that the impact of e-business on firm performance is a consequence of technological, organizational, and environmental factors. Their results were significantly affected by these factors. Based on the foregoing, this research uses the TOE framework in determine factors that may influence e-business use and performance among Nigerian financial firms.

This research model (see Figure 1) depicts nine factors within three contexts of the TOE framework: competitive intensity, regulatory

Figure 1. The Conceptual Model of Key Determinants of E-business

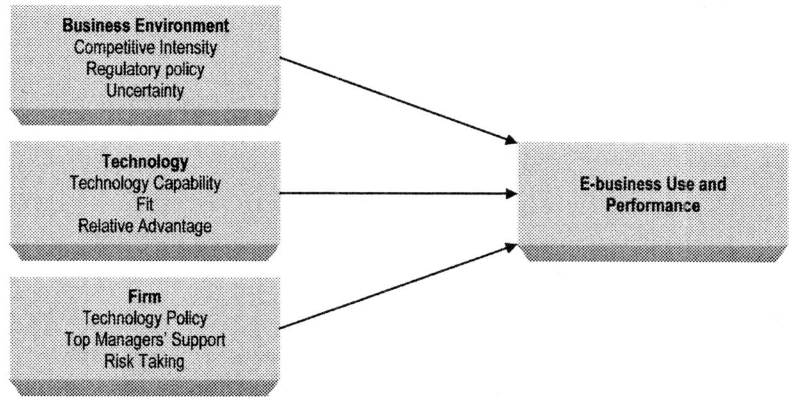

policy, uncertainty, technology capability, fit, relative advantage, technology policy, top managers' support, and risk taking. These factors are proposed to influence e-business use and performance among financial firms (Zhu et al. 2004; Kohli and Devaraj, 2003). As indicated earlier, literature identifies several factors that could influence technology implementation. The choice of the specific factors for this study, particularly the environment factors, was informed by the contextual and innovation deployment dynamics specific to Nigeria, the industry and the region. Nigerian political and economic conditions are quite different from those of the developed societies such as the United States, Canada, United Kingdom, Finland, Australia and Singapore where e-business frameworks are much stronger, developed and are of strong international standard. The next section describes the TOE factors in this paper as they pertain to the Nigerian financial services firms including the hypotheses development.

BUSINESS ENVIRONMENT FACTORS

Environmental factors entail those that create opportunities as well as threats to a firm. These factors are typically beyond the control of management, where the firm has to adjust to adapt to any change in the environment. This paper identifies three environmental factors that could influence e-business use and performance: the intensity of competition in the industry; regulatory policy; and uncertainty in the environment.

Competitive Intensity: Researchers perceive competitive intensity as an important factor in the innovation adoption literature (Grover, 1993, Iacovou at el. 1995). It refers to the extent that competitors in the marketplace affect the firm. Porter and Miller (1985) analyzed the strategic rationale underlying the relationship between competition intensity and IS innovations. A business environment characterized by an increase in competition fuels the tendency for IS deployment in firms (Zhu et al., 2004, Zhu et al., 2002; Utterback and Abernathy, 1975; Rogers, 1995). Dependence on IS can evolve with time as firms use diverse IS to differentiate products, create switching costs to retain suppliers and customers, impose barriers to market entry, create competitive advantage, or counter other firms' competitive advantage, and develop new business models (Porter, 2001, 1985). These systems can generate cost reductions, support competitive positioning, and allow the firm to compete in new and dynamics fashions. As firms become more dependent on them to process daily activities, they become critical to the success and survival of the firm (Raven, Huang, & Kim, 2007).

For Nigerian financial firms to be competitive in the region, it is imperative they venture into emerging economies such as Southern Africa, China, and India, for new markets, because international experience and knowledge would create core competencies for development, expansion and growth for Nigerian financial services firms. Nonetheless, the costs and risks firms encounter in foreign investments are extremely high. Therefore, using e-business to reach, coordinate, and integrate those markets would save enormous resources. In this regard, the e-business offers a critical low-cost option for a firm to reach wider markets, collaborate with partners and possibly remain competitive in, at least, the region. Hence, the hypothesis:

H1: The greater the competition intensity in the industry, the more likely a firm will deploy e-business.

Regulatory policy is recognized as a critical country factor affecting innovation use and performance. For example, Williamson (1983) summarized two ways in which government regulation could affect innovation deployment: (1) by taking specific actions to increase or decrease payoffs via tax and other measures to encourage research and development (R&D), (2) by taking actions that influence innovations via altering the climate in which they are received. Accordingly, Nigerian government could help financial firms to implement strong e-business by regulating the Internet to make it more a trustworthy business platform and establishing supportive business and legal frameworks to protect e-business activities. Recent empirical studies indicate that effective regulatory environment plays an important function in e-business performance (Kraemer et al., 2002). Given that financial transactions often involve very sensitive information, adequate regulatory and institutional frameworks are very critical for the Nigerian FSI. Sato et al. (2001) posited that regulatory environments, together with communication platform, are supporting infrastructures critical for e-business, highlighting the importance of regulatory environment for efficient e-business use and performance in the financial services industry. These theoretical assertions and empirical evidence leads us to the next hypothesis:

H2: *An enabling regulatory environment will positively influence the capacity of a firm to use* e-business.

Uncertainty describes the condition in which reasonable knowledge regarding risks, benefits, or the future is not available. It is a measure of how poorly we understand or can predict something such as a parameter or future behavior; in some cases, this is the same as a lack of precision. It also entails failure to know anything that may be relevant for an economic decision, such as future variables, details of a technology, or sales. Uncertainty creates the need for greater innovation and product differentiation requiring a higher level of dependence on IS resources (Sabherwal and King, 1989). Uncertainty places time constraints upon decision-making and forces firms to invest in technology for coping mechanisms. Decision-makers, for example, may turn to more sophisticated information analysis such as group collaborative systems, data mining and knowledge management systems (Holsapple and Whinston, 1996). As firms deploy more IS to support core activities and improve performance in unstable business environment, IS dependence increases. Extending this argument to e-business indicates that the rate of change in global business activities increases with every improvement in technology innovation. This paper, therefore, posits that e-business will enable Nigerian financial services firms to obtain more and richer information at faster pace compared to, say, 20-30 years ago. E-business also tends to enable better functional area visibility and better access to customers and suppliers. Experience from practitioners indicates that firms are able to manage the level of uncertainty in the business environment when they deploy e-business systems for coordination and collaborative business processes improvement. Hence the following hypothesis:

H3: *High uncertainty in the environment will positively influence* e-business *use and performance among firms.*

Technology Factors

Technological factors delineate the perceived features of the innovation. Teo et al. (1997) and Tornatzky and Klein (1982) observed that in addition to the many perceived innovation features in prior IS research, technology capability, fit, and relative advantage tend to predict innovation deployment.

Technology Capability IS literature indicates that IS capability comprises infrastructure, human resources, and knowledge (Zhu et al. (2005); Mata et al. 1995). Consistent with past research, technology capability in this paper consists of technology infrastructure and IS human resources, where technology infrastructure refers to technologies that enable Internet-related businesses (e.g., EDI, EFT, intranet, extranet), and IT human resources refer to IT professionals with the knowledge and skills to implement Internet-related applications. By this definition, technology capability is conceptualized as an integrative construct that is reflected not by the physical assets but also by human resources that are complementary to physical assets (Zhu and Kraemer, 2005; Mata et al. 1995). It refers to a firm's technical capability in the form of distributed computing power in the firm. Technology infrastructure establishes a platform on which e-business can be built; IT human resources provide knowledge and skills to develop e-business applications. Therefore, Nigerian financial firms with a higher degree of technology capability would enjoy greater capacity and readiness to use e-business in their business processes. As a result, they would be more likely to achieve a greater depth in e-business use. Hence:

H4: A firm would use e-business if it [the firm] has a strong technology capability.

Fit entails the agreement between an innovation and a firm's culture, experiences and potential needs (Rogers, 1995). It is common for a firm to deploy innovations that align closely with its specific needs, values and the business processes, which are expected to meet specific requirements of future technological developments in the industry and those that the firm deploys (Ettlie, 1983). The significance of the aforementioned approach include to ensure few changes to the firm's structure, processes and strategy, thereby enabling less resistance to e-business deployment. Further, the compatibility of new technology enables greater corporate security and minimal risk to the deploying firm, which may in turn save the firm enormous amount of investment resources for other business initiatives.

The agreement of a technology with the tasks and processes of potential users appear to be the widely researched facet of fit between technology and the firm (Tornatzky and Klein, 1982). This is because in Nigerian, as it is in some countries, firms encounter change resistance irrespective of the size, type of business and location of the firm, and therefore, would deploy an innovation when there is greater possibility of fit between the innovation and the firm culture, structure, existing processes and IS infrastructure. In contrast, Kwon and Zmud (1987) observe that inconsistency of new IS initiatives with existing business processes and IS infrastructure adversely affect users' attitudes and increases change resistance, which may impede innovation deployment. Hence the following hypothesis:

H5: The more fit e-business is with a firm's culture and existing IS infrastructure, the more likely the firm will use the innovation.

Relative Advantage: Relative advantage is the level at which potential customers observe the innovation as superior to existing alternatives. Kwon and Zmud (1987) define relative advantage as the degree to which innovation deployment provides greater firm advantage than to maintain the status quo. In essence, it describes the intensity of the reward or penalty of deploying or rejecting an innovation. The degree of relative advantage is usually denoted in economic terms: savings in time and effort, costs reduction, or better reputation in innovation application. Researchers observe relative advantage to be among key predictors of an innovation's rate of use (Rogers, 1995). Grover (1993) posits that, generally, a positive relationship exists between relative advantages of an innovation and the behavior of firms deploying new technologies. E-business gives firms competitive edge in business activities such as publicity, marketing/advertising; direct selling; research and development; and global marketing, distribution, production, communication with vendors and customers (Cockburn and Wilson, 1996). The celebrated potential benefits of using e-business include creating a worldwide clientele base, rapid information access, global scale information dissemination, cost-effective document transfer, global market reach, and groundbreaking business models (Rawn, 1994). Nigerian financial firms will derive greater value from conducting business transaction via e-business systems compared to using traditional approaches, such as the telephone. Hence, the following hypothesis:

H6: A firm will likely use EB if it [the firm] perceives greater relative advantage accruing from the innovation.

Firm Factors

Firm factors include those variables that affect the firm structure, and which the firm could adjust or change to fit its changing environmental needs. Three factors this research discusses include technology policy, top managers' support and risk taking.

Technology Policy: Ideally, technology policy is a type of long-range environmental scanning and coping mechanism, and entails actions a firm takes to outperform other firms on technology (Ettlie, 1983). These actions may involve top management's willingness to commit time, effort and financial resources to IS projects, an unflagging pursuit of new technological innovations and development in the market as well as close monitoring of competitors' e-business deployment performance and positions. This is even more critical in the financial industries in developing countries like Nigeria, where huge gaps still exist between firms' policy initiatives and policy execution.

Ettlie and Bridges (1982) observed that perceived environmental uncertainty promotes an aggressive technology policy (a long-range strategy for the deployment and production of process, product and service innovations), and that such a policy may be a means of coping with dynamic conditions. Ettlie (1983) observed that firms with aggressive technology policies constantly experiment with available processing systems; actively recruiting the best possible technical, production and marketing personnel, and often showing strong commitment to technological forecasting. This suggests that these firms tend to have specialized groups who evaluate new process innovations, which may lead to the implementation of key initiatives. Therefore, firms with aggressive technology policies would consider using e-business in response to the technological trends and market conditions; consistent with arguments on the contexts of information systems management. Thus, this paper proposes:

H7: The more aggressive a firm's technology policy is, the more it will use e-business.

Top Managers' Support: Lederer and Mendelow (1988) observed that top management influence and commitment during the assessment of an innovation and the deployment of any IS initiative is often significant. Top management support for innovation initiatives tends to enable a firm's commitment in investment needs and accelerate project-planning processes (Teo, et al., 1997). Without strong top managers' support for an innovation, it becomes difficult to secure the interest and motivation of the Chief Executive Officer (CEO), which may lead to possible rejection of the initiative. Top management influence is also vital to overcome corporate barriers and resistance to innovations and corporate changes in Nigerian financial firms, which will enable efficient overall change management.

In addition, the global connectivity and accessibility of e-business permits easy exchange of ideas and knowledge across firms, industrial and geographical boundaries. A financial firm in Nigeria deploying e-business would have to ensure that the system is used to maximum corporate advantage, particularly, in boosting productivity, creating and sustaining business relationships and in building new business models (Eze and Kam, 2001). To achieve these, top management influence is crucial. Hence, the hypothesis:

H8: The greater the extent of top managers' support, the more likely the firm will use e-business

Risk Taking entails a firm's willingness and ability to take risks associated with business including changes in the firm's structure, processes, and strategy (Eze and Kam, 2001). When a firm deploys e-business, it faces complex and dynamic changes in communication culture and reporting patterns (Lederer and Mendelow, 1988). Developing e-business capabilities means that the existing communication hierarchies in the firm may become irrelevant. With e-business, the demarcated walls of knowledge tend to dissipate, allowing information within the firm to become more easily available and accessible. This new pattern of communication and relationships may affect the conventions and culture of the firm in distinct fashions.

Management of risk delineates management's willingness to accept novel innovations. Commercial use of Internet-related e-business technologies dates back to over fifteen years ago, yet some firms still struggle with linking B2B applications to strategic significance that could enable competitive advantages (Grover and Lederer, 2004). In the context of Nigeria, a related issue is that the relevant government agencies have not fully addressed the issue of transaction security and regional standardization, which inhibits the potential corporate benefits accruing from e-business. In addition, if e-business is readily accessible, firms must also take into account privacy and intellectual property rights of firms' assets, service and product information.

Financial risk entails firm top executives' willingness to commit adequate resources into new IS projects. Although the cost of e-business deployment may be significant, a major concern for firms is the accompanying costs of training and educating firm employees in the application of e-business (Teo, et al., 1997). In the context Nigerian financial firms, a related concern is productivity; if employees are hooked on improper use of the systems, then expensive firm time and other vital assets may be used for less productive ends. Therefore, if top management is willing to accept firm restructuring, deploy novel technologies and invest resources in IS applications then the firm may be likely to use e-business. Hence the following hypothesis:

H9: The higher the risk-taking propensity of top management, the more likely the firm will use e-business.

METHODOLOGY

The data collection for this paper was conducted using survey questionnaire. The questionnaire was designed based on the research conceptual model (see Figure 1). Previous empirical studies on innovation deployment and diffusion provide key reference for the research variables and items (Zhu et al., 2005; Eze, 2003; Zhu et al., 2002; Chau and Hui, 2001; Teo, et al., 1997). Table 1 illustrates the operationalization of each independent variable, the items and their sources. The independent variable, EB use and performance was operationalized by asking the respondents to assess the improvement in their EB use and performance. Responses to the survey questions on the variables were entered on a five-point Likert-type scale as follows: **1** = Strongly Disagree, **2** = Disagree, **3** = Neutral, **4** = Agree, and **5** = Strongly Agree. Respondents were also requested to identify whether they were users with in-house EB systems, users with outsourced e-business systems or non- e-business users. In addition, the survey questionnaire includes data on participants' profile including firm type [commercial banks, finance houses, merchant banks, and insurance firms], number of employees and annual revenues.

Several steps were taken to ensure data validity and reliability. Initially, the questionnaire was pre-tested with two academics and two business executives selected from the IS industry. The questionnaire was then revised for any potentially confusing items. Before the administration of the final survey, a random subset of 10 respondents was used for pilot testing to determine further areas in the questionnaire that may need improvement. Six questionnaires were returned, and their recommendations for improvement and thus the content validity of the questionnaire were incorporated into the survey questionnaire. As noted earlier this research benefited from previously validated measurements items (see above) to help ensure the validity of the measures; multiple-item measures were used for most constructs to enhance content coverage.

For purposes of reliability, the CIO is considered most appropriate position to respond to survey questions. Directing questions to the most knowledgeable respondent often helps to overcome single source limitations (Premkumar and King 1992). The questionnaires were then mailed to CIOs of the 400 firms from the Nigerian financial industry. In the cover letter, the author assured confidentiality and offered an executive summary of the findings as an incentive for participation. Pre-paid reply envelopes were provided for the return of the responses. Follow-up calls were made to the participants who had not replied to the survey after the specified period.

To establish the absence of non-response bias, it is desirable to collect data from a set of non-respondents and compare it to data supplied willingly. For a meaningful number of surveys and for all survey items, this method is rarely achievable. A practical preference, that has been argued to provide reliable results, is to compare the mean values of responses for earlier returns with the means from later returns (Compeau 1995). This approach has the capacity to reveal any differences between early and late responders who required prompting. The assumption is that late responders share similarities with non-responders, and if no significant differences exist, the probability is strong that non-response bias does not exist (Armstrong and Overton 1977). Tests were conducted on first week responses and responses after five weeks, and the differences between the two groups were insignificant [two-tailed *t*-test $P<0.05$], indicating that time had no apparent effect on the perceptions and that non-response bias was remote.

ANALYSIS

There were 148 completed responses in this research, producing a 37% response rate. The

Table 1. Variable Operationalization

Variable	Items	Descriptions	Sources
Environmental Variables			
Competitive Intensity (CI)	V1 V2 V3	There is stiff competition based on costs; There is stiff competition based on product quality and novelty; There is increasing number of competitors in the industry.	Grover (1993); Utterbach & Abernathy (1975). Zhu et al. (2002)
Regulatory Policy (RP)	V4 V5 V6	Government provides incentive Legal protection for online purchases Required for government businesses	Eze (2006) Teo et al. (1997).
Uncertainty (U)	V7 V8 V9	There is enormous diversity in customers' buying habits There is enormous diversity in the nature of competition There is enormous diversity in product designs	Toh & Low (1993); Teo et al. (1997).
Technological Variables			
Technology Capability (TC)	V10 V11 V12	There are adequate IT professionals with requisite skills in the firm There are functional ICTs in the firm There are Internet and W3 capability in the firm	Mata et al., (1995)Zhu and Kraemer (2005)
Fit (F)	V13 V14 V15	E-business use consistent with firm's culture; Favorable attitudes towards e-business deployment present in the firm; E-business use consistent with firm's IT infrastructure	Reich & Benbasat (1990); Rogers (1995). Kendall et al. (2001)
Relative Advantage (RA)	V16 V17 V18 V19 V20 V21 V22	Convenient access to worldwide information; Creation of worldwide electronic presence; Selling and buying products through e-business; Extend global market reach; New business models and opportunities; Improve customer service; Lower operational costs using e-business	Cockburn & Wilson (1996); Rawn (1994); Rogers (1995). Kendall et al. (2001
Firm Variables			
Technology Policy (TP)	V23 V24 V25 V26	Firm's tradition of being the first to try out new technologies; Firm's expenditure on developing new products as compared to competitors; Firm's recruitment of technical personnel; Firm's awareness of the latest technological developments	Teo and Pian, 2003 Zmud (1984); Eze(2006)
Top Mangers' Support (TMS)	V27 V28 V29	Top management interested in e-business use; Top management considers e-business use as important; Top management has shown support for e-business use.	Grover (1993), Premkumar & Ramamurthy (1995) Teo et al. (1997).
Risk Taking (RT)	V30 V31 V32	Top management's willingness to accept organizational changes; Top management's willingness to accept unfamiliar technologies; Top management's willingness to invest funds in IT	Clemons & McFarlan (1986); Teo et al. (1997).

profile of participants is shown in Table 2. About 83% of the participants are CIOs and they have an average working experience of 6.5 years in their firms and 12.5 years in their respective industries. More than 65 percent of the participants are large, with more than 600 employees and average annual revenue ranging from $101 million to more than $500 million.

N = 148

Out of 148 firms, 67.6% have sophisticated e-business systems while 32.4% do not possess such e-business systems. Over 80% of the participants have been using e-business systems for about 8 years [This includes firms that use EDI and EDI over the Internet for business transactions]. Those firms with e-business systems are regarded as e-

Table 2. Participants Profile

Items	*Value*	*Percent of Respondents*
Number of Employees	***>600***	***65.76***
Annual Revenue	***101- 500 and above***	***65.76***
Duration of EC Use	***8 year and above***	***>80***
Annual Expenditure on EC	***6% and above***	***57.56***

business users while those without the systems are regarded as non-users. Out of the 100 users, 57% have in-house e-business solutions while the remaining 43% outsourced theirs. Among the non-users, about 69% intend to deploy e-business within the next one year.

RESULTS: MODEL GOODNESS-OF-FIT

There is no universally accepted indicator of goodness-of-fit, so it is usual to present several statistics as collective indicators. The Chi Square (χ^2) statistic is a fundamental measure but because it is sensitive to sample size, it is advisable to complement this measure with other measures of fit. The indicators chosen are ones used in previous IS research. These are the Non-Normed Fit Index (NNFI), the Comparative Fit Index (CFI), the Root Mean Square-Error of Approximation (RMSEA), the Root Mean Square Residual (RMSR), and the Tucker-Lewis Index (TLI). The χ^2 statistic, also a measure fit, is subject to distortion and is often replaced with the ratio of χ^2/d.f. Indexes that exceed 0.09 are acceptable for the NNFL, CFI, and TLI. Values of less than 0.06 for RMSEA and less than 0.10 for the RMSR are acceptable. The preferred value for the χ^2/d.f. ratio is below 2 (Nunnally, 1979). The probability for the χ^2 for the final measurement model was marginal ($P=0.05$). Nonetheless, all other goodness-of-fit measures for the final measurement model suggest a strong fit of the study data to the proposed model. Since the ratio of χ^2 to d.f. was low (1.47) and all other indexes were well within the prescribed range, the results are accepted as supporting overall goodness-of-fit (see Table 3).

CONSTRUCT VALIDITY AND RELIABILITY

To assess the extent to which a particular empirical indicator represents a given theoretical concept, it is pertinent to assess the validity and reliability

Table 3. Goodness-of-Fit for the Final Measurement Model

Item	**Suggested Range**	**Measurement Model Value**
χ^2	$P>0.05$	P =0.05
χ^2/d.f.	<2.00	1.47
Non-Normed Fit Index	>0.90	0.99
Comparative Fit Index	>0.90	0.96
Tucker-Lewis Index	>0.90	0.98
RMSEA	<0.06	0.04
RMSR	<0.10	0.07

of the indicators. Therefore, to ensure rigorous research protocol, construct validity and reliability assessments were conducted (see Table 4).

Construct validity focuses on the extent to which a specific measure relates to other measures in accordance with theoretically derived propositions. Construct validity may be determined in terms of convergent and discriminant validities. Convergent validity assesses the consistency across multiple operationalizations. While convergent validity assesses whether all the items measuring a construct cluster together to form a single construct, discriminant validity measures the extent to which a construct diverge from other

Table 4. Rotated Factor Matrix of the Variables

Items	Factor 1	Factor 2	Factor 3
Organizational Variables			
V1	0.2543	**0.7032**	0.2165
V2	0.2465	**0.7327**	0.2843
V3	0.1762	**0.8320**	0.0903
V4	**0.8241**	0.2120	0.1968
V5	**0.9043**	0.3198	0.1023
V6	**0.9397**	0.2921	0.1760
V7	-0.3263	0.1094	**0.7948**
V8	0.1202	0.2761	**0.8012**
V9	-0.1234	0.1032	**0.7123**
Eigenvalue	3.7	2.1	1.1
Variance (%)	32.6	16.9	13.9
Cronbach alpha	0.89	0.79	0.72
Technological Variables			
V10	0.1342	**0.7328**	0.1283
V11	0.1245	**0.8123**	0.3425
V12	0.1912	**0.7911**	0.0023
V13	**0.8453**	0.0766	0.4322
V14	**0.6124**	0.1981	0.1553
V15	**0.7078**	0.1678	-0.1122
V16	**0.8192**	0.1232	-0.3222
V17	**0.8212**	0.2154	0.2553
V18	**0.8872**	0.3456	0.5442
V19	**0.7834**	0.2198	0.1253
V20	0.3031	0.3112	**0.7322**
V21	-0.2667	0.3442	**0.8390**
V22	0.1334	-0.1425	**0.7890**
Eigenvalue	4.3	2.3	3.1
Variance (%)	41.1	17.1	21.7
Cronbach alpha	0.82	0.78	0.83
Environmental Variables			
V23	0.2342	0.2198	**0.8911**
V24	0.1211	0.1875	**0.7881**
V25	-0.2122	0.0117	**0.7119**
V26	0.2218	0.2865	**0.7295**
V27	**0.7932**	-0.0043	0.2111
V28	**0.8333**	-0.1004	0.1003
V29	**0.9133**	0.3110	-0.1212
V30	0.0111	**0.9001**	-0.3110
V31	0.2122	**0.8778**	0.3004
V32	-0.2111	**0.8279**	0.1232
Eigenvalue	3.5	2.4	1.9
Variance (%)	19.5	18.5	14.9
Cronbach alpha	0.75	0.81	0.87

constructs, and is indicated by a measure having low correlation with other measures from which it should theoretically differ (Gerbing and Anderson, 1988). Factor analysis can be used to evaluate both convergent and discriminant validities.

Considering the sample size and the number of items in this paper, individual factor analyses were performed on the firm, technological, and environmental factors to ensure that the ratio of variable items to the sample size is maintained at 1:10. This rule-of-thumb suggested by Kerlinger (1986) has also been used previously (Grover, 1993; King and Sabherwal, 1992; Premkumar and Ramamurthy, 1995; Teo at el, 1997). Separate factor analysis was performed to ensure stability of the factor loadings of the research constructs. During factor analysis, items were retained according to the following criteria: (i) factor loadings greater than 0.5 and (ii) no cross-loading of items. In other words, items were dropped where they have a loading of less than 0.5 or where their loadings are greater than 0.5 on two or more factors (King and Teo, 1996).

The results of two rounds of factor analyses for the firm, technological and environmental variables are shown in Table 4. Both convergent and discriminant validities were satisfied as the items measuring each factor clustered together to form distinct factors and there were cross-loading of items. With respect to reliability, Cronbach alpha was computed for each variable. Nunnally (1978) suggested that average reliability scores of 0.7 are sufficient for basic research. From Table 4, we see that the Cronbach alpha scores range from 0.68 to 0.89. Since the lowest score is fairly above 0.70, all the constructs are considered to have adequate reliability.

Table 5 illustrates the correlation matrix for the factors. The table indicates that the factors – top managers' support (TMS) and fit (F) relatively correlated highly. A probable explanation is that it is unlikely that top management would support the adoption of EB, if its' use is considered inconsistent with the firm's culture, business processes and existing IS infrastructure.

In testing the hypotheses, multiple regression analysis was conducted using hierarchical regression procedure. The first regression performed was on firm and e-business value variables, and the second regression was on environment and e-business variables. The level of significance to enter the framework was set at 0.20 and the level of significance to stay in the framework was set at 0.10. Being a directional hypothesis, these values are equal to the one-tailed test with values of 0.10 and 0.05, respectively. The research data supported all the hypotheses (see Table 6). However, the strengths of the coefficient of determination vary among the factors. The data supported H1 indicating that industry features characterized by competitive intensity tend to enable firms to generate value from e-business, which confirms Grover and Lederer (2004) findings. Also, the data supported H2, which indicates that efficient regulatory environment would enable more possibilities for firms to generate greater value from e-business. The findings also supported H3, suggesting that environmental uncertainty tend to enable greater e-business and performance.

The research indicates that firms operating strong technology policy have the tendency to use e-business for greater business performance. This result underscores the importance of a solid framework for IS systems planning in firms to compete effectively in this 21st century. H8 was supported as well by the analysis and suggests that, as in most business decisions, top manager's support for the deployment and use of e-business is very important to enable firms derive the best performance from an e-business initiative. This is consistent with the findings of Zhu et al. (2004). The data also supported H9, indicating that for firms that are proactive and less risk averse, the chance of deriving greater advantage from e-business use and performance is would be very high (see Eze and Kam, 2001).

Table 5. Correlation Matrix

SOURCE	CI	RP	U	TC	F	RA	TP	TMS	RT
CI	1.0000								
RP	0.0878	1.0000							
U	0.6734	0.1100	1.0000						
TC	0.5678	0.2295*	0.0924	1.0000					
F	0.0899	0.1119	0.2231	0.6233**	1.0000				
RA	0.1100	0.1299	0.0310	0.4320**	0.3782	1.0000			
TP	0.3991	0.0098	0.1269	0.7560**	0.6512**	0.3782	1.0000		
TMS	0.0764	0.1287	0.0876	0.4829**	0.8611**	0.4102	0.7600**	1.0000	
RT	0.0986	0.1199	0.0564	0.5673**	0.0328	0.0437	0.5253**	0.0277	1.0000

*p < 0.05; **p < 0.01

The data support for H9 was stronger compared to the H7 and H8.

ANOVA for EB Use Behavior

Table 7 presents the results of one-way Analysis of Variance (ANOVA) used to compare the three categories of participants along the 9 factors. Comparisons were made between three groups of firms (users with in-house e-business systems, users with outsourced e-business systems, and non- users).

Post-hoc analysis using Turkey procedure was carried out to determine which groups are significantly different from the others. Based on Table 7, the analysis indicates significant difference with respect to the respective factors with the exception of *regulatory policy and uncertainty*. This result suggests that environment factors tend to affect equally the financial services firms in Nigeria [discounting the surprise result for *competitive intensity*], irrespective of the category of e-business participants.

DISCUSSIONS

The results reveal significant difference among the three groups in terms of *competitive intensity*. All three groups perceive that they operate in

Table 6. Results of Regression Analysis

Hypotheses	Standard Error	E-business Use Coefficient of Determination
H1. Competitive Intensity	0.13	0.32**
H2. Regulatory Policy	0.17	0.36**
H3. Uncertainty	0.21	0.43**
H4. Technology Capability	0.16	0.35**
H5. Fit	0.11	0.23 *
H6. Relative Advantage	0.13	0.27*
H7. Technology Policy	0.11	0.25 *
H8. Top Managers' Support	0.17	0.24 *
H9. Risk Taking	0.09	0.42**

p<0.05; ***p*<0.01

Table 7. Testing

Factors	Users with In-house Systems Mean (S.D.)	Users with Outsourced Systems Mean (S.D.)	Non- Users Mean (S.D.)	F-Ratio	Tukey Test
CI	4.09 (0.61)	3.93 (0.68)	2.67 (0.57)	12.87**	[1]>>[2]>>[3]
RP	3.87 (0.62)	3.91 (0.72)	3.74 (0.83)	1.66	Not significant
U	2.99 (0.87)	2.84 (0.91)	2.81 (0.87)	1.52	Not significant
TC	3.98 (0.63)	3.67 (0.61)	2.76 (0.58)	11.37**	[1]>>[2]>>[3]
F	3.91 (0.65)	3.71 (0.59)	3.32 (0.66)	16.36**	[1]>>[2]>>[3]
RA	4.11 (0.51)	3.79 (0.63)	3.69 (0.77)	4.54*	[1]>>[2]>>[3]
TP	3.63 (0.54)	3.37 (0.59)	2.91 (0.73)	12.33**	[1]>>[2]>>[3]
TMS	3.87 (0.71)	3.70 (0.78)	3.21 (0.71)	14.65**	[1]>>[2]>>[3]
RT	3.34 (0.49)	3.11 (0.62)	3.13 (0.80)	7.91**	[1]>>[2]>>[3]

*$p<0.05$; **$p<0.01$

relatively different competitive sectors. Hence, the analysis rejects H7. This result is consistent with the findings of King et al. (1989), who found that pressure from competition is an important facilitator for a firm to deploy strategic information systems. However, Thong and Yap (1995) and Teo at el. (1997) observe competitiveness of the environment to be insignificant in the decision to deploy IS. This paper establishes that firms deploying B2B e-commerce do so, partly, because of the business environment they operate in, since both users and non-users of e-business may have different strategies and orientation regarding their business environments. Hence, the competitive intensity in an environment is a key determinant for firms deploying e-business. Interviewing a bank manager in Lagos, he noted, "...the competition among banks in Lagos is getting more intense and every bank wants to be perceived as the best customer-centric company and have invested heavily in e-business systems to provide quality services to customers..."

Regulatory policy is a significant determinant for e-business use among financial services firms in Nigeria, but it is not a significant dedifferentiating factor among e-business users and non-users. The relevant authorities in Nigeria such as the Nigeria Internet Group and the Information and Communication Technology Agency, should work very closely with the industries to develop comprehensive information technology frameworks that could support both national and global e-business activities. This could be done by using benchmarks from, say, Singapore, United States, Hong Kong and Finland. Similarly, the results indicate that *uncertainty* is a significant factor in firms' e-business use, which illustrates the growing necessity for financial firms to employ electronic systems to enable business operations in unstable emerging markets, such as Nigeria. However, uncertainty was not a significant factor determining differences among financial firms use of e-business. This may be due to the relatively stable political structure and governance in the country for the past 10 years following the inception of civilian government in Nigeria.

Technology capability emerged as a significant determinant of e-business deployment and performance in the firms. This indicates the prominent position advances in electronic technologies takes in business dealings. New e-business applications such as the Internet and wireless applications are driving significant changes in Nigerian business landscape and direction for future business opportunities. Technology capability also emerged as a differentiating factor among the business groups. The stronger and advanced the technology used by the firms the greater the tendency for better performance. According to an investment banker in Lagos, "…there are wider ranges of financial services in Nigeria today compared to 10 years ago, and this is made possible because of the growing impact of Internet and related technologies on the capacity of banks to provide e-business –enabled services..."

The results depict that *fit* is a strong determinant of e-business system deployment. Post-hoc analysis indicates significant differences among the three groups. Users with in-house e-business view e-business as an innovation more consistent with their firms' culture and information system infrastructures, compared with users with outsourced e-business. Similarly, users without in-house solutions perceive e-business to be more consistent with their firms' culture and information system infrastructure compared to non-users. Therefore, the analysis supports H4. This result reveals the significance of fit between e-business and the firm, which is consistent with the findings of Tornatzky and Klein (1982), that compatibility of a technology with the tasks and experiences of potential users would be a key factor in the deployment decision processes. The result also corroborates the contention that firms would deploy technologies with features that relatively match the firms' business processes and experiences (Ettlie, 1983).

Relative advantage of the innovation is a significant determinant in e-business deployment. The results in Table 6 indicate a significant difference among the three groups. This implies that users with in-house solution perceive their firms as having a relative advantage, by way of the benefits of e-business deployment. Similarly, users with outsourced systems gain relative advantage using e-business. The significant difference between users and non- users of e-business indicates support for H5. This result buttresses the assertion that relative advantage is a key factor critical to e-business deployment, which is consistent with findings of related IS researches (Zhu et al., 2002; Rogers, 1995).

The results in Table 6 illustrate that *technology policy* is a significant determinant in the deployment of e-business. Further, Post-hoc analysis suggests that all the three groups of firms are significantly different from one another. Essentially, e-business users with in-house e-business systems tend to have more aggressive technology policy compared to users without the same systems. In addition, e-business users without in-house solutions tend to have more aggressive technology policy compared to non-users of e-business, which supports H1, consistent with prior findings by Teo et al. (1997), and Ettlie and Bridges (1982), who emphasized the significance of an aggressive technology policy as a means to cope with an uncertain business environment. Early users of innovation will thus consider technology policy to be a key component of their corporate strategy formulation (He & Mykytyn, 2007).

Top managers' support is a significant e-business use determinant in Nigerian financial firms. In addition, Post-hoc analysis reveals that all three groups of firms are significantly different from one another. Users with in-house e-business tend to have the highest level of top management influence, followed closely by users with outsourced e-business systems, and non- users. Hence, the data supports H2. This result is consistent with prior related research (Grover, 1993; Premkumar and King, 1992; Premkumar and Ramamurthy, 1995), which revealed top management influence as a key determinant in technology deployment. Potential e-business users should consider gaining support and commitment from top management to reduce organizational resistance, and to ensure sufficient allocation of resources for smooth implementation of e-business.

The results illustrate that *risk taking* is a significant determinant in e-business use and performance. This implies that the risk-taking propensity of top management has enormous impact on the implementation of e-business. Hence, the data supports H3. This result is consistent with the findings of Grover (1993), who concluded that the risk-taking propensity of top management was a significant determinant of inter-organizational systems deployment. The participants in this research perceive e-business deployment as risky since it commands substantial financial investment and often, requires changes in the firm and management structure, strategy and practices, but are willing to take the challenges because of anticipated intensity competition. In addition, security and standardization issues are currently being resolved, and with the commitment from the government, commercial activities involving e-business will grow even stronger.

Implications for Practice

This paper illustrates that all three TOE constructs are equally important in e-business use and performance among Nigerian financial firms. This paper identifies 9 significant determinants of e-business use in firms within Nigerian context. They are technology policies, top managers' support, risk seeking, fit with firm culture, beliefs and IS infrastructure, relative advantage, technology capability, competitive intensity, uncertainty, and regulatory policy.

The outcome of this paper would be critical for the financial services industry in Nigeria as they plan further investments in their e-business systems. By applying the TOE framework, firms could successfully identify and address key factors in the environment, in the enterprise and related e-business issues necessary to enable effective and functional financial firms in Nigeria and at the international level. The intensity of competition in the financial industry increases with every new technological application and therefore, it is imperative for financial practitioners in Nigeria to be cognizance of the fact and prepare their firms for the business challenges of the next decade of 21st century. In essence, although the competitive use of e-business is permeating all businesses today, firms that are more sophisticated and aggressive in deploying and using advanced e-business systems would be able to keep up with the competition if not gain competitive advantage. The pressure to provide unique and acceptable services and products at reduced costs means increased deployment of e-business innovations even in traditionally low information and knowledge intensive sectors.

The outcome of this paper will also be useful to the Nigerian government and its agencies. All the economic sectors of Nigeria recognize government's role in facilitating e-business deployment, however, existing regulatory policies are at best inadequate and less comprehensive to enable the complex dynamic legal and structural frameworks needed to, successfully allow maximum business use of e-business innovation. In this regard, the Nigerian government, as a matter of urgency, needs to work closely with the private sector to determine the gaps in ICT policy implementations and initiate concrete directions for stronger e-business processes.

Implications for Research

Findings in this paper are consistent with traditional academic models of technology-organization-environment [TOE] framework (Zhu et al., 2002; Eze and Kam, 2001) in a new substantive area, Lagos, Nigeria. In this regard, it extends the field of electronic communication systems and e-business, including the diffusion and TOE perspective of the firm. In addition, this research builds on previous innovation research, and corroborates some of the findings of these studies indicated earlier (e.g., the importance of top management influence), and raises some vexing questions as well. For example, in the context of Nigeria, are uncertainty and regulatory policy still significant factors differentiating financial firms deploying e-business from those that do not, given the complexities in the Nigerian economy? This puzzle is one already being addressed in a follow-up research within the African region to determine the extent environmental uncertainty, regulatory policy, and competitive intensity influence ICT deployments in firms operating in West Africa. These findings, based on the perception of users and non-users of e-business, suggest that although the external environment appears to push firms to deploy e-business, the internal environment (firm and technological factors) is still important in determining innovation systems deployment. This indicates that while the external environment are critical key factors in the deployment of e-business, the firm's internal environment complexities tend to be the dominant significant considerations in IS deployment decision processes among Nigerian financial firms.

Finally, this paper contributes immensely to the e-business literature and creates additional poll of resources practitioners and academics could use to further enrich and extend our knowledge of the evolving phenomenon. Empirical data on e-business development and growth in Nigeria is a small step, a step nonetheless, towards enhancing and extending the discussion on e-business as a global platform for business, economic and industrial activities.

Limitations of this Research

There were some study limitations. First, the conclusions drawn from the data, while theoretically sound, are based upon perceptions of single informant. While CIOs are expected to be knowledgeable, study results would have been more reliable if paired with a second informant outside the IT area. Although all responses were anonymous, it is still possible that the CIO responses are biased in favour management information systems benefits. Second, the survey instrument relied primarily on perceptual data. Although, Venkatraman and Ramanujam (1987) and Dess and Robinson (1984) argue that subjective data correlate with objective data, further studies may consider a mix of both subjective and objective data. Third, the sample composition was, on average, large firms and as a result, inferences on the findings would be most appropriate in related cases such as South Africa and Malaysia with respect the firms that participated in this investigation.

AGENDA FOR FUTURE RESEARCH

Possible further research areas from this paper include; first, as this paper investigates the deployment of e-business in predominantly large firms from one industry, it would also be useful to gain greater understanding of e-business deployment in small firms from different multiple industries. Comparison between these categories of firms in terms of e-business implementation purposes, and usage patterns, may be useful for the firms to, properly chart the direction of the innovation implementation. Second, extending future studies to include comprehensive performance impacts of e-business, how such impacts can be evaluated, and the evolution of e-business, would be useful for firms competing in this 21st century. In addition to using a static cross-sectional approach, a longitudinal study on the e-business deployment may provide insights on the dynamics of the innovation implementation processes that may be more useful. This approach may also provide further clues to enable verification of the outcomes in this research on environmental factors. Future research may also consider a comparative study involving Asian, African and European or American firms to determine any underlying dynamics across business cultures, policy issues and entrepreneurial activities in e-business deployment. This would provide some contingency frameworks for practitioners and policymakers at an international level as e-business continues to evolve into a universal business phenomenon across the globe. Finally, future studies may consider using additional TOE factors and possibly revising some of the factors in this paper to provide for more content coverage and generalization. This, however, should be done with particular respect to the respective research circumstance.

REFERENCES

Anderson, J. C., & Gerbing, D. W. (1988). Structural equation modeling in practice: A review and recommended two-step approach. *Psychological Bulletin*, *103*, 411–423. doi:10.1037/0033-2909.103.3.411

Armstrong, J. S., & Overton, T. (1977). Estimating non-response bias in mail surveys. *JMR, Journal of Marketing Research*, *14*, 396–402. doi:10.2307/3150783

Chua, P. Y. K., & Hui, K. L. (2001). Determinants of small business EDI Adoption: An empirical Investigation. *Journal of Organizational Computing and Electronic Commerce*, *11*(4), 229–252. doi:10.1207/S15327744JOCE1104_02

Clemons, E. K., & Hitt, L. (2001). Financial services: Transparency, differential pricing, and disintermediation. In R. Litan & A. Rivlin (Eds.), *The Economic Payoff from the Internet Revolution* (pp. 87-128). Washington, D.C.: Brookings Institution Press.

Clemons, E. K., & McFarlan, W. F. (1986). Telecom: hook up or lose out. *Harvard Business Review*, *64*(4), 91–97.

Cockburn, C., & Wilson, T. D. (1996). Business use of the World Wide Web. *International Journal of Information Management*, *16*(2), 83–102. doi:10.1016/0268-4012(95)00071-2

Compeau, D. (1995). Computer Self-efficacy: Development of a Measure and Initial Test. *MIS Quarterly*, *19*, 189–211. doi:10.2307/249688

Cooper, R. B., & Zmud, R. W. (1990). Information technology research: a technological diffusion approach. *Management Science*, *36*(2), 123–139. doi:10.1287/mnsc.36.2.123

Dess, G. G., & Robinson, R. B. (1984). Measuring organizational performance in the absence of objective measures: The case of privately-held firm and conglomerate business unit. *Strategic Management Journal*, *5*, 265–273. doi:10.1002/smj.4250050306

Dun & Bradstreet. (2000/01). Key Business Directory of Singapore. *ASEAN Series* (Vol.1). Singapore.

Eccles, R. G., & Crane, D. B. (1988). *Doing Deals: Investment Banks at Work*. Boston, Massachusetts: Harvard Business School Press.

eGlobal Report, (2002). Retrieved August 12, 2002, from http://www.eMarketer.com

eGlobal Report, (2005). Retrieved August 12, 2006, from http://www.eMarketer.com

Ettlie, J. E. (1983). Organizational policy and innovation among suppliers to the food processing sector. *Academy of Management Journal*, *26*(1), 27–44. doi:10.2307/256133

Ettlie, J. E. (1986). *Implementing Manufacturing Technologies: Lessons from Experience. Managing Technological Innovation*. San Francisco, CA: Jossey-Bass Publishers.

Ettlie, J. E., & Bridges, W. P. (1982). Environmental uncertainty and organizational technological policy. *IEEE Transactions on Engineering Management*, *EM-29*, 2–10.

Eze, U. C. (2006). The Impact of Firm and Business Environment Contexts on E-Business Value. In *Proceedings of MWAIS 2006 Conference held on May 5-6 in Grand Rapids, Michigan, USA*.

Eze, U. C., & Gilbert, A. L. (2004). The significance of e-Commerce to firms' operations in relation to business location. *Australiasian Journal of Information Systems, 12*, 75–93.

Eze, U. C., & Kam, T. S. (2001). The impact of enterprise strategy and enterprise structure on the level of e-commerce deployment in enterprises in Singapore. *Journal of Global Information Technology Management, 4*(2), 38–54.

Fan, M., Stallaert, J., & Whinston, A. B. (2000). The Internet and the future of financial markets. *Communications of the ACM, 43*, 83–88. doi:10.1145/353360.353368

Floyd, S. (1988). A micro level model of information technology use by managers. In U.E. Gattiker (Ed), *Studies in Technological Innovation and Human Resources (1): Managing Technology Development* (pp. 123-142). Berlin & New York: Walter de Gruyter.

Galbraith, J. R. (1973). *Designing complex organization*. Reading, MA: Addison-Wesley.

Gilbert, A. L., & Eze, U. C. (2008). A Tale of Two Cities: Economic Geography in the Age of the Internet. Accepted for publication in *Nanyang Business Review, Singapore*.

Glazer, R. (1993). Marketing in turbulent environments: Decision processes and the time-sensitivity of information. *JMR, Journal of Marketing Research, 30*(4), 509–521. doi:10.2307/3172694

Grover, S. K., & Lederer, A. L. (2004). The Impact of Industry Contextual Factors on IT Focus and the Use of IT for Competitive Advantage. *Information & Management, 41*(7), 899–919. doi:10.1016/j.im.2003.08.018

Grover, V. (1993). An empirically derived model for the adoption of customer-based interorganizationl systems. *Decision Sciences, 24*(3), 603–640. doi:10.1111/j.1540-5915.1993.tb01295.x

He, F., & Mykytyn, P. P. (2007). Decision Factors for the Adoption of an Online Payment System by Customers. *International Journal of E-Business Research, 3*(4), 1–32.

Holsapple, C. W., & Whinston, A. B. (1996). *Decision Support Systems*. New York: West publishing company.

Iacovou, C. L., Benbasat, I., & Dexter, A. S. (1995). Electronic data interchange and small organizations: Adoption and impact of technology. *MIS Quarterly, 19*(4), 465–485. doi:10.2307/249629

Intel Information Technology White Paper. (May 2003). *Defining the Value of E-business: Seventeen Standard Measures* (pp. 1-14). Intel Publishing.

Kast, F., & Rosenzweig, J. (1973). *Contingency Views of Organization and Management*. Chicago: Science Research Associates.

Kerlinger, F. N. (1986). *Foundations of Behavioral Research*. CBS College Publishing.

King, W., Grover, V., & Hufnagel, E. (1989). Seeking Competitive Advantage Using Information-Intensive Strategies: Facilitators and Inhibitors. In *Information Technology and Management Strategy*. Eaglewood Cliffs, New Jersey: Prentice Hall.

King, W. R., & Sabherwal, R. (1992). The factors affecting strategic information systems applications: an empirical assessment. *Information & Management, 23*(2), 217–235. doi:10.1016/0378-7206(92)90046-I

King, W. R., & Teo, T. S. H. (1996). Key dimensions of facilitators and inhibitors for the strategic use of information technology. *Journal of Management Information Systems, 12*(4), 35–53.

Kohli, R., & Devaraj, S. (2003). Measuring information technology payoff: A meta-analysis of structural variables in firm-level empirical research. *Information Systems Research, 14*, 127–145. doi:10.1287/isre.14.2.127.16019

Kwon, T. H., & Zmud, R. W. (1987). *Unifying the Fragmented Models of Information Systems Implementation: Critical Issues in Information Systems Research.* New York: John Wiley.

Lederer, A. L., & Mendelow, A. L. (1988). Convincing top management of the strategic potential of information systems. *MIS Quarterly, 12*(4), 526–536. doi:10.2307/249127

Mata, F., Fuerst, J., & Barney, J. (1995). Information technology and sustained competitive advantage: A resource-based analysis. *MIS Quarterly, 19*(4), 487–505. doi:10.2307/249630

Nabseth, L., & Ray, G. F. (1974). *The Diffusion of New Industrial Processes*. London: Cambridge: University Press.

Nunnally, J. C. (1979). *Psychometric Theory*. New York: McGraw Hill.

Olazabal, N. G. (2002). Banking: the IT paradox. *McKinsey Quarterly, 1*, 47–51.

Porter, M. E. (1985). *Competitive Advantage* (3rd. Ed). New York: Free Press.

Porter, M. E. (2001). Strategy and the Internet. *Harvard Business Review*, (March): 62–78.

Premkumar, G., & King, W. R. (1992). An empirical assessment of information systems planning and the role of information systems in organizations. *Journal of Management Information Systems, 9*(2), 99–125.

Premkumar, G., & Ramamurthy, K. (1995). The role of inter-organizational and organizational factors on the decision mode for the adoption of inter-organizational systems. *Decision Sciences, 26*(3), 303–336. doi:10.1111/j.1540-5915.1995.tb01431.x

Ramamurthy, K., Premkumar, G., & Crum, M. R. (1999). Organizational and Interorganizational Determinants of EDI Diffusion and Organizational Performance: A Causal Model. *Journal of Organizational Computing and Electronic Commerce, 9*, 253–285. doi:10.1207/S153277440904_2

Raven, P. V., Huang, X., & Kim, B. B. (2007). E-business in Developing Countries: A Comparison of China and India . *International Journal of E-Business Research, 3*(1), 91–108.

Rawn (1994) The business of the Internet: an introduction to the Internet for commercial organizations. Retrieved from http://www.rtd.com/people/rawn/business.html

Reich, B. H., & Benbasat, I. (1990). An empirical investigation of factors influencing the success of customer-oriented strategic systems. *Information Systems Research, 1*(3), 325–347. doi:10.1287/isre.1.3.325

Rogers, E. M. (1995). *Diffusion of Innovations*. New York: The Free Press.

Sato, S., Hawkins, J., & Berentsen, A. (2001). E-finance: Recent developments and policy implications. In *Tracking a Transformation: e-commerce and the terms of competition in industries* (pp. 64-91). Brookings Institution Press, Washington DC.

Teo, T. S. H., & Pian, Y. (2003). A contingency perspective of Internet adoption and competitive advantage. *European Journal of Information Systems, 12*, 78–92. doi:10.1057/palgrave.ejis.3000448

Teo, T. S. H., Tan, M., & Buk, W. K. (1998). A contingency model of Internet adoption in Singapore. *International Journal of Electronic Commerce, 2*, 96–118.

Teo, T. S. H., & Too, B. L. (2000). Information systems orientation and business use of the Internet: An empirical study. *International Journal of Electronic Commerce*, *4*(4), 105–130.

Tornatzky, L. G., & Klein, K. J. (1982). Innovation characteristics and innovation adoption-Implementation: a meta-analysis of findings. *IEEE Transactions on Engineering Management*, *29*(1), 28–45.

Ujah, E. (2005, October 10). Electronic business and small business development in Nigeria. *Nigeria Business Times*.

Ujah, E. (2008, January 19). E-business Challenges for Nigerian Companies. *Nigeria Business Times*.

UNCTAD. (2002). E-commerce and development report 2002. *United Nations Conference on Trade and Development, New York and Geneva*.

United States Census Bureau. (2009). Quarterly Retail E-commerce Sales: 4th Quarter 2008. *United State Department of Commerce*. Retrieved from http://www.census.gov/mrts/www/data/html/08Q4.html

Utterback, J. M. (1974). Innovation in industry and the diffusion of technology. *Science*, *183*, 620–626. doi:10.1126/science.183.4125.620

Utterback, J. M., & Abernathy, J. M. (1975). A dynamic model of process and product innovation. *Omega*, *3*(6), 639–657. doi:10.1016/0305-0483(75)90068-7

Weill, P., & Olson, M. H. (1989). An assessment of contingency theory. *Journal of Management Information Systems*, *6*(1), 59–85.

Williamson, O. E. (1983). Organizational innovation: The transaction cost approach. In J. Ronen (Ed.), *Entrepreneurship* (pp. 101-133). Lexington, MA: Lexington Books.

Woodward, J. (1965). *Industrial Organization: Theory and Practice*. London: Oxford University Press.

Xu, M., Rohatgi, R., & Yangxing, D. (2007). E-business Adoption in SMEs: Some Preliminary Findings from Electronic Components Industry. *International Journal of E-Business Research*, *3*(1), 74–90.

Zhu, K., & Kraemer, K. L. (2005). Post-adoption variations in usage and value of e-business by organizations. *Information Systems Research*, *16*(1), 61–84. doi:10.1287/isre.1050.0045

Zhu, K., Kraemer, K. L., & Xu, S. (2002). A cross-country study of electronic business adoption using the technology-organization-environment framework. *Presented to the Twenty-Third International Conference on Information Systems, Barcelona* (pp. 337-348).

Zhu, K., Kraemer, K. L., Xu, S., & Dedrick, J. (2004). Information technology payoff in e-Business environments: An international perspective on value creation of e-Business in the financial industry. *Journal of Management Information Systems*, *21*(1), 17–54.

Zmud, R. W. (1984). An examination of 'push-pull' theory applied to process innovation in knowledge work. *Management Science*, *30*(6), 727–738. doi:10.1287/mnsc.30.6.727

Chapter 19
The Measurement of Electronic Service Quality: Improvements and Application

Grégory Bressolles
BEM – Bordeaux Management School, France

Jacques Nantel
HEC Montréal, Canada

ABSTRACT

Several measurement scales have been designed by both practitioners and researchers to evaluate perceptions of electronic service Quality. This chapter tests three of the main academically developed scales: Sitequal (Yoo & Donthu, 2001), Webqual 4 (Barnes & Vidgen, 2003) and EtailQ (Wolfinbarger & Gilly, 2003) and compares them against the scale ensuing from our research: NetQual (Bressolles, 2006). Based on 204 evaluations of consumers that participated in a laboratory experiment involving two Canadian Websites in travel and online insurance, NetQual best fits the data and offers the highest explanatory power. Then the impact of nature of task and success or failure to complete the task on the evaluation process of electronic service quality and attitude toward the site is examined and discussed on over 700 respondents that navigated on six different Websites.

INTRODUCTION

A relatively recent form of commerce, electronic commerce is increasingly becoming routine. Despite recent years' turbulence, electronic commerce is on the rise in Canadian commercial landscape. Representing a sales volume of $4.7 billion, about 1.3% of Canadian retail sales, e-commerce is becoming an indispensable tool for retailers Statistics. In total, 67% of Canadian households use the Internet (eMarketer 2005). While only 18.4% of consumers that use the Internet claim to purchase products on the Web, 56% of Web users report using this medium to obtain product information before purchasing it at a brick-and-mortar store. The systematic increase of Internet integration in consumers' decision making processes has created a strong impetus for retailers to go online. In 2005, 34% of Canadian retailers had a Website and 11.4% sold products on the Web. The proportions at finance and insurance sectors were 43% and 8%, respectively, compared with 51% and 14% for that of arts and culture. While

DOI: 10.4018/978-1-60566-910-6.ch019

both transactional and informational commercial activity on the Web is growing, studies did not find sites directed at consumers to always meet expectations.

A study by the e-tailing group[1] found that only 3% of consumers who visit a site to complete a purchase; more than 47% of consumers abandon their order before checking out (cart abandonment). Partly explained by Internet anonymity, such statistics could be also explained by the fact that many sites do not meet consumers' needs or poorly tailor their decision-making processes. One could argue that having an online presence and posting low prices seemed to be sufficient to succeed; neither of these conditions, however, does guarantee service quality. Inevitably, certain quality issues have appeared, such as the inability to carry out online transaction, non-compliance with delivery time, undelivered products, unanswered emails, and inaccessible or inadequate information. As at a brick-and-mortar store, the service quality of a commercial Website plays a vital role in its survival. Internet sales have particular characteristics that differentiate them from traditional sales. For these reasons, measurement instruments have been developed by practitioners and researchers to evaluate service quality in e-commerce.

While literature on service quality includes articles that compare various scales across different contexts such as health, arts, professional services, and retail stores, it do not offer any study that offer a comparison of Web-oriented scales. Results observed in tangible situations, where interpersonal contact is a key, can not be taken for granted in a virtual context (Bitner, Brown & Meuter, 2000; Dabholkar, 2000; Parasuraman & Grewal 2000). For instance, Parasuraman & Grewal (2000) posit that online and offline environments are sufficiently different to justify the development of scales specifically dedicated to the measurement of electronic service quality. Even when the same product or service was purchased, online and offline environments present different shopping experiences. Consequently, such measurement instruments of service quality became a necessity.

This article starts by defining the concept of electronic service quality, and compares it with the traditional one. Given the abundance of measures of electronic service quality put forth by practitioners and researchers, we then selected to test three of the main academically developed scales: Sitequal (Yoo & Donthu, 2001), Webqual 4 (Barnes & Vidgen, 2003) and EtailQ (Wolfinbarger & Gilly, 2003) and to compare them against the scale ensuing from our research, NetQual (Bressolles, 2006). Based on 204 evaluations of consumers who participated in a laboratory experiment that involved two Canadian Websites (travel and online insurance), we intend to determine the most relevant scale in terms of content, parsimony and explanatory power in an e-commerce context. Then we explore the impact of task nature and success or failure to complete a specific task on the electronic service quality evaluation. The link between electronic service quality and attitude toward site is studied. Discussion of limitations and future research avenues conclude the paper.

THEORETICAL BACKGROUND OF ELECTRONIC SERVICE QUALITY

Definition, Similarities and Differences with Traditional Service Quality

Whereas dimensions, variables, and other aspects of traditional service quality have received extensive study over the past two decades, the study of electronic service quality is a relatively new domain. While traditional service quality was defined as an overall evaluation or an attitude relative to the superiority of the service (Parasuraman, Zeithaml & Berry, 1988), electronic service quality were considered as "the extent to which a Web site facilitates efficient and effective shopping, purchasing, and delivery of products and services." (Zeithaml, Parasuraman & Malhotra, 2002, p.363). This transactional quality entails the evaluation of the pre- and post-service experience. Based on the last definition, we can draw a parallel with traditional service quality to elucidate the similarities and differences between these two concepts.

One of the most important and arguably one of the most evident differences between traditional service quality and electronic service quality is the replacement of interpersonal interaction with human-machine interaction. This distinction raises many questions concerning the types of dimensions that can or must be considered when assessing service quality in the e-commerce context. Online commerce specific characteristics render direct application of service quality dimensions developed in other environments inappropriate; at best, these dimensions fail to capture all of the subtleties of the evaluation of service quality of commercial Websites. The classic dimensions of traditional service quality are tangible elements, reliability, reactivity, assurance and empathy of service provider (Parasuraman, Zeithaml & Berry, 1988).

To date, no consensus exists regarding electronic service quality dimensions. Although largely anecdotal, dimensions proposed recur fairly systematically as security/privacy, Website design, ease of use, reliability and the quality of the information contained in the site. Moreover, positive feelings such as warmth and attachment expressed for traditional services are not present in the perception of electronic service quality; whereas negative feelings such as anger, irritation, and frustration are apparently less intense or noticeable on the Internet than that expressed during problems encountered with traditional services (Parasuraman, Zeithaml & Malhotra, 2002). On the other hand, if classic evaluations of traditional service quality were based on measuring the gap between expectations and perceptions, it is difficult to apply such a model to the electronic service quality measurement where respondents find it hard to formulate their expectations. Therefore, a direct measure of the perceptions of electronic service quality after delivering the service seems practical.

Methods of Measuring Electronic Service Quality

As offline, there are different methods to measure a Website quality (Cunliffe 2000); however, methods could be grouped under two broad categories:

- **Behavioural measures**. Behaviour measures focus on the measurement of commercial activity of the site; that is number of clicks, number of unique visitors or conversion rate of new visitors (Totty, 2003) and analysis of log files (Lynch & Ariely, 2000; Johnson *et al.*, 2004). Albeit useful, the previous measures do not capture consumer's cognitive and attitudinal evaluations of the site and are limited to analyzing apparent behaviors on the Internet.
- **Attitudinal measures**. These are based often on traditional measurement scales that evaluate consumers' perceptions or reflect these perceptions by soliciting expert opinions; nonetheless, attitudinal measures developed to date have some limitations. First, they do not evince the structure of perceived electronic service quality dimensions; for example, Madu & Madu (2002) listed multiple characteristics of a Website without specifying relations between these characteristics. Moreover, the measures examine the performance of Websites in general rather online commerce sites in particular (e.g. Hoffman & Novak, 1996). Hence there might be a lack of generalizability to purchasing behaviours on commercial Websites because these measures neglect attributes related to purchasing such

as order placing, financial security, respect for privacy, payment modes, delivery, etc. Finally, psychometric properties have not been consistently established and verified, particularly the one- or multi-dimensional nature of measures used. Two approaches fall into this category. The first is generally based on expert evaluations or interstitial surveys and is more common among practitioners. The second is more grounded in psychometric theory and, therefore, more prevalent among scholars.

Measurement of Electronic Service Quality by Practitioners

Practitioners have adopted various approaches to measure quality or efficacy of commercial Websites. Approaches ranges between questioning consumers after completing the purchase (bizrate.com, directpanel.com) and evaluating sites by professional experts (gomez.com). Despite the diversity of approaches applied, none of the initiatives taken alone does encompass the entire online transaction, from information search to order placing, problems with delivery and after-sale service. While they help forming the picture of important attributes of online shopping, neither practitioners' studies aid at constructs conceptualization nor they validate or check reliability of the utilized measures. To fill this gap and to better understand what consumers want during the online purchasing experience, scholars of marketing and computer science have both attempted to develop valid instruments to measure electronic service quality.

Measurement of Electronic Service Quality by Scholars

Four of the principal studies that examine the measurement of perceptions of electronic service quality reported in the academic literature have been retained for this study to determine which scale is best suited to measuring the perceptions of electronic service quality and to study the impact of scale dimensions on attitude toward the site assessment.

- **Webqual** (Barnes & Vidgen, 2003). This scale was developed based on an iterative process involving application in diverse domains such as online bookstores and auction sites. The authors identified three dimensions (a) usability of the site refers to pragmatic elements such as the way the consumer perceives and interacts with the site. Associated with the site design, these qualities include appearance, ease of navigation and image projected, (b) quality of information refers to quality of content offered on the site defined as precise pertinent information that users consider well formatted, and (c) quality of interaction refers to the quality of service interaction the users receive on the site; this includes elements of trust and empathy or more precisely information and transaction security, product delivery, personalization and communication with service provider. Webqual did not consider the entire online purchasing process because "all questions can be completed without having performed the full purchasing process" (Barnes & Vidgen, 2003, p.124). While this approach provides insight into users' perceptions, it does not take into account all aspects of the online service life cycle - navigation, selection, ordering, payment, delivery, and customer service. Therefore, Webqual can not be said to measure all the perceptions of electronic service quality because some service experience aspects are neglected. The use of students in the study to evaluate perceptions of electronic service quality is also questionable. While students might be considered frequent users of the Internet and books buyers, it would be more adequate to examine a sample of clients that have had several service experiences at the site concerned.

- **eTailQ** (Wolfinbarger & Gilly, 2003). The methodology behind this scale construction is online and offline focus groups, along with a classification task and an online survey of a panel of consumers. It includes four factors (a) site design that includes navigation, search for information, product selection, order process, and personalization, (b) customer service including online assistance, response to customers' emails, ease of returning items, empathy, and reactivity, (c) reliability/respect for commitments refers to adequate description, presentation, and delivery of products or services ordered at the promised quality level, and (d) security/privacy reflected by security of payments and confidentiality of personal data. The sample used at the study was not random, but rather comprised regular online purchasers; therefore it is deemed poorly representative of the overall population of cyber-consumers. In addition, evaluations obtained do not pertain to particular one or two sites, but rather to general evaluations of electronic service quality perceptions.
- **Sitequal** (Yoo & Donthu, 2001): The final version of this scale has nine items reflecting 4 dimensions (a) ease of use and capacity to obtain information, (b) design and creativity of site with multimedia content and colors, (c) speed of the order process and reactivity to consumers' requests, and (d) security of financial and personal information. The authors focused exclusively on the elements of Website experience and did not empirically verify results on a sample of cyber-users; they also used a sample of students where each student navigated three different sites.
- **NetQual** (Bressolles, 2006). The final scale tested in this study includes 18 items (see Appendix I for a list of items) distributed along 5 dimensions (a) quality and quantity of information available, (b) ease of site use, (c) design or aesthetic aspect of the site, (d) reliability or respect for commitment, (e) security/privacy of personal and financial data. The dimensions retained refer to the functional characteristics of both the site and transaction. Following a series of semi-structured interviews, the scale was developed and refined on a sample of over 1,200 online consumers who were customers of five commercial Websites representing different online sales sectors - travel, insurance, digital products and energy.

A fifth scale, the *E .S. Qual* proposed by Parasuraman et al. (2005) initially considered in this study, was not retained for further analysis. Unlike other scales, the *E .S. Qual* does not focus on the quality of the site *per se* but rather on quality of the e-services inherent in navigation, such as logistics, possibility to speak with someone, etc... Because present article proposes an optimal way to measure the quality of a Website, we decided to limit analyses to major scales proposed to measure consumers' evaluations of Websites. While scales identified above exhibit similarities in content or dimensions, each has distinctive features (table 1).

METHOD

A series of studies was conducted in concert with six Canadian companies that each has an e- commerce site. The studies were conducted on over 700 consumers between the months of July 2004 and May 2005. The companies evaluated were (a) Destina.ca, a travel site owned by Air Canada, (b) ING/Bélair Direct, a well-known Canadian

Table 1. Summary of the principal measurement scales for electronic service quality

Scale	Author	Number of items	Dimensions	Dependent variables	Sample
Sitequal	Yoo and Donthu (2001)	9	- Ease of use - Aesthetic design - Processing speed - Security	- Attitude toward site - Site loyalty - Site equity - Purchase intention - Site revisit intention - Site quality	94 students who visit and interact with 3 sales Websites online
Webqual (4)	Barnes and Vidgen (2003)	22	- Usability - Information - Interaction	-	380 questionnaires on three online bookshop Websites evaluations by students
eTailQ	Wolfinbarger and Gilly (2003)	14	- Website design - Fulfillment/reliability - Security/Privacy - Customer service	- Satisfaction - Attitude toward site - Loyalty intentions - Global quality	Online survey of 1013 customer members of a panel
NetQual	Bressolles (2006)	18	- Information - Ease of use - Reliability/fulfillment - Site design - Security/privacy	- Overall quality - Satisfaction - Attitude toward site	855 customers of 2 commercial Websites (travel and electronic goods)
e-S-Qual **e-RecS-Qual**	Parasuraman, Zeithaml and Malhotra (2005)	22 11	- Efficiency - Fulfillment - System availability - Privacy - Responsiveness - Compensation - Contact	- Perceived value - Loyalty intentions	Respondents who have carried out at least 3 online purchases over the previous 3 months on amazon.com (650 people) or walmart.com (253 people)

in automobile insurance, (c) Hydro-Québec, the largest Canadian energy provider and exporter, (d) Radio-Canada, the Website of the Canadian French-speaking official public television station, (e) the National Archives of Canada, and (f) Revenu Québec, the official Website of information, services, and applications that help promote compliance with Quebec province tax and regulations in Canada.

The same procedure was used for each of the sites studied. In each case, over 100 consumers were recruited via a hyperlink placed on the site. This procedure was mainly intended to recruit actual consumers that have at least minimal experience with both the Web and the site in question. Consumer who showed an interest in our research program by registering and providing their contact information were then invited to participate at a laboratory navigation session. At this session, each participant was to perform an individual task with an average duration of one hour. Participants were offered $50 for their participation in the study. During navigation and throughout the experiment, participants were told to verbalize out loud every thought that go through their minds, regarding difficulties encountered, surprising aspects of site, or a simple description of what they are doing. Known as protocol analysis, this approach is grounded in the work of Simon (1956) and Ericsson & Simon (1993) and proved to be very useful for Website analysis (Senecal, Gharbi & Nantel, 2002; Benbunan-Fich, 2001; Li & Biocca, 2001). In each of the four studies, i.e. one study for each scale, the experiment was supervised by a project manager. Data were collected in four of the following distinct steps.

Table 2. A listing of each site's experimental task

Site	Task
Destina	Reservation and purchase of a trip for the holiday period
ING direct	Purchase of an automobile insurance policy
Hydro-Québec	Enrolling for direct payment mode
Radio Canada	Finding information on the site
National Archives	
Revenu Québec	

Step 1: Warm-Up Task

After the welcome and a brief description of the assigned tasks, participants were invited to go to a general portal in order to read their horoscope. They were then asked to consult the movie listings on the site to find a film. This task was very important to enable the participants get familiar with navigation and with the method, i.e. verbalizing aloud thoughts and actions. The warm-up task ended when the participants finished the tasks, i.e. reading the horoscope and selecting a film on the site.

Step 2: Experimental Task

Each participant was informed of the task that they were to perform. Table 2 shows the required task that corresponds to the Website studied. Although the task was prescribed, participants were told of their ability to interrupt the navigation at any time, for any reason. Ongoing concurrent verbalization was, however, a requirement throughout the navigation process. For each task, navigation data were recorded in video and sound sequence, i.e. AVI format, using CAMTASIA software.

Step 3: The Questionnaire

After completing navigation, participants were asked to complete an online evaluation questionnaire that included items from the four electronic service quality scales under study. The questionnaire responses were saved and updated online. The questionnaire, comprising seven point Likert-type scaling, was intended to measure participants' site perceptions, i.e. electronic service quality, attitude toward site, etc.... The questionnaire also included a series of sociodemographic-oriented questions.

Step 4: The Interview

In the final step, participants were interviewed for ten minutes by the project manager with regard to completion of the experimental task and general site evaluation. This semi-structured interview consisted of three sections: (a) completion of the task, (b) general site evaluation, and (c) a series of specific questions regarding the appearance of the site under study. The interview was recorded in audio format using MP3 recording software.

RESULTS

A Comparison of Four Electronic Service Quality Measurement Scales

To determine which of the four scales selected (Webqual 4, Sitequal, EtailQ and NetQual) best reflects electronic service quality perceptions, the data gathered by the online questionnaire from 204 consumers that visited the two first partner sites of the study, travel and online insurance, were analyzed. Scales were compared to determine

Table 3. Internal reliability of measures, Cronbach's alpha and Rhô coefficients

Webqual	α	ρ	Sitequal	α	ρ	EtailQ	α	ρ	NetQual	α	ρ
Usability	.93	.93	Ease of use	.94	.94	Website design	.83	.81	Information	.93	.92
Information	.93	.93	Aesthetic design	.79	.86	Customer service	.80	.80	Ease of use	.95	.95
Interaction	.81	.81	Processing Speed	.63	.64	Fulfilment/ reliability	.77	.70	Site design	.90	.90
			Security	.85	.85	Security/ Privacy	.89	.87	Security/ Privacy	.86	.85

reliability, quality of fit and explanatory power. To verify reliability for each dimensions of scales compared, Cronbach's alpha coefficients and Jöreskog's Rhô were calculated. Table 3 summarizes findings with regard to both of the previous measures for each scale.

All coefficients obtained satisfy the criteria that Nunnally (1978) recommended for confirmatory research, over .80 for established or over .70 for new measures. The internal reliability of the dimensions for each scale is thus verified. The only dimension with a low reliability coefficient was Sitequal's "Processing speed" (Yoo & Donthu, 2001). This is probably the case because the previous scale dimension is composed of only two items with relatively low correlation, i.e. 0.46.

Confirmatory factor analysis (C.F.A.) using EQS 6 software (Bentler & Wu, 2002) were performed on each of the four scales to check fit with collected data and to determine the scale that best fit the data. Appendix 2 presents the C.F.A. results for the four scales tested. The 'Reliability' dimension of the NetQual scale, which refers to the precision and speed of delivery and after-sale service quality, was not taken into account because this dimension was not applicable to the experiment nature, specifically for the information task. Main fit indices, i.e. absolute, incremental, and parsimony, were calculated to verify the quality of measures fit and are provided in Table 4. The indices GFI, AGFI of Joreskog, NFI and NNFI of Bentler and Bonnet, and CFI of Bentler should be close to 0.95 and greater if possible. As for the RMSEA of Steiger, it is recommended that a score less than 0.05 be obtained to be acceptable (Hu and Bentler 1998). Appendix 3 presents the definition of the different fit indices retained for comparing the four scales.

While the four scales fit the data well, NetQual could be said to present the best fit index. Specifically, its RMSEA was the lowest (.084). The parsimony indices of NetQual are also lower than those of the other three scales; one should notice that this scale contains 18 items distributed along four dimensions. The other indices, the absolute and the incremental, are higher than those of the

Table 4. Fit indices of each of the four scales

	Parsimony indices		*Absolute indices*			*Incremental indices*	
	AIC	Chi² / df	RMSEA	GFI	AGFI	NFI	NNFI
Sitequal	9,66	51,66 / 21 = 2,46	.085 [.056; .114]	.94	.87	.95	.94
Webqual	15,98	531,38 / 186 = 2,85	.096 [.086; .105]	.75	.69	.83	.87
EtailQ	93,50	235,50 / 71= 3,32	.113 [.103; .123]	.85	.78	.86	.87
NetQual	36,19	175,68/72= 2,44	.084 [.07; .09]	.90	.90	.93	.94

Table 5. Explanatory power of each scale

Mean of R^2			
Sitequal	Webqual	EtailQ	NetQual
.67	.58	.57	**.73**

other scales, particularly Webqual and EtailQ. However, direct comparison of indices among the four scales is not indicative because scales do not use the same model, i.e. they are not nested models; the conclusion one can draw is that given the commonly accepted criteria, NetQual could be said to best fits the data.

To verify the superiority of NetQual over the three other scales, the means of the R^2 coefficients, based on standardized solution, were calculated. The mean provides an indication of the explanatory power of the scale and is identical to the R^2 coefficient of the regression multiple determination. Results are presented in Table 5.

Based on these results, we can conclude that NetQual is superior to the three other scales considered, Webqual 4, Sitequal, and EtailQ, in predicting electronic service quality perceptions because it best fits the data and posses the highest explanatory power. To further assess the qualities of NetQual, we want to try to determine whether the nature of the task performed by a consumer, i.e. transactional versus informational, has any impact on site evaluation. In addition, the potential of NetQual to predict the overall site performance (Chen & Wells, 1999) was evaluated.

Impact of Nature of the Task on the Evaluation Process of Electronic Service Quality and Attitude toward the Site

In order to determine the impact of the nature of the task on the evaluation process of electronic service quality and attitude toward the site, over 700 respondents on six Websites completed an online questionnaire after performing a specific task (see table 2). The questionnaire measures electronic service quality using the NetQual scale and attitude toward the site (Chen & Wells 1999). According to Chen & Wells (1999), attitude toward the site (ATS) could be considered to represent the predisposition of Internet users to respond favourably or unfavourable to a Website during a situation of particular exposure. This global measure of site evaluation includes measures concerning the capacity of the site to create a relationship with the consumer, intention to revisit the site, satisfaction with service, comfort with navigation, and the judgement about the fact that surfing the site is a good way to spend time. Although the measure of attitude toward the site is reliable, Chen & Wells (1999) conceded that it cannot supply a complete picture of consumers' judgements related to their online purchasing experience.

Second order confirmatory factor analysis was performed on a global sample of 704 observations on 6 sites to determine the fit of the model. The conditions to create a second order factor for electronic service quality were met. First, theoretical foundations are with favour to this decision, given that electronic service quality is a multidimensional construct. Second, first order dimensions are sufficiently correlated to justify the creation of a second order factor that would demonstrate construct convergent validity (Chin, 1998). Based on first order factor analysis, correlation between dimensions varies from 0.30 to 0.57. Third, to prove the importance of each dimension in the second order construct, when the second order factor is created, regression coefficients with each of the dimensions should be high. According to Hair et al. (1998), these regression coefficients should be greater than 0.50. In our case, regression

Table 6. Fit of the model for the global sample

χ^2	GFI	AGFI	RMSEA [interval] (90%)	SRMR	NFI	NNFI	CFI	χ^2/df	AIC (AICo)
630,44	.90	.90	.56 [.045; .062]	.060	.91	.92	.93	630,44/146 = 4,3	338,44 (6595,72)

coefficients varied from 0.587 for the 'security/privacy' dimension to 0.881 for the 'information' dimension. Table 6 demonstrates that fit indices of model are satisfactory.

The contribution of electronic service quality to the different dimensions, and its impact on attitude toward the site, were then further measured by differentiating the sites on which Internet users had to perform a transactional task, i.e. Destina, ING, and Hydro-Quebec, or an informational task, i.e. National Archives, Radio Canada, and Revenu Québec. We posit that the nature of the task, whether transactional or informational, influences the evaluation process of electronic service quality. To test this hypothesis, multi-group confirmatory factor analysis was performed. Both the results and the fit of the global model presented in Table 7 were found satisfactory.

Before extending the analysis, we determined whether any variations observed between the two sub-samples, transactional versus informational task, were statistically significant. To achieve this, a χ^2 test was performed between a model where the parameters are constrained to be equal between the two groups and a model where the parameters are freed. This test was significant at the .00 level (χ^2 constrained - χ^2 free = 843.61 – 824.47 = 19.14 [df = 298 – 292 = 6]). This result confirms that any differences between the two sub-samples are statistically significant and not simply due to chance.

Several results can be inferred from Table 8. First, it appears that regardless of the nature of the task, i.e. transactional versus informational, the impact of service quality on attitude toward the site is relatively strong (factor contribution > 0.90). Moreover, the global model has good predictive power (mean $R^2 > 0.65$ for the two sub-samples). NetQual therefore seems to be a good measure to predict perceptions of electronic service quality and their impact on attitude toward the site. Lastly, NetQual four dimensions (Information, Ease of use, Design and Security/Privacy) contribute, in varying degrees according to nature of task, to the formation of second order judgement of electronic service quality. The 'Reliability' dimension of the NetQual scale, which refers to the precision and speed of delivery and after-sale service quality, was not included in this analysis because this dimension was not applicable to the experiment nature, specifically the information task.

The weight of each of NetQual dimensions can be further clarified according to nature of task performed on the site. Consumers that performed a transactional task on the site (57%) placed more importance on the quality and quantity of information presented (factor contribution of the 'Information' dimension is 0.809), along with the design of

Table 7. Fit indices of a second order multi-group model

χ^2	GFI	AGFI	RMSEA [interval] (90%)	SRMR	NFI	NNFI	CFI	χ^2/df	AIC (AICo)
824,47	.90	.90	.061 [.056; .066]	.075	.90	.91	.94	824,47/292 = 2,82	240,47 (6257,93)

Table 8. Factor contributions of NetQual dimensions and their impact on attitude toward the site by type of task

	NetQual → Information	NetQual → Ease of use	NetQual → Design	NetQual → Security/ Privacy	NetQual → ATS	mean R^2
Transactional task	.809	.716	.734	.635	.987	.716
Informational task	.730	.831	.487	.398	.959	.652

the Website (0.734) and the security of personal and financial data (0.635). The transactional nature of the task users were asked to perform, along with the intangible orientation of the goods sold on the site (trips, insurance and electricity) might explain the relative importance of pervious dimensions. Inversely, for users that were asked to perform an informational task (43%), ease of use appeared to be the most important dimension (0.831). ease of navigation and information on the site are apparently prerequisites for task completion.

The evaluation process of electronic service quality varies according to the type of task performed on the site. Internet users that consult a site to find information place greater importance on ease of use of the site, whereas consumers that want to perform a transaction focus on information presented, both textual and visual, along with the security aspect of the purchase.

Impact of Completion of the Task on Attitude toward the Site and the Evaluation Process of Electronic Service Quality

To extend this analysis, we studied the impact of the respondent's ability to successfully complete the specified task on the evaluation process of the service quality of the site. We postulate that individuals that failed to complete the task will evaluate service quality more poorly than participants that successfully completed the task. To achieve this, a confirmatory multi-group factor analysis was performed for over 400 respondents that navigated four of the six sites previously studied; Responses concerning the Radio-Canada and National Archives sites were disregarded because the percentages of successful completion of task were very high and would result in a bias to the analysis. To determine whether any variations observed between the two sub-samples, i.e. success or failure, are statistically significant, a χ^2 test was performed between a model at which parameters were constrained to be equal between the two groups and a model at which the parameters were freed. Test result was significant at .00 (χ^2 constrained - χ^2 free = 773.23 – 754.75 = 18.48 [df = 297 – 292 = 5]), showing that any differences observed between the two sub-samples are statistically significant and not simply due to chance. The fit indices of the global model were satisfactory and are included in Table 9.

The contributions of electronic service quality to different dimensions, along with service quality impact on attitude toward site when the task was completed or not are illustrated in Table 10. These contributions do not seem to vary sharply between individuals that completed

Table 9. Fit indices of the second order multi-group model

χ^2	GFI	AGFI	RMSEA [interval] (90%)	SRMR	NFI	NNFI	CFI	χ^2/df	AIC (AICo)
754,75	.89	.89	.060 [.054; .065]	.067	.90	.91	.93	754,75/292 = 2,58	170,75 (5694,78)

Table 10. Factor contributions of NetQual dimensions and their impact on attitude toward the site according to whether the task was completed

	NetQual → Information	NetQual → Ease of use	NetQual → Design	NetQual → Security/ Privacy	NetQual → ATS	mean R^2
Success	.739	.765	.675	.498	.990	0.696
Failure	.801	.763	.710	.595	.971	0.704

the designated task (49%) and those that did not (51%). Regardless of the outcome of navigation, electronic service quality had a strong positive impact on attitude toward the site. In descending order of importance, dimensions in evaluation of electronic service quality are: information, ease of use, design, security/privacy, and whether the individual had completed the task. This result is not surprising and it affirms the structural stability of the NetQual scale.

Even though scale structure is stable regardless of whether the task was completed or not, a significant difference exist in dimensions' mean scores between individuals that successfully completed the task and those that did not. Individuals that completed the task tend to evaluate the site more positively (mean = 4.888) than individuals that did not successfully complete the task (mean = 4.391). This difference is significant at the .00 level (t=4.723). This applies to each scale dimensions. Table 11 summarizes the results test of equality of means for each NetQual' scale dimensions. Notably, the differences in means were higher for the "Information" ($mean_{success}$ -$mean_{failure}$ = .837) and "Security/Privacy" ($mean_{success}$ - $mean_{failure}$ = .728) dimensions, suggesting an important role for information and security/privacy on task completion.

CONCLUSION, LIMITS AND RESEARCH AVENUES

This article provides an empirical test for four academically developed scales that measure electronic service quality: Sitequal (Yoo & Donthu, 2001), Webqual4 (Barnes & Vidgen, 2003) EtailQ (Wolfinbarger & Gilly, 2003), and NetQual (Bressolles, 2006). The results educe the relative superiority of the scale developed as part of our research, NetQual, over the other three scales considered at measuring perceptions of electronic service quality. NetQual best fits the data and exhibits the strongest explanatory power; in addition, the scale has an apparent positive, significant impact on attitude toward site, regardless of the nature of the task performed, i.e. transactional versus informational, or task outcome, i.e. success or failure.

Results demonstrate that the contributions of service quality to different dimensions vary

Table 11. t-test of equality of means for dimensions of electronic service quality according to whether the task was completed

Dimension	Success	Failure	t-Student	Sig.
Information	5,475 (n = 234)	4,638 (n = 252)	6,412	.000
Ease of use	4,424 (n = 244)	4,155 (n = 258)	1,990	.000
Design	4,650 (n = 243)	4,189 (n = 257)	3,871	.000
Security/Privacy	5,470 (n = 222)	4,742 (n = 242)	5,318	.000
NetQual	4,888 (n = 218)	4,391 (n = 240)	4,723	.000

according to the type of task performed on the site. Individuals that performed a transactional task considered information presented on the site, i.e. textual and visual, along with the security/privacy aspect of the transaction to be of particular importance; on the other hand, individuals that completed an informational task placed more importance on site ease of use.

Regarding the impact of task completion on electronic service quality evaluation, success or failure to complete a task has no real impact on NetQual structure; contributions of service quality to each dimension were similar in both situations. Nevertheless, significant difference existed regarding evaluation. Respondents that did not successfully complete the task evaluate each dimension of service quality more poorly than those who successfully completed the task. Respondents' inability to complete the task, therefore, could be said to manifest itself in overall evaluation of site quality by affecting each dimension.

This study has a number of limitations that represent research avenues. To validate results, research should replicate findings on various commercial Websites, particularly e-commerce sites. Notably sites considered in this research exclusively offer services, i.e. travel, insurance, electricity, information, etc... The 'reliability' dimension of the scale which pertains to precision, speed of delivery, and after-sale service among other elements was thus removed. In addition, investigating the quality of electronic service in different cultural consumption contexts (Tsikriktsis 2002) is an interesting research area. Research should attempt to understand the impact of nature of task performed on the site, and its success or failure outcome on electronic service quality evaluation.

Rigorous attention must be paid to the nomological validity of the concept of service quality delivered by commercial Websites. This implies attentive examination of both antecedents and consequences of service quality. Antecedents of electronic service quality refer to concrete elements such as specific characteristics of design, possibility of ordering in a limited number of clicks, logos, and other signs of reassurance among other factors. Consumer perceptions of these attributes shape evaluation of service quality on a site. However, the consequences of electronic service quality have not been adequately studied and conceptualized. Such consequences would probably include both intentions, i.e. to revisit the site, to repurchase, along with actual behaviour, i.e. positive word-of-mouth, purchase volume, etc. (Yoo & Donthu 2001; Swinder *et al.* 2002). While this study demonstrates the impact of electronic service quality on attitude toward site and nuances this impact according to the type of task performed and completion of the task, electronic service quality could be better conceptualized by examining its links with other concepts such as satisfaction after purchase, perceived value, perceived control, and perceived convenience.

In addition, the study of electronic service quality evaluation suggests specific dimensions or attributes of a Website, i.e. ease of use and security, to pose particular influence on consumer decision making. Nonetheless, no published work was found to explore the influence of individual or situational characteristics on electronic service quality. Consideration of these variables may explain context variations in the importance of electronic service quality dimensions. Taking into account socio-economic, motivational and attitudinal criteria such as age, income, innovativeness, impulsiveness, propensity to seek variety, risk aversion, attitude toward advertising and direct marketing, and involvement with IT seem pertinent and necessary for the study of online purchasing (Donthu & Garcia, 1999) and perceived quality of this purchase.

Expectations of Internet users that adopt utilitarian navigational behaviour seem to differ from those users engaged in experiential hedonic behaviour (Hoffman & Novak, 1996). Visitors with goal-directed behaviour, i.e. cognitive, extrinsic motivation, usually visit sites to find specific information about a product/service, solve a particular problem, or purchase a certain product; whereas visitors engaged in hedonic navigational behaviour want to be entertained and would seek an intrinsically gratifying experience. Hoffman & Novak (1996) conclude that "the optimal design of a Website differs according to whether the behaviour is goal-directed or experiential" (p.62). Considering the orientation, utilitarian or hedonic, of the Internet user's navigation behaviour, therefore, might provide more insight into differences of the importance placed on the dimensions of perceived quality evaluation of an online purchase. On the other hand, studying the role of consumer's familiarity with the Internet and expertise at online shopping is inadequate to date (Szymanski & Hise, 2000). Extant research, notably Novak, Hoffman & Yung (2000), considers the level of user expertise as an important factor in studying online behaviour and differentiates between search and navigation habits of experts and novices. User familiarity with the Internet and expertise in online commerce are elements that might modify the importance placed on different dimensions of electronic service quality.

From a managerial standpoint, and similar to the case of traditional service quality in Parasuraman, Zeithaml & Berry (1988), the NetQual scale of electronic service quality would be useful for managers and decision makers such as Webmasters designing and upgrading commercial Websites and would help managers evaluate and monitor changes in perceptions of service quality of retail and service sites. The scale can also be used to set performance objectives in terms of electronic service quality. In addition, online merchants can use the scale to perform a competitive analysis of their sector and highlight main strengths and weakness of site in terms of electronic service quality compared with competitors. Measures of site efficiency such as the analysis of log files, visit-purchase conversion rate, and rate of retention can be determined and analyzed. The scale can then be used to refine analysis of reasons for success or failure of a particular site in terms of electronic service quality. Further studies of these themes are therefore necessary.

REFERENCES

Barnes, S. J., & Vidgen, R. T. (2003). An Integrative Approach to the Assessment of E-Commerce Quality. *Journal of Electronic Commerce Research*, *3*(3), 114–127.

Benbunan-Fich, R. (2001). Using Protocol Analysis to Evaluate the Usability of a Commercial Web Site. *Information & Management*, *39*, 151–163. doi:10.1016/S0378-7206(01)00085-4

Bentler, P. M., & Wu, E. J. C. (2002). *EQS 6 for Windows User's Guide*. Encino, CA: Multivariate Software, Inc.

Bitner, M. J., Brown, S. W., & Meuter, M. L. (2000). Technology Infusion in Service Encounters. *Journal of the Academy of Marketing Science*, *28*(1), 138–149. doi:10.1177/0092070300281013

Bressolles, G. (2006). La qualité de service électronique: NetQual. Proposition d'une échelle de mesure appliquée aux sites marchands et effets modérateurs [Electronic service quality: NetQual – Proposition of a measurement scale to commercial Websites and moderating effects]. *Recherche et Applications en Marketing*, *21*(3), 19–45.

Chen, Q., & Wells, W. D. (1999). Attitude Toward the Site. *Journal of Advertising Research*, *39*(5), 27–37.

Chin, W. W. (1998). Issues and opinion on structural equation modelling. *MIS Quarterly*, (March): 7–16.

Cunliffe, D. (2000). Developing Usable Web Sites - A Review and Model. *Internet Research*, *10*, 295–308. doi:10.1108/10662240010342577

Dabholkar, P. A. (2000). Technology in Service Delivery: Implications for Self-Service and Service Support. In Swartz T.A. and Iacobucci D. (Eds.), *Handbook of Services Marketing* (pp. 103-110). New York: Sage.

Donthu, N., & Garcia, A. (1999). The Internet Shopper. *Journal of Advertising Research, 39*(3), 52–58.

eMarketer (2005). North-America on Line: Demographics and Usage.

Ericsson, K. A., & Simon, H. A. (1993). *Protocol Analysis: Verbal reports as data* (2nd ed.). Cambridge, MA: MIT Press.

Hair, J. F., Anderson, R. E., Tatham, R. L., & Black, W. C. (1998). *Multivariate Data Analysis with Readings* (5th ed.). Upper Saddle River, NJ: Prentice-Hall.

Hoffman, D. L., & Novak, T. P. (1996). Marketing in Hypermedia Computer-Mediated Environment: Conceptual Foundations. *Journal of Marketing, 60*, 50–68. doi:10.2307/1251841

Johnson, E. J., Moe, W. W., Fader, P. S., Bellman, S., & Lohse, G. L. (2004). On the Depth and Dynamics of Online Search Behaviour. *Management Science, 50*(3), 299–309. doi:10.1287/mnsc.1040.0194

Li, H., Daugherty, T., & Biocca, F. (2001). Characteristics of Virtual Experience in Electronic Commerce: A Protocol Analysis. *Journal of Interactive Marketing, 15*(3). doi:10.1002/dir.1013

Lynch, J. G., & Ariely, D. (2000). Wine Online: Search Cost and Competition on Price, Quality and Distribution. *Marketing Science, 19*(1), 83–103. doi:10.1287/mksc.19.1.83.15183

Madu, C. N., & Madu, A. A. (2002). Dimensions of E-quality. *International Journal of Quality & Reliability Management, 19*(3), 246–258. doi:10.1108/02656710210415668

Novak, T. P., Hoffman, D. L., & Yung, Y.-F. (2000). Measuring the Customer Experience in Online Environments: A Structural Modeling Approach. *Marketing Science, 19*(1), 22–42. doi:10.1287/mksc.19.1.22.15184

Nunnally, J. (1978). *Psychometric Theory* (2nd ed.). McGraw-Hill.

Parasuraman, A., & Grewal, D. (2000). The Impact of Technology on the Quality-Value-Loyalty Chain: A Research Agenda. *Journal of the Academy of Marketing Science, 28*(1), 168–174. doi:10.1177/0092070300281015

Parasuraman, A., Zeithaml, V. A., & Berry, L. L. (1988). SERVQUAL: A Multiple-Item Scale for Measuring Consumer Perceptions of Service Quality. *Journal of Retailing, 64*(1), 12–40.

Parasuraman, A., Zeithaml, V. A., & Malhotra, A. (2005). E-S-QUAL: A Multiple-Item Scale for Assessing Electronic Service Quality. *Journal of Service Research, 7*(3), 213–234. doi:10.1177/1094670504271156

Senecal, S., Gharbi, J.-E., & Nantel, J. (2002). The Influence of Flow on Hedonic and Utilitarian Shopping Values. In S. Broniarczyk & K. Nakamoto (Eds.), *Advances in Consumer Research*, 29.

Simon, H. A. (1956). Models of Thought. New Haven, CT: Yale University Press.

Swinder, J., Trocchia, P. J., & Gwinner, K. P. (2002). Consumer Perceptions of Internet Retail Service Quality. *International Journal of Service Industry Management, 13*(5), 412. doi:10.1108/09564230210447913

Szymanski, D., & Hise, R. T. (2000). e-Satisfaction: An Initial Examination. *Journal of Retailing, 76*(3), 309–322. doi:10.1016/S0022-4359(00)00035-X

Totty, M. (2003, June 16). E-Commerce (A Special Report): Selling Strategies - Business Solutions. *Wall Street Journal*, p. 4.

Tsikriktsis, N. (2002). Does Culture Influence Web Site Quality Expectations? An Empirical Study. *Journal of Service Research*, *5*(2), 101–112. doi:10.1177/109467002237490

Wolfinbarger, M., & Gilly, M. C. (2003). eTailQ: Dimensionalizing, Measuring and Predicting Etail Quality. *Journal of Retailing*, *79*(3), 183–198. doi:10.1016/S0022-4359(03)00034-4

Yoo, B., & Donthu, N. (2001). Developing a Scale to Measure the Perceived Quality of Internet Shopping Sites (SITEQUAL). *Quarterly Journal of Electronic Commerce*, *2*(1), 31–47.

Zeithaml, V. A., Parasuraman, A., & Malhotra, A. (2002). Service Quality Delivery through Web Sites: A Critical Review of Extant Knowledge. *Journal of the Academy of Marketing Science*, *30*(4), 362–375. doi:10.1177/009207002236911

APPENDIX 1

List of Items on the NetQual Scale (Bressolles, 2006)

Information:

Info1: This site provides relevant information
Info2: This site provides accurate information
Info3: This site provides in-depth information about the product(s) or service(s) proposed

Ease of Use:

Eofu1: This site is easy to use
Eofu2: It is easy to search for information
Eofu3: This site is easy to navigate
Eofu4: The organization and layout of this site facilitate the search for information
Eofu5: The layout of this site is clear and simple

Site Design:

Design1: This site is colorful
Design2: This site is creative
Design3: This site has an attractive appearance

Reliability:

Relia1: The product or service is delivered by the time promised by the company
Relia2: You get what you ordered from this site
Relia3: You get your merchandise quickly when you order
Relia4: After-sale support on this site is excellent

Security/Privacy:

Secu1: I am confident in the security on this site
Secu2: I feel like my privacy is protected on this site
Secu3: I trust the web site administrators will not misuse my personal information

APPENDIX 2

C.F.A. Results for the Four Scales Tested (Figures 1, 2, 3, and 4)

Figure 1. Sitequal C.F.A.

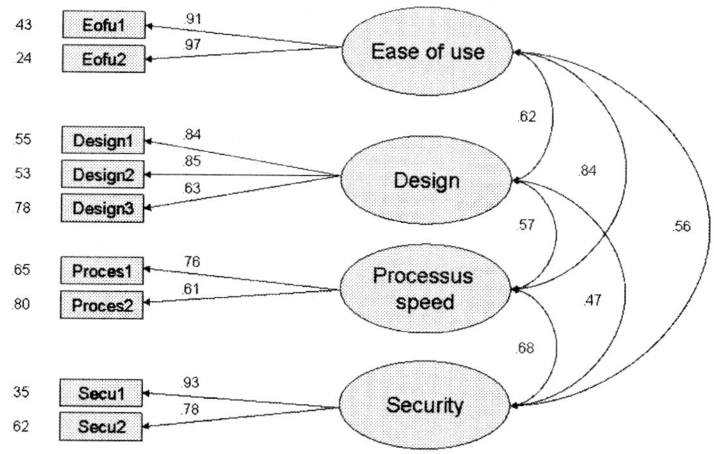

Figure 2. EtailQ C.F.A.

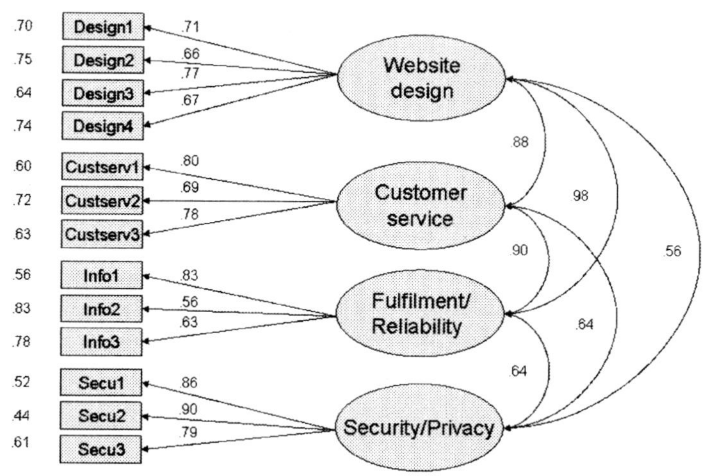

Figure 3. Webqual C.F.A

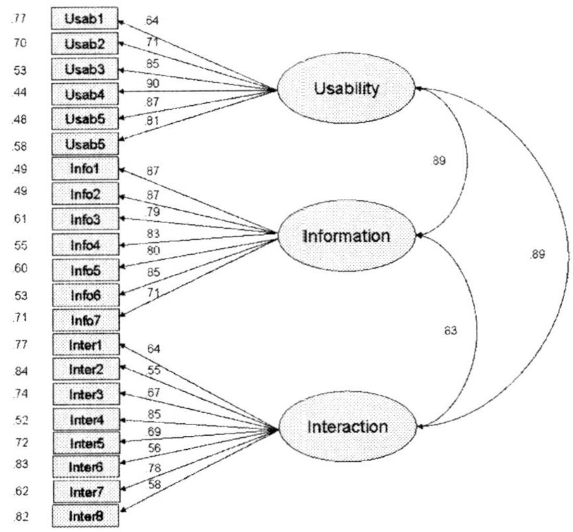

Figure 4. NetQual C.F.A

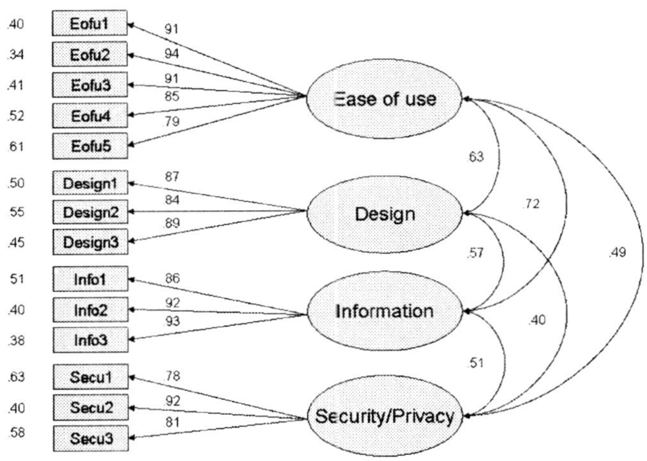

APPENDIX 3

Definition of the Different Fit Indices Retained for Comparing the Four Scales

Parsimony Indices

AIC (Akaike Information Criterion): take into account both the measure of fit and model complexity. AIC has become quite popular in SEM applications, particularly for purposes of examining competitions models.

Chi^2 / df: The Chi-square value represents a test statistic of the goodness of fit of the model, and it is used when testing the null hypothesis that the model fits the analysed covariance matrix perfectly. When the proposed model is fit to the data using a SEM program, the program will judge the obtained Chi-square value in relation to the model's degrees of freedom (df), and output associated is *p* value.

Absolute Indices

RMSEA (Root Mean Square Error of Approximation): take into account model complexity, as reflected in the degree of freedom. Some researchers have suggested that a value of the RMSEA of less than .05 is indicative of the model being a reasonable approximation to the data.

GFI (Goodness-of-Fit Index): measure the proportion of variance and covariance that the proposed model is able to explain (similar to R^2 in regression analysis).

AGFI (Adjusted Goodness-of-Fit Index): similar to the GFI but take into account the number of parameters.

Incremental Indices

NFI (Normed Fit Index): is computed by relating the difference of the Chi-square value for the proposed model to the Chi-square value for the independent or null model.

NNFI (Non Normed Fit Index): is a sample variant of the NFI that take into account the degrees of freedom of the proposed model (model complexity).

Chapter 20
Exploratory Study on the Perceived Importance of Various Features of the Internet Service as Influenced by the Perceived Necessity of the Internet and the Size and Type of Small Businesses

Minh Q. Huynh
Southeastern Louisiana University, USA

Avinash M. Waikar
University of Oklahoma, USA

ABSTRACT

In the new era of e-commerce, small businesses have emerged as the driving force because these firms comprise a significant proportion of economic activity. The spending of small businesses on IT activities continues to grow as they rely more and more on the Internet to be competitive. All these indicate a potential lucrative market for Internet Service Providers (ISPs) to serve small businesses. But how to do so? This study attempts to identify Internet service features that are important to small businesses as a way for the ISPs to exploit this potential lucrative market. It explored how various features of the Internet service were associated with the "perceived necessity" of the internet and the "size" and "type" of small businesses. Understanding these associations might help the ISPs better package their service and more successfully serve their small business clients.

DOI: 10.4018/978-1-60566-910-6.ch020

INTRODUCTION

Small Businesses represent a significant part of the economy around the world. According to In-Stat/MDR (Cyber Atlas, 2002), the small business market in the US alone grew from 2.2 million in 2001 to 2.3 million by the end of 2002, representing more than one-fifth of U.S. businesses. In-Stat forecasted a small but steady increase in the number of organizations with 5 to 99 employees growing as high as 2.7 million by 2006, and employing approximately 42 million workers. In Hong Kong, small and medium-sized enterprises represent 98% of business establishments and 50% of total employment according to Hong Kong Government statistics circa 2007. (Chan and Chao, 2008)

Small businesses around the world are also spending more today on information technology (IT) activities. According to research firm AMI Partners, businesses with fewer than 100 employees spent more than $12 billion on network and telecom equipment in 2004. Furthermore, small businesses are expected to account for 24% of all IT hardware and software spending this year (Hochmuth, 2005). According to a recent study of Latin America Wireless & Mobile Solutions 2008, it reported that Latin American small and medium-sized companies are increasingly adopting new mobile software technologies with 31% of companies having plans to increase investment on mobile CRM and ERP. For these companies mobility is a key factor as over 50% of their work force spends 10% of their time outside the office. (Latin American, 2008) CEO of "Vendio" Mr. Rodrigo Sales said, "We are entering a new era in e-commerce—one increasingly driven by smaller businesses and merchants." (Kooser, 2003). Indeed, the Web hosting industry is waking up to small business needs. Yet, to our surprise, there have not been many studies on Internet Service Providers (ISPs) and the needs of small businesses. Realizing this gap, this study attempts to explore what the small business market's needs are in regard to the services provided by ISPs.

Our broad intention for this study is to look at the potential small business market and explore possible ways for ISPs to exploit this potentially lucrative market. One of the keys to successfully exploit this market, we believe, rests upon the Internet service providers' ability to understand the market needs and to fulfill these needs effectively. Although there are many possibilities, for the scope of this study, we focus mainly on how an ISP could match its many different features to the needs of small businesses. This is essential because how effectively small businesses use the Internet would depend on "features of the Internet service". In this context, this study explored what features of Internet service are important to small businesses and to ascertain whether these preferences are related to certain organizational characteristics such as "Type of small business" and "Size of small business". Specifically, this paper looks for some insights into three aspects of the issue: (1) Is "Size of small business" related to the perceived importance for various features of the Internet service? (2) Do different "Types of small businesses" have different levels of perceived importance for various features provided in the internet service by ISPs? If so, what are the differences? And (3) Is there an association between the perceived importance for features and the perceived necessity for internet service by small businesses?

MARKET BACKGROUND

Before we can explore the small business market potential, we need to assess a few important indicators of this market. We will review available literature to address the following questions. Is there a need for ISPs among small businesses? If so, what are some of the applications that small businesses are using? How much do these small businesses spend on IT?

Let us first examine briefly the computing and web access needs of small businesses. According to Dun & Bradstreet's 20th Annual Small Business Survey (2001) on computer and Internet usage, eighty percent of U.S. small businesses have at least one computer on site used for business purposes, and in some sectors computer usage has almost reached saturation. The survey, which measured attitudes, behaviours and trends in the U.S. small business market, also found out that two-thirds of all small businesses and approximately 85 percent of small business computer owners have Internet access, more than half of those have a Web site and the number is rising. However, only 27 percent of those with a Web site sell on the Internet and average less than three Web-based orders per month. This survey shows that the need for the internet access and web presence is strong regardless of the different characteristics among many disparate companies. (Dun & Bradstreet, 2001)

Regarding the type of transactions that small businesses do, a study conducted by Celent (Moore et al., 2001) stated that small businesses infrequently use the Internet for basic transactions such as online banking. Only two percent of businesses indicated that they regularly paid bills online. Despite the fact that small businesses do not currently utilize the Internet for banking transactions, the study found that the majority of small businesses would like more services automated online, suggesting that the propensity of small businesses to use online services is high, but small businesses are either unaware of what is currently available online or frustrated by the current lack of robust solutions. (Moore et al., 2001)

Regarding the amount of IT spending by small businesses, it is interesting to observe a gap between the commercial small businesses vs. the home office small businesses. According to the results from the COMDEX Small Business Survey by Key3Media Group, Inc., home office small businesses ' IT needs are quite different from that of commercial office small businesses. The differences were reflected in terms of IT investment priorities, IT budgets and preferred purchasing channels. However, the survey did find one similarity between the two groups in that both consider the high cost of implementation to be the major impediment in their acquisition of new technologies. Regardless, small-business spending on information technology and other such functional support areas is going to be significant.

Like any other technology, the pace of changes in IT field is rapid. Any of the technological changes will likely impact small businesses and Internet service Providers. According to ISP-Market LLC report (Taking the Pulse, 2003), wireless broadband and web hosting were two new fast growing business areas. The worldwide market for Web hosting services was expected to grow from some $10.3 billion in 2001 to more than $46.9 billion and more than 40 percent of small businesses were expected to subscribe to broadband in 2006 (Pastore, 2002). More recently, many new tools come out to the market. Microsoft Windows Terminal Server that allows business owners to operate their computer remotely and Webinars that use desktop sharing software to help small business owners reduce waste are two specific solutions that could make life easier for many ISPs, while wireless connectivity and free conference calling are among hot the technologies that could save costs for small businesses. (Marks, 2008) Another promising technology is groupware. According the study conducted by Mrenoño-Cerdán (2008), it was found that groupware can positively affect organizational performance in small-medium size businesses.

Based on these observations of key indicators and trends, the small business market is potentially lucrative to ISPs. The key to successfully exploit this market will rest upon their ability to understand the market needs and to fulfil these needs effectively. Therefore, this paper focuses on what features of internet service are important to small businesses and to ascertain whether these preferences are related to certain organizational characteristics of small businesses such as "Type of small business" and "Size of small business". Specifically, the paper looks for some insights into two aspects of the issue: (1) Does different "Type of small businesses" have different levels of perceived importance for various features provided in the internet service by ISPs? If so, what are the differences? (2) Is the "Size of the small businesses" related to the perceived importance for various features of the internet service? And finally (3) Does perceived necessity of internet service by small business owner make any difference in the selection of features?

Figure 1. The association between "Size" and "Type" of the business, perceived necessity for internet service and the perceived importance of various Internet service features offered by Internet Service Provider

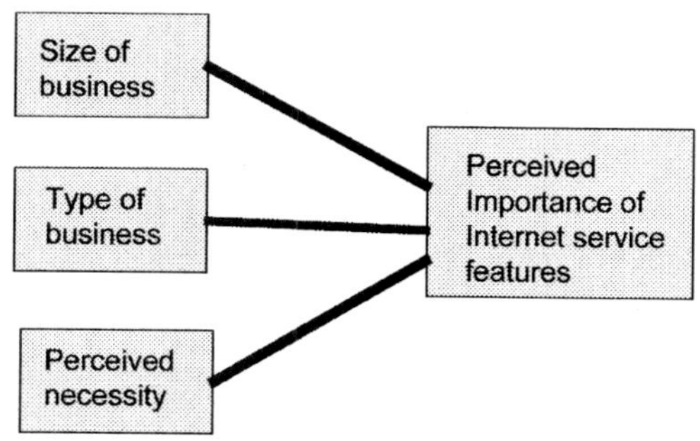

CONCEPTUAL CONTEXT

Figure 1 shows a research diagram used to seek answers to the questions in this study. It consists of four major constructs: "Size of business", "Type of business", "perceived necessity for internet service by small businesses", and the "perceived importance of various Internet service features". In this section, we will discuss some of the justifications for these variables including their descriptions and significance.

Type of Business

The use of internet may not be embraced by all firms (Daniel et al. 2002; Feindt et al . 2002). Past studies on small businesses tended to ignore subtle differences among various types of small businesses. For example, in many studies, all small businesses were considered as one homogeneous entity (Magal and Lewis, 1995; Mirchandandi, et al 2001; Cragg, et al. 1993). However, in reality, the small businesses are not uniform, but they include a wide range of business types such as construction, retailing, manufacturing, health care, services, etc. Based on their review of studies related to Information, Communication Technologies (ICT), Barton and Bear (1999) found that not all enterprises have the same ICT needs, because not al all microenterprises and small firms face similar challenges. We posit that different types of small businesses may have different needs in terms of Internet features. Hence, "Type of business" was employed as one of the major constructs in this study. Our second research question is "How does the perceived importance for various Internet features differ according to "Type of business"?

Size of Business

Prior studies showed the relevance of the construct "Size" to the firm's adoption and use of IT. For instance, the size of the firm, in particular, has been found to be related to adoption rates (DTI 2000; Haynes et al. 1998; Riquelme 2002) as well as the intensity and level of information and communication technology (ICT) activity (Fallon and Moran 2000; Palvia and Palvia 1999). In their literature review, Chen, et al. (2006) reported that the size of the business along with age, education, and computer knowledge are the most important variables related to information systems (IS) in small businesses. Furthermore, it is also reflected in Radovilsky and Hedge (2004)'s study of the variables influencing web commerce implementation, in which, they found that company size was positively correlated with the adoption of web commerce as a business strategy.

While the business size is readily recognized and well defined in the previous studies, the term "small businesses" nevertheless shows much diverse interpretations. According to the Small Business Administration (SBA), in the United States, businesses with less than 500 employees are considered small businesses. Taking the context beyond the U.S., the definition of small businesses becomes little more complicated. Instead of using the term "small businesses", various descriptive terms such as micro-, small- and medium-sized enterprises are used in Europe and their interpretations may vary widely. For instance, the term MSE includes "micro-enterprises" and "small enterprises" and these enterprises are considered the smallest. However, most numerous businesses within the larger group of enterprises are called SMEs (small-and medium-sized enterprises), (Donner 2004). Not having one common term with precise description is just one problem in research involving small businesses. Another difficulty is how to determine the threshold for the size. What size makes a business a small business? There is no agreement on an absolute number. While SBA set 500 employees as a drawing line, this arbitrary number is impractical because it is found that a large number of US small businesses have fewer than 20 employees (Zimmerer and Scarborough 2003). Therefore, studies (such as Zank and Vokurka 2003, Boyer and Olson 2002; Barua et al. 2001) of 'small business' that track firms with fewer than 100 employees may not fully capture the nuances in usage within very small businesses. This means that small business population is not homogeneous but diverse in terms of size. To account for the non-homogeneity, this study differentiates various small businesses based on the number of employees using four size groups: (a) 1-25 employees, (b) 26-50 employees, (c) 51-100 employees, and (d) more than 100 employees.

Perceived Necessity

Although the Internet is quite common and its use in businesses has been pervasive, it is interesting to find out how actual small businesses perceive the necessity of internet service. Based on their responses, this study explored further whether there is an association between this perceived necessity and the preferences regarding various features provided by internet service providers.

Perceived Importance of Internet Service Features/Benefits

When considering different features/benefits offered by various ISPs in the Internet service, it may not be simple to identify a list/subset of the most important features. Almost two decades ago, simple capabilities such as file transfer, terminal emulation, speed of the dial up would be sufficient. Over the years, the choices of features have greatly expanded. They now range from ease of installation to connection reliability and from quality of customer support to multi-protocol operability. Recently, security and privacy have also become important factors to consider. Because of such a wide scope of features and capabilities, it is necessary to categorize them with a broader notion. Our literature review revealed very limited number of studies on ISPs regarding this aspect. As a result, there is no existing framework that we could adopt for our study.

Therefore, we developed a list of features and benefits of the Internet service that we believed are relevant and important to small businesses. This list was refined and modified after informally consulting a number of small business owners. The final list of features in shown Table 1 was adopted and used in this study. To organize these features further, we classified them into three categories of technical capability, technical support, and non-technical features.

Table 1. List of categories, important features, and their descriptions

CATEGORY	FEATURE	DESCRIPTION
Technical capability	Multiple User Capability	Permits multiple users to connect to network and share information at the same time without breakdowns.
	Speed of Connection	Speed with which Internet data is transmitted over telephone lines. Speed is dependent on the type of connection. Different connections work at different speed such as Dial up (56 Kbs), ISDN (64 Kbps-128 Kbps), DSL (128 Kbs - 768 Kbs), Satellite (400 Kbps) and Cable (200 Kbs - 2 Mbs).
	Remote Access Capability	Access through a terminal to a computer that is geographically removed from the terminal.
	Web presence (also called Server space)	File storage space that is available to anyone on the World Wide Web through ISPs. Web space is typically used for storing personal Web pages. Web hosting is a typical service provided by many Internet Service Providers to design web sites for small businesses and help them maintain Internet presence on a continuous basis.
Technical support	Repair/Quality of service	Quality of after sale and repair services offered by an Internet Service Provider.
	Accessible Help Line	Telephone help line that can be easily/readily accessed 24 hours.
	Ease of Installation/Set–up	Ease with which configuration of hardware and connectivity to Internet is established.
Non technical	Company's Name Recognition	Reputation and standing of Internet Service Provider.
	Small Business Incentives	Special incentives offered to small businesses.
	Cost per month of Service	Fee and other costs paid per month by user for Internet Service

RESEARCH QUESTIONS AND HYPOTHESES

Research Design

This study used the survey method to collect data. A questionnaire was designed to determine the small businesses' preferences regarding various features provided by internet service provider. Though the questionnaire consisted of three major segments, this study focused mainly on the first segment that consisted of demographic questions such as type of industry, number of employees, years of existence, gender. Hence, the scope of this study focused mainly on the relationship between the level of importance of various features and the type of small businesses, their size, and their perceived necessity of internet service.

As shown in Table 1, the selected 10 features/ benefits included in this study are grouped into three major categories as described in the previous section. The first category is technical capability that consisted of the following features: multiple user capability, speed of connection, remote access capability, web space presence. The second category is technical support with features including repair/quality of service, accessible help line, ease of installation/set-up. The final category is other non-technical factors including company's name recognition, small business incentives, and cost per month of service.

Research Hypotheses

In this study, we wanted to examine the relationship between the preference of various features/ capabilities provided by ISPs and the type and size of small businesses. Understanding how the preferences varied according to different type and size of small businesses and how important specific features are to different small businesses would be very beneficial to ISPs. Such knowledge would help in guiding their market campaign and also in developing appropriate package of features to meet the needs of various small businesses.

As described in the earlier section, the focus of this study was to explore the following three questions specifically. First, is there any difference among various small business types in selecting features provided by ISPs? Second, does the size of small businesses influence the preference on certain features provided by ISPs? Finally, does the perceived necessity for internet service by small business owner affect the selection of features?

Based on the general research questions, we proposed the following hypotheses for this study.

H1: ***The perceived level of importance regarding the technical capability features is independent of the size of small businesses. The technical capabilities include specifically the following features:****multiple user capability, speed of connection, remote access capability, web space presence.*

H2: ***The perceived level of importance regarding the technical support features is independent of the size of small businesses. The technical supports include specifically the following features:****repair/quality of service, accessible help line, and ease of installation/set-up.*

H3: ***The perceived level of importance regarding the non-technical features is independent of the size of small businesses. The other non-technical factors include specifically the following:****company's name recognition, small business incentives, and cost per month of service.*

H4: ***The perceived level of importance regarding the technical capability features is independent of the type of small businesses. The technical capabilities include specifically the following features:****multiple user capability, speed of connection, remote access capability, web space presence.*

H5: ***The perceived level of importance regarding the technical support features is independent from the type of small businesses. The technical supports include specifically the following features:****repair/quality of service, accessible help line, and ease of installation/set-up.*

H6: ***The perceived level of importance regarding the non-technical features is independent of the type of small businesses. The other non-technical factors include specifically the following:****company's name recognition, small business incentives, and cost per month of service.*

H7: ***The perceived level of importance regarding the technical capability features is independent of the perceived necessity for internet service by small businesses. The technical capabilities include specifically the following features:****multiple user capability, speed of connection, remote access capability, web space presence.*

H8: ***The perceived level of importance regarding the technical support features is independent of the perceived necessity for internet service by small businesses.. The technical supports include specifically the following features:****repair/quality of service, accessible help line, and ease of installation/set-up.*

H9: ***The perceived level of importance regarding the non-technical features is independent of the perceived necessity for internet service by small businesses. The other non-technical factors include specifically the following:****company's name recognition, small business incentives, and cost per month of service.*

Survey Sample and Instrument

The population for the study consisted of small businesses from the region of Southeastern Louisiana. A convenience sample of 500 companies was selected from Southeastern Louisiana University's Small Business Development Center clients database. 500 questionnaires were sent out and 120 questionnaires were completed and returned. The response rate was 24%. The participants (n=120) consisted of businesses from seven types of industries: retailing (R), services (S), real estate (RE), housing (H), construction (C), manufacturing (M) and others (O). They were categorized into four different groups based on the number of employees: (a) 1-25, (b) 26-50, (c) 51-100, (d) more than 100 employees. All of the respondents included in the analysis indicated that they had subscribed to the Internet service. To ensure this, the questionnaire included the question "How long has your company had an Internet service provider?" Respondents who did not have Internet service were eliminated from the analysis.

A questionnaire was developed containing a list of features and benefits (See Table 1) that the small businesses would like to see in their Internet service. Our literature review indicated a very limited number of academic studies on this topic. Hence, no instrument was available for adoption in our study. Drawing from related trade journals, we developed a list of features and benefits of the Internet service that we believed was relevant and important to small businesses. The content validity of the initial list was evaluated by a number of small business owners. They were asked to judge each item and to comment on the relevance as well as provide suggestions for improving the list. Their input allowed us to develop the final list of features as shown in Table 1.

The questionnaire along with a letter explaining the purpose of the study was sent to the businesses in the selected sample. Participants completed the questionnaire, which requested them to provide demographic information and to respond to items related to the importance of various features of Internet service. The demographic questions included information on type of industry, number of employees, years of existence, and gender of the respondent. For instance, "What industry is your company in?" Possible responses included: Construction, Retail, Real Estate, Manufacturing, Healthcare, Service, and Other. Finally, a question asked participants to rate the importance of the features/benefits when selecting an Internet Service Provider. Ten features shown in Table 1 were presented to respondents for important ratings. The importance ratings reflecting preferences were to be based on Likert rating scale of 1 to 5 where, 1 = Very Unimportant, 2 = Unimportant, 3 = Neutral, 4 = Important, and 5 = Very Important. The importance ratings reflected their preferences for each of the features/benefits.

Data Analysis

The research questions assess how the perceived importance of various features/benefits of Internet service is related to different types and sizes of small businesses. We employed the frequency analysis in this study. Specifically, we used Chi-square to test independence to examine the association between the type and size of small businesses and the perceived importance of various features on Internet service provided by ISPs. The questionnaire responses were analyzed using statistical software SPSS-Version 10. Although Chi-square analysis is a rather simple statistical technique, it is an appropriate tool for the purpose of our study; namely, to assess the association between variables rather than to determine the cause and effect relationship. Since Chi-square test is a common tool for frequency analysis, the data was coded and analyzed according to the standard practices outlined in the SPSS manual and statistical textbook such as Anderson, Sweeney, and Williams' Statistics for Business and Economics, 2005 edition. Past studies (Mahony, et al., 2006; Bennett, et al., 2006, and Harman, et al., 2005/2006) had used the same technique for similar purpose to examine the association between two variables and another. Thus, Chi square test of independence was applied in this study because the data obtained from the survey were measured at ordinal levels. Contingency tables were used to examine the pattern between importance ratings for features offered by ISPs and organizational variables for significance. Categories were collapsed because while computing χ^2 test statistic it is necessary to ensure that none of the cells have expected frequencies of less than five. Categories were collapsed as necessary for computing χ^2 test statistic to ensure that none of the cells had expected frequencies of less than five. Table 2 is an illustration of a contingency table with collapsed categories for both factors: size of business (number of employees) and the

Table 2. The relationship between features & benefits and number of employees

Features / Benefits		**Number of employees**					
		Unimportant	Unimportant	Important	Important	χ^2 test statistic	Significant
Multiple user capability	Observed Frequency	36	2	59	19		
	Expected Frequency	31.12	6.88	63.88	14.12		
		0.77	3.46	0.37	1.69	6.28	
Company's name recognition	Observed Frequency	52	7	42	13		
	Expected Frequency	48.65	10.35	45.35	9.65		
		0.23	1.08	0.25	1.16	2.73	Y
Speed of connection	Observed Frequency	21	1	74	20		
	Expected Frequency	18.02	3.98	76.98	17.02		
		0.49	2.23	0.12	0.52	3.37	Y
Remote access capability	Observed Frequency	39	5	56	16		
	Expected Frequency	36.03	7.97	58.97	13.03		
		0.24	1.1	0.15	0.67	2.17	Y
Repair/quality of service	Observed Frequency	18	1	76	20		
	Expected Frequency	15.53	3.47	78.47	17.53		
		00.39	1.76	0.08	0.35	2.58	Y
Accessible help line	Observed Frequency	26	2	69	19		
	Expected Frequency	22.93	5.07	72.07	15.93		
		0.41	1.86	0.13	0.59	2.99	Y
Ease of installation	Observed Frequency	28	2	67	18		
	Expected Frequency	24.78	5.22	70.22	14.78		
		0.42	1.98	0.15	0.7	3.25	Y
Small business incentives	Observed Frequency	29	6	66	15		
	Expected Frequency	28.66	6.34	66.34	14.66		
		0.0	0.02	0	0.01	0.03	Y
Cost per month	Observed Frequency	21	4	74	16		
	Expected Frequency	20.65	4.35	74.35	15.65		
		0.01	0.03	0	0.01	0.04	Y
Web presence	Observed Frequency	35	5	50	12		
	Expected Frequency	33.33	6.67	51.67	10.33		
		0.08	0.42	0.05	0.27	0.82	Y

importance rating. The initial five categories for importance ratings (Very Unimportant, Unimportant, Neutral, Important, and Very Important) were collapsed into two categories of Unimportant and Important. The initial four categories for Size of business (1-25, 26-50, 50-100, more than 100) were collapsed into two categories (<50 and >50). Table 3 shows an illustration of a contingency table with collapsed categories for both factors: type of business and the importance rating. Seven

Table 3. Features and benefits preferred by type of industry (C=Construction, R=Retail, RE=Real estate, M=Manufacturing, H=Health, S=Service, and O=Others)

Features / Benefits		**Type of industry**					
		C+M+RE+O	R+H+S	C+M+RE+O	R+H+S		
		Unimportant	Unimportant	Important	Important	χ^2 test statistic	Significant
Multiple user capability	Observed Frequency	13	24	29	47		
	Expected Frequency	13.75	23.25	28.25	47.75		
		0.04	0.02	0.02	0.01	0.10	Y
Company's name recognition	Observed Frequency	23	35	19	35		
	Expected Frequency	21.75	36.25	20.25	33.75		
		0.07	0.04	0.08	0.05	0.24	Y
Speed of connection	Observed Frequency	8	14	34	57		
	Expected Frequency	8.18	13.82	33.82	57.18		
		0	0	0	0	0.01	Y
Remote access capability	Observed Frequency	14	30	28	41		
	Expected Frequency	16.35	27.65	25.65	43.35		
		0.34	0.2	0.22	0.13	0.88	Y
Repair/quality of service	Observed Frequency	8	11	33	60		
	Expected Frequency	6.96	12.04	34.04	58.96		
		0.16	0.09	0.03	0.02	0.30	Y
Accessible help line	Observed Frequency	9	18	33	53		
	Expected Frequency	10.04	16.96	31.96	54.04		
		0.11	0.06	0.03	0.02	0.22	Y
Ease of installation	Observed Frequency	8	21	34	49		
	Expected Frequency	10.88	18.13	31.13	51.88		
		0.76	0.46	0.27	0.16	1.64	Y
Small business incentives	Observed Frequency	13	21	29	50		
	Expected Frequency	12.64	21.36	29.36	49.64		
		0.01	0.01	0	0	0.02	Y
Cost per month	Observed Frequency	7	17	35	53		
	Expected Frequency	9	15	33	55		
		0.44	0.27	0.12	0.07	0.91	Y
Web presence	Observed Frequency	13	27	22	37		
	Expected Frequency	14.14	25.86	20.86	38.14		
		0.09	0.05	0.06	0.03	.24	Y

categories of business type were collapsed into two categories as shown in Table 3. A similar operation was applied to generate Table 4 to display the perceived necessity and the importance rating. A high value of the Chi-square test statistic indicates that the observed frequencies significantly deviated from the expected frequencies, leading to the rejection of the "null" hypothesis. This implies dependence or association between the two factors. A low value indicates no significant

Table 4. The relationship between features & benefits and perceived necessity for Internet service by small businesses

Features / Benefits		**Is Internet service a necessity for small businesses?**					
		Unimportant YES	Unimportant NO	Important YES	Important NO	χ^2 test statistic	Significant
Multiple user capability	Observed Frequency	21	16	66	12		
	Expected Frequency	27.99	9.01	59.01	18.99		
		1.75	5.43	0.83	2.57	10.57	Y
Company's name recognition	Observed Frequency	41	17	45	10		
	Expected Frequency	44.14	13.86	41.86	13.14		
		1.91	5.95	0.45	1.41	9.72	
Speed of connection	Observed Frequency	11	11	76	17		
	Expected Frequency	16.64	5.36	70.36	22.64		
		1.91	5.95	0.45	1.41	9.72	Y
Remote access capability	Observed Frequency	26	17	61	11		
	Expected Frequency	32.53	10.47	54.47	17.53		
		1.31	4.07	0.78	2.43	8.6	Y
Repair/quality of service	Observed Frequency	10	9	77	18		
	Expected Frequency	14.5	4.5	72.5	22.5		
		1.4	4.5	0.28	0.9	7.08	Y
Accessible help line	Observed Frequency	19	8	68	20		
	Expected Frequency	20.43	6.57	66.57	21.43		
		0.1	0.31	0.03	0.09	0.53	
Ease of installation	Observed Frequency	22	8	64	20		
	Expected Frequency	22.63	7.37	63.37	20.63		
		0.02	0.05	0.01	0.02	0.1	
Small business incentives	Observed Frequency	24	10	63	18		
	Expected Frequency	25.72	8.28	61.28	19.72		
		0.12	0.36	0.05	0.15	0.67	
Cost per month	Observed Frequency	22	3	64	25		
	Expected Frequency	18.86	6.14	67.14	21.86		
		0.52	1.61	0.15	0.45	2.73	
Web presence	Observed Frequency	22	15	51	11		
	Expected Frequency	28.96	10.04	46.04	15.96		
		0.85	2.45	0.53	1.54	5.38	Y

deviation between the observed and expected frequencies, implying independence, that is no relationship between two factors.

RESULTS

Demographic Profile

Several demographic variables were explored in the study such as type of business, number of employees, years in existence, the respondent's position in the organization, and gender. The respondents were requested to classify their business in one of the seven categories as shown in Figure 2. Thus, the sample represented companies from seven different industries namely retailing (R), services (S), real estate (RE), housing (H), construction (C), manufacturing (M) and others (O). About 29 percent of the companies were from retailing, 27 percent from services, 6 percent from construction, 8 percent from real estate, 4 percent from manufacturing, 7 percent from housing and 19 percent were from other industries.

As seen in Figure 3, 68 percent of the companies in the sample had fewer than twenty five employees, 15 percent had a workforce between twenty-five and fifty, 8 percent between fifty-one and hundred, only 9 percent had more than hundred employees.

As can be seen in figure 4, of all the respondents, 75 percent considered Internet as an important tool and a necessity for the company and 58 percent acknowledged that the Internet service has been beneficial for the company. While 21 percent felt that the service has not been beneficial for the company and another 21 percent remained neutral.

Hypothesis Testing Results

The results for the hypotheses were summarized in Table 5. Table 5 shows various null hypotheses along with the conclusion whether the null hypotheses were supported or not supported. ***Sup** meant the null hypothesis was supported because p-value showed no significance. On the other hand, **Not Sup** meant that the null hypothesis was rejected. With p-value showing significance, this implies an association between constructs in the hypothesis.

Findings for H1, H2, and H3 and Result Interpretations

The first research question explored whether there is a relationship between the perceived level of importance regarding various features provided by Internet service providers and the size of small business or not. Recall that size was measured in terms of number of people that a business employs. It has four ranges: 1-25, 26-50, 51-100, and more than 100. This question was addressed by H1, H2, and H3.

Figure 2. Businesses in the sample belonging to different industries

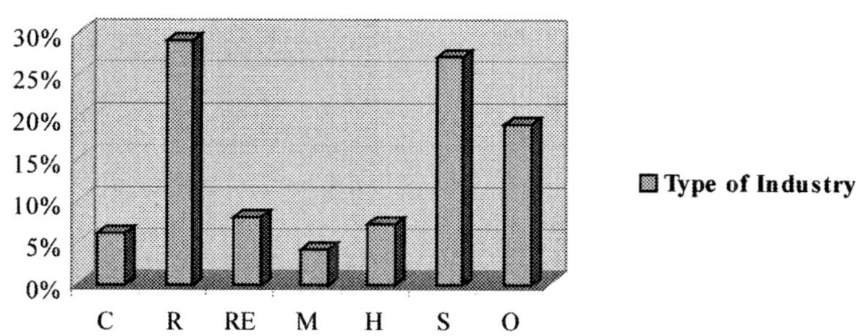

Figure 3. Proportion of industries in the sample by the number of employees

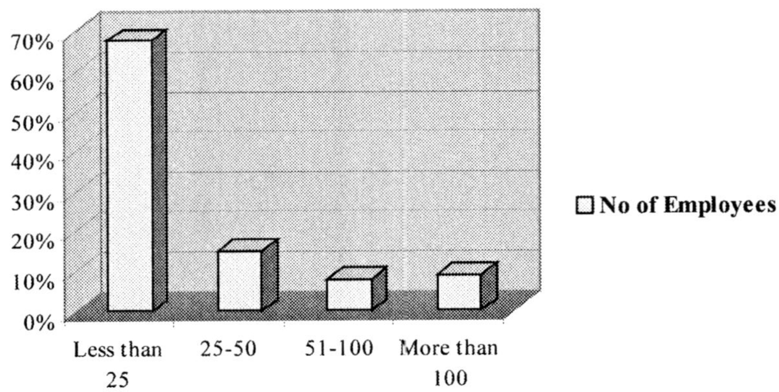

For H1, the specific question is whether the perceived importance of different technical capability features independent of the size of the business. The results showed that H1 was partially supported because the size as a variable was found to be significantly related to the technical capability features of multiple user capability ($p < 0.01$) and speed of connection ($p < 0.07$).

This is expected because more the number of employees a small business has, more likely it has to rely on a multi-user environment to support its operation rather than on a cluster of standalone systems. Then, network connection becomes the key to provide shared resources such as files, documents, printers, and database. With more users accessing the system via the network, the speed of connection also becomes critical. Hence, it is consistent that the result showed a relationship between importance of both multiple user capability and speed of connection with the size of small business.

The relationships between the importance for remote access capability and web space presence and the size of small business were found to be not statistically significant. This could be explained in terms of small businesses' scope of operations. Many small businesses tend to cater their products and services to local needs. Hence, their scope is local or regional. As a result, their employees are not scattered at many different locations as in

Figure 4. Internet as a necessity and whether beneficial or not

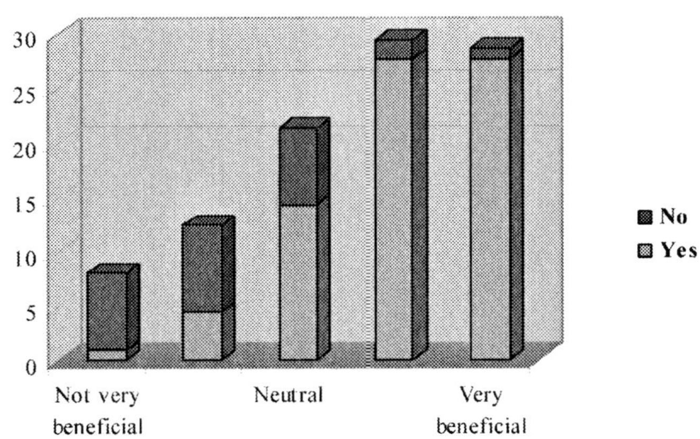

Table 5. Summary of results

Hypotheses	
H1: ***The perceived level of importance regarding the technical capability features is independent of the size of small businesses.***	*Sup – Multiple user capability
	*Sup – Speed of connection
	Not Sup – Remote access capability
	Not Sup – Allow web space
H2: ***The perceived level of importance regarding the technical support features is independent of the size of small businesses.***	*Sup – Accessible help line
	*Sup – Ease of installation / setup
	Not Sup – Repair / quality of service
H3: ***The perceived level of importance regarding the non-technical features is independent of the size of small businesses.***	*Sup – Company's name recognition
	Not Sup – Small business incentives
	Not Sup – Cost per month of service
H4: ***The perceived level of importance regarding the technical capability features is independent of the type of small businesses.***	Not Sup - Multiple user capability, Speed of connection, Remote access capability, Allow web space
H5: ***The perceived level of importance regarding the technical support features is independent of the type of small businesses.***	Not Sup - Accessible help line, Ease of installation / setup, Repair / quality of service
H6: ***The perceived level of importance regarding the non-technical features is independent of the type of small businesses.***	Not Sup - Company's name recognition, Small business incentives, Cost per month of service
H7: ***The perceived level of importance regarding the technical capability features is independent of the perceived necessity for internet service by small businesses.***	Not Sup - Multiple user capability, Speed of connection, Remote access capability, Allow web space
H8: ***The perceived level of importance regarding the technical support features is independent of the perceived necessity for internet service by small businesses.***	Not Sup - Repair / quality of service *Sup - Accessible help line, Ease of installation / setup,
H9: ***The perceived level of importance regarding the non-technical features is independent of the perceived necessity for internet service by small businesses.***	*Sup - Company's name recognition, Small business incentives, Cost per month of service

the case of large businesses. Furthermore, their reach is also limited to local rather than global community. These are perhaps reasons why remote access capability and web space presence are perceived to be lower in the level importance than other technical capability features.

For H2, the specific question is whether the perceived importance of different technical support features independent of the size of the business. The results showed that H2 was also partially supported with he statistically significant relationship between the size of small businesses and two technical support features specifically accessible help line ($p < 0.08$) and ease of installation/set up ($p < 0.07$).

The increase in number of employees does lead to an increase in the perceived preference for accessible help line and ease of installation/set up. These results are consistent with those in H1. As the size of small businesses increases, so does the complexity especially with the reliance on a multi-user environment. Most of small businesses are generally not tech savvy. As a result, easy access to tech support and simple installation/set up are important. However, an expected finding is that the increase in the size of small businesses does not increase the preference for repair/quality of service. This may be explained in terms of service contract. To save costs, many small businesses do not maintain a service contract with their ISPs as in the case of their counterpart large businesses. Subsequently, for them, the level of importance for repair/quality of service was perceived to be lower than those of others.

For H3, the specific question is whether the perceived importance of non-technical features independent of the size of the business. The results showed that H3 was weakly supported because only company's name recognition was found statistically significant to the size of small businesses. As expected, the ISP's reputation is important in the selection of service by small businesses. This is perhaps related to the small businesses' lack of technical knowledge and limited exposure of various choices. They tend to pick services from those that they are familiar and recognized. What is interested in the results are the small businesses' perceived low level of importance on both incentives and monthly service charge by ISPs. It is expected that most small businesses are generally cost conscious. For them, discount incentive and low service charge were thought to be highly important. Yet, they were not. One possible explanation is there are not enough incentives offered by ISPs to small businesses as well as various cost structures available for small business to pick from. The lack of those incentives and one set of charge for all cause small businesses to perceive their importance to be lower.

Findings for H4, H5, and H6 and Result Interpretations

The hypotheses H4, H5, and H6 focused on the possibility of whether there is a relationship between the perceived level of importance regarding various features provided by ISPs and the type of small businesses. The type was specified according to the following categories: construction, retail, real estate, manufacturing, healthcare, service, and others.

All H4, H5, and H6 were not supported. Although the relationships between the type of industry and all three categories of features were found to be statistically insignificant, these results are interesting in a number of ways. Common notion suggests that different type of businesses should have different preferences regarding features because of differences in their operations. Perhaps, this is true with larger businesses. Yet, unexpectedly for small businesses in this study, the "type of business" construct doesn't seem to play a significant role. The perceived preference on features does not vary according to the type of small business. This could be taken as good news for ISPs. Instead of custom creating packages for various types of small businesses, ISPs should pay attention to different needs based on various sizes of small businesses. In other words, ISPs should offer few packages with a subset of features that meet the preference of certain size group. This could simplify greatly the design of Internet service package for small businesses given their characteristics of highly fragmented market segment. Also, ISPs could concentrate on the size of small businesses and cater their services to meet the needs of small businesses based on their size rather than their type.

Findings for H7, H8, and H9 and Result Interpretations

The hypotheses H7, H8, and H9 were concerned with the question whether there is an association between the perceived necessity for internet service by small businesses and importance of various features provided by ISPs.

H7 was supported because the perceived necessity for internet service by small businesses was found to be significantly related to importance of all of the technical capability features. This shows that the technical capability features are important to small businesses. As they rely more on the internet service to do their businesses, they desire reliable and robust technical capability.

For H8, the only feature that was supported was the preference for Repair/Service Quality while both the preference for Accessible Help Line and the preference for Ease of Installation showed no statistical significance for the perceived necessity for internet service by small businesses. These results are quite interesting because they seem to be contrary to those results in H2 where Accessible Help Line and Ease of Installation were found to be important. One possible explanation is that those small businesses who expressed the necessity for internet service were more technically savvy. As computers become more pervasive and internet becomes a necessity, small businesses also become more familiar with basic set up and troubleshooting. This might explain the reason why these small businesses were less reliant on help line and installation assistance. However, when it comes to repair and service, it is interesting to observe that small businesses definitely rely on the support from the service providers.

H9 indicated that there were no statistical significance between the perceived necessity for internet service by small businesses and the non-technical features such as the incentives offered, cost per month, and name recognition. This is an interesting result because it might cause the ISPs to rethink about their marketing strategy. Under a normal circumstance, measures such as incentive and brand promotion are important means to retain existing businesses and lure new customers. This might not be the case with small businesses using the internet service. As a result, the ISPs might be better off in focusing more on technical features rather than non-technical features to attract small businesses. Preferences for features like multiple user capability, speed of connection, remote access capability, repair/quality of service, web space allowed are not independent of perceived necessity for internet service, in other words companies that considered Internet as a necessity are also likely to rank these benefits higher than other benefits like accessible help line, ease of installation/set-up, small business incentives, company's name recognition, and cost per month of service.

DISCUSSION

Internet is an important source of competitiveness for small businesses as more and more small businesses rely on it for customer service and communication. Small businesses also use the Internet as proactive customer relations and marketing tool. They exchange email with customers and use websites fro promotion. Internet usage brings benefits to small businesses. Among the most important benefits is perhaps the positive impact of Internet usage on their bottom line. Therefore, small businesses are embracing the Internet and recognizing the value in using it for their competitive advantage.

These observations are consistent with the evidence from the literature. Many studies showed that Internet is changing the way many businesses are operating. According to Cottrill (1997), Power and Sohal (2002), the Internet is the enabler that ultimately revolutionizes the way business is conducted in some industries. Research shows that the use of Internet by businesses has led to improvement in lead times (Hauguel and Jackson 2001; Power and Sohal 2002), better communication and collaboration within and across organizations (Garcia-Dastugue and Lambert 2003), and significant gain in achieving greater cost efficiencies (Barnes et al. 2003).

Given these findings, it appears that Internet may be assuming a transforming role in helping small businesses achieve their business objectives. Again, the literature show that the Internet was a powerful tool enabling small businesses to 'level the playing field' when competing with larger firms. (Levenburg 2005) There is growing evidence that the Internet has indeed changed the way many small businesses are conducting their operations. (Barnes et al. 2003) Perhaps, the use of Internet has penetrated much deeper into the small business market. To gain competitiveness in producing and selling products or delivering services effectively and efficiently, small businesses must rely on the use of Internet. This in turn shows a critical connection between small businesses and their Internet service providers.

As seen, there is an ever-increasing competition between Internet service providers, largely because there are no restrictions to market entries. The gap between a local and a national player is waning, as all ISPs compete on the same platform and try to lure as many customers as they possibly can. They could possibly use the results of this study. Findings from this study suggests that the size rather the type of small business is a relevant variable in assessing what features in Internet service are perceived to be important by small businesses. Perhaps, this is why more and more ISPs have recently begun to woo "smaller" businesses with tailor-made online services, (Kooser, 2003).

According to research company AMI-Partners, small and midsize businesses in 2006 will spend $175 billion on IT (Pereira, 2005). Hence, the small businesses in part represent most growth and potential lucrative market for ISPs. A number of big online service providers already take position in carving out this market. That's why companies such as AOL, Covad, Microsoft, Vendio, and Yahoo are actively wooing small businesses with many offerings designed specifically to meet their unique needs. (Kooser, 2003) This is certainly an awakening call for many smaller ISPs. To survive, they need to right-set their strategy and engage small businesses to utilize their services. They have to be competitive and proactive in developing their own niche.

Traditional view often assumes that service charge such as cost per month of service should be perceived to be important, the result for this study shows the contrary. What does this mean? This should not be taken to mean that small businesses do not prefer low cost internet service. Rather, the result reveals the need for differentiated pricing structure. In increasingly significant segment of the Internet, market is moving away from simple cost-driven undifferentiated commodity market into a market where there is a serious attempt to provide differentiation based on a specific feature of the delivered services. The result of low level of importance placed on non-technical feature "small business incentives" provides another interesting evidence for an emerging opportunity that ISPs could pursue. What this means is that ISPs should develop more differentiated incentives to attract their potential small business clients. Verizon is the case in point. In order to temp small businesses, Verizon offer the "Open Hosted Speech Services" (OHSS) for speech applications, a new interface for road warriors to locate wireless fidelity access points, and FiOS TV service to differentiate itself from other competitors. (Verizon tempts..., 2008)

An observation that is most characterized of the challenges facing ISPs is the readily available alternative for today's small business customers. For instance, small business customer awareness has increased manifolds over the years. Consumers have become much more demanding than before and make their choices after considering all relevant factors. Due to the availability of wide choice, switching from one provider to another is not overruled at any time. As per a report published by J.D. Power and Associates (AT&T Ranks, 2000), 14% of ISP subscribers are likely to change provider in a single year and another 22% consider making a change at some point in the future. The study also cited that connection speed is the most common reason for changing providers (79%).

Selection of a service provider is also affected by word of mouth. A bad experience of one customer may lead to a loss of another ten potential customers thereby making it essential for Internet Service Providers to gauge customers' needs from time to time so as to provide them competent and valuable services.

CONCLUSION

Technologies like the Internet have permanently altered the nature of competition in certain markets and accelerated the flow of information, new products and services worldwide. As a result small businesses must embrace these changes quickly, no matter what industry they are in and no matter what product or service they sell.

This need creates a tremendous market for ISPs to tap in. ISPs serve as an electronic gateway to the outside world. It plays a critical role in the operation of many small businesses. Thus, the needs of the small businesses go beyond the level of technical expertise. It is important for ISPs to have what it takes to survive and thrive in the good as well as bad time. One of the critical success factors is that the ISPs must be reliable, service-oriented, and capable of offering packages that are relevant and desirable to specific size of small businesses. Furthermore, with the right strategies, the ISPs not only can maintain their position but also can move up the ladder in terms of consumer satisfaction. In this dynamic market, it is important for the ISPs to upgrade their technology and to launch effective marketing strategies. While to be an innovator could help, there is always a danger of being copied. However, differentiation on the basis of services could help an ISP succeed, but to differentiate they must first have a deep insight of the segment they intent to serve and dedicate enough time and effort to understand their needs.

Practical Implications

There are two major implications that could be drawn from this study. The first one is that small businesses are a potentially lucrative market for ISPs. As indicated earlier, big companies such as AOL and Yahoo have started offering services targeted specifically to small business (Luhn, 2003). Despite such moves by a few big ISPs, many parts of the small business market opportunity remains untapped especially for the smaller ISPs. The reason is the fragmented nature of the small business market itself that is made up of many disparate entities and companies. Hence, there are many potential niches that small ISPs could carve into this billion-dollar-a-year market. Another implication is that while there are many potential niches, the cost of capitalizing these niches might not be as high as many small ISPs believe. The reason is because different types of small businesses show little or no variation in perceived importance of various features packaged in the service as shown in this study. Furthermore, most of today's small businesses do not want to confine themselves to traditional ways of doing business. An increasing number of them are relying and using Internet technology for carrying operations and are demanding services specifically designed to satisfy their business needs. More and more firms realize that the use of Internet-based applications and platforms such as electronic mail, instant messaging helps promote improved communications among employees, customers, and partners, while reducing associated costs and enables them to establish a national or even global presence without having to invest in physical infrastructure. What this means is a strong growing demand for good, reliable Internet service for small businesses. In turn, this demand will continue to drive the growth of ISPs in the years to come.

Another implication is the importance of adaptability. Like any other technology, the pace of changes in IT field is rapid. Any of the technological changes will likely impact small businesses and Internet service Providers. For instance, one such study by ISP-Market LLC (Taking the Pulse, 2003) reported that wireless broadband and web hosting are two new fast growing business areas. The worldwide market for Web hosting services is expected to grow from some $10.3 billion in 2001 to more than $46.9 billion and more than 40 percent of small businesses are expected to subscribe to broadband in 2006 (Pastore, 2002). What this means is that to be successful, today's ISPs must be flexible and adaptable to changes, so that they can respond to the small businesses' rapidly changing preferences of Internet service features.

Limitations

Findings of this study should be interpreted in context of the following limitations. The first major limitation of this study is the sample. It included small businesses only from Southeastern Louisiana. Any generalization should be done with caution. The second limitation is the rapid change of technology. It is difficult to include in an academic study the most up-to-date developments in the world of ISPs. What we did here is to provide a snapshot picture of the whole industry and point out some of the opportunities within the small business market segment. Another limitation is the list of Internet service features. We acknowledged that our list is quite narrow. Due to our time and resource constraints, a number of features such as business log service, script support, data backup, scalability, virtual or independent server, security control, etc could not be included although these are relevant to today's small businesses. Today's typical offering for businesses may include the followings: ADSL fast access, dedicated access, and web hosting solutions. The typical package may include static IP, 20 Email addresses, backup dialup account, DNS hosting with customer's domain name, and 24 hour technical support. Other possible emerging features to be included are good pricing model for connectivity, disaster recovery services, single bill, single service and support standard, single point of responsibility. The scope of future studies could be expanded to examine these features thoroughly. Finally, because of the lack of previous literature, we proposed and adopted variables and constructs that have not been tested before. Furthermore, most of our measures in our survey were based on the input from a few small business owners and our synthesis from literature. An adoption of this survey for future study should be noted with this limitation.

Despite these limitations, this study is one of the few that has looked at an under exposed areas of research related ISPs and small businesses' needs. One contribution that it makes is to open up some issues and encourage many more future studies.

REFERENCES

AT&T WorldNet Ranks Highest in Customer Satisfaction Among Largest National Subscription Internet Service Providers. (2000, September). J.D. Power and Associates Report. Retrieved from http://www.jdpa.com/pdf/2053.pdf.

Barnes, D., Hinton, M., & Mieczkowska, S. (2003). Competitive Advantage through e-operations. *TQM & Business Excellence*, *14*(6), 659–675. doi:10.1080/1478336032000053816

Barton, C., & Bear, M. (1999, March). Information and communications technologies: Are they the key to viable business development services for micro and small enterprises? Retrieved from http://www.mip.org/pdfs/mbp/ict.pdf.

Barua, A., Konana, P., Whinston, A. B., & Yin, F. (2001). Driving e-business Excellence. *MIT Sloan Management Review*, *43*(1), 36–44.

Bennett, G., Sagas, M., & Dees, W. (2006). Media Preferences of Action Sports Consumers: Differences Between Generation X and Y. *Sport Marketing Quarterly*, *15*(1), 40–49.

Boyer, K. K., & Olson, J. R. (2002). Drivers of Internet Purchasing Success. *Production and Operations Management*, *11*(4), 480–498.

Chan, I., & Chao, C. (2008, April). Knowledge Management in Small and Medium-Sized Enterprises. *Communications of the ACM*, *51*(4), 83–88. doi:10.1145/1330311.1330328

Chen, J., Melinda, H., & Zapatero, E. (2006, February). Understanding Web Commerce Adoption By Very Small Businesses. *Proceedings of ASBBS*, Las Vegas, NV, *13*(1), 268-277.

Cottrill, K. (1997). The Supply Chain of the Future. *Distribution, 96*(11), 52–54.

Cragg, P. B., & King, M. (1993). Small-firm Computing: Motivators and Inhibitors. *MIS Quarterly, 7*(1), 47–60. doi:10.2307/249509

Cyber Atlas Staff. (2002, October). Small Biz, Small Growth. Retrieved May 12, 2006, from http://ISP-Planet.com/research/2002/smallbiz_021009.html

Daniel, E., Wilson, H., & Myers, A. (2002). Adoption of ecommerce by SMEs in the UK. *International Small Business Journal, 20*(3), 253–268. doi:10.1177/0266242602203002

Donner, J. (2004, Fall). Microentrepreneurs and Mobiles: An Exploration of the Uses of Mobile Phones by Small Business Owners in Rwanda. *Information Technologies & International Development, 2*(1), 1–21. doi:10.1162/1544752043971198

DTI. (2000). International Bench Marking Survey for ICT Use by UK Department of Trade and Industry. Retrieved from http://www.ukonline.gov.uk

Dun & Bradstreet's Small Business Survey Reveals Eight in Ten Small Businesses Now Have Computers and Two-Thirds Have Internet Access. (2001, October). Retrieved May 12, 2006, from www.itaa.org/isec/pubs/e200110-11.pdf

Fallon, M., & Moran, P. (2000). Information Communication Technology (ICT) and Manufacturing SMEs'. In *Proceedings of the 2000 Small Business and Enterprise Development Conference* (pp. 100-109). Manchester University.

Feindt, S., Jeffcoate, J., & Chappell, C. (2002). Identifying Success Factors for Rapid Growth in SME E-commerce. *Small Business Economics, 19*(1), 51–62. doi:10.1023/A:1016165825476

Garcı'a-Dastugue, S. J., & Lambert, D. M. (2003). Internet enabled Coordination in the Supply Chain. *Industrial Marketing Management, 32*(3), 251–263. doi:10.1016/S0019-8501(02)00269-9

Harman, K., & Koohang, A. (2005/2006, Winter). Frequency of Publication and Topical Emphasis of Knowledge Management Books Versus Doctoral Dissertations: 1983-2005. *Journal of Computer Information Systems, 46*(2), 64–68.

Hauguel, P., & Jackson, N. (2001). Outward-looking Supply-chain Strategy. *European Business Journal, 13*(3), 113–118.

Haynes, P. J., Becherer, R. C., & Helms, M. M. (1998). Small and Mid-sized Businesses and Internet Use: Unrealized Potential? *Internet Research, 8*(3), 229–235. doi:10.1108/10662249810217786

Hochmuth, P. (2005, November 11). Cisco service targets small business. *New World (New Orleans, La.), 22*(45), 8–9.

Kooser, A. (2003, November). They Want You. *Entrepreneur, 31*(11), 34.

Latin American SMEs Increasingly Adopting Mobile CRM, ERP Technologies IDC Mexico. (2008, July 31). *EmergingMarketsNOW*, 1.

Levenburg, N. M. (2005, June). Does Size Matter? Small Firms' Use of E-Business Tools in the Supply Chain. *Electronic Markets, 15*(2), 94–105. doi:10.1080/10196780500083746

Luhn, R. (2003, June). CNET review. Retrieved from http://reviews.cnet.com/4520-6536_7-726606-2.html?tag=st.is.3762-8-%250920777559-1.arrow.3762-8-20777559-2

Magal, S. R., & Lewis, C. (1995, Spring). Determinants of Information Technology Success in Small Businesses. *Journal of Computer Information Systems*, 75–83.

Mahony, D., Riemer, H., Breeding, J., & Hums, M. (2006, April). Organizational Justice in Sport Organizations: Perceptions of College Athletes and Other College Students. *Journal of Sport Management*, *20*(2), 159–188.

Marks, G. (2008, February 6). Worthwhile Small Business Technologies. *Business Week Online*, 26.

Meroño-Cerdán, A. (2008, Summer). Groupware Uses and Influence on Performance in SMEs. *Journal of Computer Information Systems*, *48*(4), 87–96.

Mirchandandi, D. A., & Motwani, J. (2000, Spring). Understanding Small Business Electronic Commerce Adoption: An Empirical Analysis. *Journal of Computer Information Systems*, 70–73.

Moore, A., & Outwater, M. (2001, December). Small Business Survey: Understanding the Business Opportunity, Retrieved June 06, 2003, from http://www.celent.com/ PressReleases/20011217/ SmallBusSurvey.htm.

Palvia, P., & Palvia, S. (1999). An Examination of IT Satisfaction of Small-business Users. *Information & Management*, *35*(3), 127–137. doi:10.1016/ S0378-7206(98)00086-X

Pastore, M. (2002, January 30). *The Broadband and Wireless Revolution*. Retrieved June 06, 2003, from http://ISP-Planet.com/research/2002/ne

Pereira, P. (2005, November 16). SMBs will lead channel. *eWeek, 22*(45A), 33.

Power, D. J., & Sohal, A. S. (2002). Implementation and Usage of Electronic Commerce in Managing the Supply Chain: A Comparative Study of Ten Australian Companies, Benchmarking. *International Journal (Toronto, Ont.)*, *9*(2), 190–208.

Radovilsky, Z., & Hedge, V. (2004, March). Factors influencing e-commerce implementation: Analysis of survey results. *Journal of Academy of Business and Economics*, *4*(1), 29.

Revolutions driving the ISP Business in 2003 (2003). ISP-Market LLC.

Riquelme, H. (2002). Commercial Internet Adoption in China: Comparing the Experience of Small, Medium and Large Businesses. *Internet Research*, *12*(3), 276–286. doi:10.1108/10662240210430946

Taking the Pulse of the Internet Service Provider Industry: Evolutions and

Verizon Tempts Enterprises, SMBs With New Apps Laundry List. (2008, July 10), TelecomWeb News Break, 1.

Zank, G. M., & Vokurka, R. J. (2003). The Internet: Motivations, Deterrents, and Impact on Supply Chain Relationships. *SAM Advanced Management Journal*, *68*(2), 33–40.

Zimmerer, T. W., & Scarborough, N. M. (2002). *Essentials of Entrepreneurship and Small Business Management* (3rd ed.). Upper Saddle River, NJ: Pearson Education, Inc.

Chapter 21
Towards a Contextual and Policy-Driven Method for Service Computing Design and Development

Zakaria Maamar
Zayed University, UAE

Djamal Benslimane
CNRS & Université de Lyon, France

Youakim Badr
INSA de Lyon, France

ABSTRACT

Today, Web services are of interest to both academia and industry. However, little has so far been accomplished in terms of design and development methods to assist those who are responsible for specifying and deploying applications based on Web services in compliance with service computing principles. For this purpose, the authors developed in this chapter a method based on Context and Policy for Web Services known as CP4WS. In this method, policies manage various aspects related to Web services such as participation in composition scenarios and adjustment in response to environmental changes, and context provides the necessary information that permits for instance to trigger the appropriate policies and to regulate the interactions between Web services with respect to the current state of the environment. CP4WS consists of several steps such as the identification of user needs and the behavioral specification of Web services. Each step has a specific graphical notation that facilitates the representation, description, and validation of the composition operations of Web services. A case study that illustrates and highlights the use and originality of CP4WS, respectively, is provided in this chapter.

DOI: 10.4018/978-1-60566-910-6.ch021

INTRODUCTION

For the World Wide Web Consortium (W3C), a Web service "*is a software application identified by a URI, whose interfaces and binding are capable of being defined, described, and discovered by XML artifacts and supports direct interactions with other software applications using XML-based messages via Internet-based applications.*" In a short period of time, the development pace of Web services has proved quite spectacular (Papazoglou et al., 2007; Maurice et al. 2007; Dustdar and Papazoglou, 2008; Dustdar and Schreiner, 2005). Several Web services standards have been developed such as WSDL, SOAP, and UDDI (Curbera et al., 2002) and a good number of Web services projects have been initiated (Maamar et al., 2009a; Maamar et al., 2009b; Agarwal and Sprick, 2005; Mrissa et al., 2007; Tolk and Diallo, 2005; Bentahar et al., 2008). The efforts of these standards and projects are primarily geared towards the development of solutions that would overcome the automatic-composition problem of Web services. Composition handles the situation of a user's request that cannot be satisfied by any single, available Web service, whereas a composite Web service obtained by combining available Web services may be used.

In this chapter we shed the light on the scarcity of design and development methods for Web services. This is due to a certain extent to the lack of campaigns that would raise the awareness level of the research community to the challenges associated with Web services design and development. Roughly speaking, methods provide some kind of roadmap to assist designers and developers in delivering Web services-based information systems as per end-users' needs and requirements. Nowadays, designers and developers are put on the front line of delivering a new generation of Business-to-Business information systems that would across organizations' boundaries. Generally, a method comprises a set of steps to carry out according to a certain chronology and adopts a specific notation to comply with during graphical modeling. A graphical notation proves important as it facilitates discussions and validation exercises among design team members and with end-users, respectively. Throughout this chapter, we propose our design and development method namely CP4WS, which stands for Context and Policy for Web Services. CP4WS stems out of our previous works on Web services like those reported in (Maamar et al., 2005; Maamar et al., 2006a; Maamar et al., 2008; Mrissa et al., 2007), and stresses two major concepts which extend the Web services concept. These two concepts are Policy and Context. Policies are introduced to manage various aspects related to Web services such as participation in composition, semantic mediation, and adjustment in response to environmental changes, whereas context denotes the necessary information that permits, for instance, to trigger the appropriate policies and to regulate the interactions between Web services according to the current environmental state.

In CP4WS, an additional element, which we refer to as resource, is part of the design and development exercise of Web services-based information systems. A resource identifies a computing platform upon which a Web service operates. Since resources schedule the execution requests that Web services submit, Web services have to be constantly aware of the capabilities of and constraints on their designated resources. Resource assignment for long periods of time is by far not acceptable as the number of available Web services continues to grow, so the use of resources will become intensive (Limthanmaphon and Zhang, 2004).

The rest of this chapter proceeds as follows. The following section introduces the concepts of policy and context, discusses the rationale of adopting both concepts in CP4WS, and continues with a presentation of a case study. The subsequent section highlights the various steps that make up CP4WS. Some steps rely on graphical notations, which are illustrated through the case study as part of the modeling exercise of a composite Web service. In addition, the prototype that implements the design of the case study is introduced afterwards. Related work and concluding remarks are finally presented and drawn in the last two sections, respectively.

PRELIMINARY ELEMENTS

Definitions

Context "[...] *is not simply the state of a predefined environment with a fixed set of interaction resources. It is part of a process of interacting with an ever-changing environment composed of reconfigurable, migratory, distributed, and multiscale resources.*" (Coutaz et al., 2005). In the field of Web services, context supports the development of adaptable Web services (Keidl, 2004; Maamar et al., 2006c). These Web services should now take into account the environment in which they operate in terms of users (stationary, mobile, expert, novice, etc.), time of day (in the afternoon, in the morning, etc.), physical locations (meeting room, cafeteria, etc.), and so on. As a result, Web services should be more responsive to their surrounding environment as theybecome flexible, stable, and autonomous (Maamar et al., 2006c).

A policy is defined as "*information which can be used to modify the behavior of a system*" (Lupu and Sloman, 1999). Our definition of policy in (Maamar et al., 2006b) goes beyond behavior modification and considers policies as external, dynamically-modifiable rules and parameters that are used as input to a system. This allows the system to adjust to administrative decisions and changes in the execution environment. Policies have been applied in multiple domains, from telecommunication-devices control features (Reiff-Marganiec and Turner, 2003) to conversation regulation between intelligent components (Kagal and Finin, 2005).

Rationale of Context and Policies

The use of context and policies in CP4WS is backed by the following two points:

- Context collects information on the environment in which the composition of Web services takes place. Such information help track the composition progress, which makes it possible to identify for instance which policies to trigger and which interactions to initiate. Composition progress must occur in accordance with the current state of the environment and current constraints on Web services such as performance and reliability.
- Policies distinguish guidelines for conducting composition from guidelines for defining Web services. Guidelines for Web services composition focus in part on how to integrate user preferences into Web services, how to guarantee the satisfaction of these preferences subject to resource availability, and how to track the execution progress of Web services. Guidelines for Web services definition deals in part with how to announce Web services to potential users, how to deal with Web services non-reliability, and how to suspend Web services execution due to risks of behavior alteration or information interception.

Case Study

Our case study concerns Amin who is visiting Melissa in Oslo. They both agree to meet in a coffee shop. Amin has two options to reach the meeting place, either by taxi or by bus. For illustration purposes, we identify some potential Web services that would implement this scenario. One of the expected outcomes of CP4WS is to identify Web services according to the studied case.

At his hotel, Amin browses some Web sites about transport in Oslo. A site has **ItineraryWS**

Figure 1. CP4WS Method's principles integrating context and policies

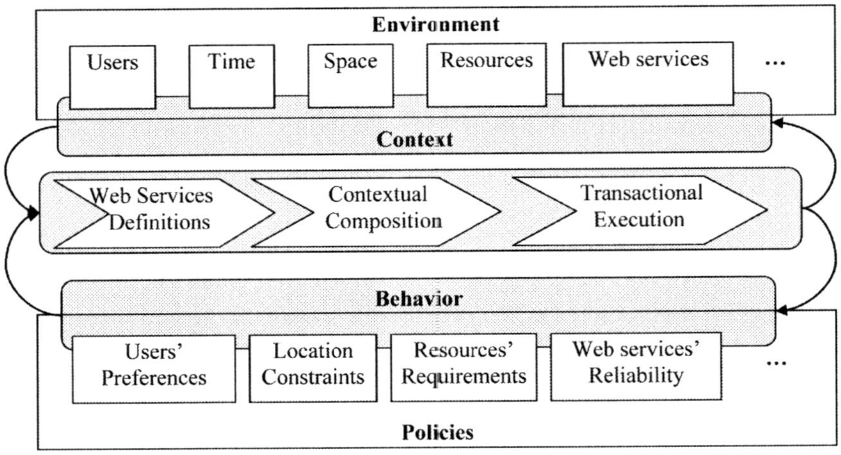

that proposes routes between the hotel of Amin and the coffee shop. The proposed routes are subject to weather forecasts and traffic. In case of cold weather, **ItineraryWS** recommends taxis, otherwise public transport including tramways and buses are recommended. Parallel to consulting **WeatherWS**, **ItineraryWS** requests details on the origin and destination places using **LocationWS**. In case **WeatherWS** returns bad weather, a taxi booking is made using **TaxiWS** upon Amin's approval. Otherwise, Amin uses public transport. The location of the hotel of Amin and the coffee shop are submitted to **BusScheduleWS**, which returns the bus numbers that Amin will ride. Potential traffic jams in Oslo force **BusScheduleWS** to regularly interact with **TrafficWS**, which monitors the transportation network. This monitoring outcome is submitted to **BusScheduleWS** so that changes in the suggested bus numbers are made.

DESIGN AND DEVELOPMENT STEPS IN CP4WS

Web services emerge as a standard-based computing approach for designing and deploying flexible distributed software applications that collaborate and evolve without major changes in their core architectures. Web services are self-contained components that can be advertised, discovered, and composed on demand. Despite the wide academic and industrial activities related to Web services use, design and deployment, no systematic end-to-end method exists to analyze and design contextual and transactional applications based on dynamic behavior and policies. Combining context and policies would be a powerful way to build complex evolving systems. A clear need of methods in service computing applications rises. The adoption of method principles helps analysts and developers provide a systematic way to identify, specify, and deploy customized service-oriented applications. In Figure 1, we illustrate our method based on Context and Policy for Web Services, known as CP4WS. The method emphasizes context elements and dynamic behavior in order to identify Web services and perform transactional execution with respect to context and policies.

Like other design and development methods in the field of information systems (presented in the Related Work Section), CP4WS tackles identical issues. Firstly, methods focus on identifying user requirements to ensure the usability of the developed system. Secondly, methods suggest guidelines during design and development. Finally, methods adopt graphical notations to

Figure 2. Design and development steps in the CP4WS method

User Needs Identification and Specification Step
Major actions: identify user needs, validate user needs, collection of information on similar applications
Representation formalism: UML use-case.
Web Services Orchestration Step
Major actions: identify types of Web services, identify interaction between Web services, compose Web services.
Representation formalism: service chart diagram and UML state chart diagram.
Web Services Contextualization Step
Major actions: identify parameters of context, specify update operations between contexts.
Representation formalism: null (not a specific one).
Web Services Behavior Specification Step
Major actions: identify behavior types of Web service, specify behaviors using policies.
Representation formalism: WSPL.
Web services Transactional Definition Step
Major actions: identify appropriate transactional properties, assign them to Web services if needed.
Representation formalism: UML state chart diagram
Web Services Deployment Step
Major actions: define behaviors of Web service towards resources, specify behaviors using policies.
Representation formalism: WSPL.

support abstract modeling and facilitate validation exercises. In this section, we detail the six steps of CP4WS and the rationale and expected outcome of each step. Some illustrations per step are also given based on Amin scenario. Figure 2 summarizes these six steps including the major actions and representation formalisms (such as graphical notation) for each step.

User Needs Identification and Specification Step

The first step in CP4WS consists of identifying and specifying user needs. Traditional software-engineering techniques to identify user needs and requirements are adopted in CP4WS such as interviewing users, studying current practices, reviewing forms and collecting information on similar applications. Regarding the specification technique of user needs, CP4WS adopts UML use-cases (Booch et al., 1998). A use case describes the system's behavior as it responds to a request initiated by an actor from outside of that system. This interaction captures functional requirement through sequences of simple activities. This adoption has several advantages: (i) most information system designers and developers are familiar with use cases, (ii) identified use cases could be mapped onto potential Web services, and (iii) interactions between use cases are an excellent indicator of composition-based collaboration between Web services. It should be noted that the identification of the appropriate actors and use cases during this step in CP4WS takes advantage of the experience and familiarity of designers with the application domain.

Figure 3 presents a part of the use case model we developed for Amin scenario. Several actors are shown such as *Amin*, *Weather Forecast agency*, and *Transportation agency*. In addition, several use cases are identified including *Itinerary Development*, *Weather Forecast Establishment*, and *Traffic Monitoring*. The same figure shows the way some use cases (using dashed lines in Figure 3) interact with one other. For instance, *Itinerary Development* use-case requires details on the traffic situation that are obtained by *Traffic Monitoring* use-case.

Web Services Orchestration Step

The second step in CP4WS consists of specifying the orchestration of the Web services that will constitute the future composite Web service. In terms of functionality, the types of the required Web services are already identified based on the outcome of User-Needs Identification and Specification Step. In general a use case is a po-

Figure 3. Part of UML use cases of Amin scenario

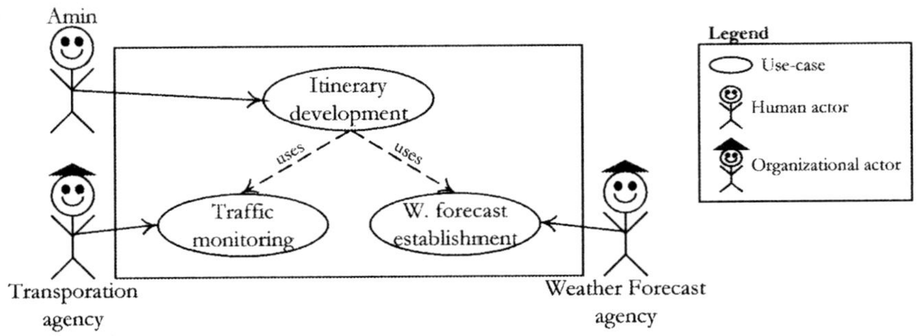

tential candidate to be mapped onto a Web service although the mapping is not always one-to-one. Several use cases can be gathered into one Web service, one use case could also be associated with several Web services, etc. It is the designer's responsibility to come up with the appropriate number of Web services.

The Web Services Orchestration Step relies on the concept of service chart diagram (Maamar et al., 2003). A service chart diagram enhances an UML state chart diagram (Booch et al., 1998), putting emphasis on the elements affecting the execution of a Web service rather than on the states that a Web service takes alone (Figure 4). As a result, the states of a Web service are wrapped into five perspectives:

- The state perspective corresponds to a regular UML state chart diagram of the Web service.
- The flow perspective corresponds to the chronology of the composite Web service in which the Web service participates.
- The business perspective identifies the various organizations that make the Web service available.
- The information perspective identifies the data exchanged between the Web service and its peers in the same composite Web service.
- The performance perspective illustrates how the Web service can be triggered, either remotely or locally.

Since a composite Web service consists of several component Web services, the business process-model that underpins the orchestration of this composite Web service is specified as an UML state chart diagram. In this diagram, states are associated with service chart diagrams and transitions are labeled with events, conditions, and variable assignment operations. Figure 5 illustrates the orchestration of the composite Web service developed for Amin scenario. Six component Web services are listed: ITineray (*IT*), WEather (*WE*), Location (*LO*), Taxi (*TA*), Bus Schedule (*BS*), and TraffiC (*TC*). We recall that ITinerayWS is associated with the Itinerary Development use-case as previously defined.

Web Services Contextualization Step

The third step in CP4WS consists of specifying contexts of the participants that could take part in Web services composition. In addition to Web

Figure 4. Service chart diagram of a component web service

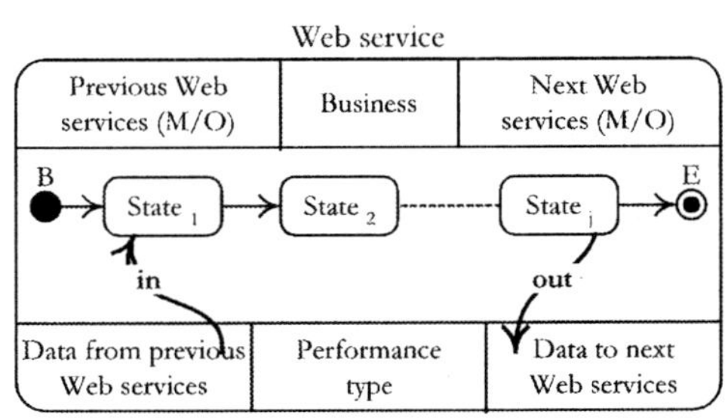

Figure 5. Composite web service for Amin scenario

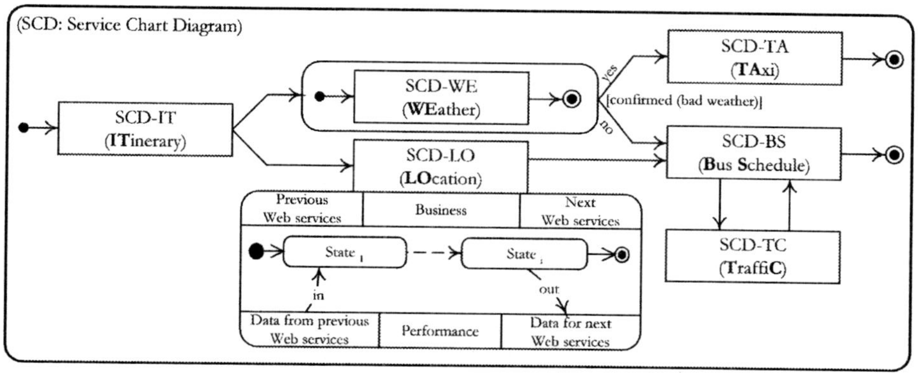

services as a default participant, CP4WS suggests the following list of participants: composite Web service, resource, and user. Additional types of participants are added as per the specific requirements of the studied application domain and business case. CP4WS associates each participant type with a dedicated context to be labeled using this participant's name, for example, W-context for context of Web service, C-context for context of Composite Web service, R-context for context of Resource, and U-context for context of User.

The next step in CP4WS demonstrates how context affects the process of loading and triggering policies. The following briefly discusses the expected role of each context type:

- W-context of a Web service returns details on the participations of this Web service in different compositions. These participations occur in accordance with the Web services instantiation principle (Maamar et al., 2005).
- C-context of a composite Web service is built upon the W-contexts of its component Web services and permits to oversee the progress of a composition.
- U-context of a user monitors her current status and reflects personal preferences such as Web services' execution times.
- R-context of a resource oversees the execution of the Web services that operate on top of a resource before this resource can accept additional Web services allocation. A resource has computational capabilities that continuously change depending on the number of Web services that are currently under execution and the execution duration of each Web service.

When the different context types are known and their expected role set, CP4WS proceeds with the specification of the internal structure of each context type. This comprises working out the relevant arguments that will populate this structure. It is expected that these arguments will include various types such as Web services execution order, forthcoming execution of Web services, current user location, and subsequent period of resource unavailability. In the following, some arguments are suggested for W/C/R/U-contexts. It should be noted that these arguments are not randomly selected, but based on our previous research on context-aware Web services (Maamar et al., 2006a; Maamar et al., 2005). To keep the focus of this chapter self-contained, just the internal structure of the W-context is discussed.

W-context, C-context, U-context and R-context encompass the arguments depicted in Table 1 whereas Table 2 illustrates how some arguments get instantiated according to Amin scenario.

Table 1. Internal structures of various context types

Participants	Context Description
Web Service Context	Label Time stamp Date Maximum Number of Active Participations Number of Active Participations Next Possibility of Participation Resource and State Per Active Participation Previous Web Services Per Active Participation Current Web Services Per Active Participation Next Web Services Per Active Participation Regular Actions Reasons of Failure Per Active Participation Corrective Actions Per Failure Type Corrective Actions Per Active Participation
Composite Web Service Context	Label Date Time stamp Previous component Web services Current component Web services Next component Web services Status per component Web service
User Context	Date Time stamp Previous locations per component services Current location per component services Next locations/component services Previous periods of time/component services Current period of time/component services Next periods of time/component services
Resource Context	Label Time Stamp Date Maximum number of component Web service Number of active component Web services Next acceptance of component Web services Previous component Web services per active composition Current component Web services per active composition Consumption and state per component Web service Consumption and state per active composition Next component Web services per active composition

Web Services Behavior Specification Step

When the steps of user needs identification and specification, Web services orchestration, and Web services contextualization are completed, the fourth step in CP4WS consists now of specifying the behavior of the component Web services that were identified and assembled together in the second step. A behavior is exposed and specified with a set of attributes and policies, respectively. The role of policies is to make a Web service bind to a certain behavior as it will be shown in this section. CP4WS suggests the following attributes to define the behavior of a Web service based on our previous research on policies for Web services (Maamar et al., 2008): permission, restriction, and dispensation. Moreover CP4WS suggests adopting the Web Services Policy Language (WSPL) to specify policies (Anderson, 2004), although other

Table 2. Internal structure of W-Context

Argument and Description
Label: corresponds to the identifier of the Web service. *Maximum Number of Active Participations*: corresponds to the maximum number of compositions in which the Web service can participate at a time. *Number of Active Participations*: corresponds to the number of active compositions in which the Web service is currently participating. *Next Possibility of Participation*: indicates when the Web service can participate in a new composition. This is related to the successful termination of the current active participations. *Resource and State Per Active Participation*: corresponds to the identifier of the selected resource and the state of the Web service in each active composition. The State can include various types namely in-progress, suspended, aborted, or completed, and will be obtained out of the state argument of R-context of this resource. *Previous Web Services Per Active Participation*: indicates the Web services that were successfully completed before the Web service per active composition (null if there are no predecessors). *Current Web Services Per Active Participation*: indicates the Web services that are concurrently being performed with the Web service per active composition (null if there is no concurrent processing). *Next Web Services Per Active Participation*: indicates the Web services that will be executed after the Web service successfully completes its execution per active composition (null if there are no successors). *Regular Actions*: illustrates the actions that the Web service normally performs. *Reasons of Failure Per Active Participation*: explains the reasons behind the failure of the execution of the Web service per active composition. *Corrective Actions Per Failure Type and Per Active Participation*: illustrates the actions that the Web service has performed due to execution failure per active composition. *Date*: identifies the time of updating the arguments above. Why there is a huge gap here? I couldn't fix it?
Application to Amin Scenario
(*Previous Web Services Per Active Participation*: WeatherWS) - Weather Web service is executed after Itinerary Web service according to the specification in Figure 5. (*Number of Active Participations*: WeatherWS(4)) - Weather Web service is currently involved in four active composite Web services. This participation number is constrained by the maximum number of active participations.

policy specification languages (Damianou et al., 2001; Horrocks et al., 2004; Moses, 2003; Schlimmer, 2004) could be integrated into CP4WS.

In the following (Figure 6), we describe first, how the attributes that define the behavior of a Web service are interpreted and secondly, how a Web service binds to a specific behavior following policy triggering. Details on the use of policies in Amin scenario are reported with some snapshots from the prototype.

- Permission: a Web service accepts the invitation to participate in a composite Web service upon validation of its current engagements in other concurrent composite Web services. The validation process relies on the content of W-context (Table 2).
- Restriction: a Web service cannot be orchestrated with some peers in the same composite Web service as these peers do not satisfy this Web service's requirements. These requirements could be related to non-functional details.
- Dispensation: a Web service breaks some policies related to either permission or restriction; in case of permission, this Web service will not participate in a composition despite the positive permission of participation. In case of restriction, this Web service will be involved in an orchestration with some peers despite the existence of restrictions.

Permission Policy. Permission authorizes a Web service to be part of a composite Web service. The following is a permission policy in WSPL. It states that a Web service participates in a composition subject to evaluating <Condition> (line 04) to true. This condition refers to some arguments such as the number of current active

Figure 6.

```
01:     Policy (Aspect="Permission")
02:     <Rule xmlns="urn:oasis:names:tc:xacml:3.0:generalization:policy:schema:wd:01"
03:     RuleId="PermissionWS">
04:      <Condition>
05:     <Apply FunctionId="and">
06:             <Apply FunctionId="integer-less-than" DataType="Boolean">
07:               <SubjectAttributeDesignator AttributeId="CurrentNumberOfParticipations"
08:                                           DataType="integer"/>
09:               <SubjectAttributeDesignator AttributeId="MaximumNumberOfParticipations"
10:                                           DataType="integer"/>
11:             </Apply>
12:               <SubjectAttributeDesignator AttributeId="Availability" DataType="Boolean"/>
13:      </Apply>
14:      </Condition>
15:      <Conclusions>
16:             <TrueConclusion Permission = "Permit"/>
17:             <FalseConclusion Permission = "Deny"/>
18:      </Conclusions>
19:      </Rule>
```

participations of the Web service in compositions (line 07) and the next possibility of participation of the Web service in additional compositions (line 12). Both arguments are part of the context of a Web service (Table 2). In the policy given below, <TrueConclusion> (line 16) shows the permission of participation, whereas <FalseConclusion> (line 17) shows the contrary.

Restriction Policy. Restrictions prevent a Web service from participating in composite Web services. They could be in relation to the QoS (e.g., time of response, throughput) (Menascé, 2002) of the component Web services with which a Web service interacts. It occurs that a Web service's provider is only interested in the Web services that have a "good" QoS record. The following illustrates a restriction policy in WSPL. It states that a Web service can be restricted from participation subject to evaluating <Condition> to true. This condition verifies that a positive permission of participation exists (line 06), a no-dispensation from participation exists as well, and finally the assessment level of the QoS of the Web services is low (line 08). In this policy, QoSAssessment (line 09) is an integer value resulting from the evaluation of the QoS of a Web service, and QoSThreshold (line 10) is the minimum QoS assessment value that is acceptable for a composition to happen (Figure 7).

Dispensation Policy. Dispensation allows a Web service to break policies related to permission or restriction. Thus, dispensation is decom-

Figure 7.

```
01:     Policy (Aspect="Restriction")
02:     <Rule xmlns="urn:oasis:names:tc:xacml:3.0:generalization:policy:schema:wd:01"
03:     RuleId="RestrictionWS"
04:      <Condition>
05:     <Apply FunctionId="and">
06:             <SubjectAttributeDesignator AttributeId="YesPermission" DataType="Boolean"/>
07:             <SubjectAttributeDesignator AttributeId="NoDispensationP" DataType="Boolean"/>
08:             <Apply FunctionId="integer-great-than-or-equal">
09:               <SubjectAttributeDesignator AttributeId="QoSAssessment" DataType="integer"/>
10:               <SubjectAttributeDesignator AttributeId="QoSThreshold" DataType="integer"/>
11:             </Apply>
12:     </Apply>
13:      </Condition>
14:      <Conclusions>
15:             <TrueConclusion Restriction = "No"/>
16:             <FalseConclusion Restriction = "Yes"/>
17:      </Conclusions>
18:     </Rule>
```

posed into permission-driven dispensation and restriction-driven dispensation. For illustration reasons, we only describe the first type. Although a Web service has been granted a permission to join in a composite Web service, a dispensation from participation can override this permission. One of the criteria that supports the dispensation is the composite Web service's invocation period to the Web service. If this period was within the Web service's peak-time period of receiving invocation requests, the Web service could cancel the permission of participation. This could affect the QoS this Web service guarantees to composite Web services. The following illustrates a permission-driven dispensation policy in WSPL. It states that a Web service cancels a permission of participation subject to evaluating <Condition> to true. The latter verifies that such a permission exists (line 06) and the invocation-request time does not fall in the peak time-period of the Web service (lines 07, 08, 09 in Figure 8).

Web Services Transactional Definition Step

The purpose of this step is to identify the constraints that can be put on the performance of Web services included in the specification of a composite Web service. Prohibiting repetition or authorizing cancellation of executions are examples of constraints that Web services need to comply with. The specification of the transactional definition in CP4WS is built upon a set of transactional properties usually referred to as *retriable*, *pivot*, and *compensatable* (Bhiri et al, 2005; Younas et al., 2006) Firstly, pivot denotes that the execution effects of a Web service remain unchanged and cannot be undone (or revoked). Secondly, retriable refers to the fact that the execution of a Web service can be repeated a certain number of times in case of failure. Thirdly, compensatable indicates that the execution side effects of a Web service can be undone. The rationale of these three properties is due to the unsuitability of the Atomicity, Consistency, Isolation, and Durability (ACID) model to Web services as reported in (Verma and Deswal, 2003): transactions may take place over long periods of time (sometimes hours, days, or even more), participants may not allow their resources to be locked for long durations of time, and a transaction may succeed even if only some of the participants (sub-transactions) choose to confirm and others choose to cancel.

Transactional properties enforce the acceptable behavior of a Web service at run-time. If a Web service is declared as pivot, then the side effects of its execution will not (and cannot) be undone or revoked. The transactional flow of this Web service is shown using a state chart (Figure 9 (a)). For a retriable/compensatable Web service, a dif-

Figure 8.

```
01:  Policy (Aspect="DispensationPermission")
02:  <Rule xmlns="urn:oasis:names:tc:xacml:3.0:generalization:policy:schema:wd:01"
03:  RuleId="DispensationPermissionWS">
04:   <Condition>
05:          <Apply FunctionId="and">
06:             <SubjectAttributeDesignator AttributeId="YesPermission" DataType="Boolean"/>
07:              <Apply FunctionId="equal">
08:               <SubjectAttributeDesignator AttributeId="PeakTime" DataType="integer"/>
09:               <SubjectAttributeDesignator AttributeId="RequestTime" DataType="integer"/>
10:              </Apply>
11:          </Apply>
12:   </Condition>
13:   <Conclusions>
14:          <TrueConclusion Permission = "Deny"/>
15:          <FalseConclusion Permission = "Permit"/>
16:   </Conclusions>
17:  </Rule>
```

Figure 9. Transactional definition of a Web service per transactional property

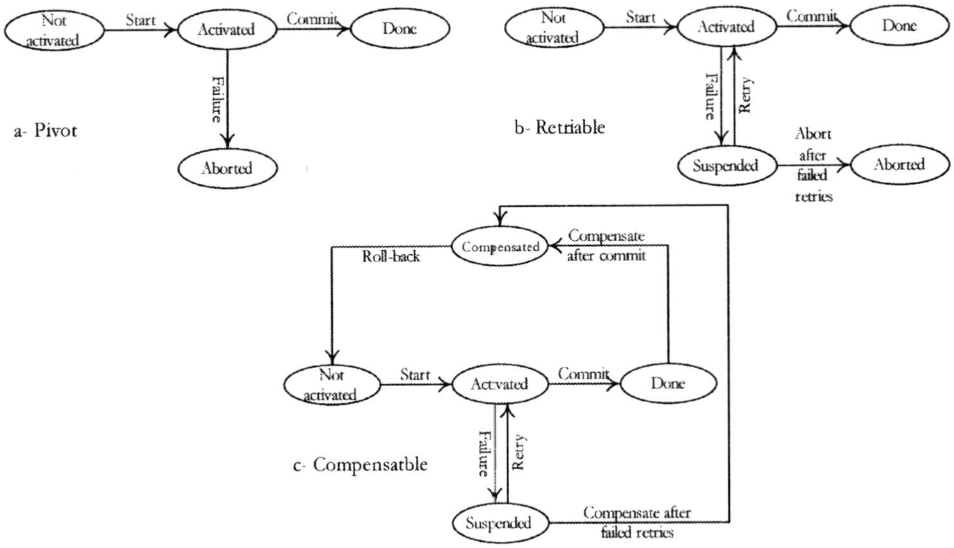

ferent transactional flow is developed (Figure 9 (b,c)). It should be noted that the UML state charts in Figures 5 and 9, respectively have different purposes. The state chart in Figure 4 indicates the business logic implementing LocationWS, while the state chart in Figure 9 illustrates the transactional behavior of this Web services.

In addition to pivot, retriable, and compensatable properties, the transactional definition of Web services in CP4WS is enriched with details that identify transactional dependencies between Web services. These dependencies are directed-oriented relationships and of the following types (Figure 10):

- Activation dependency exists from WS_x to WS_y: Upon completion of WS_x, the WS_y is executed or activated.
- Abortion dependency exists from WS_x to WS_y. In the case that the failure, cancellation, or abortion of WS_x occurs, the abortion of WS_y is triggered.
- Compensation dependency exists from WS_x to WS_y Upon failure or compensation of WS_x the WS_y compensation is triggered.

The transactional definition of Web services makes it possible to work out how to handle the exceptions that Web services raise in reaction to failures. This handling is usually dependent on the business logic that underpins the functioning of a Web service and the composition scenario in which this Web service takes part. In Amin's scenario if the planning of the meeting is aborted due to resource failure, then the developer of this scenario will have to come up with plans to compensate and probably abort the already undertaken activities such as a taxi reservation. In addition to the business logic that guides the type of exception handling strategy to set-up, transactional properties and dependencies have an impact on this strategy as well. For example, if a Web service is declared as pivot, then there will be no need to retry its execution although its business logic permits it and suggests appropriate means to carry out this retrial. The pivot property withholds any tentative of re-execution.

Web Services Deployment Step

The sixth and final step in CP4WS consists of managing the performance of Web services on

Figure 10. Dependency and transaction diagram

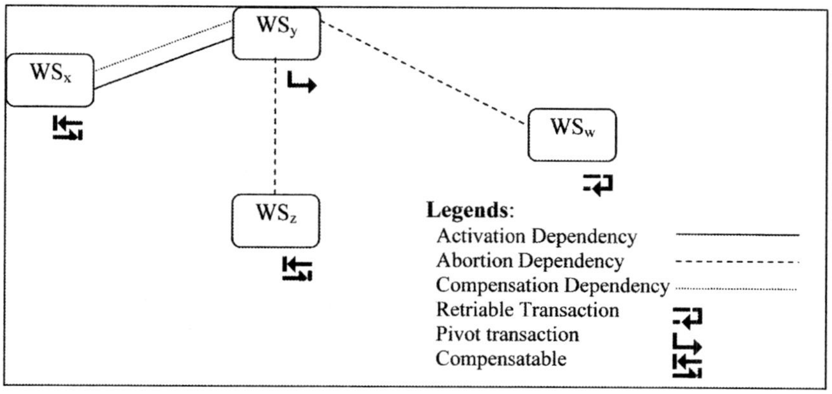

top of computing resources. The rationale of this step is to satisfy the needs of bodies that have interests and/or obligations in strengthening or restricting the execution of Web services on top of their resources. For instance, a Web service does not obtain the necessary authorizations to operate on a resource since this Web service does not comply with this resource's requirements such as permitted execution-time. CP4WS manages the deployment of Web services to resources with policies. For standardization purposes with the policies associated with Web services' behaviors, WSPL is also adopted to specify deployment policies (Figure 11).

Deployment Policy for Permission. A Web service receives the necessary execution authorizations from a resource. These authorizations are based on the state of the resource which manifests itself using its R-context. The following illustrates a deployment policy for permission in WSPL. It states that a resource accepts the execution request of a Web service subject to evaluating <Condition>. This condition refers to some arguments such as the number of active component Web services (line 07) that the resource supports their execution and next acceptance of the resource to additional component Web services (line 12). In the policy, <TrueConclusion> (line 17) shows the permission of execution, whereas <FalseConclusion> (line 20) shows the contrary. In case of a positive permission of execution, the yes-permission-deployment procedure is executed (line 18), which results in updating the following arguments: resource and state per active participation of W-context of the Web service (Table 2) and number of active component Web services of R-context of the resource.

Deployment Policy for Restriction. This policy consists of preventing a Web service form being executed on a resource. A part from the example of resource failure, restrictions could be geared towards the reinforcement of the execution clauses that are agreed upon between a Web service and a resource. For example, a Web service binds a resource for execution before the scheduled time. The following illustrates a deployment policy for restriction in WSPL. It states that a Web service can be restricted from execution subject to evaluating <Condition> to true. This condition verifies that a positive execution permission (line 04) has been issued (line 06) and the agreed execution time is valid. The execution time of a Web service is identified using next component Web services per active participation argument of R-context of the resource (Figure 12).

Figure 11.

```
Policy (Aspect="PermissionDeployment")
<Rule xmlns="urn:oasis:names:tc:xacml:3.0:permission:policy:schema:wd:01"
xmlns:proc="permission-deployment" RuleId="PermissionDeploymentWS">
 <Condition>
          <Apply FunctionId="and">
            <Apply FunctionId="integer-less-than" DataType="Boolean">
             <SubjectAttributeDesignator AttributeId="NumberofActiveComponentWebServices"
                                         DataType="integer">
             <SubjectAttributeDesignator AttributeId="MaximumNumberofComponentWebServices"
                                         DataType="integer">
            </Apply>
          <SubjectAttributeDesignator AttributeId="NextAcceptanceofComponentWebServices"
                                         DataType="Boolean">
          </Apply>
 </Condition>
 <Conclusions>
          <TrueConclusion>
              <proc:do> yes-permission-deployment </proc:do>
          </TrueConclusion>
          <FalseConclusion>
              <proc:do> no-permission-deployment </proc:do>
          </FalseConclusion>
 </Conclusions>
</Rule>
```

Figure 12.

```
Policy (Aspect="RestrictionDeployment")
<Rule xmlns="urn:oasis:names:tc:xacml:3.0:generalization:policy:schema:wd:01"
RuleId="RestrictionDeploymentWS">
 <Condition>
          <Apply FunctionId="and">
              <SubjectAttributeDesignator AttributeId="YesPermissionDeployment"
                                       DataType="Boolean">
              <Apply FunctionId="equal" DataType="Boolean">
                <SubjectAttributeDesignator AttributeId="ExecutionTime" DataType="String">
              </Apply>
          </Apply>
 </Condition>
 <Conclusions>
          <TrueConclusion RestrictionDeployment = "No">
          <FalseConclusion RestrictionDeployment = "Yes">
 </Conclusions>
</Rule>
```

CP4WS TO IMPLEMENT AMIN SCENARIO

We present the work carried out following the use of CP4WS to design the system implementing Amin scenario. For compatibility purposes, Sun Microsystems's tools are used: J2EE 1.4 to develop Web services and XACML Open Source to develop policies. Figure 13 illustrates the architecture of this system that comprises four types of managers. *It should be noted that the role of each manager is associated with a specific step in* CP4WS. Figure 14 shows some snapshots related to the use of the system implementing Amin scenario.

The **Specification Manager** supports designers during the specification of composite Web services. This calls for identifying the appropriate component Web services. The specification work is carried out through a **composition environment**, which is a set of integrated tools that assist designers in the creation and editing of new and existing specifications of composite Web services, respectively. We use a composition environment that we developed in previous research. This environment, further, supports translating composite

Figure 13. Architecture of the system CB4WS framework

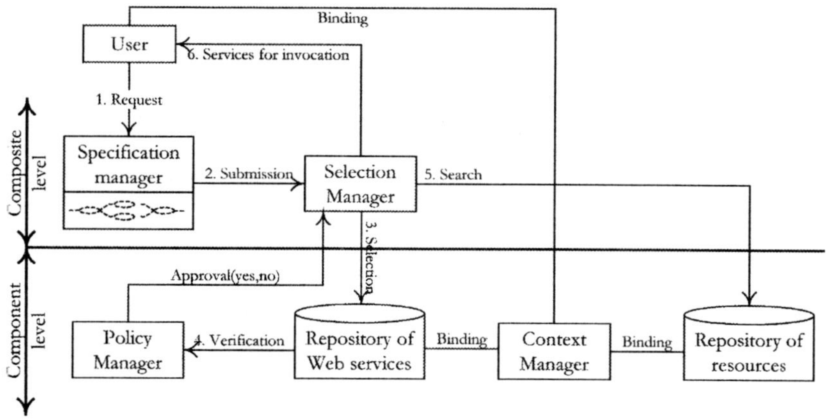

Web service specifications, such as the one shown in Figure 5, to BPEL specification.

The **Selection Manager** is responsible for identifying the component Web services to satisfy user needs. This manager is triggered upon the user's request and identification of the appropriate composite Web service specification. In the current system, the selection is not solely driven by the resulting functionality of the composition that the user needs (such as reaching a meeting place by taxi or by bus according to weather conditions). It also considers Web services QoS parameters that affect the selection process such as response time, performance and throughput. These constraints are expressed with WSPL policies. The **Policy Manager** makes Web services bind to appropriate behaviors according to the progress of a composition.

Finally, the **Context Manager** keeps track of the contexts of users, Web services, and resources. There are different types of context arguments and their values change over time. Figure 15 illustrates the specification of **W-context** of **Weather WS** using the prototype. Therefore, the context manager is supported with a real-time triggering mechanism that feeds context parameters with fresh values. Details of contexts are structured as XML files. Before sending the selected Web services' addresses to the user for invocation, the **Policy Manager** ensures that these Web services comply with the policies previously reported. Upon approval by the **Policy Manager**, the **Selection Manager** initiates a search of the resources on which the Web services will operate. Figure 16 shows parts of the policy that restricts the execution of **Weather WS** on a resource.

RELATED WORK

Several initiatives attempt to develop methods to Web services design and deployment such as SOAD (SOAD, 2004), SOMA (SOMA, 2004), MSOAM (Erl, 2007), and OASIS (Bate and Mulholland, 2005). To our knowledge, few projects have aimed to suggest design and development methods for Web services based on context and driven by policies. In the following we present some projects that have helped shape the steps and annotations of CP4WS. These projects mainly target design and development methods.

In (Baresi et al., 2005), the authors propose a policy-based approach to monitor the functional (e.g., constraints on exchanged data) and non-functional (e.g., security and reliability) requirements of Web services. In this approach, Baresi et al. report on the different types of policies that can be defined along the life cycle of a Web ser-

Figure 14. Snapshots from the prototype

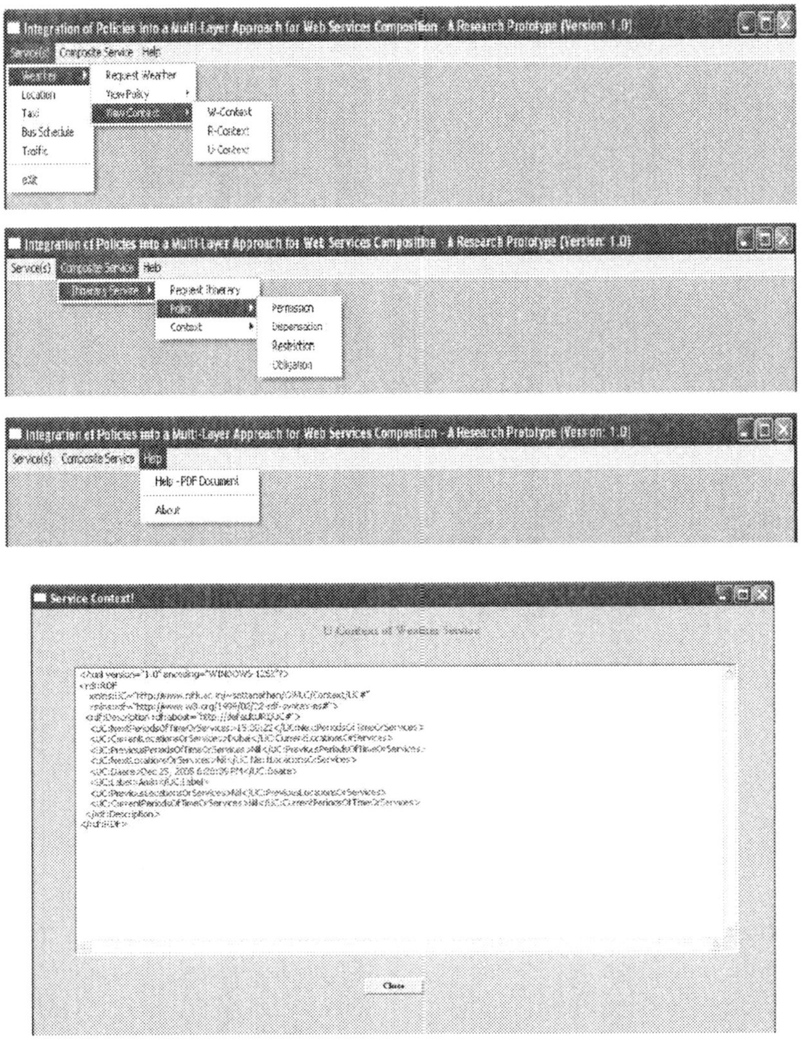

vice (Mukhi et al., 2004). These types of policies include service policies, server policies, supported policies, and requested policies.

In (UWA, 2002), the Ubiquitous Web Applications (UWA) Consortium proposes a framework that comprises several methodologies, meta-models, and tools for designing ubiquitous Web applications. Organization and execution models are used to design the business processes that implement the business logic of such applications. The organization model expresses the hierarchical relations between activities of the same business process. These relations are described with an UML class diagram. In addition, some properties (such as ACID properties) can be associated with each activity. The execution model describes the possible execution flows of the different activities that compose the organizational model. These flows constitute the dynamic aspect of the business process and are expressed with an UML activity diagram.

In (Brambilla et al., 2006), the authors propose a Web engineering method for the high-level specification of processes and Web services-based applications. This method relies on extending the

Figure 15. W-context of Weather Web Service

```
<rdf:RDF
    xmlns:WC="http://www.nitk.ac.in/ sattanathan/OWLC/Context/WC#"
    xmlns:rdf="http://www.w3.org/1999/02/22-rdf-syntax-ns#" >
  <rdf:Description rdf:about="http://defaultURI/WC#">
        <WC:CorrectiveActionsPerFailureType>Nil</WC:CorrectiveActionsPerFailureType>
        <WC:ResourceAndStatePerActiveParticipation>Resource1/Suspended
        </WC:ResourceAndStatePerActiveParticipation>
        <WC:MaximumNumberOfParticipations>10</WC:MaximumNumberOfParticipations>
        <WC:ReasonsOfFailurePerActiveParticipation>Nil</WC:ReasonsOfFailurePerActiveParticipation>
        <WC:RegularActions>Book Taxi</WC:RegularActions>
        <WC:PrevisousWebServicesPerActiveParticipation>Nil
                </WC:PrevisousWebServicesPerActiveParticipation>
        <WC:NextWebServicePerActiveParticipation>TAxi or Bus Schedule
                </WC:NextWebServicePerActiveParticipation>
        <WC:Label>WEather</WC:Label>
        <WC:Date>Dec 25, 2005 6:28:39 PM</WC:Date>
        <WC:CurrentWebServicesPerActiveParticipation>Location
                /WC:CurrentWebServicesPerActiveParticipation>
        <WC:NextPossibiliyOfParticipation>Possible</WC:NextPossibiliyOfParticipation>
        <WC:NumberOfActiveParticipations>7</WC:NumberOfActiveParticipations>
  </rdf:Description>
</rdf:RDF>
```

Figure 16. Restriction policy on Weather Web Service

```
<Policy PolicyId="Restriction"
  RuleCombiningAlgId="urn:oasis:names:tc:xacml:1.0:rule-combining-algorithm:ordered-permit-overrides">
  <Description>Restriction policy on WEather Web Service</Description>
  <Target>
   <Subjects>
    <Subject>
     <SubjectMatch MatchId="urn:oasis:names:tc:xacml:1.0:function:rfc822Name-match">
     <AttributeValue DataType="http://w3.org/...#string">users.example.com</AttributeValue>
     <SubjectAttributeDesignator AttributeId="urn:oasis:names:tc:xacml:1.0:subject:subject-id"
     DataType="urn:oasis:names:tc:xacml:1.0:data-type:rfc822Name"/>
   </SubjectMatch>
  </Subject>
  </Subjects>
 <Resources>
  <Resource>
    <ResourceMatch MatchId="urn:oasis:names:tc:xacml:1.0:function:anyURI-equal">
    <AttributeValue DataType="http://w3.org/...#anyURI">http://webservice.weather.com/</AttributeValue>
    <ResourceAttributeDesignator AttributeId="urn:oasis:names:tc:xacml:1.0:resource:resource-id"
    DataType="http://www.w3.org/2001/XMLSchema#anyURI"/>
    </ResourceMatch>
   </Resource>
  </Resources>
 <Actions>
   <AnyAction/>
 </Actions>
 </Target>
<Rule RuleId="RestrictionWS" Effect="No">
 <Target>
```

Web Modeling Notation (WebML (Ceri et al., 2000)) with standard process modeling concepts and Web services-based application distribution primitives. Process modeling concepts expressed in the Business Process Modeling Notation (BPMN 2008) are added to WebML. By doing so, process requirements in terms of interactions over the Web are expressed. Web services are used as a means to deal with process distribution requirements. It is interesting to note that this method has been implemented as a case tool through the use of WebRatio (http://www.WebRatio.com).

In (Detroyer and Casteleyn, 2003), the authors work on an adaptation of the existing Web Site Design Method (WSDM). Their objective is to model complex business processes in the context of Web applications and support some advanced features such as transactions and process persistence. Task modeling and navigational design, which are the core steps in the conceptual design of WSDM, have been changed to accommodate the requirements of modeling complex processes. The ConcurTaskTrees (CTT) notation (Paterno et al., 1997) is modified in order to consider three task categories: application, interaction, and abstract. Abstract task refers to the task that needs to be decomposed, while application and interaction tasks refer to the tasks executed by applications and users, respectively. The navigation structure that describes how a user should perform tasks is then automatically generated for different users from the CTT augmented with temporal relationships between the different tasks.

In (Cappiello and Pernici, 2006), the authors suggest a method to solve run-time data quality problems in self-healing Web services environments. Failures due to run-time data quality problems are detected and the proposed method produces a list of recovery actions for quality improvements. The main step in the method focuses on the warning message generator, which is in charge of monitoring the system, detecting all the anomalies that occur in data management, and identifying the sources of data quality problems.

In (Distante et al., 2007), a framework for analyzing and comparing Web application design methods is presented. The authors discuss business/user/system requirements to design business process. Business requirements correspond to the identification of the different activities included in a given Web transaction and their semantic associations, and the logical/temporal order in which such activities must be executed by the user. User requirements include the specification of the set of activities, which can be suspended and resumed in case of long-lived transactions and the description of how an activity can be customized depending on the state of the ongoing transaction. System requirements refer to the definition of the information objects affected by the execution of an activity and the definition and the management of the states of a Web transaction. These design requirements are then used to evaluate and compare two Web application design methodologies, namely OOHDM (Schmid and Rossi, 2004) and UWA (UWA, 2002).

In (Iacovelli at al., 2008), the authors highlight that the use of method engineering approaches is limited to those who provide these approaches and sharing them with other peers continues to be a challenge. To overcome such a limitation, Iacovelli et al. adopt the paradigm of service-oriented architecture for method engineering and by analogy to Saas (Software as a service) promote the approach of Maas (Method as a Service). This approach aims at increasing the accessibility of method services, facilitating their dynamic composition, and allowing an easy execution of method services composition.

CONCLUSION

We have presented the Context and Policy for Web Services method (CP4WS) that targets those who are responsible for the design and development of information systems based on Web Services. We illustrated how the composition of Web services makes it possible to address the situation of a user's request that cannot be satisfied by any single, available Web service, whereas a composite Web service obtained by combining a set of available Web services might be used. The core concepts of the CP4WS methods are context, policy, service chart diagram, state chart diagram, and resources. The CP4WS methods consists of six steps: user needs identification and specification, Web services orchestration, Web services contextualization, Web services behavior specification, Web services transactional definition, and last but not least Web services deployment. By using a scenario from real world we have explained how the core concepts enable the accomplishment of the CP4WS steps: The first step relies on UML use-cases to identify and specify user needs. The second step uses service chart diagrams and state chart diagrams to specify the orchestration of the component Web services that constitute a composite Web service. The third step defines the arguments that form the context of the component Web services and of other participants such as users and resources that interact with these component Web services. The fourth step uses WSPL to specify the policies that manage the behavior that the component Web services expose to the external environment. The fifth step relies again on WSPL to manage the performance of the component Web services on top of the computing resources. The sixth step deals with the transactional properties of Web service behavior at run-time. Finally, the scenario illustrates the prototype architecture which demonstrates the benefits of CP4WS method to manage various aspects related to Web services such as participation in composition, semantic mediation, and adjustment due to environmental changes and trigger appropriate policies according to the current environmental state.

REFERENCES

Agarwal, S., & Sprick, B. (2005). Specification of Access Control and Certification Policies for Semantic Web Services. In *Proceedings of the 6th International Conference on Electronic Commerce and Web Technologies (EC-Web'2005), Copenhagen, Denmark.*

Anderson, A. H. (2004). An Introduction to the Web Services Policy Language (SWPL). In *Proceedings of the 5th IEEE International Workshop on Policies for Distributed Systems and Networks (POLICY'2004), New York, USA.*

Baresi, L., Guinea, S., & Plebani, P. (2005). WS-Policy for Service Monitoring. In *Proceedings of the 6th International Workshop on Technologies for E-Services (TES'2005), in conjunction with the 31st International Conference on Very Large Data Bases (VLDB'2005), Trondheim, Norway.*

Bate, C., & Mulholland, A. (2005). A Methodology for Service Architectures Version 1.0 – August 2005. *OASIS technical report*. Capgemini, UK.

Bentahar, J., Maamar, Z., Wan, W., Benslimane, D., Thiran, Ph., & Sattanathan, S. (2008). Agent-based Communities of Web Services: An Argumentation-driven Approach. *Service Oriented Computing and Applications, 2*(4). doi:10.1007/s11761-008-0033-4

Bhiri, S., Perrin, O., & Godart, C. (2005). Ensuring Required Failure Atomicity of Composite Web Services. In *Proceedings of the 14th International World Wide Web Conference (WWW'2005), Chiba, Japan.*

Booch, G., Rumbaugh, J., & Jacobson, I. (1998). *The Unified Modeling Language User Guide*. Addison-Wesley Professional.

BPMN. (2008). Business Process Modeling Notation. Retrieved December 2008 from http://www.bpmn.org

Brambilla, M., Ceri, S., & Fraternali, P. (2006). Process Modeling in Web Applications. *ACM Transactions on Software Engineering and Methodology, 15*(4). doi:10.1145/1178625.1178627

Breener, M., & Schiffers, M. (2003). Applying Web Services Technologies to the Management of Context Provisioning. In *Proceedings of the 10th Workshop of the OpenView University Association (OVUA'2003), Geneva, Switzerland.*

Cappiello, C., & Pernici, B. (2006). A Methodology for Information Quality Management in Self-Healing Web Services. In *Proceedings of the 11th International Conference on Information Quality (ICIQ'2006), Massachusetts, USA.*

Ceri, S., Fraternali, P., & Bongio, A. (2000). Web Modeling Language (WebML): A Modeling Language for Designing Web Sites. *Computer Networks, 33*, 1–6. doi:10.1016/S1389-1286(00)00040-2

Consortium, U. W. A. (2002). Ubiquitous Web Applications. In *Proceedings of eBusiness and eWork Conference*.

Coutaz, J., Crowley, J. L., Dobson, S., & Garlan, D. (2005). Context is Key. *Communications of the ACM, 48*(3). doi:10.1145/1047671.1047703

Curbera, F., Duftler, M., Khalaf, R., Nagy, W., Mukhi, N., & Weerawarana, S. (2002). Unraveling the Web Services Web: An Introduction to SOAP, WSDL, and UDDI. *IEEE Internet Computing, 6*(2). doi:10.1109/4236.991449

Damianou, N., Dulay, N., Lupu, E., & Sloman, M. (2001). The Ponder Policy Specification language. In *Proceedings of the 2nd IEEE International Workshop on Policies for Distributed Systems and Networks (POLICY'2001), Bristol, UK.*

De Troyer, O., & Casteleyn, S. (2003). Modeling Complex process for Web application with WSDM. In *Proceedings of the International Workshop on Web Oriented Software Technology (IWWOST'2003), Asurias, Spain.*

Distante, D., Rossi, G., & Canfora, G. Modeling Business Processes in Web Applications: An Analysis Framework. In *Proceedings of the 2006 ACM Symposium on Applied Computing (SAC'2006), Dijon, France.*

Dustdar, S. & Papazoglou, M. (2008). Services and Service Composition – An Introduction. *Information Technology Journal, 50*(2).

Dustdar, S., & Schreiner, W. (2005). A Survey on Web Services Composition. *International Journal on Web and Grid Services, 1*(1).

Erl, T. (2007). *SOA Principles of Service Design* (1st ed.). Prentice Hall PTR.

Horrocks, I., Patel-Schneider, P. F., Boley, H., Tabet, S., Grosof, B., & Dean, M. (2004). *SWRL A Semantic Web Rule Language Combining OWL and RuleML*. W3C Submission, 21 May 2004. Retrieved from http://www.w3.org/Submission/2004/SUBM-SWRL-20040521/

Iacovelli, A., Souveyet, C., & Rolland, C. (2008) Method as a Service (MaaS). In *Proceedings of the IEEE Second International Conference on Research Challenges in Information Science (RCIS'2008), Marrakech, Morocco.*

Kagal, L., & Finin, T. (2005). Modeling Communicative Behavior using Permissions and Obligations. In F. Dignum, R. van Eijk, & M. P. Huget (Eds.), *Developments in Agent Communication*.

Keidl, M., & Kemper, A. (2004). A Framework for Context-Aware Adaptable Web Services. In *Proceedings of the 9th International Conference on Extending Database Technology (EDBT'2004), Crete, Greece.*

Limthanmaphon, B., & Zhang, Y. (2004). Web Service Composition Transaction Management. In *Proceedings of the 14th Australasian Database Conference (ADC'2004), Dunedin, New Zealand.*

Lupu, E., & Sloman, M. (1999). Conflicts in Policy-Based Distributed Systems Management. *IEEE Transactions on Software Engineering, 25*(6). doi:10.1109/32.824414

Maamar, Z., Benatallah, B., & Mansoor, W. (2003). Service Chart Diagrams - Description & Application. In *Proceedings of the Alternate Tracks of the 12th International World Wide Web Conference (WWW'2003), Budapest Hungary.*

Maamar, Z., Benslimane, B., Mrissa, M. & Ghedira, C. (2006a). CooPS - Towards a Method for Coordinating Personalized Services. *Software and System Modeling Journal Special Section on Service-Based Software and Systems Engineering, 5*(2).

Maamar, Z., Benslimane, D. & Anderson, A. (2006b). Using Policies to Manage Composite Web Services. *IEEE IT Professional, 8*(5).

Maamar, Z., Benslimane, D., Kouadri Mostefaoui, G., Sattanathan, S., & Mahmoud, Q. (2008). Toward Behavioral Web Services Using Policies. *IEEE Transactions on Systems, Man, and Cybernetics. Part A, Systems and Humans, 38*(6). doi:10.1109/TSMCA.2008.2003460

Maamar, Z., Benslimane, D., & Narendra, N. C. (2006c). What Can Context do for Web Services? *Communications of the ACM, 49*(12). doi:10.1145/1183236.1183238

Maamar, Z., Narendra, N. C., Benslimane, D., & Sattanathan, S. (2007). Policies for Context-Driven Transactional Web Services. In *Proceedings of the 19th International Conference on Advanced Information Systems Engineering (CAISE'2007), Trondheim, Norway.*

Maamar, Z., Sheng, Q. Z., Tata, S., Benslimane, D. & Sellami, M. (2009b). Towards An Approach to Sustain Web Services High-Availability Using Communities of Web Services. *International Journal of Web Information Systems.*

Maamar, Z., Tata, S., & Belaid, D. (2009a). Towards An Approach to Defining Capacity-Driven Web Services. In *Proceedings of the23rd International Conference on Advanced Information Networking and Applications (AINA'2009), Bradford, UK.*

Maurice, H., Bucchiarone, A., & Gnesi, S. (2007). Web Service Composition Approaches: From Industrial Standards to Formal Methods. In *Proceedings of the International Conference on Internet and Web Applications and Services (ICIW'2007), Mauritius.*

Menascé, D. A. (2002). QoS Issues in Web Services. *IEEE Internet Computing, 6*(6). doi:10.1109/MIC.2002.1067740

Moses, T. (2003). *XACML profile for Web services.* Working Draft 04, 29 September 2003, Retrieved from http://www.oasis-open.org/committees/download.php/3661/draft-xacml-wspl-04.pdf

Mrissa, M., Ghedira, C., Benslimane, D., Maamar, Z., Rosenberg, F., & Dustdar, S. (2007). A Context-based Mediation Approach to Compose Semantic Web Services. *ACM Transactions on Internet Technology, 8*(1). doi:10.1145/1294148.1294152

Mukhi, N., Plebani, P., Silva-Lepe, I., & Mikalsen, T. (2004). Supporting Policy-Driven Behaviors in Web Services: Experiences and Issues. In *Proceedings of the 2nd International Conference on Service Oriented Computing (ICSOC'2004), New-York, USA.*

Papazoglou, M. P., Traverso, P., Dustdar, S. & Leymann, L. (2007). Service-Oriented Computing: State of the Art and Research Challenges. *IEEE Computer, 40*(11).

Paterno, F., Mancini, C., & Meniconi, S. (1997) *ConcurTaskTrees: A Diagrammatic Notation for Specifying Task Models.* Paper presented at the INTERACT Conference, Sydney, Australia.

Reiff-Marganiec, S., & Turner, K. J. (2003). A Policy Architecture for Enhancing and Controlling Features. In *A Policy Architecture for Enhancing and Controlling Features.* IOS Press.

Schlimmer, J. (2004). Web Services Policy Framework (WS-Policy). http://www-128.ibm.com/developerworks/webservices/library/specification/ws-polfram/

Schmid, H. A., & Rossi, G. (2004). Modeling and Designing Processes in E-Commerce Applications. *IEEE Internet Computing, 8*(1).

SOAD. (2004). *Service Oriented Analysis and Design Methodology.* Retrieved September 26, 2008 from http://www.ibm.com/developerworks/webservices/library/ws-soad1/

Tolk, A., & Diallo, S. Y. (2005). Model-based Data Engineering for Web Services. *IEEE Internet Computing*, *9*(4). doi:10.1109/MIC.2005.87

Verma, M., & Deswal, P. (2003). Approaching Web Services Transactions (Tech. Rep.). Second Foundation Inc. Retrieved February 2005 from, http://www-128.ibm.com/developerworks/webservices/library/ws-tranart [Visited February 2005].

Younas, M., Chao, K. M., Lo, C. C., & Li, Y. (2006). An Efficient Transaction Commit Protocol for Composite Web Services. In *Proceedings of the IEEE 20th International Conference on Advanced Information Networking and Applications (AINA'2006), Vienna, Austria.*

Chapter 22
Implementation and Modeling of Enterprise Web Services:
A Framework with Strategic Work Flows

Mabel T. Kung
California State University, USA

Jenny Yi Zhang
California State University, USA

ABSTRACT

Recent years have seen a dramatic increase in business processes and research in distributed computing environments. Applications today can be composed of very heterogeneous components: some involve having the user in the loop; some deal with streaming data; while some require high-performance resources for their execution. This chapter examines the performance of a series of process-based models for the development of e-Business using enterprise software applications. Merging management technology in workflow systems is a critical step to provide service-oriented architecture and on-demand business. The authors propose a value-oriented process technique as a strategic alignment to improve investment value. The framework focuses on the guidelines for traditional users to identify the structural conflicts in integrating web services. A comparative study of workflow models for intra-and inter-organizational process control is presented. This chapter identifies the current progress in the adaptability in the design of process models coupled with structural changes of workflow views. The study provides a resource list of successful implementations for practitioners in organizational management. The research highlights the motivation of market facilitation, expert sharing and collaboration that enable commercial applications to support complex heterogeneous, autonomous and distributed information systems.

INTRODUCTION

The relationship between information technology investments and the value of the enterprise has sustained interest among researchers in workflows, cross-enterprise processes governed by business logic and rules, internet protocols, as well as web services. More and more companies collaborate with each other in a virtual way powered by the internet networking, from supplier to end-use customers. The information with the infrastructure can be expanded simultaneously available to all those involved in

DOI: 10.4018/978-1-60566-910-6.ch022

enterprise processes, such as product data, quality, costs and delivery requirements, quantity quotations, process plan efficiency, and interactions for meta-, macro-, and micro- distributed process planning (Siller, Estruch, Vila, Abellan, & Romero, 2008; Kuechler & Vaishnzvi, 2008).

Business process modeling is a significant activity in enterprises as e-Business and enterprise integration drive the need to deploy business processes online (Aissi, Malu, & Srinivasan, 2002; Weiss & Amyot, 2005; Sewing, Rosemann & Dumas, 2006; Chen, Zhang & Zhou, 2007). Most business process modeling efforts are knowledge-intensive and require organizations to formalize a large number of complex inter and intra-organizational processes to facilitate their ensuing deployment in large-scale workflow systems in enterprise planning. (Tagg, 2001) These management systems need to be integrated with the tools of a process to perform within it: productivity tools, specialized technical support systems, such as CAD systems, graphic packages, enterprise-wide integrated software applications, such as ERP (Enterprise Resource Planning), CRM (Customer Relationship Management), mail systems and other communication systems. When the applications become more modulated and service oriented, there will no longer be work-alone software. (Cimatti, Clarke, Giunchiglia & Roveri, 2000).

The most common application for process modeling, control and management is Workflow Management Systems (WfMSs). The technology has become readily available (van der Aalst, Desel, & Oberwies, 2000; van der Aalst & van Hee, 2002; van der Aalst & Jablonski, 2000; Fischer, 2001; van der Aalst & van Dongen, 2002; Grigori, Casati, Dayal, & Shan, 2001; Herbst & Karagiannis, 2000; Cook & Wolf, 1999). Commercial workflow management systems (WfMSs) such as Staffware, IBM MQSeries, and COSA offer generic modeling and enactment capabilities for structured business processes. Besides stand-alone systems, WfMSs are becoming integral components of many enterprise-wide information systems (Leymann & Roller, 2000). Consider for example Enterprise Resource Planning (ERP) systems such as SAP, PeopleSoft, Baan and Oracle, Customer Relationship Management (CRM) software, Supply Chain Management (SCM) systems, Business to Business (B2B) applications which embed workflow technology.

The introduction of large scale systems such as the ERP system changes the structure of the organization of software applications. This moves from numerous independent software development procedures to an integrated web based software framework with components for different purposes. Although ERP system can improve organization's performance, standardized ERP system from the vendor such as SAP, need to be customized to be deployed in an organization. It has to be customized to fit the business goals of the company. This customization needs the continuous input of end user involvement. In order for the dynamics of the web services to succeed, the deployment team needs to understand the business processes of the company that can be incorporated into a workflow design. The design layout can then be used for discussions with the management and end users to provide better understanding of the processes during changes.

Currently, "Eighty percent of the software that needs to be written has already been done collaboratively." (McKendrick, 2006). It was estimated that in U.S. alone, there would be 55 million user developers compared to 2.75 million professional software developers (Sutcliffe & Mehandjiev, 2004). Since the user developed software may affect the entire organization's system, more challenges and conflict issues arise in a more dynamic state (Bergeron & Berube, 1990). Although the centralized Information Technology (IT) department provides the traditional support of the enterprise-wide system, integration and workflow design are far from trivial. Without appropriate policies and control mechanisms, user development cost can be higher than the benefit it brings in (Galletta & Hufnagel, 1992). The cost related to such user software includes poor security, incompatible hardware and software, inadequate documentation (Davis, 1988), insufficient validation and testing (Alavi & Weiss, 1986; Davis, 1988; O'Donnell & March, 1987), ill-defined policies regarding access to corporate databases (Alavi & Weiss, 1986). Existing workflow systems provide general-purpose scripting languages and platforms with tool-automation features (McPhillips, Bowers, Zinn, & Ludascher, 2008; Glatard, Montagnat, Emsellem & Lingrand, 2008). Without a structured process control flow or policy, collaborative computing in enterprise wide systems can hardly be effective.

This paper presents the components of a standardized business process with the emphasis from the user point of view. The individuals involved in each process have the skills and human capital that complement one another. Next, a value-oriented framework is proposed as a benchmark for economic assessment. The synthesis and the process-based approach are discussed to align with organizational strategic goals. In this context, the types of inputs, the nature of tasks needed to perform the activity, the sorts of coordination required among the various tasks, and the intended scale of output depend on the internal top management reactions to assess values in the production level as well as the social level among the employees. Lastly, a summary of current research in workflow models provides the technological and managerial issues involving the current designs in organizations. By combining these different streams of research, our objective is to provide guidelines and structural designs to enable evaluations of process goals to improve the overall value of enterprise web services (Kung & Zhang, 2008).

BUSINESS PROCESS

A business process is composed of a series of continuous actions or operations that are performed upon a commodity (Childe, Maull, & Bennett, 1994). It is usually initiated by a customer. It must provide results directly to a customer, who may be internal or external to the company. CIM-OSA Standards Committee (1989) has sub-divided processes into three main areas: manage, operate and support. The CIM-OSA framework regards manage processes as those which are concerned with strategy and direction setting as well as with business planning and control. Operate processes are viewed as those which are directly related to satisfying the requirements of the external customer, for example the logistics supply chain from order to delivery. They are sometimes referred to as core processes. Support processes typically act in support of the management and operate processes. They include the financial, personnel, facilities management and information systems provision (IS) activities.

Information systems (technologies) make an impact at different levels (Brancheau & Brown, 1993; DeLone & McLean, 1992; Harris, 2000; Powell & Moore, 2002; Seddon, 1997; Andreescu, 2006): system or information level, individual level, group level, organization level, and system or information quality level. When user develops application changes in organizational information system, these modifications will also affect all four different levels. Without careful coordination at all levels, the organization will not obtain the possible benefit that the user intends to bring. Consider the coordination from the process perspective. The collaboration begins in these stages: (1) manage processes are related to organizational level, (2) operational processes are related to system and information level and individual level, and (3) support processes are related to group level influence.

Traditionally, project initiatives begin from the top management level and filter down to the lower level. However, sometimes a project may start from the bottom level with an idea to modify the existing system to improve effectiveness or efficiency of their job. This initial idea will then be

Figure 1. A hierarchical process perspective (or top-down and bottom-up perspective)

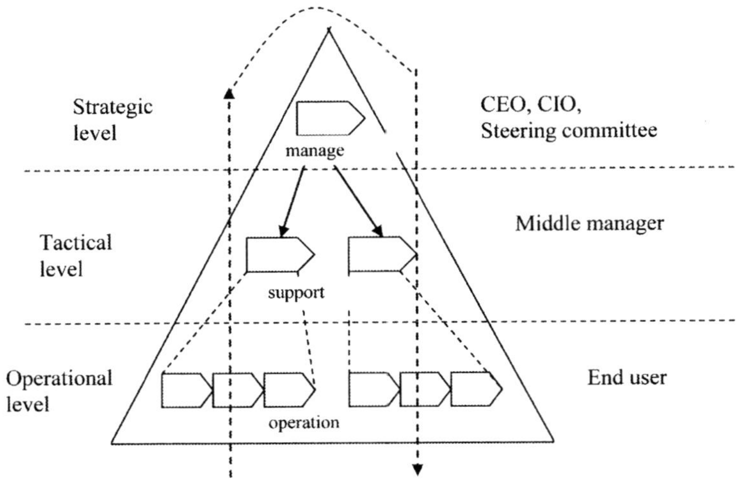

presented to the tactical middle management level to convert it into higher-level business requirement and redirected back to upper management. Upper management will then judge it based on its strategic direction. If the proposal is approved, it will be sent back to tactical level where it will be converted to a detailed plan on how to perform the project. The project will be divided into sub processes and activities, which will then be transferred to operational level where the implementation takes place. Figure 1 illustrates the hierarchical processes within the infrastructure.

Strategic Level: The top level process mainly deals with direction setting – high-level strategic planning activities. One common problem with many improvement initiations being less successful is the lack to the organizational strategy – the big picture (Rummler & Brache, 1995; Hacker & Brotherton, 1998). Sometimes, an initiation might be beneficial in local operation level, but might be malicious for the enterprise-wide strategy. In most IS projects, manage process acts as an overall management that takes ideas about direction based upon business requirement reported from the operational level, decides whether or not to proceed it based on its alignments with the company's overall direction, and sets the high-level goal for the project. Competitive advantage requires the learning, change and adaptation processes over the time horizon in terms of the availability of resources and the capabilities of the users involved. (Helfat & Peteraf, 2003; Adner & Helfat, 2003). The CEO and CIO's office has the responsibility to define the strategic enterprise architecture that provides the infrastructure for activities in each of the business units, the tactical level and the operational levels. Strategic level will decide the organizational level impact from the project.

Tactical level: This level serves as a converting or communication level. It performs several critical roles:

- Transfers the high level strategy into plans on how to convert an initiated idea into a finished product.
- Serves as a bridge between business perspective and technical perspective
- Provides guidance and support at operation level.

This level is the most important among all three levels. It serves as a glue to combine strategic and operational process all together. It encourages the creativity of user, and at the same time guarantees the quality and integrity of user development. The level does all the managerial tasks that directly

Table 1.

Level	Responsibility
Strategic	Strategic alignment Sustainable advantage Benefits measurement Evolution of resources over time
Tactical	Cost control Quality control Connectivity Control Planning Coordination with other department
Operational	Functionality/Capability Development Unit test Documentation

related to the project, for instance, agreement of requirement changes, monitoring project timetable, and quality control. Meeting customer and partner demands with proactive and efficient services, special attention to preferred business alliances such as joint market launches, analyst briefings, technical collaboration, or premium customer support, help increase custom satisfaction. Customers are provided a set of collaborative tools and relationship portals for interacting with core partner services and become stakeholders for the income generated by applying this value-oriented model. In many instances, the success of the implementation of a project depends on this level's function.

Operational Level: This is the level where an idea of implementing a system starts and also where it changes into a final product. With the trend of deploying enterprise-wide information systems, the most important issue at this level is to consider how the final product be integrated into the company's existing IS infrastructure, and be re-used by other departments. The web services and user relationships can be measured by the value of the final product. The user satisfaction may take into effect greater creative freedom and channel the skills in an innovative fashion for career growth, such as obtaining deeper specialization in an area, taking broader responsibility in serving the area services, acquiring decision making skills to meet enterprise transformation and culture changes (Table 1).

A Value-Oriented Framework

The value of the computing depends largely on the quality of the use to which it is put. Pressure to decrease costs has led to downsizing of data centers, reductions in programming staff for development, fewer resources for maintenance and an overall push to move computing power out to the users (Livari & Livari, 2006; Melville, Kraemer, & Gurbaxani, 2004; Fischer, Giaccardi, Ye, Sutcliffe, & Mehandjiev, 2004; Tallon, Kraemer, & Gurbaxani, 2000; Boynton, Zmud, & Jacobs, 1994; Guimaraes, Gupta & Rainer, 1999; Hitt & Brynjolfsson, 1997). Instead, these managers should be asking: *"How much more value can the organization create, in a given period of time, with the user computing as opposed to without it?"*

What follows is a value-oriented framework (Table 2) that addresses interrelated problems through the value of information technology. The assessment starts with Level I as the lowest stage with each level dependent on all lower levels.

Table 2.

Level	Issues
IV Transformation	New Business Doing Business Differently Sustainable Advantage
III Business Linkage	Alignment / Organization Service Level / Support Flexibility / Responsiveness Benefits Measurement
II Economic	Applications Development Production Support / Maintenance Overheads
I Mechanical / Physical	Functionality / Capability Reliability Connectivity

Level I: Mechanical/ Physical Level

Naturally, the first and most elementary question is: Do the hardware, software, and communications do what they are supposed to? The significant problems at this level today have to do not so much with individual systems, but more with integrating multiple systems from a variety of vendors. Software vendors have responded to this dilemma by aggressively developing products and services to meet the demands of both the systems personnel and users. They have embraced new technologies such as distributed processing, client/server architectures and relational database technology. But more importantly, they have developed new systems that provide users with the ability to maintain, enhance and run their own systems without constant involvement and assistance from the systems department.

Level II: Economic Level

Once the user computing works as it is intended, it must do so at an acceptable cost. Of course, purchasing managers do have up-to-the-minute information on how much the next system will cost to acquire. The problem is, once the systems have been acquired, nobody keeps track of the user computing assets as a separate category, aside from office equipment or sometimes, furniture. This often has to do with the structure of the company's chart of accounts. In the scheme of things, these technologies are still relatively new, and accountants absolutely hate to change the chart of accounts (Hitt & Brynjolfsson, 1997).

Even where all the hardware and software costs are tracked regularly and careful, companies often stumble on the scope of costs associated with the user computing. Most important, training and support costs are almost always underestimated or, worse still, ignore. These costs, combined with the cost of the users' own time, turn out to be much larger than the initial cost of the hardware and software.

Level III: Business Linkage Level

The third layer address whether the user computing is being used in ways, which promote the company's objectives. The chief question here is whether the user computing is deployed most intensively, where it can best leverage the organization's ability to create value. Too frequently, user computing gets deployed strictly along organizational lines, without much regard for the fact that user computing, like any other tool, has more value in the hands of certain groups or individuals (Compeau, Higgins & Huff, 1999; Shah, 2001; Talon, Kraemer & Gurbaxani, 2000).

The issue of business linkage also involves hardware and software selection, as well as training and support. In particular, there is a strong, natural tendency for central information systems organizations to limit the user computing choices from which user organizations can make selections and receive support. While this approach helps the information/ systems managers hold down their budgets, if taken too far it can seriously reduce the user computing leverage for a particular business or unit.

Level IV: Transformation Level

The final layer focuses on whether the business strategy has been conceived and implemented in ways that take advantage of the opportunities provided by the user computing. Doing things the same old way is comfortable, and often seems to entail less risk. As fundamental changes in the corporate environment have given rise to the wave of restructuring, dramatic advances in user computing technologies are enabling fundamental changes in the ways in which work is structured. Because some managers are unwilling to embrace substantial change to the internal culture, many of these companies are overlooking opportunities to eliminate vast amounts of paperwork, along with the associated costs and risks to quality. In retrospect, user computing enables new organizational reporting relationships. In this age of restructuring, user computing supports the need to move away from conventional hierarchical structures to more relational organizations, with less management filters (Biazzo, 1998; Davenport, 1993; Hammer, 1996).

Synthesis with Workflow

Business linkage and transformation are the most crucial levels. As the most successful deployers of user computing have found, not moving up from the lower two levels of the model is analogous to manufacturing a product, shipping it to a warehouse, and then waiting for potential customers to notice it is there (Strassmann, 1997). Put simply, products alone do not deliver value, customers do.

In any business, it is the customer who eventually determines the product value, sets a reasonable price and establishes marketplace demand. It is the customer who controls cash flow into the organization and, thereby, drives shareholder value.

For information systems management, the customer is the user, and the marketplace may be the business divisions or functional groups within the enterprise. It is the organization's users who will, therefore, ultimately determine the value and return on the user computing investments.

Consequently, the focus of management must shift from the traditional comfort zones of Level I and II – technical standards and acquisition control – toward the user and the organization's business strategy. Of course, the Level I and II issues must also be addressed appropriately if the benefits of Level III and IV are to be realized. A key element for evaluation process has been holding sessions with users to get their input on how well the current computing environment meets the real needs of their businesses. User may hold the technical expertise of their information systems organizations in high esteem, but consider it of limited value when that expertise is not applied to their business in ways that not only work technically, but also add significant value.

Today's dynamic environment has no place for the drawn-out, form-driven, bureaucratic planning process and thick planning documents. Instead, the alignment process is based on frequent, structured dialogues between the information systems management and users, and great care is taken to ensure that the discussion gets beyond the "gripe session" level. Findings from these meetings and subsequent surveys have been somewhat surprising and enlightening for the information systems management. As a direct of these efforts, some businesses can be identified as "under-served"; others as "over-served." Adjustments can be made to both resource allocation and support levels, resulting in both cost saving and revenue enhancement. Further, emerging opportunities for high-value applications of the user computing can be uncovered, particularly in the areas of marketing and sales.

This approach enables the information systems organization to enter into a partnership with the users. Equally important, by focusing on Business Linkage (Level III) and Transformation (Level IV) issues, the connection between the user computing and shareholder value has been clearly established. Finally, user managers now have responsibility for both their business unit's performance and shared responsibility for the users who support it. For their part of the partnership, the information systems managers play a supportive role with respect to user computing applications, and serve as the keepers of the corporate standards (Keen & Knapp, 1996).

In fact, coordination support is the key component that distinguishes task-oriented from process-oriented technologies. When it comes to currently available computer-supported process coordination, workflow technology has been widely recognized as the leading process-oriented coordination tool (Workflow Management Coordination, 2006). Workflows are designed to specify, execute, manage, monitor and streamline business processes that span the functional boundaries in an organization. Figure 2 shows the technology that offers effective coordination support by allocating the right task to the right person at the right point of time along with the resources needed to perform the assigned task.

In Figure 2, Interface 1 is used at build-time to define the workflow process. Interface 2 defines the standard mechanism for interacting with the user of the WfMSs – the worklists that appear on user screens. Interface 3 is the API through which the WfMS interacts with other user applications such as ERP or CRM systems. Interface 4 is the standard API through which WfMSs provided by different vendors can interoperate. Interface 5 is the API through which administrators gather information from the log maintained by the WfMSs. Facilities such as e-meetings with electronic white-boards, instant messaging, webcasts, and task-oriented community tools supplement the existing synchronous communication facilities, such as teleconferences. Asynchronous communication is supported by specialized team rooms, project databases, interactive team portals and forums, and e-mail.

The strategic level of the organization should establish a demonstrable connection (direct or indirect) to one or more critical business issues (Process Definition). Accountability related to process modeling requires a clear specification and has to be adapted with changes in the objectives, scope or size of the modeling initiative (Administration and Monitoring Tools). The true economies of scale and synergies occur when communications among the end-users at the operational level are open and are willing to migrate the wide range of purposes to one common platform (Workflow Engines). The managerial tasks at the tactical level are to identify the required skills by means of educational training, to locate expert process representatives, to explain to the users the holistic picture, and to facilitate the process visibility across the heterogeneous group of stakeholders, the strategic, tactical and operational levels (Workflow Client Application and Interchange).

The workflows or processes within a single organization can be extended to multiple, geographically distributed locations over wide-area communications networks. (Basu & Kumar, 2002; Sewing, Rosemann, & Dumas, 2006). Applying the value-oriented process framework and workflow perspectives, the user can create digital interface by means of common platform, such as Java 2JEE, Java Servlets, or using JSP, a process that requires minimal development time. (van der

Figure 2. Workflow Reference Model (Workflow Management Coalition, 2006)

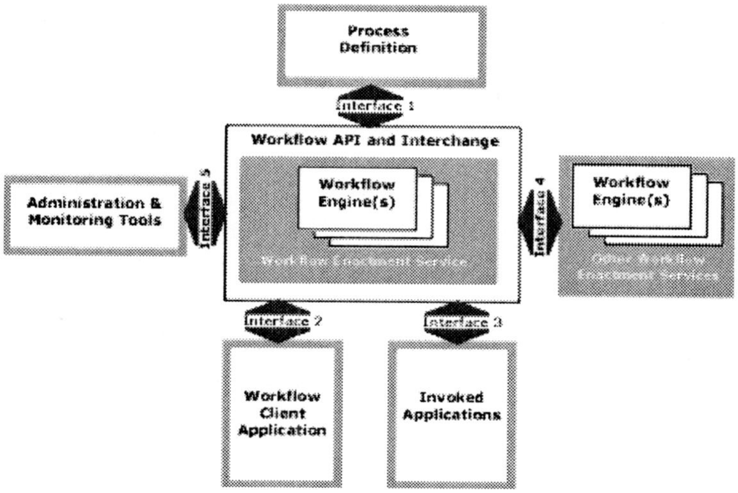

Aalst, Weske, & Grunbauer, 2005). Workflow management systems such as Ensemble (FileNet) and InConcert (InConcert) support workflows by the end-user of the system under unexpected undesirable events. (van der Aalst & Jablonski, 2000). Many enterprises select standardized commercial workflow management systems, COSA, Visual Workflow, Forte Conductor, Lotus Domino Workflow, Meteor, Mobile, MQSeries/Workflow, Staffware, Verve Workflow, I-Flow, InConcert, Changengine, SAP R/3 Workflow, Eastman, and FLOWer. (van der Aalst, ter Hofstede, Kiepuszewski, & Barros, 2003).

Mediation to link service requestors, providers and end users is supported by middleware such as the Enterprise Service Bus (ESB). (Robinson, 2004; Schmidt & Kalyana, 2004). The ESB is the infrastructure which integrates the user roles involved in creating and managing the solutions, describing service endpoint requirements, capabilities, and relationships, including information describing the specific details of interaction contracts. The service registry assembles the runtime entities, dynamic adaptation components, multiple crosscutting configuration, connection, matchmaking, channel structures and event application domain for users. (Kon, Costa, Blair & Campbell, 2002) These ESB usage patterns are realized through large-scale retail and brokerage applications. The ESB plays a central role in the implementation of the architecture for the IBM On Demand Operating Environment. (Cox & Kreger, 2005; Schmidt, Hutchison, Lambros, & Phippen, 2005; Sadtler, Cotignola, Crabtree, & Michel, 2004).

In both intra-organizational and inter-organizational WfMSs, traditional workflow systems have limitations in support of flexibility, adaptability, these limitations result in restraint control, delegation, and coordination of processes and tasks for mid-level managers. (van der Aalst, Weske, & Wirtz, 2003). In the next section new developments in WfMSs will be presented to overcome the limitations and to support workflow control over multiple organizations.

COMPARISON OF DESIGNS IN PROCESS-BASED OPERATIONS

In order for a process model to operate coherently, not only the users need to know how each activity works, but they have to *manage* the dynamic changes in the processes so that the flow of work and information between participants is reasonable and efficient (Basu & Blanning, 2000; Bolton & Davis, 2000; Stohr & Zhao, 2001). Traditionally, WfMSs support process control within one organization – Intra-organizational WfMSs. (Hevner, March, Park & Ram, 2004). However, with the evolution of the commercial internet, the trends for virtual corporations and e-commerce, increased global networking of economies is accelerating. Work has also shifted from creation of tangible goods from one organization to the flow of information through the value chain that across multiple organizations. The research in WfMSs has also shifted to define, analyze, and management the flow of information-intensive work (Basu & Kumar, 2002). This extension allows the users on the operational level to communicate and refine the process as these web-based systems move toward an open environment. Such open processed-based systems enable the employees at the operational level of companies to implement their ideas in the form of inherently distributed and inter-organizational design (Verbeek, Basten & van der Aalst, 2001; Basu & Kumar, 2002). The flexibility as an interaction agent over the internet pushes more control of the middle managers and their subordinates to perform more market-based solutions. Implementations using the Unified Modeling Language (UML) serve to be a useful technique in integrating this design (Fowler & Scott, 1997). Another analytic tool that users can develop company's views of the process is the bridging of the eXtensible Markup Language (XML) and supply chain modeling that define data elements in business documents.

Intra-Organizational Process Control

Intra-organizational WfMSs are implemented to support the modeling, analysis, and performance of routine business processes. With the trend of companies going global and joining e-commerce, many business processes are subject to change. However, the traditional WfMSs typically fail to allow for unexpected or developmental changes occurring in the business practices and processes they model (Casati & Pozzi, 1999; Borgida & Murata, 1999; Heinl, Horn, Jablonski, Need, Stein, & Teschke, 1999; Van der Aalst, 1999). They also have limited support to emergent processes which is a challenge to the coordination-related tasks that an end-user is likely to perform (Alvai & Leidner, 2001). Most available commercial workflow systems rely on a monolithic, single-schema architecture, which makes it difficult to fully capture the business process to be supported (Bichler, Preuner, & Schrefl, 1997). This has been recognized as a major limitation in the uptake of WfMSs (Heinl, Horn, Jablonski, Need, Stein, & Teschke, 1999). Also, these WfMSs provide little support for exception handling at the process-conceptual and instance-execution layers (Casati & Pozzi, 1999). These limitations restrain the control of the end users on the WfMSs, which makes it less efficient.

Recently, research in intra-organizational WfMSs has been focused on providing solutions to the above problems so that WfMSs can offer the automation of the routine tasks, and help users deal with exceptional situations, breakdowns, or emerging new processes in a secured manner. Van der Aalst (1999) presented a generic model which can provide management information at the right aggregation level and also offer adaptability. Kumar and Zhao (1999) proposed a general framework to implement dynamic routing and operational controls in WfMSs. Faustmann (2000) proposed an approach to configure parts of a detailed process model with different ways of assigning tasks to a worker, which they call support strategies. These support strategies allow changes if the situation requires. In this approach, the end user can have different ways to accomplish on a task. They can do it directly, or delegate sub-tasks to other workers. Kumar, Van der Aalst, and Verbeek (2002) proposed an approach to dynamically distribute work in order to create a balance between quality and performance. Wang and Wang (2006) used a cognitive approach to take real-time decisions on activities into consideration so that the system is more adaptable. Adams, Edmond, and Hofstede (2003) proposed an approach of handling flexibility by deriving principles for work practice from "Activity Theory". Klein and Dellarocas (2000) presented a notation – Ariadne to support different dimensions of process modeling to achieve adaptability. Hagen and Alonso (2000) presented an algorithm for improving fault tolerance of WfMSs based on exception handling from programming languages. Klein and Dellarocas (2000) proposed to use a knowledge management system for exception handling.

Another concern in WfMSs is security. Workflow authorization models were proposed in late 1990s (Atluri & Huang, 1996; Atluri & Huang, 1997; Casati, Ceri, Pernici, & Poss, 1995). In recent years, Wainer, Barthelmass and Kumar (2003) proposed security models for WfMSs with Role-Based Access Control (RBAC) model. With the advances of the internet technology, companies are becoming distributed and multinational. An extensive array of functions across the organization is being performed through the web services. The security concern is also moved to the cyberspace (Gudes, Olivier, & Riet, 1999; Gudes & Tubman, 2002). Several studies address the organizational structure changes due to the decentralization and globalization of the companies (Tan & Harker, 1999; Klarmann, 2001; Muehlen, 2004). Other assessments in WfMSs involve monitoring business process performance (Thomas, Redmond, Yoon, & Singh, 2005), using incentive mechanisms to formulate organizational modeling (Raghu, Jayaraman, & Rao, 2004).

Inter-Organizational Process Control

Compared to intra-organizational workflows, inter-organizational workflow has its unique issues. Among them, the most important ones are *heterogeneity* which consists of the hardware, software, automation level and workflow control policies, and *autonomy* of the local systems which result in a lack of cross-company access to workflow resources and the missing of a complete view of the whole workflow (Zhao, 2002). In this area, research focus is in developing techniques for ensuring semantic integrity of the information and rules for mapping it correctly between any two partners. Currently, XML and Web services gain popularity across enterprise systems and infrastructures. These services sustain major roles in inter-organizational workflow management. A major challenge in achieving the goal of Web services composition for process management is semantic interoperability. Communication among heterogeneous, independently developed Web services demands a well-defined mechanism for semantic description of services and their properties so as to make services semantically understandable by business process. Security is also a concern (Zhang, 2005).

Van der Aalst (1999) presented two possible process-oriented architectures for inter-organizational workflow systems. Several research studies in this area focus on defining languages or schemas to support inter-organizational workflow (Van der Aalst & Kumar 2003; 2005; Workflow Management Coalition, 2006). Chiu, Cheung, Till, Karlapalem, Li, & Kafeza (2004) used workflow views for interoperability of multiple workflows across business organizations. Web services present another popular topic in inter-organizational process control. Zhang (2005) evaluated the roles of web services in cross-organization process management. Cardoso and Sheth (2003) developed ways to discover web services in inter-organizational WfMSs. Kumar and Wainer (2005) explored the exception handling problem in inter-organizational setting. They used XML defined meta-workflow knowledge for control and coordination. Singh and Salam (2006) discussed the security aspect of inter-organizational process control. They deployed ontology analysis to identify central concept for e-Business process modeling.

Table 3 gives a summary of the current workflow models for both intra- and inter-organizations.

CONCLUSION

Using the value-oriented framework from a systems perspective, the user is typically working with some sort of task. Ideally this task is adding value to an activity. This activity should add value to a process output – a product, and the process and product should be vital to the organizational strategy.

While the user tend to have excellent knowledge of day-to-day operations and what is needed in these operations, he/she may not have full understanding of the process goals and how different activities together add value to the process output. On the other hand, management tends to be withdrawn from day-to-day operations and may not fully understand the details at each task level.

Ideally, the holistic big picture should be understood by the personnel in the organization, yet few seem to do. Problems potentially occur when user recognizes a clear need for some sort of improvement or development, yet since he/she may not have knowledge of the bigger picture, this improvement may cause sub-optimization of the system. In other words, the improvement may help at task level but not at process level. If users understand the systemic picture and the connection between strategy, processes and operations, then user development of activities as well as systems can be a valuable tool for organizational improvement and efficiency. This form of development will not cause sub-optimization of processes. On the contrary, organization can use standard forms of software, and still create unique solutions at user/task level. Similarly, ideas to improve operations can be implemented. WfMSs can be used to help both managers and end-user understand the business process better. It also helps managers to control and delegate tasks more efficiently and effectively.

The management of the new information based company is the entrepreneurial spirit through user computing. Exchanging and distributing knowledge allow people at the line levels more aggressively setting their own direction and objectives. Individual managers feel more control and satisfaction with the end product to resolve issues on their own. Mentoring opens wider communications between the systems personnel and levels of employees closest to operations, customers, and their associated problems. Such strategy continues to encourage the creativity and team cooperation in the business functions of the company and ultimately the systems achieve usability from the people who design and develop themselves – the users.

Although WfMSs provide us a promising solution to help understand and control processes and motivate communications among different levels of personnel in an organization, current commercial WfMSs still have limitations in supporting flexibility and adaptation, and lack of interoperability to support B2B workflow control. As the gap between academic and industry standards reduces, the above weakness can be overcome.

This paper aims to address a guide to practitioners through a series of well-defined structural steps necessary to make informed, consistent and efficient changes to business processes. The research has also contributed to the new knowledge in web-based services with the collaborative workflow applications. The mechanisms of inter-organizational workflows coupled with the performance incentives of the process framework enable the users to integrate enterprise applications in a distributed environment. In order to gain a sustainable competitive advantage in the wide spectrum of e-services, workflow technologies coupled with cross-functional business processes offer fully automated coordination support. Future debates include the standardization to bridge between systems with an organizational boundary,

Table 3. A summary of current progress in workflow models

Author	Aspect	Solutions	Technology used
Van der Aalst (1999)	Capture management information; Adaptation	Present a generic model inspired by the techniques used in product configuration to aggregate management information and also support dynamic changes	Product configuration
Kumar & Zhao (1999)	Flexibility; Exception handling	A general framework to implement dynamic routing and operational controls	Workflow control tables; Sequence constraints; Event-based workflow management rules
Hagen & Alonso (2000)	Exception Handling	An algorithm for implementing more reliable processes based on exception handling in programming languages, and atomicity	Exception handling in programming languages; atomicity
Agostini & Michelis (2000)	Flexibility; Adaptation	Present the MILANO system which is highly flexible and adaptable. The system is built on the principle that workflow models must be as simple as possible	Elementary Net System
Faustmann (2000)	Flexibility; Adaptation	Proposed an approach that configures parts of a detailed process model with different support strategies (how a system assign tasks to a worker). The explicit modeling of these support strategies allows them to be changed if demanded by the situation.	Used in the WAM approach (Wide Area Multimedia Group Interaction)
Klein & Dellarocas (2000)	Exception Handling	Proposed an approach for exception handling that is based on exploiting a generic and reusable body of knowledge concerning what kinds exceptions can occur in collaborative work processes, and how these exceptions can be handled.	Artificial Intelligence
Divitini & Simone (2000)	Adaptability	The paper claims that adaptability involves different dimensions of process modeling. These dimensions concern the possibility to flexibly combine a rich set of basic categories in order to obtain the most suitable language for modeling the target business process and the work practices around it.	Ariadne, a notation providing a set of linguistic features suitable to model processes and their evolutions.
Kumar et al. (2002)	Dynamic work distribution	A systematic approach to dynamically create a balance between quality and performance issues.	Use metrics to represent work distribution
Adams, et al. (2003)	Flexibility; Exception Handling	Derive a set of principles for work practice from "Activity Theory" to create a set of criteria to provide adequate support for flexible work methods.	Activity Theory
Wang & Wang (2006)	Adapt to change	A cognitive approach to help manage complex business activities based on continuous awareness of situations and real-time decisions on activities.	Cognitive Process
Tan & Harker (1999)	Organizational structure: centralized vs. de-centralized	Use of mathematical modeling to compare the total expected costs of decentralized and centralized organizational designs. Coordinate the flows of information and work.	Mathematical modeling
Klarmann (2001)	Changes in organizational structure	Existing systems cannot cope with frequent structural change of organizational and process structure. Use of an organizational meta-model that describes meta information about organizational structures.	Meta Model
Muehlen (2004)	Organizational Management	Provide an overview of the organizational aspects of workflow technology in the context of the workflow life cycle	Meta model

continued on the following page

Table 3. continued

Author	Aspect	Solutions	Technology used
Thomas, et al.(2005)	Monitor business process performance	A loosely coupled semantic architecture overlaid upon a business process, where agents communicate and monitor business process performance. The descriptive power of semantic languages can be used by agents to provide input for process reconfiguration decisions based on process performance measures.	BPEL, Web Ontology Language
Raghu et al. (2004)	Economic incentives	An approach to organizational modeling that integrates both agent-centric and activity-centric approaches using incentive mechanisms.	Combine agent-centric and activity-centric to model organizational process
Gudes (1999)	Security	Present a three-level framework: modeling, specification and implementation. The participation of an Alter-ego in each message enables the complete authentication and some specific individual-based checks that are required in such an environment.	Alter-ego: one object in which all relevant of an individual person is kept and which can execute actions.
Gudes & Tubman (2002)	Security	A system AutoWF is presented for secured WfMS over the Web.	Autonomous objects
Wainer et al. (2003)	Security	Present a pair of role-based access control models for workflow systems known as W-RBAC models	The Role-Based Access Control (RBAC) model
Van der Aalst (1999)	Process-oriented architecture verification (across organizations)	Evaluate two approaches of inter-organizational workflow architecture with the concern of possibility to verify correctness of inter-organizational workflows	Case transfer architecture; Loosely coupled architecture
Singh & Salam (2006)	Security aspect of inter-organizational Business process (across organizations)	An ontological analysis of an eBusiness process and identify a set of central concepts that are essential to model the eBusiness process. Utilize this eBusiness process to develop a semantic architecture.	OWL-DL (description logics)
Alast & Kumar (2003, 2005)	Inter-organizational information exchange (across organizations)	Develop process models of inter-organizational workflows and their coordination structures. Design an eXchangeable Routing Language (XRL) using XML	Petri nets, XML XRL
Kumar & Wainer (2005)	Exception handling (across organizations)	Control and coordination of inter-organizational workflow systems using meta-workflow knowledge of inter-organizational eBusiness processes	XML
Zhang (2005)	Inter-organizational process management (across organizations)	Discuss the role of web services in process management. Propose an architecture for process workflow via web services composition.	Web services
Chiu et al. (2004)	Interoperability (across organizations)	Use of workflow views as a fundamental support mechanism for the interoperability of multiple workflow across business organizations.	XML, Web services
Cardoso & Sheth (2003)	Interoperability (across organizations)	Develop ways to efficiently discover web services – based on functional and operational requirements and to facilitate the interoperability of heterogeneous web services in e-services. Use of ontology to achieve service discovery and interoperability functions more efficiently.	Web services, Ontology-based systems

continued on the following page

Table 3. continued

Author	Aspect	Solutions	Technology used
Zeng, et al. (2008)	Error detection and correction	Develop a conflict verification and resolution approach for workflow constrained by resources and non-determined duration.	Petri net
Sun et al. (2008)	Merging workflows	Describe the concept of merging workflows with business processes. Four types of merges with the corresponding algorithms, and issues related to the merge region to avoid invalid merges are discussed.	Petri net
Julia, et al. (2008)	Real time scheduling of WfMS with hybrid resources	Propose an approach based on a p-time Petri net model with hybrid resources to solve the real time scheduling problem of workflow management systems	P-time Petri net, Activity diagram, Token player algorithm
Tretola & Zimeo (2008)	Grid workflow (across organizations)	Use a technique that generates fine-grained concurrency with asynchronous invocation of services and continuation of execution to ensure pre-scheduling activities at run-time to improve performances.	Asynchronous calls and symbolic reference in concurrent language
Glatard, et al. (2008)	Service oriented architecture for optimizing distributed workflow execution	Use the service-oriented approach based on grid architecture to help dynamically group services which reduce gird overhead on the execution and help optimizing the application execution time.	Grid technologies, Web services
Deelman, et al. (2008)	Use workflows in scientific settings	Provide a summary of features and examples that scientists apply the existing workflow systems.	Grid computing, Web services
Cao et al. (2008)	Service-oriented workflow mapping	Use Contract Net Protocol (CNP) to implement service-oriented workflow mapping in ShanghaiGrid workflow management system, including Abstract workflow, Concrete workflow, and Executable workflow	Grid technology, Contract Net Protocol, Belief-Desire-Intension agent technology
De Roure, et al. (2008)	Social sharing of workflows	Use social web approach to design and build a system for scientists to share and collaborate through workflows.	Social web approach, Experiment Virtual Research environment
Truong, et al. (2008)	Performance metrics for Grid workflows using ontology	Analyze performance metrics for evaluating the performance of Grid workflows. Develop an ontology to describe performance data of workflows and use it for analyzing and monitoring purpose.	Grid systems, Ontology
Dang, et al. (2008)	Ontology for workflow system in healthcare	Provide an ontology knowledge framework for healthcare domain applying to machine intelligence.	Semantic Web, SOA, Ontology
Ly, et al. (2008)	Integration and verification of semantic constraints in adaptive PMS	Present a framework for defining constraints over processes which takes into consideration of both real-world domain knowledge and the maintenance and semantic process verification effort.	Adaptive PMS, Semantic correctness, Semantic constraints, Semantic process verification
Kuechler Jr. & Vaishnavi (2008)	Dynamically re-coordinate distributed workflows (across organizations)	Present a model using expert system to dynamically re-coordinate workflow processes when one organization changes its process based on workflow goals.	Expert system, Distributed workflows
Hwang, et al. (2007)	QoS of web-service-based workflows	Propose a unified probabilistic model for describing QoS values web-service-based workflows.	Sample-space reduction technique: dynamic programming and greedy method
Mok et al. (2006)	Computability of workflow problems	Use theoretical computer and statecharts to prove that many workflow problems do not have pure computer-based solutions. Human intervention is inevitable for many problems.	Agent-based method, statechart

continued on the following page

Table 3. continued

Author	Aspect	Solutions	Technology used
Siller et al. (2008)	Modeling workflows for collaborative process planning (across organization)	Propose a workflow model for collaborative process planning with the support of Product LifeCycle Management and CAD/CAM tools.	CAPP and PLM systems CSCW groupware tool CAD/CAM tools
Eshuis & Grefen, (2008)	Customized process views (across organization)	Describe an approach to construct customized process views on process models to hide confidential or irrelevant for the partners.	Logistics
Hauser, et al. (2008)	Data structure for workflow analysis and transformation	The concept of region tree (RT) is introduced as the central data structure for both workflow analysis and transformation. RT performs the analysis and transformation incrementally.	Region tree, Region-growing rules
Montagut, et al. (2008)	Pervasive workflow	Propose an adaptive transactional protocol for the pervasive workflow model to support executing business processes in the pervasive setting.	Web services, Transactional model, Composition algorithm
Choi, et al. (2008)	Terminality and compensatibility of cycles in business processes	Present an approach to seek the terminability of a cycle introduced by BPTrigger model. Determine whether a cycle is allowable in terms of compensatibility.	BPTrigger, Transactional workflow
Zheng, et al. (2008)	Workflow simulation	Propose a workflow simulation method with the support of interactive events mechanism. An event sub-model is introduced in the workflow meta-model, then the simulation engine performs the event-based interaction at the run time	SOA, Workflow meta model, Workflow sub-model
Lee & Suh (2008)	Estimating the duration of stochastic workflow for product development process	Propose a method to estimate the completion time for a stochastic workflow in the product development process with both random activity durations and predefined resource constraint using Markovain model	Markovian model

where the internal systems meet the external web systems and other ways of using value-oriented patterns to improve performances.

REFERENCES

Adams, M., Edmond, D., & Hofstede, A. H. M. (2003). *The Application of Activity Theory to Dynamic Workflow Adaptation Issues.* Paper presented at the 7th Pacific Asia Conference on Information Systems, Adelaide, South Australia.

Adner, R., & Helfat, C. (2003). Corporate Effects and Dynamic Managerial Capabilities . *Strategic Management Journal*, *24*, 1011–1025. doi:10.1002/smj.331

Agostini, A., & de Michelis, G. (2000). A Light Workflow Management System Using Simple Process Models . *Computer Supported Cooperative Work: The Journal of Collaborative Computing*, *9*(3-4), 335–363. doi:10.1023/A:1008703327801

Aissi, S., Malu, P. & Srinivasan. (2002). E-business Process Modeling: The Next Big Step, *IEEE Computer, 35*(5), 55-62.

Alavi, M., & Leidner, D. E. (2001). Review: Knowledge Management and Knowledge Management Systems: Conceptual Foundations and Research Issues . *MIS Quarterly*, *25*(1), 107–136. doi:10.2307/3250961

Alavi, M., & Weiss, I. R. (1986). Managing the risks associated with end-user computing . *Journal of Management Information Systems, 2*(3), 5–20.

Andreescu, F. (2006). Organisational Challenges of Implementing E-Business in the Public Services: The Case of Britain's National Mapping Agency. *International Journal of E-Business Research, 2*(4), 39–60.

Atluri, V., & Huang, W. K. (1996). An authorization model for workflows. In H. K. E. Bertino, G. Martella & E. Montolivo (Ed.), *Computer Security - ESORICS* (pp. 44-64). Springer.

Atluri, V., & Huang, W. K. (1997). An extended petri net model for supporting workflow in a multilevel secure environment. In P. S. a. R. S. Sandhu (Ed.), *Database Security X: Status and Prospects*. Chapman and Hall.

Basu, A., & Blanning, R. B. (2000). A formal Approach to Workflow Analysis . *Information Systems Research, 11*(1), 17–36. doi:10.1287/isre.11.1.17.11787

Basu, A., & Kumar, A. (2002). Research Commentary: Workflow Management Issues in e-Business . *Information Systems Research, 13*(1), 1–14. doi:10.1287/isre.13.1.1.94

Bergeron, F. & Berube, C. (1990). End-users Talk Computer Policy. *Journal of Systems Management*, December, 14-16.

Biazzo, S. (1998). A Critical Examination of the Business Process Re-engineering Phenomenon . *International Journal of Operations & Production Management, 18*(9/10), 1000–1016. doi:10.1108/01443579810225586

Bichler, P., Preuner, G., & Schrefl, M. (1997). *Workflow Transparency.* Paper presented at the 9th International Conference for Advanced Information Systems Engineering (CAiSE'97), Barcelona, Spain.

Bolton, C., & Davies, J. (2000). Activity graphs and Processes. In W. Grieskamp, T. Santen, and B. Stoddart, (eds.), *Proceedings Integrated Formal Methods 2000* (LNCS 1945).

Borgida, A., & Murata, T. (1999). Tolerating Exceptions in Workflows: A Unified Framework for Data and Processes. In D. Georgakopoulos, W. Prinz and A.L. Wolf (Eds.), *Proceedings of the International Joint Conference on Work Activities Coordination and Collaboration, WACC 99* (pp. 59-68). ACM Press, February

Boynton, A. C., Zmud, R. W., & Jacobs, G. C. (1994). The Influence of IT Management Practice on IT Use in Large Organizations . *MIS Quarterly*, (September): 299–318. doi:10.2307/249620

Brancheau, J. C., & Brown, C. V. (1993). The management of end-user computing: Status and directions . *ACM Computing Surveys, 25*(4), 450–482. doi:10.1145/162124.162138

Cao, L., Li, M. L., Cao, J., & Huang, J. S. (2008). CNP-based implementation of service-oriented workflow mapping in SHGWMS. *World Wide Web-Internet and Web Information Systems, 11*(2), 199–226.

Cardoso, J., & Sheth, A. (2003). Semantic E-Workflow Composition. *Journal of Integrated Design and Process Science, 21*(3), 191–225.

Casati, F., Ceri, B., Pernici, B., & Pozz, G. (1995). *Conceptual Modelling of Workflows.* Paper presented at the Object-oriented and Entity-Relationship Conf., Australia.

Casati, F., & Pozzi, G. (1999). *Modeling Exceptional Behaviours in Commercial Workflow Management Systems.* Paper presented at the 1999 IFCIS International Conference on Cooperative Information Systems, Edinburgh, Scotland.

Chen, M., Zhang, D., & Zhou, L. (2007). Empowering Collaborative Commerce with Web Services enabled Business Process Management Systems. *Decision Support Systems*, *43*, 530–546. doi:10.1016/j.dss.2005.05.014

Childe, S. J., Maull, R. S., & Bennett, J. (1994). Frameworks for Understanding Business Process Re-engineering . *International Journal of Operations & Production Management*, *14*(2), 22–34. doi:10.1108/01443579410072346

Chiu, D. K. W., Cheung, S. C., Till, S., Karlapalem, K., Li, Q., & Kafeza, E. (2004). Workflow View Driven Cross-Organizational Interoperability in a Web Service Environment . *Information Technology and Management*, *5*, 221–250. doi:10.1023/B:ITEM.0000031580.57966.d4

Choi, I., Jung, J., Mannino, M., & Park, C. (2008). Terminability and compensatibility of cycles in business processes with a process-oriented trigger. *Data & Knowledge Engineering*, *66*(2), 243–263. doi:10.1016/j.datak.2008.03.002

CIM-OSA. (1989). *CIM-OSA Reference Architecture*. CIM-OSA Standards Committee, AMICE ESPRIT.

Cimatti, A., Clarke, E., Giunchiglia, F., & Roveri, M. (2000). NUSMV: A New Symbolic Model Checker. *International Journal on Software Tools for Technology Transfer*, *2*(4), 410–425. doi:10.1007/s100090050046

Compeau, D., Higgins, C. A., & Huff, S. (1999). Social Cognitive Theory and Individual Reactions to Computing Technology: A Longitudinal Study . *MIS Quarterly*, (June): 145–158. doi:10.2307/249749

Cook, J. E., & Wolf, A. L. (1999). Software Process Validation: Quantitatively Measuring the Correspondence of a Process to a Model . *ACM Transactions on Software Engineering and Methodology*, *8*(2), 147–176. doi:10.1145/304399.304401

Cox, D., & Kreger, H. (2005). Management of the Service-Oriented-Architecture Life Cycle . *IBM Systems Journal*, *44*, 709–726.

Dang, J., Hedayati, A., Hampel, K., & Toklu, C. (2008). An ontological knowledge framework for adaptive medical workflow . *Journal of Biomedical Informatics*, *41*(5), 829–836. doi:10.1016/j.jbi.2008.05.012

Davenport, T. (1993). *Process Innovation – Reengineering Work Through Information Technology*. Boston, MA: Harvard Business School Press.

Davis, G. B. (1988). The Hidden Costs of End-user Computing. *Accounting Horizons*, *2*(4), 103–106.

De Roure, D., Goble, C., & Stevens, R. (2008). (in press). The design and realisation of the Experiment Virtual Research Environment for social sharing of workflows . *Future Generation Computer Systems*. doi:.doi:10.1016/j.future.2008.06.010

Deelman, E., Gannon, D., Shields, M., & Taylor, I. (2008). (in press). Workflow and e-Science: An overview of workflow system features and capabilities. *Future Generation Computer Systems*. doi:.doi:10.1016/j.future.2008.06.012

DeLone, W. H., & McLean, E. R. (1992). Information Systems Success: The Quest for the Dependent Variables . *Information Systems Research*, *3*(1), 60–95. doi:10.1287/isre.3.1.60

Divitini, M., & Simone, C. (2000). Supporting different dimensions of adaptability in workflow modeling . *Computer Supported Cooperative Work: The Journal of Collaborative Computing*, *9*(3-4), 365–397. doi:10.1023/A:1008751210054

Eshuis, R., & Grefen, P. (2008). Constructing customized process views. *Data & Knowledge Engineering*, *64*(2), 419–438. doi:10.1016/j.datak.2007.07.003

Faustmann, G. (2000). Configuration for Adaptation - A Human-centered Approach to Flexible Workflow Enactment. *Computer Supported Cooperative Work: The Journal of Collaborative Computing, 9*(3-4), 413–434. doi:10.1023/A:1008719831436

Fischer, G., Giaccardi, E., Ye, Y., Sutcliffe, A. G., & Mehandjiev, N. (2004). Meta-Design: A Manifesto for End-User Development . *Communications of the ACM, 47*(9), 33–37. doi:10.1145/1015864.1015884

Fischer, L. (Ed.). (2001). *Workflow Handbook 2001, Workflow Management Coalition, Future Strategies*. Lighthouse Point, Florida.

Fowler, M., & Scott, K. (1997). *UML Distilled.* New York: Addison-Wesley.

Galletta, D. F., & Hufnagel, E. M. (1992). A Model of End-User Computing Policy: Context, Process, Content and Compliance. *Information & Management, 22*, 1–18. doi:10.1016/0378-7206(92)90002-W

Glatard, T., Montagnat, J., Emsellem, D., & Lingrand, D. (2008). A Service-Oriented Architecture enabling dynamic service grouping for optimizing distributed workflow execution . *Future Generation Computer Systems, 24*, 720–730. doi:10.1016/j.future.2008.02.011

Grigori, D., Casati, F., Dayal, U., & Shan, M. C. (2001). Improving Business Process Quality through Exception Understanding, Prediction, and Prevention. In P. Apers, P. Atzeni, S. Ceri, S. Paraboschi, K. Ramamohanarao, & R. Snodgrass (eds.), *Proceedings of 27th International Conference on Very Large Data Bases, VLDB01* (pp. 159-168). Morgan Kaufmann.

Gudes, E., Olivier, M. S., & van der Riet, R. P. (1999). Modeling, Specifying and Implementing Workflow Security in Cyberspace . *Journal of Computer Security, 7*(4), 287–315.

Gudes, E., & Tubman, A. (2002). AutoWF: A Secure Web Workflow System Using Autonomous Objects . *Data & Knowledge Engineering, 43*(1), 1–27. doi:10.1016/S0169-023X(02)00058-7

Guimaraes, T., Gupta, Y. P., & Rainer, R. K. (1999). Empirically Testing the Relationship Between End-User Computing Problems and Information Center Success Factors . *Decision Sciences*, (Spring):393–413. doi:10.1111/j.1540-5915.1999.tb01615.x

Hacker, M. E. & Brotherton, P. A. (1998). Designing and installing effective performance measurement systems. *Industrial Engineering Solutions*, 18-23.

Hagen, C., & Alonso, G. (2000). Exception Handling in Workflow Management Systems. *IEEE Transactions on Software Engineering, 26*(10), 943–958. doi:10.1109/32.879818

Hammer, M. (1996). *Beyond Reengineering.* New York: Harper Business.

Harrington, J., Esseling, E., & Van Nimwegen, H. (1997). *Business Process Improvement Workbook.* New York: McGraw-Hill.

Harris, R. W. (2000). Schools of thought in research into end-user computing success. *Journal of End User Computing, 12*(1), 24–34.

Hauser, R. F., Friess, M., Kuster, J. M., & Vanhatalo, J. (2008). An incremental approach to the analysis and transformation of workflows using region trees. *IEEE Transactions on Systems, Man and Cybernetics. Part C, Applications and Reviews, 38*(3), 347–359. doi:10.1109/TSMCC.2008.919178

Heinl, P., Horn, S., Jablonski, S., Need, J., Stein, K., & Teschke, M. (1999). *A Comprehensive Approach to Flexibility in Workflow Management Systems.* Paper presented at the WACC'99, San Francisco, CA.

Helfat, C., & Peteraf, M. A. (2003). The Dynamic Resource-Based View: Capability Lifecycles. *Strategic Management Journal*, *24*, 997–1010. doi:10.1002/smj.332

Herbst, J., & Karagiannis, D. (2000). Integrating Machine Learning and Workflow Management to support Acquisition and Adaptation of Workflow Models. *International Journal of Intelligent Systems in Accounting Finance & Management*, *9*(1), 67–92. doi:10.1002/1099-1174(200006)9:2<67::AID-ISAF186>3.0.CO;2-7

Hevner, A., March, S. T., Park, J., & Ram, S. (2004). Design Science in Information Systems Research. *MIS Quarterly*, *28*(1), 75–105.

Hitt, L., & Brynjolfsson, E. (1997). Information Technology and Internal Firm Organization: An Exploratory Analysis. *Journal of Management Information Systems*, *14*(2), 81–101.

Hwang, S., Wang, H., Tang, J., & Srivastava, J. (2007). A probabilistic approach to modeling and estimating the QoS of web-services-based workflows. *Information Sciences*, *177*, 5484–5503. doi:10.1016/j.ins.2007.07.011

Julia, S., de Oliveira, F. F., & Valette, R. (2008). Real time scheduling of Workflow Management Systems based on a p-time Petri net model with hybrid resources . *Simulation Modelling Practice and Theory*, *16*(4), 462–482. doi:10.1016/j.simpat.2008.01.006

Keen, G., & Knapp, M. (1996). *Every Manager's Guide to Business Processes: A Glossary of Key Terms and Concepts for Today's Business Leader*. Boston, MA: Harvard Business School Press.

Klarmann, J. (2001). *A Comprehensive Support for Changes in Organizational Models of Workflow Management Systems*. Paper presented at the 4th International Conference on Information Systems Modeling ({ISM} '01).

Klein, M., & Dellarocas, C. (2000). A Knowledge-based Approach to Handling Exceptions in Workflow Systems. *Computer Supported Cooperative Work: The Journal of Collaborative Computing*, *9*(3-4), 399–412. doi:10.1023/A:1008759413689

Kon, F., Costa, F., Blair, G., & Campbell, R. H. (2002). The Case for Relective Middleware. *Communications of the ACM*, *45*(6), 33–38. doi:10.1145/508448.508470

Kuechler, W. L., & Vaishnavi, V. K. (2008). An expert system for dynamic re-coordination of distributed workflows . *Expert Systems with Applications*, *34*(1), 551–563. doi:10.1016/j.eswa.2006.09.014

Kumar, A., van der Aalst, W. M. P., & Verbeek, H. M. W. (2002). Dynamic Work Distribution in Workflow Management Systems: How to Balance Quality and Performance? *Journal of Management Information Systems*, *18*(3), 157–193.

Kumar, A., & Wainer, J. (2005). Meta Workflows as a control and coordination mechanism for exception handling in workflow systems. *Decision Support Systems*, *40*(1), 89–105. doi:10.1016/j.dss.2004.04.006

Kumar, A., & Zhao, J. L. (1999). Dynamic routing and operational controls in workflow management systems. *Management Science*, *45*(2), 253–272. doi:10.1287/mnsc.45.2.253

Kung, M. T., & Zhang, J. Y. (2008). Analysis of Business Process Models in Enterprise Web Services. *International Journal of E-Business Research*, *4*(2), 69–87.

Lee, H., & Suh, H. W. (2008). Estimating the duration of stochastic workflow for product development process. *International Journal of Production Economics*, *111*(1), 105–117. doi:10.1016/j.ijpe.2007.01.003

Leymann, H., & Roller, D. (2000). *Production Workflow: Concepts and Techniques*: Prentice Hall.

Livari, J., & Livari, N. (2006). *Varieties of User-Centeredness.* Paper presented at the 39th Annual Hawaii International Conference on System Sciences, Hawaii.

Ly, L. T., Rinderle, S., & Dadam, P. (2008). Integration and verification of semantic constraints in adaptive process management systems. *Data & Knowledge Engineering, 64*(1), 3–23. doi:10.1016/j.datak.2007.06.007

McKendrick, J. (2006, March 2). Will there be a 'Dell' of software? *ZDNet*, March 2, 559.

McPhillips, T., Bowers, S., Zinn, D., & Ludascher, B. (2008). (in press). Scientific workflow design for mere mortals . *Future Generation Computer Systems*. doi:.doi:10.1016/j.future.2008.06.013

Melville, M., Kraemer, K., & Gurbaxani, V. (2004). Information Technology and Organizational Performance: An Integrative Model of IT Business Value . *MIS Quarterly, 28*(2), 283–322.

Mok, W. Y., Palvia, P., & Paper, D. (2006). On the computability of agent-based workflows . *Decision Support Systems, 42*(3), 1239–1253. doi:10.1016/j.dss.2005.10.010

Montagut, F., Molva, R., & Golega, S. T. (2008). The pervasive workflow: A decentralized workflow system supporting long-running transactions . *IEEE Transactions on Systems, Man and Cybernetics. Part C, Applications and Reviews, 38*(3), 319–333. doi:10.1109/TSMCC.2008.919184

Muehlen, M. Z. (2004). Organizational Management in Workflow Applications - Issues and Perspectives . *Information Technology and Management, 5*, 271–291. doi:10.1023/B:ITEM.0000031582.55219.2b

O'Donnell, D., & March, S. (1987). End User Computing Environments - Finding a Balance Between Productivity and Control . *Information & Management, 13*(1), 77–84. doi:10.1016/0378-7206(87)90012-7

Powell, A., & Moore, J. E. (2002). The Focus of Research in End User Computing: Where Have We Come Since the 1980s? *Journal of End User Computing, 14*(1), 3–22.

Raghu, T. S., Jayaraman, B., & Rao, H. R. (2004). Towards an integration of agent and activity-centric approaches in organization process modeling: Incorporating incentive mechanisms . *Information Systems Research, 15*(4), 316–335. doi:10.1287/isre.1040.0031

Robinson, R. (2004). Understand Enterprise Service Bus Scenarios and Solutions in Servie-Oriented Architecture, IBM developerWorks, http://www-106.ibm.com/developerworks/webservices/library/ws-esbscen/.

Rummler, G., & Brache, A. (1995). Managing the White Space on the Organizational Chart . *Super Vision, 52*(5), 6–12.

Sadtler, C., Cotignola, D., Crabtree, B., & Michel, P. (2004). Patterns: Broker Interactions for Intra- and Inter-Enterprise. *IBM Redbooks*, SG24-6075, http://www.redbooks.ibm.com/redbooks.nsf/0/532fca172da15c6c85256d6d6d0046192e?open Document.

Schmidt, M. T., Hutchison, B., Lambros, P., & Phippen, R. (2005). The Enterprise Service Bus: Making Service-Oriented Architecture Real. *IBM Systems Journal, 44*, 727–744.

Schmidt, M. T., & Kalyana, S. (2004). The On Demand Operating Environment – Architectural Overview. *IBM developerWorks*, http://www-106.ibm.com/developerworks/ibm/library/i-odoe1/.

Seddon, P. B. (1997). A Respecification and Extension of the DeLone and McLean Model of IS Success . *Information Systems Research, 8*(3), 240–253. doi:10.1287/isre.8.3.240

Sewing, J., Rosemann, M., & Dumas, M. (2006). Process-Oriented Assessment of Web Services . *International Journal of E-Business Research, 2*(1), 19–44.

Shah, S.K. (2001). Improving Information System Performance through Client Value Assessment: A Case Study, *Review of Business*, Spring, 37-42.

Siller, H. R., Estruch, A., Vila, C., Abellan, J. V., & Romero, F. (2008). Modeling workflow activities for collaborative process planning with product lifecycle management tools . *Journal of Intelligent Manufacturing, 19*(6), 689–700. doi:10.1007/s10845-008-0120-6

Singh, R., & Salam, A. F. (2006). Semantic information assurance for secure distributed knowledge management: a business process perspective . *IEEE Transactions on Systems, Man, and Cybernetics, 36*(3), 472–486. doi:10.1109/TSMCA.2006.871792

Stohr, E. A., & Zhao, J. L. (2001). Workflow Automation: Overview and Research Issues . *Information Systems Frontiers, 3*(3), 281–296. doi:10.1023/A:1011457324641

Strassmann, P. A. (1997). *The Squandered Computer: Evaluating the Business Alignment of Information Technologies*, Information Press, New York.

Sun, S., Kumar, A., & Yen, J. (2008). Merging workflows: A new perspective on connecting business processes . *Decision Support Systems, 42*(2), 844–858. doi:10.1016/j.dss.2005.07.001

Sutcliffe, A., & Mehandjiev, N. (2004). End-user development: Introduction . *Communications of the ACM, 47*(9), 31–32. doi:10.1145/1015864.1015883

Tagg, R., (2001). Workflow in Different Styles of Virtual Enterprise, *Proceedings, Workshop in Information Technology for Virtual Enterprises*, 21-28.

Tallon, P. P., Kraemer, K., & Gurbaxani, V. (2000). Executives' perception of the business value of informaiton technology: A process-oriented approach . *Journal of Management Information Systems, 16*(4), 145–173.

Talon, P. P., Kraemer, K. L., & Gurbaxani, V. (2000). Executives' Perceptions of the Business Value of Information Technology: A Process-Oriented Approach . *Journal of Management Information Systems*, (Spring): 145–173.

Tan, J. C., & Harker, P. T. (1999). Designing Workflow Coordination: Centralized Versus Market-Based Mechanisms . *Information Systems Research, 10*(4), 328–342. doi:10.1287/isre.10.4.328

Thomas, M., Redmond, R., Yoon, V., & Singh, R. (2005). A Semantic Approach to Monitor Business Process Performance . *Communications of the ACM, 48*(12), 55–59. doi:10.1145/1101779.1101809

Tretola, G., & Zimeo, E. (2008). Activity prescheduling for run-time optimizationof grid workflows . *Journal of Systems Architecture, 54*, 883–892. doi:10.1016/j.sysarc.2008.01.009

Truong, H., Dustdar, S., & Fahringer, T. (2008). Performance metrics and ontologies for Grid workflows . *Future Generation Computer Systems, 23*, 760–772. doi:10.1016/j.future.2007.01.003

van der Aalst, W. M. P. (1999). *Generic Workflow Models: How to Handle Dynamic Change and Capture Management Information.* Paper presented at the the Fourth IFCIS International Conference on Cooperative Information Systems (CoopIS'99), Edinburgh, Scotland.

van der Aalst, W. M. P. (1999). Process-oriented Architectures for Electronic Commerce and Interorganizational Workflow . *Information Systems, 24*(8), 639–671. doi:10.1016/S0306-4379(00)00003-X

van der Aalst, W. M. P., Desel, J., & Oberweis, A. (Eds.). (2000). *Business Process Management: Models, Techniques, and Empirical Studies* (LNCS 1806).

van der Aalst, W. M. P., & Jablonski, S. (2000). Dealing with Workflow Change: Identification of Issues and Solutions. *International Journal of Computer Systems, Science, and Engineering, 15*(5), 267–276.

van der Aalst, W. M. P., & Kumar, A. (2003). XML based schema definition for support of inter-organizational workflow . *Information Systems Research, 14*(1), 23–46. doi:10.1287/isre.14.1.23.14768

van der Aalst, W. M. P., ter Hofstede, A. H. M., Kiepuszewski, B., & Barros, A. P. (2003). Workflow Patterns . *Distributed and Parallel Databases, 14*(3), 5–51. doi:10.1023/A:1022883727209

van der Aalst, W. M. P., & van Dongen, B. F. (2002). Discovering Workflow Performance Models from Timed Logs. In Y. Han, S. Tai, & D. Wikarski (Eds.), *International Conference on Engineering and Deployment of Cooperative Information Systems, EDCIS2002* (LNCS 2480).

van der Aalst, W. M. P., & van Hee, K. M. (2002). *Workflow Management: Models, Methods, and Systems*. MIT Press: Cambridge, MA.

van der Aalst, W. M. P., Weske, M., & Grunbauer, D. (2005). Case Handling: A New Paradigm for Business Process Support. *Data & Knowledge Engineering, 53*, 129–162. doi:10.1016/j.datak.2004.07.003

van der Aalst, W. M. P., Weske, M., & Wirtz, G. (2003). Advanced Topics in Workflow Management: Issues, Requirements, and Solutions. *Journal of Integrated Design and Process Science, 7*(3), 49–77.

Verbeek, H. M. W., Basten, T., & van der Aalst, W. M. P. (2001). Diagnosing Workflow Processes using Woflan . *The Computer Journal, 44*(4), 246–279. doi:10.1093/comjnl/44.4.246

Wainer, J., Barthelmess, P., & Kumar, A. (2003). W-RBC - A Workflow Security Model Incorporating Controlled Overriding of Constraints . *International Journal of Cooperative Information Systems, 12*(4), 455–485. doi:10.1142/S0218843003000814

Wang, M., & Wang, H. (2006). From Process logic to business logic - A cognitive approach to business process management . *Information & Management, 43*, 179–193. doi:10.1016/j.im.2005.06.001

Weiss, M., & Amyot, D. (2005). Business Process Modeling with URN. *International Journal of E-Business Research, 1*(3), 63–90.

Workflow Management Coalition. (2006). http://www.e-workflow.org/standards/index.htm

Zeng, Q., Wang, H., Xu, D., Duan, H., & Han, Y. (2008). Conflict detection and resolution for workflows constrained by resources and non-determined durations . *Journal of Systems and Software, 81*, 1491–1504. doi:10.1016/j.jss.2007.09.004

Zhang, D. (2005). Web Services Composition for Process Management in E-Business . *Journal of Computer Information Systems*, *XLV*(2), 83–91.

Zhao, J. L. (2002). Interorganizational Workflow and E-Commerce Applications. Presentation at HICSS-35, http://attila.stevens-tech.edu/sigpam/publications/tutorial/HICSS-35/Interorganizational_Workflow.pdf.

Zheng, Y. C., Fan, Y. S., & Tan, W. (2008). Towards workflow simulation in service-oriented architecture: an event-based approach . *Concurrency and Computation*, *20*, 315–330. doi:10.1002/cpe.1221

Chapter 23
Brokering Web Services via a Hybrid Ontology Mediation Approach Using Multi Agent Systems (MAS)

Saravanan Muthaiyah
Multimedia University, Malaysia

Larry Kerschberg
George Mason University, USA

ABSTRACT

This chapter introduces a hybrid ontology mediation approach for deploying Semantic Web Services (SWS) using Multi-agent systems (MAS). The methodology that the authors have applied combines both syntactic and semantic matching techniques for mapping ontological schemas so as to 1) eliminate heterogeneity; 2) provide higher precision and relevance in matched results; 3) produce better reliability and 4) achieve schema homogeneity. The authors introduce a hybrid matching algorithm i.e. SRS (Semantic Relatedness Score) which is a composite matcher that comprises thirteen well established semantic and syntactic algorithms which have been widely used in linguistic analysis. This chapter provides empirical evidence via several hypothesis tests for validating our approach. A detailed mapping algorithm as well as a Multi-agent based system (MAS) prototype has been developed for brokering Web services as proof-of-concept and to further validate the presented approach. Agent systems today provide brokering services that heavily rely on matching algorithms that at present focus mainly only on syntactic matching techniques. The authors provide empirical evidence that their hybrid approach is a better solution to this problem.

INTRODUCTION

The concept of Semantic Web was introduced by Tim Berners-Lee who is also the founder of the World Wide Web (WWW). He defines the Semantic Web as "a web of data that is directly or indirectly processed by machines" (Gruber, 1993). The **Semantic Web** is envisaged as more powerful web than the current web as it is able to process contextual information. It is a futuristic web platform where agent systems

DOI: 10.4018/978-1-60566-910-6.ch023

would winnow and sift through metadata on the Semantic Web to provide services to users. In other words, just like how human beings play the role of a travel agent to find the best travel package or deals for their clients, similarly **web agents** would be deployed in the Semantic Web as brokers to search, find and secure the best deals for their customer's. Agents would process customer requests by sifting through all types of metadata that involves data on pricing, travel packages, last-minute promotions, etc. In order for this to happen, agent systems must be provided with the ability to process, convert and reason data in useful and meaningful ways. For example if a user had to look for "accommodation" then, the agent must be able to winnow and sift for data that is related to "accommodation". This would include hotel, motel, lodging, etc. Thus the agents must be able to process contextual information as well. Currently, Web agents cannot process contextual data accurately. This is because; they have to acquire, match, reason and interpret contextual data from the Web and semantically as well as syntactically match them for achieving precise results.

Unfortunately, the status of the Semantic Web today is far from what Tim Berners-Lee had envisioned and this is because it hasn't really begun including **semantic** processing within its search methods(Resnik, 1998). In a semantically Web-enabled environment, agent systems would search the Web for "accommodation" where it is defined as "a type of lodging" or "a place to stay". In other words it would understand the contextual meaning of accommodation to locate all the relevant results pertaining to the search of "a place to stay". Agent systems would also accurately present to the user the best deals for those accommodations based on the criteria set by the user as a result of contextual analysis of search over heterogeneous sources. Such a search is also referred to as semantic search. Current search techniques used in information retrieval processes today, are mostly based on data label matching algorithms that use **syntactic** matching schema that does not perform semantic search(Ram, 1999).

Since various semantically associated meanings can be associated with the concept "accomodation", applying purely syntactic matching will only produce unfavorable results. In considering the Semantic Web vision, such a technology would rely on searching over all structured, semi-structured as well as unstructured heterogeneous data sources (Muthaiyah & Kerschberg, 2006). Therefore, **agent systems** or softbots must be enabled to access other agents to process readily available semantic information. In order to achieve these goals, semantics must be included in existing search algorithms(Silva & Rocha, 2003).

THE PROBLEM

This section highlights the interoperability problem. The main reasons for **data heterogeneity** amongst ontologies are: 1) structural heterogeneity (difference taxonomy structures); 2) semantic data heterogeneity (scale and representation conflict); 3) subjective mapping (conflicting data instances) and 4) atomic stored data (conflicting data type value) (Stuckenschmidt, Wache, & Visser). Sources for semantic heterogeneity also include differences in data-definition constructs, differences in object representations, and system-level differences in the way that atomic data (e.g., byte order for multibyte data, such as an integer) is stored in the two systems (Maedche, Motik, & Stojanovic, 2002).

Structural Heterogeneity Problem

Structural heterogeneity is a problem that is caused by different data structures (Stuckenschmidt, Wache, & Viss)(Stuckenschmidt, et al.). The differences in the lattice structures are common problems that exist in most structured data sources (Muthaiyah & Kerschberg). Figure

Figure 1. Structural heterogeneity problems

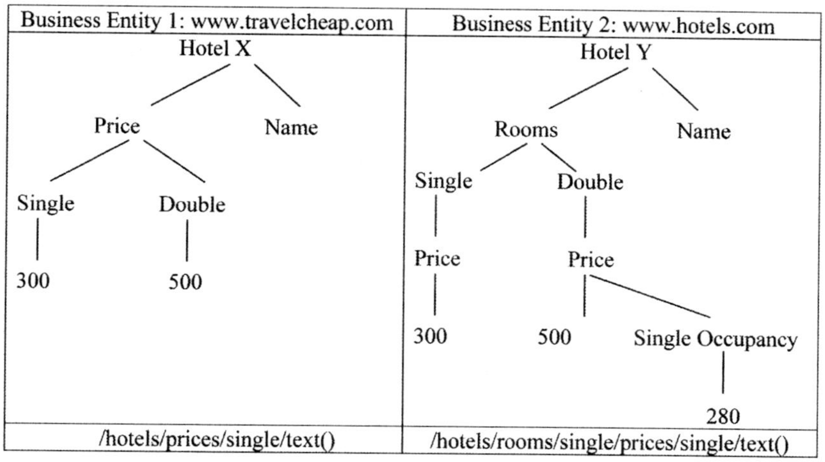

1 illustrates that Hotel Y has single and double room occupancy and the tree structure starts with "Rooms". For the same occupancy category, the tree in Hotel X starts from "Price" instead of "Rooms". Hotel Y has two separate specifications for "Price" in its tree. The difference in the lattice structure is mainly due to how the design of the **ontological structures** was made. This is usually viewed as a common problem among ontologists.

Semantic Data Heterogeneity Problem

Figure 2 depicts the **semantic data heterogeneity** problem. This happens when semantically identical data schemas are represented using different formats and scales. Thus, this type of heterogeneity is further divided into *scale* and *representation* conflict. For example the "Price" attribute is represented in "US Dollar" for Hotel X and in Hotel Y the scale used is "Euro Dollar". This is defined as, scale conflict. Hotel X denotes "Category" and Hotel Y denotes "Class" to represent ratings or rankings of the hotel (i.e. 5 indicates a 5 star hotel). Hotel Y uses a different rating scale (i.e. X which is in their definition is equivalent to 5 star). This is defined as a representation conflict (Stuckenschmidt, et al.). Suppose, Hotel Y uses the label "Quote" instead of "Price" this is again another case of representation conflict.

Subjective Mapping Problem

Figure 3 shows the **subjective mapping** conflict for class and category classification, which are

Figure 2. Semantic data heterogeneity problem

Business Entity 1 : www.travelcheap.com

Hotel X – Hotel entity

Name	Location	Category	Price
Shangri-la	City Center	5	100

Note : Scale & representation conflicts

- 5 star hotel is a category 5 (object representation conflict)
- Price in US dollar (scale conflict)

Business Entity 2 : www.hotels.com

Hotel Y–Accommodation entity

Name	Location	Class	Price
Holiday Inn	City Center	A	100

Note : Scale & representation conflicts

- Class A is a 5 star hotel (object representation conflict)
- Price in Euro dollar (scale conflict)

Figure 3. Subjective mapping problems

Business Entity 1 : www.travelcheap.com

Hotel X –Hotel entity

Name	Location	Class	Price
Shangri-La	Kuala Lumpur	Hotel	600
Marriott	Singapore	Hotel	450
Hilton	Budapest	Hotel	350
Ritz Carlton	Hamburg	Hotel	600
Holiday Inn	London	Hotel	350
Novotel	Copenhagen	Hotel	200
Sheraton	Jakarta	Hotel	450
Crown Princess	Berlin	Hotel	200
Ramada	Venice	Hotel	120

Note: Subjective mapping conflict for category. Hotel X's category (i.e. Class) refers to only hotel.

Business Entity 2 : www.hotels.com

Hotel Y–Accommodation entity

Name	Location	Category	Price
Shangri-la	Kuala Lumpur	Hotel	450
Marriott	Singapore	Hotel	340
Hilton	Budapest	Hotel	230
IBIS	Las Vegas	Apartment	450
Holiday Inn	London	Hotel	230
Radisson	Bangkok	Villa	150
Sheraton	Jakarta	Hotel	230
Schulz	Manila	Bungalow	150
Ramada	Venice	Hotel	92

Note: Subjective mapping conflict for category. Hotel Y's category refers to hotel, apartment, villa and bungalow.

both the third columns of entities Hotel X and Hotel Y. The entity names also differ in this example. Hotel X in this example classifies only hotels in its definition. Hotel Y classifies all kinds of accommodation (i.e., hotel, bungalow, apartment, villa, etc.).

Hotel Y has subjective classifications that include all kinds of accommodation but Hotel X refers to only one type of accommodation (i.e. hotels). Clearly, we can say that the subjective mapping conflict between the two hotels creates potential problems when inferences are made about them.

Atomic, Inconsistency and Redundancy Problem

System-level interoperability of **atomic data storage** is depicted in Figure 4. As mentioned earlier, atomic data storage is related to byte order-multibyte data (Stuckenschmidt, Wache, & Visser, 2003). In this example "Price" is stored on location as an integer data type for Hotel X and as a float data type for Hotel Y. We need syntactic and semantic data integration to achieve interoperability for both entities.

THE SCHEMA MEDIATION APPROACH

There are three main mediation approaches today for mapping the ontological schemas and they can be categorized as 1) domain approach; 2) hybrid approach and 3) multiple ontology approach. The approach of this chapter closely resembles the hybrid approach.

In the hybrid approach syntactic as well as semantic concept matching strategies are used to derive a similarity scores (SRS) which are then populated into a matrix. These scores are then utilized to determine ontology mappings. The motivation for our approach was based on the XMapper system architecture (Figure 5) by Lukasz Kurgan and colleagues (Kurgan, Swiercz, & Cios, 2002). They had deployed a **machine learning** technique that was used to generate the similarity of attribute pairs of XML tags which is

Figure 4. System level interoperability of atomic data storage problem

Business Entity 1 : www.travelcheap.com

Hotel X –Hotel entity

Name	Location	Class	Price
Shangri-La	Kuala Lumpur	Hotel	600

Note: Storage of atomic data is integer for price data in Hotel X.

Business Entity 2 : www.hotels.com

Hotel Y–Accommodation entity

Name	Location	Category	Price
Shangri-La	Kuala Lumpur	Hotel	600

Note: Storage of atomic data is float for price data in Hotel Y.

based on the measure for minimum distance. The values are based on attribute and selected ranking which is then normalized and populated into a distance table. Our approach is quite similar to this but the only difference is that we use "similarity" instead of distance. Our approach is based on ontologies instead of XML data. We provide a method for inter ontology concept mapping based on a normalized similarity score (SRS).

Since similarity (*s*) is easier to measure compared to distance (*d*), we used similarity instead for matching **ontological schemas**. This approach was also useful for saving time and arriving at datasets that are likely to be matched. We used Levenshtein's distance (LD) (a well established measure) string match to measure the similarity of our data labels first, that is, </price> and </quote>. Here, *distance (d)* is denoted as the *inverse* of similarity *(s) and (d=1-s)* and on a scale of 0 to 1, if *(d=0),* then has to be 1 *(s=1).* By applying LD, we first chose *price* as the source string and *quote* as the target string. Then they are syntactically matched on a scale of 1 to 10, LD = 4 denotes (*d*) = 4 and thus similarity (*s*) = 10-*d* = 6. So there is 60% similarity (i.e., 6/10). When *price* and *cost* are matched, similarity is 50% (i.e., 5/10). Only *price* and *price* resulted in 100% similarity (i.e., 10/10). Levenshtein, uses a number of deletions,

Figure 5. XMapper system architecture

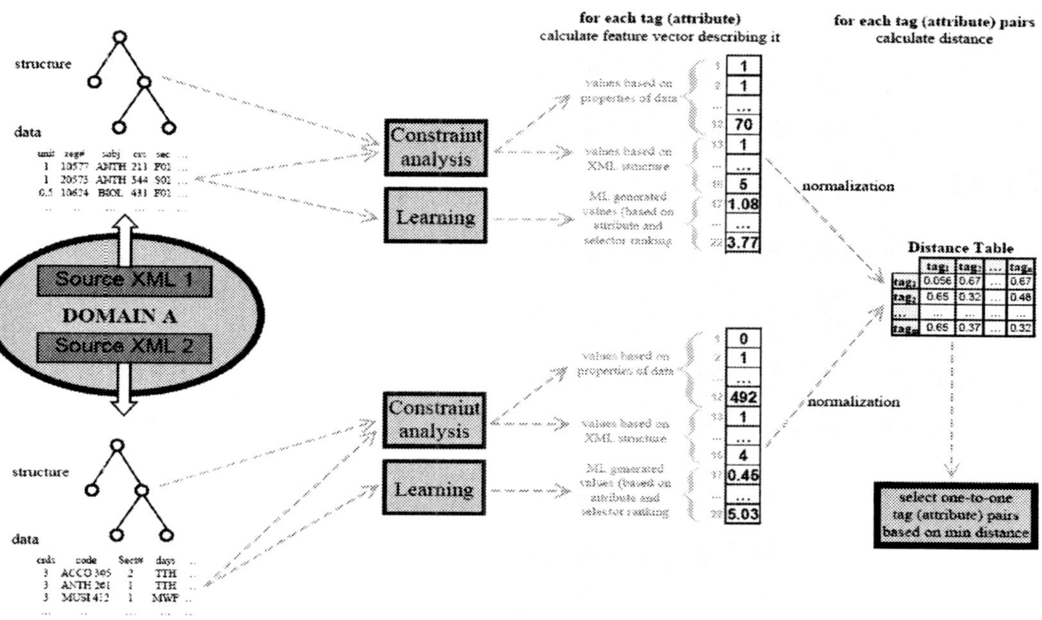

insertions, or substitutions to transform a source string into the target string. This measure has been widely used however it does not apply semantics to match data labels and this is also true for other syntactic matching systems alike. To overcome this problem we introduce SRS, which uses a six-part similarity test to check for **semantic equivalence**. The test involves concept relations and is defined as a function of equivalence (E), inclusiveness (IC), consistency (CN), semantic similarity (SEM) and syntactic similarity (SYN).

Concepts that are disjoint (D) are negated. SEM and SYN scores are aggregated to produce a unique similarity score called Semantic Relatedness Score (SRS). A unique feature of SRS is that it provides greater precision and reliability due to its hybrid nature compared to pure syntactic scores. SRS scores are used to populate and create a similarity matrix. The matrix presents the scores to an ontologist who would use them to match concepts.

The similarity matrix idea was derived from the distance table project discussed in figure 8 previously. However only high similarity scores (SRS) are used instead of minimum (*min*) distance scores. Figure 6, depicts an intuitive approach to match the data labels based on SRS. Here three ontologies i.e. hotel X, Y and Z are depicted highlighting the problem of disparate data labels. Dotted and dashed lines show the semantic correspondence achieved via SRS for mapping ontological schemas of hotels X,Y and Z. If we are matching hotel X and Y, then X can be referred to as the source ontology (SO) and Y as the target ontology (TO). However, the order really doesn't matter.

SIX PART SIMILARITY TEST

SRS matches and presents the results to the domain expert for final consideration. The domain expert's input is only required towards the final stage and this improves efficiency. The idea here is to reduce the workload of domain experts by eliminating **extraneous data** by matching source

Figure 6. Finding semantic correspondence via SRS

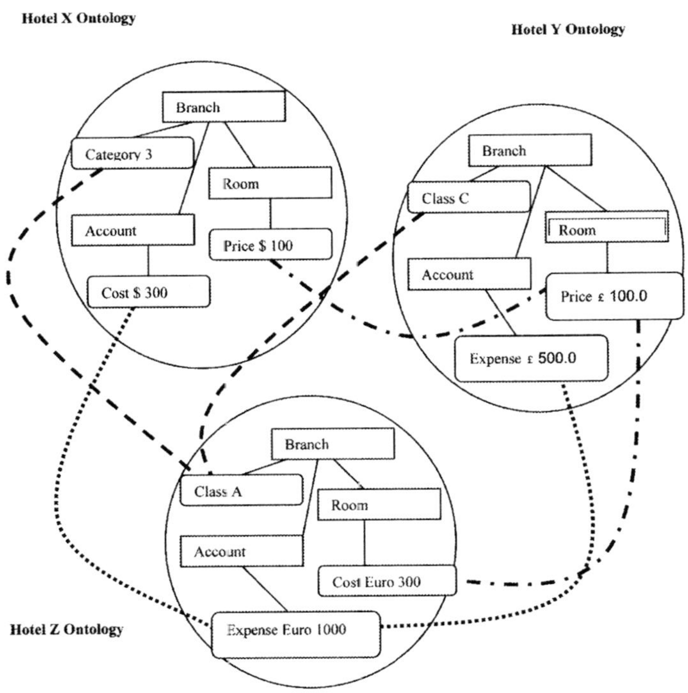

ontology (SO) and target ontology (TO) schemas faster. Parameters are entered in each execution of the algorithm and the acceptance threshold is set for SRS scores. Only classes that have SRS scores higher than the specified threshold are presented to the domain expert for selection. SRS is a highly reliable measure as it has been developed based on well established algorithms such as Lin, Gloss Vector, WordNet Vector (Resnik, 1992) and Latent Semantic Analysis (LSA) (Muthaiyah, 2008). Lin (Lin, 1998) uses the theory of similarity between arbitrary objects. This measure uses the same elements such as dist $_{JC}$ with slight changes. Where c1, c2 are synsets, sim denotes similarity, L denotes Lin, lso denotes lowest super-ordinate, len denotes length of path and p denotes the probability of encountering an instance. The formula is as follows:

$$\mathrm{sim}_{\mathrm{L}}\left(\mathrm{c1},\mathrm{c2}\right)=\frac{2x\log p(lso\ \mathrm{len}\left(\mathrm{c1},\mathrm{c2}\right))}{\log p\left(\mathrm{c1}\right)+\log p\left(\mathrm{c2}\right)}$$

The Gloss Vector (GV) measure forms a second-order co-occurrence vectors from the glosses of concept definitions. It primarily uses WordNet definitions(Resnik, 1992) to measure similarity or relatedness of two or more concepts. GV determines similarity of two concepts by determining the cosine of the angle between their gloss vectors. It augments glosses of concepts with glosses of adjacent concepts as defined by WordNet relations to resolve data sparsity problems due to extremely short glosses.LSA (Latent Semantic Analysis) was introduced by (Landauer & Dumais, 1997). It uses Singular Value Decomposition (SVD) to analyze relationships among concepts in a collection of text. It is a fully automatic computational technique for representing the meaning of text. A passage is viewed as a linear equation and it's meaning is a sum of words i.e., m (passage) = m (word_1)+m(word_2)+m(word_n). Eigenvalue is used for ordering the vector and cosine values are used to represent similarity:

$$\cos\theta_{xy} = x.y\,/|x||y| \qquad (2)$$

LSA provides better results than keyword matching; for example, *doctor-doctor* match gives a 1.0 score for both LSA and keyword match. However, *doctor-physician* results in a 0.8 score for LSA and 0 score for keyword match. This is why LSA is better. GLSA is Generalized LSA computes term vectors for vocabulary V of document collection C using corpus W (Matveeva, Levow, Farahat, & Royer, 2005). Anderson and Pirolli (Anderson & Pirolli, 1984) introduced Spreading Activation (SA), which uses a semantic network to model human memory using a Bayesian analysis. The following is their formula to measure similarity:

$$\mathrm{SA}\left(\mathrm{w1},\mathrm{w2}\right)=\log\frac{\mathrm{P(X{=}1}\,|\,\mathrm{Y{=}1)}}{\mathrm{P(X{=}1}\,|\,\mathrm{Y{=}0)}} \qquad (3)$$

This is discussed further in the fourth section. The similarity function (s) has five components, that is, (E), inclusiveness (IC), consistency (CN), syntactic similarity (SYN), and semantic similarity (SEM). The five elements within the parenthesis are independent variables that determine the dependant variable (s); thus producing the following equation:

(s) fx = { E, IC, CN, SYN,SEM } (4)

Multiple factors are considered for determining similarity including variables that have nothing in common in order to refine our results. The similarity function negates all disjoint (D) attributes between classes and the modified function is as follows:

(s) fx = { E, IC, CN, D, SYN,SEM } (5)

The first three (i.e., E, IC, and CN) tests, are iterative. These tests are based on the definitions provided in the work of (Li, Wu, & Yang, 2005).

We have expanded their definitions to include SYN and SEM which together makes up the SRS. The first three steps, specifically addresses all nodes in a hierarchical ontology structure. Then SYN and SEM tests are applied allowing full semantic matching to be computed among classes and instances (Muthaiyah & Kerschberg, 2008). All the nodes in the graph can be tested for tests in step 1, step 2, and step 3 in our algorithm. The algorithm supports Web ontology language (OWL) and resource description framework (RDF) structures and SO and TO nodes can be matched and a similarity matrix is populated.

MATCHING ALGORITHM

The process begins when two ontologies are first loaded (i.e. O_1 and O_2) and they are identified as source ontology (SO) and target ontology (TO). The taxonomies are read and translated for beginning matching. An equivalent test (E) is carried out for data labels to test their similarity in terms of three parameters, 1) test for semantically equivalent data labels, 2) test for synonyms and 3) test for similar slots or attribute names. *C* is used to refer to classes and *c* refers to attributes or slots. All the nodes in a taxonomy graph can be tested for tests in step 1, step 2 and step 3. The following are the steps involved for the matching algorithm (see figure 7):

- **Step 1—Read loaded SO and TO taxonomies.** Semantic engine reads taxonomies of the SO and TO. Prepare for detailed matching tests of data labels, go to step 2.
- **Step 2—Equivalence test.** Test for the **equivalence** of source and target classes: Test 1) do they have semantically equivalent data labels, Test 2) are they synonyms, or Test 3) do they have the same slots or attribute names. Equivalence also implies

Figure 7. Matching algorithm

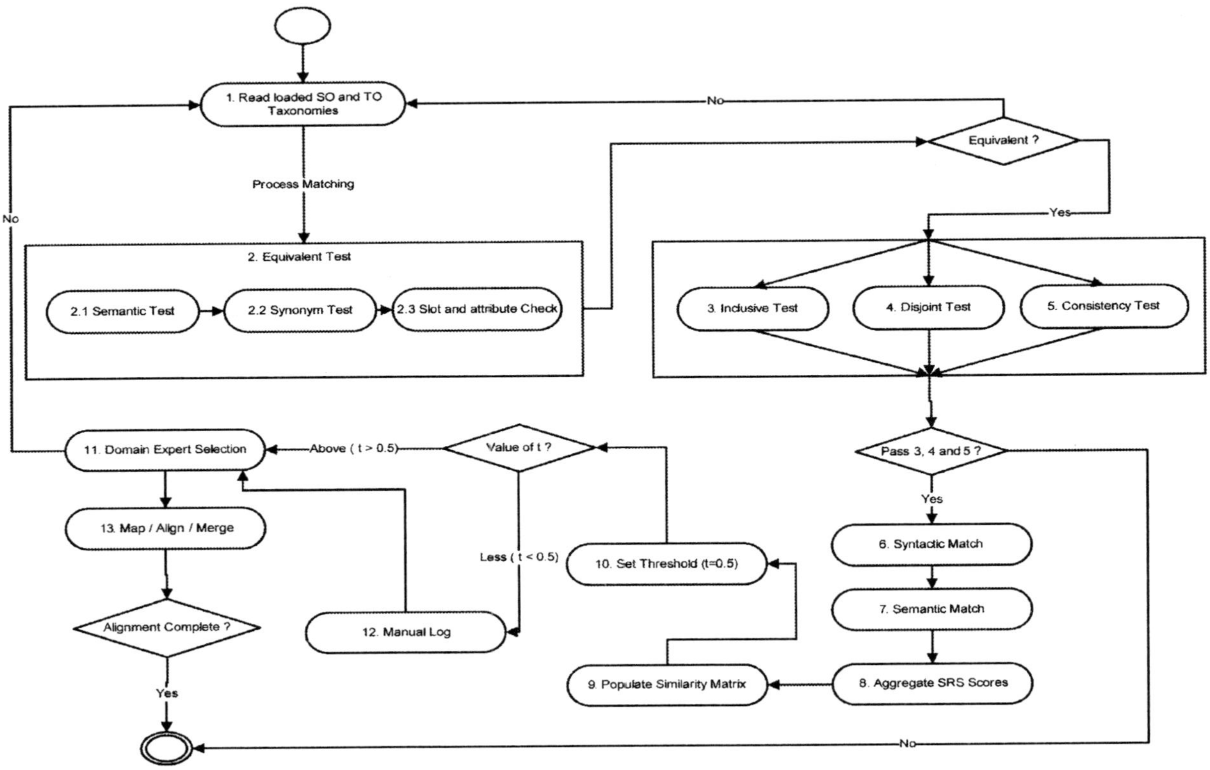

Figure 8. Comparing SRS and SYN to HCR scores

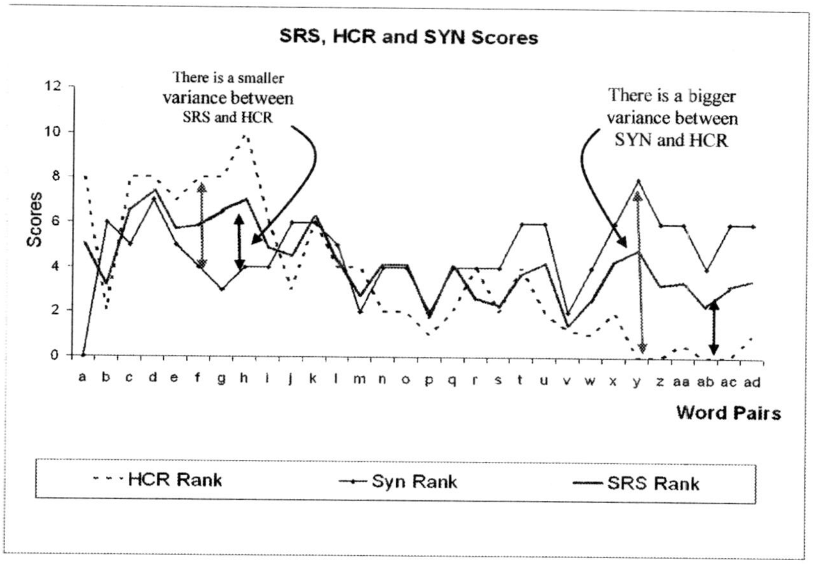

adjacent neighbors are equal. If equivalent, proceed to step 3, 4, and 5. Else go to step 1.

- **Step 3—Inclusive test.** Source and target classes or concepts (C) are **inclusive** if, the attribute (c) of one is inclusive in the other. In other words *selling price* (*ci*) is inclusive in *price* (*cj*), this is applicable to *hyponyms*. If inclusive, proceed to step 6.
- **Step 4—Disjoint test.** Source and target classes or concepts (C) are **disjoint** if, the intersection of their two attribute sets (c), *ci* and *cj* results in an empty set {} or ø. If match test is not disjoint, proceed to step 6.
- **Step 5—Consistency test.** Source and target classes or concepts (C) are **consistent** if, all the attributes or slots (i.e., c*1* and c*2*) in the class, have nothing in common s.t. $c1 \cap c2 = \{\}$. All slots must belong to class that is being tested. This can be configured with RacerPro. If consistent, proceed to step 6.
- **Step 6—Syntactic match.** Syntactic match similarity scores based on class prefix, suffix, substring matches are calculated. This calculation is performed for every class in the source and target ontology. Go to step 7.
- **Step 7—Semantic match:** Semantic match similarity scores based on cognitive measures such as LSA, Lin, Gloss Vector, and WordNet Vector are used. This calculation is done for every class in the source and target ontology. Go to step 8.
- **Step 8—Aggregate both similarity scores.** Similarity inputs from step 6 and 7 are aggregated, to produce SRS. Go to step 9.
- **Step 9—Populate similarity matrix.** The aggregated values (SRS) from step 8 of candidate labels are populated into the similarity matrix. Multiple matches are carried out. Values are to be verified against the threshold. Go to step 10.
- **Step 10—Set threshold.** Threshold value (t) is set based on scale used. For a scale between, 0 and 1 the threshold value is usually 0.5 ($t > 0.5$). Those below threshold are logged in file in step 12. If greater than the threshold value, go to step 11.
- **Step 11—Domain expert selection.** At this stage, candidates from step 10 are

presented to domain expert by the system. Input from step 12 is accepted at the discretion of the domain expert.

- **Step 12—Manual log.** Selection is made manually only for those values below threshold. The domain expert uses his own cognitive judgment. Go to step 13.
- **Step 13—Mapping/alignment/merge**: All the candidates for mapping, alignment or merge (i.e., integration) chosen from step 11 and 12 are processed. End. The matching algorithm (see figure 7) shows detailed steps before the semantic matching engine produces mappings. The process begins when two ontologies are first loaded (i.e., O1 and O2) and they are identified as SO and TO.

We provide a **matching algorithm** based on similarity of classes that uses both syntactic and semantic matching in order to determine more reliable and precise similarity scores. We have also built a prototype agent-based system using Java agent development framework (JADE), which deploys a matching agent (MA) that computes similarity using SRS (Muthaiyah, Barbulescu, & Kerschberg, 2008). Empirical evidence is provided to support this model which will be discussed in the following section.

EMPIRICAL ANALYSIS

An experiment was conducted to compare results produced by **Semantic Relatedness Score** (SRS) to that of domain experts (HCR). Respondents were asked to rank order 30 pairs of data labels for similarity (Muthaiyah, 2008). This was to simulate a real-life situation where domain experts would be called in to match data labels. Our goal here was to provide empirical validation that SRS scores did perform better than purely SYN scores which was the current practice. Fifty survey forms were distributed via e-mail, fax, regular mail and face-to-face interviews. Fifty completed survey responses were received, thus this gives the study a 100% response rate. However, twelve survey responses had to be eliminated from the analysis. Thus, the balance of thirty-eight responses was used for data analysis. The hypothesis test carried out was as follows:

(H_0): SRS Rank scores do not match HCR responses
(H_1): SRS Rank scores matches HCR responses

Table 1 shows the Pearson product-moment correlation coefficient of SRS and HCR scores. A significant positive correlation between the two scores i.e., $r = +0.919$ or (91.9%) was indicated. Significance value (p) for this 2-tailed test is <0.05 thus ($r = 0.919$, $p < 0.05$) supports the alternate hypothesis (H_1) and we therefore reject null

Table 1. Pearson product moment correlation

		SRS Rank	HCR Rank
SRSRank	Pearson Correlation	1	.919(**)
	Sig. (2-tailed)		.000
	N	30	30
HCR Rank	Pearson Correlation	.919(**)	1
	Sig. (2-tailed)	.000	
	N	30	30

hypothesis (H_0). The t-statistic was calculated to validate the hypothesis results. Given r coefficient = +0.919. The t-statistic resulted in 12.33 with the given degree of freedom of (n-2 = 28) and given α = 0.01, the critical value of *t* was 2.76. Since the t-statistic of 12.33>2.76, this clearly rejected the null hypothesis (H_0) and the alternate hypothesis (H_1) was thus accepted (Muthaiyah, 2008).

Using Kendall's tau_b and Spearman's rho, both HCR and SRS scores were tested again for nonparametric correlation. Results show a significant correlation given α = 0.01 for this 2-tailed test. Given r coefficient = +0.703, the t-statistic for Kendall's tau_b resulted in 5.23 with the given degree of freedom of (n-2 = 28) and α = 0.01, the critical value of *t* was 2.76. Since the t-statistic of 5.23>2.76, the null hypothesis (H_0) is rejected and the alternate hypothesis (H_1) is accepted. The t-statistic for Spearman's rho resulted in 8.29 with the given degree of freedom of (n-2 = 28) and α = 0.01, the critical value of *t* was 2.76. Since the t-statistic of 8.29>2.76, the null hypothesis (H_0) is rejected and the alternate hypothesis (H_1) is accepted. Figure 8, further validates that SRS has a higher proximity with HCR scores (due to a smaller variance) compared to SYN scores that has a much bigger variance (Muthaiyah, 2008).

Another hypothesis test was carried out to determine the precision and relevance of SRS. The null and alternate hypothesis is stated as:

(H_0): SRS Rank scores are more reliable than SYN scores

(H_1): SRS Rank scores are less reliable SYN scores

The results showed that SRS had a 40% score for *precision* (P_s =12/30) and 96.67% for *relevance* (R_L =29/30), however, SYN resulted only in 16.67% score for *precision* (P_s =5/30) and 73.33% for *relevance* (R_L =22/30). Therefore it was clear that SRS produced more relevant and precise scores, which is why SRS scores would be more reliable (Muthaiyah, 2008).

MULTI-AGENT SYSTEM (MAS) PROTOTYPE

A prototype using the JADE (Java Development Framework) was designed and built based on the matching algorithm and six-part similarity test for proof-of-concept (see figure 9). It was tested for the upgrade operation of a local ontology to reflect the new classes that have been added based on SRS scores. Multiple agents were deployed to perform an upgrade of the local ontology to work with a new version of a shared ontology (Muthaiyah, et al., 2008). Several agents were introduced in the prototype such as:

- **Ontology Agent (OA):**

Used by other agents to get access to the ontology. It uses OWL ontologies and stores the ontology modifications in an evolution log. It responds to queries like "what is the root class for a "accommodation"?", "what are the subclasses of "hotel"?" and performs ontology modification to create an instance as a child of an existing data label.

- **External Communication Ontology Agent (ECA):**

Acts as a proxy to talk to agents in other platforms and handles inter-platform communication via agent communication language (ACL). The protocols used for communication are Common Object Request Broker Architecture (CORBA) and Remote Method Invocation (RMI). CORBA is used for inter-platform communication for example between platform 1 and 2 (see figure 9). Whereas, RMI is used for intra-platform communication for agents to communication locally in each platform.

- **Match Agent (MA):**

Computes similarity (SRS)between two given concepts or instances. It the core of the semantic

Figure 9. MAS architecture for the upgrade process

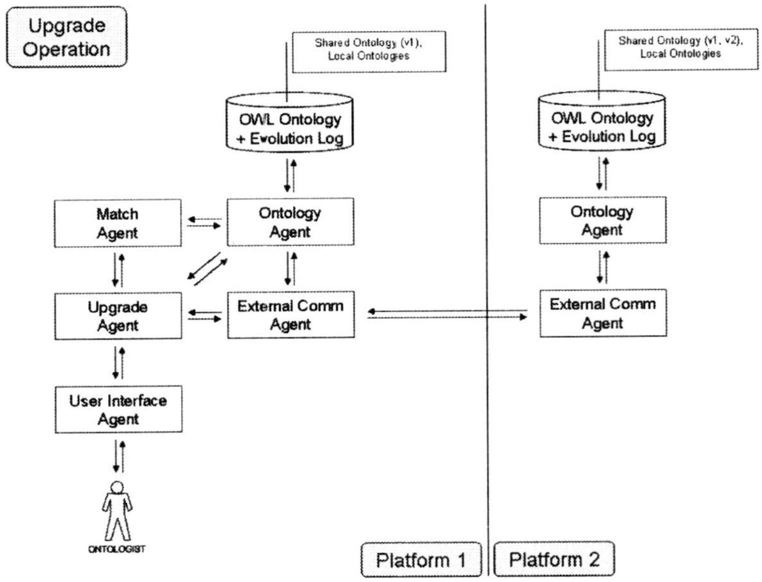

mediation process and it's based on the algorithms described earlier.

- **Upgrade Agent (UA):**

The business logic behind the upgrade process is encapsulated in this agent.

- **User Interface Agent (UIA):**

Provides interface to the ontologist and handles actual reconfiguration of existing local ontologies during the upgrade process. It also services UA.

When all agents have been invoked and are ready to receive input, the upgrade process is ready to be executed. The UA will execute to find the differences in the shared ontology with input from the ontologist. This is done by clicking the differences icon (see figure 10).

UA communicates with OA (OA2) of Platform 2 via ECA1 and ECA2 and receives the differences between the current version of the shared ontology and the latest available version. In this case the newly updated classes are those shown by the agents (Figure 11). UA will show the ontologist the differences through the UIA, that is the new definitions of ontological schemas that have been newly added for example new types of "places to stay" or "accommodation". The ontologist would then proceed with the upgrade process for these new definitions. Changes are saved in an evolution log (Muthaiyah, et al., 2008).

CONCLUSION

In this chapter we have introduced a hybrid ontology mediation approach using a hybrid matching (SRS). We presented the heterogeneity problems inherent within the travel domain to highlight the need for achieving a common data model via ontology mediation. This is in view of Service Oriented Architecture (SOA) and the Semantic Web dream for web agents to seamlessly winnow and sift metadata and to enhance search results for the user who in this case wants to make a travel reservation. Agent systems today provide brokering services that heavily rely on matching algorithms that at present focus mainly only on syntactic (SYN) matching techniques per se. We provide empirical evidence that our hybrid approach is a better solution to this problem. A detailed mapping

Figure 10. Agents loading in JADE

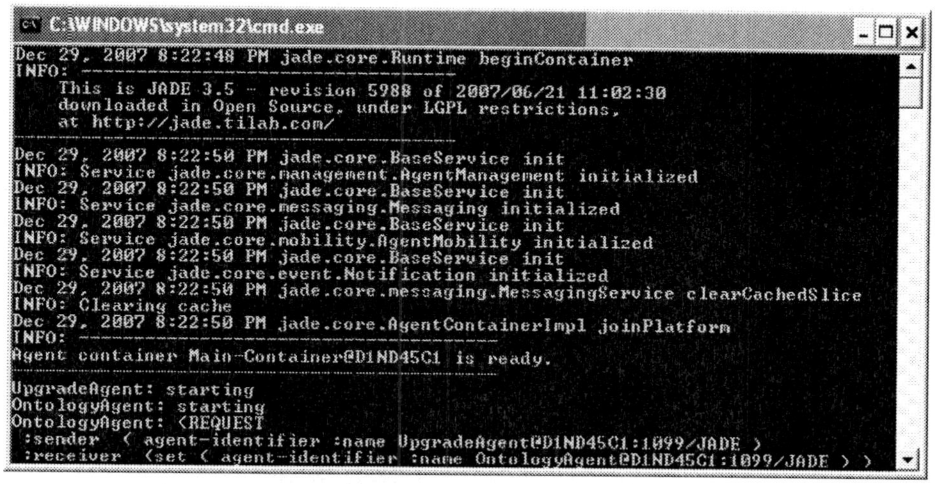

Figure 11. Agents loading in JADE (GUI)

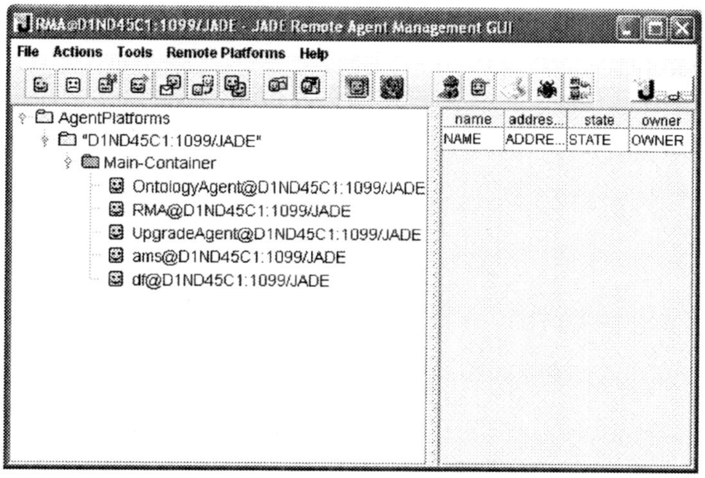

algorithm as well as a **multi-agent system** (MAS) prototype has been developed for proof-of-concept to further validate our approach. We demonstrate this by providing an example of a hotel ontology that has been updated in a shared repository. The match agent deployed in our prototype provides exact matches (via SRS) and presents the data to the ontologist for consideration. The idea here is to automate the matching process on-the-fly so that web agents can sift through the Semantic Web to find and accurately present to the user the best deals for accommodations based on the criteria set by the user as a result of contextual analysis of search over heterogeneous sources. For future work we would like to test our prototype on other domains as well besides the travel domain.

REFERENCES

Anderson, J. R., & Pirolli, P. L. (1984). Spread of Activation. *Journal of Experimental Psychology. Learning, Memory, and Cognition*, *10*(4), 791–798. doi:10.1037/0278-7393.10.4.791

Gruber, T. R. (1993). *Toward principles for the design of ontologies used for knowledge sharing.* Paper presented at the Knowledge Representation '94.

Kurgan, L. A., Swiercz, W., & Cios, K. J. (2002). *Semantic Mapping of XML Tags Using Inductive Machine Learning.* Paper presented at the ICMLA 2002.

Landauer, T. K., & Dumais, S. T. (1997). A solution to Plato's problem: The Latent Semantic Analysis theory of the acquisition, induction, and representation of knowledge. *Psychological Review, 104*, 211–140. doi:10.1037/0033-295X.104.2.211

Li, L., Wu, B., & Yang, Y. (2005). *Agent-Based Ontology Mapping Towards Ontology Interoperability.* Paper presented at the Australian Joint Conference on Artificial Intelligence (AI'05), Sydney, Australia.

Lin, D. (1998). *An information-theoretic definition of similarity.* Paper presented at the 15th International Conference on Machine Learning, Madison, WI, USA.

Maedche, A., Motik, B., & Stojanovic, N. (2002). *MAFRA - A MApping FRAmework for Distributed Ontologies.* Paper presented at the 13th International Conference, Knowledge Engineering and Knowledge Management (EKAW 2002), Siguenza, Spain.

Matveeva, I., Levow, G.-A., Farahat, A., & Royer, C. (2005). *Term Representation with Generalized Latent Semantic Analysis.* Paper presented at the Conference on recent advances in natural language processing (RANLP).

Muthaiyah, S. (2008). *A framework and methodology for ontology mediation through semantic and syntactic mapping.* George Mason University.

Muthaiyah, S., Barbulescu, M., & Kerschberg, L. (2008). Managing ontology change and evolution via a hybrid matching algorithm. *WSEAS Trans. Sys. Ctrl., 3*(9), 737–747.

Muthaiyah, S., & Kerschberg, L. (2006). *Virtual Organization Security Policies: An Ontology-based Mapping and Integration Approach.* Paper presented at the The Second Secure Knowledge Management Workshop (SKM). Brooklyn, New York.

Muthaiyah, S., & Kerschberg, L. (2006). *Dynamic Integration and Semantic Security Policy Ontology Mapping for Semantic Web Services (SWS).* Paper presented at the First IEEE International Conference on Digital Information Management (ICDIM), Bangalore, India.

Muthaiyah, S., & Kerschberg, L. (2008). Achieving interoperability in e-government services with two modes of semantic bridging: SRS and SWRL. *J. Theor. Appl. Electron. Commer. Res., 3*(3), 52–63. doi:10.4067/S0718-18762008000200005

Ram, S., et al. (1999). *Schema Integration: Past, Current and Future.* San Francisco.

Resnik, P. (1992). *WordNet and distributional analysis: A class-based approach to lexical discovery.* Paper presented at the AAAI Workshop.

Resnik, P. (1998). Semantic Similarity in a Taxonomy: An Information based Measure and its Application to Problems of Ambiguity in Natural Language. *Journal of Artificial Intelligence Research, 11*, 95–130.

Silva, N., & Rocha, J. (2003, 13-17 October). *Semantic Web Complex Ontology Mapping.* Paper presented at the IEEE WIC International Conference.

Stuckenschmidt, H., Wache, H., & Visser, U. (2003, October 20). *Information Integration on the World Wide Web* Paper presented at the 2nd International Semantic Web Conference 2003, Sanibel Island, Florida.

Compilation of References

Aarnio, A., Enkenberg, A., Heikkilä, J., & Hirvola, S. (2002). Adoption and Use of Mobile Serivces - Empirical Evidence from a Finnish Survey. In *Proceedings of the 35th Hawaii International Conference on System Sciences 2002* (pp. 1454–1463).

Adams, M., Edmond, D., & Hofstede, A. H. M. (2003). *The Application of Activity Theory to Dynamic Workflow Adaptation Issues.* Paper presented at the 7th Pacific Asia Conference on Information Systems, Adelaide, South Australia.

Adner, R., & Helfat, C. (2003). Corporate Effects and Dynamic Managerial Capabilities . *Strategic Management Journal, 24,* 1011–1025. doi:10.1002/smj.331

Agarwal, R., Arning, A., Bollinger, T., Mehta, M., Shcafer, J., & Srikant, R. (1996). *The quest data mining system.* Paper presented at the 2nd International Conference on Knowledge Discovery in Databases and Data Mining, Portland, OR.

Agarwal, S., & Sprick, B. (2005). Specification of Access Control and Certification Policies for Semantic Web Services. In *Proceedings of the 6th International Conference on Electronic Commerce and Web Technologies (EC-Web'2005), Copenhagen, Denmark.*

Agostini, A., & de Michelis, G. (2000). A Light Workflow Management System Using Simple Process Models . *Computer Supported Cooperative Work: The Journal of Collaborative Computing, 9*(3-4), 335–363. doi:10.1023/A:1008703327801

Ahadi, H. R. (2004). An examination of the role of organizational enablers in business process reengineering and the impact of information technology. *Information Resources Management Journal, 17*(4), 1–19.

Ahmad, R., & Buttle, F. (2002). Customer retention management: A reflection of theory and practice. *Marketing Intelligence and Planning, 20*(3), 149-161.

Aissi, S., Malu, P. & Srinivasan. (2002). E-business Process Modeling: The Next Big Step, *IEEE Computer, 35*(5), 55-62.

Ajzen, I. (1985). From Intentions to Actions: A Theory of Planned Behavior. In J. Kuhl & J. Beckmann (Eds.), *Action Control: From Cognition to Behavior.* New York: Springer.

Ajzen, I. (1991). The theory of planned behavior. *Organizational Behavior and Human Decision Processes, 50,* 179–211. doi:10.1016/0749-5978(91)90020-T

Ajzen, I., & Fishbein, M. (1977). Attitude-behavior relations: A theoretical analysis and review of empirical research. *Psychological Bulletin, 84*(5), 888-918.

Ajzen, I., & Fishbein, M. (1980). *Understanding attitudes and predicting social behavior.* Englewood Cliffs, NJ: Prentice-Hall.

Akyildiz, F. (2002). A Survey on Wireless Sensor Networks. *IEEE Communications, 40*(8), 102–114. doi:10.1109/MCOM.2002.1024422

Al Moajel, N. (2007). *Fostering Trust for Middle-Eastern B2C e-Commerce"An Investigation of the Effects of Voice, Accent and Culture an English e-commerce website on Trust perception among Middle Eastern Consumers".* PhD thesis in Bradford University, Bradford, UK.

Alanen, J., & Autio, E. (2003). Mobile business services: A strategic perspective. In B. E. Mennecke & T. J. Strader

(Eds.), *Mobile Commerce: Technology, Theory and Applications* (pp. 162-184). Hershey, PA: Idea Group Publishing.

Alavi, M., & Leidner, D. E. (2001). Review: Knowledge Management and Knowledge Management Systems: Conceptual Foundations and Research Issues . *MIS Quarterly, 25*(1), 107–136. doi:10.2307/3250961

Alavi, M., & Weiss, I. R. (1986). Managing the risks associated with end-user computing . *Journal of Management Information Systems, 2*(3), 5–20.

Albers-Miller, N. D., & Gelb, B. D. (1996). Business advertising appeals as a mirror of cultural dimensions: a study of eleven countries. *Journal of Advertising, 25*(4), 57–70.

Al-Diri, K., Hobbs, D., & Qahwaji, R. (2006). *Effects of Social Presence and Cultural Representation on Websites for Promoting Trust in B2C E-Commerce - A Saudi Arabian Study.* Paper presented at the Proceedings of the IADIS e-Commerce 2006 International Conference, Barcelona, Spain.

Al-Diri, K., Hobbs, D., & Qahwaji, R. (2007). *E-Commerce: An exploratory study of on-line shopping in Saudi Arabia.* Paper presented at the Proceedings of the Saudi Innovation Conference (SiC) 2007, Newcastle upon Tyne, UK.

Allen, J. P. (2000). Information systems as technological innovation. *Information Technology & People, 13*(3), 210–221. doi:10.1108/09593840010377644

Alonso, G., et al. (2003). *Web Services.* Springer.

Al-Qirim, N. (2003). E-commerce in the aerial mapping industry: a New Zealand case study. *Journal of Systems & Information Technology, 7*(1), 67–92.

Al-Qirim, N. (2005). An empirical investigation of an e-commerce adoption-capability model in small businesses in New Zealand. *Electronic Markets, 15*(4), 418–437. doi:10.1080/10196780500303136

Amit, R., & Zott, C. (2001). Value creation in e-business. *Strategic Management Journal, 22*, 493–520. doi:10.1002/smj.187

Amoako-Gyampah, K., & Salam, A. F. (2004). An extension of the technology acceptance model in an ERP implementation environment. *Information & Management, 41*, 731–745. doi:10.1016/j.im.2003.08.010

Anderson, A. H. (2004). An Introduction to the Web Services Policy Language (SWPL). In *Proceedings of the 5th IEEE International Workshop on Policies for Distributed Systems and Networks (POLICY'2004), New York, USA.*

Anderson, J. C., & Gerbing, D. W. (1982). Some methods for respecifying measurement models to obtain unidimensional construct measurement. *Journal of Marketing Research, 19,* 453-460.

Anderson, J. C., & Gerbing, D. W. (1988). Structural equation modeling in practice: A review and recommended two-step approach. *Psychological Bulletin, 103*, 411–423. doi:10.1037/0033-2909.103.3.411

Anderson, J. C., & Gerbing, D. W. (1988). Structural equation modelling in practice: A review and recommend two-step approach. *Psychological Bulletin, 103*(3), 411–423. doi:10.1037/0033-2909.103.3.411

Anderson, J. R., & Pirolli, P. L. (1984). Spread of Activation. *Journal of Experimental Psychology. Learning, Memory, and Cognition, 10*(4), 791–798. doi:10.1037/0278-7393.10.4.791

Anderson, P., & Anderson, E. (2002). The New E-commerce Intermediaries. *Sloan Management Review, 43*(4), 53–62.

Anderson, S., & Palma, A. (1992). The Logit as a Model of Product Differentiation. *Oxford Economic Papers, 44*, 51–67.

Andreassen, T. W., & Lindestad, B. (1998). The effect of corporate image in the formation of customer loyalty. *Journal of Service Research, 1*(1), 82-92.

Andreescu, F. (2006). Organisational Challenges of Implementing E-Business in the Public Services: The Case of Britain's National Mapping Agency. *International Journal of E-Business Research, 2*(4), 39–60.

Angeles, R., Corritore, C., Choton, S., & Nath, R. (2001). Success factors for domestic and international electronic data interchange (EDI) implementation for US firms. *Information & Management, 21*, 329–347. doi:10.1016/S0268-4012(01)00028-7

Anon [A] (2002). Converting Web users to Web shoppers. *Strategic Direction* 18(4) 10–12.

Anon [B] (2002). Internet users weigh in worldwide: a new report finds that even though more Internet users plan to buy online this year, security is the highest concern for Web shoppers worldwide. *Direct Marketing* 65(6), 28–36.

Anonymous (2004). At the drop of a hammer. *Economist, 371*(8375), 10.

Ansari, A., Essegaier, S., & Kohli, R. (2000). Internet Recommendation Systems. *JMR, Journal of Marketing Research, 37*(3), 363–375. doi:10.1509/jmkr.37.3.363.18779

Ansoff, H. I. (1965). *Corporate Strategy.* New York: McGraw Hill.

Anton Martin, C., Garrido Samaniego, M. J., & Rodriguez Escudero, A. I. (1998). El Grado de Elaboracion del Proceso de Eleccion de Compra como Refuerzo de la Lealtad de Marca. X Encuentro de Profesores Universitarios de Marketing, Santander (pp. 435-448).

Arbuckle, J. L. (1989). AMOS: Analysis of moment structures. *The American Statistician, 43,* 66–67. doi:10.2307/2685178

Ariely, D., & Simonson, I. (2003). Buying, bidding, playing, or competing? Value assessment and decision dynamics in online auctions. *Journal of Consumer Psychology, 13*(1&2), 113-123.

Armstrong, J. S., & Overton, T. (1977). Estimating non-response bias in mail surveys. *JMR, Journal of Marketing Research, 14,* 396–402. doi:10.2307/3150783

Asher, A. (2007). Developing a B2B e-commerce implementation framework: a study of EDI implementation for procurement. *Information Systems Management, 24*(4), 373–390. doi:10.1080/10580530701586144

AT&T WorldNet Ranks Highest in Customer Satisfaction Among Largest National Subscription Internet Service Providers. (2000, September). J.D. Power and Associates Report. Retrieved from http://www.jdpa.com/pdf/2053.pdf.

Atluri, V., & Huang, W. K. (1996). An authorization model for workflows. In H. K. E. Bertino, G. Martella & E. Montolivo (Ed.), *Computer Security - ESORICS* (pp. 44-64). Springer.

Atluri, V., & Huang, W. K. (1997). An extended petri net model for supporting workflow in a multilevel secure environment. In P. S. a. R. S. Sandhu (Ed.), *Database Security X: Status and Prospects.* Chapman and Hall.

Ba, S., & Pavlou, P. A. (2002). Evidence of the effect of trust building technology in electronic markets: Price premiums and buyer behavior. *MIS Quarterly, 26*(3), 243-269.

Ba, S., Stallaert, J., & Whinston, A. B. (2001). Optimal Investment in Knowledge within a Firm using a Market Mechanism . *Management Science, 47*(9), 225–239. doi:10.1287/mnsc.47.9.1203.9781

Ba, S., Whinston, A. B., & Zhang, H. (2003). Building trust in online auction markets through an economic incentive mechanism. *Decision Support Systems, 35*(3), 273-286.

Baatz, E. (1999). Online auctions start to pick up steam. *Purchasing, 127*(6), 46-56.

Badre, A. (2001). The Effects of Cross Cultural Interface Design Orientation on World Wide Web User Performance. *Georgia Institute of Technology.*

Bagozzi, R. (1994). *Structural equation model in marketing research.* Basic Principles, Principles of Marketing Research. Oxford: Blackwell Publishers.

Bagozzi, R. P., Wong, N., Abe, S., & Bergami, M. (2000). Cultural and situational contingencies and the theory of reasoned action: Application to fast food restaurant consumption. *Journal of Consumer Psychology, 9*(2), 97–106.

Bagozzi, R., & Yi, Y. (1988). On the evaluation of structural equation models. *Academy of Marketing Science, 16*(1), 74–94. doi:10.1007/BF02723327

Baier, A. (1986). Trust and Antitrust. *Ethics, 96,* 231–260.

Bajari, P., & Hortacsu, A. (2003). The winner's curse, reserve prices, and endogenous entry: Empirical insights from eBay auctions. *Rand Journal of Economics, 34*(2), 329-355.

Bakalis, S., Abeln, M., & Mante-Meijer, E. (1998) *Adoption and use of mobile telephony in Europe.* Cost 248 Report. Telia, Farsta, Sweden.

Baker, W., Marn, M., & Zawada, C. (2001). Price smarter on the net. *Harvard Business Review, 79*, 122-127.

Bakos, J. Y. (1991). A strategic analysis of electronic marketplaces. *MIS Quarterly, 15*(3), 295–310. doi:10.2307/249641

Bakos, J. Y. (1991). Information Links and Electronic Marketplaces: Implications of Interorganizational Information Systems in Vertical Markets. *Journal of Management Information Systems, 8*(2), 1991.

Bakos, J. Y. (1997). Reducing buyer search costs: Implications for electronic marketplaces. *Management Science, 43*(12), 1676–1692. doi:10.1287/mnsc.43.12.1676

Bakos, J. Y. (1998). The Emerging Role of Electronic Marketplaces on the Internet. *Communications of the ACM, 41*(8), 35–42. doi:10.1145/280324.280330

Bakos, Y. (1997). Reducing Buyer Search Costs: Implications for Electronic Marketplaces. *Management Science, 43*(12), 1676–1692. doi:10.1287/mnsc.43.12.1676

Balasubramanian, S. (1998). Mail versus mall: A strategic analysis of competition between direct marketers and conventional retailers. *Marketing Science, 17*(3), 181–195.

Baldinger, A. L., & Rubinson, J. (1996). Brand loyalty: The link between attitude and behavior. *Journal of Advertising Research, 36*(6), 22-34.

Banker, R. D., Kauffman, R. J., & Morey, R. C. (1990). Measuring gains in operational efficiency from information technology: A study of the positran deployment at Hardee's Inc. *Journal of Management Information Systems, 7*(2), 29–54.

Bapna, R., Goes, P., & Gupta, A. (2000). A theoretical and empirical investigation of multi-item on-line auctions. *Information Technology and Management, 1*(1-2), 1-23.

Bapna, R., Goes, P., & Gupta, A. (2001). Comparative analysis of multi-item online auctions: Evidence from the laboratory. *Decision Support System.*

Bapna, R., Goes, P., & Gupta, A. (2001). Insights and analyses of online auctions. *Association for Computing Machinery. Communications of the ACM, 44*(11), 42-50.

Bapna, R., Goes, P., & Gupta, A. (2003). Analysis and design of business-to-consumer online auctions. *Management Science, 49*(1), 85-102.

Bapna, R., Goes, P., Gupta, A., & Karuga, G. (2002). Optimal design of the online auction channel: Analytical, empirical, and computational insights. *Decision Sciences, 33*(4), 557-578.

Barber, W., & Badre, A. N. (1998). *Culturability: The merging of culture and usability.* Paper presented at the 4th Conference on Human Factors and the Web, Baskin, Ridge, New Jersey, US.

Baresi, L., Guinea, S., & Plebani, P. (2005). WS-Policy for Service Monitoring. In *Proceedings of the 6th International Workshop on Technologies for E-Services (TES'2005), in conjunction with the 31st International Conference on Very Large Data Bases (VLDB'2005), Trondheim, Norway.*

Barich, H., & Kotler, P. (1991). A framework for marketing image management. *Sloan Management Review, 32*(2), 94-104.

Barkham, R., Gudgin, G., Hart, M., & Hanvey, E. (1996). *The determinants of small firm growth: An inter-regional study in the United Kingdom, 1986-1990.* London: Jessica Kingsley.

Barnes, D., Hinton, M., & Mieczkowska, S. (2003). Competitive Advantage through e-operations. *TQM & Business Excellence, 14*(6), 659–675. doi:10.1080/1478336032000053816

Barnes, S. J., & Vidgen, R. T. (2003). An Integrative Approach to the Assessment of E-Commerce Quality. *Journal of Electronic Commerce Research, 3*(3), 114–127.

Barnes, S., & Vidgen, R. (2001). Assessing the quality of auction Web sites. In *Proceedings of the 34th Annual Hawaii International Conference on System Sciences* (Vol. 7).

Barton, C., & Bear, M. (1999, March). Information and communications technologies: Are they the key to viable business development services for micro and small enterprises? Retrieved from http://www.mip.org/pdfs/mbp/ict.pdf.

Barua, A., Konana, P., Whinston, A. B., & Yin, F. (2001). Driving e-business Excellence. *MIT Sloan Management Review, 43*(1), 36–44.

Barua, A., Konana, P., Whinston, A. B., & Yin, F. (2001a). Measures for e-business value assessment. *IT Pro,* 35-39.

Barua, A., Kriebel, C. H., & Mukhopadhyay, T. (1991). An economic analysis of strategic information technology investments. *MIS Quarterly, 15*(3), 313–331. doi:10.2307/249643

Barua, A., Kriebel, C. H., & Mukhopadhyay, T. (1995). Information technologies and business value: An analytic and empirical investigation. *Information Systems Research, 6*(1), 3–23. doi:10.1287/isre.6.1.3

Basu, A., & Blanning, R. B. (2000). A formal Approach to Workflow Analysis . *Information Systems Research, 11*(1), 17–36. doi:10.1287/isre.11.1.17.11787

Basu, A., & Kumar, A. (2002). Research Commentary: Workflow Management Issues in e-Business . *Information Systems Research, 13*(1), 1–14. doi:10.1287/isre.13.1.1.94

Bate, C., & Mulholland, A. (2005). A Methodology for Service Architectures Version 1.0 – August 2005. *OASIS technical report.* Capgemini, UK.

Bauer, H. H., Grether, M., & Leach, M. (2002). Building Customers Relations over the Internet. *Industrial Marketing Management, 31*(2), 55–63. doi:10.1016/S0019-8501(01)00186-9

Bauer, R. (1960). Consumer behavior and risk taking. In R. Hancock (Ed.), *Proceedings of the American Marketing Association* (pp. 389-398).

Bearden, W. O., & Teel, J. E. (1983). Selected determinants of consumer satisfaction and complaint reports. *Journal of Marketing Research, 20,* 21-28.

Belanger, F., Hiller, J., & Smith, W. (2002). Trustworthiness in electronic commerce: the role of privacy, security, and site attributes. *The Journal of Strategic Information Systems, 11*(3-4), 245–270.

Benbasat, I., Goldstein, D. K., & Mead, M. (1987). The case research strategy in studies of information systems. *MIS Quarterly, 11*(3), 369–386. doi:10.2307/248684

Benbunan-Fich, R. (2001). Using Protocol Analysis to Evaluate the Usability of a Commercial Web Site. *Information & Management, 39,* 151–163. doi:10.1016/S0378-7206(01)00085-4

Benjamin, R. I., & Levinson, E. (1993). A framework for managing IT enabled change. *Sloan Management Review, 1993*(Summer), 22–33.

Benjamin, R. I., & Scott, M. S. S. (1988). Information technology, integration, and organizational change. *Interfaces, 18*(3), 86–98. doi:10.1287/inte.18.3.86

Bennett, G., Sagas, M., & Dees, W. (2006). Media Preferences of Action Sports Consumers: Differences Between Generation X and Y. *Sport Marketing Quarterly, 15*(1), 40–49.

Bennett, R., Härtel, C., & McColl-Kennedy, J. R. (2005). Experience as a moderator of involvement and satisfaction on brand loyalty in a business-to-business setting 02-314R. *Industrial Marketing Management, 34,* 97–107. doi:10.1016/j.indmarman.2004.08.003

Benou, P., & Bitos, V. (2008). Developing mobile commerce applications. *Journal of Electronic Commerce in Organizations, 6*(1), 63–78.

Bentahar, J., Maamar, Z., Wan, W., Benslimane, D., Thiran, Ph., & Sattanathan, S. (2008). Agent-based Communities of Web Services: An Argumentation-driven Approach. *Service Oriented Computing and Applications, 2*(4). doi:10.1007/s11761-008-0033-4

Bentler, P. M. (1995). *EQS structural equations program manual.* CA: Multivariate Software, Inc.

Bentler, P. M., & Bonett, D. G. (1980). Significance tests and goodness of fit in the analysis of covariance structures. *Psychological Bulletin, 88*(3), 588–606. doi:10.1037/0033-2909.88.3.588

Bentler, P. M., & Wu, E. J. C. (2002). *EQS 6 for Windows User's Guide.* Encino, CA: Multivariate Software, Inc.

Bergeron, F. & Berube, C. (1990). End-users Talk Computer Policy. *Journal of Systems Management,* December, 14-16.

Berne, C. (1997). Modelizacion de la Poscompra: Satisfaccion y Lealtad. In J. M. Mugica Grijalva & S. Ruiz de Maya

(Eds), *El Comportamiento del Consumidor,* Cap. 5, Ariel Economia, Barcelona, 163-180.

Berry, M., & Linoff, G. (1997). *Data mining techniques for marketing, sales, and customer support.* New York: John Wiley & Sons.

Bhatnagar, A., Misra, S., & Rao, H. R. (2000). On risk, convenience, and Internet shopping behavior. Communications of the ACM, 43(11), 98-105.

Bhatnagar, A., Misra, S., & Rao, H. R. (2000). On risk, convenience, and Internet shopping. *Communications of the ACM, 43*(11), 98-105.

Bhatt, G. D., & Emdad, A. F. (2001). An analysis of the virtual value chain in electronic commerce. *Logistics Information Management, 14*(1/2), 78–84. doi:10.1108/09576050110362465

Bhattacherjee, A. (2002). Individual trust in online firms: Scale development and initial test. *Journal of Management Information Systems, 19*(1), 211.

Bhimani, A. (1996). Securing the commercial Internet. *Communications of the ACM, 39*(6), 29-31.

Bhiri, S., Perrin, O., & Godart, C. (2005). Ensuring Required Failure Atomicity of Composite Web Services. In *Proceedings of the 14th International World Wide Web Conference (WWW'2005), Chiba, Japan.*

Biazzo, S. (1998). A Critical Examination of the Business Process Re-engineering Phenomenon . *International Journal of Operations & Production Management, 18*(9/10), 1000–1016. doi:10.1108/01443579810225586

Bichler, M. (2001). *The Future of E-Markets.* Cambridge University Press.

Bichler, P., Preuner, G., & Schrefl, M. (1997). *Workflow Transparency.* Paper presented at the 9th International Conference for Advanced Information Systems Engineering (CAiSE'97), Barcelona, Spain.

Bickmore, T. W., & Picard, R. W. (2005). Establishing and maintaining long-term human-computer relationships. *ACM Transactions on Computer-Human Interaction, 12*(2), 293–327.

Bitner, M. J. (1992). Servicescapes: The impact of physical surroundings on customers and employees. *Journal of Marketing, 56*(2), 57-71.

Bitner, M. J., Brown, S. W., & Meuter, M. L. (2000). Technology Infusion in Service Encounters. *Journal of the Academy of Marketing Science, 28*(1), 138–149. doi:10.1177/0092070300281013

Blili, S., & Raymond, L. (1993). Information technology – threats and opportunities for small and medium sized enterprises. *International Journal of Information Management, 13*, 439–448. doi:10.1016/0268-4012(93)90060-H

Bloemer, J. M., & Kasper, H. (1993). Brand loyalty and brand satisfaction: The case of buying audio cassettes anew in the Netherlands. In *Proceedings of the 22nd European Academy Conference*, Barcelona, Spain.

Bloemer, J. M., & Kasper, H. (1995). The complex relationship between consumer satisfaction and brand loyalty. *Journal of Economic Psychology, 16*(2), 311-329.

Bloemer, J. M., & Lemmink, J. G. (1992). The importance of customer satisfaction in explaining brand and dealer loyalty. *Journal of Marketing Management, 8*(4), 351-364.

Bockem, S. (1994). A Generalized Model of Horizontal Product Differentiation. *The Journal of Industrial Economics, 42*, 287–298. doi:10.2307/2950571

Bollen, K. A. (1989). *Structural equations with latent variables.* New York: Wiley.

Bolton, C., & Davies, J. (2000). Activity graphs and Processes. In W. Grieskamp, T. Santen, and B. Stoddart, (eds.), *Proceedings Integrated Formal Methods 2000* (LNCS 1945).

Bolton, R. N., Kannan, P. K., & Bramlett, M. D. (2000). Implications of loyalty program membership and service experiences for customer retention and value. *Journal of Academy of Marketing Science, 28*(1), 95-108.

Booch, G., Rumbaugh, J., & Jacobson, I. (1998). *The Unified Modeling Language User Guide.* Addison-Wesley Professional.

Borgida, A., & Murata, T. (1999). Tolerating Exceptions in Workflows: A Unified Framework for Data and Processes. In D. Georgakopoulos, W. Prinz and A.L. Wolf (Eds.), *Proceedings of the International Joint Conference on Work Activities Coordination and Collaboration, WACC 99* (pp. 59-68). ACM Press, February

Bouchard, L. (1993). Decision criteria in the adoption of EDI. In J. I. DeGross, R. P. Bostrom & D. Robey (Eds.), *Proceedings of the Fourteenth International Conference on Information Systems* (pp. 365-376). Orlando FL: ICIS.

Boulding, W., Kalra, A., Staelin, R., & Zeithaml, V. A. (1993). A dynamic process model of service quality: From expectations to behavioral intentions. *Journal of Marketing Research, 30,* 7-27.

Bowen, J. T., & Shoemaker, S. (1998). Loyalty: A strategic commitment. *Cornell and Hotel Restaurant Administration Quarterly, 39*(1), 12-25.

Boyer, K. K., & Olson, J. R. (2002). Drivers of Internet Purchasing Success. *Production and Operations Management, 11*(4), 480–498.

Boynton, A. C., Zmud, R. W., & Jacobs, G. C. (1994). The Influence of IT Management Practice on IT Use in Large Organizations . *MIS Quarterly*, (September): 299–318. doi:10.2307/249620

BPMN. (2008). Business Process Modeling Notation. Retrieved December 2008 from http://www.bpmn.org

Brambilla, M., Ceri, S., & Fraternali, P. (2006). Process Modeling in Web Applications. *ACM Transactions on Software Engineering and Methodology, 15*(4). doi:10.1145/1178625.1178627

Brancheau, J. C., & Brown, C. V. (1993). The management of end-user computing: Status and directions. *ACM Computing Surveys, 25*(4), 450–482. doi:10.1145/162124.162138

Breener, M., & Schiffers, M. (2003). Applying Web Services Technologies to the Management of Context Provisioning. In *Proceedings of the 10th Workshop of the OpenView University Association (OVUA'2003), Geneva, Switzerland.*

Breiman, L., Friedman, J., Olshen, R., & Stone, C. (1984). *Classification and regression trees.* Belmont, CA: Wadsworth International Group.

Bressolles, G. (2006). La qualité de service électronique: NetQual. Proposition d'une échelle de mesure appliquée aux sites marchands et effets modérateurs [Electronic service quality: NetQual – Proposition of a measurement scale to commercial Websites and moderating effects]. *Recherche et Applications en Marketing, 21*(3), 19–45.

Brint, A. T. (2003). Investigating buyer and seller strategies in online auctions. *Journal of the Operational Research Society, 54*(11), 1177-1197.

Brint, A. T. (2003). Investigating buyer and seller strategies in online auctions. *The Journal of the Operational Research Society, 54*(11), 1177-1188.

Brosig, J., Ockenfels, A., & Weimann, J. (2002). The effects of communication media on cooperation. *German Economic Review, 4*(2), 217–241.

Brown, C. V. (2003). Performance metrics for IT human resource alignment. *Information Systems Management, 20*(4), 36–42. doi:10.1201/1078/43647.20.4.20030901/77291.6

Brown, D., Oleksik, G., & Bisdee, D. (2007). Consumer attitudes review, Internet shopping – Annexe E. *Prepared for the Office of Fair Trading, Crown Copyright.*

Bruce, N., Haruvy, E., & Rao, R. (2004). Seller rating, price, and default in online auctions. *Journal of Interactive Marketing, 18*(4), 37-51.

Brynjolfsson, E. (2003). IT Productivity Gap. *Optimize, 21*(7).

Brynjolfsson, E., & Hitt, L. (1993). Is information technology spending productive? New evidence and new results. In *Proceedings of the 14th Annual International Conference on Information Systems.*

Brynjolfsson, E., & Hitt, L. (1998). Beyond the productivity paradox. *Communications of the ACM, 41*(8), 41–55. doi:10.1145/280324.280332

Brynjolfsson, E., & Kahim, B. (Eds.). (2000). *Understanding the Digital Economy.* Cambridge: MIT Press.

Brynjolfsson, E., & Smith, M. D. (2000). Frictionless commerce? A comparison of Internet and conventional retailers. *Management Science, 46*(4), 563-585.

Budish, E., & Takeyama, L. N. (2001). Buy prices in online auctions: Irrationality on the Internet? *Economics Letters, 72*(3), 325-334.

Burn, J. M., & Szeto, C. (2000). A comparison of the views of business and IT management on success factors for strategic alignment. *Information & Management, 37*(4), 197–216. doi:10.1016/S0378-7206(99)00048-8

Burton-Jones, A., & Hubona, G. S. (2005, Spring). Individual Difference and Usage Behavior: Revisiting a Technology Acceptance Model Assumption. *The Data Base for Advances in Information Systems, 36*(2), 75.

Cahill, K. (2007). Worth the Price? Virtual Reference, Global Knowledge Forums, and the Demise of Google Answers. *Journal of Library Administration, 46*(3/4), 73–86.

Cameron, D., & Galloway, A. (2005). Consumer motivations and concerns in online auctions. *International Journal of Consumer Studies, 29*(3), 181-193.

Campbell, & Fiske (1959). Convergent and discriminant validation by the multitrait-multimethod matrix. *Psychological Bulletin, 56*(2), 81-105.

Cao, L., Li, M. L., Cao, J., & Huang, J. S. (2008). CNP-based implementation of service-oriented workflow mapping in SHGWMS. *World Wide Web-Internet and Web Information Systems, 11*(2), 199–226.

Cao, M., Zhang, Q., & Seydel, J. (2005). B2C e-commerce web site quality: an empirical examination. *Industrial Management & Data Systems, 105*(5), 645–661. doi:10.1108/02635570510600000

Cappiello, C., & Pernici, B. (2006). A Methodology for Information Quality Management in Self-Healing Web Services. In *Proceedings of the 11th International Conference on Information Quality (ICIQ'2006), Massachusetts, USA.*

Carayannis, E. G., & Turner, E. (2006). Innovation diffusion and technology acceptance: The case of PKI technology. *Technovation, 26*(7), 847–855. doi:10.1016/j.technovation.2005.06.013

Cardoso, J., & Sheth, A. (2003). Semantic E-Workflow Composition. *Journal of Integrated Design and Process Science, 21*(3), 191–225.

Carr, N. (2003). IT doesn't matter. *Harvard Business Review*, (May): 5–12.

Casati, F., & Pozzi, G. (1999). *Modeling Exceptional Behaviours in Commercial Workflow Management Systems.* Paper presented at the 1999 IFCIS International Conference on Cooperative Information Systems, Edinburgh, Scotland.

Casati, F., Ceri, B., Pernici, B., & Pozz, G. (1995). *Conceptual Modelling of Workflows.* Paper presented at the Object-oriented and Entity-Relationship Conf., Australia.

Cassar, K., Swerdlow, F. S., Kim, H., & Johnson, M. (2000). Local commerce: Internet threat mandates brick-and-mortar reconstruction. *Jupiter Vision Report,* (11), January 3.

Cattani, K. D., Gilland, W. G., & Swaminathan, J. M. (2006). Boiling frogs: Pricing strategies for a manufacturer adding a direct channel that competes with the traditional channel. *Production and Operations Management, 15*(1), 40-57.

Celsi, R., Rose, R., & Leigh, T. (1993). An exploration of high-risk leisure consumption through skydiving. *Journal of Consumer Research, 20*(1), 1-23.

Ceri, S., Fraternali, P., & Bongio, A. (2000). Web Modeling Language (WebML): A Modeling Language for Designing Web Sites. *Computer Networks, 33,* 1–6. doi:10.1016/S1389-1286(00)00040-2

Chaffey, D. (2004). *E-business and e-commerce management* (2nd ed.). Prentice Hall.

Chamberlin, J. R., Cohen, J. L., & Coombs, C. H. (1984). Social choice observed: Five presidential elections of the American Psychological Association. *The Journal of Politics, 46*(2), 479-502.

Chan, E. Y., & Reich, H. B. (2007). IT alignment: What have we learned? *Journal of Information Technology, 22,* 297–315. doi:10.1057/palgrave.jit.2000109

Chan, E. Y., Huff, L. S., Barclay, W. D., & Copeland, G. D. (1997). Business strategic orientation, information systems strategic orientation, and strategic alignment. *Information Systems Research, 8*(2), 125–150. doi:10.1287/isre.8.2.125

Chan, E. Y., Huff, S. L., & Copeland, G. D. (1998). Assessing realized information systems strategy. *Strategic*

Information Systems, 6(4), 273–298. doi:10.1016/S0963-8687(97)00005-X

Chan, I., & Chao, C. (2008, April). Knowledge Management in Small and Medium-Sized Enterprises. *Communications of the ACM, 51*(4), 83–88. doi:10.1145/1330311.1330328

Chan, Y. E. (2000). IT value: The great divide between qualitative and quantitative and individual and organizational measures. *Journal of Management Information Systems, 16*(4), 225–261.

Chatterjee, D. & Ravichandran, T. (2004). Beyond exchange models: understanding the structure of B2B information systems. *Information Systems and e-Business Management, 2*, 169-186.

Chau, P. Y. K., & Hui, K. L. (2001). Determinants of small business EDI adoption: an empirical investigation. *Journal of Organizational Computing and Electronic Commerce, 11*(4), 229–252. doi:10.1207/S15327744JOCE1104_02

Chau, P. Y. K., & Tam, K. Y. (1997). Factors affecting the adoption of open systems: an exploratory study. *MIS Quarterly, 21*(1), 1–24. doi:10.2307/249740

Chau, P. Y. K., & Tam, K. Y. (2000). Organizational adoption of open systems: a 'technology-push, need-pull' perspective. *Information & Management, 37*(5), 229–239. doi:10.1016/S0378-7206(99)00050-6

Chau, Y. K. (1997). Reexamining a model for evaluating information center success using a structural equation modeling approach. *Decision Science, 28*(2), 309-334.

Chaudhury, A., Mallick, D. N., & Rao, H. R. (2001). Web channels in e-commerce. *Communications of the ACM, 44*(1), 99–104. doi:10.1145/357489.357515

Chen, J., Melinda, H., & Zapatero, E. (2006, February). Understanding Web Commerce Adoption By Very Small Businesses. *Proceedings of ASBBS*, Las Vegas, NV, *13*(1), 268-277.

Chen, M., Zhang, D., & Zhou, L. (2007). Empowering Collaborative Commerce with Web Services enabled Business Process Management Systems. *Decision Support Systems, 43*, 530–546. doi:10.1016/j.dss.2005.05.014

Chen, Q., & Wells, W. D. (1999). Attitude Toward the Site. *Journal of Advertising Research, 39*(5), 27–37.

Chen, S. C., & Dhillon, G. S. (2003). Interpreting Dimensions of Consumer Trust in E-Commerce. *Information Technology and Management, 4*(2-3), 303.

Chen, Y., Chen, G., & Wu, S. (2005). Issues and opportunities in E-Business research: A Simonian perspective. *International Journal of E-Business Research, 1*(1), 37–56.

Cheng, H., & Schweitzer, J. C. (1996). Cultural values reflected in Chinese and U.S television commercials. *Journal of Advertising Research*, (May/June): 27–45.

Cheskin, & Sapient. (1999). *eCommerce Trust Study*, from http://www.studioarchetype.com/cheskin

Chiang, W. Y., Chhajed, D., & Hess, J. D. (2003). Direct marketing, indirect profits: A strategic analysis of dual-channel supply chain design. *Management Science, 49*(1), 1-20.

Childe, S. J., Maull, R. S., & Bennett, J. (1994). Frameworks for Understanding Business Process Re-engineering. *International Journal of Operations & Production Management, 14*(2), 22–34. doi:10.1108/01443579410072346

Chin, W. W. (1998). Issues and opinion on structural equation modelling. *MIS Quarterly*, (March): 7–16.

Chiou, J. S. (2004). The antecedents of customers' loyalty toward Internet service providers. *Information and Management, 41*(6), 685-695.

Chircu, A. M., & Kauffman, R. J. (2000). Limits to value in electronic commerce-related IT investments. *Journal of Management Information Systems, 17*(2), 59–80.

Chiu, D. K. W., Cheung, S. C., Till, S., Karlapalem, K., Li, Q., & Kafeza, E. (2004). Workflow View Driven Cross-Organizational Interoperability in a Web Service Environment. *Information Technology and Management, 5*, 221–250. doi:10.1023/B:ITEM.0000031580.57966.d4

Cho, V. (2006). Factors in the adoption of third-party B2B portals in the textile industry. *Journal of Computer Information Systems, 46*(3), 18–31.

Choi, I., Jung, J., Mannino, M., & Park, C. (2008). Terminability and compensatibility of cycles in business processes with a process-oriented triggrer. *Data & Knowledge Engineering*, *66*(2), 243–263. doi:10.1016/j.datak.2008.03.002

Choi, K. S., Lee, H., Kim, C., & Lee, S. (2005). The service quality dimensions and patient satisfaction relationships in South Korea: Comparisons across gender, age and types of service. *Journal of Services Marketing, 19*(3), 140-149.

Chong, B., & Wong, M. (2005). Crafting an effective customer retention strategy: A review of halo effect on customer satisfaction in online auctions. *International Journal of Management and Enterprise Development, 2*(1), 12-26.

Chow, S., & Holden, R. (1997). Toward an understanding of loyalty the moderating role of trust. *Journal of Managerial Issues, 9*(3), 275–298.

Christensen, C. M. (1997). *The Innovator's Dilemma*. Harvard Business School Press.

Chua, P. Y. K., & Hui, K. L. (2001). Determinants of small business EDI Adoption: An empirical Investigation. *Journal of Organizational Computing and Electronic Commerce, 11*(4), 229–252. doi:10.1207/S15327744JOCE1104_02

Chuang, M. C., Chang, C. C., & Hsu, S. H. (2001). Perceptual factors underlying user preferences toward product form of mobile phones. *International Journal of Industrial Ergonomics, 27*, 247–258. doi:10.1016/S0169-8141(00)00054-8

Churchill, J. R. (1979). A paradigm for developing better measures of marketing construct. *JMR, Journal of Marketing Research*, (February): 64–73. doi:10.2307/3150876

Chwelos, P., Benbasat, I., & Dexter, A. S. (2001). Research report: empirical test of an EDI adoption model. *Information Systems Research, 12*(3), 304–321. doi:10.1287/isre.12.3.304.9708

Cimatti, A., Clarke, E., Giunchiglia, F., & Roveri, M. (2000). NUSMV: A New Symbolic Model Checker. *International Journal on Software Tools for Technology Transfer, 2*(4), 410–425. doi:10.1007/s100090050046

CIM-OSA. (1989). *CIM-OSA Reference Architecture*. CIM-OSA Standards Committee, AMICE ESPRIT.

Clark, D. D., Lehr, W., Bauer, S. J., Faratin, P., Sami, R., & Wroclawski, J. (2006). Overlay Networks and Future of the Internet. *Journal of Communications and Strategies, 3*(63), 1–21.

Clemons, E. K., & Hitt, L. (2001). Financial services: Transparency, differential pricing, and disintermediation. In R. Litan & A. Rivlin (Eds.), *The Economic Payoff from the Internet Revolution* (pp. 87-128). Washington, D.C.: Brookings Institution Press.

Clemons, E. K., & McFarlan, W. F. (1986). Telecom: hook up or lose out. *Harvard Business Review, 64*(4), 91–97.

Clemons, E. K., Row, M. C., & Reddi, S. P. (1993). The Impact of Information Technology on the Organization of Economic Activity: The 'Move to the Middle' Hypothesis. *Journal of Management Information Systems, 10*(2), 9–35.

Clemons, E., & Wang, Y. (2000). Special Issue: Technology Strategy for Electronic Marketplaces. *Journal of Management Information Systems, 17*(2), 5–7.

Cockburn, C., & Wilson, T. D. (1996). Business use of the World Wide Web. *International Journal of Information Management, 16*(2), 83–102. doi:10.1016/0268-4012(95)00071-2

Cohen, L., Manion, L., & Morrison, K. (2000). *Research methods in education* (5th ed.). London: Routledge Falmer.

Cohen, M., & Whang, S. (1997). Competing in Product and Service: A Product Life-Cycle Model. *Management Science, 43*, 535–545. doi:10.1287/mnsc.43.4.535

Cohen, W. M., & Levinthal, D. A. (1990). Absorptive capacity: A new perspective on learning and innovation. *Administrative Science Quarterly, 35*(1), 1–14. doi:10.2307/2393553

Colgate, M., & Lang, B. (2001). Switching barriers in consumer markets: An investigation of the financial services industry. *Journal of Consumer Marketing, 18*(4/5), 332-347.

Compeau, D. (1995). Computer Self-efficacy: Development of a Measure and Initial Test. *MIS Quarterly, 19*, 189–211. doi:10.2307/249688

Compeau, D. R., & Higgins, C. A. (1995). Computer self-efficacy: development of a measure and initial test. *MIS Quarterly*, *19*(2), 189–211. doi:10.2307/249688

Compeau, D., Higgins, C. A., & Huff, S. (1999). Social Cognitive Theory and Individual Reactions to Computing Technology: A Longitudinal Study . *MIS Quarterly*, (June): 145–158. doi:10.2307/249749

Consensus. (n.d.). Retrieved from http://essential-facts.com/primary/ethics/Consensus.html

Consortium, U. W. A. (2002). Ubiquitous Web Applications. In *Proceedings of eBusiness and eWork Conference*.

Constat. 2001. ConStat's Technology Adoption Model. The Framework for the WISDM Program (White Paper). Retrieved November 26, 2006 from http://www.constat.com/wisdm/ConStat_Tech_Adoption_Model.pdf

Cook, C., Heath, F., & Thompson, R. (2000). A meta-analysis of response rates in Web- or Internet-based surveys. *Education and Psychological Measurement, 60*(6), 821-836.

Cook, D. P., & Luo, W. (2003). The role of third-party seals in building trust online. *E-Service Journal, 2*(3), 71-84.

Cook, D., & Das, S. (2004). *Smart Environments: Technology, Protocols, and Applications*. Wiley Pub.

Cook, J. E., & Wolf, A. L. (1999). Software Process Validation: Quantitatively Measuring the Correspondence of a Process to a Model . *ACM Transactions on Software Engineering and Methodology*, *8*(2), 147–176. doi:10.1145/304399.304401

Cooper, R. B., & Zmud, R. W. (1990). Information technology research: a technological diffusion approach. *Management Science*, *36*(2), 123–139. doi:10.1287/mnsc.36.2.123

Corritore, C. L., Kracher, B., & Wiedenbeck, S. (2003). Online trust: concepts, evolving themes, a model. *International Journal of Human-Computer Studies*, *58*(6), 737–758.

Cottrill, K. (1997). The Supply Chain of the Future. *Distribution*, *96*(11), 52–54.

Couper, M. (2000). Web surveys: A review of issues and approaches. *The Public Opinion Quarterly, 64*(4), 464-494.

Coutaz, J., Crowley, J. L., Dobson, S., & Garlan, D. (2005). Context is Key. *Communications of the ACM*, *48*(3). doi:10.1145/1047671.1047703

Cox, D., & Kreger, H. (2005). Management of the Service-Oriented-Architecture Life Cycle . *IBM Systems Journal*, *44*, 709–726.

Cragg, P. B., & King, M. (1993). Small firm computing – motivators and inhibitors. *MIS Quarterly*, *17*, 47–60. doi:10.2307/249509

Cragg, P., King, M., & Hussin, H. (2002). IT alignment and firm performance in small manufacturing firms. *The Journal of Strategic Information Systems*, *11*(1), 109–132. doi:10.1016/S0963-8687(02)00007-0

Crockett, R. O. (1999). Going, going...richer. *Business Week* (3659), EB16.

Cron, W., & Sobol, M. (1983). The relationship between computerisation and performance: a strategy for maximising the economic benefits of computerisation. *Information & Management*, *6*, 171–181. doi:10.1016/0378-7206(83)90034-4

Cronin, J. J., Brady, M. K., & Hult, G. T. M. (2000). Assessing the effects of quality, value and customer satisfaction on consumer behavioral intentions in service environments. *Journal of Retailing, 76*(2), 193-218.

Cronin, J. J., Jr., & Taylor, S. A. (1992). Measuring service quality: A re-examination and extension. *Journal of Marketing, 56*(3), 55-68.

Croteau, A., & Bergeron, F. (2001). An information technology trilogy: Business strategy, technological deployment and organization performance. *The Journal of Strategic Information Systems*, *10*(2), 77–99. doi:10.1016/S0963-8687(01)00044-0

Cunliffe, D. (2000). Developing Usable Web Sites - A Review and Model. *Internet Research*, *10*, 295–308. doi:10.1108/10662240010342577

Cunningham, R. M. (1961). Customer loyalty to store and brand. *Harvard Business Review, 39*(6), 127-137.

Cunningham, S. M. (1967). The major dimensions of perceived risk. In *Risk taking and information handling in consumer behavior* (pp. 82-108). Boston: Harvard University Press.

Curbera, F., Duftler, M., Khalaf, R., Nagy, W., Mukhi, N., & Weerawarana, S. (2002). Unraveling the Web Services Web: An Introduction to SOAP, WSDL, and UDDI. *IEEE Internet Computing, 6*(2). doi:10.1109/4236.991449

Customer Care - Best Practices @ Cisco. (2001). Retrieved 2001, from http://www.cisco.com/warp/public/779/ibs/solutions/customer/practices/

Cyber Atlas Staff. (2002, October). Small Biz, Small Growth. Retrieved May 12, 2006, from http://ISP-Planet.com/research/2002/smallbiz_021009.html

Cyr, D., & Trevor-Smith, H. (2004). Localization of web design: An empirical comparison of German, Japanese, and United States web site characteristics. *Journal of the American Society for Information Science and Technology, 55*(13), 1199–1206.

Dabholkar, P. A. (2000). Technology in Service Delivery: Implications for Self-Service and Service Support. In Swartz T.A. and Iacobucci D. (Eds.), *Handbook of Services Marketing* (pp. 103-110). New York: Sage.

Dahlen, M. (2002). Learning the web: Internet user experience and response to web marketing in Sweden. *Journal of Interactive Advertising, 3*(1).

Dai, Q., & Kauffman, R. J. (2002). B2B E-Commerce Revisited: Leading Perspectives on the Key Issues and Research Directions. *Electronic Markets, 12*(2).

Dai, Q., & Kauffman, R. J. (2002). Business models for internet-based B2B electronic markets. *International Journal of Electronic Commerce, 6*(4), 41–72.

Damianou, N., Dulay, N., Lupu, E., & Sloman, M. (2001). The Ponder Policy Specification language. In *Proceedings of the 2nd IEEE International Workshop on Policies for Distributed Systems and Networks (POLICY'2001), Bristol, UK.*

Damsgaard, J., & Lyytinen, K. (2000). The dynamics of factors explaining EDI diffusion in Hong Kong in the late 1990s. In J. Thong, P. Chau & K. Y. Tam (Eds.), *Proceedings of the Fourth Pacific Asia Conference on Information Systems* (pp. 1061-1074). Hong Kong: AIS.

Dang, J., Hedayati, A., Hampel, K., & Toklu, C. (2008). An ontological knowledge framework for adaptive medical workflow . *Journal of Biomedical Informatics, 41*(5), 829–836. doi:10.1016/j.jbi.2008.05.012

Daniel, E., Wilson, H., & Myers, A. (2002). Adoption of ecommerce by SMEs in the UK. *International Small Business Journal, 20*(3), 253–268. doi:10.1177/0266242602203002

Davenport, H. T., & Prusak, L. (1998). *Working Knowledge: how organizations manage what they know.* Boston, MA: Harvard Business School Press.

Davenport, T. (1993). *Process Innovation – Reengineering Work Through Information Technology.* Boston, MA: Harvard Business School Press.

Davern, M. J., & Kauffman, R. J. (2000). Discovering potential and realizing value from information technology investments. *Journal of Management Information Systems, 16*(4), 121–143.

Davis, F. D. (1989). Perceived Usefulness, Perceived Ease of Use, and User Acceptance of Information Technology. *MIS Quarterly, 13*, 319–340. doi:10.2307/249008

Davis, F. D., Bagozzi, R. P., & Warshaw, P. R. (1989). User Acceptance of Computer Technology: A Comparison of Two Theoretical Models. *Management Science, 35*, 982–1003. doi:10.1287/mnsc.35.8.982

Davis, F., Baggozi, R., & Warshaw, P. (1989). User acceptance of computer technology: A comparison of two theoretical models. *Management Science, 35*(8), 982–1002. doi:10.1287/mnsc.35.8.982

Davis, G. B. (1988). The Hidden Costs of End-user Computing. *Accounting Horizons, 2*(4), 103–106.

De Mooij, M. (2006). *Global marketing and advertising: Understanding cultural paradoxes*: Sage Publications: London.

De Roure, D., Goble, C., & Stevens, R. (2008). (in press). The design and realisation of the Experiment Virtual Research Environment for social sharing of workflows . *Future Generation Computer Systems.* doi:.doi:10.1016/j.future.2008.06.010

De Troyer, O., & Casteleyn, S. (2003). Modeling Complex process for Web application with WSDM. In *Proceedings of the International Workshop on Web Oriented Software Technology (IWWOST'2003), Asurias, Spain.*

Dechter, D. (2003). *Constraint Processing.* Morgan Kaufmann.

Dedrick, J., & West, J. (2004). An exploratory study into open source platform adoption. In *Proceedings of the 37th Hawaii International Conference on System Sciences* (pp. 1-10). Hawaii: IEEE.

Deelman, E., Gannon, D., Shields, M., & Taylor, I. (2008). (in press). Workflow and e-Science: An overview of workflow system features and capabilities. *Future Generation Computer Systems.* doi:.doi:10.1016/j.future.2008.06.012

Dehning, B., & Richardson, V. J. (2002). Returns on investments in information technology: A research synthesis. *Journal of Information Systems, 16*(1), 7–30. doi:10.2308/jis.2002.16.1.7

Dekleva, S. (2000). Electronic commerce: A half-empty glass. *Communications of the Association for Information Systems, 3*(18).

Del Galdo, E., & Neilson, J. (1996). *International user interfaces.* New York: John Wiley & Sons.

DeLone, W. H., & McLean, E. R. (1992). Information systems success: The quest for the dependent variable. *Information Systems Research, 3*(1), 60–95. doi:10.1287/isre.3.1.60

Denscombe, M. (2006). Web-based questionnaires and the mode effect. *Social Science Computer Review, 24*(2), 246-254.

DePietro, R., Wiarda, E., & Fleischer, M. (1990). The context for change: organization, technology, and environment. In L. G. Tornatzky & M. Fleischer (Eds.), *The Processes of Technological Innovation* (pp. 151-175). Lexington MA: Lexington Books.

Deshpande, R., Hoyer, W. D., & Donthu, N. (1986). The intensity of ethnic affiliations: A study of the sociology of Hispanic consumption. *The Journal of Consumer Research, 13,* 214–220.

Desouza, K. C., & Awazu, Y. (2003). Constructing internal knowledge markets: Considerations from mini cases. *International Journal of Information Management, 23,* 345–353. doi:10.1016/S0268-4012(03)00056-2

Dess, G. G., & Robinson, R. B. (1984). Measuring organizational performance in the absence of objective measures: The case of privately-held firm and conglomerate business unit. *Strategic Management Journal, 5,* 265–273. doi:10.1002/smj.4250050306

Desuoza, K., Awazu, Y., Yamakawa, S., & Umezawa, M. (2005). Facilitating knowledge market mechanism. *Knowledge and Process Management, 12*(2), 99–107. doi:10.1002/kpm.226

Dewar, R. D., & Dutton, J. E. (1986). The adoption of radical and incremental innovation: An empirical analysis. *Management Science, 32,* 1422–1433. doi:10.1287/mnsc.32.11.1422

Dhillon, G. (2008). Organizational competence for harnessing IT: A case study. *Information & Management, 45,* 297–303. doi:10.1016/j.im.2008.01.008

Dholakia, U. (2005). Concept discovery, process explanation, and theory deepening in e-marketing research: The case of online auctions. *Marketing Theory, 5*(1), 117-124.

Dholakia, U. (2005). The usefulness of bidders' reputation ratings to sellers in online auctions. *Journal of Interactive Marketing, 19*(1), 31-40.

Dholakia, U. M. (2005a). Concept discovery, process explanation, and theory deepening in e-marketing research: The case of online auctions. *Marketing Theory, 5*(1), 117-124.

Dholakia, U. M. (2005b). The usefulness of bidders' reputation ratings to sellers in online auctions. *Journal of Interactive Marketing, 19*(1), 31-41.

Dholakia, U. M., & Soltysinski, K. (2001). Coveted or overlooked? The psychology of bidding for comparable listings in digital auctions. *Marketing Letters, 12*(3), 225-237.

Dholakia, U., & Simonson, I. (2005). The effect of explicit reference points on consumer choice and online bidding behavior. *Marketing Science, 24*(2), 206-217.

Dholakia, U., Bagozzi, R., & Klein, L. (2004). A social influence model of consumer participation in network and small-group-based virtual communities. *International Journal of Research in Marketing, 21*, 241–263. doi:10.1016/j.ijresmar.2003.12.004

Dholakia, U., Basuroy, S., & Soltysinski, K. (2002). Auction or agent (or both)? A study of moderators of the herding bias in digital auctions. *International Journal of Research in Marketing, 19*(2), 115-130.

Dichter, E. (1985). What's an image? *Journal of Consumer Marketing, 2*(1), 75-81.

Dick, A. S., & Basu, K. (1994). Customer loyalty: Toward an integrated conceptual framework. *Journal of the Academy of Marketing Science, 22*(2), 99-113.

Dillman, D. A. (1978). *Mail and Telephonic Surveys: The Total Design Method.* New York: Wiley.

Ding, M., Elishaberg, J., Huber, J., & Saini, R. (2005). Emotional bidders - An analytical and experimental examination of consumers' behavior in a Priceline-like reverse auction. *Management Science, 51*(3), 352-365.

Dishaw, M., & Strong, D. (1999). Extending the technology acceptance model with task-technology fit constructs. *Information & Management, 36*(1), 9–21. doi:10.1016/S0378-7206(98)00101-3

Distante, D., Rossi, G., & Canfora, G. Modeling Business Processes in Web Applications: An Analysis Framework. In *Proceedings of the 2006 ACM Symposium on Applied Computing (SAC'2006), Dijon, France.*

Divitini, M., & Simone, C. (2000). Supporting different dimensions of adaptability in workflow modeling . *Computer Supported Cooperative Work: The Journal of Collaborative Computing, 9*(3-4), 365–397. doi:10.1023/A:1008751210054

Donner, J. (2004, Fall). Microentrepreneurs and Mobiles: An Exploration of the Uses of Mobile Phones by Small Business Owners in Rwanda. *Information Technologies & International Development, 2*(1), 1–21. doi:10.1162/1544752043971198

Donthu, N., & Garcia, A. (1999). The Internet Shopper. *Journal of Advertising Research, 39*(3), 52–58.

Doolin, B., & Troshani, I. (2007). Organizational adoption of XBRL. *Electronic Markets, 17*(3), 199–209. doi:10.1080/10196780701503195

Doolin, B., McLeod, L., McQueen, B., & Watton, M. (2003). Internet strategies for established retailers: four New Zealand case studies. *Journal of Information Technology Cases and Applications, 5*(4), 3–20.

Dos Santos, B. L., & Sussman, L. (2000). Improving the return on IT investment: the productivity paradox. *International Journal of Information Management, 20*, 429–440. doi:10.1016/S0268-4012(00)00037-2

Dos Santos, B. L., Peffers, K., & Mauer, D. C. (1993). The impact of information technology investment announcements on the market value of the firm. *Information Systems Research, 4*(1), 1–23. doi:10.1287/isre.4.1.1

DTI. (2000). International Bench Marking Survey for ICT Use by UK Department of Trade and Industry. Retrieved from http://www.ukonline.gov.uk

Dun & Bradstreet. (2000/01). Key Business Directory of Singapore. *ASEAN Series* (Vol.1). Singapore.

Dun & Bradstreet's Small Business Survey Reveals Eight in Ten Small Businesses Now Have Computers and Two-Thirds Have Internet Access. (2001, October). Retrieved May 12, 2006, from www.itaa.org/isec/pubs/e200110-11.pdf

Dustdar, S. & Papazoglou, M. (2008). Services and Service Composition – An Introduction. *Information Technology Journal, 50*(2).

Dustdar, S., & Schreiner, W. (2005). A Survey on Web Services Composition. *International Journal on Web and Grid Services, 1*(1).

Easley, R. F., & Tenorio, R. (2004). Jump bidding strategies in Internet auctions. *Management Science, 50*(10), 1407-1420.

Eastin, M. S. (2002). Diffusion of e-commerce: An analysis of the adoption of four e-commerce activities. *Telematics and Informatics, 19*, 251–267. doi:10.1016/S0736-5853(01)00005-3 eBay (2006). eBay Inc. announces fourth quarter and full year 2005 financial results. Retrieved April 3, 2006, from www.investor.ebay.com/news/Q105/ebay0420-2932ld.pdf

Eccles, R. G., & Crane, D. B. (1988). *Doing Deals: Investment Banks at Work*. Boston, Massachusetts: Harvard Business School Press.

Edelman, B. (2004). *Earnings and Ratings at Google Answers* (Working draft). Harvard Law School, Harvard University.

Edwards, W., & Fasolo, B. (2001). Decision Technology. *Annual Review of Psychology, 52*(1), 581–606. doi:10.1146/annurev.psych.52.1.581

Egger, F. N. (2001, June 27-29). *Affective design of e-commerce user interface: How to maximize perceived trustworthiness.* Paper presented at the the International Conference on Affective Human Factors Design, Singapore.

eGlobal Report, (2005). Retrieved August 12, 2006, from http://www.eMarketer.com

Eisenhardt, K. M., & Martin, J. A. (2000). Dynamic capabilities: what are they? *Strategic Management Journal, 21*(10), 1105–1121. doi:10.1002/1097-0266(200010/11)21:10/11<1105::AID-SMJ133>3.0.CO;2-E

Elliot, S. (2002). Research model and theoretical implications. In S. Elliot (Ed.), *Electronic Commerce B2C Strategies and Models* (pp. 291-326). Brisbane: Wiley.

El-Shinnawy, M., & Markus, M. (1998). Acceptance of communication media in organizations: richness orfeatures? *Professional Communication . IEEE Transactions on, 41*(4), 242–253.

eMarketer (2005). North-America on Line: Demographics and Usage.

e-Marketer (June, 2008). *June 2008: U.S. Retail E-Commerce.* Retrieved 2008 from http://www.iab.net/insights_research/530422/1675/334589.

Emmelhainz, M. A. (1993). *EDI. A total management guide* (2nd ed.). Van Nostrand Reinhold, New York.

Engelbrecht-Wiggans, R. (1980). Auctions and bidding models: A survey. *Management Science, 26*(2), 119-143.

Epstein, M. J. (2000). Organizing your business for the internet evolution. *Strategic Finance, 82*(1), 56–60.

Ericsson, K. A., & Simon, H. A. (1993). *Protocol Analysis: Verbal reports as data* (2nd ed.). Cambridge, MA: MIT Press.

Erl, T. (2005). *Service Oriented Architecture (SOA): Concepts, Technology, and Architecture.* Prentice Hall.

Erl, T. (2007). *SOA Principles of Service Design* (1st ed.). Prentice Hall PTR.

Ernst, & Young. (2002). *Gulf Cooperation Council eCommerce Study*: Gulf Cooperation Council.

Eshuis, R., & Grefen, P. (2008). Constructing customized process views. *Data & Knowledge Engineering, 64*(2), 419–438. doi:10.1016/j.datak.2007.07.003

Ettlie, J. E. (1983). Organizational policy and innovation among suppliers to the food processing sector. *Academy of Management Journal, 26*(1), 27–44. doi:10.2307/256133

Ettlie, J. E. (1986). *Implementing Manufacturing Technologies: Lessons from Experience. Managing Technological Innovation.* San Francisco, CA: Jossey-Bass Publishers.

Ettlie, J. E. (1990). What makes a manufacturing firm innovative? *Academy of Management Executive, 4*(4), 7–20.

Ettlie, J. E., & Bridges, W. P. (1982). Environmental uncertainty and organizational technological policy. *IEEE Transactions on Engineering Management, EM-29*, 2–10.

Eze, U. C. (2006). The Impact of Firm and Business Environment Contexts on E-Business Value. In *Proceedings of MWAIS 2006 Conference held on May 5-6 in Grand Rapids, Michigan, USA.*

Eze, U. C., & Gilbert, A. L. (2004). The significance of e-Commerce to firms' operations in relation to business location. *Australiasian Journal of Information Systems, 12*, 75–93.

Eze, U. C., & Kam, T. S. (2001). The impact of enterprise strategy and enterprise structure on the level of e-commerce deployment in enterprises in Singapore. *Journal of Global Information Technology Management, 4*(2), 38–54.

Fallon, M., & Moran, P. (2000). Information Communication Technology (ICT) and Manufacturing SMEs'. In *Proceedings of the 2000 Small Business and Enterprise Development Conference* (pp. 100-109). Manchester University.

Fan, M., Stallaert, J., & Whinston, A. B. (2000). The Internet and the future of financial markets. *Communications of the ACM, 43*, 83–88. doi:10.1145/353360.353368

Fasolo, B., McClelland, G. H., & Lange, K. A. (2005). The effect of site design and interattribute correlations on interactive web-based decisions. In C.P. Haugtvedt, K. Machleit, & R. Yalch (Eds.), *Online consumer psychology: understanding and influencing behavior in the virtual world* (pp. 325-344). Lawrence Erlbaum Associates, Inc.

Faustmann, G. (2000). Configuration for Adaptation - A Human-centered Approach to Flexible Workflow Enactment. *Computer Supported Cooperative Work: The Journal of Collaborative Computing, 9*(3-4), 413–434. doi:10.1023/A:1008719831436

Feeny, D. F., & Ives, B. (1990). In search of sustainability: Reaping long-term advantage from investments in information technology. *Journal of Management Information Systems, 7*(1), 27–46.

Feindt, S., Jeffcoate, J., & Chappell, C. (2002). Identifying Success Factors for Rapid Growth in SME E-commerce. *Small Business Economics, 19*(1), 51–62. doi:10.1023/A:1016165825476

Feng, J., Lazar, J., & Preece, J. (2004). Empathy and online interpersonal trust: A fragile relationship. *Behaviour & Information Technology, 23*(2), 97–106.

Ferguson, G. A. (1989). *Statistical analysis in psychology and education* (5 ed.): New York; London: McGraw-Hill

Ferle, C., Edwards, S., & Mizuno, Y. (2002). Internet diffusion in Japan: Cultural consideration. *Journal of Advertising Research*, (March-April): 65–79.

Fernandes, T. (1995, May). *Global User Interface Design.* Paper presented at the ACM Conference on Human Factors in Computing Systems—CHI '95 Mosaic of Creativity, New York.

Fichman, R. G. (2004). Going beyond the dominant paradigm for information technology innovation research: emerging concepts and methods. *Journal of the Association for Information Systems, 5*(8), 314–355.

Fillis, I., Johannson, U., & Wagner, B. (2004). Factors impacting on e-business adoption and development in the smaller firm. *International Journal of Entrepreneurial Behaviour & Research, 10*(3), 178–191. doi:10.1108/13552550410536762

Finextra (2006). Security fears scare off US customers from online banking, shopping. Retrieved July 17, 2007, from: http://www.finextra.com/fullstory.asp?id=16204

Fischer, G., Giaccardi, E., Ye, Y., Sutcliffe, A. G., & Mehandjiev, N. (2004). Meta-Design: A Manifesto for End-User Development . *Communications of the ACM, 47*(9), 33–37. doi:10.1145/1015864.1015884

Fischer, L. (Ed.). (2001). *Workflow Handbook 2001, Workflow Management Coalition, Future Strategies.* Lighthouse Point, Florida.

Fishbein, M., & Ajzen, I. (1975). *Belief, attitude, intention and behavior: an introduction to theory and research.* Reading, MA: Addison-Wesley.

Flanagin, A. J. (2000). Social pressures on organizational Website adoption. *Human Communication Research, 26*(4), 618–646. doi:10.1111/j.1468-2958.2000.tb00771.x

Flanagin, A. J., & Metzger, M. J. (2001). Internet use in the contemporary media environment. *Human Communication Research, 27*, 153–181. doi:10.1093/hcr/27.1.153

Fletcher, W. (1999). Buying the groceries via home shopping is not a fresh concept. *Marketing* (May), 16–17.

Floyd, S. (1988). A micro level model of information technology use by managers. In U.E. Gattiker (Ed), *Studies in Technological Innovation and Human Resources (1): Managing Technology Development* (pp. 123-142). Berlin & New York: Walter de Gruyter.

Fock, H. (2000). *Cultural influences on marketing communication on the World Wide Web.* Paper presented at the Multicultural Marketing Conference, Hong Kong, September.

Fodness, D., & Murray, B. (2007). Passengers' expectations of airport service quality. *Journal of Services Marketing, 21*(7), 492-506.

Fogg, B. J. (2002). *Stanford Guidelines for Web Credibility A Research Summary from the Stanford Persuasive Technology Lab*, from http://www.webcredibility.org/guidelines

Fogg, B. J., Marshall, J., Kameda, T., Solomon, J., Rangnekar, A., Boyd, J., et al. (2001). *Web credibility research: A method for online experiments and early study results.* Paper presented at the CHI '01 extended abstracts on Human factors in computing systems, Seattle,WA.

Ford, W., & Baum, M. S. (1997). *Secure electronic commerce*. Upper Saddle River, NJ: Prentice Hall.

Fornell, C. (1992). A national customer satisfaction barometer: The Swedish experience. *Journal of Marketing, 56*(1), 6-21.

Fornell, C., & Larcker, D. (1981). Structural equation models with unobserved variables and measurement error. *JMR, Journal of Marketing Research, 36*(3), 39–50. doi:10.2307/3151312

Fornell, C., & Larker, D. F. (1981). Evaluating structural equation models with unobservable variables and measurement error. *JMR, Journal of Marketing Research, 18*(1), 39–50. doi:10.2307/3151312

Fornell, C., Johnson, M. D., Anderson, E. W., Cha, J., & Bryant, B. E. (1996). The American customer satisfaction index: Nature, purpose, and findings. *Journal of Marketing, 60*(4), 7-18.

Forsythe, S.M., & Shi, B. (2003). Consumer patronage and risk perceptions in Internet shopping. *Journal of Business Research* 56, 867–875.

Fournier, S., & Yao, J. L. (1997). Reviving brand loyalty: A conceptualization within the framework of consumer-brand relationships. *International Journal of Research in Marketing, 14*(5), 451-472.

Fowler, M., & Scott, K. (1997). *UML Distilled.* New York: Addison-Wesley.

Fram, E.H., & Grady, D.B. (1997). Internet shoppers: is there a surfer gender gap?. *Direct Marketing* 59 (9), 46 – 51.

Frambach, R. T., & Schillewaert, N. (2002). Organizational innovation adoption: a multi-level framework of determinants and opportunities for future research. *Journal of Business Research, 55*(2), 163–176. doi:10.1016/S0148-2963(00)00152-1

Franklin, M., & Reiter, M. (1996). The design and implementation of a secure auction service. *In IEEE Transactions on Software Engineering, 22*(5), 302-312.

Frolick, N. M., & Ariyachandra, R. T. (2006). Business performance management: one truth. *Information Systems Management, 2006*(Winter), 41–48. doi:10.1201/1078.10580530/45769.23.1.20061201/91771.5

Fromkin, V., & Rodman, J. (1983). An Introduction to Language. New York, NY: CBS College Publishing.

Fulk, J., Schmitz, J., & Power, G. J. (1987). A Social Information Processing Model of Media Use in Organizations. *Communication Research, 14*, 520–552.

Funk, J. L. (2004). *Mobile Disruption - the technologies and applications driving the mobile internet.* Wiley.

Furrer, O., Thomas, H., & Goussevskaia, A. (2008). The structure and evolution of the strategic management field: A context analysis of 26 years of strategic management research. *International Journal of Management Reviews, 10*(1), 1–23. doi:10.1111/j.1468-2370.2007.00217.x

Galbraith, J. R. (1973). *Designing complex organization.* Reading, MA: Addison-Wesley.

Gale, B. T. (1994). *Managing customer value.* New York: The Free Press.

Galletta, D. F., & Hufnagel, E. M. (1992). A Model of End-User Computing Policy: Context, Process, Content and Compliance. *Information & Management, 22*, 1–18. doi:10.1016/0378-7206(92)90002-W

Gallivan, M. J. (2001). Organizational adoption and assimilation of complex technological innovations: development and application of a new framework. *The Data Base for Advances in Information Systems, 32*(3), 51–85.

Gao, M., Hyytinen, A., & Toivanen, O. (2006). Demand for and Pricing of Mobile Internet: Evidence from a Real-World Pricing Experiment. *Helsinki Center of Economic Research Discussion Paper No. 122* (October 2006).

Gao, S. (2005). Designing a limited supply online auction. In *Proceedings of The Best Conference Ever '05*, Vancouver.

Garcı'a-Dastugue, S. J., & Lambert, D. M. (2003). Internet enabled Coordination in the Supply Chain. *Industrial Marketing Management, 32*(3), 251–263. doi:10.1016/S0019-8501(02)00269-9

Gardner, R. (2003). *Games for business and economics, 2nd ed.* Hoboken, NJ: John Wiley and Sons Publishers.

Garrity, E. J., Glassberg, B., Kim, Y. J., Sanders, G. L., & Shin, S. K. (2005). An experimental investigation of Web-based information systems success in the context of electronic commerce. *Decision Support Systems, 39*(3), 485–503. doi:10.1016/j.dss.2004.06.015

Gatignon, H., & Xuereb, J. (1997). Strategic orientation of the firm and new product performance. *Journal of Business Research, 41*, 161–174.

Gefen, D. (2000). E-commerce: the role of familiarity and trust. *Omega: The International Journal of Management Science, 28*(6), 725–737.

Gefen, D. (2002). Customer loyalty in e-commerce. *Journal of the Association for Information Systems, 3*(1), 27-51.

Gefen, D. (2002). Reflections on the dimensions of trust and trustworthiness among online consumers. *The Data Base for Advances in Information Systems, 33*(3), 38.

Gefen, D., & Straub, D. W. (1997). Gender differences in the perception and use of e-mail: an extension to the technology acceptance model. *MIS Quarterly, 21*(4), 389–400. doi:10.2307/249720

Gefen, D., & Straub, D. W. (2000). The relative importance of perceived ease of use in IS adoption: A study of e-commerce adoption. *Journal of the Association for Information Systems, 1*(8).

Gefen, D., & Straub, W. D. (2004). Consumer trust in B2C e-Commerce and the importance of social presence: experiments in e-Products and e-Services. *Omega.Oxford, 32*(6), 407.

Gefen, D., Karahanna, E., & Straub, D. W. (2003). Trust and TAM in online shopping: An integrated model. *MIS Quarterly, 27*(1), 51-90.

Geng, X., Stinchcombe, M., & Whinston, A. B. (2001). Radically new product introduction using on-line auctions. *International Journal of Electronic Commerce, 5*(3), 169-189.

Gentry, C. R. (2003). Recycled returns. *Chain Store Age, 79*(1), 72-73.

Gerbing, D. W., & Anderson, J. C. (1988). An updated paradigm for scale development incorporating unidimensionality and its assessment. *JMR, Journal of Marketing Research, 15*, 186–192. doi:10.2307/3172650

Gerpott, T., Rams, W., & Schindler, A. (2001). Customer retention, loyalty, and satisfaction in the German mobile cellular telecommunications market. *Telecommunication Policy, 25*(4), 249-269.

Geyskens, I., Gielens, K., & Dekimpe, M. G. (2002). The market valuation of Internet channel additions. *Journal of Marketing, 66*, 102-119.

Gibbs, J., Kraemer, K. L., & Dedrick, J. (2003). Environment and policy factors shaping global E-Commerce diffusion: A cross-country comparison. *The Information Society, 19*, 5–18. doi:10.1080/01972240309472

Giddens, A. (1991). *The Consequences of Modernity.* Cambridge, UK: Polity Press.

Gilbert, A. L., & Eze, U. C. (2008). A Tale of Two Cities: Economic Geography in the Age of the Internet. Accepted for publication in *Nanyang Business Review, Singapore.*

Gilkeson, J., & Reynolds, K. (2003). Determinants of Internet auction success and closing price: An exploration study. *Psychology and Marketing, 20*(6), 537-566.

Gillespie, T. (2000). Trust and Risk in Internet Commerce. *Library Journal, 125*(10), 186.

Girard, T. Korgaonkar, P. , & Silverblatt, R. (2003). Relationship of type of product, shopping orientations and demographics with preference for shopping on the Internet. *Journal of Business and Psychology* 18(1), 101–120.

Glatard, T., Montagnat, J., Emsellem, D., & Lingrand, D. (2008). A Service-Oriented Architecture enabling dynamic service grouping for optimizing distributed workflow execution . *Future Generation Computer Systems, 24*, 720–730. doi:10.1016/j.future.2008.02.011

Glazer, R. (1993). Marketing in turbulent environments: Decision processes and the time-sensitivity of information. *JMR, Journal of Marketing Research, 30*(4), 509–521. doi:10.2307/3172694

Gommans, M., Krishan, K. S., & Scheddold, K. B. (2001). From brand loyalty to e-loyalty: A conceptual framework. *Journal of Economic and Social Research, 3*(1), 43–58.

Grandon, E., & Pearson, J. M. (2004). Electronic commerce adoption: An empirical study of small and medium US businesses. *Information & Management, 42*, 197–216.

Grant, R. M. (1996). Towards a knowledge-based theory of the firm. *Strategic Management Journal, 17*, 109–122. doi:10.1002/(SICI)1097-0266(199602)17:2<109::AID-SMJ796>3.0.CO;2-P

Green, N., Harper, R. H. R., Murtagh, G., & Cooper, G. (2001). Configuring the mobile user: sociological and industry views. *Personal and Ubiquitous Computing, 5*, 146–156. doi:10.1007/s007790170017

Gregg, D., & Walczak, S. (2003). E-commerce auction agents and online-auction dynamics. *Electronic Markets, 13*(3), 242-251.

Gregor, S., Hart, D., & Martin, N. (2007). Enterprise architecture: Enablers of business strategy and IS/IT alignment in government. *Information Technology & People, 20*(2), 2007. doi:10.1108/09593840710758031

Gregory, G., & Munch, J. (1997). Cultural values in international advertising: An examination of familial norms and roles in Mexico. *Psychology and Marketing, 14*(2), 99–119.

Grenci, R. T., & Todd, P. A. (2002). Solutions-Driven Marketing. *Communications of the ACM, 45*(2), 64–71. doi:10.1145/504729.504730

Grewal, D., Monroe, K. B., & Krishnan, R. (1998). The effects of price comparison advertising on buyers' perceptions of acquisition value and transaction value. *Journal of Marketing, 62*(2), 46-59.

Grewal, G., Munger, J. L., Iyer, G. R., & Levy, J. (2003). The influence of Internet retailing factors on price expectations. *Psychology and Marketing, 20*(6), 477–493.

Griffin, J. (1996). The Internet's expanding role in building customer loyalty. *Direct Marketing, 59*(7), 50-53.

Griffin, K., Ladd, P., & Whitehead, R. (1997). Internet commerce: Security is still a concern. *The Review of Accounting Information Systems, 2*(1), 67-71.

Grigori, D., Casati, F., Dayal, U., & Shan, M. C. (2001). Improving Business Process Quality through Exception Understanding, Prediction, and Prevention. In P. Apers, P. Atzeni, S. Ceri, S. Paraboschi, K. Ramamohanarao, & R. Snodgrass (eds.), *Proceedings of 27th International Conference on Very Large Data Bases, VLDB01* (pp. 159-168). Morgan Kaufmann.

Gronroos, C. (1984). A service quality model and its market implications. *European Journal of Marketing, 18*(4), 36-44.

Grover, S. K., & Lederer, A. L. (2004). The Impact of Industry Contextual Factors on IT Focus and the Use of IT for Competitive Advantage. *Information & Management, 41*(7), 899–919. doi:10.1016/j.im.2003.08.018

Grover, V. (1993). An empirically derived model for the adoption of customer-based interorganizationl systems. *Decision Sciences, 24*(3), 603–640. doi:10.1111/j.1540-5915.1993.tb01295.x

Grow, B. (2002). Excess inventory? EBay to the rescue. *Business Week*, 8.

Gruber, T. R. (1993). *Toward principles for the design of ontologies used for knowledge sharing.* Paper presented at the Knowledge Representation '94.

Gudes, E., & Tubman, A. (2002). AutoWF: A Secure Web Workflow System Using Autonomous Objects . *Data & Knowledge Engineering, 43*(1), 1–27. doi:10.1016/S0169-023X(02)00058-7

Gudes, E., Olivier, M. S., & van der Riet, R. P. (1999). Modeling, Specifying and Implementing Workflow Security in Cyberspace . *Journal of Computer Security, 7*(4), 287–315.

Guimaraes, T., Gupta, Y. P., & Rainer, R. K. (1999). Empirically Testing the Relationship Between End-User Computing Problems and Information Center Success Factors. *Decision Sciences*, (Spring): 393–413. doi:10.1111/j.1540-5915.1999.tb01615.x

Gunasekaran, A., Okko, P., Martikainen, T., & Yli-Olli, P. (1996). Improving productivity and quality in small and medium enterprises: Cases and analysis. *International Small Business Journal, 15*(1), 59–72. doi:10.1177/0266242696151004

Gurbaxani, V., & Mendelson, H. (2001). An Integrative Model of Information Systems Spending Growth. *Information Systems Research, 1*(1), 23–46. doi:10.1287/isre.1.1.23

Hacker, M. E. & Brotherton, P. A. (1998). Designing and installing effective performance measurement systems. *Industrial Engineering Solutions*, 18-23.

Haddon, L. (2001). *Domestication and mobile telephony.* Presented at the conference "Machines that Become Us," Rutgers University, April 18-19, New Jersey.

Hagel, J., & Brown, J. S. (2002). *Out of the Box: Strategies for Achieving Profits Today & Growth Tomorrow Through Web Services.* HBS Press.

Hagen, C., & Alonso, G. (2000). Exception Handling in Workflow Management Systems. *IEEE Transactions on Software Engineering, 26*(10), 943–958. doi:10.1109/32.879818

Hagmann, C., & McCahon, C. (1993). Strategic information systems and competitiveness. *Information & Management, 25*, 183–192. doi:10.1016/0378-7206(93)90067-4

Hair, J. F., Anderson, R. E., Tatham, R. L., & Black, W. C. (1998). *Multivariate Data Analysis, Fifth edition.* Delhi: Pearson Education.

Halstead, D., Hartman, D., & Schmidt, S. L. (1994). Multi-source effects on the satisfaction formation process. *Journal of the Academy of Marketing Science, 22,* 114-129.

Hammer, M. (1996). *Beyond Reengineering.* New York: Harper Business.

Harman, K., & Koohang, A. (2005/2006, Winter). Frequency of Publication and Topical Emphasis of Knowledge Management Books Versus Doctoral Dissertations: 1983-2005. *Journal of Computer Information Systems, 46*(2), 64–68.

Harrington, J., Esseling, E., & Van Nimwegen, H. (1997). *Business Process Improvement Workbook.* New York: McGraw-Hill.

Harris, R. W. (2000). Schools of thought in research into end-user computing success. *Journal of End User Computing, 12*(1), 24–34.

Harrison, J. G. (1997). *Enhancements to the data mining process.* Unpublished Ph.D., Stanford University, CA.

Härtel, C. E. (1999). Development and test of a two-stage model of performance appraisal. *Asia Pacific Journal of Human Resources, 37*(2), 76–91. doi:10.1177/103841119903700207

Hassanein, K., & Head, M. M. (2004). Manipulating social presence through the Web interface and its impact on consumer attitude towards online shopping. Paper presented at the McMaster eBusiness Research Centre (MeRC)Working Paper Series, Hamilton, ON, Canada.

Haubl, G., & Trifts, V. (2000). Consumer decision making in online shopping environments: The effects of interactive decision aids. *Marketing Science, 19*(1), 14–21. doi:10.1287/mksc.19.1.4.15178

Hauble, G., & Murray, K. (2003). Preference Construction and Persistence in Digital Marketplaces: The Role of Electronic Recommendation Agents. *Journal of Consumer Psychology, 13*(1), 75–91. doi:10.1207/153276603768344807

Hauguel, P., & Jackson, N. (2001). Outward-looking Supply-chain Strategy. *European Business Journal, 13*(3), 113–118.

Hauser, R. F., Friess, M., Kuster, J. M., & Vanhatalo, J. (2008). An incremental approach to the analysis and transformation of workflows using region trees. *IEEE Transactions on Systems, Man and Cybernetics. Part C, Applications and Reviews*, *38*(3), 347–359. doi:10.1109/TSMCC.2008.919178

Haynes, P. J., Becherer, R. C., & Helms, M. M. (1998). Small and Mid-sized Businesses and Internet Use: Unrealized Potential? *Internet Research*, *8*(3), 229–235. doi:10.1108/10662249810217786

He, F., & Mykytyn, P. P. (2007). Decision Factors for the Adoption of an Online Payment System by Customers. *International Journal of E-Business Research*, *3*(4), 1–32.

Heck, E., & Vervest, P. (1998). How should CIOs deal with Web-based auctions?. *Communications of the ACM*, *41*(7).

Heijden van der, H. & Ogertschnig, M. & van der Gaast, L. (2005). Effects of Context Relevance and Perceived Risk on User Acceptance of Mobile Information Services. *ECIS 2005.*

Heinl, P., Horn, S., Jablonski, S., Need, J., Stein, K., & Teschke, M. (1999). *A Comprehensive Approach to Flexibility in Workflow Management Systems.* Paper presented at the WACC'99, San Francisco, CA.

Heinonen, K., & Pura, M. (2006). *Classifying Mobile Services.* Paper Presented at Helsinki Mobility Roundtable, 1-2 June 2006, Helsinki.

Helfat, C., & Peteraf, M. A. (2003). The Dynamic Resource-Based View: Capability Lifecycles. *Strategic Management Journal*, *24*, 997–1010. doi:10.1002/smj.332

Henderson, J. C., & Venkatraman, N. (1993). Strategic alignment: Leveraging information technology for transforming organizations. *IBM Systems Journal*, *32*(1), 4–16.

Henriksen, H. Z. (2006). Motivators for IOS adoption in Denmark. *Journal of Electronic Commerce in Organizations*, *4*(2), 25–39.

Herbst, J., & Karagiannis, D. (2000). Integrating Machine Learning and Workflow Management to support Acquisition and Adaptation of Workflow Models. *International Journal of Intelligent Systems in Accounting Finance & Management*, *9*(1), 67–92. doi:10.1002/1099-1174(200006)9:2<67::AID-ISAF186>3.0.CO;2-7

Hevner, A., March, S. T., Park, J., & Ram, S. (2004). Design Science in Information Systems Research. *MIS Quarterly*, *28*(1), 75–105.

Hewitt-Dundas, N., & Roper, S. (1999). Overcoming customer dependency: A case study of the strategies of small consumer food producers in Northern Ireland. *International Small Business Journal*, *17*(4), 49–65. doi:10.1177/0266242699174003

Heyman, J. E., Orhun, Y., & Ariely, D. (2004). Auction fever: The effect of opponents and quasi-endowment on product evaluations. *Journal of Interactive Marketing*, *18*(4), 7-21.

Hildebrandt, L. (1987). Consumer retail satisfaction in rural areas: A reanalysis of survey data. *Journal of Economic Psychology*, *8*, 19–42. doi:10.1016/0167-4870(87)90004-3

Hillier, M. (2003). The role of cultural context in multilingual website usability. *Electronic Commerce Research and Applications*, *2*, 2–14.

Hitt, L. M., & Brynjolfsson, E. (1994). The Three Faces of IT Value: Theory and Evidence. In *Proceedings of the International Conference on Information Systems* (pp. 263-277).

Hitt, L., & Brynjolfsson, E. (1997). Information Technology and Internal Firm Organization: An Exploratory Analysis. *Journal of Management Information Systems*, *14*(2), 81–101.

Hochmuth, P. (2005, November 11). Cisco service targets small business. *New World (New Orleans, La.)*, *22*(45), 8–9.

Hoffman, D. L., & Novak, T. P. (1996). Marketing in Hypermedia Computer-Mediated Environment: Conceptual Foundations. *Journal of Marketing*, *60*, 50–68. doi:10.2307/1251841

Höflich, J. R., & Rössler, P. (2001). Mobile schriftliche Kommunikation oder: E -Mail für das Handy. *Medien & Kommunikationswissenschaft*, *49*, 437–461.

Hofstede, G. (2005). *Cultures and Organizations*: Harper Collins Business, London.

Hogarth, R. (1987). *Judgment and Choice* (2nd ed). New York: John Wiley and Sons.

Holsapple, C. W., & Whinston, A. B. (1996). *Decision Support Systems*. New York: West publishing company.

Horn, D. B., Olson, J. S., & Karasik, L. (2002). The effects of spatial and temporal video distortion on lie detection performance. *Conference on Human Factors in Computing Systems - Proceedings*, 714-715.

Horrocks, I., Patel-Schneider, P. F., Boley, H., Tabet, S., Grosof, B., & Dean, M. (2004). *SWRL A Semantic Web Rule Language Combining OWL and RuleML*. W3C Submission, 21 May 2004. Retrieved from http://www.w3.org/Submission/2004/SUBM-SWRL-20040521/

Hsieh, C., & Lin, B. (2004). Impact of standarization on EDI in B2B development. *Industrial Management & Data Systems*, *104*(1/2), 68–77. doi:10.1108/02635570410514106

http://www.blonnet.com/catalyst/2002/03/07/stories/2002030700120200.htm.

Hu, X., Lin, Z., & Zhang, H. (2003). Trust promoting seals in electronic markets: An exploratory study of their effectiveness for online sales promotion. *Journal of Promotion Management*, *9*(1/2), 163-180.

Hu, X., Lin, Z., Whinston, A. B., & Zhang, H. (2004). Hope or hype: On the viability of escrow services as trusted third parties in online auction environments. *Information Systems Research*, *15*(3), 236-250.

Huang, C. D., & Hu, Q. (2007). Achieving IT – business strategic alignment via enterprise-wide implementation of balanced scorecards. *Information Systems Management*, *24*, 173–184. doi:10.1080/10580530701239314

Huang, Z., Chen, L., & Frolick, M. N. (2002). Integrating web-based data into a data warehouse. *Information Systems Management*, *19*(1), 23–34. doi:10.1201/1078/43199.19.1.20020101/31473.4

Huselid, M. A., & Becker, B. E. (1997). The impact high performance work systems, implementation effectiveness, and alignment with strategy on shareholder wealth. *Academy of Management Proceedings* (pp. 144-148).

Hwang, S., Wang, H., Tang, J., & Srivastava, J. (2007). A probabilistic approach to modeling and estimating the QoS of web-services-based workflows. *Information Sciences*, *177*, 5484–5503. doi:10.1016/j.ins.2007.07.011

Iacovelli, A., Souveyet, C., & Rolland, C. (2008) Method as a Service (MaaS). In *Proceedings of the IEEE Second International Conference on Research Challenges in Information Science (RCIS'2008), Marrakech, Morocco.*

Iacovou, C. L., Benbasat, I., & Dexter, A. S. (1995). Electronic data interchange and small organizations: Adoption and impact of technology. *MIS Quarterly*, *19*(4), 465–485. doi:10.2307/249629

Igbaria, M., Zinatelli, N., Cragg, P., & Cavaye, A. L. M. (1997). Personal computing acceptance factors in small firms: A structural equation model. *MIS Quarterly*, *21*(3), 279–302. doi:10.2307/249498

Iivari, J. (1992). The organizational fit of information systems. *Journal of Information Systems*, *2*, 3–29. doi:10.1111/j.1365-2575.1992.tb00064.x

Im, K. S., Dow, K. E., & Grover, V. (2001). Research report: A reexamination of IT investment and the market value of the firm - An event study methodology. *Information Systems Research*, *12*(1), 103–117. doi:10.1287/isre.12.1.103.9718

Intel Information Technology White Paper. (May 2003). *Defining the Value of E-business: Seventeen Standard Measures* (pp. 1-14). Intel Publishing.

Internet Crime Complaint Center. (2007). *The Internet crime complaint center hits 1 million!* Retrieved February 13, 2008, from http://www.ic3.gov/media/2007/070613.htm

Internet Fraud Watch. (2005). *Internet fraud statistics report*. Retrieved September 24, 2007, from http://www.fraud.org/internet/intstat.htm

Internet Fraud Watch. (2007). *Internet fraud tips*. Retrieved February 13, 2008, from http://www.fraud.org/tips/internet/general.htm

Jacoby, J., & Kyner, D. B. (1973). Brand loyalty vs repeat purchasing behavior. *Journal of Marketing Research, 10,* 1-9.

Jai-Yeol, S., Narasimhan, S., & Riggins, F. J. (2005). Effects of relational factors and channel climate on EDI usage in the customer-supplier relationship. *Journal of Management Information Systems, 22*(1), 321.

Jamieson, S. (2004). Likert scales: How to (ab)use them. *Medical Education, 38,* 1217-1218.

Jarvenpaa, S. L., & Tractinsky, N. (1999). Consumer trust in an internet store a cross-cultural validation. *Journal of Computer-Mediated Communication, 5*(2), 1–35.

Jarvenpaa, S., Tractinsky, L. N., & Vitale, M. (2000). Consumer trust in an Internet store. *Information Technology and Management, 1*(1/2), 45-71.

Jedetski, J., Adelman, L., & Yeo, C. (2002). How Web Site Decision Technology Affects Consumers. *IEEE Internet Computing, 6*(2), 72–79. doi:10.1109/4236.991446

Jeyaraj, A., Rottman, J. W., & Lacity, M. C. (2006). A review of the predictors, linkages, and biases in IT innovation adoption research. *Journal of Information Technology, 21*(1), 1–23. doi:10.1057/palgrave.jit.2000056

Jiménez, J., & Polo, Y. (1998). International diffusion of a new tool: The case of electronic data interchange in the retailing sector. *Research Policy, 26*(7-8), 811–827. doi:10.1016/S0048-7333(97)00045-0

Jinling, C., & Guoping, X. (2005). Comprehensive Evaluation of E-commerce Websites Based on Concordance Analysis. In *Proceedings of the 2005 IEEE International Conference on e-Business Engineering* (pp. 179-182).

Johannes, L. (2000). Competing online, drugstore chains virtually undersell themselves --- their Web sites charge less than their own stores, with some strings attached. *Wall Street Journal,* (January 10, 2000) B1.

Johnson, C. A., et al. (2003). *US eCommerce Overview: 2003 - 2008.* Cambridge, MA: Forrester Research, Inc.

Johnson, E. J., & Payne, J. W. (1985). Effort and accuracy in choice. *Management Science, 31*(4), 394–414. doi:10.1287/mnsc.31.4.395

Johnson, E. J., Moe, W. W., Fader, P. S., Bellman, S., & Lohse, G. L. (2004). On the Depth and Dynamics of Online Search Behaviour. *Management Science, 50*(3), 299–309. doi:10.1287/mnsc.1040.0194

Johnson, J., Boucher, K. D., Conners, K., & Robinson, J. (2001). The Criteria for Success. *Software Magazine, 21*(1), s3–s11.

Johnson, J., Boucher, K., Connors, K., & Robinson, J. (2001). Collaboration: Development and Management Collaborating of Project Success. *Software Magazine.* Retrieved from http://www.softwaremag.com/archive/2001feb/CollaborativeMgt.html

Johnson-George, C., & Swap, W. C. (1982). Measurement of specific interpersonal trust: Construction and validation of a scale to assess trust in a specific other. *Journal of Personality and Social Psychology, 43*(6), 1306–1307.

Johnston, H. R., & Vitale, M. R. (1988). Creating competitive advantage with interorganizational information systems. *MIS Quarterly,* ▪▪▪, 152–165.

Jones, D. R., & Brown, D. (2003). The division of labor between human and computer in the presence of decision support system advice. *Decision Support Systems, 33*(4), 375–388. doi:10.1016/S0167-9236(02)00005-2

Jones, J.M., & Vijayasaratgy L.R. (1998). Internet consumer catalog shopping: findings from an exploratory study and directions for future research. *Internet Research: Electronic Networking Applications and Policy* 8(4) 322 – 330.

Jones, M. A. (1999). Entertaining shopping experiences: An exploratory investigation. *Journal of Retailing and Consumer Services,* 6.

Jones, M. C., & Beatty, R. C. (1998). Towards the development of measures of perceived benefits and compatibility of EDI: a comparative assessment of competing first order factor models. *European Journal of Information Systems, 7*(3), 210–220. doi:10.1057/palgrave.ejis.3000299

Jones, T. O., & Sasser, W. E. (1995). Why satisfied customers defect? *Harvard Business Review, 73*(6), 88-99.

Jöreskog, K. (1971). Statistical analysis of sets of cogeneric tests. *Psychometrika, 36,* 109–133. doi:10.1007/BF02291393

Jöreskog, K., & Sörbom, D. (1993). LISREL 8 structural equation modelling with the simplis comand language. *Scientific software International*. Chicago-Illinois.

Josefek, R. A. Jr, & Kauffman, R. J. (1997). Dark pockets and decision support: The information technology value cycle in efficient markets. *Electronic Markets*, *7*(3), 36–42. doi:10.1080/10196789700000036

Julia, S., de Oliveira, F. F., & Valette, R. (2008). Real time scheduling of Workflow Management Systems based on a p-time Petri net model with hybrid resources. *Simulation Modelling Practice and Theory*, *16*(4), 462–482. doi:10.1016/j.simpat.2008.01.006

Julien, P., Joyal, A., Deshaies, L., & Ramangalahy, C. (1997). A typology of strategic behavior between small and medium sized exporting businesses: A case study. *International Small Business Journal*, *15*(2), 33–50. doi:10.1177/0266242697152002

Kacen, J., & Lee, J. A. (2002). The influence of culture on consumer impulsive buying behavior. *Journal of Consumer Psychology*, *12*(2), 163–176.

Kaefer, F., & Bendoly, E. (2004). Measuring the impact of organizational constraint on the success of business e-commerce efforts: A transactional focus. *Information & Management*, *41*, 529–541. doi:10.1016/S0378-7206(03)00088-0

Kagal, L., & Finin, T. (2005). Modeling Communicative Behavior using Permissions and Obligations. In F. Dignum, R. van Eijk, & M. P. Huget (Eds.), *Developments in Agent Communication*.

Kagel, J. H. (1995). A survey of experimental research. In J. H. Kagel & A. E. Roth (Eds.), *The handbook of experimental economics* (pp. 501-586). Princeton, NJ: Princeton University Press.

Kaiser, H. F. (1970). A second-generation little jiffy. *Psychometrika*, *35*, 401–415. doi:10.1007/BF02291817

Kakati, P. R., & Dhar, U. R. (2002). Competitive strategies and new venture performance. *Vikalpa*, *27*(3), 13–26.

Kalakota, R., & Robinson, M. (2001). *e-Business 2.0: Roadmap for Success* (2nd ed.). Boston, MA: Addison-Wesley.

Kalakota, R., & Robinson, M. (2002). *M-Business: The Race to Mobility*. New York: McGraw-Hill.

Kalakota, R., Robinson, M., & Gundepudi, P. (2003). Mobile applications for adaptive supply chains: A landscape analysis. In E.-P. Lim & K. Siau (Eds.), *Advances in Mobile Commerce Technologies* (pp. 298-311). Hershey, PA: Idea Group Publishing.

Kamins, M., Dreze, X., & Folkes, V. S. (2004). Effects of seller-supplied prices on buyers' product evaluations: Reference prices in an Internet auction context. *Journal of Consumer Research*, *30*(4), 622-628.

Kammerer, M. (2000). Die Bestimmung von Vertrauen in Internetangebote. Zurich: Lizenziatsarbeit der Philisophischen Fakultaet der Universitaet Zurich.

Kandampully, J., & Suhartanto, D. (2000). Customer loyalty in the hotel industry: The role of customer satisfaction and image. *International Journal of Contemporary Hospitality Management*, *12*(6), 346-351.

Kang, K. S., & Corbitt, B. (2001). Effectiveness of graphical components in web site e-commerce application- A cultural perspective. *The Electronic Journal Systems in Developing Countries*.

Kannan, P. K., & Kopalle, P. (2001). Dynamic pricing on the Internet: Importance and implications for consumer behavior. *International Journal of Electronic Commerce*, *5*(3), 63-83.

Kao, D., & Decou, J. (2003). A strategy-based model for e-commerce planning. *Industrial Management & Data Systems*, *103*(4), 238–252. doi:10.1108/02635570310470638

Kaplowitz, M., Hadlock, T., & Levine, R. (2004). A comparison of Web and mail survey response rates. *Public Opinion Quarterly*, *68*(1), 94-101.

Karahanna, E., Straub, D. W., & Chervany, N. L. (1999). Information technology adoption across time: a cross-sectional comparison of pre-adoption and post-adoption beliefs. *MIS Quarterly*, *23*(2), 183–213. doi:10.2307/249751

Karlson, B., Bri, A., Link, J., Lönnqvist, P., & Norling, C. (2003). *Wireless Foresight: Scenarios of the Mobile World in 2015*. John Wiley and Sons.

Kasper, J. D. (1988). On problem perception, dissatisfaction and brand loyalty. *Journal of Economic Psychology, 9*(3), 387-397.

Kast, F., & Rosenzweig, J. (1973). *Contingency Views of Organization and Management.* Chicago: Science Research Associates.

Kau, A. K., Tang, Y. E., & Ghose, S. (2003). Typology of online shoppers. *The Journal of Consumer Marketing, 20*(2/3), 139-156.

Kauffman, R. J., & Kriebel, C. H. (1988). Modeling and measuring the business value of information technologies. In P.A. Strassman, P. Berger, E.B. Swanson, C.H. Kriebel and R.J. Kauffman (Eds.), *Measuring the Business value of Information Technologies.* Washington, D.C.: ICIT Press.

Kauffman, R. J., & Walden, E. A. (2001). Economics and electronic commerce: A survey and directions for research. *International Journal of Electronic Commerce, 5*(4), 5.

Kauffman, R., & Wood, C. (2005). The effects of shilling on final bid prices in online auctions. *Journal of Electronic Commerce Research and Applications.*

Kearney, A. T. (2001). *Satisfying the experienced on-line shopper: Global e-shopping survey.* London: A.T. Kearney Ltd., Inc.

Kearns, G. S., & Sabhewarl, R. (2007). Antecedents and consequences of information systems planning integration. *IEEE Transactions on Engineering Management, 54*(4), 628–643. doi:10.1109/TEM.2007.906848

Keen, G., & Knapp, M. (1996). *Every Manager's Guide to Business Processes: A Glossary of Key Terms and Concepts for Today's Business Leader.* Boston, MA: Harvard Business School Press.

Keeney, R. L. (1999). The value of internet commerce to the customer. *Management Science, 45*(4), 533–542. doi:10.1287/mnsc.45.4.533

Keevil, B. (1998, September 24 to 26, 1998). *Measuring the Usability Index of Your Web Site.* Paper presented at the the annual Special Interest Group on Systems Documentation conference SIGDOC98, Quebec City, Province of Quebec, Canada.

Keidl, M., & Kemper, A. (2004). A Framework for Context-Aware Adaptable Web Services. In *Proceedings of the 9th International Conference on Extending Database Technology (EDBT'2004), Crete, Greece.*

Kenney, A. R., McGovern, N. Y., Martinez, I. T., & Heidig, L. J. (2003). Google Meets eBay-What Academic Librarians Can Learn from Alternative Information Providers. *D-Lib Magazine, 9*(6). doi:10.1045/june2003-kenney

Kerlinger, F. N. (1986). *Foundations of Behavioral Research.* CBS College Publishing.

Khandwalla, P. N. (1977). *The Design of Organization.* New York: Harcourt Brace Jovanovich.

Kickul, J., & Gundry, L. K. (2001). Breaking through boundaries for organizational innovation: New managerial roles and practices in e-commerce firms. *Journal of Management, 27,* 347–361. doi:10.1016/S0149-2063(01)00095-2

Kikuchi, H., Harkavy, M., & Tygar, J. (1998). *Electronic auctions with private bids.* In *Proceedings of the 3rd USENIX Workshop on Electronic Commerce,* (pp. 61-73).

Kim, C. S., & Peterson, D. K. (2001). Developers' perceptions of information systems success factors. *Journal of Computer Information Systems, 41*(2), 29–36.

Kim, D. J., Ferrin, D. L., & Rao, H. R. (2003). A Study of the Effect of Consumer Trust on Consumer Expectations and Satisfaction: The Korean Experience. *Proceedings of the ACM Conference on Electronic Commerce, 5,* 310-315.

Kim, E., & Tadisina, S. (2005). Factors Impacting Customers' Initial Trust in E-Businesses: An Empirical Study. *Proceedings of the Hawaii International Conference on System Sciences, Big Island, HI, January 2005,* 1-10.

Kim, J. (2001). *Analyzing mobile Internet users.* Paper Presented at the CHI 2001 Workshop on mobile communications.

Kim, J. W., Lee, B. H., Shaw, M. J., Chang, H. -L., & Nelson, M. (2001). Application of decision-tree induction techniques to personalized advertisements on Internet storefronts. *International Journal of Electronic Commerce, 5*(3), 45-62.

Kim, N., & Srivastava, R. K. (1998). Managing intraorganizational diffusion of technological innovations. *Industrial Marketing Management, 27*, 229–246. doi:10.1016/S0019-8501(97)00059-X

Kim, N., Pae, J. H., Han, J. K., & Srivastava, R. K. (in press). Utilization of business technologies: Managing relationship-based benefits for buying and supplying firms. *Industrial Marketing Management.*

Kimery, K. M., & McCord, M. (2006). Signals of trustworthiness in e-commerce: Consumer understanding of third-party assurance seals. *Journal of Electronic Commerce in Organizations, 4*(4), 52-74.

Kincaid, J. W. (2003). *Customer Relationship Management: Getting it Right!* Upper Saddle River, NJ: Prentice Hall PTR.

King, R. C., & Gribbins, M. L. (2002). Internet technology adoption as an organisational event: an exploratory study across industries. In *Proceedings of the 35th Annual Hawaii International Conference on System Sciences* (Vol. 7, pp. 201). Washington DC: IEEE Computer Society.

King, R. C., Sen, R., & Xia, M. (2004). Impact of Web-based e-commerce on channel strategy in retailing, *International Journal of Electronic Commerce, 8*(3), 103-130.

King, W. R., & Sabherwal, R. (1992). The factors affecting strategic information systems applications: an empirical assessment. *Information & Management, 23*(2), 217–235. doi:10.1016/0378-7206(92)90046-I

King, W. R., & Teo, T. S. H. (1996). Key dimensions of facilitators and inhibitors for the strategic use of information technology. *Journal of Management Information Systems, 12*(4), 35–53.

King, W., Grover, V., & Hufnagel, E. (1989). Seeking Competitive Advantage Using Information-Intensive Strategies: Facilitators and Inhibitors. In *Information Technology and Management Strategy.* Eaglewood Cliffs, New Jersey: Prentice Hall.

Klarmann, J. (2001). *A Comprehensive Support for Changes in Organizational Models of Workflow Management Systems.* Paper presented at the 4th International Conference on Information Systems Modeling ({ISM} '01).

Klarreich, E. (2003). Sold to the latest bidder. *SIAM News, 36*(2).

Klein, B., & Leffler, K. B. (1981). The role of market forces in assuring contractual performance. *The Journal of Political Economy, 89*(4), 615-641.

Klein, M., & Dellarocas, C. (2000). A Knowledge-based Approach to Handling Exceptions in Workflow Systems. *Computer Supported Cooperative Work: The Journal of Collaborative Computing, 9*(3-4), 399–412. doi:10.1023/A:1008759413689

Klemperer, P. (1995). Competition when consumers have switching costs: An overview with applications to industrial organization macroeconomics and international trade. *Review of Economics Studies, 62*, 515-539.

Klemperer, P. (1999). Auction theory: A guide to the literature. *Journal of Economic Surveys, 13*(3), 227-286.

Klemperer, P. (Ed.). (2000). *The economic theory of auctions.* Cheltenham, UK: Edward Elgar.

Kline, R. B. (1998). *Principles and Practice of Structural Equation Modeling.* New York: Guilford Press.

Kohli, R., & Devaraj, S. (2003). Measuring information technology payoff: A meta-analysis of structural variables in firm-level empirical research. *Information Systems Research, 14*, 127–145. doi:10.1287/isre.14.2.127.16019

Kon, F., Costa, F., Blair, G., & Campbell, R. H. (2002). The Case for Relective Middleware. *Communications of the ACM, 45*(6), 33–38. doi:10.1145/508448.508470

Kooser, A. (2003, November). They Want You. *Entrepreneur, 31*(11), 34.

Korgaonkar, P., & Wolin, L. (1999). A multivariate analysis of Web usage. *Journal of Advertising Research, 39*(2), 53-68.

Korgaonkar, P., Silverblatt, R., & Girard, T. (2006). Online retailing, product classifications, and consumer preferences. *Internet Research, 16*(3), 267-287.

Kotler, P. (1982). *Marketing for nonprofit organizations* (2nd ed.). Englewood Cliffs, NJ: Prentice Hall.

Kotler, P. (2002). *Marketing Management.* Englewood Cliffs, NJ: Prentice Hall.

Kovar, S. E., Burke, K. G., & Kovar, B. R. (2000). Consumer responses to the CPA WEBTRUST assurance. *Journal of Information Systems, 14*(1), 17-35.

Kowtha, N. R., & Choon, T. W. I. (2001). Determinants of website development: a study of electronic commerce in Singapore. *Information & Management, 39*(3), 227–242. doi:10.1016/S0378-7206(01)00092-1

Koyuncu, C., & Bhattacharya, (2004). The impacts of quickness, price, payment risk, and delivery issues on on-line shopping. *Journal of Socio-Economics 33(2), 241-251.*

Kraemer, K. L., & Dedrick, J. (2002). Strategic use of the Internet and e-commerce: Cisco Systems. *The Journal of Strategic Information Systems, 11,* 5–29. doi:10.1016/S0963-8687(01)00056-7

Krauter, S. G., Kaluscha, E. A., & Fladnitzer, M. (2006). *Perspectives of Online Trust and Similar Constructs –A Conceptual Clarification.* Paper presented at the ICEC'06, Fredericton, Canada.

Krishna, V. (2002). *Auction theory.* San Diego, CA: Academic Press.

Kuan, K. K. Y., & Chau, P. Y. K. (2001). A perception-based model for EDI adoption in small businesses using a technology-organization-environment framework. *Information & Management, 38*(8), 507–521. doi:10.1016/S0378-7206(01)00073-8

Kuechler, W. L., & Vaishnavi, V. K. (2008). An expert system for dynamic re-coordination of distributed workflows . *Expert Systems with Applications, 34*(1), 551–563. doi:10.1016/j.eswa.2006.09.014

Kumar, A., & Wainer, J. (2005). Meta Workflows as a control and coordination mechanism for exception handling in workflow systems. *Decision Support Systems, 40*(1), 89–105. doi:10.1016/j.dss.2004.04.006

Kumar, A., & Zhao, J. L. (1999). Dynamic routing and operational controls in workflow management systems. *Management Science, 45*(2), 253–272. doi:10.1287/mnsc.45.2.253

Kumar, A., van der Aalst, W. M. P., & Verbeek, H. M. W. (2002). Dynamic Work Distribution in Workflow Management Systems: How to Balance Quality and Performance? *Journal of Management Information Systems, 18*(3), 157–193.

Kumar, M., & Feldman, S. (1998). Internet auctions. In *Proceedings of the 3rd USENIX Workshop on Electronic Commerce.*

Kumar, N., & Benbasat, I. (2002). Para-social presence and communication capabilities of a Web site: A theoretical perspective. *e-Service Journal, 1*(3).

Kumar, S., & Zahn, C. (2003). Mobile communications: Evolution and impact on business operations. *Technovation, 23*(6), 515–520. doi:10.1016/S0166-4972(02)00120-7

Kung, M. T., & Zhang, J. Y. (2008). Analysis of Business Process Models in Enterprise Web Services. *International Journal of E-Business Research, 4*(2), 69–87.

Kurgan, L. A., Swiercz, W., & Cios, K. J. (2002). *Semantic Mapping of XML Tags Using Inductive Machine Learning.* Paper presented at the ICMLA 2002.

Kurnia, S., & Johnston, R. (2005). The Dynamic Interactional Model of Inter-Organizational System Adoption: The Case of Category Management Adoption in Australia, *Asia Pacific . Management Review, 10*(1).

Kwak, H., Fox, R. J., & Zinkhan, G. M. (2002). What products can be successfully promoted and sold via the Internet? *Journal of Advertising Research, 2,* 23-38.

Kwon, T. H., & Zmud, R. W. (1987). *Unifying the Fragmented Models of Information Systems Implementation: Critical Issues in Information Systems Research.* New York: John Wiley.

LaBarbera, P. A., & Mazursky, D. (1983). A longitudinal assessment of consumer satisfaction/dissatisfaction. *Journal of Marketing Research, 20,* 393-404.

Lal, R., & Sarvary, M. (1999). When and how is the Internet likely to decrease price competition? *Marketing Science, 18*(4), 485-503.

Lancaster, K. (1990). The Economics of Product Variety: A Survey. *Marketing Science, 9,* 189–206. doi:10.1287/mksc.9.3.189

Landauer, T. K., & Dumais, S. T. (1997). A solution to Plato's problem: The Latent Semantic Analysis theory of the acquisition, induction, and representation of knowledge. *Psychological Review, 104,* 211–140. doi:10.1037/0033-295X.104.2.211

Larrick, R. P. (2004). Debiasing. In D.J. Koehler & N. Harvey (Eds.), *Blackwell Handbook of Judgment and Decision Making.* Oxford: Blackwell Publishing.

Lasica, J. D. (1999). *The confidence game.* Retrieved September 24, 2007, from http://www.jdlasica.com/idrive/TIS/confidence.html

Latin American SMEs Increasingly Adopting Mobile CRM, ERP Technologies IDC Mexico. (2008, July 31). *EmergingMarketsNOW,* 1.

Lau, H. C. W., Lee, C. K. M., Ho, G. T. S., Ip, W. H., Chan, F. T. S., & Ip, R. W. L. (2006). M-commerce to support the implementation of a responsive supply chain network. *Supply Chain Management, 11*(2), 169–178. doi:10.1108/13598540610652564

Lederer, A. L., & Mendelow, A. L. (1988). Convincing top management of the strategic potential of information systems. *MIS Quarterly, 12*(4), 526–536. doi:10.2307/249127

Lederer, A. L., Maupin, D. J., Sena, M. P., & Zhuang, Y. (2000). The technology acceptance model and the world wide web. *Decision Support Systems, 28,* 269–282. doi:10.1016/S0167-9236(00)00076-2

Lederer, A. L., Mirchandani, D. A., & Sims, K. (2001). The Search for Strategic Advantage from the World Wide Web. *International Journal of Electronic Commerce, 5*(4), 117–133.

Lee, B., & Barua, A. (1999). An integrated assessment of productivity and efficiency impacts of information technology investments: Old data, new analysis and evidence. *Journal of Productivity Analysis, 12,* 21–43. doi:10.1023/A:1007898906629

Lee, H. G., & Clark, T. H. (1997). Market process reengineering through electronic market systems: Opportunities and challenges. *Journal of Management Information Systems, 13*(3), 113–136.

Lee, H., & Suh, H. W. (2008). Estimating the duration of stochastic workflow for product development process. *International Journal of Production Economics, 111*(1), 105–117. doi:10.1016/j.ijpe.2007.01.003

Lee, M. K. O., & Turban, E. (2001). A trust model for consumer Internet shopping. *International Journal of Electronic Commerce, 6*(1), 75-91.

Lee, S. C., Pak, B. Y., & Lee, H. G. (2003a). Business value of B2B electronic commerce: The critical role of inter-firm collaboration. *Electronic Commerce Research and Applications, 2,* 350–361. doi:10.1016/S1567-4223(03)00003-6

Lee, Y., Kozar, K. A., & Larsen, K. R. T. (2003b). The technology acceptance model: past, present, and future. *Communications of the Association for Information Systems, 12,* 752–780.

Lefebvre, L. A., Langley, A., Harvey, J., & Lefebvre, E. (1992). Exploring the strategy-technology connection in small manufacturing firms. *Production and Operations Management, 1*(3), 1–17.

Legner, C. (2008). The evolution of B2B e-services from first generation e-commerce solutions to multichannel architectures. *Journal of Electronic Commerce in Organizations, 6*(2), 58–77.

Lehmann, D. R., Grupta, S., & Steckel, J. H. (1999). *Marketing Research.* New York: Addison-Wesley.

Lehtinen, U., & Lehtinen, J. R. (1982). *Service quality—A study of dimensions.* Working Paper. Service Management Institute, Helsinki, Finland.

Leonard, K. (2005) .It's not shopping cart abandonment, it's comparison shopping. Retrieved July 17, 2007, from Internet Retailer Web site: http://www.internetretailer.com.

Leung, L. (2001). College student motives for chatting on ICQ. *New Media & Society, 3*(4), 483–500. doi:10.1177/14614440122226209

Leung, L., & Wei, R. (2000). More than just talk on the move: Uses and gratifications of the cellular phone. *J& MC Quarterly, 77*, 308–320.

Levary, R., & Mathieu, R. G. (2000). Hybrid retail: Integrating e-commerce and physical stores. *Industrial Management, 42*(5), 6-13.

Levenburg, N. M. (2005, June). Does Size Matter? Small Firms' Use of E-Business Tools in the Supply Chain. *Electronic Markets, 15*(2), 94–105. doi:10.1080/10196780500083746

Levy, M., & Powell, P. (2000). Information systems strategy for small and medium sized enterprises: An organizational perspective. *The Journal of Strategic Information Systems, 9*(1), 63–84. doi:10.1016/S0963-8687(00)00028-7

Lewicki, R. J., & Bunker, B. B. (1995). Trust in Relationships: A Model of Trust Development and Decline. In I. B. B. B. J. Z. Rubin (Ed.), *Conflict, cooperation and justice* (pp. 133-173). San Francisco.

Lewis, J. D., & Weigert, A. (1985). Trust as a social reality. *Social Forces, 63*(4), 967–985.

Lewis, R. C., & Booms, B. H. (1983). The marketing aspects of service quality. In L. L. Berry, G. Shostack, & G. Upah (Eds), *Emerging perspectives in service marketing* (pp. 99-107). Chicago: American Marketing Association.

Leymann, H., & Roller, D. (2000). *Production Workflow: Concepts and Techniques*: Prentice Hall.

Lhose, G. L., Biolsi, K., Walker, N., & Reuter, H. H. (1994). A classification of visual representations. *Communications of the ACM, 37*(12), 36-49.

Li, H., Daugherty, T., & Biocca, F. (2001). Characteristics of Virtual Experience in Electronic Commerce: A Protocol Analysis. *Journal of Interactive Marketing, 15*(3). doi:10.1002/dir.1013

Li, L., Wu, B., & Yang, Y. (2005). *Agent-Based Ontology Mapping Towards Ontology Interoperability.* Paper presented at the Australian Joint Conference on Artificial Intelligence (AI'05), Sydney, Australia.

Liang, T., & Huang, J. (1998). An empirical study on consumer acceptance of products in electronic markets: A transaction cost model. *Decision Support Systems, 24*, 29-43.

Liljander, V., Polsa, P., & Forsberg, K. (2007). Do Mobile CRM Services Appeal to Loyalty Program Customers? *International Journal of E-Business Research, 3*(2), 24.

Limthanmaphon, B., & Zhang, Y. (2004). Web Service Composition Transaction Management. In *Proceedings of the 14th Australasian Database Conference (ADC'2004), Dunedin, New Zealand.*

Lin, D. (1998). *An information-theoretic definition of similarity.* Paper presented at the 15th International Conference on Machine Learning, Madison, WI, USA.

Ling, R. (2001). *It is 'in.' It doesn't matter if you need it or not, just that you have it. Fashion and the domestication of the mobile telephone among teens in Norway.* Oslo: Working Paper, Telenor R&D, Norway.

Ling, R., & Yttri, B. (2002). Hyper-coordination via mobile phone in Norway. In J. E. Katz & M. Aakhus (Eds.), *Perpetual contact.* New York: Cambridge University Press.

List, J. A., & Lucking-Reiley, D. (2000). Demand reduction in multiunit auctions: Evidence from a sportscard field experiment. *The American Economic Review, 90*(4), 961-982.

Liu, C., Arnett, K. P., & Litecky, C. (2000). Design quality of websites for electronic commerce: fortune 1000 webmasters' evaluations. *Electronic Markets, 10*(2), 120–129. doi:10.1080/10196780050138173

Livari, J., & Livari, N. (2006). *Varieties of User-Centeredness.* Paper presented at the 39th Annual Hawaii International Conference on System Sciences, Hawaii.

Lucking-Reiley, D. (1999). Using field experiments to test equivalence between auction formats: Magic on the Internet. *The American Economic Review, 89*(5), 1063-1081.

Lucking-Reiley, D. (2000a). Auctions on the Internet: What's being auctioned, and how? *The Journal of Industrial Economics, 48*(3), 227-253.

Lucking-Reiley, D. (2000b). Vickrey auctions in practice: From nineteenth-century philately to twenty-first-century e-commerce. *The Journal of Economic Perspectives, 14*(3), 183-193.

Luftman, J. N., Lewis, P. R., & Oldach, S. H. (1993). Transforming the enterprise: The alignment of business and IT strategies. *IBM Systems Journal, 32*(1), 198–221.

Luhmann, N. (1979). *Trust and power*: Wiley, London.

Luhn, R. (2003, June). CNET review. Retrieved from http://reviews.cnet.com/4520-6536_7-726606-2.html?tag=st.is.3762-8-%250920777559-1.arrow.3762-8-20777559-2

Luna, D., Peracchio, L. A., & de Juan, M. D. (2002). Cross-cultural and cognitive aspects of web site navigation. *Journal of the Academy of Marketing Science, 30*(4), 397–410.

Lupu, E., & Sloman, M. (1999). Conflicts in Policy-Based Distributed Systems Management. *IEEE Transactions on Software Engineering, 25*(6). doi:10.1109/32.824414

Ly, L. T., Rinderle, S., & Dadam, P. (2008). Integration and verification of semantic constraints in adaptive process management systems. *Data & Knowledge Engineering, 64*(1), 3–23. doi:10.1016/j.datak.2007.06.007

Lynch, J. G., & Ariely, D. (2000). Wine Online: Search Cost and Competition on Price, Quality and Distribution. *Marketing Science, 19*(1), 83–103. doi:10.1287/mksc.19.1.83.15183

Maamar, Z., Benatallah, B., & Mansoor, W. (2003). Service Chart Diagrams - Description & Application. In *Proceedings of the Alternate Tracks of the 12th International World Wide Web Conference (WWW'2003), Budapest Hungary.*

Maamar, Z., Benslimane, B., Mrissa, M. & Ghedira, C. (2006a). CooPS - Towards a Method for Coordinating Personalized Services. *Software and System Modeling Journal Special Section on Service-Based Software and Systems Engineering, 5*(2).

Maamar, Z., Benslimane, D. & Anderson, A. (2006b). Using Policies to Manage Composite Web Services. *IEEE IT Professional, 8*(5).

Maamar, Z., Benslimane, D., & Narendra, N. C. (2006c). What Can Context do for Web Services? *Communications of the ACM, 49*(12). doi:10.1145/1183236.1183238

Maamar, Z., Benslimane, D., Kouadri Mostefaoui, G., Sattanathan, S., & Mahmoud, Q. (2008). Toward Behavioral Web Services Using Policies. *IEEE Transactions on Systems, Man, and Cybernetics. Part A, Systems and Humans, 38*(6). doi:10.1109/TSMCA.2008.2003460

Maamar, Z., Narendra, N. C., Benslimane, D., & Sattanathan, S. (2007). Policies for Context-Driven Transactional Web Services. In *Proceedings of the 19th International Conference on Advanced Information Systems Engineering (CAISE'2007), Trondheim, Norway.*

Maamar, Z., Sheng, Q. Z., Tata, S., Benslimane, D. & Sellami, M. (2009b). Towards An Approach to Sustain Web Services High-Availability Using Communities of Web Services. *International Journal of Web Information Systems.*

Maamar, Z., Tata, S., & Belaid, D. (2009a). Towards An Approach to Defining Capacity-Driven Web Services. In *Proceedings of the 23rd International Conference on Advanced Information Networking and Applications (AINA'2009), Bradford, UK.*

Machlis, S. (1998). Online auction services thriving. *Computerworld, 32*(24).

Madu, C. N., & Madu, A. A. (2002). Dimensions of E-quality. *International Journal of Quality & Reliability Management, 19*(3), 246–258. doi:10.1108/02656710210415668

Maedche, A., Motik, B., & Stojanovic, N. (2002). *MAFRA - A MApping FRAmework for Distributed Ontologies.* Paper presented at the 13th International Conference, Knowledge Engineering and Knowledge Management (EKAW 2002), Siguenza, Spain.

Magal, S. R., & Lewis, C. (1995, Spring). Determinants of Information Technology Success in Small Businesses. *Journal of Computer Information Systems*, 75–83.

Mahajan, & Peterson, R. A. (1985). *Models for Innovation Diffusion.* SAGE Publications, Inc.

Mahmood, M. A., & Mann, G. J. (1993). Measuring the organizational impact of information technology investment:

An exploratory study. *Journal of Management Information Systems, 10*(1), 97–122.

Mahmood, M. A., & Mann, G. J. (2000). Impacts of information technology investment on organizational performance. *Journal of Management Information Systems, 17*(1), 3–10.

Mahmood, M. A., & Sniezek, J. A. (1989). Defining Decision Support Systems: An Empirical End-User Assessment. *INFOR -- Canadian Journal of Operational Research and Information Processing*, 27, 253-271.

Mahmood, M. A., Gemoets, L., Lopez, F., & Hall, L. (2008). Measuring E-Commerce Technology Enabled Business value: An Exploratory Research. *International Journal of E-Business Research, 4*(2), 48–68.

Mahmood, M. A., Hall, L., & Swanberg, D. (2001). Factors affecting information technology usage: A meta-analysis of the empirical literature. *Journal of Organizational Computing and Electronic Commerce, 11*(2), 107–130. doi:10.1207/S15327744JOCE1102_02

Mahony, D., Riemer, H., Breeding, J., & Hums, M. (2006, April). Organizational Justice in Sport Organizations: Perceptions of College Athletes and Other College Students. *Journal of Sport Management, 20*(2), 159–188.

Mallat, N. (2006a). *Consumer and merchant adoption of mobile payments.* Doctoral dissertation. Helsinki School of Economics.

Mallat, N. (2006b). *Exploring Consumer Adoption of Mobile Payments.* Paper Presented at Helsinki Mobility Roundtable, 1-2 June 2006, Helsinki.

Malone, T. W., Yates, J., & Benjamin, R. I. (1989). The Logic of Electronic Markets. *Harvard Business Review, 67*(5), 166–170.

Mano, H., & Oliver, R. L. (1993). Assessing the dimensionality and structure of the consumption experience: Evaluation, feeling, and satisfaction. *Journal of Consumer Research, 20*(3), 451-66.

Marcoux, A. (2003). Snipers, stalkers, and nibblers: Online auction business ethics. *Journal of Business Ethics, 46*(2), 163-173.

Marcus, A., & Gould, E. W. (2000). Cultural dimensions and global web user interface design. *Interaction*, (July/August): 33–46.

Marks, G. (2008, February 6). Worthwhile Small Business Technologies. *Business Week Online*, 26.

Mata, F., Fuerst, J., & Barney, J. (1995). Information technology and sustained competitive advantage: A resource-based analysis. *MIS Quarterly, 19*(4), 487–505. doi:10.2307/249630

Matsumoto, D. (2000). *People: Psychology from a Cultural Perspective.* CA: Brookes/Cole.

Matveeva, I., Levow, G.-A., Farahat, A., & Royer, C. (2005). *Term Representation with Generalized Latent Semantic Analysis.* Paper presented at the Conference on recent advances in natural language processing (RANLP).

MÄuller. R. M., Spiliopoulou, M., & Lenz, H.-J. (2002). Electronic Marketplaces of Knowledge: Characteristics and Sharing of Knowledge Assets. In *Proceedings of the International Conference on Advances in Infrastructure for e-Business, e-Education, e-Science, and e-Medicine on the Internet (SSGRR 2002w).*

Maurice, H., Bucchiarone, A., & Gnesi, S. (2007). Web Service Composition Approaches: From Industrial Standards to Formal Methods. In *Proceedings of the International Conference on Internet and Web Applications and Services (ICIW'2007), Mauritius.*

Mayer, R. C., Davis, J. H., & Schoorman, F. D. (1995). An integrative model of organizational trust. *Academy of Management Review, 3*(20), 709–734.

Mayer, R.N. (2002). Shopping from a list: international studies of consumer online experiences. *Journal of Consumer Affairs* 36(1), 115–127.

McAfee, A. (2005). Will Web Services Really Transform Collaboration? *Sloan Management Review.*

McAfee, R. P., & McMillan, J. (1987). Auctions and bidding. *Journal of Economic Literature, 25*(2), 699-739.

McAllister, D. J. (1995). Affect- and cognition-based trust as foundations for interpersonal cooperation in organizations. *Academy of Management Journal, 38*(1), 24–59.

McDonald, C. G., & Slawson, V. C. (2002). Reputation in an Internet auction market. *Economic Inquiry, 40*(4), 633-650.

McDonald, R. (1981). The dimensionality of test and items. *The British Journal of Mathematical and Statistical Psychology, 34*, 110–117.

McIlvaine, B. (2000). John Chambers. *Electronic Buyers' . News, 1242*, 42.

McIntyre, K., & Perlman, E. (2000). Nike - Channel conflict. Stanford GSB Case, EC-9B.

McKendrick, J. (2006, March 2). Will there be a 'Dell' of software? *ZDNet*, March 2, 559.

McKnigh, D. H., Choudhury, V., & Kacmar, C. (2002). Developing and validating trust measures for e-commerce: An integrative typology. *Information Systems Research, 13*(3), 334.

McKnight, D. H., Choudhury, V., & Kacmar, C. (2002). The impact of initial consumer trust on intentions to transact with a Web site: A trust building model. *Journal of Strategic Information Systems, 11*, 297-323.

McKnight, D. H., Cummings, L. L., & Chervany, N. L. (1998). Initial trust formation in new organizational relationships. *AMR, 3*(23), 472–490.

McKnight, D., & Chervany, N. (2001). What trust means in e-commerce customer relationships: An interdisciplinary conceptual typology. *International Journal of Electronic Commerce, 6*(2), 35–59.

McPhillips, T., Bowers, S., Zinn, D., & Ludascher, B. (2008). (in press). Scientific workflow design for mere mortals . *Future Generation Computer Systems*. doi:.doi:10.1016/j.future.2008.06.013

Mehrabian, A., & Epstein, N. (1972). A measure of emotional empathy. *Journal of Personality, 40*, 525–543.

Mehrtens, J., Cragg, P. B., & Mills, A. M. (2001). A model of Internet adoption by SMEs. *Information & Management, 39*(3), 165–176. doi:10.1016/S0378-7206(01)00086-6

Mehta, K. (2002). *Understanding revenue drivers of electronic auctions*. Unpublished Ph.D., University of Illinois at Chicago.

Mehta, M., Agarwal, R., & Rissanen, J. (1996). *SLIQ: A fast, scalable classifier for data mining*. Paper presented at the Fifth International Conference on Extending Database Technology, Avignon, France.

Meister, D. B., & Compeau, D. R. (2002). Infusion of Innovation Adoption: An Individual Perspective. In *Proceedings of the ASAC, Winnipeg, Manitoba*.

Melville, M., Kraemer, K., & Gurbaxani, V. (2004). Information Technology and Organizational Performance: An Integrative Model of IT Business Value . *MIS Quarterly, 28*(2), 283–322.

Menascé, D. A. (2002). QoS Issues in Web Services. *IEEE Internet Computing, 6*(6). doi:10.1109/MIC.2002.1067740

Mentzer, J. T., DeWitt, W., Keebler, J. S., Min, S., Nix, N. W., & Smith, C. D. (2001). Defining supply chain management. *Journal of Business Logistics, 22*(2), 1–25.

Meroño-Cerdán, A. (2008, Summer). Groupware Uses and Influence on Performance in SMEs. *Journal of Computer Information Systems, 48*(4), 87–96.

Merrilees, B., & Fenech, T. (2007). From catalog to Web: B2B multichannel marketing strategy. *Industrial Marketing Management, 37*(1), 44–49. doi:10.1016/j.indmarman.2006.06.014

Metro Stores (2006). *Innovative technologies at the METRO Group*. Retrieved July 28, 2007, from: http://www.future-store.org/servlet/PB/show/1008042/off-Publikationen-InnovativeTechnologien_engl.pdf.

Metzger, M. J. (2006). Effects of site, vendor, and consumer characteristics on Web site trust and disclosure. *Communication Research, 33*(3), 155-179.

Milgrom, P. (1985). The economics of competitive bidding: A selective survey. In L. Hurwicz, D. Schmeidler, and H. Sonnenschein (Ed.), *Social goals and social organization*. Cambridge: Cambridge University Press.

Milgrom, P. (1986). Auction theory. In T. E. Bewley (Ed.), *Advances in economic theory*. Cambridge: Cambridge University Press.

Milgrom, P. R., & Weber, R. J. (1982). A theory of auctions and competitive bidding. *Econometrica, 50*(5), 1089-1122.

Miller, D. (1987). Strategy making and structure: Analysis and implications for performance. *Academy of Management Journal, 30*(1), 7–32. doi:10.2307/255893

Minges, M. (2005). Is the Internet Mobile? Measurements from the Asia-Pacific region. *Telecommunications Policy, 29*, 113–125. doi:10.1016/j.telpol.2004.11.002

Mintel (2000). *UK vs. US Online Shopping.* Mintel, London.

Mintel (2003). *Home Shopping, UK.* Mintel, London

Mintzberg, H. (1988). Generic strategies: Toward a comprehensive framework. *Advances in Strategic Management, 5*, 1–67.

Mirchandandi, D. A., & Motwani, J. (2000, Spring). Understanding Small Business Electronic Commerce Adoption: An Empirical Analysis. *Journal of Computer Information Systems*, 70–73.

Mittal, V., & Kamakura, W. A. (2001). Satisfaction, repurchase intent and repurchase behavior: Investigating the moderating effect of customer characteristics. *Journal of Marketing Research, 38*, 131-142.

Miyazaki, A. D., & Fernandez, A. (2001). Consumer perceptions of privacy and security risks for online shopping. *The Journal of Consumer Affairs, 35*(1), 27-44.

Mok, W. Y., Palvia, P., & Paper, D. (2006). On the computability of agent-based workflows. *Decision Support Systems, 42*(3), 1239–1253. doi:10.1016/j.dss.2005.10.010

Molla, A., & Licker, P. S. (2001). E-Commerce Systems Success: An Attempt to Extend and Respecify the Delone and Maclean Model of IS Success. *Journal of ECommerce Research, 2*(4).

Montagut, F., Molva, R., & Golega, S. T. (2008). The pervasive workflow: A decentralized workflow system supporting long-running transactions. *IEEE Transactions on Systems, Man and Cybernetics. Part C, Applications and Reviews, 38*(3), 319–333. doi:10.1109/TSMCC.2008.919184

Montgomery, A. L., Hosanagar, K., Krishnan, R., & Clay, K. B. (2004). Designing a Better Shopbot. *Management Science, 50*(2), 189–206. doi:10.1287/mnsc.1030.0151

Moon, J. W., & Kim, Y. G. (2001). Extending the TAM for a world-wide-web context. *Information & Management, 38*, 217–230. doi:10.1016/S0378-7206(00)00061-6

Moore, A., & Outwater, M. (2001, December). Small Business Survey: Understanding the Business Opportunity, Retrieved June 06, 2003, from http://www.celent.com/PressReleases/20011217/SmallBusSurvey.htm.

Moore, G. C., & Benbasat, I. (1991). Development of an instrument to measure the perceptions of adopting an information technology innovation. *Information Systems Research, 2*(3), 192–222. doi:10.1287/isre.2.3.192

Morganosky, M.A., & Cude, B.J. (2000). Consumer response to online grocery shopping *International Journal of Retail and Distribution Management* 28(1), 17–26.

Moses, T. (2003). *XACML profile for Web services.* Working Draft 04, 29 September 2003, Retrieved from http://www.oasis-open.org/committees/download.php/3661/draft-xacml-wspl-04.pdf

Mrissa, M., Ghedira, C., Benslimane, D., Maamar, Z., Rosenberg, F., & Dustdar, S. (2007). A Context-based Mediation Approach to Compose Semantic Web Services. *ACM Transactions on Internet Technology, 8*(1). doi:10.1145/1294148.1294152

Muehlen, M. Z. (2004). Organizational Management in Workflow Applications - Issues and Perspectives . *Information Technology and Management, 5*, 271–291. doi:10.1023/B:ITEM.0000031582.55219.2b

Mukherjee, A., & Nath, P. (2005). An empirical assessment of comparative approaches to services quality measurement. *Journal of Services Marketing, 19*(3), 174-84.

Mukhi, N., Plebani, P., Silva-Lepe, I., & Mikalsen, T. (2004). Supporting Policy-Driven Behaviors in Web Services: Experiences and Issues. In *Proceedings of the 2nd International Conference on Service Oriented Computing (ICSOC'2004), New-York, USA.*

Mukhopadhyay, T., & Kekre, S. (1995). Business value of information technology: A study of electronic data interchange. *MIS Quarterly, 19*(2), 137–156. doi:10.2307/249685

Munshi, J. (1990). A method for constructing Likert scales. Retrieved June 2004, from http://www.munshi.4t.com

Murphy, T. B., & Martin, D. (2003). Mixtures of distance-based models for ranking data. *Computational Statistics & Data Analysis, 41,* 645-655.

Murray, K. B. (1991). A test of services marketing theory: Consumer information acquisition activities. *Journal of Marketing, 55*(1), 10-25.

Muthaiyah, S. (2008). *A framework and methodology for ontology mediation through semantic and syntactic mapping.* George Mason University.

Muthaiyah, S., & Kerschberg, L. (2006). *Dynamic Integration and Semantic Security Policy Ontology Mapping for Semantic Web Services (SWS).* Paper presented at the First IEEE International Conference on Digital Information Management (ICDIM), Bangalore, India.

Muthaiyah, S., & Kerschberg, L. (2006). *Virtual Organization Security Policies: An Ontology-based Mapping and Integration Approach.* Paper presented at the The Second Secure Knowledge Management Workshop (SKM). Brooklyn, New York.

Muthaiyah, S., & Kerschberg, L. (2008). Achieving interoperability in e-government services with two modes of semantic bridging: SRS and SWRL. *J. Theor. Appl. Electron. Commer. Res., 3*(3), 52–63. doi:10.4067/S0718-18762008000200005

Muthaiyah, S., Barbulescu, M., & Kerschberg, L. (2008). Managing ontology change and evolution via a hybrid matching algorithm. *WSEAS Trans. Sys. Ctrl., 3*(9), 737–747.

Nabseth, L., & Ray, G. F. (1974). *The Diffusion of New Industrial Processes.* London: Cambridge: University Press.

Namiki, N. (1988). Export Strategy for small business. *Journal of Small Business Management, 26*(2), 33–37.

Naor, M., Pinkas, B., & Sumner, R. (1999). Privacy preserving auctions and mechanism design. In *The 1st ACM Conference on Electronic Commerce.*

Nass, C., Moon, C., Fogg, Y., Reeves, B. J., & Dryer, D. C. (1996). Can computer personalities be human personalities? *International Journal of Human-Computer Studies, 43,* 223–239.

Nass, C., Moon, Y., & Carney, P. (1999). Are respondents polite to computers? Social desirability and direct responses to computers. *Journal of Applied Social Psychology, 5*(29), 1093–1110.

Naylor, J. B., & Williams, J. (1994). The successful use of IT in SMEs on Merseyside. *European Journal of Information Systems, 3*(1), 48–56. doi:10.1057/ejis.1994.6

Ndede-Amadi, A. A. (2004). What strategic alignment, process redesign, enterprise resource planning, and e-commerce have in common: enterprise-wide computing. *Business Process Management Journal, 10*(2), 184–199. doi:10.1108/14637150410530253

Neo, B. S. (1992). The implementation of an electronic market for pig trading in Singapore. *Journal of Strategic Management Systems, 1*(5), 278-288.

Net4now (2007). *De Montfort to make technology trustworthy.* Retrieved July 17, 2007, from: http://www.net4now.com/isp_news/news_article.asp?News_ID=3883.

News Blaze (2007). *Online privacy and security: The fear factor.* Retrieved July 17, 2007, from: http://newsblaze.com/story/20060314050502000001.mwir/newsblaze/LEGAL-LAW/Legal

Ngai, E. W. T. (2005). Customer Relationship Management Research, (1999-2002) an Academic Literature Review and Classification. *Marketing Intelligence & Management, 23*(6), 582–605. doi:10.1108/02634500510624147

Ngai, E. W. T., & Wat, F. K. (2002). A literature review and classification of electronic commerce research. *Information & Management, 39,* 415–429. doi:10.1016/S0378-7206(01)00107-0

Nguyen, N., & LeBlanc, C. (1998). The mediating role of corporate image on customers' retention decisions: An investigation in financial services. *International Journal of Bank Marketing, 16*(2), 52-65.

Nguyen, N., & LeBlanc, G. (2001). Corporate image and corporate reputation in customers' retention decisions in services. *Journal of Retailing and Consumer Services, 8*(4), 227-236.

Niederman, F. (1998). The diffusion of electronic data interchange technology. In T. J. Larsen & E. McGuire (Eds.), *Information Systems Innovation and Diffusion: Issues and Directions* (pp. 141-160). London: Idea Group.

Nielsen, J. (1996). *Top Ten Mistakes in Web Design*, from http://www.sun.com/columns/alertbox/9605.html. date of download.02-08-2006

Nielsen, J. (2004). *Designing for Web Usability*. Indianapolis: New Riders Publications.

Nikitkov, A. (2006). Information assurance seals: How they impact consumer purchasing behavior. *Journal of Information System. 20*(1), 1-17.

Nokia Telecommunications. (1999). *The Demand for Mobile Value Added Services: A Market Study*, Retrieved from http://www.nokia.com/press/background/pdf/study-vas.pdf

Nonaka, I., Toyama, R., & Nagata, A. (2000). A firm as a knowledge-creating entity: A new perspective on the theory of the firm. *Industrial and Corporate Change, 9*(1), 1–20. doi:10.1093/icc/9.1.1

Noori, B., & Salimi, M. H. (2005). A decision-support system for business-to-business marketing . *Journal of Business and Industrial Marketing, 20*(4-5), 226–236. doi:10.1108/08858620510603909

Normann, R. (1991). *Service management: Strategy and leadership in service business.* New York: John Wiley and Sons.

Noteberg, A., Christiaanse, E., & Wallage, P. (2003). Consumer Trust in Electronic Channels. *E- Service Journal, 2*(2), 46-67.

Novak, T. P., Hoffman, D. L., & Yung, Y.-F. (2000). Measuring the Customer Experience in Online Environments: A Structural Modeling Approach. *Marketing Science, 19*(1), 22–42. doi:10.1287/mksc.19.1.22.15184

Nunnally, J. (1978). *Psychometric Theory.* New York: McGraw-Hill.

Nurosis, M. (1993). *Estatistical Data Analysis.* SPSS Inc.

Nysveen, H., Pedersen, P. E., & Thorbjørnsen, H. (2005a). Intentions to Use Mobile Services: Antecedents and Cross-Service Comparisons. *Journal of the Academy of Marketing Science, 33*(3), 330–346. doi:10.1177/0092070305276149

Nysveen, H., Pedersen, P. E., & Thorbjørnsen, H. (2005b). Explaining intention to use mobile chat services: Moderating effects of gender. *Journal of Consumer Marketing, 22*(5), 247–256. doi:10.1108/07363760510611671

O'Callaghan, R., Kaufmann, P. J., & Konsynski, B. R. (1992). Adoption correlates and share effects of electronic data interchange systems in marketing channels. *Journal of Marketing, 56*, 45–56. doi:10.2307/1252041

O'Donnell, D., & March, S. (1987). End User Computing Environments - Finding a Balance Between Productivity and Control . *Information & Management, 13*(1), 77–84. doi:10.1016/0378-7206(87)90012-7

Ockenfels, A., & Roth, A. (2002). The timing of bids in Internet auctions: Market design, bidder behavior and artificial agents. *AI Magazine*, (Fall), 79-88.

Odom, M. D., Kumar, A., & Saunders, L. (2002). Web assurance seals: How and why they influence consumers' decisions. *Journal of Information Systems, 16*(2), 231-250.

Office of the Telecommunications Authority Web site. Retrieved from www.ofta.gov.hk.

Oh, W. (2002). C2C versus B2C: A comparison of the winner's curse in two types of electronic auctions. *International Journal of Electronic Commerce, 6*(4), 115-138.

Okamoto, T. (2002). Analysis and design of an online "discreet" Vikery auction protocol. *Technical Project.*

Olazabal, N. G. (2002). Banking: the IT paradox. *McKinsey Quarterly, 1*, 47–51.

Oliva, T. A., Oliver, R. L., & MacMillan, I. C. (1992). A catastrophe model for developing services satisfaction strategies. *Journal of Marketing, 56*(3), 83-95.

Oliver, R. L. (1980). A cognitive model of the antecedents and consequences of satisfaction decisions. *Journal of Marketing Research, 17*, 460-469.

Oliver, R. L. (1997). *Satisfaction—A behavioral perspective on the consumer*. New York: McGraw-Hill.

Oliver, R. L. (1999). Whence consumer loyalty? *Journal of Marketing, 63*(4), 33-44.

Olson, E. L., & Widing, R. E. (2002). Are interactive decision aids better than passive decision aids? A comparison with implications for information providers on the Internet. *Journal of Interactive Marketing, 16*(2), 22–33. doi:10.1002/dir.10007

Omote, K., & Miyaji, A. (2001). A practical English auction with one-time registration. *In Proceedings of the Sixth Australiasian Conference on Information Security and Privacy*, (pp. 221-234).

Orman, L. V. (2008). Service Semantics, Structure, and Design. *E-Service Journal, 6*(2), 57–75. doi:10.2979/ESJ.2008.6.2.58

Orman, L. V. (2008a). Browsing Databases with Constraint Hierarchies. *Electronic Commerce Research, 8*(1), 29–56. doi:10.1007/s10660-008-9016-y

Orson, L., & Donald, J. (2004). Critical success factors for online auction Web sites. *Bulletin of applied computing and information technology* (Vol. 2).

Ostrowski, P. L., O'Brien, T. V., & Gordon, G. L. (1993). Service quality and customer loyalty in the commercial airline industry. *Journal of Travel Research, 32,* 16-24.

Ottaway, T. A., Bruneau, C. L., & Evans, G. E. (2003). The impact of auction item image and buyer/seller feedback rating on electronic auctions. *The Journal of Computer Information Systems, 43*(3), 56-60.

Palvia, P., & Palvia, S. (1999). An Examination of IT Satisfaction of Small-business Users. *Information & Management, 35*(3), 127–137. doi:10.1016/S0378-7206(98)00086-X

Papazoglou, M. P., Traverso, P., Dustdar, S. & Leymann, L. (2007). Service-Oriented Computing: State of the Art and Research Challenges. *IEEE Computer, 40*(11).

Parasuraman, A., & Grewal, D. (2000). The Impact of Technology on the Quality-Value-Loyalty Chain: A Research Agenda. *Journal of the Academy of Marketing Science, 28*(1), 168–174. doi:10.1177/0092070300281015

Parasuraman, A., Zeithaml, V. A., & Berry, L. L. (1985). A conceptual model of service quality and its implication for future research. *Journal of Marketing, 49*(2), 41-50.

Parasuraman, A., Zeithaml, V. A., & Berry, L. L. (1988). SERVQUAL: A multiple-item scale for measuring consumer perceptions of service quality. *Journal of Retailing, 64*(1), 12-40.

Parasuraman, A., Zeithaml, V. A., & Berry, L. L. (1994). Reassessment of expectations as a comparison standard in measuring service quality: Implications for further research. *Journal of Marketing, 58*(1), 111-124.

Parasuraman, A., Zeithaml, V. A., & Malhotra, A. (2005). E-S-QUAL: A Multiple-Item Scale for Assessing Electronic Service Quality. *Journal of Service Research, 7*(3), 213–234. doi:10.1177/1094670504271156

Parente, D., Venkataraman, R., Fizel, J., & Milett, I. (2001). B2B online reverse auctions: What's new?. *Decision Line.*

Park, H., Ribiere, V., & Schulte, W. D. (2004). Critical attributes of organizational culture that promote knowledge management technology implementation success. *Journal of Knowledge Management, 8*(3), 106–117. doi:10.1108/13673270410541079

Parkinson, T. L. (2002). The role of seals and certifications of approval in consumer decision-making. *Journal of Consumer Affairs, 9,* 1-14.

Parsons, C. (2007). Web-based surveys: Best practices based on the research literature. *Visitor Studies, 10*(1), 1064-5578. Retrieved July 15, 2007, from http://www.informaworld.com/10.1080/10645570701263404

Pastore, M. (2002, January 30). *The Broadband and Wireless Revolution*. Retrieved June 06, 2003, from http://ISP-Planet.com/research/2002/ne

Paterno, F., Mancini, C., & Meniconi, S. (1997) *ConcurTaskTrees: A Diagrammatic Notation for Specifying Task Models*. Paper presented at the INTERACT Conference, Sydney, Australia.

Pather, S., Erwin, G., & Remenyi, D. (2003). Measuring e-Commerce effectiveness: a conceptual model. In J. Eloff,

A. Engelbrecht, P. Kotzé, & M. Eloff (Eds.), *Proceedings of the 2003 Annual Research Conference of the South African institute of Computer Scientists and information Technologists on Enablement Through Technology* (September 17-19, 2003) (pp.143-152). ACM Press.

Patterson, P. G., & Smith, T. (2003). A cross-cultural study of switching barriers and propensity to stay with service providers. *Journal of Retailing, 79*(2), 107-120.

Patterson, P. G., Johnson, L. W., & Spreng, R. A. (1997). Modelling the determinants of customer satisfaction for business-to-business profesional service. *Journal of the Academy of Marketing Science, 25*(1), 4–17. doi:10.1007/BF02894505

Pavlou, P. A. (2002). Trustworthiness as a source of competitive advantage in online auction markets. *Academy of Management Proceedings*, A1-A7.

Peak, D., & Guynes, C. S. (2003). Improving information quality through IT alignment planning: a case study. *Information Systems Management, 20*(4), 22–29. doi:10.1201/1078/43647.20.4.20030901/77289.4

Peak, D., & Guynes, C. S. (2003). The IT alignment planning process. *Journal of Computer Information Systems, 44*(1), 9–15.

Pedersen, P. E. (2001). An adoption framework for mobile commerce. In *Proceedings of the 1st. IFIP Conference of E-Commerce*, Zürich, Switzerland, October 3-5, 2001.

Pedersen, P. E. (2005). Adoption of mobile Internet services: An exploratory study of mobile commerce early adopters. *Journal of Organizational Computing and Electronic Commerce, 15*(3), 203–221. doi:10.1207/s15327744joce1503_2

Pedersen, P. E., & Thorbjørnsen, H. (2003). *Adoption of Mobile Services.* Model development and Cross-Service study. Norwegian School of Economics and Business Administration.

Pedersen, P. E., Methlie, L. B., & Thorbjørnsen, H. (2001). *Understanding mobile commerce end-user adoption: a triangulation perspective and suggestions for an exploratory service evaluation framework.* Paper presented at HICSS-35, Hawaii, US, Jan 7-10, 2002

Pell, G. (2005). Use and misuse of Likert scales. *Medical Education, 39,* 970.

Pereira, P. (2005, November 16). SMBs will lead channel. *eWeek, 22*(45A), 33.

Pereira, R. E. (2001). Influence of query-based decision aids on consumer decision making in electronic commerce. *Information Resources Management Journal, 14*(1), 31–48.

Peter, J. P. (1981). Construct Validity: A review of basic issues and marketing practices. *JMR, Journal of Marketing Research, 18*(5), 133–145. doi:10.2307/3150948

Peter, S., & Smeeton, N. C. (2007). *Applied Nonparametric Statistical Methods*: CRC Press.

Peteraf, M. A. (2000). The cornerstones of competitive advantage: A resource-based view. *Strategic Management Journal, 14*(3), 179–191. doi:10.1002/smj.4250140303

Peterson, R. A., Balasubramanian, S., & Bronnenberg, B. J. (1997). Exploring the implications of the Internet for consumer marketing. *Journal of the Academy of Marketing Science, 25*(4), 329-346.

Pew Internet and American Life. Internet Penetration and Impact (2006). Retrieved November 9, 2007 from http://www.pewinternet.org/PPF/r/182/report_display.asp

Phau, I., & Poon, S.M. (2000). Factors influencing the types of products and services purchased over the Internet. *Internet Research: Electronic Networking Applications and Policy* 10(2), 102–113.

Plakoyiannaki, E., & Tzokas, N. (2002). Customer relationship management: a capabilities portafolio perspective. *Journal of Database Marketing, 9*(3), 228–237. doi:10.1057/palgrave.jdm.3240004

Poel, D., & Leunis, J. (1999). Consumer acceptance of the Internet as a channel of distribution. *Journal of Business Research, 45.*

Pollalis, Y. A. (2003). Patterns of co-alignment in information-intensive organizations: business performance through integration strategies. *International Journal of Information Management, 23*(6), 469–492. doi:10.1016/S0268-4012(03)00063-X

Poon, S. (2000). Business environment and Internet commerce benefit – a small business perspective. *European Journal of Information Systems*, *9*(2), 72–81. doi:10.1057/palgrave/ejis/3000361

Poon, S. (2000). Business environment and Internet commerce benefit – a small business perspective. *European Journal of Information Systems* 9, 72 – 81.

Poon, S., & Swatman, P. M. C. (1998). A combined-method study of small business Internet commerce. *International Journal of Electronic Commerce*, *2*(3), 31–46.

Porter, M. (2001). Strategy and the Internet. *Harvard Business Review*, *79*(3), 63–78.

Porter, M. E. (1980). *Competitive strategy – technologies for analyzing industries and competitors*. New York: Free Press.

Porter, M. E. (1985). *Competitive Advantage* (3rd. Ed). New York: Free Press.

Porter, M. E. (2001). Strategy and the Internet. *Harvard Business Review*, (March): 62–78.

Porter, M. E. (2004). Building the microeconomic foundations of prosperity: Findings from the business competitiveness index. In *The Global Competitiveness Report 2003-2004*. World Economic Forum, Rajendran, R. (1999). *A study on computer based information systems of small-scale industries in Coimbatore, India*. Unpublished master thesis, Bharathiar University, Coimbatore.

Porter, M. E., & Millar, V. E. (1985). How information gives you competitive advantage. *Harvard Business Review*, *63*(4), 149–160.

Porter, R., & Shoham, Y. (2004). On cheating in sealed bid auctions. *Decision Support Systems*, (39), 41-54.

Porter, S., & Whitcomb, M. (2005). E-mail subject lines and their effect on Web survey viewing and response. *Social Science Computer Review*, *23*(3), 380-387.

Powell, A., & Moore, J. E. (2002). The Focus of Research in End User Computing: Where Have We Come Since the 1980s? *Journal of End User Computing*, *14*(1), 3–22.

Power, D. (2005). Implementation and use of B2B-enabling technologies: five manufacturing cases. *Journal of Manufacturing Technology Management*, *16*(5/6), 554–572. doi:10.1108/17410380510600527

Premkumar, G., & King, W. R. (1992). An empirical assessment of information systems planning and the role of information systems in organizations. *Journal of Management Information Systems*, *9*(2), 99–125.

Premkumar, G., & Potter, M. (1995). Adoption of computer aided software engineering (CASE) technology: an innovation adoption perspective. *The Data Base for Advances in Information Systems*, *26*(2/3), 105–124.

Premkumar, G., & Ramamurthy, K. (1995). The role of inter-organizational and organizational factors on the decision mode for the adoption of inter-organizational systems. *Decision Sciences*, *26*(3), 303–336. doi:10.1111/j.1540-5915.1995.tb01431.x

Premkumar, G., & Roberts, M. (1999). Adoption of new information technologies in rural small business. *OMEGA-The International Journal of Management Science*, *27*, 467–484. doi:10.1016/S0305-0483(98)00071-1

Premkumar, G., Ramamurthy, K., & Crum, M. (1997). Determinants of EDI in the transportation industry. *European Journal of Information Systems*, *6*(2), 107–121. doi:10.1057/palgrave.ejis.3000260

Premkumar, G., Ramamurthy, K., & Nilakanta, S. (1994). Implementation of electronic data interchange: an innovation diffusion perspective. *Journal of Management Information Systems*, *11*(2), 157–186.

Prescott, M. B., & Conger, S. A. (1995). Information technology innovations: a classification by IT locus of impact and research approach. *The Data Base for Advances in Information Systems*, *26*(2/3), 20–41.

Puto, C., Patton, W., & King, R. (1985). Risk handling strategies in industrial vendor selection decisions. *Journal of Marketing*, *49*(1), 89-98.

Quinlan, J. R. (1990). Induction of decision trees. In J. W. Chavlik, W. Jude & T. G. Dietterich (Eds.), *Readings in machine learning* (pp. 57-69). San Mateo, CA: Morgan Kauffman.

Radovilsky, Z., & Hedge, V. (2004, March). Factors influencing e-commerce implementation: Analysis of survey results. *Journal of Academy of Business and Economics, 4*(1), 29.

Rafaeli, S., Raban, D. R., & Ravid, G. (2007). How social motivation enhances economic activity and incentives in the Google Answers knowledge sharing market. *International Journal of Knowledge and Learning, 3*(1), 1–11. doi:10.1504/IJKL.2007.012598

Raghu, T. S., Jayaraman, B., & Rao, H. R. (2004). Towards an integration of agent and activity-centric approaches in organization process modeling: Incorporating incentive mechanisms. *Information Systems Research, 15*(4), 316–335. doi:10.1287/isre.1040.0031

Ram, S., et al. (1999). *Schema Integration: Past, Current and Future*. San Francisco.

Ramamurthy, K., Premkumar, G., & Crum, M. R. (1999). Organizational and Interorganizational Determinants of EDI Diffusion and Organizational Performance: A Causal Model. *Journal of Organizational Computing and Electronic Commerce, 9*, 253–285. doi:10.1207/S153277440904_2

Ranaweera, C., & Neely, A. (2003). Some moderating effects on the service quality-customer retention link. *International Journal of Operations and Production Management, 23*(2), 230-248.

Rangone, A., & Renga, F. M. (2006). B2e mobile internet: An exploratory study of Italian applications. *Business Process Management Journal, 12*(3), 330–343. doi:10.1108/14637150610667999

Raphel, M. (1996). How supermarkets capture customers with their net. *Direct Marketing* 59(1), 14–17.

Rask, M., & Dholakia, N. (2001). Next to the customer's hart and wallet: Frameworks for exploring the emerging m-commerce arena. *AMA Winter Marketing Educator's Conference*,12, 372-378.

Raven, P. V., Huang, X., & Kim, B. B. (2007). E-business in Developing Countries: A Comparison of China and India . *International Journal of E-Business Research, 3*(1), 91–108.

Ravichandran, T., & Lertwongsatien, C. (2002). Impact of information systems resources and capabilities on firm performance: A resource-based perspective. In *Proceedings of 23rd International Conference on Information System*, 577-582.

Rawn (1994) The business of the Internet: an introduction to the Internet for commercial organizations. Retrieved from http://www.rtd.com/people/rawn/business.html

Raymond, L., Pare, G., & Bergeron, F. (1995). Matching information technology and organizational structure: An empirical study with implication for performance. *European Journal of Information Systems, 4*(1), 3–16. doi:10.1057/ejis.1995.2

Rayport, J., & Sviokla, J. (1994). *From marketplace to market space*. Ed. Materials, Harvard University.

Reginaldo, R., Braga, R., & Masiero, P. (2001). A pattern language for online auctions management. In *Proceedings of the PLoP 2001 Conference*.

Reich, B. H., & Benbasat, I. (1990). An empirical investigation of factors influencing the success of customer-oriented strategic systems. *Information Systems Research, 1*(3), 325–347. doi:10.1287/isre.1.3.325

Reich, B. H., & Benbasat, I. (2000). Factors that influence the social dimensions of alignment between business and information technology objectives. *MIS Quarterly, 24*(1), 81–113. doi:10.2307/3250980

Reichheld, F. (1996). *The loyalty effect*. Boston: Harvard Business School Press.

Reichheld, F. F. (2001). Lead for Loyalty. *Harvard Business Review, 79*(7), 76–84.

Reichheld, F. F., & Schefter, P. (2000). E-loyalty: your secret weapon on the web. *Harvard Business Review, 78*(4), 105–113.

Reid, G. C. (1993). *Small business enterprise: An economic analysis*. London: Routledge.

Reiff-Marganiec, S., & Turner, K. J. (2003). A Policy Architecture for Enhancing and Controlling Features. In *A Policy Architecture for Enhancing and Controlling Features*. IOS Press.

Resnik, P. (1992). *WordNet and distributional analysis: A class-based approach to lexical discovery.* Paper presented at the AAAI Workshop.

Resnik, P. (1998). Semantic Similarity in a Taxonomy: An Information based Measure and its Application to Problems of Ambiguity in Natural Language. *Journal of Artificial Intelligence Research, 11*, 95–130.

Revolutions driving the ISP Business in 2003 (2003). ISP-Market LLC.

Rhee, B., & Park, S. (1999). *Online store as a new direct channel and emerging hybrid channel system* (Working Paper). Hong Kong University of Science and Technology, Clear Water Bay, Kowloon, Hong Kong.

Ribeiro, J. (2004, March 2). Researcher compiles auction data for online merchants. Retrieved June 7, 2005, from http://www.computerworld.com/managementtopics/ebusiness/story/0,10801,90668,00.html

Richard, C., & Halstead, D. (2004). Characteristics and Internet marketing strategies of online auction sellers. *International Journal of Internet Marketing and Advertising, 1*(1).

Riegelsberger, J. (2006). *Big brands must actively build trust on line.* Retrieved July 17, 2007, from: http://www.e-consultancy.com/about/press-centre/e-consultancy-press-releases.asp?id=1492.

Riegelsberger, J., & Sasse, M. A. (2002). *Face it - photos don't make a web site trustworthy.* Paper presented at the CHI '02 extended abstracts on Human factors in computing systems, Minneapolis, Minnesota, USA.

Riegelsberger, J., Sasse, M. A., & McCarthy, J. D. (2002). *Eye-catcher or Blind Spot? The Effect of Photographs of Faces on E-commerce Sites.* Paper presented at the Proceedings of the 2nd IFIP Conference on E-commerce, E-business, E-government (i3e), Boston: Kluwer.

Riegelsberger, J., Sasse, M. A., & McCarthy, J. D. (2003). *Shiny happy people building trust? Photos on e-commerce websites and consumer trust.* Paper presented at the Proceedings of CHI 2003, New York,US.

Riegelsberger, J., Sasse, M. A., & McCarthy, J. D. (2005). *Do people trust their eyes more than ears?: media bias in detecting cues of expertise.* Paper presented at the CHI '05 extended abstracts on Human factors in computing systems, Portland, OR, USA.

Riemenschneider, C., Harrison, D. A., & Mykytyn, P. P. J. (2003). Understanding ICT adoption decisions in small business: Integrating current theories. *Information & Management, 40*(4), 269–285. doi:10.1016/S0378-7206(02)00010-1

Riggins, F. J., & Mukhopadhyay, T. (1994). Interdependent benefits from interorganizational systems: Opportunities for. *Journal of Management Information Systems, 11*(2), 37–57.

Riggins, F. J., & Rhee, H. S. (1999). Developing the learning network using extranets. *International Journal of Electronic Commerce, 4*, 65–84.

Riquelme, H. (2002). Commercial Internet Adoption in China: Comparing the Experience of Small, Medium and Large Businesses. *Internet Research, 12*(3), 276–286. doi:10.1108/10662240210430946

Rivard, S., Raymond, L., & Verreault, D. (2006). Resource-based view and competitive strategy; An integrated model of the contribution of information technology to firm performance. *The Journal of Strategic Information Systems, 156*, 29–50. doi:10.1016/j.jsis.2005.06.003

Robbins, S., & Stylianou, A. (2003). Global corporate web sites: an empirical investigation of content and design. *Information & Management, 40*(3), 205–212.

Roberts, M. Xu, M.X., & Mettos, N. (2003). Internet Shopping: the supermarket and consumer perceptions. *Journal of Electronic Commerce in Organizations* 1 (2) 32–43.

Robinson, J. P., Shaver, P. R., & Wrightsman, L. S. (1991). *Criteria for scale selection.* In *Measures of Personality and Social Psychological Attitudes.* San Diego: Academic Press.

Robinson, R. (2004). Understand Enterprise Service Bus Scenarios and Solutions in Servie-Oriented Architecture, IBM developerWorks, http://www-106.ibm.com/developerworks/webservices/library/ws-esbscen/.

Rodgers, S. and Harris, M.A. (2003). Gender and e-commerce an exploratory study. Journal of Advertising Research 43(3), 322–330.

Rogers, E. M. (1962). *Diffusion of innovation.* New York, The Free Press.

Rogers, E. M. (2003). *Diffusion of Innovations* (5th ed.). New York: Free Press.

Rogers, M. E. (1983). *Diffusion of innovations* (3rd ed.). New York: Free Press.

Rose, G. M., & Straub, D. W. (1999). The Effect of Download Time on e-Commerce: The Download Time Brand Impact Model. In *Proceedings of the Ninth Workshop on Information Technologies and Systems* (pp. 167-172).

Roth, A., & Ockenfels, A. (2001). Last-minute bidding and the rules for ending second-price auctions: Evidence from eBay and Amazon auctions on the Internet. *American Economic Review.*

Rothkopf, M. H. (1969). A model of rational competitive bidding. *Management Science, 15*(7), 362-374.

Rothkopf, M. H., & Harstad, R. M. (1994). Modeling competitive bidding: A critical essay. *Management Science, 40*(3), 364-384.

Rousseau, D. M., Sitkin, S. B., Butt, R. S., & Camerer, C. (1998). Not so different after all: a cross-discipline view of trust. *Academy of Management Review, 23*(3), 393–404.

Rovenpor, J. (2003). Explaining the e-commerce shakeout: Why did so many Internet-based businesses fail? *E-Service Journal, 3*(1), 53-76.

Rowley, J. (1996). Retailing and shopping on the Internet. *International Journal of Retail and Distribution Management* 24(3), 26–38.

Rowley, J. (1998). Internet food retailing: the UK in context. *British Food Journal* 100 (2), 85–95.

Ruhi, U., & Turel, O. (2006). Enabling the glass pipeline: the infusion of mobile technology applications in supply chain management. In Y. Lan & B. Unhelkar (Eds.), *Global Integrated Supply Chain Systems* (pp. 291-309). Hershey, PA: Information Science Publishing.

Rummler, G., & Brache, A. (1995). Managing the White Space on the Organizational Chart . *Super Vision, 52*(5), 6–12.

Russell, D. M., & Hoag, A. M. (2004). People and information technology in the supply chain: social and organizational influences on adoption. *International Journal of Physical Distribution & Logistics Management, 34*(2), 102–122. doi:10.1108/09600030410526914

Rust, R. T., & Kannan, P. K. (2003). E-Service: A New Paradigm for Business in the Electronic Environment. *Communications of the ACM, 46*(6), 36–42. doi:10.1145/777313.777336

Rust, R. T., Zeithaml, V. A., & Lemon, K. N. (2004). Customer-Centered Brand Management. *Harvard Business Review, 82*(9), 110–118.

Ruyter, K., Wetzels, M., & Bloemer, J. (1996). On the relationship between perceived service quality and behavioral intentions: A cross-sectional perspective. In J. Beracs, A. Bauer, & J. Simons (Eds), *Proceedings 25th EMAC Conference*, Budapest, Hungary.

Saarikoski, V. (2006). *The Odyssey of the Mobile Inter*net. Doctoral Dissertation. University of Oulu, Finland.

Sadtler, C., Cotignola, D., Crabtree, B., & Michel, P. (2004). Patterns: Broker Interactions for Intra- and Inter-Enterprise. *IBM Redbooks*, SG24-6075, http://www.redbooks.ibm.com/redbooks.nsf/0/532fca172da15c6c85256d6d6d0046192e?openDocument.

Sako, K. (1999). Universal verifiable auction protocol which hides losing bids. In *Proceedings of the SCIS'99*, (pp. 35-39).

Sanders, N. R., & Premus, R. (2002). IT applications in supply chain organizations: a link between competitive priorities and organizational benefits. *Journal of Business Logistics, 23*(1), 65–83.

Sanderson, B. (2000). Cyberspace retailing a threat to traditionalists. *Retail World, 53*(14), 6-7.

Sandholm, T. (2000). Issues in computational vickrey auctions. *International Journal of Electronic Commerce, 4*(3), 107.

Sapienza, H. J., Smith, K. G., & Gannon, M. J. (1988). Using subjective evaluation of organizational performance in small business research. *Entrepreneurship Theory and Practice, 1988*(Winter), 45–53.

Sarkar, M. B., Butler, B., & Steinfield, C. (1995). Intermediaries and Cybermediaries: A Continuing Role for Mediating Players in the Electronic Marketplace. *Journal of Computer-Mediated Communication, 1*(3).

Sato, S., Hawkins, J., & Berentsen, A. (2001). E-finance: Recent developments and policy implications. In *Tracking a Transformation: e-commerce and the terms of competition in industries* (pp. 64-91). Brookings Institution Press, Washington DC.

Sawhney, M., Balasubramanian, S., & Krishnan, V. V. (2004). Creating Growth with Services. *Sloan Management Review, 45*(2), 33–43.

ScanAlert (2005). *New report shows online consumers window shopping for days before buying.* Retrieved July 17, 2007 from: http://www.marketwire.com/2.0/release.do?id=661502&sourceType=1.

Schlimmer, J. (2004). Web Services Policy Framework (WS-Policy). http://www-128.ibm.com/developerworks/webservices/library/specification/ws-polfram/

Schmid, H. A., & Rossi, G. (2004). Modeling and Designing Processes in E-Commerce Applications. *IEEE Internet Computing, 8*(1).

Schmidt, J., Calantone, R., Griffin, A., & Montoya-Weiss, M. (2005). Do certified mail third-wave follow-ups really boost response rates and quality? *Marketing Letters, 16*(2), 129-141.

Schmidt, M. T., & Kalyana, S. (2004). The On Demand Operating Environment – Architectural Overview. *IBM developerWorks*, http://www-106.ibm.com/developerworks/ibm/library/i-odoel/.

Schmidt, M. T., Hutchison, B., Lambros, P., & Phippen, R. (2005). The Enterprise Service Bus: Making Service-Oriented Architecture Real. *IBM Systems Journal, 44*, 727–744.

Schumacker, R. E., & Lomax, R. G. (2004). *A Beginner's Guide to Structural Equation Modeling* (2nd ed.). Mahwah, NJ: Lawrence Erlbaum Associates.

Seddon, P. B. (1997). A Respecification and Extension of the DeLone and McLean Model of IS Success . *Information Systems Research, 8*(3), 240–253. doi:10.1287/isre.8.3.240

Segars, A. H., & Grover, V. (1998). Strategic information systems planning success: An investigation of the construct and its measurement. *MIS Quarterly*, (June): 139–163. doi:10.2307/249393

Segev, A., Beam, C., & Shanthikumar, J. G. (2001). Optimal design of Internet-based auctions. *Information Technology and Management, 2*(2), 121-163.

Selnes, F. (1993). An examination of the effect of product performance on brand reputation, satisfaction and loyalty. *European Journal of Marketing, 27*(9), 19-35.

Senecal, S., Gharbi, J.-E., & Nantel, J. (2002). The Influence of Flow on Hedonic and Utilitarian Shopping Values. In S. Broniarczyk & K. Nakamoto (Eds.), *Advances in Consumer Research*, 29.

Serva, M. A., Benamati, J., & Fuller, M. A. (2005). Trustworthiness in B2C E-Commerce: An Examination of Alternative Models. *The Data Base for Advances in Information Systems, 36*(3), 89.

Seth, A., & Thomas, H. (1994). Theories of the firm: Implications for strategy research. *Journal of Management Studies, 31*, 165-191.

Sewing, J., Rosemann, M., & Dumas, M. (2006). Process-Oriented Assessment of Web Services . *International Journal of E-Business Research, 2*(1), 19–44.

Seybold, P. (2002). *Shopping online at Tesco.* Retrieved October 10, 2007 from: Business Line, Financial Daily from the Hindu Group of Publications Web Site:

Shah, S.K. (2001). Improving Information System Performance through Client Value Assessment: A Case Study, *Review of Business*, Spring, 37-42.

Shang, R., Chen, Y., & Shen, L. (2005). Extrinsic versus intrinsic motivations for consumers to shop on-line. *In-*

formation & Management, 42(3), 401–413. doi:10.1016/j.im.2004.01.009

Shankar, V., & O'Driscoll, T. (2002). How wireless networks are reshaping the supply chain. *Supply Chain Management Review, 6*(4), 44–51.

Shao, B. B. M., & Lin, W. T. (2001). Measuring the value of information technology in technical efficiency with stochastic production frontiers. *Information and Software Technology, 43*, 447–456.doi:doi:10.1016/S0950-5849(01)00150-1

Shao, B. B. M., & Lin, W. T. (2002). Technical efficiency analysis of information technology investments: a two-stage empirical investigation. Information & Management, 39, 391–401. doi:10.1016/S0378-7206(01)00105-7

Shapiro, C. (1983). Premiums for high quality products as returns to reputations. *The Quarterly Journal of Economics, 98*(4), 659-680.

Shelat, B., & Egger, F. (2002). *What makes people trust online gambling sites?* Paper presented at the Conference on Human Factors in Computing Systems.

Sheskin, D. J. (2004). *Handbook of Parametric and Nonparametric Statistical Procedures* (2004 ed.): CRC Press.

Sheth, A., Verma, K., & Gomadam, K. (2006). Services Science: Semantics to Energize the Full Services Spectrum. *Communications of the ACM, 49*(7). doi:10.1145/1139922.1139949

Shih, H. (2004). An empirical study on predicting user acceptance of e-shopping on the web. *Information & Management, 41*, 351–368. doi:10.1016/S0378-7206(03)00079-X

Shim, S. Eastlick, M.A. Lotz, S.L., & Warrington (2001). An online pre-purchase intentions model: the role of intentions to search. *Journal of Retailing* 77, 397–416.

Shin, N. (2001). Strategies for competitive advantage in electronic commerce. *Journal of Electronic Commerce Research, 2*(4), 34–41.

Shirley, D., Stanley, W., & Daniel, C. (2004). *Statistics for research* (3 ed.): New York; Chichester: Wiley-Interscience

Shneiderman, B. (2000). Designing trust into online experiences. *Communications of the ACM, 43*(12), 57–59.

Short, J., Williams, E., & Christie, B. (1976). *The Social Psychology of Telecommunications.* London: Wiley.

Siau, K. (2003). Interorganizational systems and competitive advantages - lessons from history. *Journal of Computer Information Systems, 44*(1), 33–39.

Siau, K., Lim, E.-P., & Shen, Z. (2003). Mobile commerce: Current states and future trends. In E.-P. Lim & K. Siau (Eds.), *Advances in Mobile Commerce Technologies* (pp. 1-17). Hershey, PA: Idea Group Publishing.

Sigurdson, J. (2001). *WAP OFF - Origin, Failure and Future.* Prepared for the Japanese-European Technology Studies (JETS), October 9, 2001, University of Edinburgh.

Siller, H. R., Estruch, A., Vila, C., Abellan, J. V., & Romero, F. (2008). Modeling workflow activities for collaborative process planning with product lifecycle management tools . *Journal of Intelligent Manufacturing, 19*(6), 689–700. doi:10.1007/s10845-008-0120-6

Silva, N., & Rocha, J. (2003, 13-17 October). *Semantic Web Complex Ontology Mapping.* Paper presented at the IEEE WIC International Conference.

Simon, H. A. (1955). A Behavioral Model of Rational Choice. *The Quarterly Journal of Economics, 69*(1), 99–118. doi:10.2307/1884852

Simon, H. A. (1956). Models of Thought. New Haven, CT: Yale University Press.

Simon, S. J. (2001). The impact of culture and gender on web sites: An empirical study. *The Data Base for Advances in Information Systems, 1*(32), 18–37.

Singh, N., Xhao, H., & Hu, X. (2003). Cultural adaptation on the Web: A study of American companies' domestic and Chinese websites. *Journal of Global Information Management, 11*(3), 63–80.

Singh, R., & Salam, A. F. (2006). Semantic information assurance for secure distributed knowledge management: a business process perspective . *IEEE Transactions on Systems, Man, and Cybernetics, 36*(3), 472–486. doi:10.1109/TSMCA.2006.871792

Sinha, A. R., & Greenleaf, E. A. (2000). The impact of discrete bidding and bidder aggressiveness on sellers' strategies in open English auctions: Reserves and covert shilling. *Marketing Science,* (19), 244-266.

Skog, B. (2000). The mobile phone as symbolic capital in teenage cultures (in Norwegian). In R. Ling & K. Thrane (Eds.), *Social consequences of mobile telephony* (in Norwegian), Proceedings from a seminar on society, children, and mobile telephony, Telenor FoU - Notat nr. 38/2000.

Skog, B. (2002). Mobiles and the Norwegian teen: identity, gender and class. In J. E. Katz & M. Aakhus (Eds.), *Perpetual contact.* New York: Cambridge University Press.

Slater, S. F. (1977). Developing a customer value-based theory of the firm. *Journal of the Academy of Marketing Science, 25*(2), 162-167.

Sledgianowski, D., & Luftman, J. (2005). IT-business strategic alignment maturity: A case study. *Journal of Cases on Information Technology, 7*(2), 102–120.

Smith, M. D., Bailey, J., & Brynjolfsson, E. (2000). Understanding digital markets: Review and assessment. In E. Brynjolfsson & B Kahin (Eds.), *Understanding digital economy.* Cambridge, MA: MIT Press.

SOAD. (2004). *Service Oriented Analysis and Design Methodology.* Retrieved September 26, 2008 from http://www.ibm.com/developerworks/webservices/library/ws-soad1/

Soliman, K., & Janz, B. (2004). An exploratory study to identify the critical factors affecting the adoption of internet-based interorganizational information systems. *Information & Management, 41,* 697–706. doi:10.1016/j.im.2003.06.001

Solomon, D. (2001). Conducting Web-based surveys. *Practical Assessment Research and Evaluation, 7*(19).

Song, J., Jones, D., & Gudigantala, N. (2007). The Effect of Incorporating Compensatory Choice Strategies in Web-based Consumer Decision Support Systems. *Decision Support Systems, 43*(2), 359–374. doi:10.1016/j.dss.2006.10.007

Sorensen, E. H., Miller, K. L., & Ooi, C. K. (2000). The decision tree approach to stock selection. *Journal of Portfolio Management, 27*(1), 42-52.

Sougata, R. (2004). Environment-strategy-performance linkages: A study of Indian firms during economic liberalization. *Vikalpa, 29*(2), 9–23.

Southern, A., & Tilley, F. (2000). Small firms and information and communication technologies: Toward a typology of ICT usage. *New Technology, Work and Employment, 15*(2), 115–123. doi:10.1111/1468-005X.00070

Stafford, M. R., & Stern, B. (2002). Consumer bidding behavior on Internet auction sites. *International Journal of Electronic Commerce, 7*(1), 135-150.

Stafford, M. R., & Stern, B. (2002). Consumer bidding behavior on Internet auction sites. *International Journal of Electronic Commerce, 7*(1), 135-150.

Standifird, S. S. (2001). Reputation and e-commerce: EBay auctions and the asymmetrical impact of positive and negative ratings. *Journal of Management, 27*(3), 279-295.

Standifird, S. S., Roelofs, M. R., & Durham, Y. (2004). The impact of eBay's buy-it-now function on bidder behavior. *International Journal of Electronic Commerce, 9*(2), 167-176.

Stanton, J. (1998). An empirical assessment of data collection using the Internet. *Personal Psychology, 51*(3), 709-725.

Stark, R., & Rothkopf, M. H. (1979). Competitive bidding: A comprehensive bibliography. *Operations Research, 27*(2), 364-391.

Steenkamp, J. P., & Van Trijp, H. C. M. (1991). The use of Lisrel in validating marketing constructs. *International Journal of Research in Marketing, 8,* 283–299. doi:10.1016/0167-8116(91)90027-5

Steinberg, D., & Colla, P. (1997). *CART: Interface and documentation.* San Diego, CA: Salford Systems.

Steinbruck, U., Schaumburg, H., Kruger, T., & Duda, S. (2002). *A picture says more than a thousand words photographs as trust builders in e-commerce websites.* Paper presented at the Conference on Human Factors in Computing Systems.

Steinfield, C., Marcus, M. L., & Wigand, R. (2005). Exploring Interorganizational Systems at the Industry Level of

Analysis: Evidence from the U.S. Home Mortgage Industry. *Journal of Information Technology, 20*(4). doi:10.1057/palgrave.jit.2000051

Stewart, K. J. (2003). Trust transfer on the World Wide Web. *Organization Science, 14*(1), 5-17.

Stohr, E. A., & Zhao, J. L. (2001). Workflow Automation: Overview and Research Issues . *Information Systems Frontiers, 3*(3), 281–296. doi:10.1023/A:1011457324641

Storey, D. J. (1994). *Understanding the small business sector.* London: Routledge.

Strassman, P. (1990). *The Business value of Computers.* The Economic Press.

Strassmann, P. A. (1997). *The Squandered Computer: Evaluating the Business Alignment of Information Technologies,* Information Press, New York.

Stratopoulos, T., & Dehning, B. (2000). Does successful investment in information technology solve the productivity paradox? *Information & Management, 38,* 103–117. doi:10.1016/S0378-7206(00)00058-6

Straub, D. E., Hoffman, D. L., Weber, B. W., & Steinfield, C. (2002). Measuring e-Commerce in net-enabled organizations: An introduction to the special issue. *Information Systems Research, 13*(2), 115–124. doi:10.1287/isre.13.2.115.86

Straub, D. W. (1994). The effect of culture on IT diffusion: E-mail and FAX in Japan and the U.S. *Information Systems Research, 5,* 23–47.

Strnadl, F. C. (2006). Aligning business and IT: The process driven architecture model. *Information Systems Management, 2006*(Fall), 67–77. doi:10.1201/1078.10580530/46352.23.4.20060901/95115.9

Stubblebine, S., & Syverson, P. (1999). Fair online auctions without special trusted parties. In *Proceedings of the Third Annual Conference on Financial Cryptography,* (pp. 230-240).

Stuckenschmidt, H., Wache, H., & Visser, U. (2003, October 20). *Information Integration on the World Wide Web* Paper presented at the 2nd International Semantic Web Conference 2003, Sanibel Island, Florida.

Subramani, M. (2004). How do suppliers benedif from information technology use in supply chain relationships? *MIS Quarterly, 28*(1), 45–73.

Subramani, M., & Walden, E. (2001). The impact of e-commerce announcements on the market value of firms. *Information Systems Research, 12*(2), 135–154. doi:10.1287/isre.12.2.135.9698

Subramaniam, R., Mittal, V., & Inman, J. J. (2004). Revenue equivalence in online auctions: The case of Yahoo and eBay. In P. T. L. Popkowski Leszczyc (Ed.), *Advances in consumer research* (Vol. 31, pp. 1-2).

Sun, H. (2001). Building a culturally-competent corporate web site: An explanatory study of cultural markers in multilingual web design. SIGDOC, 1(October 21-24), 95-102.

Sun, H., & Hong, C. (2002). The alignment between manufacturing and business strategies: its influence on business performance. *Technovation, 22*(11), 699–705. doi:10.1016/S0166-4972(01)00066-9

Sun, S., Kumar, A., & Yen, J. (2008). Merging workflows: A new perspective on connecting business processes . *Decision Support Systems, 42*(2), 844–858. doi:10.1016/j.dss.2005.07.001

Sutcliffe, A., & Mehandjiev, N. (2004). End-user development: Introduction . *Communications of the ACM, 47*(9), 31–32. doi:10.1145/1015864.1015883

Swatman, P. M. C., & Swatman, P. A. (1992). EDI system integration: a definition and literature survey. *The Information Society, 8,* 169–205.

Swerts, M., Krahmer, E., Barkhuysen, P., & van de Laar, L. (2003). *Audiovisual cues to uncertainty.* Paper presented at the ISCA Workshopon Error Handling in Spoken Dialog Systems, Switzerland.

Swinder, J., Trocchia, P. J., & Gwinner, K. P. (2002). Consumer Perceptions of Internet Retail Service Quality. *International Journal of Service Industry Management, 13*(5), 412. doi:10.1108/09564230210447913

Szymanski, D., & Hise, R. T. (2000). e-Satisfaction: An Initial Examination. *Journal of Retailing, 76*(3), 309–322. doi:10.1016/S0022-4359(00)00035-X

Tagg, R., (2001). Workflow in Different Styles of Virtual Enterprise, *Proceedings, Workshop in Information Technology for Virtual Enterprises*, 21-28.

Taking the Pulse of the Internet Service Provider Industry: Evolutions and

Talon, P. P., Kraemer, K. L., & Gurbaxani, V. (2000). Executives' Perceptions of the Business Value of Information Technology: A Process-Oriented Approach . *Journal of Management Information Systems*, (Spring): 145–173.

Tan, J. C., & Harker, P. T. (1999). Designing Workflow Coordination: Centralized Versus Market-Based Mechanisms . *Information Systems Research, 10*(4), 328–342. doi:10.1287/isre.10.4.328

Tan, M., & Teo, T. S. H. (1998). Factors influencing the adoption of the Internet. *International Journal of Electronic Commerce, 2*(3), 5–18.

Tastle, W. J., & Wierman, M. J. (2005). Consensus and dissention: A new measure of agreement. *North American Fuzzy Information Processing Society (NAFIPS) Conference,* Ann Arbor, MI.

Tastle, W. J., & Wierman, M. J. (2006a). An information theoretic measure for the evaluation of ordinal scale data. *Behavioral Research Methods, 3*, 487-494.

Tastle, W. J., & Wierman, M. J. (2006b). Consensus and dissension: A new measure of agreement. In *NAFIPS 2006,* Montreal, Canada.

Tastle, W. J., & Wierman, M. J. (in press-b). The development of agreement measures: From general to targeted. *International Journal of Approximate Reasoning.*

Tastle, W. J., & Wierman, M. J. (in press-c). Determining risk assessment using the weighted ordinal agreement measure. *Journal of Homeland Security.*

Taylor, S., & Todd, P. A. (1995). Understanding information technology usage: a test of competing models. *Information Systems Research, 6*, 144–176. doi:10.1287/isre.6.2.144

Teece, D. J. (2000), *Managing intellectual capital.* UK: Oxford University Press.

Teece, D. J., Pisano, G., & Shuen, A. (1997). Dynamic capabilities and strategic management. *Strategic Management Journal, 18*(7), 509–533. doi:10.1002/(SICI)1097-0266(199708)18:7<509::AID-SMJ882>3.0.CO;2-Z

Teo, T. S. H., & Ang, J. S. K. (1999). Critical success factors in the alignment of IS plans with business plans. *International Journal of Information Management, 19*, 173–185. doi:10.1016/S0268-4012(99)00007-9

Teo, T. S. H., & Ang, J. S. K. (2000). How useful are strategic plans for information systems? *Behaviour & Information Technology, 19*(4), 275–282. doi:10.1080/01449290050086381

Teo, T. S. H., & Ang, J. S. K. (2001). An examination of major IS planning problems. *International Journal of Information Management, 21*, 457–470. doi:10.1016/S0268-4012(01)00036-6

Teo, T. S. H., & King, W. R. (1996). Assessing the impact of integrating business planning and IS planning. *Information and Management*, 30, 309-321.

Teo, T. S. H., & King, W. R. (1997). Integration between business planning and information systems planning: An evolutionary-contingency perspective. *Journal of Management Information Systems, 14*(1), 185–214.

Teo, T. S. H., & Liu, J. (2005). Consumer trust in e-commerce in the United States, Singapore and China. *Omega, 35*(1), 22–38.

Teo, T. S. H., & Pian, Y. (2003). A contingency perspective of Internet adoption and competitive advantage. *European Journal of Information Systems, 12*, 78–92. doi:10.1057/palgrave.ejis.3000448

Teo, T. S. H., & Too, B. L. (2000). Information systems orientation and business use of the Internet: An empirical study. *International Journal of Electronic Commerce, 4*(4), 105–130.

Teo, T. S. H., Tan, M., & Buk, W. K. (1997). A contingency model of Internet adoption in Singapore. *International Journal of Electronic Commerce, 2*(2), 95–118.

Teo, T. S. H., Tan, M., & Buk, W. K. (1998). A contingency model of Internet adoption in Singapore. *International Journal of Electronic Commerce, 2*, 96–118.

Teo, T., & Ranganathan, C. (2004). Adopters and non-adopters of business-to-business electronic commerce in Singapore. *Information & Management*, *42*, 89–102.

Teo, T.S.H. (2002). Attitudes towards online shopping and the Internet. *Behaviour and Information Technology* 21(4), 209–271.

Thomas, D. (2003). E-shopping continues to boom. *Computer Weekly* 27(February), 3.

Thomas, D. (2003). Supermarket boosts efficiency tenfold by integrating systems. *Computer Weekly*, 2/20/2003, 12.

Thomas, M., Redmond, R., Yoon, V., & Singh, R. (2005). A Semantic Approach to Monitor Business Process Performance . *Communications of the ACM*, *48*(12), 55–59. doi:10.1145/1101779.1101809

Thong, J. Y. L. (1999). An integrated model of information systems adoption in small businesses. *Journal of Management Information Systems*, *15*(4), 187–214.

Thorelli, H. B. (1986). Networks: Between markets and hierarchies. *Strategic Management Journal*, *7*, 37–51. doi:10.1002/smj.4250070105

Tian, R. G., & Emery, C. (2002). Cross-cultural issues in Internet marketing. *Journal of American Academy of Business*(March), 217-224.

Tiessen, J., Wright, R., & Turner, I. (2001). A model of e-commerce use by internationalizing SMEs. *Journal of International Management*, *7*, 211–233. doi:10.1016/S1075-4253(01)00045-X

Todd, P., & Benbasat, I. (1994). The influence of decision aids on choice strategies: An experimental analysis of the role of cognitive effort. *Organizational Behavior and Human Decision Processes*, *60*(1), 36–65. doi:10.1006/obhd.1994.1074

Tolk, A., & Diallo, S. Y. (2005). Model-based Data Engineering for Web Services. *IEEE Internet Computing*, *9*(4). doi:10.1109/MIC.2005.87

Tornatzky, L. G., & Klein, K. J. (1982). Innovation characteristics and innovation adoption-Implementation: a meta-analysis of findings. *IEEE Transactions on Engineering Management*, *29*(1), 28–45.

Totty, M. (2003, June 16). E-Commerce (A Special Report): Selling Strategies - Business Solutions. *Wall Street Journal*, p. 4.

Tretola, G., & Zimeo, E. (2008). Activity pre-scheduling for run-time optimizationof grid workflows . *Journal of Systems Architecture*, *54*, 883–892. doi:10.1016/j.sysarc.2008.01.009

Trevathan, J. (2005). *Electronic auctions literature review* (JCU Tech. Rep.).

Truman, G. (2000). Integration in electronic exchange environments. *Journal of Management Information Systems*, *17*(1), 209–244.

Truong, H., Dustdar, S., & Fahringer, T. (2008). Performance metrics and ontologies for Grid workflows . *Future Generation Computer Systems*, *23*, 760–772. doi:10.1016/j.future.2007.01.003

Tsai, D. H. A. (2005). Knowledge spillovers and high-technology clustering: evidence from Taiwan's Hsinchu Science-Based Industrial Park. *Contemporary Economic Policy. Huntington Beach*, *23*(1), 116–139.

Tsai, L. H., & Koehler, G. J. (1993). The accuracy of concepts learned from induction. *Decision Support Systems*, *10*(2), 161-172.

Tsay, A., & Agrawal, N. (2004). Channel Conflict and Coordination in the E-Commerce Age. *Production and Operations Management*, *13*(1), 93–110.

Tseng, F. C., Teng, C. I., & Chiang, D. M. (2007). Delivering Superior Customer Perceived Value in the Context of Network Effects. *International Journal of E-Business Research*, *3*(1), 41.

Tsikriktsis, N. (2002). Does Culture Influence Web Site Quality Expectations? An Empirical Study. *Journal of Service Research*, *5*(2), 101–112. doi:10.1177/109467002237490

Tung, F.-C. (2007). Using the e-CRM information system in the high-tech industry: predicting salesperson intentions. *Journal of American Academy of Business*, *11*(2), 131–136.

Turban, E., King, D., Lee, J., & Viehland, D. (2004). *Electronic commerce: A managerial perspective*. Pearson Education.

Turban, E., King, D., Lee, J., Warkentin, M., & Chung, H. M. (2002). *Electronic Commerce 2002: A Managerial Perspective*. Upper Saddle River, NJ: Pearson Education.

TwoCrows (1999). *Introduction to data mining and knowledge discovery*. Potomac, MD: Two Crows Corporation.

U.S. Census Bureau. (2007). *Quarterly retail e-commerce sales*. Retrieved February 13, 2008, from http://www.census.gov/mrts/www/ecomm.html

U.S. Department of Commerce Report. (2004). *A Nation Online, Entering the Broadband Age*. Retrieved November 9, 2007 from http://www.ntia.doc.gov/reports/anol

U.S. Department of Justice. (2007). *Internet and telemarketing fraud*. Retrieved February 13, 2008, from http://www.usdoj.gov/criminal/fraud/internet/#howtodeal

Ueltschy, L. C., Laroche, M., Eggert, A., & Bindl, U. (2007). Service quality and satisfaction: An international comparison of professional services perceptions. *Journal of Services Marketing, 21*(6), 410-423.

Ujah, E. (2005, October 10). Electronic business and small business development in Nigeria. *Nigeria Business Times*.

Ujah, E. (2008, January 19). E-business Challenges for Nigerian Companies. *Nigeria Business Times*.

UNCTAD. (2002). E-commerce and development report 2002. *United Nations Conference on Trade and Development, New York and Geneva*.

UNCTAD. *(2004)*. United Nations Conference on Trade and Development E-commerce and Development Report 2004.

United States Census Bureau. (2009). Quarterly Retail E-commerce Sales: 4th Quarter 2008. *United State Department of Commerce*. Retrieved from http://www.census.gov/mrts/www/data/html/08Q4.html

Urban, G. L., Fareena, S., & Qualls, W. (1999). *Design and Evaluation of a trust based advisor on the internet*. Retrieved 20-January, 2006, from http://ebusiness.mit.edu/research/papers/123%20Urban,%20Trust%20Based%20Advisor.pdf. date of download.11-10-2006

Urban, G. L., Sultan, F., & Qualls W. J. (2000). Placing trust at the center of your Internet strategy. *Sloan Management Review, 42*(1), 39-48.

Utterback, J. M. (1974). Innovation in industry and the diffusion of technology. *Science, 183*, 620–626. doi:10.1126/science.183.4125.620

Utterback, J. M., & Abernathy, J. M. (1975). A dynamic model of process and product innovation. *Omega, 3*(6), 639–657. doi:10.1016/0305-0483(75)90068-7

Van den Poel, D., & Leunis, J. (1999). Consumer acceptance of the Internet as a channel of distribution. *Journal of Business Research, 45*, 249-256.

van der Aalst, W. M. P. (1999). *Generic Workflow Models: How to Handle Dynamic Change and Capture Management Information*. Paper presented at the the Fourth IFCIS International Conference on Cooperative Information Systems (CoopIS'99), Edinburgh, Scotland.

van der Aalst, W. M. P. (1999). Process-oriented Architectures for Electronic Commerce and Interorganizational Workflow. *Information Systems, 24*(8), 639–671. doi:10.1016/S0306-4379(00)00003-X

van der Aalst, W. M. P., & Jablonski, S. (2000). Dealing with Workflow Change: Identification of Issues and Solutions. *International Journal of Computer Systems, Science, and Engineering, 15*(5), 267–276.

van der Aalst, W. M. P., & Kumar, A. (2003). XML based schema definition for support of inter-organizational workflow . *Information Systems Research, 14*(1), 23–46. doi:10.1287/isre.14.1.23.14768

van der Aalst, W. M. P., & van Dongen, B. F. (2002). Discovering Workflow Performance Models from Timed Logs. In Y. Han, S. Tai, & D. Wikarski (Eds.), *International Conference on Engineering and Deployment of Cooperative Information Systems, EDCIS2002* (LNCS 2480).

van der Aalst, W. M. P., & van Hee, K. M. (2002). *Workflow Management: Models, Methods, and Systems*. MIT Press: Cambridge, MA.

van der Aalst, W. M. P., Desel, J., & Oberweis, A. (Eds.). (2000). *Business Process Management: Models, Techniques, and Empirical Studies* (LNCS 1806).

van der Aalst, W. M. P., ter Hofstede, A. H. M., Kiepuszewski, B., & Barros, A. P. (2003). Workflow Patterns . *Distributed and Parallel Databases, 14*(3), 5–51. doi:10.1023/A:1022883727209

van der Aalst, W. M. P., Weske, M., & Grunbauer, D. (2005). Case Handling: A New Paradigm for Business Process Support. *Data & Knowledge Engineering, 53*, 129–162. doi:10.1016/j.datak.2004.07.003

van der Aalst, W. M. P., Weske, M., & Wirtz, G. (2003). Advanced Topics in Workflow Management: Issues, Requirements, and Solutions. *Journal of Integrated Design and Process Science, 7*(3), 49–77.

van Heck, E., & van Damme, E. (1997). *New entrants and the role of IT: The tele-flower auctions in the Netherlands.* Paper presented at the 30th HICSS, Hawaii.

Van Mulken, S., Andre, E., & Müller, J. (1999). An empirical study on the trustworthiness of life-like interface agents. In (pp. 152-156). Mahwah, NJ: Lawrence Erlbaum.

Varki, S., & Colgate, M. (2001). The role of price perceptions in an integrated model of behavioural intentions. *Journal of Service Research, 3*(3), 232-240.

Venkatesh, V., & Davis, F. D. (2000). A theoretical extension of the Technology Acceptance Model: four longitudinal field studies. *Management Science, 46*(2), 186–204. doi:10.1287/mnsc.46.2.186.11926

Venkatraman, N. (1989). Strategic orientation of business enterprises: the construct, dimensionality and measurement. *Management Science, 35*(8), 942–962. doi:10.1287/mnsc.35.8.942

Venkatraman, N. (1994). IT-enabled business transformation: from automation to business scope redefinition. *Sloan Management Review, 35*(2), 73–87.

Venkatraman, N., & Ramanujam, V. (1986). Measurement of business performance in strategy research: A comparison of approaches. *Academy of Management Review, 11*, 801–814. doi:10.2307/258398

Verbeek, H. M. W., Basten, T., & van der Aalst, W. M. P. (2001). Diagnosing Workflow Processes using Woflan . *The Computer Journal, 44*(4), 246–279. doi:10.1093/comjnl/44.4.246

Verizon Tempts Enterprises, SMBs With New Apps Laundry List. (2008, July 10), TelecomWeb News Break, 1.

Verkasalo, H. (2005). *Handset-Based Monitoring of Mobile Customer Behavior.* Master's Thesis Series. Networking Laboratory. Department of Electrical and Telecommunications Engineering. Helsinki University of Technology.

Verkasalo, H. (2006a). *Mobile Data Service Evolution - Empirical Implications from Europe and the USA.* Paper presented at the 3rd International CICT Conference, November 30 - December 1, 2006, Denmark, 2006.

Verkasalo, H. (2006b). *Empirical Observations on the Emergence of Mobile Multimedia Services in the U.S. and Europe.* Paper presented at the 5th International Conference on Mobile and Ubiquitous Multimedia. December 4-6, 2006, Stanford University, California, 2006.

Verkasalo, H. (2007a). A Cross-Country Comparison of Mobile Service and Handset Usage. Licentiate's thesis, Helsinki University of Technology, Networking Laboratory, Finland.

Verkasalo, H. (2007b). Empirical Insights on the Evolution of the Finnish Mobile Market. Paper presented at Conference on Telecommunication Techno-Economics (CTTE) 2007, 14-15 June 2007, Helsinki, Finland.

Verkasalo, H. (2007c). *Empirical Findings on the Mobile Internet and E-Commerce.* Paper presented at 20th Bled e Conference: eMergence: Merging and Emerging Technologies, Processes, and Institutions. Bled, Slovenia, June 4 - 6, 2007.

Verkasalo, H. (2007d). *Contextual Usage-Level Analysis of Mobile Services.* Accepted for publication at The 4th Annual International Conference on Mobile and Ubiquitous Systems: Computing, Networking and Services (MOBIQUITOUS 2007). August 6-10, 2007 - Philadelphia, PA, USA.

Verkasalo, H., & Hämmäinen, H. (2006). *Handset-Based Monitoring of Mobile Subscribers*. Paper presented at Mobility Roundtable, 1-2 June, 2006, Helsinki, Finland.

Verkasalo, H., & Hämmäinen, H. (2007). A Handset-Based Platform for Measuring Mobile Service Usage. *INFO: The Journal of Policy, Regulation and Strategy, 9*(1).

Verma, M., & Deswal, P. (2003). Approaching Web Services Transactions (Tech. Rep.). Second Foundation Inc. Retrieved February 2005 from, http://www-128.ibm.com/developerworks/webservices/library/ws-tranart [Visited February 2005].

Vesa, J. (2005). *Mobile Services in the Networked Economy*. Hershey, PA: IRM Press.

Vickrey, W. (1961). Counter speculation, auctions and sealed tenders. *Journal of Finance, 16*, 8-37.

Vijayasarathy, L.R. (2003). Psychographic Profiling of Online Shopper. *Information Science Publishing* 1(3), 48-74.

Vilagines. (1994). *Marketing Relocional: Como Obtener Clientes Leales y Rentables*. Ed., Gestion S. A., Barcelona.

Viswanathan, K., Boyd, C., & Dawson, E. (2000). A three phased schema for sealed bid auction system design. In *ACISP*.

Vivekanandan, K., & Rajendran, R. (2005). Adoption of electronic commerce: An exploratory study of Tirupur knitwear apparel exporters. *Organisational Management, 21*(3), 5–11.

Vivekanandan, K., & Rajendran, R. (2006). Export marketing and the World Wide Web: perceptions of export barriers among Tirupur knitwear apparel exporters - An empirical analysis. *Journal of Electronic Commerce Research, 7*(1), 27–40.

Vogelstein, F., Boyle, M., Lewis, P., & Kirkpatrick, D. (2004). 10 tech trends to bet on. *Fortune, 149*(4), 75-81.

Von Krogh, G., Ichijo, K., & Nonaka, I. (2000). *Enabling knowledge creation: How to unlock the mystery of tacit knowledge and release the power of innovation*. New York: Oxford University Press.

Voss, G., Parasuraman, A., & Grewal, D. (1998). The role of price and quality perceptions in prepurchase and post-purchase evaluation of services. *Journal of Marketing, 62*(4), 46-61.

Wade, M., & Hulland, J. (2004). Review: The resource-based view and information systems research: Review, extension and suggestions for future research. *MIS Quarterly, 28*(1), 107–142.

Wainer, J., Barthelmess, P., & Kumar, A. (2003). W-RBC - A Workflow Security Model Incorporating Controlled Overriding of Constraints . *International Journal of Cooperative Information Systems, 12*(4), 455–485. doi:10.1142/S0218843003000814

Wakefield, R. L., & Whitten D. (2006). Examining user perceptions of third-party organization credibility and trust in an e-retailer. *Journal of Organizational and End User Computing, 18*(2), 1-19.

Waldfogel, J., & Chen, L. (2006). Does Information Undermine Brand? Information Intermediary Use and Preference for Branded Web Retailers. *The Journal of Industrial Economics, 54*(4), 425–449. doi:10.1111/j.1467-6451.2006.00295.x

Wallis, G. (2006). Internet spending: measurement and recent trends. *Economic Trends, 628*(MARCH,2006).

Wang, C., & Leung, H. (2004). Anonymity and security in continuous double auctions for Internet retails market. In *Proceedings of the 37th Hawaii International Conference on System Sciences*.

Wang, M., & Wang, H. (2006). From Process logic to business logic - A cognitive approach to business process management . *Information & Management, 43*, 179–193. doi:10.1016/j.im.2005.06.001

Wang, S., Beatty, S. E., & Foxx, W. (2004). Signaling the trustworthiness of small online retailers. *Journal of Interactive Marketing, 18*(1), 53-69.

Wang, Y. S., Wang, Y., Lin, H. H., & Tang, T. I. (2003). Determinants of user acceptance of Internet banking: An empirical study. *International Journal of Service Industry Management, 14*(5), 501–519. doi:10.1108/09564230310500192

Wang, Y., & Emurian, H. (2005). An overview of online trust: Concepts, elements, and implications. *Computers in Human Behavior*, *21*(1), 105–125.

Ward, M., & Lee, M. (2000). Internet shopping, consumer search and product branding. *Journal of Product and Brand Management*, *9*(1), 6–20. doi:10.1108/10610420010316302

Ward, S. G., & Clark. J. M. (2002). Bidding behavior in on-line auctions: An examination of the eBay Pokemon card market. *International Journal of Electronic Commerce*, *6*(4), 139-155.

Watson, R., & McKeown, P. (1999). Manheim auctions: transforming interorganizational relationships with an extranet. *International Journal of Electronic Commerce*, *3*(4), 29–46.

Webb, K. (2002). Managing channels of distribution in the age of electronic commerce. *Industrial Marketing Management*, *31*, 95–102. doi:10.1016/S0019-8501(01)00181-X

Weick, K. (1995). What theory is not, theorizing is. *Administrative Science Quarterly*, *40*(3), 385-390.

Weill, P. (1990). Strategic investment in information technology - an empirical study. *Information Age*, *12*(3), 141–147.

Weill, P., & Olson, M. H. (1989). An assessment of contingency theory. *Journal of Management Information Systems*, *6*(1), 59–85.

Weiss, M., & Amyot, D. (2005). Business Process Modeling with URN. *International Journal of E-Business Research*, *1*(3), 63–90.

Wernerfelt, B. (1991). Brand loyalty and market equilibrium. *Marketing Science*, *10*(3), 229-245.

West, J. (2001). Customer relationship management and you. *IIE Solutions*, *33*(4), 34–37.

West, J. (2002). Information for Sale: My Experience with Google Answers. *Searcher - The Magazine for Database Professionals*, *10*(9), 14.

Wheeler, C. B. (2002). NEBIC: A dynamic capabilities theory for assessing net-enablement. *Information Systems Research*, *13*(2), 125–146. doi:10.1287/isre.13.2.125.89

Whetten, D. (1989). What constitutes a theoretical contribution. *Academy of Management Review*, *14*(4), 490-495.

Whinston, B. A., & Geng, X. (2004). Operationalising the essential role of the information technology artifact in information systems research: Gray area, pitfalls, and the importance of strategic ambiguity. *MIS Quarterly*, *28*(2), 149–159.

Widing, R. E., & Talarzyk, W. W. (1993). Electronic Information Systems for Consumers: An Evaluation of Computer-Assisted Formats in Multiple Decision Environments. *JMR, Journal of Marketing Research*, *30*(2), 125–141. doi:10.2307/3172823

Wierman, M. J., & Tastle, W. J. (2005). Consensus and dissention: Theory and properties. *North American Fuzzy Information Processing Society (NAFIPS) Conference*, Ann Arbor, MI.

Wilcox, R. T. (2000). Experts and amateurs: The role of experience in Internet auctions. *Marketing Letters*, *11*(4), 363-374.

Williamson, O. E. (1983). Organizational innovation: The transaction cost approach. In J. Ronen (Ed.), *Entrepreneurship* (pp. 101-133). Lexington, MA: Lexington Books.

Wilson, R. B. (1969). Communications to the editor: Competitive bidding with disparate information. *Management Science*, *15*(7), 446-449.

Wilson, R. B. (1987). Auction theory. In J. Eatwell, M. Milgate & P. Newman (Eds.), *The new Palgrave: A dictionary of economic theory*. London: Macmillan.

Wingfield, N. (2001). EBay watch: Corporate sellers put the online auctioneer on even faster track --- Goods from IBM, Disney help dot-com pioneer post a surge in profits --- Why mom and pop are mad. *Wall Street Journal*, A.1.

Witkowski, M., Neville, B., & Pitt, J. (2003). Agent mediated retailing in the connected local community. *Interacting with Computers*, *15*(1), 5–32.

Witt, L. (2005). Building sales on eBay. Retrieved May 10, 2005, from http://www.fortune.com/fortune/smallbusiness/answercentral/0,15704,601929

Wold, H. (1989). Introduction to the Second Generation of Multivariate Analysis. In H. Wold (Ed.), *Theoretical Empiricism* (pp. vii-xl). New York: Paragon House.

Wolfe, R. A. (1994). Organisational innovation: review, critique and suggested research directions. *Journal of Management Studies*, *31*(3), 405–431. doi:10.1111/j.1467-6486.1994.tb00624.x

Wolfinbarger, M., & Gilly, M. C. (2003). eTailQ: Dimensionalizing, Measuring and Predicting Etail Quality. *Journal of Retailing*, *79*(3), 183–198. doi:10.1016/S0022-4359(03)00034-4

Wolfingarger, M., & Gilly, M. C. (2001). Shopping online for freedom, control, and fun. *California Management Review*, *43*(2), 34-55.

Wong, C. C., & Hiew, P. L. (2005). Correlations between factors affecting the diffusion of mobile entertainment in Malaysia. *ICEC*, *2005*, 615–621. doi:10.1145/1089551.1089661

Woo, C. Y., & Willard, G. E. (1983). *Performance representation in business policy research: discussion and recommendation*. Paper presented at the 23rd Annual National Meetings of the Academy of Management, Dallas.

Wood, S. T. (2001). Remote purchase environments: The influence of return policy leniency on two-state decision processes. *Journal of Marketing Research, 38*(2), 156-169.

Woodward, J. (1965). *Industrial Organization: Theory and Practice*. London: Oxford University Press.

Woolley, S. (1998). E-muscle. *Forbes*, *161*(5).

Workflow Management Coalition. (2006). http://www.e-workflow.org/standards/index.htm

Wright, S. (1921). Correlation and causation. *Journal of Agricultural Research*, *20*, 557–585.

Wright, S. (1934). The method of path coefficients. *Annals of Mathematical Statistics*, *5*, 161–215. doi:10.1214/aoms/1177732676

Wright, S. (1960). Path coefficients and path regression: Alternative or complementary concepts? *Biometrics*, *16*, 189–202. doi:10.2307/2527551

Wymer, S. A., & Regan, E. A. (2005). Factors influencing e-commerce adoption and use by small and medium businesses. *Electronic Markets*, *15*(4), 438–453. doi:10.1080/10196780500303151

Xiao, B., & Benbasat, I. (2007). E-Commerce Product Recommendation Agents: Use, Characteristics, and Impact. *MIS Quarterly*, *31*(1), 137–209.

Xu, M., Rohatgi, R., & Yangxing, D. (2007). E-business Adoption in SMEs: Some Preliminary Findings from Electronic Components Industry. *International Journal of E-Business Research*, *3*(1), 74–90.

Xu, Y., Yen, D. C., Lin, B., & Chou, D. C. (2002). Adopting Customer Relationship Management Technology. *Industrial Management & Data Systems*, *102*(8/9), 442–452. doi:10.1108/02635570210445871

Yang, J., & Papazoglu, M. P. (2000). Interoperation support for electronic business. *Communications of the ACM*, *43*(6), 39–47. doi:10.1145/336460.336473

Yao, D. Q., & Liu, J. (2005). Competitive pricing of mixed retail and e-tail distribution channels. *Omega, 33*(25), 235-247.

Yin, R. K. (2003). *Case Study Research: Design and Methods* (3rd ed.). Thousand Oaks, CA: Sage.

Yoo, B., & Donthu, N. (2001). Developing a Scale to Measure the Perceived Quality of Internet Shopping Sites (SITEQUAL). *Quarterly Journal of Electronic Commerce*, *2*(1), 31–47.

Yoo, Y., & Alavi, M. (2002). Media and group cohesion: Relative influences on social presence, task participation, and group consensus. *MIS Quarterly*, *25*, 371–390.

Yoon, S. J. (2002). The antecedents and consequences of trust in online-purchase decisions. *Journal of Interactive Marketing, 16*(2), 47-63.

Younas, M., Chao, K. M., Lo, C. C., & Li, Y. (2006). An Efficient Transaction Commit Protocol for Composite Web Services. In *Proceedings of the IEEE 20th International Conference on Advanced Information Networking and Applications (AINA'2006), Vienna, Austria.*

Yuen, W., Sung, W., & Wong, W. (2003). Optimal price decremental strategy for Dutch auctions.

Yuille, J. C., & Catchpole, M. J. (1977). The role of imagery in models of cognition. *Journal of Mental Imagery, 1*(1), 171-180.

Zajonc, R. (1980). Feeling and thinking: Preferences need no inferences. *The American Psychologist, 35*(2), 151–175.

Zanakis, S., & Becerra-Fernandez, I. (2005). Competitiveness of nations: A knowledge discovery examination. *European Journal of Operational Research, 166*(2), 185-211.

Zank, G. M., & Vokurka, R. J. (2003). The Internet: Motivations, Deterrents, and Impact on Supply Chain Relationships. *SAM Advanced Management Journal, 68*(2), 33–40.

Zeithaml, V. A., Parasuraman, A., & Malhotra, A. (2002). Service Quality Delivery through Web Sites: A Critical Review of Extant Knowledge. *Journal of the Academy of Marketing Science, 30*(4), 362–375. doi:10.1177/009207002236911

Zeithaml, V., Berry, L. L., & Parasuraman, A. (1996). The behavioral consequences of service quality. *Journal of Marketing, 60*(2), 31-46.

Zeng, Q., Wang, H., Xu, D., Duan, H., & Han, Y. (2008). Conflict detection and resolution for workflows constrained by resources and non-determined durations . *Journal of Systems and Software, 81,* 1491–1504. doi:10.1016/j.jss.2007.09.004

Zeng, X., Zeng, J., & Guo, Q. (2005). *Research of trust on B2C electronic commerce.* Paper presented at the the Proceedings of the 7th international conference on Electronic commerce, Xi'an, China.

Zeng, Y. E., Wen, H. J., & Yen, D. C. (2003). Customer-relationship management (CRM) in business to business e-commerce. *Information Management & Computer Security, 11*(1), 39–44. doi:10.1108/09685220310463722

Zhang, D. (2005). Web Services Composition for Process Management in E-Business. *Journal of Computer Information Systems, XLV*(2), 83–91.

Zhang, M., & Tansuhaj, P. (2007). Organizational culture, information technology capability, and performance: The case of born global firm. *The Multinational Business Review, 15*(3), 43–77.

Zhang, P., & Von Dran, G. M. (2001). Expectations and rankings of website quality features: Results of two studies on user perceptions. In *Proceedings of the 34th Hawaii International Conference on System Sciences,* HICSS-34(7), 7019.

Zhang, P., Von Dran, G. M., Small, R. V., & Barcellos, S. (1999). Websites that Satisfy Users: A Theoretical Framework for Web User Interface Design and Evaluation. In *Proceedings of the 32nd Hawaii International Conference on System Sciences,* HICSS-32(2), 2016.

Zhang, P., Von Dran, G. M., Small, R. V., & Barcellos, S. (2000). A Two-Factor Theory for Website Design. In *Proceedings of the 33rd Hawaii International Conference on System Sciences,* HICSS-33(6), 6026.

Zhang, X., & Zhang, Q. (2005). *Online trust forming mechanism: approaches and an integrated model.* Paper presented at the Proceedings of the 7th international conference on Electronic commerce, Xi'an, China.

Zhang, Y., & Gelb, B. (1996). Matching Advertising Appeals to Culture: The Influence of Products' Use Conditions. *Journal of Advertising, 25*(3).

Zhao, J. L. (2002). Interorganizational Workflow and E-Commerce Applications. Presentation at HICSS-35, http://attila.stevens-tech.edu/sigpam/publications/tutorial/HICSS-35/Interorganizational_Workflow.pdf.

Zheng, Y. C., Fan, Y. S., & Tan, W. (2008). Towards workflow simulation in service-oriented architecture: an event-based approach . *Concurrency and Computation, 20,* 315–330. doi:10.1002/cpe.1221

Zhu, K. (2004). The Complementarity of Information Technology Infrastructure and E-Commerce Capability: A Resource Based Assessment of their Business value. *Journal of Management Information Systems, 21*(1), 187–202.

Zhu, K., & Kraemer, K. L. (2002). e-Commerce metrics for net-enhanced organizations: Assessing the value of e-Commerce to firm performance in the manufacturing sector. *Information Systems Research, 13*(3), 275–295. doi:10.1287/isre.13.3.275.82

Zhu, K., & Kraemer, K. L. (2005). Post-adoption variations in usage and value of e-business by organizations. *Information Systems Research, 16*(1), 61–84. doi:10.1287/isre.1050.0045

Zhu, K., Kraemer, K. L., & Xu, S. (2002). A cross-country study of electronic business adoption using the technology-organization-environment framework. *Presented to the Twenty-Third International Conference on Information Systems, Barcelona* (pp. 337-348).

Zhu, K., Kraemer, K. L., Xu, S., & Dedrick, J. (2004). Information technology payoff in e-Business environments: An international perspective on value creation of e-Business in the financial industry. *Journal of Management Information Systems, 21*(1), 17–54.

Zhu, K., Kraemer, K., & Xu, S. (2003). Electronic business adoption by European firms: a cross-country assessment of the facilitators and inhibitors. *European Journal of Information Systems, 12*(4), 251–268. doi:10.1057/palgrave.ejis.3000475

Zimmerer, T. W., & Scarborough, N. M. (2002). *Essentials of Entrepreneurship and Small Business Management* (3rd ed.). Upper Saddle River, NJ: Pearson Education, Inc.

Zins, A. H. (2001). Relative attitudes and commitment in customer loyalty models—Some experiences in the commercial airline industry. *International Journal of Service Industry Management, 12*(3), 269-294.

Zmud, R. O., Lind, M., & Young, F. (1990). An attribute space for organizational communication channels. *Information Systems Research, 14*, 440–457.

Zmud, R. W. (1984). An examination of 'push-pull' theory applied to process innovation in knowledge work. *Management Science, 30*(6), 727–738. doi:10.1287/mnsc.30.6.727

Zott, C., Amit, R., & Donlevy, J. (2000). Strategies for value creation in e-commerce: Best practice in europe. *European Management Journal, 18*(5), 463–475. doi:10.1016/S0263-2373(00)00036-0

Zuboff, S. (2002). *The Support Economy: Why Corporations are failing Individuals and the Next Episode of Capitalism.* Viking Books.

About the Contributors

In Lee is a professor in the Department of Information Systems and Decision Sciences in the College of Business and Technology at Western Illinois University. He received his Ph.D. from University of Illinois at Urbana-Champaign. He is a founding editor-in-chief of the International Journal of E-Business Research. He has published his research in such journals as Communications of the ACM, IEEE Transactions on Systems, Man, and Cybernetics, IEEE Transactions on Engineering Management, International Journal of Production Research, Computers and Operations Research, Computers and Industrial Engineering, Knowledge and Process Management, Business Process Management Journal, Journal of E-Commerce in Organizations, and International Journal of Simulation and Process Modeling. His current research interests include e-commerce technology development and management, investment strategies for computing technologies, and telecommunications planning and management.

* * *

Khalid Aldiri is currently a visiting researcher at the electronic imaging and media communications department at the Informatics school at University of Bradford. His main research interests lie in the area of the effect media cues as a form of social presence on trust in B2C e-commerce, and he has many publications in this field. He has publications in the fields of, virtual environments, e-commerce and multimedia systems.

Eman Ibrahim Al Haj Ali graduated from Auckland University of Technology, New Zealand, with a Master of Information Technology degree in 2005. Her thesis was on mobile commerce adoption across the supply chain in New Zealand businesses. Her research has been presented at the Americas Conference on Information Systems. She is now at United Arab Emirates University.

Jeff Baker (jeff.baker@ttu.edu) is a doctoral student in information systems and quantitative sciences at the Rawls College of Business Administration at Texas Tech University. His research interests include electronic commerce, visualization of multidimensional datasets, and information systems strategy. His research appears in the Journal of Electronic Commerce and Electronic Commerce Research.

Djamal Benslimane is a full professor of computer science at Lyon University, Lyon, France. His research interests include databases interoperability, Web services, and ontologies. He received his PhD in computer science from Blaise Pascal University. He's member of the LIRIS-CNRS lab.

Amit Bhatnagar is an associate professor of marketing at the University of Wisconsin-Milwaukee. He received his PhD in marketing from the State University of New York at Buffalo, MS in aeronautical engineering from IIT, Kanpur, India and BS in mechanical engineering from IIT, Kanpur, India.

Michael Bourlakis is currently a senior lecturer at Brunel University (UK). Michael has more than 10 years experience dealing with marketing, distribution channels and supply chain management. Michael graduated in business administration at the Athens University of Economics and Business (Greece) and obtained both his MBA and PhD degrees from the University of Edinburgh (UK). Michael worked as a research associate at the Management Centre, University of Leicester (UK) and at the Oxford Institute of Retail Management, Templeton College, University of Oxford (UK) and as a lecturer at Newcastle University. He has published in various logistics, supply chain management and marketing journals.

Carl S. Bozman, PhD teaches courses in principles of marketing, research methods, business planning, and new product development. His research interests include demand estimation methods and the new product development process. Professor Bozman also consults in the areas of new product development and demand assessment.

Grégory Bressolles is Professor of Marketing at BeM — Bordeaux Management School. He holds a Ph.D from the University of Social Sciences of Toulouse (France). He is responsible for the e-Commerce and Retail Chair of BeM. His research interests included consumer behaviour, the Internet, e-commerce, retail and services marketing. He has published his research in many international journals like International Journal of e-Business Research, Journal of Customer Behavior, Recherche et Applications en Marketing and he has presented his works in many international conferences.

T. C. Edwin Cheng is chair professor of management in the Department of Logistics, The Hong Kong Polytechnic University. He obtained a BSc(Eng)(Hons) from the University of Hong Kong, an MSc from the University of Birmingham, UK, and a PhD and an ScD from the University of Cambridge, UK. He has previously taught in Canada, England and Singapore. His expertise is in operations management; in particular quality management, business process re-engineering and supply chain management. An active researcher, he has co-authored four books and published over 380 academic papers in such journals as Management Science, Operations Research, and MIS Quarterly, among others. A registered professional engineer and a seasoned management consultant, Prof. Cheng regularly advises business and industry and provides management training and executive development to public and private organizations.

Bill Doolin is Microsoft Professor of E-Business at Auckland University of Technology, New Zealand. His research focuses on the organizational implications of information technology and the adoption of electronic business technologies. His work has appeared in journals such as Electronic Markets, Information Systems Journal, Journal of Global Information Management and Journal of Information Technology.

Uchenna Cyril Eze is a faculty at Multimedia University, Malaysia. He holds a Ph.D. in e-commerce strategic value from NTU, Singapore. Earlier, he worked in First Bank PLC Nigeria, after which he joined Lawdon Maritime Ltd. Nigeria. He is the Chair of Center for e-Services Entrepreneurship and

Marketing [CESEM] in Multimedia University, where he drives research activities among members and external partners. He is also involved in extensive research collaborations with academics and practitioners from several parts of the globe including Malaysia, Nigeria, Singapore, Hong Kong, India, and USA. He serves on the editorial boards of several journals and seats in the committee of advisors of several international and regional conferences. He has published several papers in international conferences and journals including Americas Conference on Information Systems, Australasian Conference on Information Systems, Journal of Global Information Technology Management, International Journal of E-Business Research, International Journal of Electronic Business, and Australiasian Journal of Information Systems. His research interests include knowledge management, IS, e-business, strategic management, entrepreneurship and development. His current research involves ISP service quality and customer satisfaction and IT governance.

Helen Fox has graduated from the School of Agriculture, Food and Rural Development at Newcastle University.

Daniel Friesner, PhD is an assistant professor in the Gonzaga University School of Business. His areas of expertise are applied econometrics and health economics. Professor Friesner teaches economics and statistics at both the undergraduate and MBA levels. His research interests lie primarily in the area of management science, and in particular, efficiency measurement and process improvement issues in health care.

Leopoldo Gemoets is an Associate Professor in the Department of Information and Decision Sciences at the University of Texas at El Paso. He is the founding director of UTEP's CEDARS (Center for Entepreneurial Development, Advancement, Research and Support). A former NASA faculty fellow with the California Institute of Technology, he has over 20 years experience in developing and implementing information technology for industry and government. He is also the faculty coordinator to the Institute for community-based teaching and learning. Dr. Gemoets has published in various international and national journals in the area of international information technology. Other areas of expertise include: E-Commerce, Smart Communities, Strategic use of IT, ERPs and CRMs.

Naveen Gudigantala is a PhD Candidate in MIS in the Rawls College of Business at Texas Tech University. He received his MS in MIS from Texas Tech University. His research interests include Web-based decision support systems, Systems analysis and design, and Information security. His work has appeared in Decision Support Systems, International Journal of E-Business Research, International Journal of Information and Operations Management Education, and Twenty-Sixth International Conference on Information Systems.

Laura L. Hall is an Associate Professor in the Department of Information and Decision Sciences, in the College of Business at the University of Texas at El Paso. Dr. Hall's major research interests are in e-Commerce with emphasis on supply chain management, distance education development, and minority workforce training. She has published in the Journal of Organizational Computing and Electronic Commerce, the Journal of Behavior and Information Technology, Information Resource Management, the International Online Conference on Teaching Online in Higher Education, the Proceedings of the Decision Sciences Institute, and the Proceedings on Human Computer Interaction.

Ezz Hattab is an associate professor in e-business. He developed several software and e-commerce applications. He is an active member of numerous professional and scientific societies, including the Arab Society of Computers (ASC), and Jordanian Society of Computers (JSC). Dr. Hattab has 25 publications and 4 books in the area of Information Technology, e-business, and web applications. He is a member of the technical committee of The International Arab Journal of Information Technology (IAJIT), EBEL, e- Jordan and an associate editor of the International Journal of Mobile Learning and Organization (IJMLO). He is a member of the high supervision committee of Management Information Stream at the Ministry of Education in Jordan. Dr. Hattab has been awarded a scholarship from the EU to pursue his PhD in information technology. His PhD research was part of the Europe research project "The Webminer", in which he proposed numerous algorithms and techniques that handle web information retrieval and search.

Blanca Hernandez holds a Ph.D. in Business Administration and is lecturer in the Department of Marketing and Business Studies at the University of Zaragoza (Spain). Her research interests include the acceptance of new technologies, knowledge management systems, and e-commerce. Ms. Hernandez's work has been published in journals such as Industrial Marketing Management, Internet Research, Technovation, Interacting with Computers and Online Information Review.

David Hobbs was a Senior Lecturer at the Electronic Imaging and media communications department at the University of Bradford. He is a Fellow of the Higher Education academy (FHEA), and a member of The Institution of Electrical and Electronic Engineer (IEEE), British Computing Society (MBCS). His main research interests lie in the area of virtual environments, particularly in the representation and conveying of emotion through facial expressions and gestures presented on virtual humans. Application areas include e-learning, e-commerce and collaborative work environments (CWE). This arises out of long-standing interests in computer-based self-instructional systems, particularly multimedia systems, and in the application of artificial intelligence techniques within such systems. He has publications in the fields of, virtual environments, e-learning, e-commerce and collaborative work environments (CWE), multimedia systems, and in the application of artificial intelligence techniques.

Xiaorui Hu is an associate professor of decision sciences and information technology management at the John Cook School of Business at Saint Louis University. She received her PhD in economics from the University of Texas at Austin. Her research focuses on trust related issues in electronic commerce, B2B markets, telecommunication market evolution, information security, and culture impact on international business. Her work has appeared in Information Systems Research, Decision Support Systems, IEEE Computer, Journal of Organizational Computing and Electronic Commerce, International Marketing Review, Journal of Global Information Management, and Journal of Promotion Management, and other outlets.

Minh Q. Huynh is an Assistant Professor at Southeastern Louisiana University. He received his Ph.D. from State University of New York at Binghamton. His teaching expertise are in the areas of E-commerce, Telecommunications, Decision Support Systems. Research interests include E-learning, computer supported collaborative learning, and IS outsourcing. His publications appear in such journals as the Communications of ACM, Journal of AIS, Communications of AIS, European Journal of IS, Journal of Electronic Commerce in Organizations. Prior to his academic career, he had worked in the areas related to computer programming, systems management, technical support, and network security.

Sajjad M. Jasimuddin is a Lecturer at the School of Management and Business, Aberystwyth University, UK, and is also an Associate Professor of Management, University of Dhaka, Bangladesh. He holds his MPhil in International Business from the Judge Business School, University of Cambridge, and PhD in Knowledge Management from the School of Management, University of Southampton. His articles have been published, among others, in Journal of Knowledge Management, International Journal of E-Business Research, Knowledge and Process Management, Management Decision, Management Decision, Knowledge Management Review, International Journal of Organizational Analysis, Journal of Information and Knowledge Management, Advances in Doctoral Research in Management, Journal of Business and Industrial Marketing, Management Research News, The Encyclopaedia of Knowledge Management, Business Strategy Series (Formerly Handbook of Business Strategy), Encyclopaedia of Mobile Computing & Commerce, Encyclopaedia of Portal Technology and Applications, Asian Affairs, Journal of Management, Journal of Air Transport Management, and Journal of Internet Banking and Commerce. His current research interests are in the areas of knowledge management, international business environment, linkage between human resource and strategic management.

Julio Jimenez holds a Ph.D. in Business Administration and is professor in the Department of Marketing and Business Studies at the University of Zaragoza (Spain). His research in adoption and diffusion of innovations has been published in several journals, such as Research Policy, Industrial Marketing Management, Internet Research, Interacting with Computers, Online Information Review and Technovation.

Donald R. Jones is an associate professor and James C. Wetherbe Professor of MIS in the Rawls College of Business at Texas Tech University. He received his PhD from the University of Texas at Austin. His research interests focus on the effects of DSS features on decision making, use of database queries in problem solving, and human behavior in online environments. His research has appeared in Decision Support Systems, Organization Behavior and Human Decision Processes, The Journal of Information Systems, Information & Management, among others.

Larry Kerschberg is professor of computer science, at George Mason University, Fairfax, VA 22030 USA. He is director of E-Center for E-Business and directs the MS in E-Commerce Program. He is past chairman of the information and software engineering department at Mason. During 1998 he was a Fellow of the Japan Society for the Advancement of Science at Kyoto University. He holds a BS in engineering from Case Institute of Technology, and MS in Electrical Engineering from the University of Wisconsin—Madison, and a PhD in engineering from Case Western Reserve University. He is editor-in-chief of the Journal of Intelligent Information Systems, published by Springer. He recently served as an editor and contributor of the book The Functional Approach to Data Management: Modeling, Analyzing and Integrating Heterogeneous Data, published at Heidelberg, Germany, Springer, 2004. His areas of expertise include expert database systems, intelligent integration of information, knowledge management, and agent-based semantic search. His recent papers have focused on ontology-driven semantic search in knowledge sifter, knowledge representation using topic maps, and methodologies for the creation and management of semantic Web services.

Mabel T. Kung is professor of information systems and decision sciences department at the Mi-Haylo College of Business and Economics, California State University at Fullerton. Her PhD is from the McCombs School of Business at The University of Texas at Austin. Her research publications and industrial experience include the fields of e-commerce, corporate training, strategic planning and benchmark technology, web component development and enterprise database systems.

Lawrence C. F. Lai is a general manager of New World TMT Limited responsible for technology, media and telecommunication investments. He has more than 15 years experience in serving senior positions in private equity investment and telecommunication operations in Hong Kong. He received a BA from the University of Western Ontario, Canada, an MBA from Heriot-Watt University, UK, and a Doctor of Business Administration from The Hong Kong Polytechnic University.

Francisco J. Lopez received a college degree in actuarial sciences from Universidad Anahuac, in Mexico City, in 1984. He enrolled in the Ph.D. program in Business at the University of Mississippi in August 1996 and obtained his doctoral degree in August 1999. Teaching experience includes a visiting assistant position at the University of Mississippi between August 1999 and May 2000; assistant and associate professor at the University of Texas at El Paso between August 2000 and July 2007; and associate professor at Macon State College since August 2007. He was also a visiting professor at the Helsinki School of Economics, campus Mikkeli during Fall 2002. His research interest includes convex analysis (linear and positive hulls), DEA, linear programming, scheduling, and the development of theory and algorithms in these areas.

Zakaria Maamar received his PhD in computer sciences from Laval University (Canada) in 1998. Currently, he is an associate professor in the College of Information Technology at Zayed University, Dubai, United Arab Emirates. His research interests lie in the areas of context-aware computing, Web services, and software agents.

M. Adam Mahmood is a Professor of Computer Information Systems in the Department of Information and Decision Sciences at the University of Texas at El Paso. He also holds the Ellis and Susan Mayfield Professorship in the College of Business Administration. He was a visiting faculty at the Helsinki School of Economics and Business Administration in Finland and a visiting Erskine Scholar at the University of Canterbury in New Zealand. Dr. Mahmood's research interest centers on economics of information systems, electronic commerce, strategic and competitive information systems, group decision support systems, and organizational and end-user computing. On this topic and others, he has published over 85 technical research papers in some of the leading journals and conference proceedings, including the Journal of Management Information Systems, MIS Quarterly, Decision Sciences, European Journal of Information Systems, INFOR—Canadian Journal of Operation Research and Information Processing, Information and Management, Journal of Organizational and End User Computing, Data Base, and others. Dr. Mahmood presently serves as the editor of the Journal of Organizational and End User Computing. As a Governor's appointee, he also serves as a member of the Texas Department of Information Resources Board of Directors. He has also served as president of the Information Resources Management Association.

Maria José Martín holds a Ph.D. in Business Administration and is senior lecturer in the Department of Marketing and Business Studies at the University of Zaragoza (Spain). Her main research line is online consumer behavior and e-commerce. Ms. Martín de Hoyos' work has been published in several journals, such as Industrial Marketing Management, Internet Research, Interacting with Computers, Online Information Review and Technovation.

Matthew Q. McPherson, PhD is an assistant professor in the Gonzaga University Jepson School of Business. His areas of expertise include international economics, development, and health economics. Professor McPherson teaches international economics and global economic issues at both the undergraduate level. His research interests include international trade and finance, management science, and issues in health care. Professor McPherson is also the managing editor of the Journal of Economic Development and Business Policy.

Saravanan Muthaiyah is a senior lecturer at Multimedia University, Cyberjaya, Malaysia. He holds a bachelors degree in accounting and finance, a Msc in Information Technology and a doctoral degree in Information Technology from George Mason University, Fairfax, VA. He is also a Fulbright scholar and fellow under the auspicious of thr graduate research exchange program sponsored by the US Department of State. His research interests include semantic web, ontology mapping, systems integration, systems engineering, topic maps, knowledge management and enterprise architectures. His recent papers have focused on ontology mapping and mainly solving heterogeneity issues for Semantic Web.

Jacques Nantel is Professor of Marketing at H.E.C. Montréal. He holds a Ph.D from Indiana University. He is the founder and the first director of the RBC Financial Group Chair of e-commerce. Since March 2007, he is General Secretary (Senior Associate Dean) of HEC Montréal. Concerning research, He has also published several articles in journals such as the Journal of Retailing, The Journal of Interactive Marketing, The International Journal of e-commerce, The Journal of Business Research, the European Journal of Marketing, the Journal of Social Behaviour and Personality, and the Journal of Business Ethic. His research interests included the impact of Information Technologies (I.T.) on consumers' behaviour and on marketing strategies of organisations.

Levent V. Orman is Professor of Information Systems at Cornell University. He received Ph.D. in Information Systems from Northwestern University. He has taught courses and written articles on electronic commerce, database management, and decision support systems. His articles appeared in a variety of journals ranging from Information Systems, Acta Informatica, and IEEE Transactions on Knowledge and Data Engineering, to MIS Quarterly, Journal of MIS, and Electronic Commerce Research. He serves on the editorial board of the Journal of Information Technology and Management. Professor Orman is currently working on electronic commerce and electronic services.

M. A. Otair is an instructor at the Arab Academy for Banking and Financial Sciences. He received his BSc in computer science from IU-Jordan and his MSc and PhD in 2000, 2004, respectively, from the Department of Computer Information Systems -Arab Academy. His major interests are neural network learning paradigms, mobile-computing, e-learning and e-business. Dr. Otair has 19 publications (International journals and conferences) in the area of neural networks, information technology, m-computing, and e-applications.

Savvas Papagiannidis graduated from the physics department of the University of Newcastle upon Tyne. Upon completion of his PhD he joined the eBusiness Group at the business school in the same University. Savvas has started a number of eBusiness ventures and also worked as a freelance Internet developer. His research interests include management of Internet and emerging technologies, high-technology related entrepreneurship and e-business models.

Rami Qahwaji is currently a Senior Lecturer at the Electronic Imaging and media communications department at the University of Bradford. He is a member of the Higher Education academy (HEA), The Institution of Electrical Engineer (IEE), International Society for Computers and Their Applications (ISCA) and the American Geophysical Union (AGU). His research expertise include: image processing, pattern recognition and machine learning and the design of machine vision systems. He has publications in the fields of solar imaging, medical imaging, biometrics and face recognition, morphological transforms, statistical classifiers and neural networks, security and watermarking.

R. Rajendran is a professor in the Department of Management at Sri Ramakrishna Institute of Technology, Coimbatore, India. He holds a Ph.D. and M.Phil. in management from the Bharathiar University and an M.B.A. and B.E.(Hons.) degree in Mechanical Engineering from the Madras University. He has prior business and industry experience. He served as a senior engineer for Bharat Electronics Limited, a public sector undertaking of Government of India under administrative control of Ministry of Defense, for ten years. Prior to joining academic in the year 2000, he was managing a small manufacturing partnership firm for six years. His research focuses on information systems strategic management, electronic commerce, electronic business and small business management. He has published his findings in journals such as Digital Business Review, Journal of Electronic Commerce Research, International Journal of E-Business Research and Organization Management. He is a member of the international editorial review board of International Journal of E-Business Research.

Jaeki Song (jaeki.song@ttu.edu) is an associate professor of information systems and quantitative sciences at the Rawls College of Business Administration at Texas Tech University. His research interests include electronic commerce, web design, information systems strategy, and technology adoption. His research appears in Management Science, Decision Support Systems, Information and Management, IEEE Transactions on Professional Communication, and other leading journals. He received his PhD from the University of Wisconsin-Milwaukee.

William J. Tastle received his PhD in advanced technology with specialization in systems science from the Thomas J. Watson School of Engineering and Applied Science of the State University of New York, University Center at Binghamton in 1994 and an MBA in MIS from the School of Management at Binghamton. He is a professor of information systems in the business school at Ithaca College and is active in the IS community having served as the president of the Association for Information Technology Professionals, Education Special Interest Group (EDSIG), and on many other committees. He is the managing editor of the International Journal of General Systems. Currently he is organizing the 2008 conference of the North American Fuzzy Information Processing Society and serves on its board of directors as treasurer. He is a consultant for numerous organizations including Los Alamos National Laboratory. Dr. Tastle's current areas of interest involve measures of consensus, dissention, agreement and disagreement, requirements analysis elicitation, and outsourcing.

Hannu Verkasalo acts as a project manager at the national research project MoMI (modeling the usage and business of the mobile Internet) at the Helsinki University of Technology in Finland. Hannu Verkasalo has actively published on the evolution of the mobile service market. He is currently finishing his doctoral dissertation on the emergence of the mobile Internet and new mobile service concepts. Verkasalo has particularly focused on the empirical modeling of mobile services, utilizing the recently developed handset-based research method. In addition to research work and consulting in the mobile industry, Verkasalo gives lectures on network economics, operator business and telecommunication regulation.

K. Vivekanandan is a reader of Bharathiar School of Management and Entrepreneurship Development at Bharthiar University, India. He received his Ph.D. in computer simulation from the same university in the year 1996. His research interest includes electronic commerce, electronic business, and data mining. His publications have appeared in Journal of Electronic Commerce Research, Digital Business Review, and other national and international research journals.

Avinash M. Waikar, Ph.D., University of Oklahoma, is a registered professional engineer with the Louisiana State Board of Professional Engineers. He has approximately ten years of industrial and manufacturing experience in operations management. He has been a consultant in the areas of operations management and statistics to businesses and has published numerous journal articles in the areas of operations management and other business disciplines. His research interests are in the area of operations management and MIS. He has been a recipient of Outstanding Paper Award. Currently he teaches operations management and statistics at Southeastern Louisiana University.

Mark J. Wierman is an assistant professor of computer science at Creighton University in Omaha, Nebraska. He received his PhD in Systems Science from Binghamton University. His current research interests are the application of the mathematics of uncertainty to political science and generalized information theory. He has published over 30 papers and his second book, Applying Fuzzy Mathematics to Formal Models in Comparative Politics (with TD Clark, JM Larson, JN Mordeson, and JD Potter) will be published in spring 2008.

Yuhong Wu was an assistant professor of marketing at Christos M. Cotsakos College of Business, William Paterson University. She received her PhD degree in marketing from the University of Texas at Austin. Her research interests lie in the areas of internet marketing and e-commerce, strategies in network market, and new product development and management. Her works has appeared in Journal of Marketing and other outlets.

Ruiliang Yan is an assistant professor of marketing at the Virginia State University. He received his PhD in marketing from the University of Wisconsin, Milwaukee, MS from Sichuan University, China and a bachelor's degree from Southwest Agricultural University, China.

Andy C. L. Yeung received an MSc from the University of Bradford, UK, and a PhD from the University of Hong Kong. He is currently an Associate Professor in the Department of Logistics, The Hong Kong Polytechnic University. His research interests include quality management, operations management and supply chain management. He has published papers in Production and Operations

Management, Journal of Operations Management and IEEE Transactions on Engineering Management, among others.

Jenny Yi Zhang is associate professor of information systems and decision sciences department at the MiHaylo College of Business and Economics, California State University at Fullerton. She earned her PhD in information systems from the New Jersey Institute of Technology. Her research areas are mainly focused on e-business, knowledge management, virtual communities, virtual teams, and business intelligence.

Zuopeng (Justin) Zhang is an Assistant Professor of Management Information Systems at School of Business and Economics in State University of New York at Plattsburgh. He obtained his Ph.D. in Business Administration from Smeal College of Business, The Pennsylvania State University. His research interests include economics of information systems, knowledge management, electronic business, and workflow systems. His research has recently appeared at Knowledge and Process Management, International Journal of E-Business Research, Encyclopaedia of Portal Technology and Applications, Encyclopaedia of Mobile Computing and Commerce, and in the proceedings of major conferences in Information Systems.

Index

D

E

F

G

H

I

K

L

M

N

O

P

Q

R

S

T

U

V

W

X